II International Conference on Computational Methods in Engineering Science (CMES'17)

ITM Web of Conferences Volume 15 (2017)

Lublin, Poland
23 - 25 November 2017

Editors:

Magdalena Borys
Zbigniew Czyz
Katarzyna Falkowicz
Justyna Kujawska
Monika Kulisz
Miroslaw Szala

ISBN: 978-1-5108-5379-9

Printed from e-media with permission by:

Curran Associates, Inc.
57 Morehouse Lane
Red Hook, NY 12571

Some format issues inherent in the e-media version may also appear in this print version.

Printed by Curran Associates, Inc. (2018)

For additional information, please contact EDP Sciences – Web of Conferences at the address below.

EDP Sciences – Web of Conferences
17, Avenue du Hoggar
Parc d'Activité de Courtabœuf
BP 112
F-91944 Les Ulis Cedex A
France

Phone: +33 (0) 1 69 18 75 75
Fax: +33 (0) 1 69 28 84 91

contact-edps@webofconferences.org

Additional copies of this publication are available from:

Curran Associates, Inc.
57 Morehouse Lane
Red Hook, NY 12571 USA
Phone: 845-758-0400
Fax: 845-758-2633
Email: curran@proceedings.com
Web: www.proceedings.com

TABLE OF CONTENTS

Statement of Peer review

In submitting conference proceedings to ITM Web of Conferences, I certify to the Publisher that I adhere to the **Policy on Publishing Integrity** of the journal in order to safeguard good scientific practice in publishing.

1. All articles have been subjected to peer review administered by the proceedings editors.
2. Reviews have been conducted by expert referees, who have been requested to provide unbiased and constructive comments aimed, whenever possible, at improving the work.
3. Proceedings editors have taken all reasonable steps to ensure the quality of the materials they publish and their decision to accept or reject a paper for publication has been based only on the merits of the work and the relevance to the journal.

Title, date and place of the conference

II International Conference of Computational Methods in Engineering Science CMES'17, 23-25 November 2017, Lublin University of Technology, Lublin, Poland

Proceedings editor(s):

Magdalena Borys
Zbigniew Czyż
Katarzyna Falkowicz
Justyna Kujawska
Monika Kulisz
Mirosław Szala

Date and editor's signature

14.11.2017

17, avenue du Hoggar - PA de Courtabœuf – BP 112 - 91944 Les Ulis Cedex A (France)
Tél. : 33 (0)1 69 18 75 75 - Fax : 33(0)1 69 07 45 17 – www.edpsciences.org

 DOI: 10.1051/itmconf/20171501001

An inexpensive environmental monitoring system with IoT agents

Dariusz Czerwinski[1,*], and *Marek* Milosz[1]

[1]Lublin University of Technology, Institute of Computer Science, Nadbystrzycka 38A, 20-618 Lublin, Poland

Abstract. Air quality is of great importance for human health and life expectancy. It becomes crucial to monitor atmospheric dust in the air of cities. In connection with the development of mobile networks and low-cost sensory agents, it has become possible to create inexpensive environmental monitoring systems. The paper presents results of studies on the system monitoring dust concentration in city air. The system consists of moving IoT agents placed on vehicles (taxies, busses, private cars) and measure the dust concentration. Agents, using a wireless connection, are sending the data to the recording server. The server application collects the data and visualises them on the map in a certain colour, depending on the dust concentration in the air and the values acceptable by standards. The system architecture, the algorithm of measurements and the agent-server data exchange protocol were presented in the article, as well as the example of data visualisation.

1 Introduction

Environmental pollution is a big problem all over the world. Nowadays people pay more attention to the physical health. Clean and healthy environment becomes more important to all. Pollution incidents and smog in the cities cause healthiness problems and create welfare threads [1-3].

In developing countries more and more attention is paid on building up air quality monitoring systems. Some of them give pretty accurate air quality measurement results with the use of monitoring systems and reference measurement verification [4-6].

Current research and development of such monitoring systems can be divided mainly into two solutions: interactive and real time. In the interactive systems the pollutant is chosen and after this the query to the measurement database is performed. After this the related pollutant information is presented [7]. In the real time monitoring systems the information on pollutant or dust level is presented online after receiving the measurement data. This kind of systems are the mainstream, however there are various methods of implementation [4, 8, 9].

With the development of new technologies the new area is open for building the monitoring systems. Such promises technology is Internet of Things (IoT), which is the global network of devices which are allowed to connect to Internet and interact with available web services.

The IoT is significant because a device with the unique ID can represent itself digitally and therefore becomes so called smart thing [10-12].

Sensor technologies prices are quickly decreasing and the area of IoT systems is expected to expand very fast. The most popular application areas of IoT include: energy savings and management [13], production management [14], environment protection [15], smart home applications [16], security and monitoring systems [17, 18].

Fig. 1. Location of dust monitoring stations in Poland [21].

In Poland and other European countries dust level measuring systems use expensive equipment at fixed locations or dedicated mobile environment monitoring stations. Data obtained in such measurements is used for calculations in dispersion models to get the environmental pollution maps. This kind of system is the system with the coarse sensors network where the measurement points are few and far in-between. The advantage of this system type is the high measurement accuracy due to the used expensive equipment. The main disadvantage is that access to such data is limited only to persons who are well informed on the pollution sources and levels [19, 20].

* Corresponding author: d.czerwinski@pollub.pl

DOI: 10.1051/itmconf/20171501001

In consequence, measurement stations are very few (Fig. 1) and the results of measurements are available with daily delay. For example, in Lublin province of Poland is the only one air dust concentration monitoring station (Fig. 1).

The other solution is the system with the dense sensors network which can provide more frequent and spatially dense measurements. These types of measurements are essential for monitoring spatially inhomogeneous environments with variable emission sources like in urban areas [23,24].

There are also some related works on low cost air quality monitoring systems. The mobile electrochemical sensors (usually CO, NO and NO 2) designed in highly compact way and carry out by volunteers were shown in [23]. Data from sensors were sent directly to Apache server and visualized using Google Maps API.

Brienza et al. in [9] presented low cost sensing system for measuring concentrations of polluting gases in various areas of the city. The users can deployed the nodes in different areas of interest and can change location of the node whenever they want to monitor a different place. Usually nodes were placed on users houses and made use of WiFi connection with user access point. System contained database and web servers which allow users to register sensing nodes and set their geographical coordinates locating them on a map. The work out website uses the Google Maps API to visualize sensor nodes and Air Quality Index levels.

Low-cost wireless monitoring system, which performs air quality (dust, CO and CO2) referential parameters measurements, was presented by Fuertes et al. in [25]. The agents were based on Arduino prototyping board and placed in chosen places of the city. Measurements were transferred through the wireless connection to the server. Server web application presents measured values using Google Maps and brief description of point with measured parameters.

Castell et al. in [24] present the novel approach followed in the Citi-Sense-MOB project. The idea of the system consists of the use of lower-cost sensors for monitor-ing air quality at street level and web system offering the user-friendly and visually informative layout. The authors proposed that sensors will be mounted on mobile platforms, such as buses and cars.

In this article, the authors described the system of the big data sensing type, which consists of three main layers: mobile agent, middleware and application. The purpose of the presented work is the collection and analysis of dust level and temperature data from IoT agents. The middleware and application layers are implemented by the dedicated application. This solution offers the scalability, platform independence and real-time data processing and visualization.

The remainder of the paper is organized as follows. Section 2 presents the information on air quality measurements standards. In the Section 3 the layered architecture of the proposed system is described. The details of the implementation of mobility IoT agents are presented in Section 4. Server side application is described in Section 5. Finally, a brief conclusion is made in Section 6.

2 Air quality measurements standards

The measurement of air quality is carried out by state services based on standards acceptable levels of various substances in the air. Polish and EU standards limit the usual average levels of these pollutants. Polish standard [26] defines maximum levels for dust pollution for two sizes: PM2.5 and PM10. PM2.5 is the concentration of particulate matter with an aerodynamic diameter of grains to 2.5 microns, and PM10 – to 10 microns. Acceptable levels of dust concentrations are determined for the average daily (PM10, the permissible level is 50 µg/m3), or an average annual (PM2.5, the permissible level is 25 µg/m3). The standard [26] was defined two additional levels of pollution (only for PM10):
- the level of informing the public (value - 200 µg/m3),
- the alarm level (300 µg/m3).

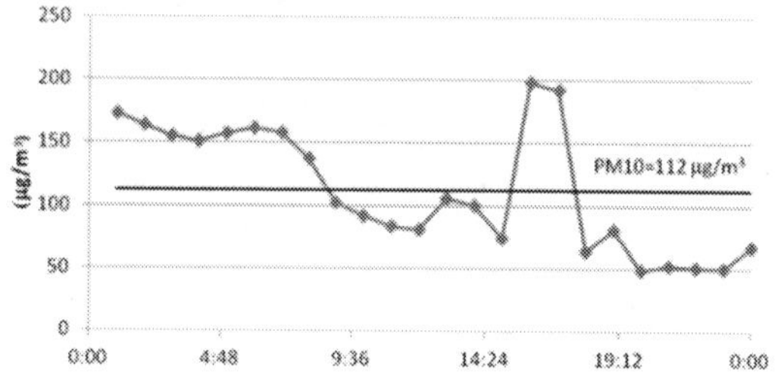

Fig. 2. The results of the daily measurement of the dust concentration (PM10; Dietla str., Krakow, Poland; March 18, 2016) with the marked daily average value (Source: [27]).

Both of these levels are defined in the standard for the average daily. Short-term increases in the level of pollution do not therefore constitute formal causes for the alarms (Fig. 2). Such information may, however, be useful in the choice time and places of jogging or resting by citizens. For example, afternoon jogging would be better start after 18 and even better after 20 in the neighbourhood of the measuring station with the data from Fig. 2. The concentration at this time was four times less than 16-17 hours.

3 Inexpensive environmental monitoring system architecture

Inexpensive Environmental Monitoring system consists of two basic components (Fig. 3):
- mobile measuring stations,
- server-side software.

Many mobile measuring stations, placed on the public transport vehicles (buses, taxi) or private (volunteers) during motion collect data on the concentration of dust in the air. They used algorithms to optimize energy consumption and make sense of measurement and act as IoT agents.

Measuring stations transfer the data to the server in a situation where they will in the field of public WiFi network (e.g. accessible to the public in buses or other free Hot Spots located in different places of the city). This solution reduces the operating costs of the system. In addition, the algorithms of transfer the measurement

DOI: 10.1051/itmconf/20171501001

data to the server optimize energy consumption and transmission time.

On the server side are implemented three basic features:
- mobile measuring stations (IoT agents) registration,
- measurement data acquisition,
- data presentation in various forms, layouts and profiles.

Server-side part of the system uses:
- HTTP server (Apache)
- PHP parser,
- database server (MySQL)
- external services (Google Maps)
- dedicated software (HTML, CSS, JavaScript, PHP).

The software has been developed using the responsive web design (RWD) approach and model-view-controller (MVC) architecture.

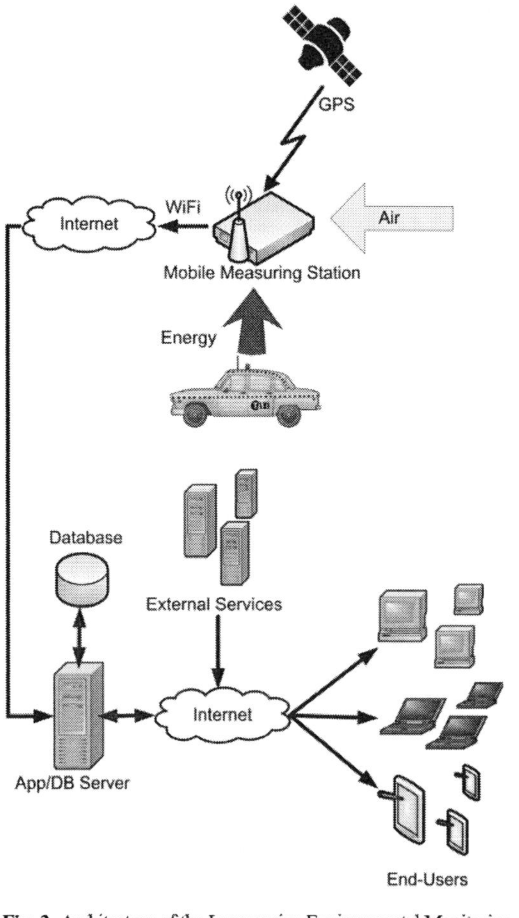

Fig. 3. Architecture of the Inexpensive Environmental Monitoring System.

4 The moving IoT agent for dust concentration measurements

The measuring station, which also acts as IoT agent, is designed in the compact way to fulfil the mobility requirements. The overall architecture of the station is shown in Fig. 4.

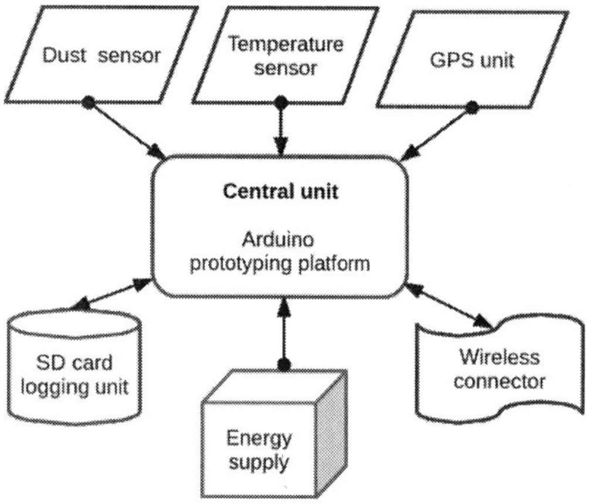

Fig. 4. Architecture of the moving measuring station of dust concentration.

Station consists of the few sensors. The main one is the dust sensor for measuring the dust concentration, which gives a good indication of the air quality in an environment. However due to the specific way of dust sensor automatic measurement, this values cannot be assumed as reference values for dust concentration.

The other two sensors are: GPS for determining the position and temperature sensor for acquire working temperature. The range of working temperatures is important due to the dust sensor operation temperatures which are from -10 °C to +65 °C.

The heart of the IoT agent is the Arduino prototyping platform with Atmega chipset onboard. Arduino platform allows for programming all components together and act as single IoT agent. The measurement data is written to SD memory card for later transmission. In the designed system the WiFi module connected to the Arduino board acts as wireless connector. Wiring of the designed measurement station is shown in Fig. 5.

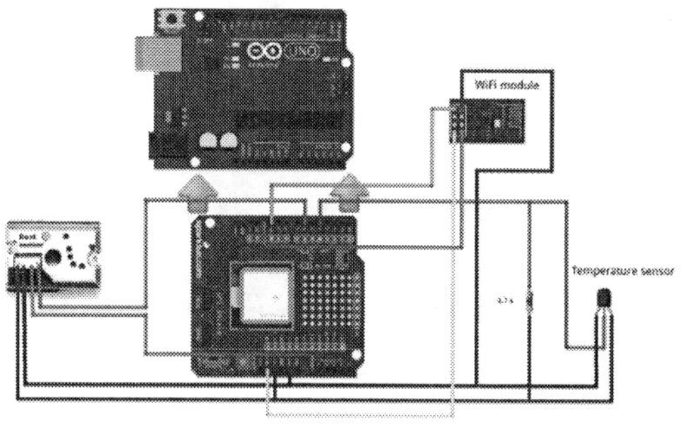

Fig. 5. Wiring of the dust measuring IoT agent.

On the top of Arduino Uno board the GPS shield is set. This shield is also equipped in SD card slot and additionally allows for transparent wire connection of external devices to Arduino board. All other devices of an IoT agent i.e. dust sensor, temperature sensor and WiFi module are properly connected to the supplying pins (GND, +5V, +3.3V) and digital and analogue input pins (Fig. 5). An agent is supplied with the electrical system of the car or the USB power bank of about

10000 mAh capacity. That allows for approximately 6 hours measurement and communication session.

The hardware parameters of an IoT mobile dust level measuring station are shown in Table 1. Total cost of the station with additional hardware (cables, resistor, and casing) is about 125 €.

The agent performs measurements according to flowchart shown in Fig. 6. The series of measured data are averaged (10 measurements of dust level and temperature per second) and the averaged value is written to the file. That was assumed, that with allowed speed limit in the cities in Europe which is 50 km/h the vehicle changes its position by the 13.8 m during one second.

Table 1. Hardware parameters of IoT mobile dust concentration measuring station.

Hardware element	Description
Arduino programmable board	Arduino Uno board which is used as a main platform for sensors and dedicated agent software. Maximum current consumption is 46.5 mA with supplying voltage equals to 5V. Price 25€
Optical Dust Sensor	The sensor GP2Y1010AU0F has a very low current consumption (20mA maximum), supplying voltage up to 7V of direct current. The output of the sensors is an analogue voltage proportional to the measured dust density, with a sensitivity of 0.5V/0.1mg/m3. Price 14€
WiFi module	Wireless ESP8266 module, WiFi 802.11 b/g/n, working frequency 2.4 GHz, PCB antenna, current consumption 140 mA during transmission. Price 3€
GPS and SD card shield	The shield is SparkFun GPS Logger Shield - GPS GP3906-TLP with SD Card slot. It allows for GPS data recording and storing data on SD Card. Current consumption is 30 mA. Price 50€
Power bank	Polymos 10 AIR - 10000mAh power bank. Output current 2.1 A. Price 25€

However according to the [28] average speed varies from 19 to 35 km/h, which translates into a travelled distance from 5.3 to 9.7 meters per one second.

During the movement of the vehicle the IoT agent tries to connect to predefined access points or Hot Spots. Shared by smartphone and other WiFi connection can also be used. It agent succeeds in connecting to Internet, it tries to send to server stored data. The flowchart of data sending is shown in Fig. 7. Agent during sending data is working according to FIFO algorithm. The structure of sending to the server data frames is as follow:

Frame := MAC, Number of measurements, {GPS position, DataTime, Type, Measurement;}

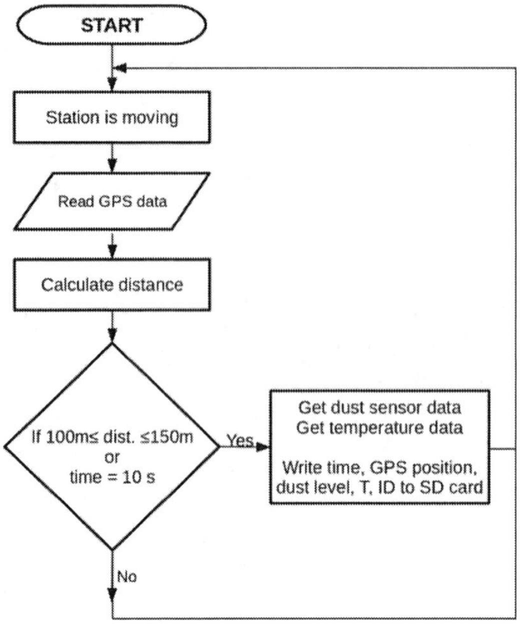

Fig. 6. Dust level measurement flowchart.

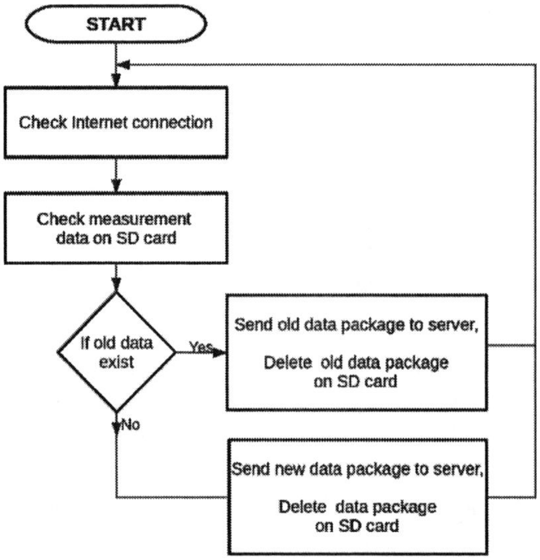

Fig. 7. Flowchart of the sending the measurement data to server protocol.

5 Server-side application

Server application was developed using technology WAMP (Windows + Apache + MySQL + PHP). Developed software for server-side part of the system has functionalities shown in Table 2.

The systems stores all collected data in the database. Table *Users* contains the information on registered in system users.

DOI: 10.1051/itmconf/20171501001

Table 2. Functionalities of server-side part of the system.

Area	Functionality	Description
Mobile measuring stations registration	Can register a station	The system use WiFi MAC to identification a station
	Can deactivate the station	Temporary deactivate measuring station
	Can activate the station	Re-activate the measuring station
Measurement data acquisition	Can receive and store the measured data in database	Measurement data are received and stored in a database of the control unit and using a special data transmission protocol
Data presentation	Can present measurement data in various forms, systems and sections	Various filters and external services (geographic) will be used for data presentation

The data from mobile stations are stored in *MMStations* and *Measurements* tables. The structure of system's data base is presented in Fig. 8.

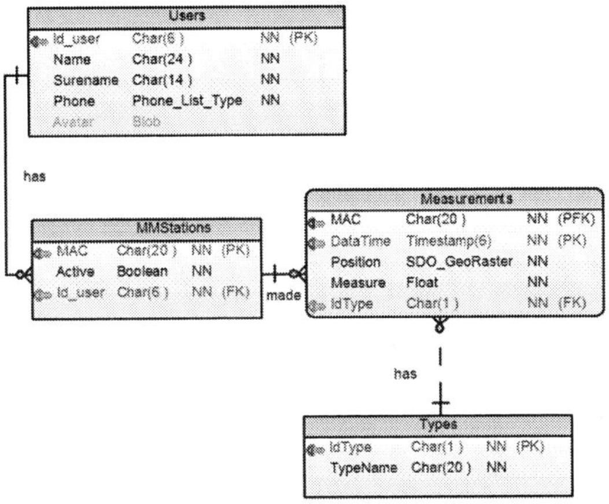

Fig. 8. Structure of the system database.

Server side application can present in graphic way conducted measurements. They are represented as spots in a colour corresponding to the intensity of the dust level. The examples of the server-side application screens are shown in Fig. 9.

6 Conclusions and future works

The paper presents the results of the studies on inexpensive system to monitor the dust level in the air by IoT agents. The system can be implemented in industry due to its reliability, modularity and low implementation costs. The independent scaling of each layer was applied in the project. This allows for expansion of this solution de-pending on the needs. Designed moving IoT agents allow for dust level, temperature and position recording as also as sending the data to server system.

On the basis of data obtained during the operation of the system will be possible to implement different data mining algorithms, including:

- Search incorrectly operating measurement agents.
- Mapping of pollution (temporary and average) by averaging and approximation of data.
- Analysis of the dynamics of the propagation of dust pollution with the ability to detect sources (backward approximation and modelling of the pollutant propagation).

Fig. 9. Data filtering and visualizations screen.

References

1. M. Kampa, E. Castanas, *Environmental Pollution*, **151**, 362–367. Elsevier (2008)
2. J. Anderson, J. Thundiyil, A. Stolbach, J. Med. Toxic., **8**, 166–175. Springer (2012)
3. S. Steinle, S. Reis, C.E. Sabel, *Science of The Total Environment*, **443**, 184–193. Elsevier (2013)
4. S. Devarakonda, P. Sevusu, H. Liu, R. Liu, L. Iftode, B. Nath, *Proceedings of the 2nd ACM SIGKDD International Workshop on Urban Computing (UrbComp '13)*, 8 p. ACM, New York (2013)
5. Y. Cheng, X. Li, Z. Li, S. Jiang, Y. Li, J. Jia, X. Jiang, *Proceedings of the 12th ACM Conference on Embedded Network Sensor Systems (SenSys '14)*, 251–265. ACM, New York (2014)
6. W. Jiao, G. Hagler, R. Williams, R. Sharpe, L. Weinstock, J. Rice, *Environmental Science & Technology*, **49**, 6085–6092. American Chemical Society (2015)
7. J. H. Liu, Y. F. Chen, T. S. Lin, C. P. Chen, P. T. Chen, T. H. Wen, C-H. Sun, J-Y. Juang, J. A. Jiang, Intern. J. on Smart Sens. and Intel. Syst., vol. **5**, 191–214 (2012)
8. M.F. Ramadhani, B. Yuliarto, Nugraha, *Proceedings of the 3Rd International Conference*

DOI: 10.1051/itmconf/20171501001

On Instrumentation Control & Automation (ICA), 270–273. IEEE (2013)

9. S. Brienza, A. Galli, G. Anastasi, P. Bruschi, Sens., **15**, 12242–12259 (2015)
10. S. Agrawal, M.L. Das, Proceedings of the 2011 *Nirma University International Conference On Engineering*, 1-7. IEEE (2011)
11. H. Kopetz, *Design Principles for Distributed Embedded Applications*, 307–323. (Springer 2011)
12. J. Gubbi, R. Buyya, S. Marusic, M. Palaniswami, Fut. Gen. Comp. Syst., **29**, 1645–1660. (Elsevier, 2013)
13. R. Wójcicki, *Studia Informatica*, **36**, 59–72 (2015)
14. X. Jia, Q. Feng, T. Fan, Q. Lei, *Proceedings of the 2012 2nd International Conference On Remote Sensing, Environment & Transportation Engineering*, 1282–1285 (IEEE, 2012)
15. L. Dan, C. Xin, H. Chongwei, J. Liangliang, In *Proceedings of the 2015 International Conference On Intelligent Transportation, Big Data & Smart City*, 487–490. (IEEE 2015)
16. S.T. Kelly, N.K. Suryadevara, S.C. Mukhopadhyay, IEEE Sens. J., **13**, (IEEE 2013)
17. K. Cai, Adv. in Inf. Sci. and Serv. Sci., **4**, (2012)
18. Z. Marszalek, R. Sroka, M. Stencel, Metr. Meas. Syst., **18**, 69-76 (2011)
19. M. Joly, V. Peuch, Atmos. Env., **47**, 111–123 (Elsevier, 2012)
20. M. Viana, J. Pey, X. Querol, A. Alastuey, F. de Leeuw, A. Lükewille, Sci. Tot. Env., **472**, 825–833. (Elsevier, 2014)
21. Chief Inspectorate for Environmental Protection, http://powietrze.gios.gov.pl/pjp/maps/measuringstation/W?lang=en
22. The Regional Inspectorate of Environmental Protection in Lublin. Monitoring System of Air Quality, http://envir.wios.lublin.pl/?par=2&l=
23. M. Mead, O. Popoola, G. Stewart, P. Landshoff, M. Calleja, M. Hayes, J.J. Baldovi, M.W. McLeod, T.F. Hodgson, J. Dicks, A. Lewis, J. Cohen, R. Baron, J.R. Saffell, R. Jones, Atm. Env., **70**, 186-203 (Elsevier 2013)
24. N. Castell, M. Kobernus, H. Liu, P. Schneider, W. Lahoz, A.J. Berre, J. Noll, Urb. Clim., **14**, 370-382. (Elsevier, 2015)
25. W. Fuertes, D. Carrera, C. Villacis, T. Toulkeridis, F. Galarraga, E. Torres, H. Aules, *Proceedings of the 2015 IEEE/ACM 19Th International Symposium On Distributed Simulation & Real Time Applications (DS-RT)*, 58-67 (IEEE 2015)
26. Regulation of the Minister of the Environment of 24 August 2012 on the levels of certain substances in the air, http://isap.sejm.gov.pl
27. MAŁOPOLSKI Region. The air quality monitoring. The measurement data. http://monitoring.krakow.pios.gov.pl/dane-pomiarowe/automatyczne/stacja/149/parametry/1725-1723-1724-1726/dzienny/18.03.2016
28. Statista. Average speed in Europe's 15 most congested cities, http://www.statista.com/statistics/264703/average-speed-in-europes-15-most-congested-cities/

 DOI: 10.1051/itmconf/20171501002

Decomposition of business process models into reusable sub-diagrams

Piotr Wiśniewski[1,*]

[1]AGH University of Science and Technology, Department of Applied Computer Science, al. Mickiewicza 30, 30-059 Krakow, Poland

Abstract. In this paper, an approach to automatic decomposition of business process models is proposed. According to our method, an existing BPMN diagram is disassembled into reusable parts containing the desired number of elements. Such elements and structure can work as design patterns and be validated by a user in terms of correctness. In the next step, these component models are categorised considering their parameters such as resources used, as well as input and output data. The classified components may be considered a repository of reusable parts, that can be further applied in the design of new models. The proposed technique may play a significant role in facilitating the business process redesign procedure, which is of a great importance regarding engineering and industrial applications.

1 Introduction

Business Process Management stands for a set of techniques which aim to design, analyse, implement and improve organisational processes. Process models can help organisations in visualising processes of the company and give opportunity for optimisation of the process structure. Owing to this, the company may achieve business goals in a more efficient way.

BPMN (Business Process Model and Notation) [1], maintained by the OMG group, is the most widely adopted notation for modelling business processes. BPMN 2.0 is quite complex, as there are many additional documents explaining the notation, such as handbooks and papers devoted exclusively to the pragmatics of the notation usage. In the paper, a brief overview of the guidelines for process modelling as well as various strategies of process design were provided.

According to the research based on industrial applications [2], most of the manually created process models are badly affected by quality issues. Therefore, an automated support in the phase of process design is highly appreciated. The proposed method may be seen as a business process modelling technique based on sub-diagrams generated as a result of a decomposition of existing models. The workflow specification begins with process elements and sub-processes of which the reusable sub-diagrams are created. These component parts are then combined to form a comprehensive process. An overview of the decomposition method is presented in Figure 1.

This paper is organised as follows. In Section 2, an overview of the existing solutions in the area of process model design, as well as their applications, is presented. Section 3 describes in detail the idea of business process modelling and their mathematical representation. The proposed decomposition algorithm is presented in Section 4. Section 5 shows the idea of a component process repository, while in Section 6 the main guidelines of sub-diagrams merge were described.

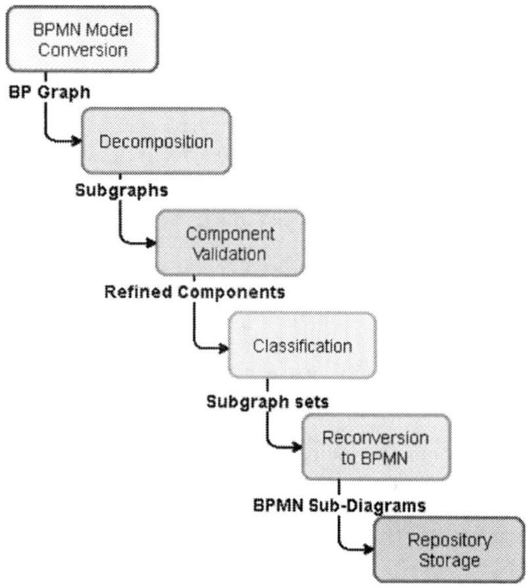

Fig. 1. Overview of the decomposition method.

2 Related Works

Decomposition of business process models is commonly used in the literature to redesign processes or create new models from existing components. One of the approaches consists in identifying maximal repeating fragments, called exact clones, which appear in multiple

* Corresponding author: wpiotr@agh.edu.pl

process models [3]. This is conducted by identifying SESE (Single Entry, Single Exit) structures in a process tree. Milani et al. [4] discuss criteria and heuristics for model decomposition. They also provide an evaluation regarding the influence of these criteria on the understandability of selected models. The decomposition procedure may be supported by detecting names of non-uniformly specified process element in order to enable a faster creation of consistent complex models [5]. Another solution to the discussed problem was presented in [6] where business processes were presented as a hierarchical ontology of business tasks apt for re-composition. Such a description may simplify the dynamic creation of models of collaborative business processes.

This paper is also related to structured and formal process representations which facilitate process modelling, as well as their reconstruction. One of the user-friendly approaches is the spreadsheet-based representation [7] where the set of activities is listed in a table with specified branches and routing conditions. Process models apt for re-composition can be also described in a declarative specification [8], which focuses on constraints and dependencies between activities instead of determining an explicit sequence flow. They can be also represented as a randomly-ordered list of tasks, along with their input and output conditions, which are then used for a constraint-based process composition [9]. Another approach is a semantic model for BPMN specified in the CSP process algebra [10]. This is especially useful for specifying time constraints between process activities.

Process model repositories are an important tool in the area of business process redesign. They can be developed in a form of a database storing relevant process fragments with a reference to the process metadata [11]. This approach leads to an automated generation of complex models. In case of such a composition, the correctness of the generated process model must be ensured. Therefore, it is necessary to mention the identification and evaluation of issues regarding the flow consistency [12] as well as the layout of the generated business process model [13]. The automated process modelling should also include verification whether the created model contains any typical anomalies such as deadlocks, livelocks or the lack of synchronisation [14] [15] or event anomalies [16]. The analysis of the formal aspects of a BPMN model can be also conducted using Alvis Modelling Language [17].

3 Business Process Models

Highly developed software systems are present in many areas of industrial and business applications, which range from human-machine interface systems, through production management, to web services covering a large domain of implementations. To ensure the reliability and efficiency of such systems and simplify their design and development phases, advanced process engineering approaches are in use.

The engineering process usually starts with business process modelling which constitutes a set of techniques used to represent a certain workflow by providing visual diagrams depicting a business process with its dependencies. Thus, a business process model presents the way operations are performed to accomplish the planned objective of an organisation.

An example field of application for business process models is the specification of Manufacturing Execution Systems (MES) [18]. Building a complete modelling framework for the integration of a system technical and functional model, as well as production process model ensures the understandability of the specification for all the participants involved in the implementation process.

3.1 Business Process Model and Notation

Business process models in BPMN notation are represented by diagrams composed of a limited set of graphical symbols [19]. Visualisation of activity flows allows business users and developers to understand processes more clearly. It is a transparent visual tool dedicated to modelling complex processes.

BPMN model elements can be divided into four major groups: flow objects, connecting objects, swim lanes and artefacts. Flow objects represent the principal elements of the process diagram. In this set, three major object types can be distinguished [1]:
- activities (rounded rectangles) which represent a work that must be performed in the process,
- events (circles) which represent incidents occurring at the time of process execution,
- gateways (rhombi) which control the flow of tokens in the process.

There are two kinds of activities: tasks and sub-processes. A task is an atomic action that can be performed by a person or executed as an automated activity. On the other hand, a sub-process is a compound activity representing a complex work, divisible into smaller parts which are defined as a separate process at the lower level.

The set of events covers three types of elements: start events which indicate the beginning of a process with optional triggering conditions, end events that determine the end of a path in a process or subprocess, as well as intermediate events which may happen at any time within a process.

Gateways control the process flow by splitting it into branches and joining them afterwards. The most typical BPMN gateways are data-based, which means that they evaluate an expression of process data, while they cannot decide regarding the flow of the process [20]. Gateways can be exclusive (XOR) which enable only one outgoing sequence flow based on given process data, inclusive (OR) that precede potentially parallel branches, as well as parallel (AND) used to represent concurrent flows executed without any conditions.

A set of the most typically used BPMN flow objects is presented in Figure 2.

DOI: 10.1051/itmconf/20171501002

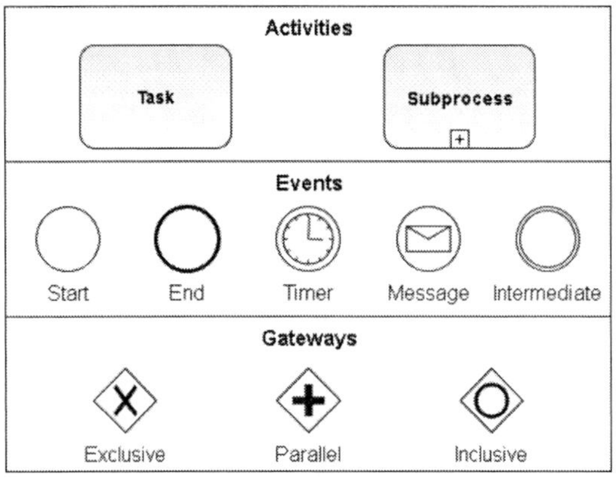

Fig. 2. Set of common BPMN flow objects. Based on [20].

All the flow objects need to be connected using so-called connecting objects of two major types:
- sequence flow which shows the execution order of a process,
- message flow which represents the exchange of messages between process participants.

To distinguish business functions or parts of the system related to certain flow objects, the concept of swimlanes is usually used in BPMN. Business process participants, which can either be entities within an organisation or different collaborators in a process [21] are represented by pools. Sub-partitions of a pool, which represent specific objects or roles, are called lanes. They organise and categorise flow objects, connecting objects and artefacts in a pool. Lanes are present in BPMN diagrams in a form of a rectangle extended either vertically or horizontally along the length of a pool.

Artefacts show additional pieces of information which enable process designers to include more details in a model. Three main types of artefacts were defined in the standard [1]: data objects, groups and annotations.

3.2 Running Example

To illustrate the proposed method, a sample supply process was selected. Its BPMN model is presented in Figure 3. The process goal is to purchase goods in case if the available inventory is below the desired value.

After the start, a warehouse employee checks the inventory: if the available quantity is above the minimal value, the process is ended; otherwise a requester creates a purchase order which is then reviewed by a supervisor. If the created order does not fulfil the requirements, it should be reprocessed by the requester, otherwise the funds are reserved by the accounting department and the supervisor sends the order to the supplier. If the product is not available at the supplier's stock, a negative response is sent to the supervisor and the purchase order is cancelled. Otherwise, the customer receives an invoice and a packing slip, which need to be recorded before releasing the funds and issuing the payment. As soon as the payment is sent to the supplier, the process is completed.

3.3 Mathematical Model of a BPMN diagram

To use the proposed process decomposition method, it is necessary to transform a BPMN model into a form of a graph. Therefore, a formal model of the process should be determined before the conversion. A simple BPMN model can be mathematically represented as a tuple $\mathbb{P} = (\mathbb{O}, \mathbb{F})$ [23] where:
- $\mathbb{O}$ is the set of flow objects, such as: activities (including task and sub-processes), events and gateways,
- $\mathbb{F} \subset \mathbb{O} \times \mathbb{O}$ is the set of sequence flows.

For the decomposition method, a task with boundary events is represented in the formal model as a single flow object having multiple outputs. In addition, the proposed approach includes neither message flows nor artefacts, as the resulting sub-diagrams are generated only for one determined process participant and additional information may differ depending on a specific process model.

3.4 Conversion to a Business Process Graph

Directed graphs are one of the common representation for business process models [24]. Since the proposed decomposition approach considers only the layout of the process model and not its execution, it was possible to generalise the existing model of atomic and complex process graphs. In the proposed structure called business process graph, the existence of loops as well as multiple start and end events is allowed. Therefore, there are no restrictions regarding the cyclicity of the graph and degrees of its vertices. In this case, the resulting graph model may serve as a precise copy of a BPMN diagram.

Based on the formal model presented in Section 3.3, a business process graph can be defined as a connected, directed graph $G_P = (V_O, E_F)$ where:
- V_O is a non-empty set of vertices representing all flow objects in a process,
- E_F is a non-empty set of edges representing all sequence flows in a process.

Transforming the BPMN model into a business process graph consists in listing all process flow objects along with the following properties:
- Object ID and Name (optional),
- Pool/Lane,
- Object Type (task, XOR split gateway, etc.),
- Condition (optional, used for OR and XOR gateways),
- Input/Output Data Entities (optional).

All the objects represented by vertices should be then connected with directed edges which correspond to the sequence flow of the process. A directed edge $e \in E_F$ is a pair:

$$(v_1, v_2) \in V_O \quad (1)$$

where v_1 represents its starting point and v_2 its endpoint. In the business process graph edges contain also optional information about a condition value, provided in case of alternative flows starting with XOR and OR gateways.

DOI: 10.1051/itmconf/20171501002

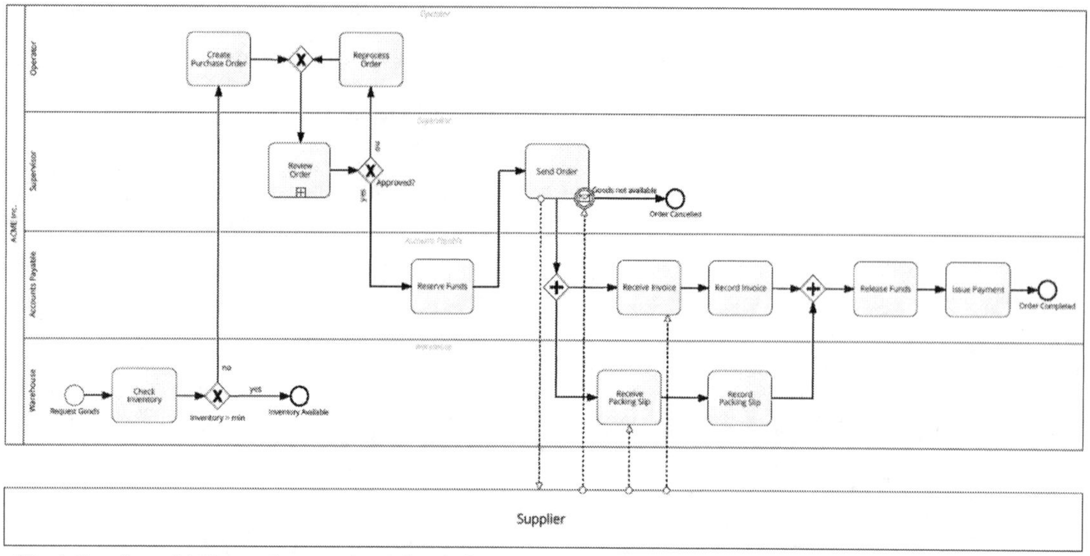

Fig. 3. Sample model of a supply process. Based on [22].

In the next step, it is necessary to define adjacency matrix D [25] of the generated business process graph. Assuming that the analysed process consists of n flow objects, D is a square matrix of size n x n, defined by Formula (2):

$$D = (d_{ij}) \tag{2}$$

where rows and column indexes $i, j = 1..n$ correspond to the IDs of process objects. Each element of matrix D is assigned to a value according to Formula (3):

$$\forall\, v_i, v_j \in V_O : d_{ij} = \begin{cases} 1, & if\ (v_i, v_j) \in E_F \\ 0, & otherwise \end{cases} \tag{3}$$

Such a matrix stores all the information regarding flow connections between objects in the analysed process.

4 Model Decomposition

The process model decomposition algorithm consists in finding the set S_k of all the subgraphs induced by k vertices, where $2 \le k < n$. In the first step, it is necessary to define appropriate subsets of k vertices. Then, an adjacency matrix of each subgraph is created by deleting all rows and columns from matrix D which do not correspond to vertices for the selected subset. The following subsections present methods for generating sub-diagrams having two and three elements, as well as the concept of a general solution.

4.1 Bi-gram Decomposition

In the most basic example, the process graph is decomposed into all the possible pairs of vertices connected by an edge.

$$S_2 = \left\{ \left(\{v_i, v_j\}, (v_i, v_j) \right) : d_{ij} = 1 \right\} \tag{4}$$

4.2 Tri-gram Decomposition

A more complex case occurs when there is a need to generate sub-diagrams consisting of three process flow objects which can be connected to multiple elements. In such a situation, a component trigram for an object v can be in one of the following forms:

1. v and its two direct predecessors,
2. v with one predecessor and one successor,
3. v and its two direct successors.

Providing the process graph $G_P = (V_O, E_F)$, having n vertices, as well as its incidence matrix D, the decomposition algorithm consists in determining a set S_3 by performing the following operation for each vertex v of graph G_P:

1. Get the sets of all direct predecessors and successors of v, denoted as $Pred(v)$ and $Succ(v)$, respectively.
2. Calculate cardinality of the determined sets, setting n_p as the number of predecessors and n_s as the number of successors.
3. If $n_p > 0$ and $n_s > 0$ then add to set S_3 all subgraphs induced by a set of vertices:

$$\{v_p, v, v_s\} \tag{5}$$

where $v_p \in Pred(v)$ and $v_s \in Succ(v)$.

4. If $n_p \ge 2$ then add to S_3 all subgraphs induced by a set of vertices:

$$\{v_{p1}, v_{p2}, v\} \tag{6}$$

where $v_{p1}, v_{p2} \in Pred(v)$ and $v_{p1} \ne v_{p2}$.

5. If $n_s \ge 2$ then add to S_3 all subgraphs induced by a set of vertices:

$$\{v, v_{s1}, v_{s2}\} \tag{7}$$

where $v_{s1}, v_{s2} \in Succ(v)$ and $v_{s1} \ne v_{s2}$.

In the analysed example presented in Figure 3, 21 flow objects were identified. After running the decomposition algorithm for tri-grams 29 different subgraphs were generated. An example subgraph is shown in Figure 4.

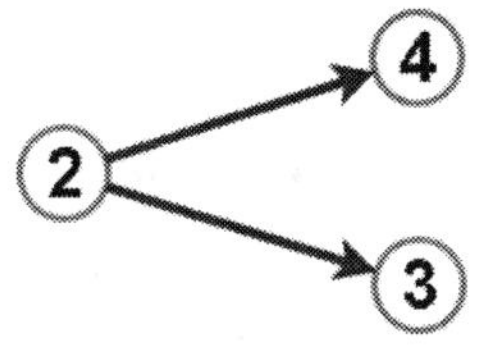

Fig. 4. Three-element subgraph generated as a result of process graph decomposition.

Properties of the flow objects corresponding to vertices of the generated subgraph are listed in Table 1 while Table 2 presents edges included in the subgraph.

Table 1. Properties of flow objects.

Object ID	2	3	4
Pool/Lane	Warehouse		Operator
Object Type	XOR split	end event	task
Object Name	--	Inventory Available	Create PO
Condition	Inventory > min	--	--

Table 2. Properties of sequence flows.

Source Object	2	2
Dest. Object	3	4
Condition Value	yes	no

4.3 K-gram Decomposition

The general problem can be solved by determining for a given number $k < n$, all k-element subsets of V_O and generating connected subgraphs of G_P induced by these subsets. A related solution was presented in [26], however in this paper, a use of constraint programming technique is proposed.

Following input data should be prepared to solve a constraint satisfaction problem:
- decision variable: k-element vector s_k representing a subset of process graph vertices,
- constraints: all elements of s_k are different and connected,
- domain: adjacency matrix of graph G_P.

4.4 Resulting Sub-Diagrams

The generated subgraphs can be retransformed into BPMN models by executing the inversed method presented in Section 3.4 – vertices of each subgraph represent flow objects and non-zero values of matrix D correspond to the sequence flow of the process.

Each of the sub-diagrams can have a different number of inputs and outputs, according to the flow objects from which it is composed. In general, flow objects can be divided into five main groups:
1. Sources: elements that can contain only output sequence flow, such as start events.
2. Sinks: elements that can contain only outputs, such as end events.
3. SESE (Single Entry Single Exit): tasks.
4. SEME (Single Entry Multiple Exit): split gateways, tasks with boundary events, subprocesses.
5. MESE (Multiple Entry Single Exit): merge gateways.

Regarding a best practice of BPMN process modelling [2], implicit splits and joins should be avoided and replaced by gateways. Therefore, if the decomposed model is constructed correctly, all its tasks are SESE elements, except those containing boundary events, as described in Section 3.3. Sub-Diagrams that contain at least one SEME and one MESE object represent the MEME type (Multiple Entry Multiple Exit).

Based on the defined groups, it is possible to determine the number of potential inputs and outputs in each of the generated sub-diagrams, by counting the number of ingoing and outgoing edges of each process flow object and comparing it to its group. This number is used to define functions $\sigma_{I/O}$, described by Formula 8:

$$\sigma_I = \begin{cases} 0, & \text{if there are no available entries} \\ 1, & \text{if there is one available entry} \\ 2, & \text{otherwise} \end{cases} \quad (8)$$

Respectively, function σ_O is defined for exits. Table 3 presents to which group a sub-diagram belongs depending on values of functions σ_I and σ_O.

Figures 5-7 present three sub-diagrams of different groups generated based on the example BPMN model presented in Figure 3:
1. SESE (Figure 5) – one available entry for "Reserve Funds" task and one for "Send Order" task.
2. SEME (Figure 6) – one available entry for split AND gateway and two available exits, one per each task.
3. MEME (Figure 7) – one available entry for "Reprocess Order" task, multiple available entries for join XOR gateway and multiple available exits for "Review Order" subprocess.

Table 3. Sub-Diagram groups depending on functions $\sigma_{I/O}$.

σ_I \ σ_O	**0**	**1**	**2**
0	subprocess	source	source
1	sink	SESE	SEME
2	sink	MESE	MEME

 DOI: 10.1051/itmconf/20171501002

Fig. 5. A SESE sub-diagram.

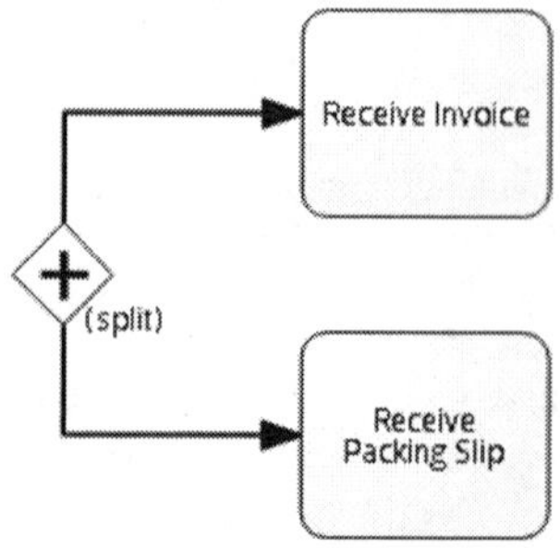

Fig. 6. A SEME sub-diagram.

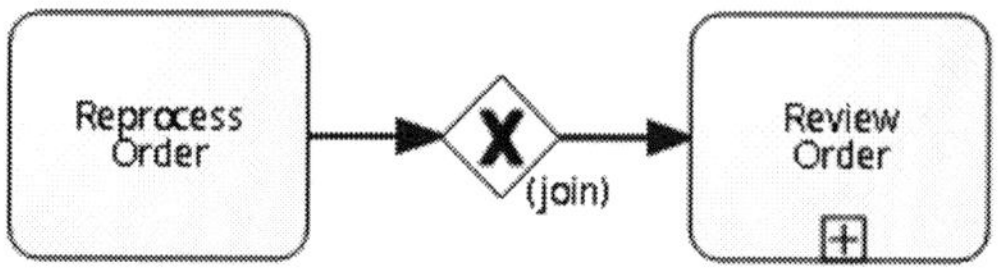

Fig. 7. A MEME sub-diagram.

5 Towards a Sub-Diagram Repository

The decomposition algorithm presented in Section 4 generates all the possible sub-diagrams of k elements. This solution ensures that every possible configuration of process elements is ready to be reused in new process models. However, some of the generated solutions may potentially cause process inconsistencies, as they do not contain enough information about the process flow.

Let us analyse one of the tri-grams generated based on the example software testing process model which is presented in Figure 8. In this case, the resulting process fragment will not be fully reusable because of the AND gateway without any output element assigned. Taking into consideration the fact, that the process flow description contains information about the gateway type, as stated in Section 3.4, such a tri-gram configuration prevents from executing only one activity after the "Sign Agreement" task which limits the component utility. At the same time, the number of branches after the AND gateway is not specified which makes the sub-diagram underdetermined.

Fig. 8. Tri-gram generated as a result of process model decomposition.

To ensure the correct repository creation without omitting any relevant process sub-diagrams, a user-based component validation was proposed. The operator who performs the decomposition is presented a set of potentially inconsistent component models. The selected sub-diagrams need to satisfy at least one of the following conditions:

1. Presence of a split gateway without any output branch.
2. Presence of a join gateway without any input branch.
3. More than two swimlanes considered.

In the next step, the component processes are classified considering two criteria: the potential number of inputs and outputs, as well as the diagram similarity. The former is used for grouping models according to the values of functions σ_I and σ_O defined in Section 4.4. The latter is performed by applying the vector model technique for business processes [27]. In this case, it is necessary to define three input entities for the algorithm:

- document collection: two sub-diagram graphs $G_1, G_2 \in S_k$,
- set of terms Θ: the union of all graph vertices with their properties listed in Section 3.4,
- weights: for each term $\theta \in \Theta$ a value of 1 is assigned if the corresponding vertex is a part of the selected subgraph and 0 otherwise.

Considering these two criteria, selecting the appropriate sub-diagram consists either in choosing its group, as shown in Table 3 or specific flow objects that should belong to the component. In the second case, the user is shown a set of similar sub-diagrams, out of which a final selection is needed.

6 Merge Guidelines Outline

Composition of the BPMN model using the repository of reusable parts should always be conducted with respect to business process modelling guidelines to avoid typical anomalies [28]. Incoherent business process models may be affected by syntactical anomalies which include incorrect usage of activities, gateways, connecting objects or swimlanes. Another group are structural anomalies, related to the improper dynamic behaviour of the process, for example the presence of deadlocks and infinite loops.

Considering the correctness of the composed process model, it is recommended that its generation is supported by a graphical tool connected to the repository. The interface should suggest recommended connection using the position-based classification technique [29]. An example of a forward model completion is presented in Figure 9.

On the other hand, the system should also prevent the user from creating links between flow objects which may lead to structural anomalies. A potential deadlock is shown in Figure 10.

7 Conclusions

The approach presented in this paper offers a set of activities which support creation of reusable component business process models. The proposed method uses

 DOI: 10.1051/itmconf/20171501002

complete BPMN diagrams as input, based on which the component models are generated.

Such a decomposition algorithm may be used to create a component model repository which facilitates process redesign as well as provides a possibility to avoid common modelling anomalies.

In the future work it is planned to include the message flow in the decomposition which will increase the support of the method for various process elements. Another valuable extension would be to allow generation of sub-diagrams representing a specific set of activities, based on their semantic description.

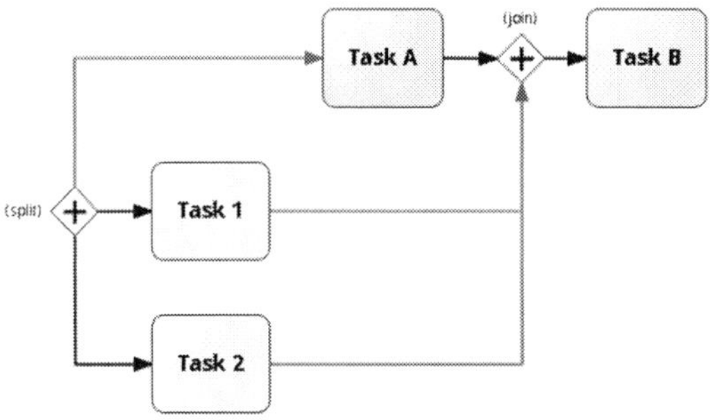

Fig. 9. BP modelling recommendation based on forward model completion technique. Recommended links were marked with green colour.

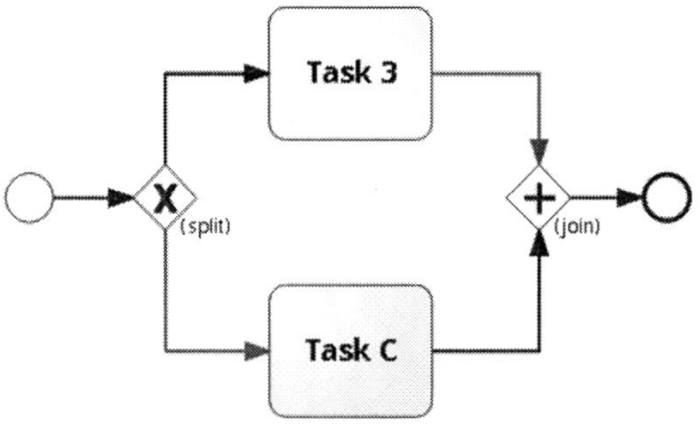

Fig. 10. Forbidden connections between sub-diagrams.

References

1. OMG, *Business Process Model and Notation (BPMN): Version 2.0 specification. Technical Report* (Object Management Group, 2011)
2. H. Leopold, J. Mendling, O. Günther, IEEE Software **33**.4, p. 26-33 (2016)
3. M. Dumas, L. García-Bañuelos, M. La Rosa, R. Uba, Information Systems **38**.4, p. 619-633 (2013)
4. F. Milani et al., Business & Information Systems Engineering **58**.1, p.7-17 (Springer, 2016)
5. A. Koschmider, E. Blanchard, User assistance for business process model decomposition, In: *RCIS*, p.445-454 (IEEE, 2007)
6. R. KL Ko, E. W. Lee, S. G. Lee., IEEE Transactions on Services Computing **5**.2, p. 246-259 (2012)
7. K. Kluza, P. Wiśniewski, Spreadsheet-based business process modelling. In: *2016 Federated Conference on Computer Science and Information Systems* p. 1355-1358 (IEEE, 2016)
8. F. M. Maggi, A. J. Mooij, W. MP van der Aalst, "User-guided discovery of declarative process models." In: *2011 IEEE Symposium on Computational Intelligence and Data Mining*, p.192-199 (2011)
9. P. Wiśniewski, K. Kluza, M. Ślażyński, A. Ligęza, Constraint-Based Composition of Business Process Models. In: BPAI Workshop held in International Conference on Business Process Management (accepted)
10. P. YH Wong, J. Gibbons, Science of Computer Programming **76**.8, p. 633-650 (2011)
11. M. Skouradaki, V. Andrikopoulos, F. Leymann, Representative BPMN 2.0 process model generation from recurring structures, In: *IEEE International Conference on Web Services*, p. 468-475 (IEEE, 2016)
12. A. Burattin, et al., Software & Systems Modelling, p. 1-22 (Springer, 2017)
13. V. Bernstein, P. Soffer, Identifying and quantifying visual layout features of business process models, In: *International Conference on Enterprise, Business-Process and Information Systems Modelling*, p. 200-213 (Springer, 2015)
14. A. Ligęza, T. Potempa, Artificial Intelligence for Knowledge Management with BPMN and Rules, In: *IFIP International Workshop on Artificial Intelligence for Knowledge Management,* p.19-37 (Springer, 2012)
15. A. Suchenia et al., Selected Approaches Towards Taxonomy of Business Process Anomalies. In: *Advances in Business ICT: New Ideas from Ongoing Research*. p. 65-85 (Springer, 2017)
16. A. Suchenia, A. Ligęza, International Journal of Computer Technology & Applications **6**.5, p. 789-797 (2015)
17. M. Szpyrka, G. J. Nalepa, A. Ligeza, K. Kluza, Proposal of Formal Verification of Selected BPMN Models with Alvis Modelling Language. In: *IDC*, p. 249-255 (2011)
18. M. Ricken, B. Vogel-Heuser, Modelling of manufacturing execution systems: An interdisciplinary challenge, *2010 IEEE Conference on Emerging Technologies and Factory Automation (ETFA)*, p.1-8 (IEEE, 2010)
19. A. Ligęza, Pomiary Automatyka Robotyka **15**.12, p. 210-212 (2011)
20. B. Silver, B. Richard, *BPMN method and style.* Vol. 2. (Cody-Cassidy Press, 2009)
21. A. Lodhi, V. Küppen, G. Saake, Scientific Journal of Riga Technical University. Computer Sciences **43**.1, p. 27-34 (2011)
22. BPM Initiative Model Collection, http://bpmai.org/

23. A. Ligęza, Decision Making in Manufacturing and Services **5**.1-2, p. 57-67 (2011)
24. I. M. Weber, *Semantic Methods for Execution-level Business Process Modelling: Modelling Support Through Process Verification and Service Composition*. Vol. 40 (Springer, 2009)
25. K. Ruohonen, *Graph theory*, Graafiteoria lecture notes, (TUT, 2013)
26. P. Vasiliu, P. Tiberiu, "Mircea cel Batran" Naval Academy Scientific Bulletin, **18**.2, p. 419 (2015)
27. B. van Dongen, R. Dijkman, J. Mendling, *Measuring similarity between business process models*, In: Seminal Contributions to Information Systems Engineering, p. 405-419, (Springer, 2013)
28. A. Suchenia, P. Wiśniewski, A. Ligęza, Overview of Verification Tools for Business Process Models. In: *2017 Federated Conference on Computer Science and Information Systems*, p. 303-310 (PTI, 2017)
29. K. Kluza, M. Baran, S. Bobek, G. J. Nalepa, Overview of Recommendation Techniques in Business Process Modelling, In: *Knowledge Engineering and Software Engineering* (Würzburg University, 2013)

 DOI: 10.1051/itmconf/20171501003

Identification, characterisation, and correction of artefacts in electroencephalographic data in study of stationary and mobile electroencephalograph

Monika Kaczorowska[1,*]

[1]Lublin University of Technology, Electrical Engineering and Computer Science Faculty, Institute of Computer Science, Nadbystrzycka 36B, 20-618 Lublin, Poland

Abstract. The present paper covers the review of artefacts of EEG recording, as well as methods to eliminate them, i.e. PCA and ICA. The reviewed artefacts were of various origins: biological (muscle, eye blinking, eyeball moving), as well as technological artefacts caused mainly by the power network with the frequency of 50 Hz. The research has been conducted on a group consisting of 8 students. The research was carried out with the use of two defined models of electroencephalographs: stationary and mobile, then results have been compared. The comparison of two amplifiers combined with various artefact elimination methods has been presented. Presented research has shown that the record performed by the mobile amplifier contains the artefacts of significantly lower amplitude.

1 Introduction

Electroencephalography (EEG) is a technique applied for registering and monitoring an electrical activity of a brain. It is a non-invasive method of measurement of specific waveforms emitted by a human brain. The special type of amplifier is dedicated to record EEG signals with a small amplitude. The signal is gathered from the human scalp by electrodes which are placed on a special cap [1]. The EEG signals are applied in medicine and in the construction of brain-computer interfaces (BCI). They may be applied in rehabilitation as well [2].

EEG signal is a resultant of all the potential electromagnetic fields that are present nearby the examined person. Most often it is disturbed by many undesirable signals called artefacts [1, 3]. They can originate from the bioelectrical activity of an examined person, as well as from the external environment [1]. The aim of the paper is to characterise and compare the process of artefact correction in EEG signals using two defined models of amplifiers, correspondingly: stationary and mobile. The comparison was made by using dedicated software (WinEEG). The task of artefact elimination is still a challenging one. This paper presents two methods applied to remove artefacts: PCA (Principal Component Analysis) and ICA (Independent Components Analysis).

The signals have been gathered from the representatives of two types of amplifiers: stationary and mobile. The mobile electroencephalograph is a comparatively new type of electroencephalograph convenient for implementing BCIs [4, 5]. Mobile EEG does not hamper the examined person due to the absence of cables connecting it with a computer. It also meets the needs of creating BCIs not only because of its mobility but also due to the fact, that this type of EEG is equipped with several auxiliary devices, such as accelerometer, which helps to correct the artefacts from lower levels of recording [6]. The goal of the research was to compare an old-type stationary electroencephalograph with a modern type mobile EEG, in terms of noise level and artefacts.

2 Related works

2.1 Artefacts

Most artefacts in an EEG recording, such as eye-blinking and eye-movement artefacts, can be seen as a sudden peak. The other type of artefacts, such as muscle ones (especially jaw muscle-related artefacts) may be observed as a local increase of the high-frequency component. What is more, the technical artefacts, related to the power network with the frequency of 50 Hz, usually appear as a permanent presence of the 50 Hz frequency component in the EEG recording. The artefacts usually have a higher amplitude than the amplitude of the brain potentials. Most often artefacts can be registered by many or even all the electrodes attached to the scalp of the examined person. Along with that, the EEG components occurring in a typical EEG recording are usually quite local [6, 7]. Artefact correction is an essential part of pre-processing data to obtain a clean EEG recording [1].

Artefacts may be divided into two groups. The first group includes technical artefacts related to signal

* Corresponding author: m.kaczorowska@pollub.pl

registration. This group includes: inappropriate electrode attachment localisation, insufficient contact of an electrode with the skin, influence of the power network's electromagnetic field, electromagnetic field of outer electronic devices, unsuitable conditions of an examination, e.g. the result of the EEG examination can be affected by walking near an examined person [8].

The second one includes the biological artefacts concerning the motions of the examined person's body. The signal picked up by the electrodes is amplified by a specialised EEG amplifier. This amplifier is a source of noise. A very important parameter of the amplifier is the value of the amplitude of the output noise related to its input [8]. This group includes participants sweating, blinking, head muscle artefacts, muscle shivering and heart's electrical activity. The most popular biological artefacts which can be observed in most EEG recordings are eye-blinking artefacts, eye-movement artefacts and muscle activity artefacts. The disturbances may be observed in one or many electrodes, in different proportions [8, 9].

Artefacts may also appear because of muscle tension caused by movement. Movement creates an electrical potential (EMG – electromyography), which is many times higher than the level of the EEG signal. This kind of artefacts is caused by movements of the limbs, neck muscles, swallowing, tongue movement, or clamping of the jaw. Blinking of the examined person may cause a potential up to an amplitude of 100 μV [8, 10].

2.2 Parameters for artefact detection

There are several possibilities to classify channels as artefactual. One of the parameter is the mean of the channel correlation coefficients with the other channels. The channel with technical artefacts related to high impedance has a low correlation with other channels. The other parameter, which describes appearance of artefacts, is the variance of the channel. Despite of quite correlation, there is a chance that due to the artefacts, the variance is high. The next parameter is the Hurst exponent. The Hurst exponent is a measure of long-range dependence within a signal. To detect artefacts, it is possible to compute the amplitude range of the epoch. The deviation from each channel's average value is the following parameter. In case if the deviation's value is large, the probability of artefact appearance is high. It is important to choose the appropriate type of measure to reveal the artefacts in the signal [11].

2.3 Methods of artefact correction

There are several methods of artefacts reduction. However, they can never be eliminated completely. The first way is to reduce the movements and muscle tension of the examined person [3].

In the past, the fragment of the signal which contained artefacts was rejected [3]. Nowadays, there are several methods to detect artefacts automatically. One of them is the method called "frequency method". The main idea of artefact identification is to determine what frequencies dominate in the recorded signal. If the signal frequencies deviate significantly from the frequency of the EEG signal, the classifier selects this fragment as the artefact [6, 12, 13].

The second type of methods of artefact correction are methods based on subtracting the interfering signal from the EEG signal. These methods may be applied in eye movement artefacts. Besides EEG recordings gathered from the scalp, electrooculography (EOG) electrodes are applied to record the electrooculography signal. There is a possibility to subtract the signal including the artefacts originating from the electrooculography activity. It may reduce artefacts in a significant way [9, 14, 15].

The most popular methods of removing artefacts are PCA and ICA [6,16]. This paper contains a comparison between removing artefacts from the signal gathered from a stationary and a mobile amplifier.

3 Applied methods

3.1 Principal Component Analysis

Two methods of correcting artefacts are reviewed in the paper. The first method is PCA, i.e. Principal Component Analysis. It is one of the Factor Analysis Methods. It is typically used to reduce the number of dimensions, eliminating low variety dimensions as they do not contain significant information [16].

The idea of the method may be described in the following way: the signal collected from a certain number of channels n within the time interval of k periods may be understood as a cloud of k points in the n-dimensional space or as a matrix with n rows and k columns (1).

$$X = \begin{bmatrix} x_{11} & x_{12} & \cdots & x_{1k} \\ x_{21} & x_{22} & \cdots & x_{2k} \\ \vdots & \vdots & \ddots & \vdots \\ x_{n1} & x_{n2} & \cdots & x_{nk} \end{bmatrix} \quad (1)$$

Clearly, the basis of PCA is matrix calculus. In accordance with it, the following steps should be taken, to obtain a principal components matrix:

1. the mean values for the columns should be calculated;

$$X_{mean} = \frac{1}{n}\left[\sum_{i=1}^{n} x_{i1} \quad \sum_{i=1}^{n} x_{i2} \quad \cdots \quad \sum_{i=1}^{n} x_{ik}\right] \quad (2)$$

2. the deviation matrix should be computed in accordance with (3):

$$\Delta X = X - \begin{bmatrix} 1 \\ 1 \\ \vdots \\ 1 \end{bmatrix}_n \cdot X_{mean} \quad (3);$$

3. based on the Singular Value Decomposition (SVD) method three matrixes: U, Σ and V^* should be calculated, where:

- U is the left singular matrix with the size of nxn, containing the eigenvectors of the matrix product $\Delta X \Delta X^T$;

- V^* is the conjugate transpose of the matrix containing the eigenvectors of the matrix product $\Delta X^T \Delta X$, its size is kxk;

- Σ is the diagonal matrix containing the singular values of the deviation matrix.

Singular values are arranged in the descending order and represent principal components of the signal. The first principal component corresponds to the most common information contained in the signal, i.e. the part that is present in all the channels. The successive principal components correspond to the progressively more individual details of the signal [17].

As mentioned above, artefacts have higher amplitude compared to EEG signal and therefore, affect all or most of the channels, while brain signals are usually local and can be detected by one electrode or small group of electrodes. Hence, the highest principal component corresponds to the artefacts contained in the signal. That is why one or two first components are usually eliminated i.e. set to zero. The corrected signal is obtained by adding means to the matrix product $U\Sigma' V^*$, where Σ' is the principal component matrix with one or several principal components set to zero [17].

3.2 Independent Component Analysis

The other method that is considered in the paper is ICA, i.e. Independent Component Analysis. The main idea of the method is to divide the signal into a series of components, statistically independent of each other, and having frequency distribution maximally different from normal distribution [16]. Kurtosis is commonly used as the measure of a distribution's normality. Kurtosis of a signal **x** may be defined as:

$$Kurt(\mathbf{x}) = E(\mathbf{x}^4) - 3(E(\mathbf{x}^2))^2 \quad (4)$$

Kurtosis may be described as a measure of probability density flatness. The value of kurtosis of a normal distribution equals zero, while for non-Gaussian distribution it may possess the following values:

1. kurtosis is positive, the probability density is more concentrated than in normal distribution,
2. kurtosis is negative, the probability density is less concentrated than in normal distribution [18].

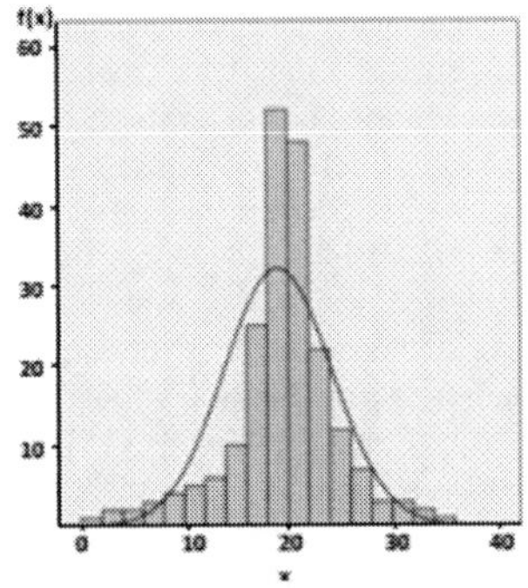

Fig 1a. Close to normal probability distribution [19].

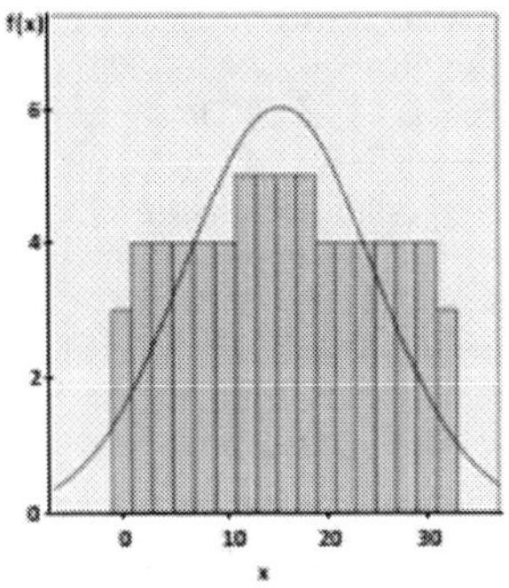

Fig 1b. Probability distribution differing significantly from normal [19].

Figure 1 presents a comparison of the probability density of two variables. Finding an independent component with a maximally non-Gaussian probability density, distribution is based on the maximisation or minimisation of the kurtosis value [19].

ICA is used in many fields, such as face and voice recognition, communication, and even stock price research [20]. Artefacts correction may be also performed by means of ICA. The main feature of the method is that the number of the components, that the signal can be divided into, is not greater than the channel number [16, 18].

3.3 Frequency based method

Another method used, was the frequency based method. The idea of this method is based on the use of various filters, such as lowpass, high-pass and band-stop (notch). The Fig.2 presents the amplitude response of band-stop filter. In the current application, the value of frequency requiring attenuation was equal to 50 Hz.

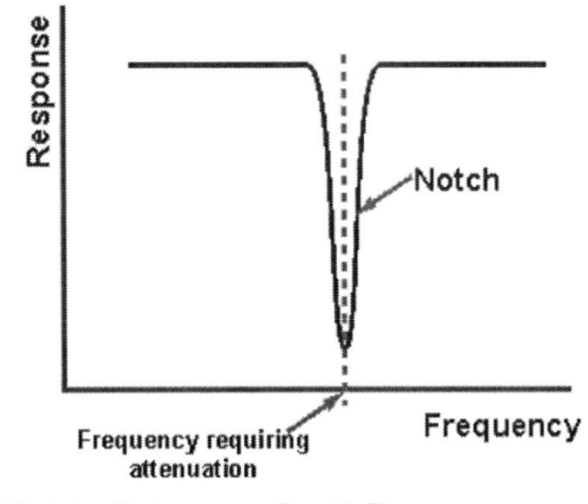

Fig 2. Amplitude response of a notch filter.

4 The case study

Eight participants took part in an experiment, in a quiet room of a dedicated laboratory. The room was illuminated with standard fluorescent light and outside light was blocked. The participants were of similar age. During the test, the person examined was left alone.

Two 21-channel amplifiers were applied during the experiments. The first of them was the stationary – Mitsar EEG 201 amplifier, and the other – mobile EEG Smart BCI. The EEG amplifiers transmitted data to a computer with native software (WinEEG). The stationary amplifier used USB protocol to communicate with computer, and the mobile amplifier used Bluetooth. Every participant was examined with stationary amplifier first, and then with the mobile one.

The samples were recorded at a frequency of 500 Hz. The signals were gathered from 19 electrodes placed in a special EEG cap, which kept them in the right position. The cap was compatible with international 10-20 EEG standard and located over an examined person's scalp.

The ground electrode was placed in the centre of the frontal lobe and two reference electrodes (A1, A2) were placed on the ears.

The participants were asked to perform the following activities:
- sitting with eyes opened and closed,
- blinking the eyes,
- moving the jaw,
- writing words on a sheet of paper,
- adding up even numbers – cognitive activity data,
- squeezing the hands,
- moving the eyeballs.

Frequency filtration was applied to remove the noisy part of the signal containing low level information. It is the signal which has frequency lower than 1 Hz or higher than 45 Hz, including signals originating from external electronic devices. EEG signals have low amplitude, so quality and high signal-to-noise ratio is very important [8].

EEG signal pre-processing covered signal filtration and artefact elimination. In the study, Principal Component Analysis and Independent Component Analysis were applied to remove the artefacts.

5 Data processing results

Tables 1 and 2 show the amplitude for the raw EEG signal and the signal after elimination of the first component, the first two components and the first three components for PCA and the selected components of ICA algorithms. The tables show also the amplitude of the EEG signal of relaxation state. This value is used as the baseline for other data in the table. It was observed that the increase in the number of the principal components eliminated caused a decrease of the artefact amplitudes. This effect may be explained by the fact that the principal components with the highest values contain the most common part of the signals recorded by the channels, so the common part, i.e. the artefacts, was removed because of removing the highest principal components. The individual information from the parts of the head closest to the corresponding electrodes was left unaffected [9, 17, 21].

Data in Tables 1 and 2 shows that the artefact amplitude is higher after using ICA, than the artefact amplitude of the signal obtained because of using PCA. The values of the amplitudes were measured with dedicated software (WinEEG).

The results show that the amplitude of artefacts decreased after correction. The obtained results, i.e. the artefact amplitude values, deviated from mean amplitude values insignificantly, so mean values are sufficient to demonstrate the results.

Table 1. Mobile amplifier. Amp - amplitude of EEG signal with artefacts, Amp. rel. - amplitude of EEG signal of relaxation state.

Type of artefacts	Amp. [μV]	PCA, amplitude [μV]			ICA, amp. [μV]	Amp. rel. [μV]
		1.	2.	3.		
Eyes blinking	150	15	4	4	30	8.5
Jaw moving	220	26	22	18	50	
Eyeballs moving	140	16	9	7.5	90	
Hands squeezing	20	14	8	4.5	17	
Writing words on the sheet of paper	14	12	6	5	9	

Table 2. Stationary amplifier. Amp - amplitude of EEG signal with artefacts, Amp. rel. - amplitude of EEG signal of relaxation state.

Type of artefacts	Amp. [μV]	PCA, amplitude [μV]			ICA, amp. [μV]	Amp. rel. [μV]
		1.	2.	3.		
Eyes blinking	170	15	13	13	90	11
Jaw moving	380	80	60	50	130	
Eyeballs moving	100	60	24	13	40	
Hands squeezing	40	20	20	9	25	
Writing words on the sheet of paper	22	18	18	8	18	

As it is known, the ICA method operates in a different way. It does not separate the signal into parts describing progressively less common information of the signal, but rather divides the signal (being a mix of independent signals) into independent components. On the basis of that feature, it is theoretically not enough to eliminate several components with the highest values. The components corresponding to particular artefacts should be found and removed [15]. The research confirmed this statement: the average number of rejected independent components was significantly higher than in the case of PCA [12, 17].

One of the measures presented in the section 2.2 was used for artefact detection. This measure was channel potential variance. The mean values of separate electrode potential variance among all the participants for all types of biological artefacts mentioned in the paper were calculated. The tables 3, 4 and 5 present the selected mean values of separate electrode potential variances before and after PCA. On the basis of decreased value of variance and performed visual inspection, it was concluded that the artefacts were successfully eliminated to a high degree. The electrodes presented in the tables were chosen due to the fact that the variance value was the highest for these electrodes during presence of specific artefacts, i.e.: eyes blinking, eyeballs moving and jaw moving. As data in the tables show, artefacts affect less the record carried out by the mobile amplifier. That conclusion was made on the basis of comparison of variance values of signals recorded by stationary and mobile amplifiers in the Tables 3-5.

Table 3. Eyes blinking.

	Before		After	
Electrodes	Stationary	Mobile	Stationary	Mobile
Fp1	2041.32	715.03	7.92	4.86
Fp2	2041.32	932.99	23.79	9.94
F7	2152.57	399.01	33.03	6.59

Table 4. Eyeballs moving.

	Before		After	
Electrodes	Stationary	Mobile	Stationary	Mobile
Fp1	867.86	512.04	12.51	7.63
Fp2	813.06	528.49	10.14	5.88
F7	3402.70	2075.65	33.33	21.33
F8	2868.24	1720.94	18.21	10.01

Table 5. Jaw moving.

	Before		After	
Electrodes	Stationary	Mobile	Stationary	Mobile
F3	1939.47	1635.90	238.72	28.43
F4	3625.23	378.58	188.29	39.33
T3	3032.03	2393.03	206.31	59.89
C3	2041.64	1138.21	222.78	28.95
T4	3155.57	1770.70	271.70	62.89

The Figures 3-7 present the EEG recordings with artefacts of various types that were registered in the process of the research.

The Fig. 3 presents the example of frequency based method: the signal spectrum obtained by the means of Fourier Transform clearly indicates presence of undesired frequency component.

The muscle and eye-related artefacts have a very characteristic shape in the records of both amplifiers, but as Tables 1 and 2 show, the artefacts in the signal recorded by the mobile EEG demonstrate a considerably lower magnitude.

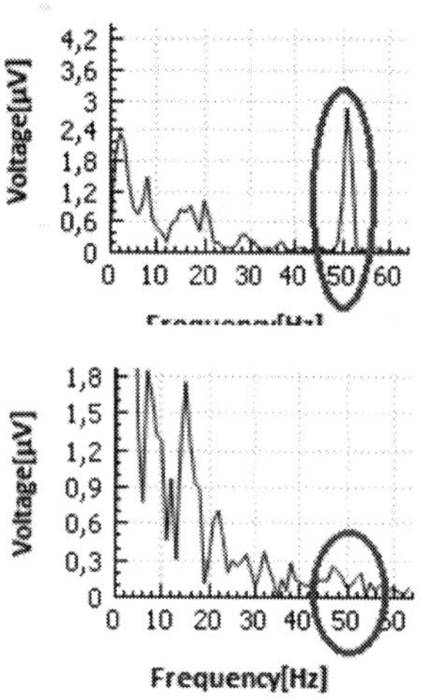

Fig 3. Technical artefacts related to a 50 Hz network (Fourier transform – frequency domain). Top – stationary, bottom – mobile.

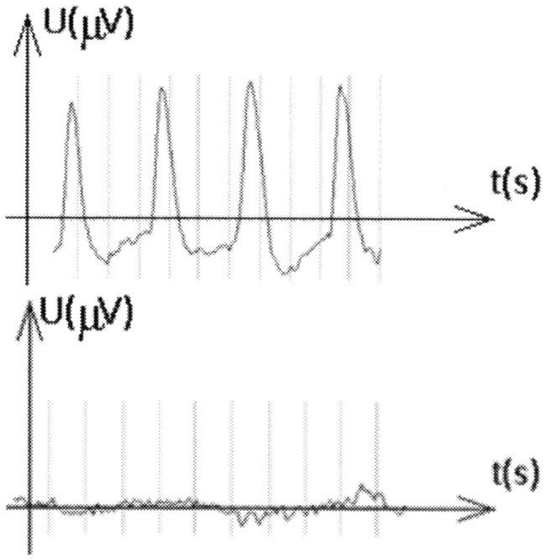

Fig 4. Eyes blinking. Top – original, bottom – corrected.

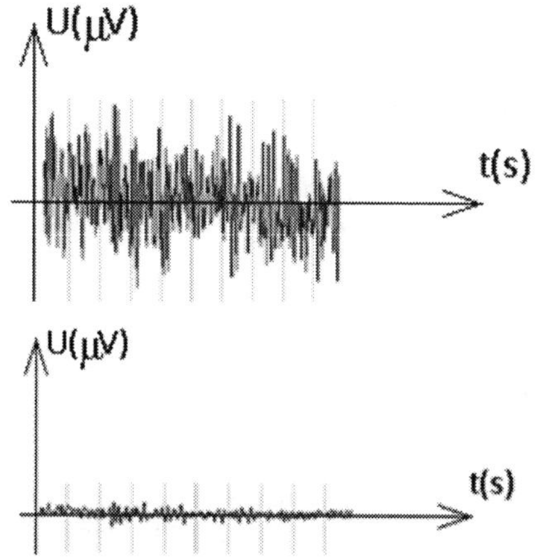

Fig 5. Jaw moving. Top – original, bottom – corrected.

 DOI: 10.1051/itmconf/20171501003

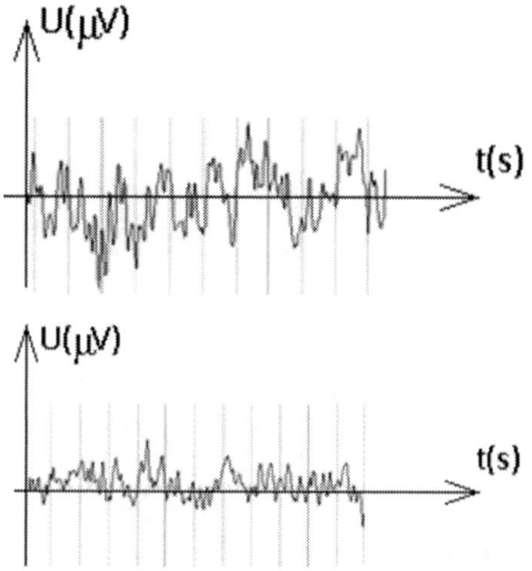

Fig 6. Hands squeezing. Top – original, bottom – corrected.

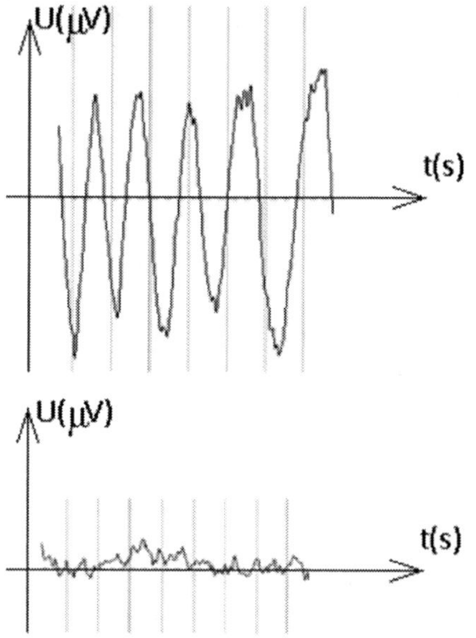

Fig 7. Eyeballs moving. Top – original, bottom – corrected.

Another type of artefacts, being worth mentioning and necessary to discuss, are technical artefacts. After comparing the records performed by the two types of electroencephalograph, it may clearly be seen that the mobile EEG is significantly less sensitive to the effect of the power network with the frequency of 50 Hz. As it is known, the technical artefacts of that kind are caused by the potentials induced in the long cables used to transmit the potential sensed by the electrodes placed on the scalp of the person examined to the EEG [8]. The length of cables used in combination with stationary EEG usually equals to 2 m, while the cables used for the mobile encephalograph are considerably shorter (20 cm), so the length of cables is not enough for the noisy 50 Hz potentials to be generated.

The artefacts in the signals gathered from standard and mobile amplifier were compared. Data are presented in Tables 1 and 2. The research has shown that jaw mowing generated the artefacts with the highest amplitude. The artefacts of the lowest amplitudes in the research were generated during the part covering hand writing. The said effects have been observed on both amplifiers. When comparing the muscle artefacts in the signals from the two amplifiers, it was noticed that the signal recorded by the mobile amplifier contained the muscle artefacts of a significantly (up to 1/3) lower amplitude. The influence of 50 Hz power network has been proven to be considerably lower on the mobile electroencephalograph.

6 Conclusions

The goal of the present paper was to compare two methods of artefact correction and two electroencephalographs. The research has shown that the PCA method is more universal and simpler in use, since the artefacts are usually visible at all the electrodes, and originate from one or two sources, so it is usually enough to eliminate one or two principal components [8]. The ICA method divides the signal into a series of independent components, their number is equal to the number of the channels [16]. However, the identification of a component as the artefact is more complicated. The localisation of the component, as well as its character, should also be analysed [15]. Another inconvenience was the fact that it was necessary to delete a comparatively larger number of components (6 IC compared to 1-2 PC).

A comparison of the two electroencephalographs has shown that the mobile encephalograph deals better with the artefacts of both biological and technological origin. The results obtained prove that the mobile EEG better meets the needs of BCI applications, as the signal recorded by this type of EEG contains a considerably lower number of biological artefacts and a significantly decreased number of technological artefacts related to the power network with the frequency of 50 Hz [5]. These features might facilitate achieving the following results:

1. the performance requirements for the system based on the mobile EEG would be lower, since it enables to process the obtained signal consuming less performance resources (simpler artefact correction method, absence of notch filter);
2. better possibilities to process data in the frequency domain in the range of higher frequencies (40+ Hz), which would be achieved thanks to the absence of the notch filter. That would allow to monitor the gamma component of brain waves, the significant part of which (the band between 45 and 65 Hz) would be thrown away if the notch filter was used.

References

1. Lopes da Silva F. H., Storm van Leeuwan W., Remond A., *Handbook of electroencephalography and clinical neurophysiology*. Revised series 1986; vol. **2**. Amsterdam: Elsevier; pp. 15-62, (1986)

2. M. Plechawska-Wojcik, M. Kaczorowska, *10th International Technology, Education and Development Conference (INTED)*, pp. 844-852, (2016)
3. C. Jutten, J. Herault, *Signal Processing*, 24: pp. 1-10,(1991)
4. M. Plechawska-Wojcik, P. Wolszczak, R. Cechowicz, *9th International Conference on Human System Interactions (HSI)*, pp. 348-354, (2016)
5. M. Plechawska-Wojcik, P. Wolszczak, *12th International Scientific Conference on Beyond Databases, Architectures and Structures (BDAS)*, pp. 485-496, (2016)
6. Z. Zhang, H. Li, D. Mandic, *2016 IEEE 13th International Conference*, pp. 177-182, (2016)
7. N. C. Rogasch, R. H. Thomson, F. Farzan, B. M. Fitzgibbon, Neil. W. Bailey, J. C. Hernandez - Pavon, Z. J. Daskalakis, Paul B. Fitzgerald, Elsevier, pp. 425-439, (2014)
8. J. A. Uriguen, B. Garcia-Zapirain, *Journal of Neural Engineering* Vol. **12**, pp. 342-348, (2015)
9. Y. Li, Z. Ma, W. Lu, Y. Li, *Physiological Measurement*, 27 (4), (2006)
10. B. Singh, H. Wagatsuma, *Computational and Mathematical Methods in Medicine*, pp. 612-620, (2017)
11. H.Nolan, R.Whelan, R.B. Reilly, Elsevier pp. 152-162, (2010)
12. A. Cichocki, S. Amari, *Learning Algorithms and Applications*, Wiley, pp. 106-111, (2003)
13. S. Motamedi-Fakhr, M. Moshrefi-Torbati, M. Hill, C. M. Hill, P. R. White, Elseiver pp. 21-33,(2014).
14. Y. Ding, I. W. Selesnick, Elsevier, pp. 236-248, (2016)
15. G. Barbati, C. Porcaro, F. Zappasodi, P. M. Rossini, F. Tecchio, *Clin. Neurophysiology*, pp. 1220-1232,(2004)
16. M. Kaczorowska, M. Plechawska-Wojcik, M.Tokovarov, R. Dmytruk, *International Symposium on Advanced Topics in Electrical Engineering (ATEE) 2017*, pp. 262-267, (2017)
17. L. I. Smith: *A tutorial of Principal Components Analysis*, pp 218-223, (2002)
18. A. J. Bell, T. Sejnowski, *Neutral Computation*,: pp. 1129-2259, (1995).
19. M. Kołodziej, Warsaw University of Technology, pp. 56-60, (2011)
20. J. Koronacki, WNT, pp. 250-256, (2005)
21. W. J. Krzanowski, *Oxford University Press*, (2000)

 DOI: 10.1051/itmconf/20171501004

Modification of Adaptive Huffman Coding for use in encoding large alphabets

Mikhail *Tokovarov*[1,*]

[1]Lublin University of Technology, Electrical Engineering and Computer Science Faculty, Institute of Computer Science, Nadbystrzycka 36B, 20-618 Lublin, Poland

Abstract. The paper presents the modification of Adaptive Huffman Coding method – lossless data compression technique used in data transmission. The modification was related to the process of adding a new character to the coding tree, namely, the author proposes to introduce two special nodes instead of single NYT (not yet transmitted) node as in the classic method. One of the nodes is responsible for indicating the place in the tree a new node is attached to. The other node is used for sending the signal indicating the appearance of a character which is not presented in the tree. The modified method was compared with existing methods of coding in terms of overall data compression ratio and performance. The proposed method may be used for large alphabets i.e. for encoding the whole words instead of separate characters, when new elements are added to the tree comparatively frequently.

1 Introduction

Efficiency and speed – the two issues that the current world of technology is centred at. Information technology (IT) is no exception in this matter. Such an area of IT as social media has become extremely popular and widely used, so that high transmission speed has gained a great importance. One way of obtaining high communication performance is developing more efficient hardware. The other one is to develop the software that would allow to compress the data in such a way that would reduce the size of data but not affect its information content. In other words, to encode the data by the method called *lossless data compression*. This term means that the methods of this type allow the original data to be perfectly reconstructed from the encoded message.

Binary or entropy encoding are the most popular branches among lossless data compression methods. The term *entropy encoding* means that the length of a code is approximately proportional to the negative logarithm of the occurrence of the character encoded with the code. Simplifying it may be said: the higher probability of the character is, the shorter is its code [1].

Several methods of entropy encoding exist, these are the most frequently used methods of this branch:
- arithmetic coding,
- range coding,
- Huffman coding,
- asymmetric Numeral Systems.

Arithmetic and Range coding are quite similar with some differences [2], but arithmetic coding is covered by patent, that is why, due to lack of patent coverage, Huffman coding is frequently chosen for implementing open source projects [3]. The present paper contains the description of the modification that may help to improve the algorithm of adaptive Huffman coding in terms of data savings.

2 Study of related works

Huffman coding has been developed in 1952 by David Huffman. He developed the method during his Sc. D. study if MIT and published in the 1952 paper "A Method for the Construction of Minimum-Redundancy Codes" [1]. Huffman coding is the most optimal among methods encoding symbols separately, but it is not always optimal compared to some other compression methods, such as e.g. arithmetic and Lempel-Ziv-Welch coding [4]. However, the last two methods, as it has been said, are patent-covered, so developers often tend to use Huffman coding [3]. Comparative simplicity and high speed due to lack of arithmetic calculations are the advantages of this method as well. Due to these reasons Huffman coding is often used for developing encoding engines for many applications in various areas [5].

The fact that billions of people are exchanging gigantic amount of data every day stimulated development of compression technologies. This topic constantly presents significant interest for researches, the following works may be presented as the examples:

Jarosław Duda et al. worked out the method called Asymmetric Numeral Systems, the method was developed on the basis of the two encoding methods: arithmetic and Huffman coding. It combined the advantages from the two methods:

* Corresponding author: m.tokovarov@pollub.pl

DOI: 10.1051/itmconf/20171501004

- near accurate symbol probabilities hence better compression ratio of arithmetic coding, and capability to fast encoding and decoding of Huffman coding [6];
- Facebook Zstandard algorithm is based on LZ77 dictionary coder and tANS – effective entropy encoding based on the Huffman method [7];
- Brotli coding algorithm which is used in most of the modern Internet Browsers, such as Chrome, Opera.

Similarly, to the Zstandard it is based on the combination of LZ77 and modified Huffman coding [8]. So, it may be clearly seen that despite its long history the Huffman encoding algorithm still presents great interest for application.

3 Huffman coding explained

3.1 Static Huffman coding algorithm

The concept of Huffman coding is based on the binary tree. The tree consists of two kinds of nodes: *intermediate nodes*, i.e. the nodes having descendant nodes and the nodes which do not have descendants. These nodes are called *leaves*. A character may be stored only in a leaf node, this condition ensures the character codes to be *prefix-free* [1]. It means, that no character has the code, that would be the initial segment of another character's code. As the example of **prefix** codes, the following bit sequences may be used: 110101 and 110. The code 110 is identical to the initial segment of the code 110101, so these two codes may not be decoded unambiguously. The tree is organized according to the following principles [4]:

a) any node in the tree may not have a single descendant: either two or none;

b) each node in the tree has the number assigned to it. This number is called *weight*. Depending on the type of the node its weight may have the following meanings:

- if the node is a leaf, the weight value is equal to the number of times the character stored in the leaf occurs in the message sent;
- if the node is an intermediate node its weight value is equal to the sum of its descendants' weights.

c) the weight of the right descendant should be not less than the weight of the left descendant.

The codes for every character are defined as the path in the binary tree from the root to the leaf containing the character (See Figure 1), e.g. the blank space character which is the most frequent character in the tree has the code 111, and the 'p' character, which occurs only once has the code 10011 [4].

The Figure 1 presents the tree constructed in accordance with static method of Huffman coding. The main feature of it is that the tree is constructed before the transmission is started on the basis of analysis of the probabilities of separate characters in the whole message [1].

The data compression ratio(DCR) is described by the formula (1):

$$DCR = \frac{numberOfBitsInCompressedMessage}{numberOfBitsInOriginalMessage} \quad (1)$$

But the compression ratio does not show actual space saving as besides of encoded message the table with the code-character pairs should be transmitted. For that reason, the other index should be introduced. The sent-to-original-bits ratio(SOBR) is described in accordance with the formula below:

$$SOBR = \frac{numOfBitsInCompMes + numOfBitsToSendTab}{numberOfBitsInOriginalMessage} \quad (2)$$

The only way to make SOBR equal to DCR is to use one coding tree for all messages, but this tree will not be optimal since the character probabilities may be different in various messages.

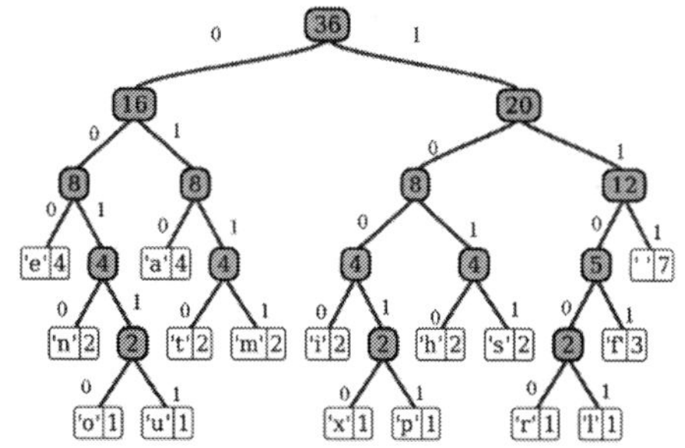

Fig. 1. Huffman tree generated based on the phrase "this is an example of a Huffman tree".

3.2 Adaptive Huffman coding algorithm

The method of adaptive Huffman Coding (AHC) reviewed in the article was proposed by Jeffrey Vitter in his paper published in 1987 [9].

The algorithm working during creating the tree may be observed in Figure 2.

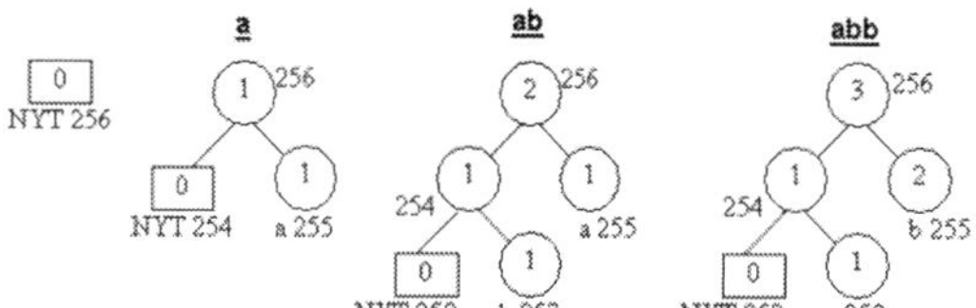

Fig. 2. The algorithm of AHC, example for encoding the word "abb".

This method is based on the same principles as the static method plus the following extensions [9]:

a) every node in the tree has its *key number*, the key numbers are arranged in the following way:

- maximum value of the key number in the tree may be calculated as:

$$keyNum = 2 \cdot maxNumOfCharInAlphabet + 1 \quad (3)$$

- the root has the largest key number;
- an ancestor has larger number than any of its descendants;
- the right descendant should have larger key number than the left descendant;

b) after input of any character the tree is updated;

c) a special leaf node called NYT (not yet transmitted) is used for both indicating the place for a new character and for signalizing that the new character is obtained, its key has the least value in the tree, its weight always equals to zero;
d) a set of nodes with equal weight values is called a block.

The encoding algorithm is described in Listing 1. The algorithm contains only the process of updating the tree and does not consider communication. If the communication algorithm based on the AHC is used, the process becomes more complicated. The idea may be presented as follows:
- transmitter and receiver have identical trees which are updated in accordance with the algorithm described in the Listing 1;
- the communication operates in the way presented in the Listing 2.

As it may be seen from the Listing 2, in case of AHC not only the codes of the characters are transmitted. Auxiliary codes such as the code of *NYTNode* and ASCII codes of the new-coming characters are being transmitted as well. The necessity to transmit auxiliary codes negatively influences SOBR causing it to increase. Due to this fact overall data saving decreases. But along with higher degree of complexity compared to static Huffman coding ACH may still present interest as lossless data compression technique after implementing the modifications described in the next chapters.

3.3 Encoding words instead of separate characters

The title of the article contains the phrase "large alphabet". But the chapter is focused at encoding entire words. What is the connection between these two facts? The idea is that in the method proposed, the words are treated as separate characters, so the leaves of the coding tree store not characters, but complete words, so these words are treated as separate characters in a large alphabet. The author does not claim, that this idea belongs to him. This technique is quite well known and was presented in many works [5, 10].

The greatest success may be achieved in the case of applying this method for encoding the words of an *analytic language*. An analytic language is the type of language where grammatical relationships are established by using strict word order, prepositions, postpositions, particles and special auxiliary words that do not have individual meaning and only indicate some grammatical categories. Analytic languages do not have extensive systems of conjugation and declension as *synthetic languages* do [11].

In many cases a word in a synthetic language may have several forms which are treated as individual words by encoding algorithm. This approach would cause the coding tree to be excessively extensive. It is worth to note that the statistics show that around 95% of all common English texts may be covered by 7000 words [11, 13]. The situation becomes even more optimistic when communication in social networks and mass media is considered.

Listing 1. Pseudocode presenting updating tree of AHC.

```
UpdateTree(character ch)
if ch is not found in the tree
   make a newLeafNode as the right descendant of the
  oldNYTNode;
   newLeafNode's keyValue := oldNYTNode's
  keyValue-1;
   make a newNYTNode as the left descendant
  of the oldNYTNode
   newNYTNode's keyValue := oldNYTNode's
  keyValue-2;
   NYTNode := newNYTNode;
end if
activeNode := node containing ch;
do
  find the nodeWithTheLargestKeyNumber
  in the activeNode's block;
  if the nodeWithTheLargestKeyNumber's keyNumber
  > activeNode's keyNumber
    swap nodeWithTheLargestKeyNumber
    and activeNode;
    swap nodeWithTheLargestKeyNumber's
    keyNumber and activeNode's keyNumber;
    //key numbers should not change the place
  end if
  activeNode's weight := activeNode's weight+1;
  activeNode := activeNode's ancestor;
while activeNode <> root
```

Listing 2. Communication with the use of data compression based on AHC.

Transmitter	Receiver

Transmit (character *ch*)
if *ch* is found in the tree
    send the code of *ch* bit by bit;
else
    send the code of *NYTNode* bit by bit;
    send the ASCII code of *ch*;
end if
UpdateTree(*ch*);
	Receiving is performed in the stream, i.e. in infinite loop.
	Receive()
	while (true)
	    receive one bit from the transmitter,
	    add received bit to *bitSequence*;
	  if *bitSequence* leads to a leaf node
	    obtain the character *ch* stored in the node;
	    add *ch* to decoded message;
	    UpdateTree(*ch*);
	    set *bitSequence* empty;
	  end if
	  if *bitSequence* leads to *NYTNode*
	    receive 8 bit;
	    convert received 8 bit to character *ch*;
	    add *ch* to decoded message;
	    UpdateTree(*ch*);
	    set *bitSequence* empty;
	  end if
	end while

 DOI: 10.1051/itmconf/20171501004

To prove the feasibility of the method the term of *information entropy* should be mentioned. This term has several definitions:

- measure of unpredictability of the state;
- expected (mean) value of information contained in a message.

The value of entropy is calculated by the following formula:

Entropy of i-th character in a message:

$$I_i = -log_m p_i \quad (4)$$

Average entropy of a message:

$$H = \sum_{i=1}^{n} -p_i log_m p_i = \sum_{i=1}^{n} -p_i log_m I_i \quad (5)$$

where: p_i - probability of i-th character in the message; n - number of unique characters in the message; m - logarithm base, usually taken equal to 2, as binary system is used in computer technics.

Practically it may be stayed, that the average entropy of a message is equal to the least possible average code length of the characters contained in the message [1].

It is well known, that amongst discrete distribution with equal number of states the uniform distribution has the maximum value of entropy [12]. Every state of the uniform distribution has the same probability equal to 1/n, where n is the number of states, which yields the entropy:

$$H_{uniform} = log_2 n \quad (6)$$

To estimate the maximum entropy, the maximum number of words encoded should be defined. It has been decided to take the max number of words equal to 16384, which yields the maximal entropy equals to 14 bits. The max word number is taken because the practical number of stored words may be higher than 7000 because of some capitalized words, abbreviations, mistakes and user-defined words. To roughly estimate the compression ratio, the average length of English word is used, its value approximately equals to 4 letters. In case if ASCII coding is taken into consideration the average length in bits equals to 32. Based on this value the average compression ratio equals to 14/32 ≈ 0.43. This estimation is fairly promising as average compression ratio for separate-character AHC is about 0.55 [4].

As it is stated in the previous chapter, sent-to-original bit ratio tends to be significantly higher than data compression ratio for complete-word AHC. This effect is especially noticeable during the phase of initial building the coding tree, when new words are coming especially frequently. There are two factors that affect SOBR in this case: sending the complete ASCII codes of the new-coming words and sending the NYT bit sequence.

Precoded dictionaries may be used to decrease the influence of first factor. This possibility is not considered in the current paper.

However, the second factor, i.e. sending the NYT bit sequence will be optimized within the frame of this research. As it is described in the chapter 2 the NYT is used for both indicating the place in the tree the new-coming word is attached to and for sending the signal meaning that the ASCII code of the new-coming signal is going to be sent. This fact provides the opportunity for optimising the algorithm.

4 Modification

4.1 Introduction of NCW node

The NYT node should only act as the pointer for the new-coming word. Its weight still should be 0 and its key number should have the least value in the tree. The new node NCW should be introduced to the tree. Its initial weight should be equal 0. This node should be used for sending the "new-coming word" signal. The introduced NCW node should be treated as a usual leaf node, i.e. after sending the bit sequence corresponding to the NCW node, the procedure described in the lines 8-17 of the Listing 1 should be carried out for the NCW node. The Listing 3 presents the modified algorithm.

Listing 3. Modified algorithm.

Transmitter	Receiver
Transmit (character *ch*) 1 if *ch* is found in the tree 2 send the code of *ch* bit by bit; 3 else **4 send the code of *NCWNode* bit by bit;** **5 updateTree(*NCWNode*);** 6 send the ASCII code of *ch*; 7 end if 8 UpdateTree(*ch*);	Receiving is performed in the stream, i.e. in infinite loop. Receive() 1 while (true) 2 receive one bit from the transmitter, 3 add received bit to *bitSequence*; 4 if *bitSequence* leads to a leaf node 5 obtain the character *ch* stored in the node; 6 add *ch* to decoded message; 7 UpdateTree(*ch*); 8 set *bitSequenc* empty; 9 end if **10 if *bitSequence* leads to *NCWNode*** **11updateTree(*NCWNode*);** 12 receive 8 bit; 13 convert received 8 bit to character *ch*; 14 add *ch* to decoded message; 15 UpdateTree(*ch*); 16 set *bitSequence* empty; 17 end if 18 end while

The Figure 3 presents the difference between modified and non-modified method. The phrase "A friend in need is a friend indeed" was used for the test. As it may be noticed the weight of the NCW node is equal to the number of the unique leaf nodes.

 DOI: 10.1051/itmconf/20171501004

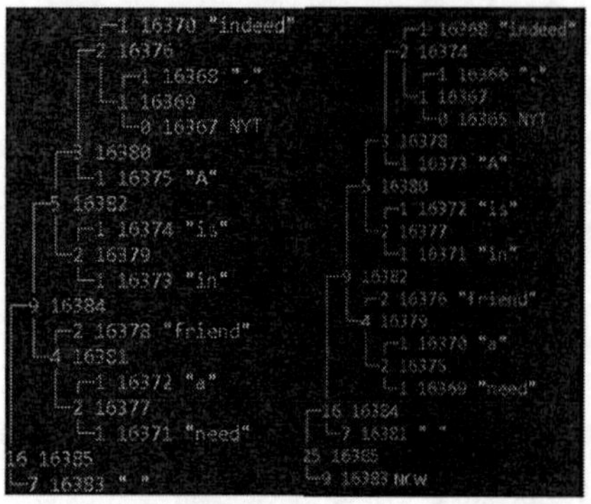

Fig. 3. Comparison of full-word AHC trees. Left – two-purpose NYT node, right – separate NCW and NYT nodes.

4.2 Forgetting

It is a quite frequent fact in real social media application, when a user makes some mistakes in the text. In the case of the complete-word Huffman algorithm it would cause the coding tree to be overgrown and not optimal, since these mistyped words are used very rarely, but keep the place in the tree, causing the entropy to be larger. The same thing may be stated about rare words. Another problem that may raise is overflowing of dynamically allocated memory. In the current application dynamically, allocated arrays are used instead of linked lists due to their better performance rates. To deal with these problem, the algorithm of forgetting should be introduced.

This algorithm is based on the estimation of the relevance function of the stored words. The relevance function may be defined as Euclidean norm:

$$WF = \sqrt{weight^2 + agingFactor^2} \quad (7)$$

where: *agingFactor* is the value characterizing how long cycles ago the word was used last time; *weight* is the number of word's appearances in the text, the more the *weight*, the more often appears the word in the text.

The forgetting function is based on the algorithm presented in Listing 4.

Listing 4. Forgetting function.

```
if numberOfStoredWords > thresholdWordNum
   Sort the array of the leaf nodes basing on WF in
   descending order;
   while numberOfStoredWords > desiredWordNum
     delete the last word from the leaf nodes array;
   end while
end if
```

It is reasonable to set *thresholdWordNum* close to the max number of stored words and *desiredWordNum* – approximately 10-15% less to ensure periodical "cleaning" of the tree.

The Figure 4 presents the example of forgetting function operation. The threshold number of words is equal to 16000 and the desired number of words is 15000.

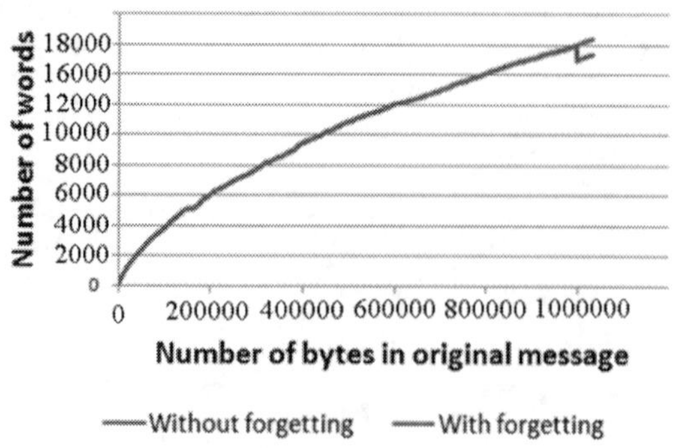

Fig. 4. Example of forgetting function operation.

5 Experiment and results

The tests were conducted on the data set containing more than million characters. The data set was composed on the basis text collected from various sources: news posts, social network comments and private correspondence. Use of forgetting function was completely feasible as many rare words, such as proper names and mistypes appeared. The results have proven that another application of complete-word AHC may be composing a dictionary for dictionary-type coding. Table 1 presents some frequent words and symbols found in the test texts.

Table 1. The most frequent words and symbols used in the text.

Word	Weight	Code
he	937	101001001
have	887	100111001
as	866	100110101
be	866	100110100
are	815	100011101
from	796	100011001
his	793	100011000
has	781	100010101
at	770	100010011
Trump	755	100000000
not	752	1111111110

The next issue that arose during the tests and comparison of the algorithms was the selection of features for analysis. The overall sent-to-original bits ratio has been chosen first. The Figure 5 presents the comparison of SOBR for the three implemented methods: separate-character AHC, unmodified complete-word AHC and modified complete-word AHC.

It may be clearly seen from the picture that even unmodified complete-word AHC demonstrates better sent-to-original bit ratio, as it was stated in the chapter 3.3. The modified method allows to decrease the SOBR even more. The end difference between overall SOBR values for modified and unmodified methods equals to 2.3 %.

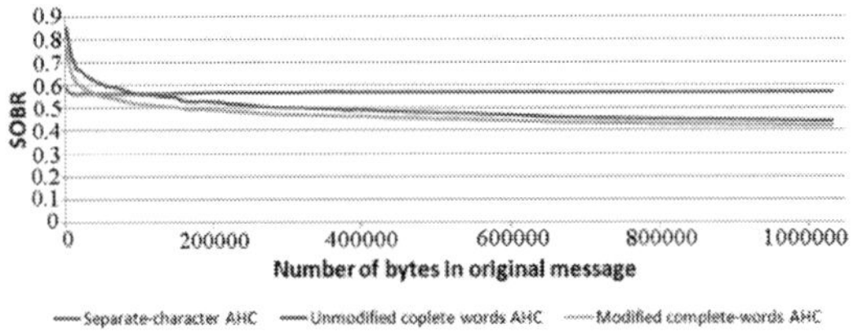

Fig. 5. Comparison of SOBR for the implemented AHC methods.

Another feature that may help to compare the two complete-words methods is specified sent-to-original bits ratio (*SSOBR*). Its value may be computed as first order divided difference of SOBR:

$$SSOBR_i = \frac{NBSM_i - NBSM_{i-1}}{NBOM_i - NBOM_{i-1}} \quad (8)$$

where: *NBSM* is the number of bits in the sent message; *NBOM* is the number of bits in the original message; *i* is the number of step.

The value of the denominator corresponds to the step or the number of bits the specified SOBR is computed per. In the current research the step was chosen to be equal to 1 Kbyte. The Figure 6 presents the differences of SSOBR of unmodified and modified AHC.

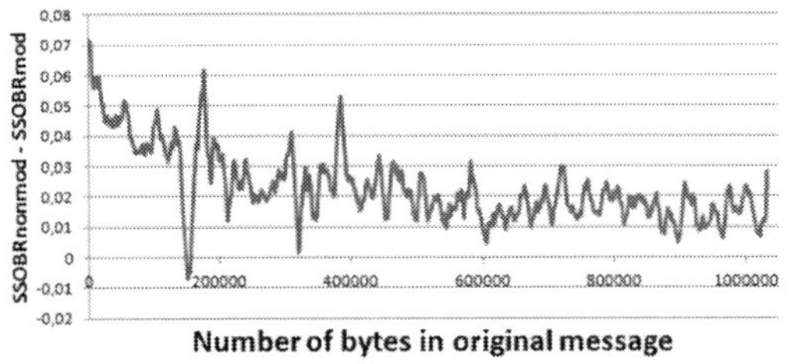

Fig. 6. Differences between SSOBR of unmodified and modified AHC methods.

The Figure 6 shows that the difference between SSOBR of the both methods tends to be more in the periods when a lot of new words is added to the tree (up to 6 percentage point). However, it might have quite low and even negative values, as in case of absence of new words unmodified AHC method becomes more optimal.

6 Conclusion

The comparison of single-character AHC and complete words AHC have been conducted, the research has proven the following:

- overall sent-to-original bits ratio of complete word AHC is approximately 12.8 percentage points lower than separate-words AHC one;
- better SOBR comes at a cost: the algorithm needs more memory to store the tree, the search of the node in the tree is slower due to its size, initial sent-to-original bits ratio is higher than this of separate-character AHC.

The implemented modification allowed to further improve the SOBR of complete-word AHC. The tests have shown that:

- sent to original bits ratio is 15.1 percentage points less compared to separate-character AHC method;
- the proposed modification proved to be more effective during the periods when large number of words is being added to the tree, and less when the number of new words is decreasing. Possible solution, that would be useful in that case, would be using the modified algorithm while the tree is being formed and then deleting the NCW node by the means of the function used for "forgetting" and using unmodified algorithm.

References

1. D.A. Huffman, *Proceedings of I.R.E,* pp. 1098–1101, (1952)
2. G. Nigel N. Martin, *Video & Data Recording Conference*, pp. 612-620, (1979)
3. B. Ryabkot, *Data Compression Coding Conference Proceedings*, pp 246-252, (2004)
4. J. Van Leeuwen, *ICALP*, pp. 382-410, (1976)
5. J. Lee, *MIT Undergraduate Journal of Mathematics*, pp. 122-130, (2007)
6. J. Duda; K. Tahboub; N.J. Gadgil; E.J. Delp, *Picture Coding Symposium (PCS)*, pp. 65-69, (2015)
7. https://code.facebook.com/posts/1658392934479273 /smaller-and-faster-data-compression-with-zstandard/ - last access 07.07.2017
8. J. Alakuijala, Z. Szabadka, *Internet Engineering Task Force* (IETF), (2016)
9. J. S. Vitter, *Journal of the ACM*, 34(4), pp 825–845, (1987)
10. M.B. Baer, *2013 IEEE International Symposium on Information Theory Proceedings (ISIT)*, pp. 1749-1753, (2013)
11. V.V. Bochkarev, A.V. Shevlyakova, V.D. Solovyev, Social *Evolution & History*, Vol.**14**, number 2, pp. 153-175 (2015)
12. H. Yokoo, *2016 International Symposium on Information Theory and Its Applications (ISITA)*, (2016)
13. http://www.ravi.io/language-word-lengths - last access 07.07.2017

ITM Web of Conferences **15**, 01005 (2017)
CMES'17
DOI: 10.1051/itmconf/20171501005

Optimisation of milling parameters using neural network

Jerzy Lipski[1,*], and *Kazimierz* Zaleski[2]

[1]Lublin University of Technology, Department of Organisation of Enterprise, Nadbystrzycka 38, 20-618 Lublin, Poland
[2]Lublin University of Technology, Mechanical Engineering Faculty, Nadbystrzycka 38, 20-618 Lublin, Poland

Abstract. The purpose of this study was to design and test an intelligent computer software developed with the purpose of increasing average productivity of milling not compromising the design features of the final product. The developed system generates optimal milling parameters based on the extent of tool wear. The introduced optimisation algorithm employs a multilayer model of a milling process developed in the artificial neural network. The input parameters for model training are the following: cutting speed v_c, feed per tooth f_z and the degree of tool wear measured by means of localised flank wear (VB3). The output parameter is the surface roughness of a machined surface R_a. Since the model in the neural network exhibits good approximation of functional relationships, it was applied to determine optimal milling parameters in changeable tool wear conditions (VB3) and stabilisation of surface roughness parameter R_a. Our solution enables constant control over surface roughness parameters and productivity of milling process after each assessment of tool condition. The recommended parameters, *i.e.* those which applied in milling ensure desired surface roughness and maximal productivity, are selected from all the parameters generated by the model. The developed software may constitute an expert system supporting a milling machine operator. In addition, the application may be installed on a mobile device (smartphone), connected to a tool wear diagnostics instrument and the machine tool controller in order to supply updated optimal parameters of milling. The presented solution facilitates tool life optimisation and decreasing tool change costs, particularly during prolonged operation.

1 Introduction

Milling processes are non-stationary operations, the realisation of whose involves the constant change of parameters crucial to productivity and quality of machining. One of such parameters, critical to the surface roughness of machined workpiece, is the condition of the cutting tool [1]. In the case in question, the relationship between tool wear and surface roughness parameter R_a is exponential. Both the pace and the functional character of these changes are relative to an array of difficult-to-define factors and exhibit non-linear dependence on the tool operation time [2]. These relationships affect production economics, particularly in lot production at constant milling parameters. The limit value of R_a specified in the product design documentation must be attained, nevertheless tool wear may hamper these efforts at certain stages of production. In the event of such an occurrence, constant monitoring of the state of machined surfaces [3,4,5] enables spotting the defect and applying the most sensible solution – cutting tool change. However, this will inevitably lead to halting the production process and generate delay of machining time of a given lot of machined products. This leads to the question, whether altering machining parameters may prolong tool life, and provide the desired surface roughness of the workpiece.

In our experimental study, we developed a model which confirmed such presumption. However, implementation of the model for adjusting machining parameters that would ensure obtaining the required surface roughness of products may compromise the productivity of the process. The optimising algorithm is therefore designed to select parameters, from the established variants, that will produce specified R_a at a minimum loss of machining productivity. This strategy proves particularly effective when the surface roughness requirements are to be maintained in high lot size production. Resolving to the conventional solution would entail frequent stops for changing the cutting tool and maintenance-work-related productivity loss. Approaching both strategies from the perspective of production economics, what must be resolved is the issue of the limit tool wear, beyond which it is pointless to search for machining parameters to obtain the required R_a of machined surfaces. Therefore, to assess the suitability of approaches we must collate the increased machining costs, resulting from decreased machining parameters, and the alternative cost of halting the machining process for tool change operations, the cost of unused machining capacity and a potential cost of setting up a new tool. If lower machining productivity, resulting from the necessary machining parameters alteration, should generate comparable or higher costs, the

* Corresponding author: j.lipski@pollub.pl

decision-making algorithm should select the cutting tool change variant.

The solution to the presented problem may be obtained by means of the algorithms, implemented by the developed decision-aiding software. The computer application may either aid the decision-making process of a machine tool operator, or alternatively, conduct the decision-making process automatically [6], given it is connected through proper interfaces with the machine tool controller. The model was built for cyclic milling of workpieces in high-sized lot production, at relatively low surface roughness parameter R_a.

The system was developed and tested following the stages listed below:

- conducting experimental milling processes for a specified area of acceptable parameters and different tool wear;
- employing the obtained experimental data into building the multilayer non-linear model defining relationships between feed per tooth, cutting speed and tool wear parameters and the target surface roughness. The model was developed by means of a designed artificial neural network;
- developing optimisation algorithms for milling process parameter selection;
- testing the system.

2 Data collection for the neural network model

The data for the neural network was obtained from the series of test runs of the milling machine tool, conducted at a specified constant milling parameters.

The study was carried out on 25 mm x 25 mm x 80 mm cuboid-shaped workpiece specimens made of 55NiCrMoV7 tool steel (hardness 240 HB). The specimens were subjected to face milling with cutting fluid, carried out on FV-580A vertical machining centre supplied by *MOC-Mechanicy*. The process was realised with R390-050C5-18060 square shoulder milling cutter (D = 50 mm, entering angle $\kappa_r = 90^0$) equipped with R390-180612H-PL coated cemented carbide cutting inserts, produced by Sandvik Coromant.

The milling process was conducted at the following constant milling parameters:

- axial depth of cut ap = 0.5 mm,
- radial depth of cut ae = 25 mm (partial symmetrical face milling).

Changeable parameters of the process were:

- cutting speed in the range of vc = 60 ÷ 450 m/min.,
- feed per tooth, in the range of fz = 0.02 ÷ 0.3 mm/tooth,
- localised flank wear at the tool-workpiece interface VB3 = 0 – 0.6 mm.

The value of VB3 was monitored immediately after each machining of a test specimen by means of Keyence VHX 5000 digital microscope.

Average surface roughness parameter R_a, the arithmetic average of the absolute values of the profile heights, measured with Surtronic 3+ surface texture measurement system produced by Rank Talyor Hobson, was specified as nominal roughness.

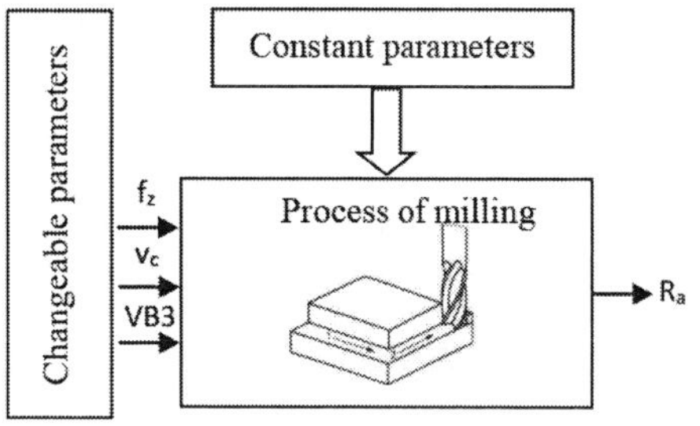

Fig. 1. Data collection for the neural network model of milling process.

3 Neural network as a model of milling process

Considering the non-linear and multilayer character of the modelling object it was resolved that the best solution for representation would be provided by the neural network [7,8]. Among its advantages, a model generated by the neural network exhibits one significantly beneficial feature, *i.e.* it is capable of calculating values of a modelled function outside of the training data set.

The neural model building process consists of three stages:

- defining the structure of a neural network,
- network learning,
- validation and testing.

The process was modelled in MatLab environment with the application of Neural Network Toolbox. Machine learning in MatLab begins with a processing function creating the network based on the values of input and output data:

$$net\_mf = newff(P,T,S) \tag{1}$$

where: *net_mf* – a variable defining the structure of the neural network, *P*- *R*-by-*Q* matrix of *Q* input vectors of *R* elements each, *T*- matrix of output vectors, *S*- number of neurons in hidden layer.

The structure of thus defined network may be modified by means of optional parameters, according to the descriptions found in Neural Network Toolbox guidebook. The network designed for solving the problem of a given process is shown in Fig. 2.

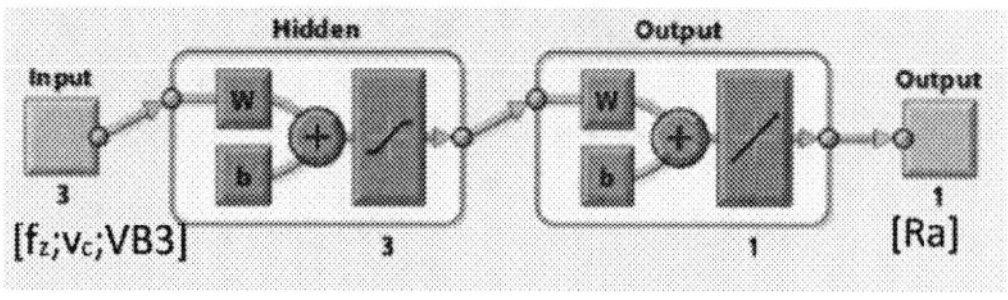

Fig. 2. Designed neural network.

The network model is adjusted to fit the experimental data at the learning stage, which in MatLab is carried out by the function *train*:

$$net\_mf = train(net\_mf, P, T) \tag{2}$$

where: *net_mf* – milling process model.

 DOI: 10.1051/itmconf/20171501005

The process of learning involves specifying weights in particular neurons and fit assessment of the model to training data and to data outside of the training data set but were obtained from the same experimental milling process. The experimental machining process in question produced 27 values of surface roughness parameter Ra of machined surfaces for different combinations of three input parameters: cutting speed vc, feed per tooth fz, and tool wear VB3. Variable P in the function of training took the form of a 3x27 matrix, whereas variable T was represented by a single-row matrix with 27 columns. The history of training and the correlation between targets and outputs may assessed from diagram in Fig. 3.

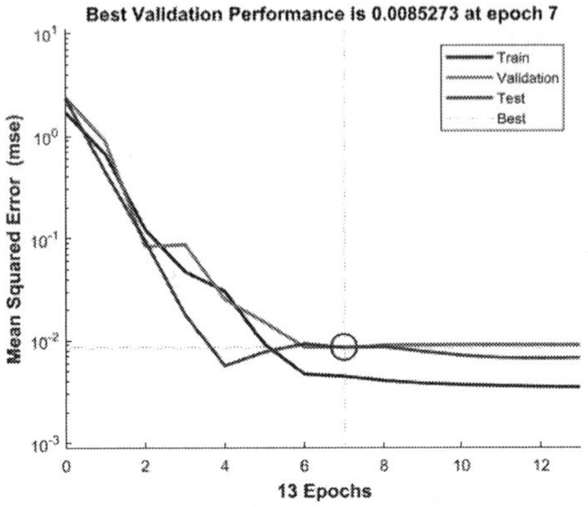

Fig. 3. Training history of neural network.

The network model *net_mf* obtained through training is a virtual model representing the process of milling for values of three selected input machining parameters from the set of training data. The statement is true if other parameters of the milling process, constant and not included in the model, are not subject to change. This condition is fulfilled for lot production processed on the same milling machine, on the same workpiece, and with the same tools and inserts.

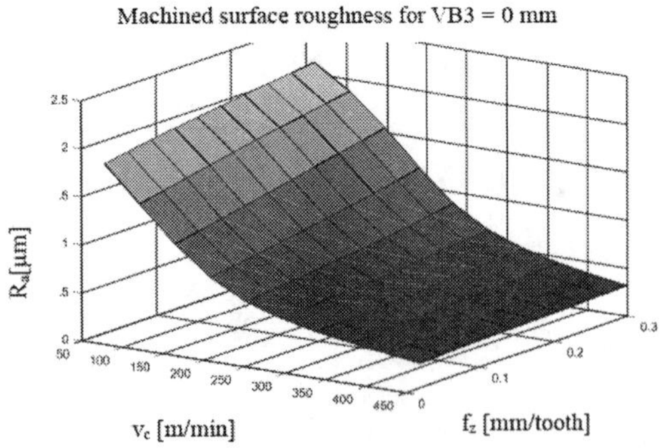

Fig. 4. Graphic representation of machined surface roughness changes depending on feed per tooth f_z and cutting speed v_c, given zero tool wear (VB3=0).

The model may be represented graphically only to a limited extent due to the fact that the function describing the architecture of the model contains three variables. Fig. 4 shows the correlation between surface roughness of the machined workpiece, feed per tooth and cutting speed at zero tool wear VB3, while Fig. 5 represents surface roughness at identical machining parameters but carried out with a worn tool, VB3 = 0.6 mm. In both diagrams the relationship exhibits a non-linear character, whereas advancing tool wear results in higher sensitivity to feed per tooth and cutting speed changes, as well as lower and more limited range of surface roughness values.

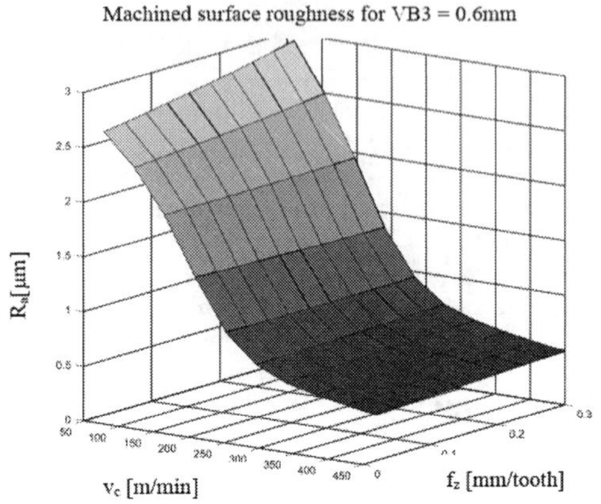

Fig. 5. Graphic representation of machined surface roughness changes depending on feed per tooth f_z and cutting speed v_c, at given tool wear of vb = 0.6 mm.

In MatLab software implementation of *net_mf* model for computing R_a of machined surfaces for any given variant of f_z, v_c and VB3 independent variables combination is carried out with the function *sim*:

$$R_a = sim\left(net\_mf, v_c, f_z, VB3\right) \quad (3)$$

4 Optimisation algorithm based on the model of milling process

The model of milling process obtained from the neural network provides us with data regarding the surface roughness of a machined surface obtained as a result of particular combination of machining parameters: feed per tooth, cutting speed and tool wear. This functionality itself is insufficient in terms of the optimisation objective: specifying particular parameters of milling that fulfil the optimisation objective function, *i.e.* maximum productivity of milling process *W*, defined as the material removal rate in a given time unit:

$$W_{max} = f_z * z * \frac{1000 * v_c}{\pi * d_f} \quad (4)$$

where: f_z – feed per tooth [mm/tooth]; z – number of cutting edges; v_c – cutting speed [m/min]; d_f – milling cutter diameter [mm].

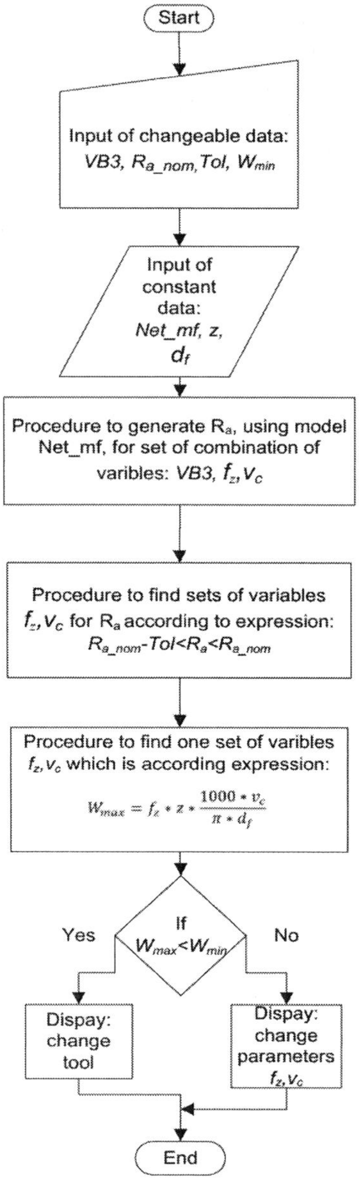

Fig. 6. Schematic representation of the algorithm for find best parameters of milling process, developed with the application of model described in the neural network.

Simultaneously, one more condition must be fulfilled: the surface roughness of a machined workpiece R_a must be within the given tolerance range *Tol*:

$$R_{a\,nom} - Tol < R_a < R_{a\,nom} \tag{5}$$

The surface roughness [9] of a machined workpiece R_a on a higher than nominal level is unacceptable, however, the range of R_a may be limited to values no lower than defined by *Tol*. Such limits may result from product design specifications. Decision-making variables are f_z and v_c for a particular localised flank wear VB3.

The developed algorithm searches the space of combination parameters generated by the trained neural model with the aim of specifying a sub-set of parameters fulfilling the inequality (4). The next step is to find a pair of parameters controlling the process of milling, f_z and v_c, which fulfil the objective function (3). The algorithm may include the minimal permitted productivity of milling W_{min} defined by the machine operator and dictated by economic analysis of the production process. Upon achieving the specified value, the algorithm suggests tool change. Fig. 6 shows a simplified representation of the optimisation algorithm for the milling process, developed with the use of the neural network model. In this algorithm has been applied invers method for identification of parameters [10, 11]. The was programmed in MatLab ver. R2017a.

The programme is composed of two modules: network learning aimed at producing the model, and application employing the developed model in controlling milling process parameters via milling machine controller. At a current level of software development the controller conducts regular tool wear measurements and updates the VB3 value in the system. The interface of the programme is a simple graphic design programmed with *Guide* tool in MatLab software.

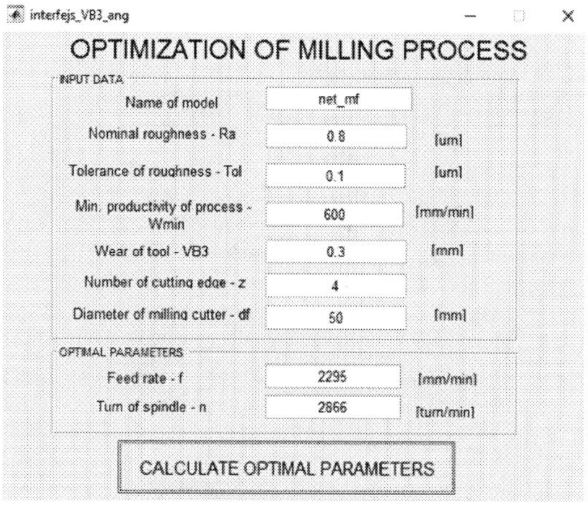

Fig. 7. Programme interface for calculation of optimal milling parameters.

Dialogue boxes are completed automatically with constant machining parameters (number of cutting edges and milling cutter diameter) specific for a given machine tool model, entered in the programme. The remaining fields must be completed by a machine tool operator, who must specify:

- nominal surface roughness - Ra;
- tolerance range of surface roughness - Tol;
- minimal permitted productivity of a milling process carried out with a worn-out tool, producing the surface of specified surface roughness – W_{min};
- regularly updated value of localised tool flank wear VB3.

Selecting „CALCULATE OPTIMAL PARAMETERS" button initiates generation of new optimal parameters of feed per tooth f_z and cutting speed v_c, which should be input in the machine tool controller in subsequent

DOI: 10.1051/itmconf/20171501005

milling operations. The parameters are automatically translated to corresponding values of feed rate f [mm/min] and turns of spindle [rev/min]. Updating the set of parameters in the milling machine controller will ensure maintaining desired surface roughness of machined surfaces at a specified level VB3, at maximal productivity of the process.

5 Testing milling parameter optimisation software

Programme testing consisted in analysing the fit of calculated parameters, f_z and v_c, and predicted productivity of machining. Table 1 collates test results, which show that the algorithm for high nominal R_a (between 1 and 3 μm) always proposed parameters close to maximal for a given cutting tool and milling machine, regardless of tool wear indication. As a consequence, high values of productivity *W*, which result from the fact that such high limit for surface roughness had little impact on decreasing the productivity. It may be furthermore seen that lower surface roughness values cause the decrease in productivity, particularly for higher values of tool wear VB3. For localised flank wear values VB3 higher than 0.15 mm it was impossible to attain a desired nominal surface roughness of 0.3 μm.

6 Conclusions and future work in the field of computer aided milling

The results obtained from the presented application tests confirmed the proposed theses. The system is capable of maintaining specified surface roughness of machined workpiece surface. Simultaneously, the generated pair of milling parameters *f* and *n* ensure maximal productivity at a constantly updated tool wear level VB3. Economic-wise, the efficiency of the proposed solution will depend on individually assessed tool change and set-up costs and the interval of tool wear measurement VB3.

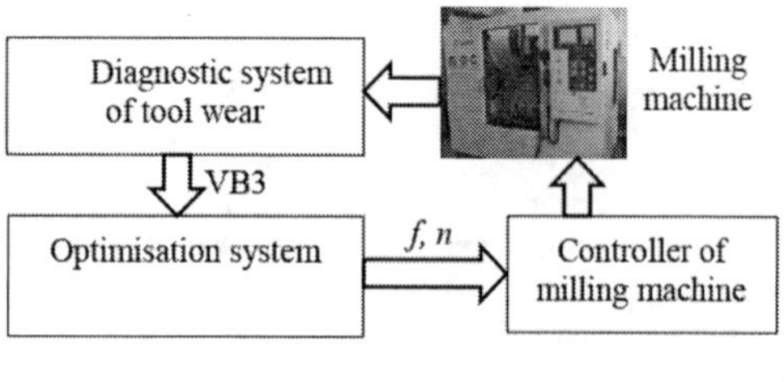

Fig.8. Automation of a milling parameter correction system.

The target application of the presented solution is automation of the localised tool flank wear monitoring procedure [12]. The value of VB3 obtained after machining of subsequent workpieces will be sent to the optimisation programme. The optimal parameters *f* and *n* will be automatically updated in the milling controller software. Our study suggests that the presented application (Fig. 8) will shorten correction time and ensure full operation of milling cutter inserts without compromising the specified design features of the final product.

Table 1 . Results of software testing.

Ra nom [μm]	VB3=0 [mm]		
	v_c [m/min]	f_z [mm/tooth]	W [mm/min]
3.0	450	0.3	3443
2.0	450	0.3	3443
1.2	450	0.24	2754
1	450	0.2	2295
0.8	450	0.2	2295
0.7	450	0.18	2066
0.5	450	0.14	1607
0.3	450	0.05	574
Ra nom [μm]	VB3=0.15 [mm]		
	v_c [m/min]	f_z [mm/tooth]	W [mm/min]
3.0	450	0.3	3443
2.0	450	0.3	3443
1.2	459	0.25	2926
1	450	0.24	2754
0.8	450	0.2	2295
0.7	363	0.17	1574
0.5	450	0.14	1607
0.3	450	0.02	230
Ra nom [μm]	VB3=0.3 [mm]		
	v_c [m/min]	f_z [mm/tooth]	W [mm/min]
3.0	450	0.3	3443
2.0	450	0.3	3443
1.2	450	0.23	2639
1	450	0.23	2639
0.8	450	0.2	2295
0.7	450	0.17	1951
0.5	450	0.11	1262
0.3	not available		0
Ra nom [μm]	VB3=0.45 [mm]		
	v_c [m/min]	f_z [mm/tooth]	W [mm/min]
3.0	450	0.3	3443
2.0	450	0.26	2984
1.2	450	0.25	2869
1	450	0.22	2525
0.8	450	0.18	2066
0.7	450	0.17	1951
0.5	450	0.08	918
0.3	not available		0
Ra nom [μm]	VB3=0.6 [mm]		
	v_c [m/min]	f_z [mm/tooth]	W [mm/min]
3.0	450	0.3	3443
2.0	450	0.3	3443
1.2	450	0.17	1951
1	450	0.17	1951
0.8	450	0.14	1607
0.7	320	0.1	816
0.5	450	0.05	574
0.3	not available		0

 DOI: 10.1051/itmconf/20171501005

References

1. L. Guoliang, Z. Bin, H. Chuanzhen, Int. J. Adv. Manuf. Tech., **83**, 257 (2016)
2. S. Fa; L. Guofu, *MATEC Web of Conferences* **75** Article Number: UNSP 02002 (2016)
3. O. Taga, Z. Kiral, K. Yaman, Int. J. Precis. Eng. Man., **17**, 579 (2016)
4. Oktem, T. Erzurumlu, F. Erzincanli, Mater. Design., **27**, 735 (2006)
5. U. Zuperl, F. Cus, Precis. Eng., **43**, 530 (2016)
6. S. Moshat, S. Datta, A. Bandyopadhyay, P. Pal, Int. J. Eng. Sci, **2**, 95 (2010)
7. I. Kandilli, Soenmez, E. Murat, M. HuseyiConference: IEEE International Conference on Mechatronics and Automation, **8**, (2007)
8. L. Xiaohong, H. Xiaochen; H. Wang et al. Ind. Lubr. Tribol., **68** , 206, (2016)
9. I. Jawahir, S. Brinksmeier, E. M'Saoubi, Cirp Ann-Manuf. Techn., **60**, 603, (2011)
10. A. Maurel, G. Michel, S. Thibaud, *10th ESAFORM Conference on Material Forming,* **18**, (2007)
11. S. Guha, T. Banerjee, A. Bandyopadhyay, *28th International Conference on CAD/CAM, Robotics and Factories of the Future (CARs and FoF),* **8**, (2016)
12. P. Stavropoulos, A. Papacharalampopoulos, E. Vasiliadis, Int. J. Adv. Manuf. Tech., **82,** 509 (2016)

ITM Web of Conferences 15, 01006 (2017)
CMES'17 DOI: 10.1051/itmconf/20171501006

Repeatability of the three dimensional kinematics of the pelvis, spine and lower limbs while performing selected exercises

Maria Skublewska-Paszkowska[1,*], *Edyta* Łukasik[1], *Jakub* Smołka[1], *Magdalena* Zawadka[2], *Mirosław* Jabłoński[3], and *Piotr* Gawda[4]

[1]Lublin University of Technology, Faculty of Electrical Engineering and Computer Science, Institute of Computer Science, Nadbystrzycka 38D, 20-618 Lublin, Poland
[2]Medical University of Lublin, Faculty of Health Sciences, Jaczewskiego 8 Street, 20-090 Lublin, Poland
[3]Medical University of Lublin, Department of Rehabilitation and Orthopedics, Jaczewskiego 8, 20-090 Lublin, Poland
[4]Medical University of Lublin, Faculty of Health Sciences, Department of Rehabilitation and Physiotherapy, Magnoliowa 2, 20-143 Lublin, Poland

Abstract. Three-dimensional optical systems are used for calculating many kinetic and kinematic parameters. The obtained data are precise; however, their repeatability is a very important aspect. The aim of this paper is to verify the range of motion repeatability of one healthy subject in the joints of the pelvis, spine and lower limbs based on the coefficient of variation. The participant performed seven exercises, repeated five times: two-leg squat, single-leg squat, forward bending, forward-step motion, step onto the stair, hip extension in a standing position and tip-toe extension while standing. Motion was recorded using Vicon motion capture system consisting of eight NIR cameras. The participant had 39 markers attached to her body according to the Plug-in Gait model. The coefficient of variation was calculated in three dimensions (X, Y and Z). The greatest repeatability, pursuant to the coefficient, was observed during the two-leg squat and forward bending in the sagittal plane (X coordinate). It was also high during the single-leg squat. The lowest repeatability was observed during the tip-toe extension while standing and the hip extension in a standing position. During the step onto the stair and the forward-step motion, a higher repeatability of measurement occurred in the open kinematic chain than in the closed chain; in the hip extension the reverse occurred. Repeatability of a range of motion is different in two types of kinematic chain and in 7 exercises. Exercises such as tip-toe extension and hip extension, which require a greater ability to balance, indicated more variability in movement.

1 Introduction

Three-dimensional analyses are commonly used in biomechanical and clinical movement research to understand pathological movement patterns. Kinematic movement analysis has contributed valuable information to the physiology of human movement and physical therapy examination. It is also used to describe specific parameters of motor functionality.

Three-dimensional (3D) computerized motion analysis systems are routinely used in gait analysis research and in clinical practice, and have been applied in evaluating lower limb and lumbopelvic motion during functional tasks [1] .The use of various 3D systems for evaluating joints' kinematics has been previously investigated [2]. The VICON system is an optical tracking system which tracks three-dimensional movement trajectories of reflective markers with up to 1mm accuracy. It is seen as the gold standard in optical gait analysis, and it is used for making comparisons with other systems [3-6].

Dynamic accuracy cannot be predicted based on static parameters. Previous studies revealed the significant influence of the system environment on the performance of video-based motion capturing systems. The number of cameras and the measurement site have the most significant influence on instrumental errors. The use of standard laboratory equipment makes the procedure widely applicable in functional error assessment [7-8]. Larger marker diameters led to a higher precision of measurement. Accuracy decreases significantly when using an optical lens filter. Optical motion capture provides precise measurement tools for biomechanical applications [9]. Marker placement, which differs from examiner to examiner, is identified as the largest source of variability in the evaluation of kinematic parameters.

Previous results demonstrate that system accuracy and reliability can be obtained allowing the collection of comparable data across different laboratories with varying configurations and equipment [10]. Vicon optical motion capture is accepted as a sensitive data research technology used for making comparisons with

* Corresponding author: maria.paszkowska@pollub.pl

DOI: 10.1051/itmconf/20171501006

new 3D motion systems [11]. The kinematic parameters of human gait measured using the Vicon system are some of most often parameters described in literature [12,13]. Motion capture systems were used for investigating running parameters [14,15]. However, there is a lack of kinematic analysis during exercises in open-closed chains.

The intra-class correlation coefficient of different methods of measuring the range of motion (goniometer, Vicon system and marker-less system) shows that the reliability of all methods is adequate. An alternative to Plug-in-Gait's model for calculating the center of a joint has also been proposed [16]. Electromagnetic motion capture systems also achieve a high level of repeatability in the analysis of pelvis and thorax kinematics in different environmental settings during a golf swing [17]. Because of motion capture usage in sport [18] and physical therapy research, the repeatability and variability evaluation of kinematic parameters are still needed to improve the quality of the physical measurement of human movement.

The purpose of this study is to use seven tests to evaluate the measurement repeatability of three-dimensional motion analysis systems (using the Vicon optical motion capture system with the Plugin-gait model as an example). Seven different functional movement tasks (two-leg squat, single-leg squat, forward bending, forward-step motion, step onto the stair, hip extension in a standing position and tip-toe extension while standing) were recorded with 3D full-body kinematics.

To define the repeatability of joint kinematics measurement, range of motion (ROM) was selected. It was calculated as the difference between the maximum value of the angle and the minimum value of the angle. The authors take into account the ankle joint, the knee joint, the hip joint, the pelvic tilt and the spine.

2 Materials and methods

One young healthy woman (age 22, body length 1.76m, body mass 61kg) participated in this study. She made five repetitions of each of the following exercises:

- single leg squat with erect posture and legs held parallel,
- forward bending with extended knees,
- stepping onto a 16cm high stair with one leg and stepping onto the stair (the participant started with toes 15cm away from the step),
- full two-leg squat with upper limps held parallel to the ground (The participant squatted as low as she could safely, and then rose back up. The heels could not lift off the floor),
- forward-step motion, 30 cm in length and return-step motion,
- hip extension in standing position without trunk bending forward,
- maximal tip-toe ascent while holding her position for 3 seconds.

Approval for this research was obtained from both the ethical committee at the Lublin University of Technology (no. 6/2015) and the ethical committee at the Medical University of Lublin (no. KE-0254/331/2015). The participant of the study was informed about the study procedures in detail. She signed a consent form prior to the measurements.

3D kinematic data was captured using a Vicon (Vicon Motion Systems Ltd., Oxford, United Kingdom) optical motion capture system equipped with eight NIR T40S cameras (two cameras mounted on each wall of the motion recording room) and two reference Bonita cameras on tripods. The Vicon Nexus 2.0 was used for data collection and post-processing. The data was sampled at 100 Hz.

During a session the participant had 39 markers placed all over her entire body as specified in the Plug-in Gait Model documentation [19]. The markers were fixed to the skin using Vicon double sided tape. The participant was measured in compliance with the Plug-in Gait Model specification (leg length, knee width, ankle width, shoulder offset, elbow width, wrist width, hand width, weight and height) so that subject created in Vicon Nexus software could be calibrated. All motion recordings were made in a room without daylight access. Ankle-joint angles were calculated using the Euler angle conventions of plantar flexion/ dorsiflexion about the y-axis and varus/ valgus about the x-axis. The Plug-in Gait Model computed the subject kinematics and kinetics data. The measurements were input into the Vicon system so that the a new subject could be created. The first participant's capture was in a motorbike pose and was used for the subject's model calibration. Additionally, the marker placements were validated.

Recorded 3D data was post-processed: (1) each marker was labelled, (2) gaps in marker trajectories were filled, (3) the Plug-in Gait dynamic model was run. This computed the model outputs, some of which were used in this analysis. Then the recorded and generated data was exported to a C3D file (a file that contains the markers' positions and the model outputs for each frame of the recording). For this purpose a custom piece of software was written in C++ using B-TK (Biomechanical Toolkit) and Eigen libraries.

The software (figure 1) is used for automatic data export and collection. It (1) sets up variables containing input data details, values being exported, trial names etc., (2) goes through data recordings of consecutive objects and extracts their names (from file paths), (3) enumerates for each object all sessions (recording days) and extracts the session names, (4) enumerates trials and their variants, constructs the input file names and clears the temporary file list, (5) enumerates the exported values if the input file exists and saves them to temporary files, adds the names of those files to the temporary file list, (6) merges files in the temporary file list (to produce a single file with all the values for a particular trial), removes the temporary files.

 DOI: 10.1051/itmconf/20171501006

```
//set data directory...
DATA_DIR ← //...
//result file prefix
PREFIX ← "angles_";
//specify exported angles array
ANGLES ← {"LabsAnkleAngle",
"RAbsAnkleAngle", "LAnkleAngles",
"RAnkleAngles", "LHipAngles",
"RHipAngles", "LKneeAngles",
"RKneeAngles", "LPelvisAngles",
"RPelvisAngles", "LSpineAngles",
"RSpineAngles"};
//specify trial names array
TRIALS ← {"squat_one_leg_left",
"squat_one_leg_right", "squat_two_legs",
"bend_over"};
//specify trial variants array
VARIANTS ← {"_1", "_1b", "_2", "_2b",
"_3", "_3b", "_4", "_4b", "_5", "_5b"};

//browse patients
for all patient_directories in DATA_DIR
 do
  begin
   name ←
    basename(patient_directory)
   //browse recording sessions of a
   //patient
   for all session_directories in
          patient_directory do
    begin
     session_name ←
      basename(session_directory)
     //browse recorded trials
     for all trials in TRIALS do
      begin
       //browse attampts/variants of
       //trials
       for all variants in VARIANTS
        do
         begin
          file ← session_directory+
           trial+ variant+ ".c3d";
          files_for_merging ← "";
          if (file_exists(file))
           begin
            //save angle data to
            //temporary files
            for all angles in ANGLES do
             begin
              out_file ← PREFIX+ trial+
               "_"+ name+ "_"+ angle+
               ".csv";
              //angles are saved as
              //markers in c3d files
              export_marker_value(angle,
               out_file);
              files_for_merging ←
               files_for_merging +
               out_file;
             end
           end
          //merge temporary files
          result_file ← PREFIX+ trial+
           variant+ "_"+ name+ ".csv";
          merge_files(files_for_merging,
           result_file);
          remove_files(
           files_for_merging);
         end
      end
    end
  end
```

Fig. 1. Pseudo-code used for data export and merging.

The coefficient of variation – CV (see formula 1) [20] was used for computing the repeatability of ROM.

$$CV = \frac{\sigma}{\mu} \cdot 100 \qquad (1)$$

where σ denotes the standard deviation and μ the mean. An examined parameter was assumed to be repeatable when its CV was lower than ten.

3 Results

The greatest repeatability was observed during the two-leg squat and the forward bending in a sagittal plane (X coordinate). It was also high during the single-leg squat. The lowest repeatability was observed during the tip-toe standing and the extension of the hip.

Coefficient of variation (CV%) was the highest in the transverse plane (Z coordinate) in most exercises except for the tip-toe standing and single-right-leg squat, where the frontal plane motion had the worst repeatability.

Repeatability of a range of motion is different in the two types of kinematic chains and in 7 exercises. Exercises such as tip-toe extension and hip extension, which require a greater ability to maintain balance, indicated more variability in movement.

Specifically high repeatability, pursuant to the coefficient, was observed during the two-leg squat (Table 5) and the forward bending (Table 6) in a sagittal plane (X coordinate). The lowest repeatability was observed during the tip-toe extension while standing (Table 6) and the hip extension in a standing position (Table 4).

A higher repeatability of measurement occurred in open rather than in closed kinematic chains during stepping onto the stair and stepping forward. The same does not hold for hip extension in a standing position (the CV% values are similar – Table 4).

However the range of motion was found to be highly repeatable in the saggital plane (X) for all joints in the closed kinematic chain during the single-leg squat (the average of CV% values for all joints is equal to 5,02% - see Table 1). The average CV% values for stepping forward and stepping onto the stair in the open kinematic chaing were slightly higher, equaling 6,64% (Table 3) and 7,08% (Table 2) respectively.

4 Discussion

Repeatability of human gait parameters measured by the Vicon motion capture system was previously investigated using the coefficient of variation and the coefficient of multiple correlation [13, 14]. Intra-subject repeatability was excellent for kinematic data in the sagittal plane both within a test day as well as between test days. In our study joint movements in the sagittal plane were more repeatable than in the frontal or transverse planes motions.

Table 1. Coefficient of variation (%) of range of motion during the single-leg squat (for each joint the upper row contains the values for the left side, the lower row contains values for the right side).

Joints	Single-leg squat					
	Open kinematic chain			Closed kinematic chain		
	x	y	z	x	y	z
Knees	6.78	4.93	13.94	6.74	3.6	14.49
	3.15	0.85	6.08	1.92	21.55	4.57
Ankles	10.8	26.92	26.4	6.04	11.68	11.57
	36.02	12.4	10.55	3.79	18.86	18.32
Hips	4.37	7.72	22.02	5.69	7.21	11.01
	10.95	5.99	6.15	5.95	7.33	14.7
Mean	12.01	9.8	14.19	5.02	11.7	12.44
SD	12.19	9.19	8.41	1.81	7.12	4.66

Table 2. Coefficient of variation (%) of range of motion during stepping onto the stair (for each joint the upper row contains the values for the left side, the lower row contains values for the right side).

Joints	Step onto the stair					
	Open kinematic chain			Closed kinematic chain		
	x	y	z	x	y	z
Knees	5.34	2.45	11.12	28.68	16.59	11.27
	8.59	5.3	5.19	26.34	0.2	11.76
Ankles	11.3	13.58	13.24	16.73	21.24	21.12
	11.85	11.54	11.42	19.48	31.57	30.65
Hips	2.14	11.88	17.45	6.56	10.66	9.94
	3.29	20.65	12.21	6.64	43.25	43.2
Mean	7.08	10.9	11.77	17.4	24.94	21.32
SD	4.11	6.42	3.96	9.44	11.57	13.32

Table 3. Coefficient of variation (%) of range of motion during forward step (for each joint the upper row contains the values for the left side, the lower row contains values for the right side).

Joints	Forward step					
	Open kinematic chain			Closed kinematic chain		
	x	y	z	x	y	z
Knees	10.63	7.87	18.16	27.76	42.32	12.95
	5.35	4.98	6.04	8.45	18.9	12.17
Ankles	3.32	9.02	8.7	4.33	19.45	18.76
	8.57	11.47	11.15	3.08	13.86	13.73
Hips	7.69	13.17	35.64	8.87	18.9	26.48
	4.28	11.83	9.12	9.2	10.75	20.92
Mean	6.64	9.73	14.8	10.28	20.7	17.5
SD	2.79	3.03	11	8.94	11.15	5.61

Table 4. Coefficient of variation (%) of range of motion during hip extension in a standing position (for each joint the upper row contains the values for the left side, the lower row contains values for the right side).

Joints	Hip extension					
	Open kinematic chain			Closed kinematic chain		
	x	y	z	x	y	z
Knees	21.12	25.02	23.59	23.74	45.4	20.61
	41.03	70.39	61.57	17.65	16.05	18.82
Ankles	9.4	36.61	36.69	43.18	26.58	26.3
	31.83	33.36	33.36	46.41	22.82	21.83
Hips	6.52	33.93	53.21	11.16	22.97	28.8
	7.67	33.55	59.71	8.16	6.09	25.34
Mean	19.6	38.81	44.69	25.05	23.32	23.62
SD	14.34	15.96	15.62	16.24	13.03	3.8

Table 5. Coefficient of variation (%) of range of motion during two-leg squat (for each joint the upper row contains the values for the left side, the lower row contains values for the right side).

Joint	Two-leg squat		
	x	y	z
Knees	2.95	3.05	7.45
	3.45	3.94	2.98
Ankles	3.68	10.19	9.94
	4.15	17.98	18.43
Hips	2	11.78	14.42
	2.06	17.47	14.72
Pelvis	8.47	13.44	13.06
Spine	1.12	13.54	16.64
Mean	3.49	11.42	12.21
SD	2.25	5.55	5.12

Table 6. Coefficient of variation (%) of range of motion during tip-toe standing and forward bending (for each joint the upper row contains the values for the left side, the lower row contains values for the right side).

Joint	Tip-toe			Forward bending		
	x	y	z	x	y	z
Knees	15.56	20.85	8	10.03	8.01	10.88
	30.11	25.04	18.08	6.88	8.79	12.13
Ankles	1.76	6.6	6.37	6.81	25.54	24.26
	2.75	11.3	11.47	7.96	9.64	9.58
Hips	6.84	5.8	7.78	2.77	11.47	14.2
	22.99	96.91	38.21	3.22	11.45	7.32
Pelvis	22.94	29.83	25.16	3.2	8.67	5.65
Spine	16.23	37.09	77.75	1.73	8.15	9.13
Mean	14.9	29.18	24.1	5.33	11.46	11.64
SD	10.34	29.55	24.24	2.98	5.85	5.75

DOI: 10.1051/itmconf/20171501006

The marker-based MoCap system has high reliability and validity as indicated by the coefficient of variation (CV%) [21]. Another study provides evidence that a marker-based and a markerless protocol report similar ranges of pelvis and lower limb angles in the single-leg squat [22]. The knee's range of motion was under special consideration because of the frequency of injuries in sport activities. The repeatability of knee motion measurement turned out to be the lowest during squat and lung exercises when compared to eleven motor tasks [23]. In our study we analyzed the hip, the ankle and the knee joints and the pelvic and the spine motion, and we observed the highest repeatability during the two-leg squat and the forward bending in a sagittal plane. Because the hip extension and the tip-toe extension require balancing and holding one's position, there is more variation in the range of motion than in the case of others exercises. More stable exercises, especially with two-leg support have the highest repeatability of movement.

The joint angle can be adjusted without incurring any changes in other joints in an open kinematic chain. There is one primary joint axis and a single dominant plane of movement during the exercise. Moreover, more isolated joint motion occurs. In a closed kinematic chain both ends of the chain are fixed and any changes in the angle of one joint reciprocally results in altered angles in the other joints. Triplanar multiple-joint movement occurs. Complex movements such, as stepping onto the stair, lunging, stepping forward, typically consist of both open and closed kinematic chains. When one lower limb is fixed to the ground, the other moves freely.

Previously, the parameters of the lower limbs and pelvis were correlated between kinematic chains during gait [24]. Our study shows that open kinematic chains have higher repeatability of movement, probably because the main part of the motion occurred in this chain. Legs in the closed kinematic chain have to maintain balance and give stable support. Because the range of motion in closed kinematic chains was small in most cases, the relative change values may seem large.

5 Conclusions

Repeatability of the joint range of motion may be affected by the type of exercise, the stability of the position and the type of kinematic chain. Further investigations of tridimensional motion are needed to explain the differences observed between closed and open kinematic chains.

References

1. A. G. Schache, P.D. Blanch, D,A. Rath, T.V. Wrigley, R. Starr, K. L. Bennell. Hum. Mov. Sci. 21, 273–293 (2002)
2. M. Kim, C. Yi, O. Kwon, S. Cho, H. Cynn, Y. Kim, S. Hwang, B. Choi, J. Hong, D. Jung. Spine 38, 1260–1267 (2013)
3. Y. Levanon, A. Gefen, Y. Lerman, U. Givon, N. Z. Ratzon. Gait Posture 32, 469–474 (2010)
4. B. Müller, W. Ilg, M. A. Giese, N. Ludolph. PLOS ONE 12, e0175813 (2017)
5. S. Springer, G. Y. Seligmann. Sensors 16, 194 (2016)
6. L. D. Duffell, N. Hope, A. H. McGregor. Proc. Inst. Mech. Eng. H (2014)
7. P. Eichelberger, M. Ferraro, U. Minder, T. Denton, A. Blasimann, F. Krause, H. Baur. J. Biomech. 49, 2085–2088(2016)
8. G. E. Gorton III, D. A. Hebert, M. E. Gannotti. Gait Posture 29, 398–402 (2009).
9. M. Windolf, N. Götzen, M. Morlock. J. Biomech. 41, 2776–2780 (2008)
10. E. Miller, K. Kaufman, T. Kingsbury, E. Wolf, J. Wilken, M. Wyatt. Gait Posture 50, 116–119 (2016)
11. K. Otte, B. Kayser, S. Mansow-Model, J. Verrel, Friedemann Paul, A. U. Brandt, T. Schmitz-Hübsch. PLoS ONE 11, (2016)
12. M. P. Kadaba, H. K. Ramakrishnan, M. E. Wootten. J. Orthop. Res. 8, 383–392 (1990)
13. M. P Kadaba, H. K. Ramakrishnan, M. E. Wootten, J. Gainey, G. Gorton, G. V. B. Cochran. J. Orthop. Res. 7, 849–860 (1989)
14. A. G. Schache, P.D. Blanch, D,A. Rath, T. V. Wrigley, R. Starr, K. L. Bennell. Gait Posture 15, 136–145 (2002)
15. D. Levine, M. A. Colston, M. W. Whittle, E. C. Pharo, D. J. Marcellin-Little. J. Athl. Train. 42, 29–34 (2007)
16. S. P. Nair, S. Gibbs, G. Arnold, R. Abboud, W. Wang. Clin. Biomech. Bristol Avon 25, 582–587 (2010)
17. VICON, 2016. Plug-in Gait Product Guide [WWW Document]. VICON. URL http://www.vicon.com/downloads/documentation/vicon-documentation/plug-in-gait-product-guide (accessed 26.06.2017)
18. J. Wasik, GB Shan. ARCHIVES OF BUDO 11, 61-67, (2015)
19. JF. Reed, F. Lynn. Meade, BD. Clin Diagn Lab Immunol 9 (6) 1235 –1239 (2002)
20. K. Evans, SA. Horan, RJ. Neal, RS. Barrett, PM. Mills. Sports Biomech. 11, 262–272 (2012)
21. M. Yazdifar, M. Yazdifar. Eur. Sci. J. ESJ 11, (2015) http://eujournal.org/index.php/esj/article/view/5979
22. M. A. Perrott, T. Pizzari, J. Cook, J. A. McClelland. Gait Posture 52, 57–61 (2017)
23. K. Desloovere, P. Wong, L. Swings, B. Callewaert, H. Vandenneucker, A. Leardini. Gait Posture 32, 597–602 (2010)
24. Z. Svoboda, M. Janura, P. Kutilek, E. Janurova. J. Hum. Kinet. 51, 37–43 (2016)

DOI: 10.1051/itmconf/20171501007

Segmentation and reconstruction of the 3D geometry of the middle and inner ear

Yanfei Lu[1,*]

[1]Warsaw University of Technology, Faculty of Production Engineering, Institute of Mechanics and Printing, Department of Construction Engineering and Biomedical Engineering, Narbutta 85, 02-524 Warsaw, Poland

Abstract. The anatomical model of the ear is of great importance in the design of ossicular prosthesis, cochlear implant electrodes, as well as for the preoperative planning and navigation of surgery. By means of micro-computed tomography (micro-CT) and technology of 3D reconstruction, an anatomical model of the middle and inner ear was built. Region of interest includes the ossicular chain (malleus, incus, and stapes), cochlea (scala vestibule-ST, scala tympani-ST, basilar membrane-BM, spiral ligament-SL and osseous spiral lamina-OSL), tympanic membrane-TM, oval window membrane-OWM, round window membrane-OWM and stapedial annular ligament-SAL. The micro-CT images of a cadaver's temporal bone were acquired by "SkyScan 1076" (Kontich, Belgium, www.skyscan.be) and then reconstructed to cross-section images by SkyScan NRecon™ (v1.6.10.4). The image processing and 3D geometry reconstruction of temporal bone were performed by software Mimics® (v14.0, Materialise NV, Leuven, Belgium). The obtained structures are measured and validated against literature data and the results are in good agreement.

1 Introduction

Middle ear and inner ear are important structures responsible for sound transduction and balance. Sound wave in the air is firstly detected by eardrum (tympanic membrane), then transformed through ossicular chain to cochlea and there transduced to bioelectric signal. Ossicular chain in the middle ear consists of malleus, incus and stapes. The three ossicles are attached to surrounding bony structures by ligaments and tendons and are connected with each other by two joints. In this way, they keep a complex spatial structure and fulfill the function of sound transmission. Malleus connects to tympanic membrane, and stapes connects to cochlea. Around the footplate of stapes, there is a ring of fibrous tissue that connects the stapes to the oval window of the inner ear, called stapedial annular ligament. The opening on the cochlea in contact with stapes is covered with a membrane, called oval window membrane.

The bony labyrinth of the inner ear is the network of passages with bony walls lined with periosteum in the temporal bone. It consists of three parts: the vestibule, semicircular canals, and cochlea. Inside these cavities, there exists a system of endolymph filled tube, membranous labyrinth, which is separated from bony labyrinth by a layer of perilymph fluid. Cochlea is a spiral shaped cavity in the bony labyrinth. In humans, it makes about 2.5 turns around its axis. There are three scalae in the cochlea: scala vestibule, scala tympani and scala media. Scala vestibule lies superior to the cochlea spiral and filled with the fluid of perilymph. Scala tympani lies inferior to cochlea spiral and filled with the fluid of endolymph. Between them, there is scala media containing also the fluid of endolymph. Normally, the membrane between SV and SM, Reissner's membrane, is very thin and hard to illustrated by means of micro-CT. The basilar membrane, which separates apart the ST and SM, has some degrees of thickness that can be detected by micro-CT. Mechanical waves excited by the OWM, which locates on the bottom end of the SV, transfer along the spiral coli to the apex of the cochlea, the helicotrema, where the SV and ST merge, then come back along another duct ST. Sensory hair cells on the BM are activated by the fluid wave in SV and ST, and convert the mechanical signal to biological signal transferred to the brain along the auditory nerve. The other two parts of the bony labyrinth, the vestibule and semicircular canals, are responsible for the sensations of balance and motion. They use the same kinds of lymph fluids and sensory cells as the cochlea uses, and detect the attitude and motion of the body.

Diseases or injuries of the middle ear or inner ear may cause severe healing loss. About 5% population suffer from disability associated with hearing problem worldwide [1]. The causes of hearing loss are different from conductive (middle ear) or sensorineural (inner ear). Ossicles prosthesis are applied to middle ear damage caused by chronic inflammatory or disease, such as otosclerosis (stiffness of stapes), to reconstruct the

* Corresponding author: y.lu@wip.pw.edu.pl

 DOI: 10.1051/itmconf/20171501007

conduction system [2, 3]. While damage of inner ear requires the application of cochlear implant.

Optimal designed ossicular prosthesis and cochlear implant electrodes, as well as their precise insertion are possible thanks to the development of computer-aided engineering (CAE) and computer-aided surgery (CAS): design of the ossicular prosthesis and cochlear implant electrodes is greatly simplified by computer-aided design (CAD) [4, 5]; afterwards verification and optimization of them are achieved by finite element method (FEM) [6, 7, 10]; due to the complexity and small scale of the structure of the middle and the inner ear, CAS is usually applied in preoperative surgery planning and navigation during the surgery [8, 9].

Effectiveness of these computer-based approaches is based on a reliable geometric model. To obtain the 3D anatomical information of the analyzed structures, several methods have been used, among them histological microgrinding [11], Magnetic Resonance Imaging (MRI) [12] and micro computed tomography (micro-CT). Histological microgrinding traditionally used and features for high accuracy but destructive procedure. MRI features for large contract for soft tissue but low spatial resolution. Therefor micro-CT is preferred for its high resolution and nondestructive procedure.

Aim of this work is to build a 3D anatomical model of the middle and inner ear by means of micro-CT and 3d reconstruction. Region of interest includes the ossicular chain (malleus, incus, and stapes), cochlea (scala vestibule-ST, scala tympani-ST, basilar membrane-BM, spiral ligament-SL and osseous spiral lamina-OSL), tympanic membrane-TM, oval window membrane-OWM, round window membrane-OWM and stapedial annular ligament-SAL.

2 Methods

2.1. Temporal bone preparation

The human cadaver's temporal bone was dissected at Warsaw Medical University from male cadaver. The temporal bone was harvested within 2 days of death and removed from soft tissues and frozen immediately. The specimen was cut to the size about 2 cm × 2 cm × 2 cm cubic, in order to minimize the X-ray absorption by surrounding bone.

2.2 Micro-CT scanning process and the image acquisition

2.2.1 Methodology of micro-CT scanning process

Micro-CT is a non-invasive and efficient technique to investigate the 3D geometry of anatomic structure. In this study, to examine the structure of the middle and inner ear, a piece of extracted calvarial specimen was scanned using a micro-CT instrument "SkyScan 1076" (Kontich, Belgium, www.skyscan.be).

The micro-CT system consists of the X-ray lamp with a high voltage power supply, one bed for the sample with a precise object positioning mechanism and a CCD camera connected to a computer with software for storing and processing of the received data. This equipment is able to emit X-ray on energy range from 20kV to 100kV. The data are obtained by rotating the X-ray tube and the CCD cameras for up to 180 degrees or 360 degrees, with adjustable increments (at least 0.02 degree) around the axis of the immovable object. The resulting individual 2D projection x-ray images are saved as 16-bit TIFF files. The rotation of the imaged sample relative to the x-ray source and camera allows the precise 3D location of the scanned objects to be calculated.

2.2.2 Parameters of the scanning process

The scan was performed with a source voltage and source current of, respectively, 100 kV and 100 μA. An aluminum filter with a thickness of 1mm was used to narrow the X-ray energy spectrum, reducing the low energy component, thus reducing beam hardening. The CT images were acquired at a step angle of 0.6° for a total circular orbit of 360°. Exposure time per frame was 620 ms. In order to improve quality of the final image file, averaging of four images was conducted. Scanning resolution in this study is set at 18 μm. All projection images were saved as 16-bit TIFF files (Fig.1).

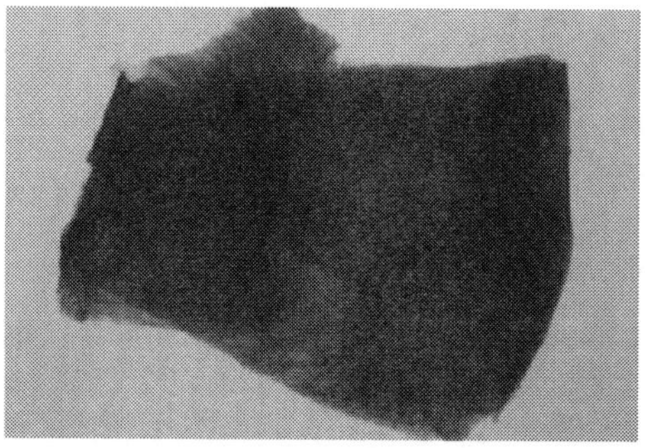

Fig. 1. Example 2D projection image of micro-CT scan.

2.2.3 2D Reconstruction

Following projection images acquisition from multiple angles, cross-section reconstruction of from CT image was performed based on the Feldkamp algorithm which consists of back-projection and convolution. Back-projection derives object position from the shadow position from multiple angles, while convolution removes a blurring halo from around the periphery of objects which back-projection alone would leave, by a selective image subtraction method. In this study, SkyScan NRecon (Version 1.6.10.4) package was used to reconstruct cross-section images of the object from tomography projection.

Some imaging artifacts are peculiar to CT, such as ring artifacts and beam hardening. Beam hardening rises from emission from the micro-focus x-ray source of a mixture of low and high energy x-rays. Ring artifacts arise from the incremental rotation of the object-camera geometry, in combination with non-uniformity of the x-ray detection performance of individual camera pixels. During the reconstruction, these artifacts could be corrected by software to some extent.

The final projection images in 16-bit TIFF format were loaded and then the following parameters of the reconstruction process were adjusted by a try-and-error fashion and defined as: Ring Artifact Correction=10, Beam Hardening Correction (%)=17. Post-alignment was set as 5.5 to reduce tails, doubling or blurring in the reconstructed images.

Region-of-interest was chosen to include the ossicular chain (malleus, incus, stapes), tympanic membrane and bony labyrinth (vestibule, semicircular canals and cochlea). Grey level threshold was adjusted according to a preview slice to make sure the structures of interest obvious and in good contrast. A total number of 600 slice images were saved in 8-bit grayscale BMP format at the resolution 2000 x 1336 pixels. Figure 2 shows an example of one slice image of the inner ear.

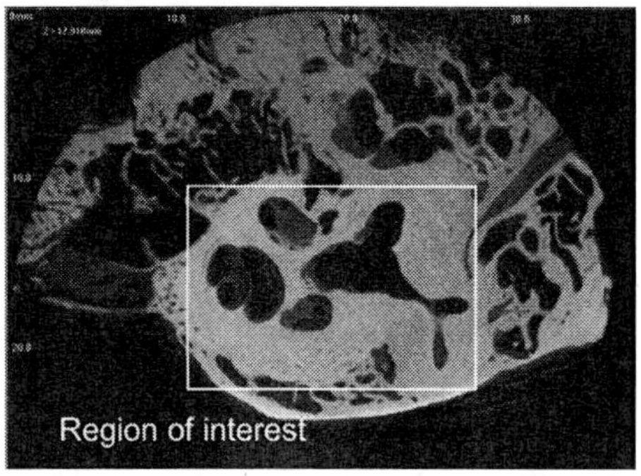

Fig. 2. Example of cross sectional image.

2.3 Image processing and three-dimensional reconstruction

In order to separate the regions of interest from the rest of the image, a series of segmentation tools include Thresholding, Region growing, Boolean and Morphology operations were used.

The first step was a separation of bone from soft tissue by Thresholding tool. The defined radiodensity for bone is from 226 to 3024 HU and for soft tissue is from -700 to 225 HU. This defined CT value well reflected the bone and soft tissue partition, despite some of the unavoidable noise points which could be removed manually in later steps. Because the ossicular chain is connected to temporal bone only by ligaments and tenders which were previously removed in thresholding step, we could separate it to a new mask by Region growing tool. However, there are some structures cannot be separated by automatic tools: structures in between small gap such as the 2 joints between ossicles, SAL and stapes; structures have similar contrast such as BM, SL and OSL; thin layers such as RWM and OWM. In this case, manually segmentation is applied.

3 Results

Due to the small space between incus-malleus joint and incus-stapes joint, the separation of the ossicular chain was performed manually by Mask edit tool. After segmentation, we got three masks of Malleus, Incus and Stapes, as shown in Fig.3.

Closely next to malleus, there is a membranous structure-tympanic membrane (TM), which is marked in light green, as shown in Fig.4. The 3D view reveals the topological relationship between TM and the ossicular chain. The head of malleus attaches to the incus; and the handle of malleus is connected to the cone-shaped membrane TM.

Due to the big contrast between bone and soft tissue, it's not too difficult to fill the bony labyrinth. Spiral structure of cochlear and spatial arrangement of three semicircular canals could be clearly observed in Fig.5. Round window niche and oval window niche could also be noticed.

The next step was cut out of scala vestibuli and scala tympani from the mask of cochlear capsule. Scala vestibuli (SV) lies superior to the cochlea spiral and scala tympani (ST) lies to the inferior. Both of the two spiral-shaped canals make about 2.5 turns about the modiolus. Between the ducts of SV and ST is scala media. Organ of corti (OC) and basilar membrane (BM) were not marked separately in this project because thickness of BM (5-13 μm) is less than 1 pixel (18 μm). SV and ST join at the helicotrema, the tip of the helix.

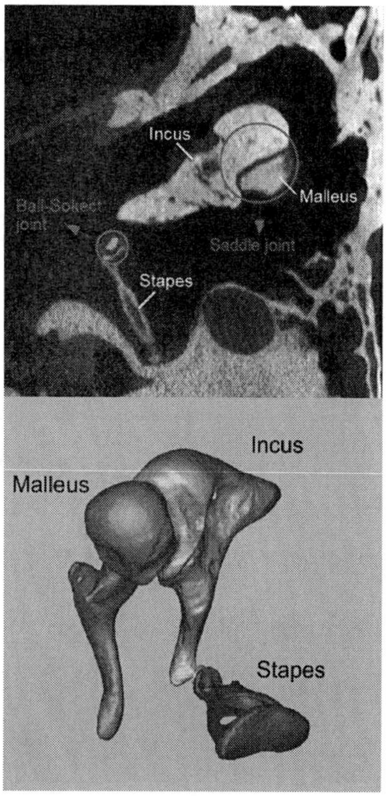

Fig. 3. Segmentation and 3D representation of ossicular chain.

DOI: 10.1051/itmconf/20171501007

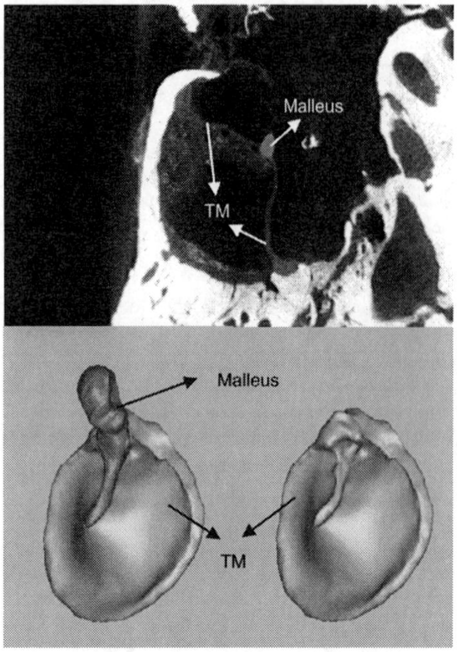

Fig. 4. Tympanic membrane (TM).

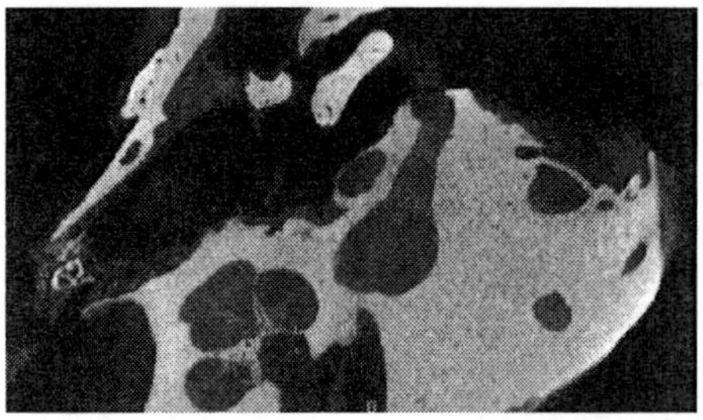

Fig. 5. Bony labyrinth.

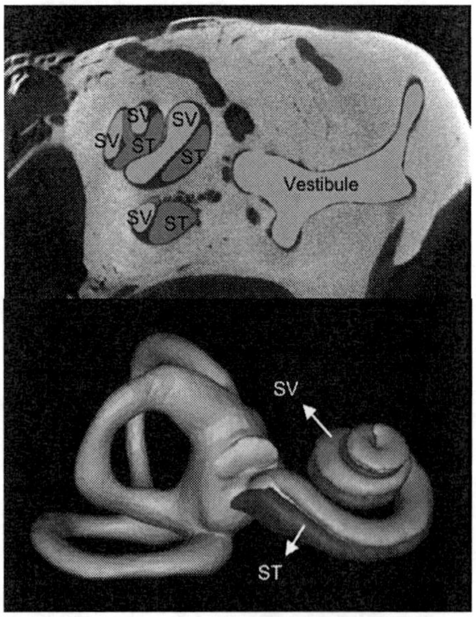

Fig. 6. Segmentation and 3D representation of scala vestibuli and scala tympani.

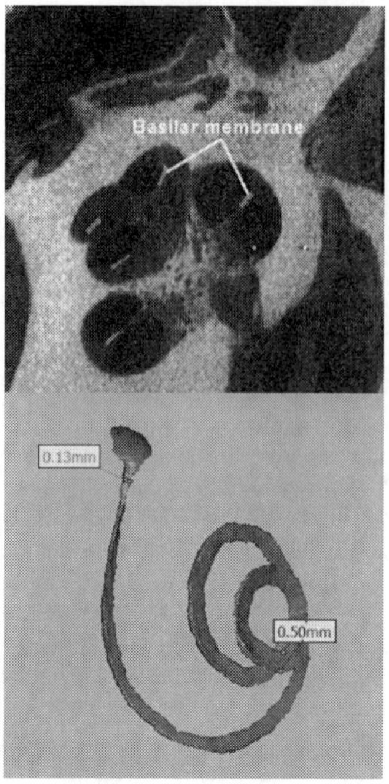

Fig.7. Basilar membrane and measurement.

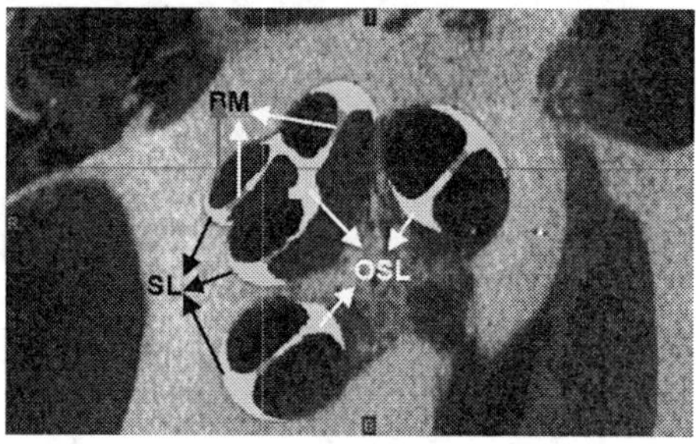

Fig.8. Segmentation of basilar membrane, spiral ligament and osseous spiral lamina.

In Fig.6, the gap between SV and ST, consisted of three parts, that are the compose of OC and BM, spiral ligament (SL) and osseous spiral lamina (OSL). Contrast between those parts was not big enough to separate them directly, therefore data reported in literature is referred, that width of BM is between 0.1 mm and 0.5 mm, narrowest at the base and widest at the apex of cochlear. Mask and reconstruction of the compose of organ of corti and basilar membrane is shown in Fig.7. SL and OSL were obtained by subtracting BM, SV and ST from the mask of bony labyrinth, as shown in Fig.8.

Round window situated at the bottom end of ST, is a membrane-covered opening between the middle ear and the cochlea. Light mask represents the RWM. This saddle-shaped membrane seals the RW niche, as shown in Fig.9.

Another tiny structure of interest is stapedial annular ligament (SAL). SAL is a ring of fibrous tissue between the base of the stapes to the oval window bony wall, which allows the vibration of stapes and seals the oval window niche. Geometry of SAL is various, the average height is 0.2 mm and thickness is between 0.04 and 0.1 mm. Mask and 3D reconstruction of SAL are shown as in Fig.10.

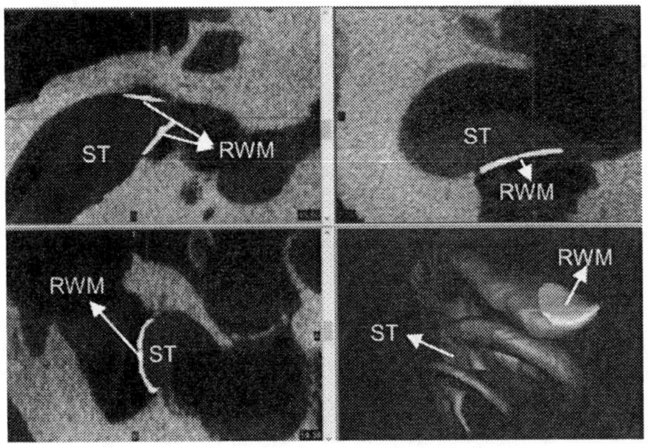

Fig.9. Segmentation of the round window membrane.

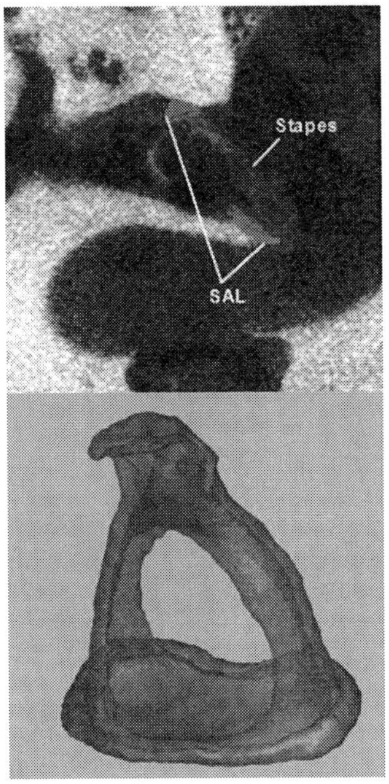

Fig.10. Segmentation of the stapes and the stapedial annular ligament.

Comparison of feature diameters are performed in order to verify the reliability of the 3D model. Firstly, the widths of 3 cochlea turns are measured and compared with data reported by Erixon [13]. The width of the first turn of cochlea is 8.17 mm (range 5.6-8.2 mm, mean width 6.8 mm), the width of the second turn is 4.25 (range 3.3-4.3 mm, mean width 3.8 mm) and the width of the third turn is 3.11 mm (range 0.6-3.6 mm, mean width 2.1 mm). Semimajor anteroposterior and semiminor supero-inferior axes of stapes footplate measured in this 3D model are 2.99 mm and 1.80 mm respectively. In vitro measurement gives the results of 2.93 mm and 1.82 mm [14]. Measurement difference between this model and in vitro experiment are 2.0% and 1.1% respectively, which are not larger than the error due to individual difference.

For membranous structure, the thickness of membranes is generally ranging between several micrometers to dozens of micrometers, which is smaller or in the same range of micro-CT resolution (18 um). Therefore, it makes no sense to compare the thickness of membranous structure. Instead, the measurement is made along the membranous surface. RWM has irregular shape, therefore the diameter is measure as the distance over the surface. Short diameter and long diameter of the RWM are 2.18 mm and 2.23 mm respectively, which are in line with the study performed by Nomura [15]. In his study, the short diameter of the round window is ranging from 1.43 to 2.38 mm (average 1.87 mm) and the long diameter is ranging from 1.83 to 3.25 mm (average 2.3 mm).

Therefore, it can be concluded that the reconstructed model reflects the real structure of the middle and inner ear. It provides reliable geometric information for anatomical structures. This model could be used for planning surgery and as well as student education in the field of otology and audiology. Furthermore, a FEM model could be generated from this model after proper modification and meshing. Application of this model could be extended to the fields of CAE and CAS.

This work is performed under the project of PBS3/B7/25/2015 founded by NCBiR.

References

1. Global Burden of Disease Study 2013 Collaborators, Lancet. **386**, 743 (2015)
2. K. Jahnke, *Middle Ear Surgery* (Thieme, Stuttgart-New York, 2004)
3. M. Neudert, T. Zahnert, N. Lasurashvili, M. Bornitz, Z. Lavcheva, C. Offergeld, Otol. Neurotol. **30**, 332 (2009)
4. Y. Song, C. Lee, Tzu Chi Medical Journal, **24**, 178 (2012)
5. R. Gan, C. Dai, X. Wang, D. Nakmali, M. Wood, Hearing Research, **263**, 138 (2010)
6. D. Kelly, P. Prendergast, A. Blayney, Otol. Neurotol. **24**, 11 (2003)
7. M. Kwacz, J. Wysocki, P. Krakowian, Biocybern Biomed Eng, **32**, 27 (2012)
8. F. Dammann, A. Bode, E. Schwaderer, M. Schaich, M. Heuschmid, M. Maassen, Radio Graphics, **21**, 183 (2001)
9. Y. Zhao, B. Dawant, R. Labadie, J. Noble, MICCAI 2014, Part I, 331
10. A. Paolis, H. Watanabe, J. Nelson, M. Bikson, M. Packer, L. Cardoso, J Biomech, **50**, 209 (2017)
11. W. Funnell, W. Decraemer, S. Khanna, J. Acoust. Soc. Am. **91**, 2082 (1992)
12. J. Lane, R. Witte, O. Henson, C. Driscoll, R. Robb, Clin Anat. **18**, 409 (2015)

 DOI: 10.1051/itmconf/20171501007

13. E. Erixon, H. Hogstorp, K. Wadin, Otol Neurotol, **30**, 14 (2009)
14. L. Aristeguieta, L. Acuna, Int J Morphol, **28**, 1059 (2010)
15. Y. Nomura, Advances in Oto-Rhino-Laryngology, **33**, 1 (1984)

ITM Web of Conferences **15**, 02001 (2017) DOI: 10.1051/itmconf/20171502001
CMES'17

Artificial neural network modelling of cutting force components in milling

Ireneusz Zagórski[1], *Monika* Kulisz[2,*], and *Aleksandra* Semeniuk[3]

[1]Lublin University of Technology, Faculty of Mechanical Engineering, Department of Production Engineering, Nadbystrzycka 36, 20-618 Lublin, Poland,
[2]Lublin University of Technology, Faculty of Management, Department of Organisation of Enterprises, Nadbystrzycka 38, 20-618 Lublin, Poland
[3]Lublin University of Technology, Faculty of Mechanical Engineering, Nadbystrzycka 36, 20-618 Lublin, Poland

Abstract. The following paper will give an account of experimental tests and simulation of the cutting force components Fx, Fy and Fz in down milling of AZ91D magnesium alloy. The milling operation employed two milling cutters with a different helix angle, λs = 20° and λs = 50°, and was conducted with changeable milling machining parameters: cutting speed, feed per tooth, axial depth of cut. The simulation part of the study was conducted in Statistica software environment with the application of Multi-Layered Perceptron neural network architecture, and using a "black box" approach, which guarantees a good fit of input and output data obtained from the experimental tests.

1 Introduction

Manufacturing innovative modern components of machines and devices for the aircraft and automotive industries requires using state-of-the-art engineering materials and effective machining methods, such as milling with advanced tools and machining centres. The ever-increasing challenges of the modern market may generate serious production costs, therefore, engineers and manufacturers must seek to increase the cost-effectiveness of production without compromising the quality of the final product. In subtractive machining optimisation may be achieved by means of analysis of the actual cutting forces and computer simulation. The combined data obtained from such analyses may improve the stability and effectiveness of machining. Artificial neural network simulation aids cost optimisation through reducing the number of actual iterations regarding the specification and optimisation of technological parameters of machining, and is moreover an effective tool for minimising machining errors.

2 The state of knowledge

Subtractive machining lies at the heart of machine building and should therefore by characterised by high productivity, efficiency, and effectiveness of subsequent machining layers removal. Milling and High Speed Milling are important representatives of subtractive machining. Qualitative description of these operations frequently employs machinability factors.

The cutting forces occurring in milling constitute the practical machinability factors. A sudden increase in the volume of cutting forces may result in undesirable machined surface errors. The total cutting force is a resultant of a number of factors, including: the material of the cutting tool and workpiece, or the cutting tool geometry; however, these are machining parameters that are of critical importance to the cutting forces. Machining with high cutting speeds, feeds or depths of cut contributes to increasing the productivity of milling [1-2]. The paper [3] presents the research results of the influence of technological parameters on the values and course of the radial force during brushing process. Based on the research it was found that the radial force increases with the increase of the brushing speed and brush stiffness. Analysis of the cutting force components is essential particularly in aircraft and medical applications, where the machined components require ultra-high precision, and involve thin-walled features of intricate geometry [4]. Excessive cutting forces may have an adverse effect on the quality of machined surfaces, which results from the increased level of vibration in the milling machine-milling cutter-workpiece-fixture system [5].

Other factors that have an impact on the total cutting force are adhesion and the built-up edge. They may lead to significant fluctuations in the values of the cutting force components in milling, deterioration of the machined surface quality and of the dimensional and shape accuracy of the workpiece [6]. Cutting speed v_c is the parameter which to a considerable extent determines the cutting force and productivity of milling. According to cutting speed milling operations may be classified into conventional and high speed milling (HSM). The range of HSM parameters may be defined as $\partial F/\partial v_c<0$ whereas $\partial F/\partial v_c>0$ determines the range of speeds of conventional

* Corresponding author: m.kulisz@pollub.pl

machining. It is accepted that what distinguishes HSM from conventional machining is that in HSM the increase in the cutting speed v_c leads to the decrease of the cutting forces [7]. Alternatively, HSM may be understood as a high-performance cutting method which produces high-quality machined surface, eliminates the need for finishing operations, traditionally realised through *e.g.* grinding and polishing.

Research studies in the field also focus on the effect of changing tool geometry on the cutting forces in milling. One of the studies analysed the changes of specific technological milling parameters (v_c, f_z, a_p) performed with the carbide milling cutters at a different contact angle (γ = 5° and γ = 30°). Lower values of cutting force components and their amplitudes, indicating greater stability of the process, were observed for the tool γ = 30°. Increasing the axial depth of cut triggered a proportional increase in the cutting force components and their amplitudes. A change in feed per tooth provoked the increase of the cutting forces and subsequently their stabilisation [8]. In machining with Kordell geometry tools, the F_x and F_y components and their amplitudes rise with the increase of feed per tooth f_z. In the case of milling with cutters of traditional tool geometry, a more significant influence on cutting forces and their amplitudes is observed when changing the feed per tooth f_z rather than cutting speed v_c. Furthermore, it ought to be remarked that the cutting force components decrease with the increase of cutting speed to v_c = 1200 m/min in traditional tool geometry [8,13].

Another factor of high importance is the impact of cutting tool coating (such as the TiB_2 and TiAlCN type) on the cutting forces in milling. The lowest values of the cutting forces (F_x, F_y) in milling Al6082 alloy with carbide tools were obtained for a tool with a TiB_2 coating. During the v_c change, the characteristic point of „transition" to the range of HSC (where v_{cgr} = 450÷600 m/min) was observed [9].

Research work [10] analysed, inter alia, the values and amplitudes of the cutting force F_x. The machining carried out as a part of the test was conducted with three types of milling cutters (two monolithic tools with high-speed steel or cemented carbide, and an indexable milling cutter R390-020B20-11L). The workpiece material was EN AlSi21CuNi aluminium alloy. In tests it was determined that machining with the indexable tool was characterised by the highest increase in the cutting forces and their amplitudes. Such phenomenon is highly undesired in milling as it has a detrimental effect on tool life, machine tool life and machined surface quality. The study showed that machining performed with the cemented carbide tool exhibited superior quality in comparison with the other cutting solutions. Lower cutting force observed in that case resulted from lower cutting resistance ("sharp" tool geometry).

The indexable tool generated comparable cutting forces to that of the high-speed steel cutter. Moreover, when milling with an indexable tool the amplitude of the cutting force F_x, which is an important indicator of stability in milling, was by far the highest among the tested tools. This predominantly results from the unfavourable cutting edge geometry of the milling tool in question, low contact angles and helix angle (λs = 5°). Such parameters are highly undesired in machining as they generate instability and resulting increase in the amplitudes of the cutting forces.

In AZ91HP alloy milling with PCD cutter and in the presence of cutting fluid cutting forces assume low values and grow linearly with the increase of feed. Lower cutting forces translate into smaller tendency of tools to overheat (smaller coefficient of friction at a tool-workpiece interface). What can be observed along with lower cutting forces, especially as far as small cross-section of a machined layer is concerned, is a significantly lower temperature in the cutting area [6].

The arguments presented in the preceding paragraphs show that there are numerous factors that contribute to the total cutting force. The character of relationships between them is frequently non-linear, therefore we may observe a growing interest in modelling machining processes by means of mathematical methods of analysis [1, 11], or artificial intelligence [12].

The main purpose of milling modelling is prediction of non-linear technological processes. Its analysis can contribute to the creation of a system which could support decision-making processes in an enterprise (for instance through optimisation of milling process focusing on the selection of suitable technological parameters of machining).

For the purpose of cutting force components simulation, artificial neural network was used. The applied software was Statistica. During testing, two networks were scrutinised: RBF (Radial Basis Function) and MLP (Multi-Layered Perceptron) [13]. The authors of the work showed a simple model of milling process of two changeable machining parameters, cutting speed v_c and feed per tooth f_z, whereas the remaining machining parameters were fixed (axial depth of cut ap, radial depth of cut a_e and helix angle λs). The obtained model enables testing different cutting speed v_c and feed per tooth f_z settings without the need for conducting any additional milling tests. In [14] the cutting force prediction model for circular end milling process was employed to analyse the curvature effects of tool path on chip thickness, along with the impact of entry and exit angles. Experimental results showed a good fit of measured and simulated results.

The presented brief review of selected literature in the field shows that analysis of cutting forces is of great importance to effects of milling processes. Simultaneously, the results of numerical simulation enable prediction of processes of a non-linear nature. Simulation of such processes may prove particularly significant when optimal process parameters are required in the face of a limited amount of input data available.

3 Methodology, aim of study and research subject

The workpiece material tested in the presented study was AZ91D magnesium alloy, whose high mechanical properties and excellent corrosion resistance determine it high applicability in different industrial applications.

DOI: 10.1051/itmconf/20171502001

The chemical composition and properties of the alloy under analysis are presented in Table 1.

Table 1. Chemical composition and mechanical properties of AZ91D magnesium alloy [15-17].

Chemical composition of AZ91D [%]							
Al	Zn	Mn	Fe	Be	Si	Cu	Ni
8.95 - 9.11	0.77 - 0.79	0.20 - 0.24	0.0025 - 0.0027	0.0009 - 0.00112	0.025 - 0.029	0.0015 - 0.0021	0.00063 - 0.00078
Mechanical properties of AZ91D alloy							
Tensile Strength [MPa]		Yield Strength (0.2%) [MPa]		Hardness [HB]		Elongation [%]	
230		160		63		3	
Physical properties of AZ91D alloy							
Density [g/cm^3]		Melting Point [°C]			Thermal Conductivity [W/mK]		
1.81		533			72.3		

The machining operations were carried out by means of AVIA VMC 800 HS vertical machining centre controlled by Heidenhain iTNC 530 control (power 25 kW, n_{max} up to 24000 rev./min and $v_{f\ max}$ up to 40 m/min).

Milling operations were carried out with the solid carbide end mill with unequal helix angle, λs = 20° and λs = 50°, diameter d = 16 mm, number of cutting teeth z = 3. The cutting tool was mounted in the SECO HSK63A SFD 16x120 tool holder with thermal shrink-fit technology. Before the tool holder was mounted in the spindle they were balanced in CIMAT CMT-15V2N balancing machine, balancing class G 2.5 at 25 000 rev./min according to ISO1940:2003, which allows to have an unbalance of 1 g mm/kg. The unbalance in our study was as follows: for the cutting tool with λs = 20° - 0.42 g mm/kg, for the helix angle λs = 50° - 0.23 g mm/kg. The cutting forces were measured with 927B dynamometer by Kistler, operating in the range of -5kN ÷ 5kN, equipped with a charge amplifier 5017B and LMS SCADAS Mobile SCM05 data acquisition hardware.

The range of changeable technological parameters of machining in tests was as follows: cutting speed v_c = 400÷1200 m/min, feed per tooth f_z = 0.05÷0.3 mm/tooth, axial depth of cut ap = 1÷6 mm. The radial depth of cut parameter a_e = 14 mm was constant in all milling operations.

Given that the milling process is a type of network design then the cutting forces F_x, F_y and F_z are output units and cutting speed v_c, feed per tooth f_z and axial depth of cut a_p along with helix angle λs are input variables. The presented network employs the black box model, which ensures good performance on the training set. The black box approach was selected as it was the best to reflect the complex character of milling. Such a model is implemented when it is difficult to determine mathematical equations that would describe the analysed process [13].

The cutting forces F_x, F_y and F_z are defined in the dynamometer's system of coordinates, and these components have their counterparts in the coordinates system of the machine tool [18]:

- F_x – feed force F_f,
- F_y – normal to the feed force F_{fn},
- F_z – reactive force F_p.

4 Results of experimental tests and cutting force modelling

The cutting forces F_x, F_y and F_z were measured throughout the milling process of AZ91D magnesium alloy with two tools of different helix angles λs (20°, 50°). Presented in Figs. 1-3 are the results of measurements for constant feed per tooth f_z = 0.15 mm/tooth and axial depth of cut a_p = 6 mm. The results are visualised with the whiskers and box plot, which shows the maximum range of the cutting forces and the difference between the minimum and maximum values.

In certain cases we may observe outliers and extremes, which result from the increase in the cutting forces on the cutting tool entry into and exit from the workpiece.

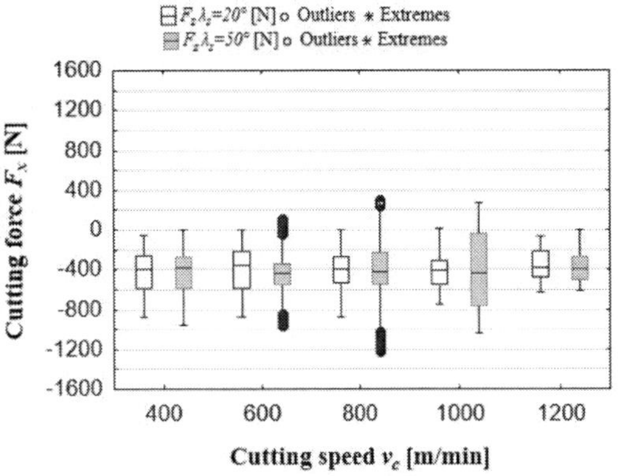

Fig. 1. Effect of cutting speed v_c on the cutting force F_x, f_z = 0.15 mm/tooth, a_p = 6 mm.

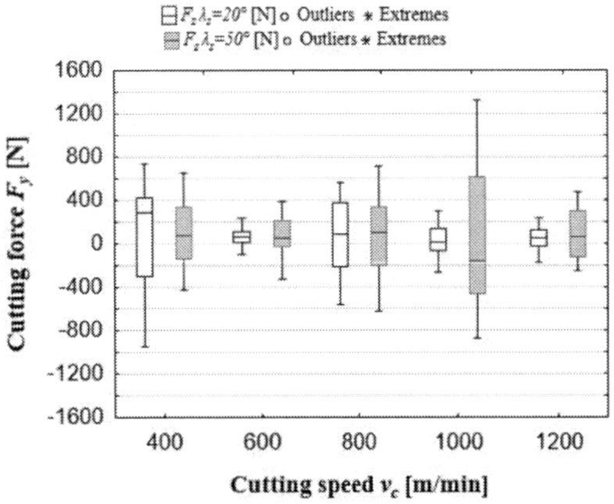

Fig. 2. Effect of cutting speed v_c on the cutting force F_y, f_z = 0.15 mm/tooth, a_p = 6 mm.

DOI: 10.1051/itmconf/20171502001

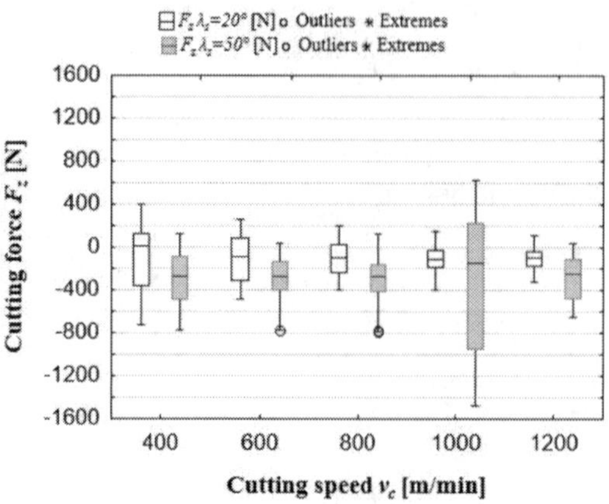

Fig. 3. Effect of cutting speed v_c on the cutting force F_z, f_z = 0.15 mm/tooth, a_p = 6 mm.

The highest amplitude/range of the F_x (Fig. 1) is observed in milling with the cutting speed of 800 m/min, by means of the tool with helix angle λs equal to 20°, which denotes the highest tool stability range. The cutting force component F_y (Fig. 2) exhibits the highest range at the cutting speed of 1000 m/min, when machined with the cutting tool of helix angle of λs = 50°. The cutting speeds which provide a better stability of milling process are v_c = 600 m/min and v_c = 1200 m/min, with milling cutter helix angle of λs = 20°. The reactive cutting force F_z (Fig. 3) shows the highest amplitude at v_c = 1000 m/min, when machined with a cutting tool with a helix angle λs = 50°. In the case of a lower helix angle, λs = 20°, the range of the cutting force F_z decreases with the decrease of cutting speed v_c.

The comparison of both tools applied in the study indicates that the cutting tool with a helix angle λs = 50° generates higher cutting forces. Machining magnesium alloys with this cutting tool is characterised by decreasing cutting force components F_x, F_y, F_z at the cutting speed of v_c = 1200 m/min. In milling with this type of tool this is the cutting force Fx which dominates. In order to decrease the Fx the magnesium alloy workpiece should be machined with the cutting speed v_c = 1200 m/min. The recommended range of feed per tooth in the case is 0.15÷0.3 [mm/tooth].

The cutting forces obtained from the milling of AZ91D magnesium alloy provided the input data for the artificial network simulation. The simulation was conducted with Statistica Neural Network software using MLP (Multi-Layered Perceptron).

The key parameters of a good fit of input/output data are: training quality, validation quality and the training error determined with the least squares method. The simulations of the cutting forces F_x, F_y *and* F_z were conducted for two tool cutting tool helix angles λs = 20° and λs = 50°. As a result 6 models were obtained. An example of the cutting force F_x simulation is shown in Fig. 4. The network architecture for other cutting force components models is analogical, and the produced output of the network is the corresponding cutting force.

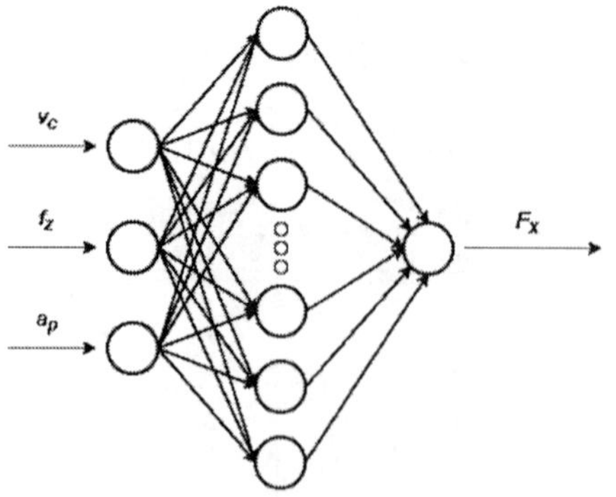

Fig. 4. Artificial neural network architecture.

It was established that the obtained neural network architecture should be simplest; hence, it contained one hidden layer. The input layer consists of three neurons (cutting speed v_c, feed per tooth f_z and axial depth of cut a_p), and the output layer is one neuron (a respective cutting force component F_x, F_y or F_z). The number of training epochs was defined experimentally in the range of 100÷200, whereas the range of hidden neurons 1÷10. The MLP network was trained with the use of BFGS (Broyden-Fletcher-Goldfarb-Shanno) technique. This is one of the recommended network training techniques as it offers the best performance in reducing the function to a desired level. The implemented activation functions were: linear, exponential, logistic and hyperbolic (tanh). The networks were trained for the total of 17 machining parameter sets. 80% of the results from milling test were used as the training data set whereas the remaining 20% built the validation set. Given the small number of data sets, the test data set was omitted [19].

In each simulation 100 networks were trained and the best three models were determined. The criteria for the selection of the best networks were mentioned earlier and included: training quality, validation quality, the lowest training error. Network parameters for the particular cutting forces F_x, F_y, F_z and the cutting tool helix angle of λs = 20° in milling of AZ91D magnesium alloy are collated in Table 2.

The analysis of the neural network architectures indicates that the best network for modelling the cutting force component F_x is network 2 (MLP 3-4-1 with 4 neutrons) as the training and validation quality amounted to 99.98% and 98.57% respectively, and the training error value amounted to approx. 14N. The best network to model the cutting force F_y is network 1 (MLP 3-7-1 with 7 neutrons), and for the cutting force component Fz it is network 2 (MLP 3-4-1 with 4 neutrons). The values of the cutting forces F_x, F_y and F_z obtained from the neural network models depending on cutting speed and feed per tooth are shown in Figs. 6-7.

The values of F_x increase with the increase of feed per tooth f_z decrease when cutting speed v_c decreases. The same relationship is observed for the cutting force F_z. The lowest values of F_y are obtained when the machining parameters are v_c = 1200 [m/min] and f_z = 0.05÷0.15 [mm/tooth], and are within the range of 150÷200 N.

Table 2. Parameters of the best MLP networks for particular cutting forces, cutting tool helix angle $\lambda s = 20°$.

Feed force F_x			
Id.	1	2	3
Network name	MLP 3-4-1	MLP 3-4-1	MLP 3-8-1
Quality (training) [%]	99.99	99.98	99.97
Quality (validation) [%]	91.1	98.57	93.8
Error (training)	12.87	13.58	22.12
Activation (hidden)	Logistic	Logistic	Exponential
Activation (output)	Exponential	Tanh	Logistic
Normal to the feed force F_y			
Id.	1	2	3
Network name	MLP 3-7-1	MLP 3-5-1	MLP 3-6-1
Quality (training) [%]	99.98	99.96	99.95
Quality (validation) [%]	99.99	99.38	99.33
Error (training)	6.41	15.01	19.46
Activation (hidden)	Logistic	Logistic	Tanh
Activation (output)	Logistic	Exponential	Logistic
Reactive force F_z			
Id.	1	2	3
Network name	MLP 3-4-1	MLP 3-4-1	MLP 3-4-1
Quality (training) [%]	99.85	99.84	99.83
Quality (validation) [%]	90.78	99.34	98.97
Error (training)	27.65	28.5	29.22
Activation (hidden)	Tanh	Logistic	Logistic
Activation (output)	Linear	Exponential	Tanh

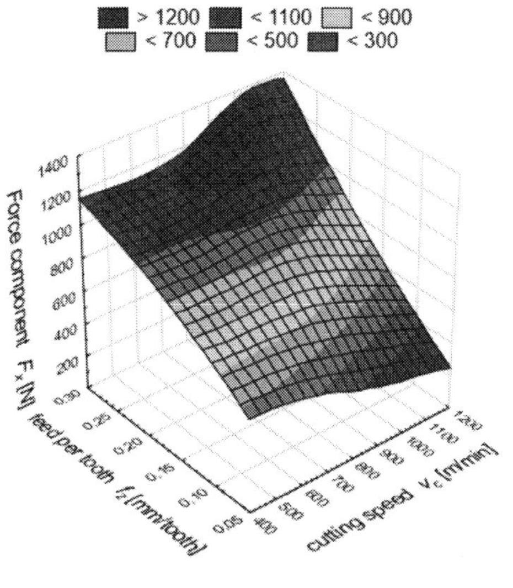

Fig. 5. Results of the cutting force F_x simulation for cutting tool helix angle $\lambda s = 20°$, in the function of cutting speed v_c and feed per tooth f_z for the MLP 3-4-1 neural network architecture.

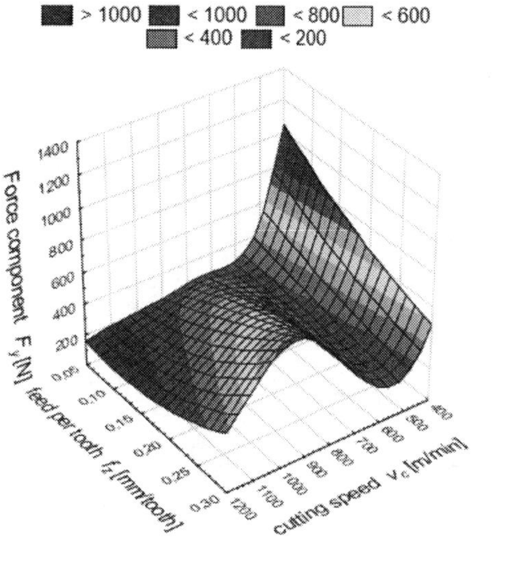

Fig. 6. Results of the cutting force F_y simulation for cutting tool helix angle $\lambda s = 20°$, in the function of cutting speed v_c and feed per tooth f_z for the MLP 3-7-1 neural network architecture.

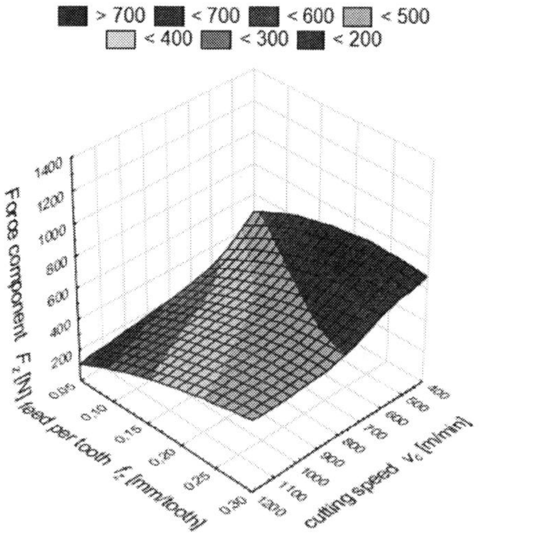

Fig. 7. Results of the cutting force F_z simulation for cutting tool helix angle $\lambda s = 20°$, in the function of cutting speed v_c and feed per tooth f_z for the MLP 3-4-1 neural network architecture.

The simulation of the cutting forces was also conducted for the milling cutter helix angle $\lambda s = 50°$. The parameters of the best MLP networks for modelling the cutting force components F_x, F_y and F_z are shown in Table 3.

The best network with regards to the cutting force Fx is network 1 (MLP 3-5-1 with 5 neutrons), as training quality was 99,99% and validation quality 96.87%, the training error was the lowest and was equal to 17.33 N. The cutting force Fy is best modelled with network 1 (MLP 3-4-1 with 4 neutrons), with training quality of 99.99%. Although the validation quality is slightly lower than in network 2, network 1 exhibits the lowest training error, 3.69 N. In the case of the cutting force component

DOI: 10.1051/itmconf/20171502001

F_z the best parameters are shown by network 1 (MLP 3-4-1 with 4 neutrons). Predicted values of the cutting forces for the milling cutter helix angle λs = 50° obtained from the analysed neural networks are shown in Figs. 8-10.

Table 3. Parameters of the best MLP networks for particular cutting forces, cutting tool helix angle λs = 50°.

Feed force F_x			
Id.	1	2	3
Network name	MLP 3-5-1	MLP 3-4-1	MLP 3-5-1
Quality (training) [%]	99.99	99.99	99.98
Quality (validation) [%]	96.87	93.56	72.44
Error (training)	17.33	19.12	30.21
Activation (hidden)	Tanh	Tanh	Tanh
Activation (output)	Linear	Logistic	Logistic
Normal to the feed force F_y			
Id.	1	2	3
Network name	MLP 3-4-1	MLP 3-7-1	MLP 3-4-1
Quality (training) [%]	99.99	99.98	99.98
Quality (validation) [%]	98.77	99.88	93.85
Error (training)	3.69	10.55	13.33
Activation (hidden)	Tanh	Exponential	Tanh
Activation (output)	Linear	Tanh	Linear
Reactive force F_z			
Id.	1	2	3
Network name	MLP 3-4-1	MLP 3-5-1	MLP 3-4-1
Quality (training) [%]	99.98	99.97	99.97
Quality (validation) [%]	99.29	74.24	82.47
Error (training)	22.76	25.4	29.24
Activation (hidden)	Tanh	Tanh	Tanh
Activation (output)	Tanh	Exponential	Tanh

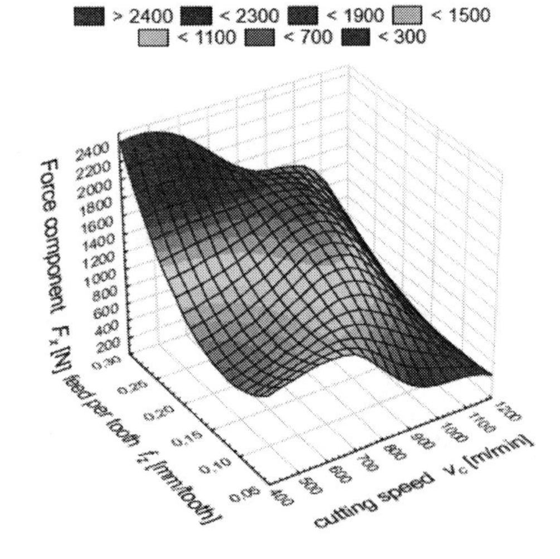

Fig. 8. Results of the cutting force F_x simulation for cutting tool helix angle λs = 50°, in the function of cutting speed v_c and feed per tooth f_z for the MLP 3-5-1 neural network architecture.

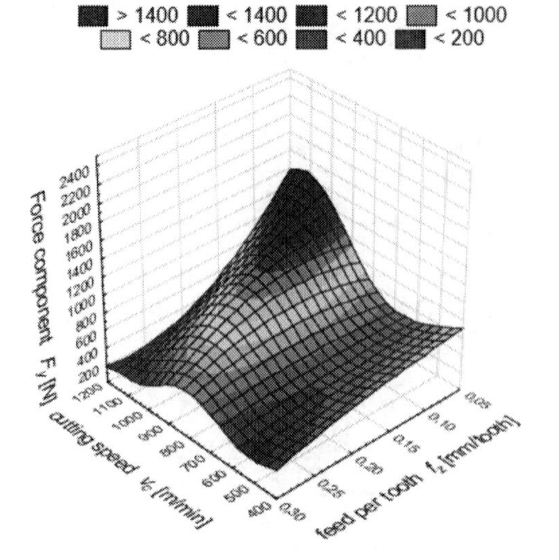

Fig. 9. Results of the cutting force F_y simulation for cutting tool helix angle λs = 50°, in the function of cutting speed v_c and feed per tooth f_z for the MLP 3-4-1 neural network architecture.

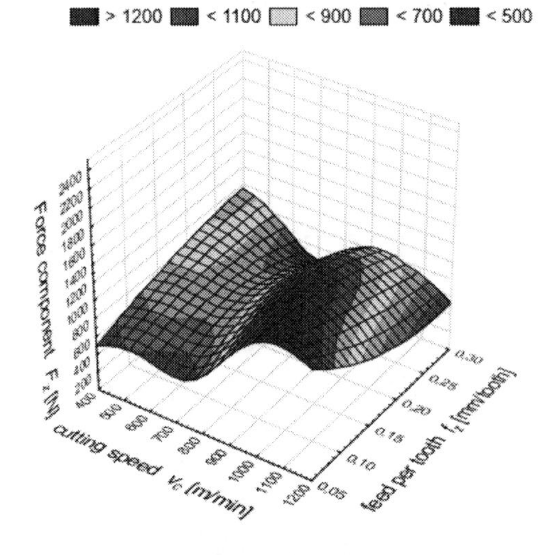

Fig. 10. Results of the cutting force F_z simulation for cutting tool helix angle λs = 50°, in the function of cutting speed v_c and feed per tooth f_z for the MLP 3-4-1 neural network architecture.

To present the quality of simulated networks the tables below collate the measured values of the cutting force F_x with the values obtained from the networks, F_{xp}. The results for the cutting tool helix angle λs = 20° are show in Table 4, while for λs = 50° in Table 5.

Table 4. The cutting force F_x and the forecasted cutting force F_{xp} for given scenarios, $\lambda s = 20°$ (Δ – absolute error, δ – relative error).

v_c [m/min]	f_z [mm/tooth]	F_x [N]	F_{xp} [N]	Δ	δ
400	0.15	824.04	824.53	0.49	0.06%
600	0.15	807.16	807.39	0.23	0.03%
800	0.15	829.83	841.65	11.82	1.42%
1000	0.15	680.05	684.28	4.23	0.62%
1200	0.15	593.98	592.13	1.86	0.31%
800	0.05	441.70	443.1	1.38	0.31%
800	0.10	609.40	619.5	10.11	1.66%
800	0.20	1114.5	1115.5	0.96	0.09%
800	0.25	1105.8	1105.6	0.25	0.02%
800	0.30	1100.6	1100.3	0.36	0.03%

Table 5. The cutting force F_x and the forecasted cutting force F_{xp} for given scenarios, $\lambda s = 50°$ (Δ – absolute error, δ – relative error).

v_c [m/min]	f_z [mm/tooth]	F_x[N]	F_{xp} [N]	Δ	δ
400	0.15	910.14	920.60	10.46	1.15%
600	0.15	901.97	890.29	11.68	1.29%
800	0.15	1176.3	1176.1	0.15	0.01%
1000	0.15	996.61	860.48	136.13	13.66%
1200	0.15	581.57	581.75	0.17	0.03%
800	0.05	800.04	800.38	0.34	0.04%
800	0.10	1208.5	1356.5	148.05	12.25%
800	0.20	1862.6	1861.4	1.20	0.06%
800	0.25	1913.7	1912.1	1.65	0.09%
800	0.30	1321.1	1321.4	0.26	0.02%

The error of the forecasted data with reference to the actual values of the cutting force F_x, for cutting tool angle helix $\lambda s = 20°$, is within the range of 0.02% ÷ 1.66%, and for helix angle $\lambda s = 50°$ the range is 0.02% ÷ 13.66%. The analysis of other cutting forces was also performed; the error of the forecasted data referred to the actual values of the cutting force component F_y and F_z, for cutting tool angle helix $\lambda s = 20°$ is within the range of 0.11% ÷ 4.44%, and 0.19% and 7.59% respectively; while for helix angle $\lambda s = 50°$, the cutting force F_y is between 0.17% ÷ 8.04%, and F_z – 0.002% ÷ 6.26%.

Simulation results exhibit acceptable error, below 15%. Therefore, the artificial neural network appears to be an excellent solution for simulation of *e.g.* the cutting forces in milling. It is possible to build effective models of the cutting force components provided that the machining process is properly characterised with such data as machining parameters

These observations imply that artificial neural networks exhibit great potential for non-linear machining processes prediction, such as milling. The presented analysis could be applied to determining the optimal set of parameters and to generate desired cutting forces in milling and provide characteristics of the process without the need for a number of technological tests.

5 Summary and conclusions

The conducted analytical and experimental works lead to the following general conclusions:

1. The change of each of the analysed technological parameters: v_c, f_z and a_p, and the milling cutter helix angle affect the cutting force components. The feed force F_x dominates over other cutting forces.
2. Higher stability and productiveness of milling, as well as lower cutting forces in milling are obtained when machining with a mill cutter of helix angle $\lambda s = 20°$. The tool $\lambda s = 50°$ generated higher cutting forces.
3. Simulations of the cutting forces and errors of forecasted data in reference to the actual values measured in tests show that the error for both cutting tool helix angles is within the acceptable range of error. The error for the tool $\lambda s = 20°$ is lower than for $\lambda s = 50°$, however the value of error in either of the cases does not exceed 15%.
4. Simulations by means of artificial neural networks enable initial specification of particular technological parameters and forecasted values of the cutting forces in milling.
5. The non-linear relationships between technological parameters and the cutting forces represented by means of neural networks may be used to calculate optimal results of machining, without the need for labour-intensive and time-consuming machining tests, which often require conducting costly actual milling operations.
6. A high-precision tool may be devised for modelling phenomena occurring in milling under conditions of incomplete data. The simulation may constitute an important decision-making tool in specifying technological parameters in machine and device component milling applications.
7. Establishing initial machining parameters of milling and forecasted values of the cutting forces by means of artificial neural network simulations increases the productivity and efficiency of the production process, and generates considerable material savings.
8. Artificial neural networks enable the "transfer" from discrete data (input data from the cutting force measurements in milling) to continuous data (simulations enable building diagrams in the form of response surface).

References

1. Z. T. Fu, W. Y. Yang, X. L. Wang, J. Leopold, *15th Cirp Conf. Model. Mach. Oper. (15th C.* **31**, 258–263 (2015)
2. J. Salguero, M. Batista, M. Calamaz, F. Girot, M. Marcos, Procedia Eng. **63**, 735–742 (2013)
3. J. Matuszak, Adv. Sci. Technol. Res. J.(to be published)
4. M. Nejman, D. Śniegulska-Grądzka, K. Jemielniak, Mechanik **8–9**, 98–104 (2015)
5. J.-D. Kim, K.-B. Lee, Engineering **2**, 788–792 (2010)
6. K. E. Oczoś, Mechanik **5–6**, 386–400 (2009)

7. W. Adamski, Adv. Manuf. Sci. Technol. **34**, 73–84 (2010)
8. O. Gziut, J. Kuczmaszewski, I. Zagórski, Adv. Manuf. Sci. Technol. **38**, 83–92 (2014)
9. J. Kuczmaszewski, P. Pieśko, Mechanik **10**, 846–854 (2013)
10. J. Kuczmaszewski, P. Piesko, Eksploat. I Niezawodn. Reliab. **16**, 37–41 (2014)
11. I. Danis, F. Monies, P. Lagarrigue, N. Wojtowicz, Int. J. Adv. Manuf. Technol. **84**, 1801–1820 (2016)
12. P. Kazemi, M. H. Khalid, J. Szlek, A. Mirtic, G. K. Reynolds, R. Jachowicz, A. Mendyk, Powder Technol. **301**, 1252–1258 (2016)
13. M. Kulisz, I. Zagórski, A. Semeniuk, Appl. Comput. Sci. **12**, 49–58 (2016)
14. B. H. Wu, X. Yan, M. Luo, G. Gao, Chinese J. Aeronaut. **26**, 1057–1063 (2013)
15. PN-EN 1753:2001Magnez i stopy magnezu – Gąski i odlewy ze stopów magnezu
16. Świadectwo kontroli jakości – stop magnezowy AZ91, www.stanchem.pl(2014)
17. https://www.dynacast.com/az91d
18. A. Skoczylas, K. Zaleski, Mechanik **8–9**, 311–319 (2015)
19. http://www.statsoft.pl/Portals/0/Downloads/Sieci%20neuronowe.pdf

DOI: 10.1051/itmconf/20171502002

Classification of user performance in the Ruff Figural Fluency Test based on eye-tracking features

Magdalena Borys[1,*], *Sara* Barakate[2], *Karim* Hachmoud[2], *Małgorzata* Plechawska-Wójcik[1], *Paweł* Krukow[3], and *Marek* Kamiński[1]

[1]Lublin University of Technology, Faculty of Electrical Engineering and Computer Science, Institute of Computer Science, Nadbystrzycka 38D, 20-618 Lublin, Poland
[2]Abdelmalek Essaâdi University, Polydisciplinary Faculty, Martil Road, Tétouan, Morocco
[3]Medical University of Lublin, Department of Clinical Neuropsychiatry, Głuska 1, 20-439 Lublin, Poland

Abstract. Cognitive assessment in neurological diseases represents a relevant topic due to its diagnostic significance in detecting disease, but also in assessing progress of the treatment. Computer-based tests provide objective and accurate cognitive skills and capacity measures. The Ruff Figural Fluency Test (RFFT) provides information about non-verbal capacity for initiation, planning, and divergent reasoning. The traditional paper form of the test was transformed into a computer application and examined. The RFFT was applied in an experiment performed among 70 male students to assess their cognitive performance in the laboratory environment. Each student was examined in three sequential series. Besides the students' performances measured by using in app keylogging, the eye-tracking data obtained by non-invasive video-based oculography were gathered, from which several features were extracted. Eye-tracking features combined with performance measures (a total number of designs and/or error ratio) were applied in machine learning classification. Various classification algorithms were applied, and their accuracy, specificity, sensitivity and performance were compared.

1 Introduction

The Ruff Figural Fluency Test (RFFT) was created by Ronald M. Ruff [1] to be a nonverbal analogue of verbal fluency measures. Fluency itself is understood as the ability to apply some strategies to generate nonverbal responses maximising response efficiency and simultaneously minimising response repetition [2]. Such effective repetition needs processes including planning, inhibition, cognitive flexibility, decision-making and self-monitoring [3, 4]. The test proved to be reliable, and the data support the thesis that the performance depended not on sex, but on age and education [5]. It includes a baseline measurement of cognitive function and executive functioning [6].

The RFFT was proved to be related to performance on the Design Fluency Test (DFT) [7]. It also has good test-retest reliability [5] and interrater reliability [2, 8]. Since figural fluency is related to the functioning of the right frontal lobe [9], the RFFT has also proved successful in distinguishing patients with frontal lobe lesions from a group of healthy control patients [10].

Originally the RFFT was designed as a paper and pencil test used to evaluate nonverbal fluency and executive functioning [2, 11]. In this research a computerised version of the RFFT was used. The test group was homogeneous, composed of seventy healthy students aged 18-31 years.

The work in this paper is novel in recognising eye activity patterns for different groups of cognitive performance during the figural fluency test and classifying those groups of cognitive performance based on eye activity patterns.

2 Related work

The popularity of figural fluency tests has increased in recent years. It is more and more often included in cognitive and neuropsychological test batteries [4]. The RFFT proved to be sensitive to the first changes in cognitive functioning in the domain of executive function [3]. Furthermore, the RFFT is useful in searching for early changes in cognitive function among young and middle-aged persons [12].

Hicks et al. [13] developed an oculomotor-driven version of the trail-making tests (TMT) to assess visual attention, psychomotor processing speed, and task-switching.

Poletti at al. [14] adopted a set of neuropsychological tests for eye-tracking (ET) control. These tests evaluated such issues as language, attentional abilities, executive functions and social cognition. This solution was checked on a sample of healthy participants. The authors applied ET in a neuropsychological battery combined with classical ''paper and pencil'' measures. The set of tests was a cognitive measure of working memory

* Corresponding author: m.borys@pollub.pl

 DOI: 10.1051/itmconf/20171502002

abilities and global cognitive efficiency. The researchers found significant correlations between "paper and pencil" screening and ET-based neuropsychological measures.

Keller et al. [15] developed a neuropsychological assessment method for patients with Amyotrophic lateral sclerosis (ALS) based on eye movement. The participants (forty-eight ALS patients and thirty-two healthy controls) were tested by two tests (D2-test and Raven's coloured progressive matrices) adapted for eye-tracking control. The results were compared to the classical paper-pencil version and similar results with high correlation were obtained.

Other studies evaluate cognitive functions using measures based on eye movements and deduce a lack of attention [16] or muscular/neuro-muscular fatigue [17]. Performance of the saccadic eye movement was found to be a sensitive marker for even minor impairment of cerebral function in various physiological and pathological conditions, including sleep deprivation [18], early Alzheimer's [19], extra-motor cerebral pathology [13] and cognitive frontal alterations [20, 21] in amyotrophic lateral sclerosis or mild frontotemporal dementia [22] [Boxer et al. 2006] and even prolonged hypobaric hypoxia monitored in climbers [23].

3 Methodology

3.1 Research questions

A total number of designs (or a total number of unique designs) drawn is the indicator of user performance in the RFFT, while the error ratio indicates correctness of performance taking into consideration a number of perseverative errors in relation to a total number of designs. Those two indices are used to asses an individual figural fluency. On the other hand, the eye-tracking measures can support or substitute some of those test indicators as research in attention and cognitive studies shows.

The primary research question that guided the study is stated as follows: "*Is there any relationship between cognitive performance (measured in a total number of designs and/or error ratio) and eye-tracking measures in the RFFT?*".

If there is any relationship between cognitive performance and eye-tracking measures, the secondary research question is: "*Is it possible to classify users with high and low cognitive performance based on eye measures in the RFFT?*".

3.2 Experiment setting

The experiment was conducted in a special testing laboratory. The room was illuminated with standard fluorescent light and outside light was blocked to ensure stable conditions for the duration of the experiments. The average illumination in the room was about 600 Lux.

Eye activity was recorded using screen-based eye tracker Tobii Pro TX300 (Tobii AB, Sweden). The Tobii TX300 performed binocular tracking with the sampling frequency of 300Hz using the dark pupil and corneal reflection method. The gaze was measured by the eye-tracker with the accuracy of 0.5° and precision of 0.01°.

The experiment was created using Tobii Studio 3.2 and a visual stimulus was presented on a separate monitor equipped with an eye tracker (23'' TFT monitor at 60 Hz). The participants were seated while working with the application. The distance between the screen and the subject was in a range from 50 to 80 cm. The distance depended on individual participant preferences (a comfortable position for working with a computer).

All participants were assessed using the same software and hardware settings (Laptop Asus G750JX-T4191H with Intel Core i7-4700HQ and 8GB of RAM).

The RFFT applied in the study was a computerised version developed on the basis of the original "paper and pencil" test. The application was developed in Java Swing and it is operated using a computer mouse.

The developed version of the RFFT covers three different dot configurations and two distractors added to the first, symmetric dot configuration (beyond the basic plain one). Each trial lasts 60 seconds. When the participant has created single design, he had to press the "Next→" button to clear the board. The accomplished designs were presented in the form of miniatures at the left part of the screen.

The computerised RFFT allows to log all user interactions, and thus to register not only the total number of designs or unique designs, but also the execution time for each individual design or part of the test.

3.3 Procedure

The eye tracker was calibrated using a 9-point built-in calibration procedure at the beginning of each session. Once calibrated, the participants were provided with the experimental instructions on the screen, in which they asked to make as many unique designs as possible by connecting the dots stimuli.

Next, the participants completed five parts of the RFFT using the computer application. Each part had different stimulus presentations (dot configurations and distractors). They covered: one part with the basic dot configuration with five equally spaced points on the white screen (without distractors), two parts with basic dot configurations and distractors added and two parts with different dot configurations on the white background (without distractors).

At the beginning of each part, a short trial, consisting of 3 designs, was run to familiarise the participants with the task. After the trial, the proper test was started.

Each part of the test was repeated three times, then another part started.

3.4 Participants

In the experiment, the exclusion criteria included having a history of significant head injury or brain disfunction and currently experiencing psychological or psychiatric problems, prescription medication use, and problems or pain related to movement. After data recording, the

DOI: 10.1051/itmconf/20171502002

further exclusion criteria included having insufficient eye-tracking ratio (less than 90%) or poor eye-tracking data quality (due to eye-glasses or excessive head movement during study).

Seventy healthy participants (male) were recruited among students at the Lublin University of Technology. From the initial group, 9 participants were excluded (1 individual had a psychiatric medical history, 3 individuals had low eye-tracking ratio and 5 individuals had poor data quality). The final group consisted of 26 Computer Science students and 35 Biomedical Engineering students ranging in age from 18 to 31, mean = 21.4 years, std. dev. =1.96. All were right-handed with normal or corrected-to-normal vision.

The study was approved by the Research Ethics Committee of the Lublin University of Technology (Approval ID 4/2015 from 12 November 2015) and all participants received verbal and written information about the study. All participants signed an informed consent. The participation was voluntary and no compensation was offered.

3.5 Signal pre-processing

The eye activities were analysed off-line using Tobii Studio 3.2 and custom programs written in MATLAB. For each test attempt, the tracking ratio were calculated (based on signal lost), the observations with eye tracking ratio lower than 90% were excluded from further analysis.

Fixations and saccades were exported from the Tobii Studio, they were detected using the Velocity-Threshold Identification (I-VT) fixation classification algorithm [24] for average eye selection. The noise was reduced by using the average moving filter with the window size of 3 samples. Gaps shorter than 75 ms were interpolated. The velocity threshold was set to 30°/s and window length to 20ms. The adjacent fixations were merged and fixations shorter than 60 ms were discarded.

Using 3D eye positions and screen gaze points from the eye-tracker, the amplitude of saccades was calculated in degrees (based on its total distance along the trajectory of the movement between start and end points). The saccades with amplitude greater than 10 degrees were classified as large ones. Also the blinks were identified as zero data (both eyes are not found) embedded in two saccadic events [25]. To avoid any confusion that the signal loss will be identified as a blink, the blinks detected automatically were manually rechecked using the video annotation technique.

The pupil diameter change was extracted for both eyes. The pupil diameter change was normalised with baseline subtraction (subtracting the mean pupil size over the first 100ms of recording) [26].

3.6 Measured feature sets

Based on the log file from the RFFT application the following features were extracted for each part of the test:
- total number of designs – the number of designs drawn by the participant;
- perseverative errors – repetitions of the same design drawn by the participant;
- error ratio – an index for assessing the participant's ability to minimise repetition while maximising unique productions calculated as percentage of errors in the number of all designs.

A variety of ET features inspired by the literature were gathered and some others were proposed. The eye-tracking features were extracted on the basis of the following eye movements and activities:
- fixations (total number of fixations, mean fixation duration, standard deviation of fixation duration, max fixation duration);
- saccades (total number of saccades, total number of small saccades, total number of large saccades, mean saccade duration, standard deviation of saccade duration, max saccade duration, mean saccade amplitude, standard deviation of saccade amplitude, max saccade amplitude);
- blinks (total number of blinks, mean blink duration, standard deviation of blink duration, max blink duration);
- pupillary response (mean left/right pupil size change, standard deviation of left/right pupil size change, max of left/right pupil size change).

All features based on duration are determined in milliseconds and based on amplitude values are in degrees.

3.7 Methods

One-way Analysis of Variance (ANOVA) [27] was applied to analyse potential differences in a continuous random variable between and among the groups. The ANOVA test was, afterwards, followed by the post-hoc test using Scheffe's method. The Kruskal-Wallis test by ranks was used as the equivalent of ANOVA for a variable which does not follow normal distribution. In this work, those methods were used in STATISTICA software.

Hierarchical Agglomerative Clusters (HAC) method [28] was used to identify the set of observations (groups) that are similar and then to differentiate between them. HAC is a bottom-up method that creates a dendrogram in the form of a tree of k-block set partitions where $1 \leq k \leq n$ and n are the number of points to be clustered. The algorithm begins with each point (observation) separately in its own cluster, and the closest pairs of clusters are joined together, resulting in reducing the k number of clusters by 1. The algorithm is repeated until k is equal to 1 and the dendrogram is formed. In this work, the HAC implementation used in Tanagra software [29] was employed, where the clusters are created using a hybrid clustering method [30].

To determine the predictors, two methods were applied: discriminant function analysis and chi-square test. Discriminant Function Analysis (DA) [31] is dedicated to identify the continuous variables best discriminating between the naturally occurring groups. In DA, the groups represent the dependent variables and the predictors are the independent variables. DA might be applied if groups are known a priori. DA is usually applied to determine if a given set of variables could be used in the task of category membership prediction. The chi-square test was also used to determine significant perdictors. This test is commonly used for testing the independence between two categorical variables, so it detects the existence of a relationship between the independent variables and the dependent variables.

Several supervised learning classification methods were used to create a model for predicting high- and low performance in the RFFT using ET features. The best results were obtained for Support Vector Machines, K-Nearest Neighbours and Bagged Decision Trees. To validate the accuracy of models the data set was split into training/testing (80%) and validating (20%) subsets based on the randomise assignment. Firstly, the model was trained and tested using 5 cross-validation on the training/testing subset, then the model was evaluated on the basis of the validating subset. The accuracy and other properties of the models were calculated on their performance on the validating subset.

Support Vector Machines (SVM) [32] is a supervised learning classification method using an appropriately designated discriminant hyperplane. If learning sub-sets are fully separable, the SVM idea is to find two parallel hyperplanes which delimit the wider area and do not contain any probe elements. Those hyperplanes are based on the probe elements called support vectors The idea of the method is to find a maximum-margin hyperplane in a transformed input space, which splits the classes and maximises the distance between them. The method deploys a quadratic programming optimisation problem to get the parameters of the hyperplane.

The K-Nearest Neighbours (KNN) algorithm [33] is also a supervised learning classification method. The algorithm consists only of storing the feature vectors in a multidimensional feature space and class labels of the training samples. To classify a new unlabelled vector it assigns the label which is most frequent among the k training samples (k-nearest neighbours in the feature space) nearest to that query point.

Bagged (or bootstrap aggregation) Decision Trees (Bagged Trees) [34] is a supervised learning classification method that combines the results of many decision trees, which reduces the effects of overfitting (like in an individual decision tree) and improves generalisation. Bagged trees use Breiman's random forests algorithm, which is a set of tree predictors in which each tree depends on the values of a random vector sampled independently and with the same distribution for all trees in the forest [35].

4 Results

The data evaluation was performed only for the first part of the RFFT test repeated by each participant three times to preserve the homogeneity of tasks and eliminate the possible influence of participant fatigue. Each repetition was treated as a single observation since eye-activity should correspond to general cognitive performance, without individual/intra-individual properties. After the exclusion based on criteria described in 3.3, there was 177 observations in the analysis.

The analysis consisted of three parts. The first part refers to the division of observations into two groups according to the median of a total number of designs. The second part refers to splitting the observations into two clusters based on the error ratio value. The last part also represents the analysis with clusters based on the total number of designs and the error ratio together. In each part, statistical analysis and classification results are presented.

4.1 A total number of designs

A total number of designs is a simple indicator of the performance in the design fluency test. Two groups based on its median were created. Since the median of a total number of design is equal to 18 designs, group 1 represents 85 observations lower than 18 and group 2 represents another 92 observations. The mean total number of designs in group 1 is 13.6 (with std. dev. = 2.6) and group 2 is 20.7 (with std. dev. = 2.3). One-way ANOVA does not show any statistically significant difference between mean values of age or education (in years) in the groups.

Table 1 shows that there is a statistically significant difference between means of five ET features in two groups created on the basis of the median of a total number of designs. While there are statistically significant difference between means of ET features in groups, the ET features can be used as predictors to classified data into groups.

Table 1. ANOVA results for ET features in two groups based on the median of a total number of designs.

Feature	Group 1 M (SD)	Group 2 M (SD)	F-test F(1,175)	p-value
Total number of fixations	147.1 (19.4)	156.0 (20.0)	9.060	0.003
Mean fixation duration	473.3 (77.8)	442.0 (67.4)	8.238	0.005
Total number of blinks	5.1 (5.1)	3.3 (3.5)	8.125	0.005
Total number of saccades	154.4 (23.7)	162.5 (23.5)	5.202	0.024

The discriminant function analysis (tolerance value = 0.05, forward stepwise procedures) with Wilks' Lambda test of significance were used to investigate the best predictors. Two ET features (total number of blinks with p-value<0.00003 and mean fixation duration with p-value<0.0002) were selected as significant predictors. Moreover, using the chi-squared test, additional features were selected as the best predictors: standard deviation of saccade duration (with p-value<0.002), the total number of small saccades (with p-value<0.0025), of fixations, of large saccades, and of saccades.

Combinations of the identified predictors were used in binary classification. To identify a high performance user in the RFFT, group 1 was seen as the negative class and group 2 as the positive class. The best accuracy (Table 2) was obtained for:

DOI: 10.1051/itmconf/20171502002

- K-NN classifier with k=15 cosine distance metric and equal distance weight for standardised data using as (2) predictors the total number of blinks and mean fixation duration;
- Bagged Trees classifier with 30 decision tree learners and random forest bag ensemble method using as (5) predictors the standard deviation of the saccade duration, the total number of small saccades, of large saccades, of fixations, and of blinks.

Table 2. The statistical measures of the performance of a binary classification.

Classifier	ACC	Sensitivity	SPC	PPV	MCC
KNN	71%	67%	74%	62%	0.4
Bagged Trees	79%	93%	70%	67%	0.62

The model with the Bagged Trees classifier has the accuracy (ACC) of 79% and detects high performance users correctly, as the sensitivity parameter indicated. The precision (positive predictive value, PPV) in both models is low because the validating subset is unbalanced in favour of the negative class (group 1). However, as the specificity (SPC) parameter shows, the Bagger Trees classifier turned out to be worse than the K-NN in classifying a low performance user. The Matthews correlation coefficient (MCC) implied that the Bagged Tress classifier is in general better in the stated binary classification.

4.2 Error ratio

The median for perseverative errors as well for the error ratio is 0, furthermore the perseverative errors are correlated with the total number of design, therefore the error ratio better describes the correctness of performance.

To find the groups (clusters) of similar observations the HAC method was used. The highest jump in the dendrogram indicated as the best clustering in two groups is presented in Table 3.

Table 3. The groups' characterisation using error ratio as similarity measure.

Feature	Group 1	Group 2	Overall
Number of observations	110 (62%)	67 (38%)	177 (100%)
Error ratio M (SD)	0.00 (0.01)	0.10 (0.06)	0.04 (0.06)

ANOVA does not find any significant differences in variables between the groups. Therefore, no further classification was conducted on the basis of the obtained clustering.

4.3 The total number of designs and error ratio

The additional clustering was proposed on the basis of two features: the total number of designs and the error ratio as indicators of performance and its correctness. The HAC method determined that the optimal number of a cluster is 3. The characteristic of those groups is presented in Table 4. In each group both features were significant and explained the created model. The visualisation of the clusters using on the axis two principal component values of principal component analysis (PCA) for a total number of designs and error ratio values is presented in Figure 1. Moreover, Kruskal-Wallis ANOVA does not show any statistically significant difference between mean values of age or education (in years) in the groups.

The size of the groups varies. Taking into consideration the small size of group 2, it was necessary to check whether the population was normally distributed according to all ET features. For features not following normal distribution the Kruskal-Wallis ANOVA was applied replacing one-way ANOVA. The following features have statistically significant difference between mean (or median) values in at least two of three groups: total number of blinks, mean fixation duration, standard deviation of fixation duration, standard deviation of saccade duration, and standard deviation of left and right pupil size change.

Table 4. The groups' characterisation using a total number of designs and error ratio values as similarity measure.

Feature	Group 1	Group 2	Group 3
Number of observations	66 (37.3%)	12 (6.8%)	99 (55.9%)
Total number of designs M (SD)	12.73 (2.26)	21.33 (3.47)	19.85 (2.48)
Error ratio M (SD)	0.04 (0.05)	0.22 (0.05)	0.02 (0.03)

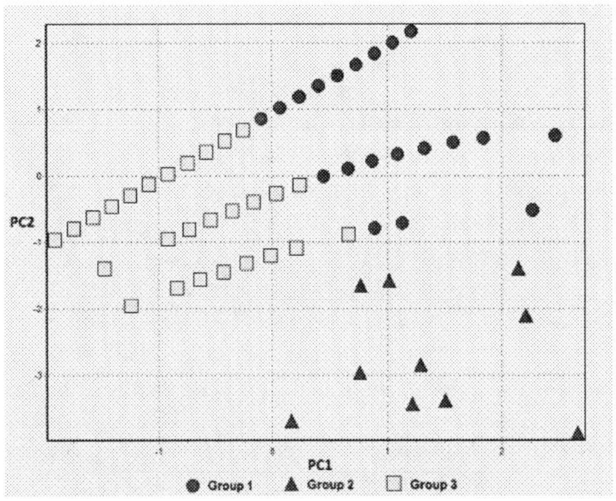

Fig. 1. The visualisation of the clustering groups using two primary components.

The chi-squared test feature selected the following predictors: standard deviation of right pupil size change (with p-value=0.004), standard deviation of saccade duration and a total number of blinks.

The identified predictors from the chi-squared test, features as significantly different between groups from ANOVA and their combinations were used in three-class classification. The group size in the validating subset was unbalanced (11 observations in group 1, 2 in group 2 and 23 in group 3) as results of the HAC cluster size.

The best classifier performances (Table 5) were obtained for:

- SVM (SVM-1) classifier with linear kernel function, kernel scale equals 4, and box constraint level equals 3, the one vs one method and standardised data; with mean fixation duration, standard deviation of saccade duration, standard deviation of the right pupil size change and a total number of blinks as (4) predictors.
- SVM (SVM-2) classifier with linear kernel function, kernel scale and box constraint level equal 4, the one vs one method and standardised data; with mean fixation duration, standard deviation of saccade duration, standard deviation of the left and right pupil size change and a total number of blinks as (5) predictors.

Table 5. The statistical measures of the performance of multiclass classification.

Classifier	ACC	Kappa
SVM-1	63%	0.244
SVM-2	67%	0.308

The classifiers are almost the same, however the additional predictor (standard deviation of left pupil size change) even is strongly correlated with the standard deviation of the right pupil size change allows for a small improvement of the model's accuracy. Both models cannot properly classify observations from group 2 (which are classified as group 3) and show more precision in classifying observations from group 3.

5 Discussion and conclusion

The research questions stated in this work were addressed. The relationship between eye-tracking features and cognitive performance (measured in a total number of designs and/or error ratio) were examined. The extracted features were explored, the best predictors were found and then the classification model was built. The ET features (total number of blinks, mean fixation duration, standard deviation of saccade duration, a total number of small saccades, a total number of fixations, a total number of large saccades, a total number of saccades) proved to be statistically significant and they were used as predictors in order to classify the data into groups.

For binary classification, where groups were created on the basis of the median value of the total number of designs, the accuracy of 79% was obtained for the Bagged Trees classifier with 5 predictors. For multiclass classification, where groups were determined by HAC using the total number of designs and error ratio values, the accuracy was 67% for SVM with linear kernel and 5 predictors. The limit size of the groups was creating a barrier for the proper multiclass training and classification, and mainly for group 2. Among all predictions, the following turned out to be the best: standard deviation of the right pupil size change, standard deviation of saccade duration and a total number of blinks. Further detailed investigation about the multiclass classification will only be achievable after increasing the size of the dataset.

The authors would like to thank all the students who participated in the experiment. This research did not receive any specific grant from funding agencies in the public, commercial, or not-for-profit sectors.

References

1. R. M. Ruff, *Ruff Figural Fluency Test professional manual* (Psychological Assessment Resources Inc., Odessa, FL, 1988)
2. P. S. Foster, J. B. Williamson, D. W. Harrison, *Arch. Clin. Neuropsychol.* **20**, 427–434 (2005)
3. M. E. A. van Eersel, H. Joosten, J. Koerts, R. T. Gansevoort, J. P. J. Slaets, G. J. Izaks, *PLoS One.* **10**, e0121411 (2015)
4. J. S. Kuiper, R. C. O. Voshaar, F. E. A. Verhoeven, S. U. Zuidema, N. Smidt, *BMC Psychol.* **5**, 15 (2017)
5. R. M. Ruff, R. H. Light, R. W. Evans, *Dev. Neuropsychol.* **3**, 37–51 (1987)
6. G. J. Izaks, H. Joosten, J. Koerts, R. T. Gansevoort, J. P. Slaets, *PLoS One.* **6**, e17045 (2011)
7. G. J. Demakis, D. W. Harrison, *Psychol. Rep.* **81**, 443–448 (1997)
8. L. C. Berning, N. C. Weed, M. S. Aloia, *Assessment.* **5**, 181–186 (1998)
9. M. Regard, E. Strauss, P. Knapp, *Percept. Mot. Skills.* **55**, 839–844 (1982)
10. J. V Baldo, A. P. Shimamura, D. C. Delis, J. Kramer, E. Kaplan, *J. Int. Neuropsychol. Soc.* **7**, 586–596 (2001)
11. E. Łojek, J. Stańczak, *Pol. standaryzacja i Norm. Pod. [The Ruff Fig. Fluen. Test. Polish Standarisation Norm. Warszawa Prac. Testów Psychol. Pol. Tow. Psychol.* (2005)
12. P. Krukow, M. Harciarek, J. Morylowska-Topolska, H. Karakuła-Juchnowicz, K. Jonak, *Cogn. Neuropsychiatry.* **22**, 391–406 (2017)
13. S. L. Hicks, R. Sharma, A. N. Khan, C. M. Berna, A. Waldecker, K. Talbot, C. Kennard, M. R. Turner, *PLoS One.* **8**, 8–13 (2013)
14. B. Poletti, L. Carelli, F. Solca, A. Lafronza, E. Pedroli, A. Faini, S. Zago, N. Ticozzi, A. Ciammola, C. Morelli, P. Meriggi, P. Cipresso, D. Lul?, A. C. Ludolph, G. Riva, *et al.*, *Neurol. Sci.* **38**, 595–603 (2017)
15. J. Keller, M. Gorges, H. T. Horn, H. E. A. Aho-Özhan, E. H. Pinkhardt, I. Uttner, J. Kassubek, A. C. Ludolph, D. Lulé, *J. Neurol.* **262**, 1918–1926 (2015)
16. B. Gaymard, C. J. Ploner, S. Rivaud, A. I. Vermersch, C. Pierrot-Deseilligny, *Exp. Brain Res.* **123**, 159–163 (1998)
17. J. J. S. Barton, A. Jama, J. A. Sharpe, *Neurology.* **45**, 2065–2072 (1995)
18. P.-A. Fransson, M. Patel, M. Magnusson, S.

Berg, P. Almbladh, S. Gomez, *J. Vestib. Res.* **18**, 209–222 (2008)

19. Q. Yang, T. Wang, N. Su, S. Xiao, Z. Kapoula, *Age (Omaha).* **35**, 1287–1298 (2013)
20. C. Donaghy, R. Pinnock, S. Abrahams, C. Cardwell, O. Hardiman, V. Patterson, R. C. McGivern, J. M. Gibson, *J. Neurol.* **256**, 420–426 (2009)
21. M. Gorges, H.-P. Müller, D. Lulé, K. Del Tredici, J. Brettschneider, J. Keller, K. Pfandl, A. C. Ludolph, J. Kassubek, E. H. Pinkhardt, *PLoS One.* **10**, e0142546 (2015)
22. A. L. Boxer, S. Garbutt, K. P. Rankin, J. Hellmuth, J. Neuhaus, B. L. Miller, S. G. Lisberger, *J. Neurosci.* **26**, 6354–6363 (2006)
23. T. M. Merz, M. M. Bosch, D. Barthelmes, J. Pichler, U. Hefti, K. U. Schmitt, K. E. Bloch, O. D. Schoch, T. Hess, A. J. Turk, U. Schwarz, *Eur. J. Appl. Physiol.* **113**, 2025–2037 (2013)
24. A. Olsen, "The Tobii I-VT Fixation Filter: Algorithm description" (2012), (available at http://www.tobii.com/)
25. K. Holmqvist, M. Nystrom, R. Andersson, R. Dewhurst, H. Jarodzka, J. van de Weijer, *Eye Tracking. A comprehensive guide to methods and measures* (Oxford University Press, 2011)
26. H. K. Wong, J. Epps, *Comput. Methods Programs Biomed.* **137**, 47–63 (2016)
27. M. G. Larson, *Circulation.* **117**, 115–121 (2008)
28. I. Davidson, S. S. Ravi, *9th Eur. Conf. Princ. Pract. Knowl. Discov. Databases, PKDD 2005*, 59–70 (2005)
29. R. Rakotomalala, *Proc. EGC.* **2**, 697–702 (2005)
30. M. A. Wong, *J. Am. Stat. Assoc.* **77 (380)**, 841–847 (1982)
31. J. Poulsen, A. French, *J. Forensic Sci.* **56**, 297–301 (1996)
32. A. Shmilovici, in *Data Mining and Knowledge Discovery Handbook*, O. Maimon, L. Rokach, Eds. (Springer New York, Second Edi., 2010), pp. 231–247
33. T. M. Mitchell, others, Machine learning. *McGraw Hill Ser. Comput. Sci.* (1997), pp. I--XVII
34. L. Breiman, *Mach. Learn.* **24**, 123–140 (1996)
35. L. Breiman, *Mach. Learn.* **45**, 5–32 (2001)

ITM Web of Conferences **15**, 02003 (2017)
CMES'17 DOI: 10.1051/itmconf/20171502003

Classifier testing for the brain-machine interface (BCI) based on Steady State Visually Evoked Potential (SSVEP)

Arkadiusz Kubacki[1,*], and *Arkadiusz* Jakubowski[1]

[1]Poznan University of Technology, Institute of Mechanical Technology, Skłodowska-Curie Square 5, 60-965 Poznań, Poland

Abstract. The paper describes the research on the classifiers for brain-computer interface (BCI) based on Steady State Visually Evoked Potential (SSVEP). Authors presented research on the checking the usability of classifiers for recognizing an EEG signal during the stimulus. Three classifiers have been checked: Support Vector Machine (SVM), Linear Discriminant Analysis (LDA) and one based on Artificial Neural Network (ANN). First part is concentrated on brain-computer interfaces and classification of them. The second part describes algorithms of all using classifiers. In the next part, authors present test stand and how the experiment is built. The last part consists of results of these tests. The best was the classifier based on Artificial Neural Network – up to 95% of correct identified. The worst results were obtained from Support Vector Machine – about 70%.

1 Introduction

Nowadays, people cannot imagine life without computers. So, it is normal that they are used to newer devices to control them. At present, two main trends are observed. These are vision system interfaces and interfaces based on human biosynthesis. In the first area, people use various cameras, such as Microsoft Kinect [1], to track and recognize some gestures or body movements. In the other area, the most popular are interfaces using electromyographic (EMG) sensors like the Myo armband [2], which can recognize hand and arm gestures using the electrical activity of forearm's muscles or brain-computer interfaces (BCI), which can recognize some activities directly from the brain [3]. There are many methods which can be used to monitor brain activity [4]:
- Electroencephalography (EEG),
- Positron emission tomography (PET),
- Functional magnetic resonance imaging (fMRI),
- Magnetoencephalography (MEG).

Owing to decreasing prices of headsets, BCIs based on EEG are developing rapidly [5]. EEG is a non-invasive method used to record the activity of brain from the examined person skull through the electrodes. Currently, the most popular and most commonly used system of electrodes placement is a "10-20" system [6]. This is an international system to describe and apply the location of scalp electrodes in the context of an EEG test or experiment. This method was created to guarantee standardized reproducibility. Thus, the EEG results of different people can be compared to each other. This system is based on a relationship between a location of an electrode and an underlying area of cerebral cortex. The name of this system refers to the fact that actual distances between adjacent electrodes are either 10% or 20% of the total front–back or right–left distance of the skull. The electrodes placement in this system is shown in Fig.1. All electrodes are labelled with a capital letter and a number. The letter refers to the area of brain e.g. F – Frontal lobe and T – Temporal lobe. Moreover, even numbers refer to the right side of the head and odd numbers the left side.

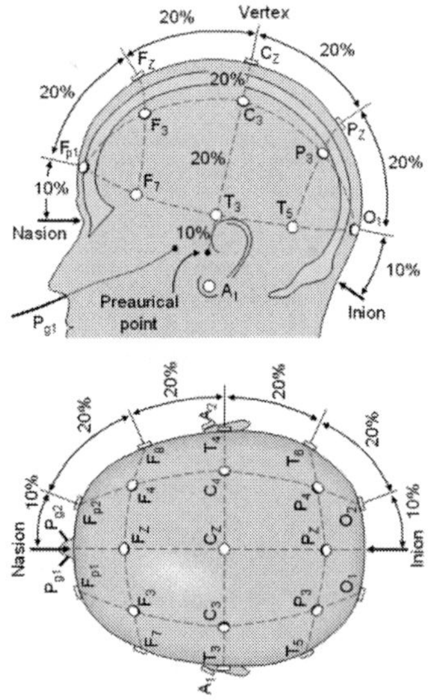

Fig. 1. Electrode placement in accordance with the "10-20" system [6].

* Corresponding author: arkadiusz.j.kubacki@doctorate.put.poznan.pl

DOI: 10.1051/itmconf/20171502003

1.1 Classification of BCIs

Currently, the basic method for brain-computer interfaces is electroencephalography, as it provides an opportunity to observe changes in brain activity in a short period of time. Unfortunately, EEG caps have poor spatial resolution and signal susceptibility to interference by artefacts from both, facial expressions and other activities of the body [7]. In the article [8], authors have shown how to find artefacts in the electroencephalogram using artificial neural network. Fig. 2 shows currently used methods to measure brain activity. Passive methods use external stimuli to induce brain activity. In active methods, human brain activity is caused through adequate imaginary [7].

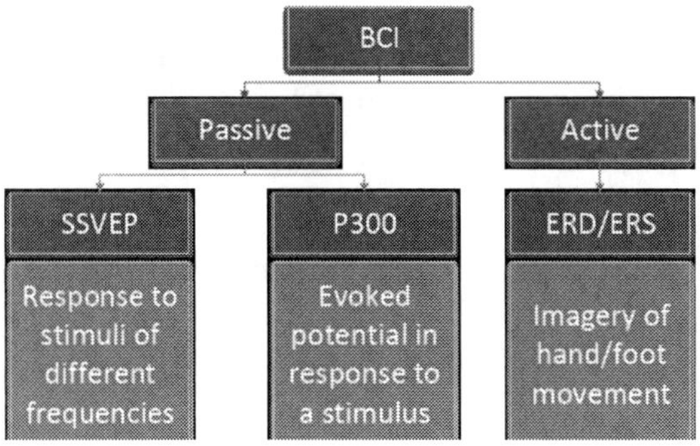

Fig. 2. Classification of BCIs [7].

Using only one method limits the user, especially in the number of possible commands and the effectiveness of their detection. It is possible to use used hybrid BCI. Their classification is shown in Fig. 3.

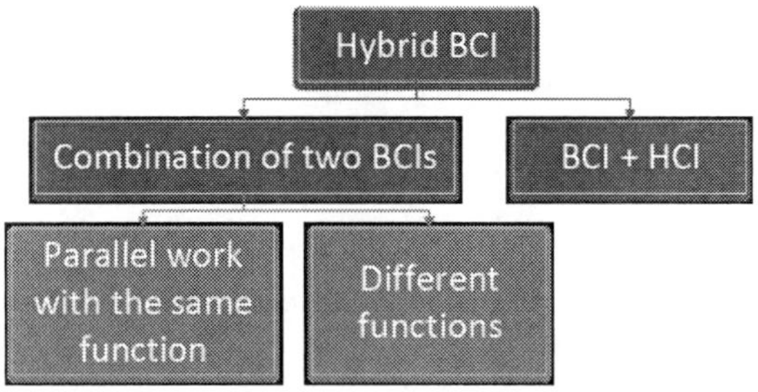

Fig. 3. Classification of hybrid BCIs [7].

Furthermore, more than only one type of interface can be used to increase the accuracy of the classification commands at parallel connection or increase the number of possible commands. BCI method can be also linked with other interfaces, such as speech recognition or eye movement observation, to make hybrid method. The aim is to have more accurate commands recognition and expanding their numbers [7].

1.2 Steady-State Visually Evoked Potential (SSVEP)

Steady-State Visually Evoked Potential (SSVEP) is a periodic call of evoked potentials through repeated visual stimulation. Usually, the frequency of blinking is more than 6Hz [9]. Flashing light is commonly used as a stimulating element. It is possible to use a LED diode or bulb or monitor with blinking point or chessboard [10]. Corresponding to the stimulation frequency and their harmonics, in an EEG signal appears the same frequency [11]. Therefore, a person without prior training can use this method for testing, which is the most important advantage of this method. Fig. 4 shows the Fast Fourier Transformation of signal taken from electrode called Oz, during initial testing of the SSVEP method. In Fig. 4a, the FFT is shown while the LED is not blinking and in Fig. 4b, the FFT of the same electrode output signal is shown when white LED is flashing with frequency 15 Hz.

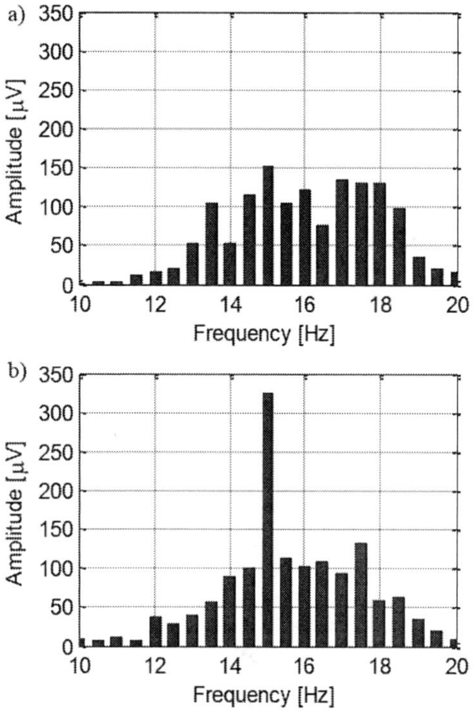

Fig. 4. Electrode Oz output signal FFT characteristics for: no blinking (a), LED blinking with 15 Hz (b).

1.3 State of art

Nowadays, brain-computer interfaces are built worldwide. In this state of art, authors present only this BCI which structure is based on Steady State Evoked Potential. Articles mentioning another two methods, shown in Fig. 2., are omitted.

In the article [11] authors created a speller based on Steady-State Visually Evoked Potential. Evoked potentials and stimulating subjects, by stimuli of frequencies from 6.66 to 8Hz, were indispensable for

 DOI: 10.1051/itmconf/20171502003

that purpose. The monitor with five squares was used. Three of them were used to divide the alphabet into 3 subgroups. On next level, a group of nine signs was divided into next 3 subgroups. On the last level, a user chose one letter. The other squares were used to delete letter and repeat the last action.

The authors of the article [12] were using the SSVEP to control prosthetic hand. Four kinds of commands can be invoked by the blinking field on the screen. Fields had a rate of 6Hz for the command "turn the hand to the left", 7Hz for the command "turn the hand to the right", 8Hz for the command "open the hand", 13Hz for the last command "close the hand".

In the article [5] authors presented the possibility of SSVEP to control a virtual car in virtual reality. Two HUD screens were used to display images of frequencies, 12 and 13Hz respectively for the command "turn left" or "turn right". Respondents got around previously designed route.

The authors of the article [13] used the SSVEP method to build hybrid BCI. They combined the SSVEP and MI task to play Tetris game. The SSVEP method was responsible for the left and right rotation of blocks and pause the game. The MI tasks were responsible for the left and right movement of blocks.

2 Classifiers

Three classifiers were considered in this paper: Linear Discriminant Analysis (LDA), Support Vector Machine (SVM) and Artificial Neural Network (ANN). The following subsections describe their algorithms.

2.1. Linear Discriminant Analysis (LDA)

Linear Discriminant Analysis (LDA) is a very important classifier, which can be used for wide variety of problems. For instance, in different machine learning problems, such as pattern or face recognition, feature extraction and data dimensionality reduction [14].

In this method, the ratio of between-class variance to the within-class variance in any data set is maximized, what guarantees maximal separability [15].

In this article, authors used LDA to two-class problem. In first step, the data sets were formulated (1).

$$set1 = \begin{bmatrix} a_{11} & a_{12} \\ a_{21} & a_{22} \\ \cdots & \cdots \\ a_{m1} & a_{m2} \end{bmatrix} \quad set2 = \begin{bmatrix} b_{11} & b_{12} \\ b_{21} & b_{22} \\ \cdots & \cdots \\ b_{m1} & b_{m2} \end{bmatrix} \quad (1)$$

The equation (2) was used to compute the mean of each data set and mean of entire data set, where μ1 was mean of set1 and μ2 mean of set2, p1 and p2 were the a priori probabilities of the classes. In the case of this simple two-class problem, the probability factor was assumed to be 0.5.

$$\mu_3 = p_1 \times \mu_1 + p_2 \times \mu_2 \quad (2)$$

Within-class and between-class scatter in LDA method were used to formulate criteria for class separability. Covariance of each of the classes was the expected for within-class scatter. The scatter measures were computed using (3).

$$S_w = 0.5 \times cov_1 + 0.5 \times cov_2 \quad (3)$$

The following equation was used to compute covariance matrix (4).

$$cov_j = (x_j - \mu_j)(x_j - \mu_j)^T \quad (4)$$

The between-class scatter was computed using the equation (5).

$$S_b = \Sigma_j(\mu_j - \mu_3) \times (\mu_j - \mu_3)^T \quad (5)$$

As defined earlier, the optimizing criterion in LDA was the ratio of between-class scatter to the within-class scatter. Maximizing this criterion defined the axes of the transformed space. The class-dependent transform of the optimizing criterion was computed using equations (4) and (5). If the LDA is a class dependent type, all lasses should have separate optimizing criterion for each class. The equation (6) was used to compute the optimizing factors in case of class dependent type.

$$criterion_j = inv(cov_j) \times S_b \quad (6)$$

The optimizing criterion for the class independent transform was computed (7).

$$criterion = inv(S_w) \times S_b \quad (7)$$

Having obtained the transformation matrices, the data sets should be transformed using the single LDA transform or the class specific transforms whichever the case may be. The decision region in the transformed space was a solid line separating the transformed data sets. For the class dependent LDA – (8) and for the class independent LDA – (9).

$$transformed\_set\_j = transform_j^T \times set\_j \quad (8)$$

$$transformed\_set = transform\_spec^T \times data^T (9)$$

When the transformations were completed, Euclidean distance or RMS distance was used to classify data points. Euclidean distance was computed using (10), where μ_{ntrans} was the mean of the transformed data set. Thus, for classes Euclidean distances were obtained for each test point.

$$dist_n = transform_{n_{spec}}^T \times x - \mu_{ntrans} \quad (10)$$

The smallest Euclidean distance among all distances classified the test vector as belonging to class n [14].

2.2 Support Vector Machine (SVM)

Support Vector Machines (SVMs) are a set of related methods for supervised learning, applicable to both classification and regression problems. SVM classifier was introduced by Vapnik in 1995. In this article,

DOI: 10.1051/itmconf/20171502003

authors presented two types of SVM – C-SVM and Nu-SVM [17].

2.2.1 C-Support Vector Classification

In first step C-SVM solves the following (11) optimization problem.

$$\min_{w,b,\xi} \frac{1}{2} w^T w + C \sum_{i=1}^{l} \xi_i \quad (11a)$$

subject to

$$y_i(w^T \phi(x_i) + b) \geq 1 - \xi_i, \xi_i \geq 0, i = 1, \ldots, l, \quad (11b)$$

In this equation $\phi(x_i)$ maps x_i into a higher-dimensional space and $C > 0$ is the regularization parameter. It is possible that the vector variable w will be high dimensionality. Usually, this problem is solved by equation (12).

$$\min_{\alpha} \frac{1}{2} \alpha^T Q \alpha - e^T \alpha \quad (12a)$$

subject to

$$y^T \alpha = 0, \; 0 \leq \alpha \leq C, i = 1, \ldots, l \quad (12b)$$

In this equation $e = [1, \ldots, l]^T$is the vector of all ones, Q is an l by l positive semidefinite matrix, $Q_{ij} \equiv y_i y_j K(x_i x_j)$, and $K(x_i x_j) \equiv \phi(x_i)^T \phi(x_j)$ is the kernel function. After solving (12), the optimal w satisfies

$$w = \sum_{i=1}^{l} y_i \propto_i \phi(x_i) \quad (13)$$

Equation (14) is a decision function.

$$sgn(w^T \phi(x) + b)\, sgn\left(\sum_{i=1}^{l} y_i \propto_i K(x_i, x) + b\right) \quad (14)$$

2.2.2 Nu-Support Vector Classification

In first step Nu-SVM solves the following (15) optimization problem.

$$\min_{w,b,\xi} \frac{1}{2} w^T w - v\rho + \frac{1}{l} \sum_{i=1}^{l} \xi_i \quad (15a)$$

subject to

$$y_i(w^T \phi(x_i) + b) \geq 1 - \xi_i, \xi_i \geq 0, i = 1, \ldots, l, \quad (15b)$$

Usually, this problem is solved by equation (16).

$$\min_{\alpha} \frac{1}{2} \alpha^T Q \alpha \quad (16a)$$

subject to

$$y^T \alpha = 0, \; 0 \leq \alpha \leq \frac{1}{l}, \; i = 1, \ldots, l, e^T \alpha \geq v \quad (16b)$$

In this equation $Q_{ij} \equiv y_i y_j K(x_i x_j)$. Problem (16) is feasible if and only if (17)

$$v \leq \frac{2min(\#y_i = +1, \#y_i = -1)}{l} \leq 1 \quad (17)$$

Equation (18) is decision function.

$$sgn\left(\sum_{i=1}^{l} y_i \propto_i K(x_i, x) + b\right) \quad (18)$$

2.3 Artificial Neural Network (ANN)

Artificial Neural Network (ANN) used in this classifier is Multilayer Perceptron (MLP). ANN consists of at least two layers, which means that in this method only one hidden layer is used. The number of neurons on this layer is changeable. Inside the hidden layer, the hyperbolic tangent is used as the activation function. Backpropagation algorithm is used to teach the ANNs. It is one of the most popular methods of teaching ANN [18]. The aim is to minimize the error E between the network outputs O and the target values T in step t, using the method of steepest descent in a network. In the equation (19) S stands for training samples and N for the number of neurons in the output layer [18, 19].

$$E(t) = \frac{1}{2} \sum_{s=1}^{S} \sum_{n=1}^{N} \left(T_s^n(t) - O_s^n(t)\right)^2 \quad (19)$$

In this method, the idea is to iterate to the moment when error E decreases. In the first step of teaching algorithm, all weights of network w are set on small random numbers and a low learning rate η is selected. In the next step, the input x is propagated forward through the ANN. The computed values for each output O are being saved. The errors are being propagated backwards through the network. For every output of the network, an error δ_k is being computed (20).

$$\delta_k = T_k - O_k \quad (20)$$

The equation (21) is used to calculate error δ_h for each hidden neuron, where i_h stands for input of the activation function from earlier units.

$$\delta_h = f'(i_h) \sum_{k \in outputs} w_{kh} \delta_k \quad (21)$$

The equation (22) is used to update the weights w_{ji} in all layers.

$$w_{ji}(t + 1) = w_{ji}(t) + \eta \delta_j x_{ji} \quad (22)$$

Designing a neural network structure, consisting of several layers and the quantity of neurons in each layer, may be convoluted. At this moment, it is worth mentioning the Kolmogorov's theorem, in accordance with every continuous real function $f(x_1, \ldots, x_n)$, which could be approximated using the function (23).

$$f(x_1, \ldots, x_n) = \sum_{j=1}^{2n+1} g_j \left(\sum_{i=1}^{n} \phi_{ij}(x_i)\right) \quad (23)$$

Corresponding with the Kolmogorov's theorem, ANN structure has two layers with n inputs, 2n+1 neurons in the hidden layer and one output neuron with a linear activation function ϕ. There are very few problems requiring more than two hidden layers.

3 Equipment and experiment

In this paper, authors presented study conducted with the use of a neuroheadset called Emotiv EPOC+, which related to a computer via Bluetooth (Fig. 5). This device consisted of 14 EEG channels + 2 reference electrodes, which were placed in accordance with the "10-20" system (described before). Emotiv EPOC+ device uses one 16-bit ADC with a sampling rate of 256 samples per second (2048 Hz internal). This device could recognize all five main brain rhythms since it received brain waves in a bandwidth of 0.16 to 43 Hz [20].

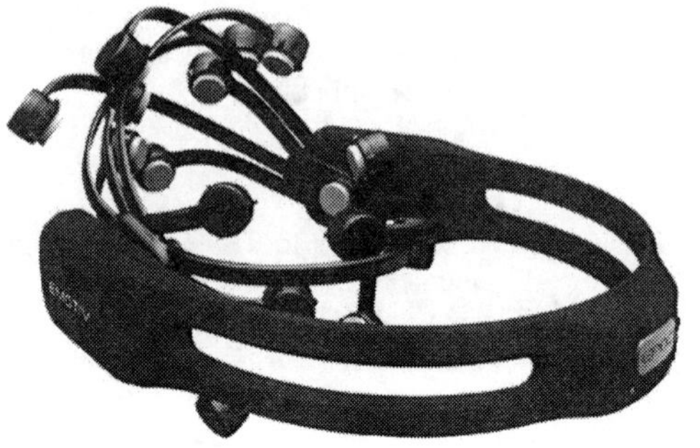

Fig. 5. Emotiv EPOC+ headset [19].

Authors have built a special light with dimensions of 45x45x45 mm. Inside of it, there was a 3W Power LED (Fig. 6). Thanks to the RGB LED, there is possible to flash a light in different colours. A control module based on the Atmega 328P microcontroller has been developed, allowing to set the diode frequency between 5 and 30 Hz, and set the brightness of the diode in the range of 0 to 100%.

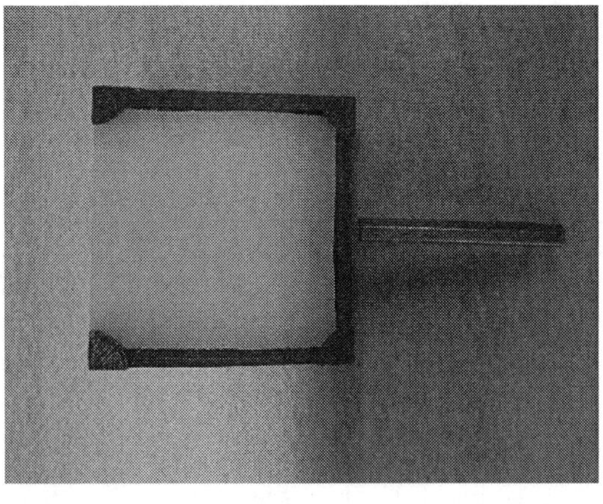

Fig. 6. Blinking module used in tests.

Authors created two scenarios in OpenVibe. The scheme of the program is shown in Fig. 7. In the first step, the input signal is filtered to the set frequency – 15Hz, 17Hz and 19Hz. Then, in the next block, an appropriate channel for the test is selected. For SSVEP, the part of the brain responsible for the sight is used; as it is placed on the back of the head, the Oz channel should be selected. Time based epoching (TBE) block was 1s.

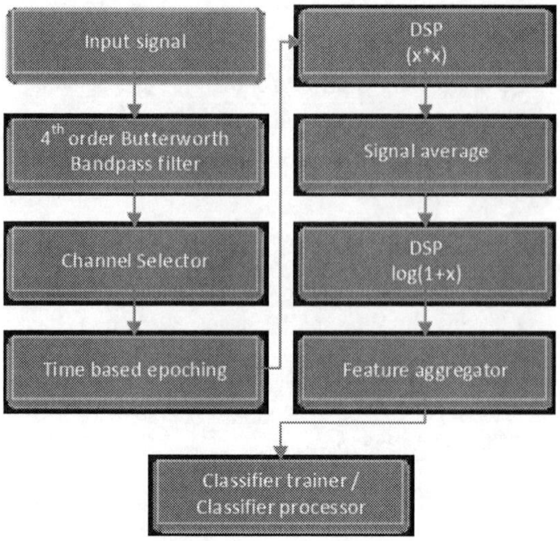

Fig. 7. Block diagram of signal processing during the experiments, made in OpenVibe system.

In the next step, the signal is raised to the power of 2, averaged, then logarithmically and outputted as Power of EEG signal. The last block before Classifier is a Feature aggregator. This block aggregates the features received on its inputs into a feature vector. Finally, signal go to the Classifier block. In the first one authors teach selected classifier with chosen settings. These settings are described in next chapter. The classifier has always been taught using the same input data. Data was collected in previous sessions of EEG test. Authors recorded data for frequency 15Hz, 17Hz and 19Hz. There was an additional scenario in OpenVibe for teaching classifiers. Authors recorded five different recorded data per one frequency per one person. Three people took part in the test. There were 3-minutes breaks between each frequency. In all tests data, which have been recorded before, were used.

4 Results

In the first test, authors checked LDA classifier. Authors taught three classifiers for 15 Hz, 17 Hz and 19Hz. To train k-fold cross validation test was set to 4 for all classifiers and frequency. In the next step, all classifiers were tested by five data per one frequency which was collected before. The results of this test are presented in Table 1.

Table 1. LDA classifier test results.

Frequency	Correctly identified
15 Hz	86 %
17 Hz	77 %
19 Hz	90 %

 DOI: 10.1051/itmconf/20171502003

The results prove that the worst score was obtained for 17 Hz. The same regularity can be seen in next tests. Most likely, this is due to noisy teaching data for this frequency. Positive predictive value of the 19 Hz LDA test is presented in Table 2.

Table 2. Positive predictive value of the 19 Hz LDA test.

	Condition positive	Condition negative
Test positive	0.96	0.04
Test negative	0.06	0.94

In first test, authors checked C-SVM and Nu-SVM classifiers. Authors taught three classifiers for 15 Hz, 17 Hz and 19Hz. To train k-fold cross validation test was set to 4, ε to 0.1, cost to 1 and cache size was set to 100 for all classifiers and frequency. Authors checked four types of kernel: Linear, Polynomial, Radial basis function and Sigmoid. In the next step, all classifiers were tested by five data per one frequency which was collected before. The results of C-SVM test are presented in Table 3 and the results of Nu-SVM test are presented in Table 4.

Table 3. C-SVM classifier test results.

Frequency	**Correctly identified**			
	Linear	Polynomial	Radial	Sigmoid
15 Hz	90 %	82 %	86 %	1 %
17 Hz	66 %	50 %	70 %	33 %
19 Hz	95 %	95 %	95 %	81 %

Table 4. Nu-SVM classifier test results.

Frequency	**Correctly identified**			
	Linear	Polynomial	Radial	Sigmoid
15 Hz	70 %	77 %	56 %	0 %
17 Hz	70 %	62 %	82 %	62 %
19 Hz	95 %	90 %	95 %	86 %

The results show that the worst score was obtained for Sigmoid kernel. For all frequencies, that result is the lowest in C-SVM and Nu-SVM. The sigmoid kernel is a special type because it has only positive values. Due to this feature, is not suitable for all applications. Probably, the fitting turned out to be very difficult, as in case. The best score is obtained with C-SVM classifier using Linear kernel. Positive predictive value of the 19 Hz C-SVM linear test is presented in Table 5.

Table 5. Positive predictive value of the 19 Hz C-SVM linear test.

	Condition positive	Condition negative
Test positive	0.95	0.05
Test negative	0.00	1.00

In the last test, authors checked ANN classifiers. Authors taught three classifiers for 15 Hz, 17 Hz and 19Hz. To train k-fold cross validation test was set to 4, learning stop condition was set to 0.000001 and Learning coefficient was set to 0.01, for all classifiers and frequency. Authors changed number of neurons on hidden layer from 1 to 5. In the next step, all classifiers were tested by five data per one frequency which was collected before. The results of this test are presented in Table 6.

Table 6. ANN classifier test results.

Frequency	**Correctly identified**				
	1	2	3	4	5
15 Hz	86 %	86 %	86 %	86 %	86 %
17 Hz	53 %	82 %	82 %	82 %	82 %
19 Hz	95 %	95 %	95 %	95 %	95 %

The results show that the score for all frequencies and numbers of neuron on hidden layer are similar. Only in case of 17 Hz this classifier needs two neurons for correct recognition of input signal. The ANN classifier has got the best results from all tested in this article classifiers. Positive predictive value of the 19 Hz ANN test is presented in Table 7.

Table 7. Positive predictive value of the 19 Hz ANN test.

	Condition positive	Condition negative
Test positive	0.99	0.01
Test negative	0.04	0.96

5 Conclusion

The paper describes the research on the classifiers for brain-computer interface (BCI) based on Steady State Visually Evoked Potential (SSVEP). Authors presented research on checking the usability of classifiers for recognizing the EEG signal during the stimulus. The results show that it is possible to recognize Steady-State Visually Evoked Potential using Linear Discriminant Analysis (LDA), Support Vector Machine (SVM) and Artificial Neural Network (ANN). Achieved results were acceptable, but not perfect. The best performance presented the classifier based on Artificial Neural Network, while Support Vector Machine presented the worst. No articles were found on comparing these classifiers, in the context of BCI based on Steady State Visually Evoked Potential (SSVEP). Therefore, it is not possible to compare the results with other articles.

In further studies, the authors take as their aim an improvement of the classifier, so that the recognition results can cause minor errors. These classifiers will be very useful to build their hybrid brain-computer interface.

The work described in this paper was funded from 02/22/DSMK/1408 (Sterowanie urządzeniami mechatronicznymi z zastosowaniem technik nieliniowych oraz bioelektrycznej aktywności mózgu).

DOI: 10.1051/itmconf/20171502003

References

1. L. Shao, J. Han, D. Xu, J. Shotton, IEEE T Cybernetics, **43**, 1314-1317, (2013)
2. Z. Arief, I. A. Sulistijono, R. A. Ardiansyah, IES, **2015**, 11-14 (2015)
3. A. Kubacki, L. Sawicki, D. Rybarczyk, P. Owczarek, Adv Intell Syst, **550**, 433–440 (2017)
4. J. R. Wolpaw, N. Birbaumer, D. J. McFarland, G. Pfurtscheller, T. M. Vaughan, Clin Neurophysiol, **113**, 767-791, (2002)
5. L. Bi, X. a Fan, K. Jie, T. Teng, H. Ding, Y. Liu, IEEE T Intell Transp, **15**, 959-966, (2014)
6. V. K. Varadan, S. Oh, H. Kwon, P. Hankins, J. Nanotechnol. Eng. Med, **1**, 3, (2010)
7. A. Cudo, E. Zabielska, B. Bałaj, KUL, (2011).
8. A. Kubacki, L. Sawicki, P. Owczarek, MMAR, **2016**, 783-787, (2016)
9. Z. Lin, C. Zhang, W. Wu, X. Gao, IEEE T Bio-Med Eng, **53**, 2610-2614, (2006)
10. D. Zhu, J. Bieger, G. G. Molina, R. M. Aarts, Intell. Neuroscience, **2010**, (2010)
11. H. Cecotti, IEEE T Neur Sys Reh, **18**, 127-133, (2010)
12. G. R. Muller-Putz and G. Pfurtscheller, IEEE T Bio-Med Eng, **55**, 361-364, (2008)
13. R.O. Duda, P.E. Hart, D.H. Stork., Pattern Classification (2nd ed.), (Wiley Interscience, 2000)
14. O. Dehzangi, Y. Zou, and R. Jafari, NER, **2013**, 1303–1306, (2013)
15. Balakrishnama, S., & Ganapathiraju, A. Linear discriminant analysis-a brief tutorial. Institute for Signal and information Processing, **18**, (1998)
16. A. Shmilovici, Springer, **2009**, 231–247, (2009)
17. C. Chih-Chung, L. Chih-Jen, (2001) http://www.csie.ntu.edu.tw/~cjlin/papers/libsvm.pdf. [Accessed: 30-Jul-2017].
18. L. Rutkowski, Methods and techniques of artificial intelligence, (Wydawnictwo naukowe PWN, 2005)
19. T. Mitchell, Machine Learning, (McGraw-Hill, 1997)
20. http://emotiv.com/

Kohonen network as a classifier of Polish emotional speech

Paweł Powroźnik[1,*]

[1]Lublin University of Technology, Faculty of Electrical Engineering and Computer Science, Institute of Computer Science, Nadbystrzycka 38D, 20-618 Lublin, Poland

Abstract. The power of speech is a main tool in human communication. There are a lot of factors as age, emotions, gender, pitch of the voice which can influence features of speech. Obviously, information conveyed by voice intonation has more than only textual meaning. The same sentence pronounced in two different ways can have two completely different meanings. This paper describes Kohonen networks as a classifier of Polish emotional speech. The usage of Discrete Wavelet Transform (DWT) as well as an innovative approach to scaleogram processing is also presented in this article. Mexican Hat Wavelet and the Haar Wavelet were used in researches. All simulations were carried out in MatLab 2016 with Neural Network Toolbar. During whole research more than 9000 simulation have been done. Three different speech databases were used in conducted researches. One of them was prepared by professional actors – four women and four men, and contains 240 wav files. Two others are results of researchers works. The structures of used Kohonen networks depend on speech signal decomposition's level and scaleogram division. During conducted researches the following emotional states were considered: anger, joy, sadness, boredom, fear and neutral state. Achieved results were between 68% and 80% depends of used wavelet, speech signal and signal decomposition's level.

1 Introduction

Recognition of speaker's emotional state based on speech signal processing is relatively new issue but its significance has been rapidly increasing. One of the reasons of such a direction of changes is burgeoning development of systems based on Brain-Computer Interface, as well as Virtual Reality (VR) environments [1].

There are a lot of factors such as: age, emotions, language or gender of a speaker which may have great influence on features of speech signal [2, 3]. Obviously, that information conveyed by intonation of voice has more than only literal meaning.

The biggest problem in emotional speech recognition systems is the number of emotional states; therefore, developing an application which correctly identifies most emotions is not trivial. So far, in researches the following emotional states are often considered: anger, joy, sadness, fear, boredom and neutral state [4, 5].

In the Figure 1, oscillograms for three emotional states are presented, including: neutral state (top), joy (middle), anger (bottom) for the same expression in the semantic sense. As it can be easily seen, the appearance of a particular emotion changed not only ranging from frequency but also to the shape of oscillogram.

In this article, an innovative system of Polish emotional speech signal processing has been described. The system based on discrete wavelet transform and scaleograms. As a classifier the Kohonen networks have been used.

The whole article was divided in four parts. The first one charactering the discussed matter. The second one described used databases of Polish emotional speech. The third part included the description of signal processing algorithm, research methods, and parameters for the Polish emotional speech and presented obtained results and suggestions for improving the adapted research methods.

2 Analysis of issues

Nowadays, the researches struggle with developing a proper and useful model of emotion. Contemporarily, two most popular models exist in the literature: James-Lange's and Plutchik's models [3, 4, 6]. The first one assumes that behavioural and somatic changes are interpreted as emotion which triggers a reaction. Based on James-Lange's model the following sentence is true: 'I am afraid of murder because I am running' instead of 'I am running because I am afraid of murder' [7]. The second model was proposed in 1960 by Robert Plutchik. He introduced eight basic emotional states which were related to behaviours essential for surviving, that is: joy, fear, trust, sadness, surprise, anger, disgust and anticipation. All other states arose from the basic ones [8].

As it was mentioned, the most complicated issue in Polish emotional speech recognition is the number of emotional states which should be detected. It should

* Corresponding author: p.powroznik@pollub.pl

 DOI: 10.1051/itmconf/20171502004

be emphasised that for an average person it is possible to recognise another person's emotional state only in 60% of all cases [9]. Some methods of emotions' detection in speech signal, based on the Polish language, have been described in the literature. These methods relate to Support Vector Machine (SVM) [10] or k-Nearest Neighbour algorithms [4]. However, all authors conclude that obtained results are not fully satisfactory and used signal processing methods need to be improved. All mentioned researches focused on six most popular emotional states: anger, joy, boredom, fear, neutral state and sadness [4, 10].

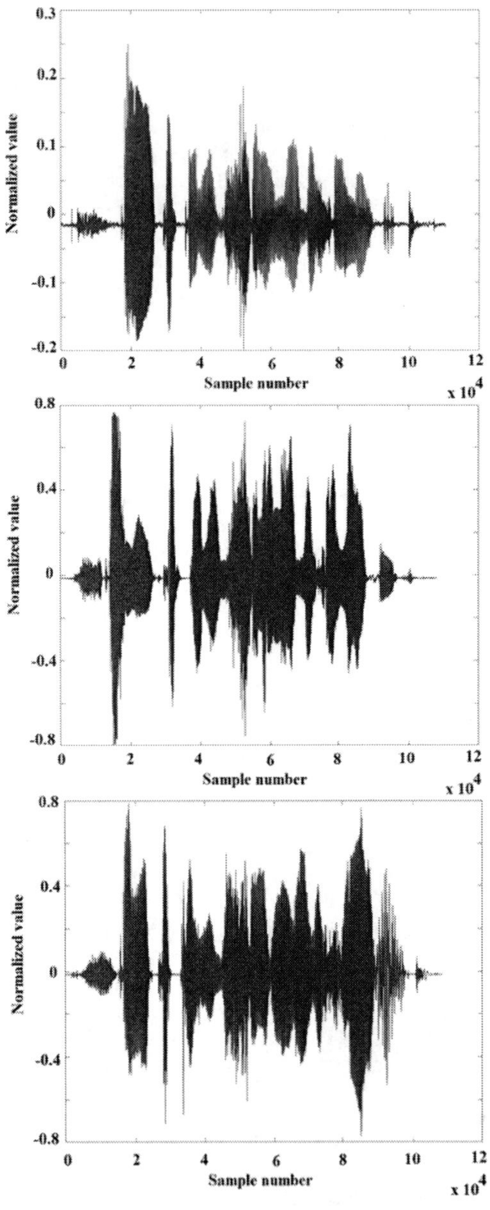

Fig. 1. Comparison of oscillograms of the same sentence in three different emotional states: neutral (top), joy (middle), fear (bottom).

Time-frequency methods are among the most popular speech signal processing tools [11]. They allow to estimate speech signal spectrum in short and finite time based on window functions and overlapping method [11]. A particular role in these cases, has Short Time Fourier Transform (STFT) and Discrete Wavelet Transform (DWT) [2, 4, 10, 12]. In this article the usage of second one has been described in detail.

3 Description of used databases

Berlin Database of Emotional Speech (BES) [13] is commonly used in many researches connected with emotional speech signal processing. However, if Polish speech is considered, a researcher would rather use their own databases or the one prepared by Medical Electronics Division of the Lodz University of Technology [14]. The base has been prepared by professional actors: four women and four men, and collected files was recorded in six emotional states. The whole database contains 240 records sampled with 44.1 kHz frequency and the bit rate of 16 bps. This database includes the following statements: 'Johnny went to the hairdresser today', 'They bought a new car today', 'I've stopped shaving from today on', 'His girlfriend is coming here by plane' and 'This lamp is on the desk today'. It was the first database used in conducted researches.

The second one was prepared in Lublin University of Technology and contained the same records as the first one. The abovementioned collection was recorded in acoustic chamber. The research involved people between 20 and 30, who were not involved in acting. The entire database contains 306 records and its structure is presented in the Table 1. Unfortunately, it was not possible to collect the same number of recordings for each emotional state for both sexes.

Table 1. Structure of recordings in second database.

Emotion	Number of recordings (women)	Number of recordings (men)
Anger	14	19
Joy	13	27
Sadness	32	30
Neutral state	28	29
Boredom	31	29
Fear	28	26

The third database was also prepared in Lublin University of Technology but the recordings was collected not in acoustic chamber but in the city environment (lecture class, street, shop). Three women and six men aged 22-31 were involved in this study. They uttered following five sentences: 'I need to talk to you', 'Is this what you really think?', 'You don't understand anything', 'Peter bought a new bike', 'This package is in Krakow already'. Whole database contains 266 records and its structure is shown in the Table 2.

 DOI: 10.1051/itmconf/20171502004

Table 2. Structure of recordings in third database.

Emotion	Number of recordings (women)	Number of recordings (men)
Anger	22	28
Joy	18	20
Sadness	24	23
Neutral state	26	24
Boredom	21	27
Fear	15	18

In case of second and third database, the affiliation of recordings to specific group of emotion was verify by 94 respondents.

4 Conducted researches

During conducted researches, following emotional states were considered: anger, joy, sadness, boredom, fear and neutral state. All simulations have been done in MatLab 2016 with Neural Network Toolbox. The number of all experiments exceeded 9000.

4.1 Wavelet Transformation

Continuous Wavelet Transform (CWT) was developed by Jean Morlet and Alex Grossman. One dimensional signal is expressed by following form [15]:

$$x(t) \epsilon L^2(\mathcal{R}) \tag{1}$$

CWT has the following form [16]:

$$w(a,b) = \frac{1}{\sqrt{a}} \int_{-\infty}^{+\infty} x(t) \psi^* \left(\frac{t-b}{a}\right) dt \tag{2}$$

where: * – indicates Conjugate complex function, a (a>0) – scale parameter, b – offset parameter, ψ – mother wavelet.

Continuous Wavelet Transform has many advantages but in speech signal processing its discrete form is more often used due to its simplicity [17]. Discrete Wavelet Transform (DWT) is defined by the following form [17]:

$$DWT(m,n) = \frac{1}{\sqrt{a^m}} \sum_n S(k) \psi(a^{-m} n - bk) \tag{3}$$

where: $S(k)$ – indicates input signal, a (a>0) – scale parameter, b – offset parameter, ψ – mother wavelet.

One of the advantages of Discrete Wavelet Transformation, compared for example to Fourier transform, is that DWT provides accurate and uninterrupted time information, which is a significant enhancement for signal processing [18].

Two types of mother wavelet were considered in conducted researches: the Haar Wavelet and the Mexican Hat Wavelet. The first one is defined as follows [19]:

$$\psi(t) = \begin{cases} 0, for\ t < 0 \\ 1, for\ 0 \le t < 0{,}5 \\ -1, for\ 0{,}5 \le t < 1 \\ 0, for\ t \ge 1 \end{cases} \tag{4}$$

The Mexican Hat Wavelet has the following form [20]:

$$\psi(t) = \frac{2}{\sqrt{3}} \pi^{\frac{-1}{4}} (1 - t^2) e^{\frac{-t^2}{2}}. \tag{5}$$

Sample graph of above-mentioned function is shown the in Figure 2.

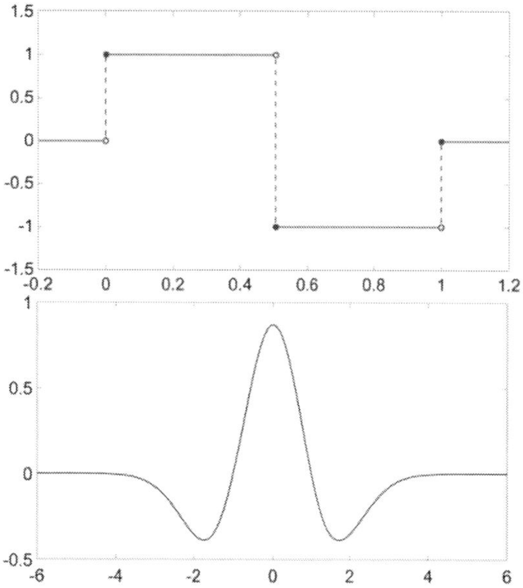

Fig. 2. Examples of wavelet functions: Haar (top) and Mexican Hat (bottom).

4.2 Speech signal processing scheme

The biggest challenge in conducted researches were preparation and processing of data process. It was divided into several steps. The first one related to noise reduction process from input signals. To fulfil this task The VOICEBOX: Speech Processing Toolbox for MATLAB was used. The second one related to speech signal values normalisation process to the range [-1,1]. These two steps constituted the pre-processing phase. In the next step the scaleogram was created. To create speech signal spectrum, the discrete wavelet transform was used. The example of signal decomposition is shown in the Figure 3.

In the next step, features were extracted from scaleograms. The main issue of this phase was the preparation of an input vector for processing by Kohonen Networks. At the beginning of features extraction process all scaleograms were transformed into grey scale images. The next step was a transformation to binary images which was as follows: all values, above specific threshold, were replaced by 1, other ones by 0. To define the threshold all values between 50 and 200 were tested. The best results were obtained at 100. The next step included scaleogram division into several subareas. The number of subfields depended on speech signal decomposition's level. The example of scaleogram division is shown in the Figure 4.

DOI: 10.1051/itmconf/20171502004

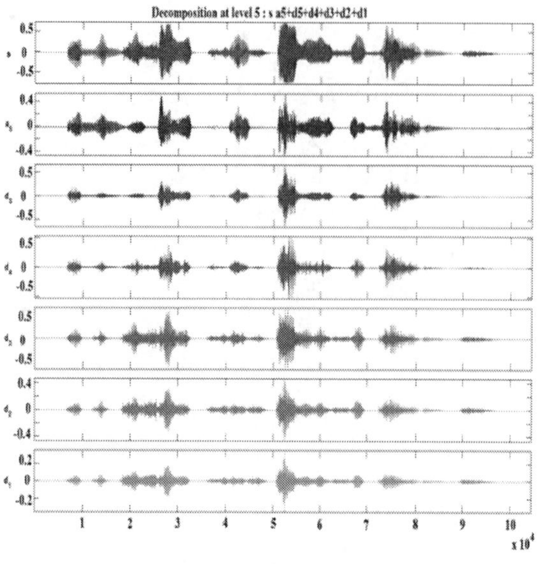

Fig. 3. Examples of speech signal decomposition using DWT.

Commonly used k-Nearest Neighbours algorithm was used also in this research, to determine the best level of signal decomposition. The fastest form of k-NN algorithm is its 1-NN version. In this algorithm, the unknown tested sample should have been assigned to the same group as its closest neighbour. Results of researches conducted during Statlog project [21], in which a several classifiers were compared, showed that for 75% of k-NN test the best results were achieved for 1-NN version. In that research the aforementioned algorithm was also used. Achieved results were shown in the Figure 5. Based on abovementioned algorithm, the 7th and 9th speech signal decomposition's level were used. It can be easily noticed that achieved results, under 7th level of decomposition, were unsatisfactory. What is more, the processing time above 9th level has been growing rapidly. The number of scaleogram subareas were multiple of level of decomposition. For 7th decomposition's level the following numbers of subfields were tested: 70, 105, 140, 175, 210, 245, 270, 315, 350 and 385 and for 9th decomposition's level: 90, 135, 180, 225, 270, 315, 360, 405.

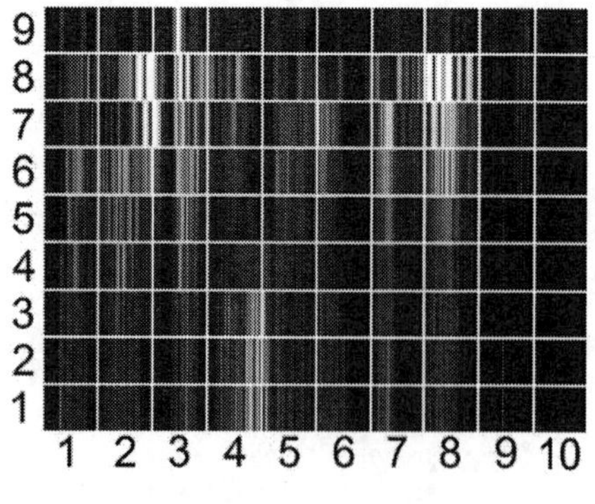

Fig. 4. Examples of scaleogram divisions.

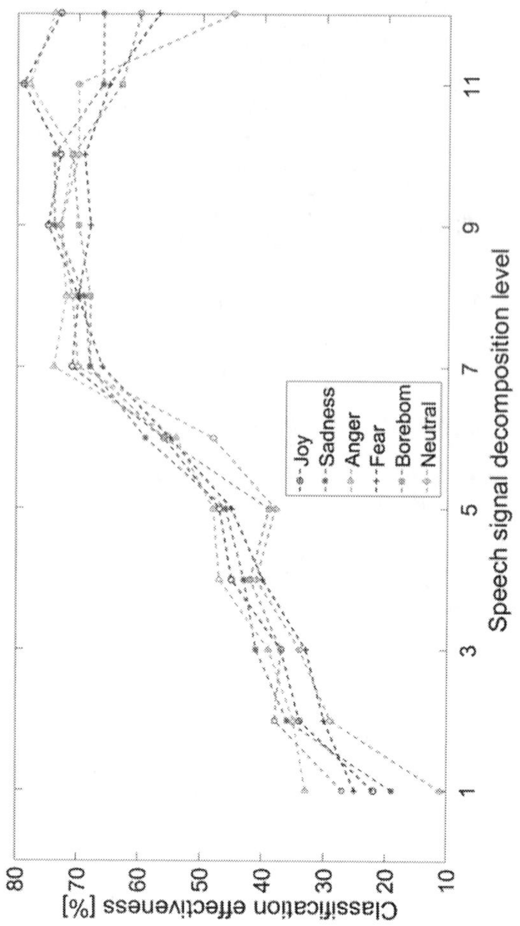

Fig. 5. Classification results using 1-NN rule depending on signal decomposition's level.

The best effectiveness was achieved from division into 210 subareas for 7th decomposition's level and into 360 for 9th level. The last step in Polish emotional speech signal processing, before classification, was summarising values in each subarea. The whole processing scheme is illustrated in the Figure 6.

4.3 Kohonen Networks Architecture and achieved results

Kohonen Networks belong to a group of Self-Organizing Maps. The main difference between this kind of neural networks and one-way artificial neural networks is that the correct output cannot be defined before training process (*a priori*). The main aim for Kohonen networks is to organise multidimensional information in such a way that it can be presented and analysed in space with smaller number of dimensions. As it was mentioned, scaleograms have been divided into 70, 105, 140, 175, 210, 245, 270, 315, 350 and 385 subfields for 7th signal decomposition's level and 90, 135, 180, 225, 270, 315, 360, 405 for 9th level of decomposition. In researches, a two-dimensional map was used, corresponding to scaleogram division size. The map in Kohonen network

was constructed in a way that for one input value there were four corresponding neurons in a map. So, the size of the map was four time bigger then input vector. The Kohonen networks work as follows:
1. Inputs relate to all nodes in map.
2. Each node stores a weight vector with identical size as an input vector.
3. Each node calculates its activation level as scalar product of vector weights and input vectors.
4. A node with the highest level of activation is the winner and can update its weight vector.
5. Nodes in the winner's neighbourhood may update their weight vectors.

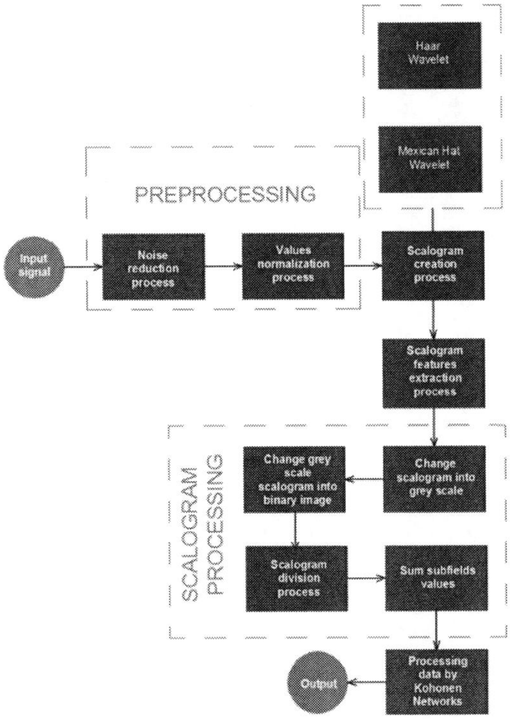

Fig. 6. Polish emotional speech signal processing scheme.

In conducted researches, a Euclidean distance weight function was used as a neighbourhood function. Number of epochs was set to 1000. As a training set 50% of recordings from the first database was used. Achieved results are presented in Tables 3 to 5.

Table 3. Achieved classification results for the first database.

Emotion	Correct classification [%]
Anger	72,5
Joy	77,4
Sadness	69,2
Neutral state	70,3
Boredom	72,7
Fear	79,8

Table 4. Achieved classification results for the second database.

Emotion	Correct classification [%]
Anger	74,7
Joy	76,5
Sadness	69,4
Neutral state	73,4
Boredom	77,1
Fear	77,6

Table 5. Achieved classification results for the third database.

Emotion	Correct classification [%]
Anger	77,4
Joy	76,6
Sadness	73,3
Neutral state	74,2
Boredom	73,5
Fear	80,2

It can be noticed that the strong emotions, such as fear or anger, were more often recognised correctly than the weak ones. What is more, regardless of used database, fear was identified with the highest efficiency and sadness with the worst. The justification for such results should be sought in the appearance of the speech signal spectrum which for sadness is quite similar as for boredom and neutral states; therefore, these three emotions were often confused. The opposite situation occurs in the case of fear, which spectrum was clearly different from spectrums of other emotional states. For all data the average effectiveness of classification was almost 76%.

5 Conclusions

Conducted researches showed that identification of emotion in speech signal is not trivial issue. There are not many publications directly related to the possibilities offered by spectrographic methods and Kohonen networks in issues connected to the identification of Polish emotional speech. Scaleogram division and determination of energy in its subfields allowed to create the appropriate input vector for Kohonen networks. Conducted researches and obtained results, as well as preliminary results of subsequent experiments, allow to assume that proposed method of speech signal processing can be universal and it is possible to uniquely identify the speaker's emotional state.

References

1. D. Ververidis, C. Kotropoulos, I. Pitas, *IEEE International Conference on Acoustics, Speech, and Signal Processing*, **149**, 1(2004)
2. P. Powroźnik, D. Czerwiński, Advances in Science and Technology Research Journal, **11**, 10, 32(2016)
3. S. P. Panda, A. K. Nayak, International Journal of Speech Technology, **2**, 9(2016)

4. D. Kamińska, A. Pelikant, IAPGOŚ, **10**, 03 (2012)
5. S. Ramakrishnan, Speech Enhancement, Modelling and Recognition – Algorithms and Applications, **7**, (2012)
6. D. Kamińska, T. Sapiński, D. Niewiadomy, A. Pelikant, Studia Informatica, **5**, 34 (2013)
7. T. Dalgleish, Nature Reviews Neuroscience, **18**, 5, 7(2004)
8. L. Zhen, G. Zhi, Affective Computing and Intelligent Interaction, **81**, (2005)
9. K. Scherer, Speech Communication, **13**, 40, 1-2(2003)
10. A. Janicki, M. Turkot, Przegląd telekomunikacyjny - wiadomości telekomunikacyjne, **98**, 8-9(2008)
11. J.H Conolly, E. A. Edmonds, J. J. Guzy, S. R. Johnson, A. Woodcock, International Journal of Man-Machine Studies, **7**, 24, 6(1986)
12. I. Józefczyk, Problemy Transportu, **7**, 4, 3(2009)
13. F. Burkhart, A. Paesche, M. Rolfes, F. Sendlmeier, B. Weiss, *A database of german emotional speech* (Proceedings of Interspeech 2005)
14. Database of Polish Emotional Speech, available: http://www.eletel.p.lodz.pl/bronakowski/med_catalog/ (Accessed 10.08.2014)
15. V. Maz, G. Schmidt, Applied and Computational Harmonic Analysis, 1, 6, 3(1999)
16. L. Zsu-Hsin Chuang, L. Wu,J. Wang, Sensors, **60**, 13, 8(2013)
17. A.Anoop Suraj, M. Francis T.S.Kavya T.M.Nirmal, Journal of Electrical Systems and Information Technology, **8**, 1, 1(2014)
18. R. Sharma, R.B. Pachori, U. R. Acharya, Entropy, **5**, 17, 8(2015)
19. U. Lepik, Applied Mathematics and Computation, **33**, 176, 1(2006)
20. Y. Hong, H. Liang, Optical Letters, **39**, 31, 3 (2006)
21. R. D. King, D. Feng, A. Southerland, Applied Artificial Inteligence, **2**, 9, 3(1995)

 DOI: 10.1051/itmconf/20171502005

Prediction of monthly electric energy consumption using pattern-based fuzzy nearest neighbour regression

Paweł Pełka[1,*], and *Grzegorz* Dudek[1]

[1]Czestochowa University of Technology, Department of Electrical Engineering, Al. Armii Krajowej 17, 42-200 Czestochowa, Poland

Abstract. Electricity demand forecasting is of important role in power system planning and operation. In this work, fuzzy nearest neighbour regression has been utilised to estimate monthly electricity demands. The forecasting model was based on the pre-processed energy consumption time series, where input and output variables were defined as patterns representing unified fragments of the time series. Relationships between inputs and outputs, which were simplified due to patterns, were modelled using nonparametric regression with weighting function defined as a fuzzy membership of learning points to the neighbourhood of a query point. In an experimental part of the work the model was evaluated using real-world data. The results are encouraging and show high performances of the model and its competitiveness compared to other forecasting models.

1 Introduction

Electric energy consumption forecasting is an essential issue in power system planning and operation. Medium-term forecasting is necessary for technical and operational purposes, such as: scheduling maintenance activities, planning of production levels and fuel purchases, and planning of network investments. From an economic viewpoint, energy consumption forecasts are fundamental for negotiating contracts between energy companies and concluding contracts with customers.

Fig. 1 shows a periodical time series representing monthly energy consumption for four European countries (the data from ENTSO-E repository – www.entsoe.eu). In this figure, seasonal variations and rising tendency can be observed, caused by the influence of the economic and technological development on the electric market. Seasonal variations reflect the annual cycle and are dependent on climatic factors, which are comparable during the same month of different years. Other factors affecting directly or indirectly the level of energy consumption are political decisions and economic policy. They can disturb general rising trend and monthly fluctuations. They include: the emergence of alternative energy sources and technologies, fluctuating economic inflation, violent change in energy prices, industrial development, and global warming issues [1], [2].

The time series of monthly electric energy demand presented in Fig. 1 differ depending on the power system size and economic development of the country. Note significant share of random component in the time series and larger amplitude of annual cycles for France than for other countries.

Two approaches have been developed for medium-term electric energy consumption forecasting [3]. The first one, called conditional modelling approach, focuses on economic analysis, management and long-term planning and forecasting of energy load and energy policies. It considers socioeconomic conditions which impact energy demands, such as economic indicators and electrical infrastructure measures. These additional inputs are introduced to the model together with historical load data and weather-related variables. Such a model can be found in [4]. It includes macroeconomic indicators, such as the consumer price index, the average salary earning and the currency exchange rate.

The second approach, called autonomous modelling approach, requires a smaller set of input information to forecast future electricity demand, primarily historical loads and weather factors. Because the economic factors are not taken into consideration, this approach is more suited for stable economies.

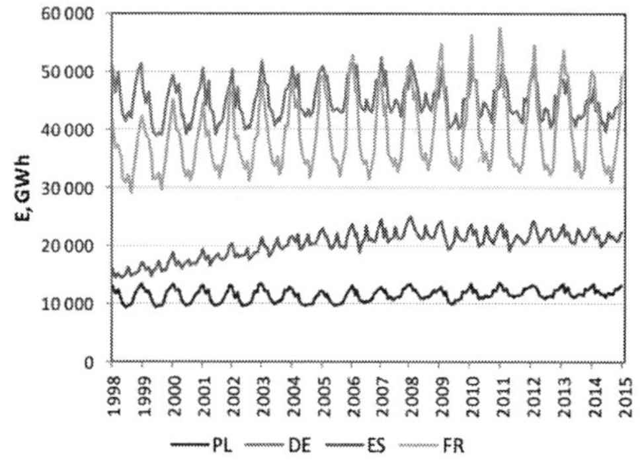

Fig. 1. Time series of monthly electric energy consumption.

* Corresponding author: pavelle50@gmail.com

Different forecasting models are used in this case such as classical autoregressive integrated moving average (ARIMA) and multiple linear regression [5], as well as computational intelligence methods, e.g. neural networks [6]. Examples of such models can be found in [7], where ARIMA, neural networks and neuro-fuzzy systems are employed to forecast future load demand based on various weather-related parameters and historical load profiles. Another example is a model presented in [8], where interval load forecasting is proposed using multi-output support vector regression. In addition, a memetic algorithm is used to select input variables among the variable candidates, which include time lagged loads and temperatures. In [9] neural network is used for forecasting load time series components extracted using digital filtering. Evolving fuzzy neural networks are proposed for monthly electricity demand forecasting in [4]. In this solution fuzzy neurons represent degree of importance of each input variable (loads, weather factors and daylight time). Different weights assigned to input variables lead to improved model accuracy and more precise prediction.

The forecasting model proposed in this work belongs to the latter category. It uses fuzzy nearest neighbour regression (FNNR), based on patterns of the time series fragments. An underlying assumption in this model is: if two fragments of the time series are similar in shape, then the fragments following them are also similar in shape [10]. This approach is especially attractive when the time series expresses seasonal pattern. In our earlier works, we proposed models from the same class of pattern similarity-based nonparametric regression models: the model based on k-nearest neighbours (k-NN) [11] and Nadaraya-Watson estimator [12]. The proposed FNNR allowed to consider similarity degree between shapes of the time series fragments using fuzzy set theory.

The remainder of the paper is organised as follows. In Section 2, a time series representation is described, using patterns of their fragments. In Section 3, forecasting model was defined, using fuzzy nearest neighbour regression. The model has been tested on real-world data in Section 4. Finally, the work was concluded in Section 5.

2 Patterns of time series fragments

In the first stage of the proposed approach load time series were pre-processed using methods presented in [10]. Input and output patterns were defined. The input pattern was an n-dimensional vector representing a time series fragment preceding the forecasted one. Let us denote the forecasted fragment by $Y_i = \{E_{i+1}\ E_{i+2}\ \dots\ E_{i+m}\}$, and the preceding fragment by $X_i = \{E_{i-n+1}\ E_{i-n+2}\ \dots\ E_i\}$, where E_k is the monthly energy consumption and k is the time index. An input pattern $\mathbf{x}_i = [x_{i,1}\ x_{i,2}\ \dots\ x_{i,n}]^T$ represented the fragment X_i. Components of that vector were pre-processed points of the sequence X_i. For example [11]:

$$x_{i,t} = E_{i-n+t} \tag{1}$$

$$x_{i,t} = \frac{E_{i-n+t}}{\overline{E}_i} \tag{2}$$

$$x_{i,t} = E_{i-n+t} - \overline{E}_i \tag{3}$$

$$x_{i,t} = \frac{E_{i-n+t} - \overline{E}_i}{D_i} \tag{4}$$

where $t = 1, 2, \dots, n$, $\overline{E}_i$ is the mean value of the points in sequence X_i, and $D_i = \sqrt{\sum_{j=1}^{n}(E_{i-n+j} - \overline{E}_i)^2}$ is a measure of their dispersion.

A pattern defined using (1) is a copy of the sequence X_i without processing. Pattern components defined using (2) are the points of the sequence X_i divided by the mean value of this sequence. Patterns (3) are composed of the differences between points and the mean sequence value. Pattern (4) is the normalised vector $[E_{i-n+1}\ E_{i-n+2}\ \dots\ E_i]^T$. All patterns defined using (4) have the unity length, mean value equal to zero and the same variance.

The output pattern $\mathbf{y}_i = [y_{i,1}\ y_{i,2}\ \dots\ y_{i,m}]^T$, representing the forecasted sequence Y_i, had the components defined similarly to the x-pattern components:

$$y_{i,t} = E_{i+t} \tag{5}$$

$$y_{i,t} = \frac{E_{i+t}}{\overline{E}_i} \tag{6}$$

$$y_{i,t} = E_{i+t} - \overline{E}_i \tag{7}$$

$$y_{i,t} = \frac{E_{i+t} - \overline{E}_i}{D_i} \tag{8}$$

In the above formulas (5)-(8) $\overline{E}_i$ and D_i are determined from the sequence X_i, and not from the sequence Y_i. This is because the sequence Y_i is not known at the moment of forecasting. To determine the forecast of the monthly energy consumption E_{i+t} on the basis of the forecasted y-pattern generated by the forecasting model, transformed equations have been used (5)-(8). For example, in the case of (8) the forecasted energy consumption is calculated as follows:

$$\hat{E}_{i+t} = \hat{y}_{i,t} D_i + \overline{E}_i \tag{9}$$

Patterns $\mathbf{x}_i$ and $\mathbf{y}_i$ are paired ($\mathbf{x}_i$, $\mathbf{y}_i$). The set of these pairs determined from the history is used for learning the forecasting model.

3 Fuzzy nearest neighbour regression

The nearest neighbour estimate $m(\mathbf{x})$ is defined as the weighted average of the y-patterns in a varying neighbourhood of the query x-pattern. Typically, this neighbourhood is defined through the x-patterns which are among the k nearest neighbours of the query pattern

[11]. The value of k determines the number of training patterns from which the regression function is constructed and controls the degree of smoothing. The k-NN estimator gives the regression function, which is discontinuous. In the points where the set of the nearest neighbours changes, the jumps on the function graph are observed. To avoid this inconvenience, a fuzzy membership of the training points to the neighbourhood of the query point was introduced [13]. In this approach, each training point belongs to the query point neighbourhood with a degree depending on the distance between these points.

The regression function $m(\mathbf{x})$ has the nonparametric form:

$$m(\mathbf{x}) = \sum_{j=1}^{N} w(\mathbf{x}, \mathbf{x}_j)\mathbf{y}_j \qquad (10)$$

where the weighting function $w(\mathbf{x},\mathbf{x}_j)$ is dependent on the similarity or distance between patterns $\mathbf{x}$ and $\mathbf{x}_j$. Usually it decreases monotonically with the distance. When using fuzzy approach, the weighting function has a form of the membership function, e.g. a Gaussian-type function:

$$\mu(\mathbf{x}, \mathbf{x}_j) = \exp\left(-\frac{d^2(\mathbf{x}, \mathbf{x}_j)}{\sigma^2}\right) \qquad (11)$$

where σ is a parameter controlling the width of the function, and $d(\mathbf{x},\mathbf{x}_j)$ is a Euclidean distance between patterns $\mathbf{x}$ and $\mathbf{x}_j$.

An estimator (10) is a linear combination of vectors $\mathbf{y}_j$ weighted by the membership degree (11) which nonlinearly maps the distance $d(\mathbf{x},\mathbf{x}_j)$. The greater the distance, the lower the weight. The width parameter σ decides about the bias-variance trade-off of the estimator. Too small σ value results in undersmoothing, whereas too large value results in over-smoothing. Thus, the selection of the width parameter is a key problem. In a training procedure the optimal value of σ is selected, as well as the optimal length of the input pattern n. These parameters are being searched using grid search method.

The training set contains pairs of patterns ($\mathbf{x}_i$, $\mathbf{y}_i$), which are historical for the forecasted sequence, i.e. these ones for which $i = n, n+1, ..., i^*-m$, where i^* is an index of the last month before the forecasted sequence. The forecasting task is to generate the forecasts for months $i^*+1, i^*+2, ..., i^*+m$.

The forecasting procedure consists of four steps:

1. Pre-processing of load time series into x- and y-patterns.
2. Calculating the weights for the training x-patterns using membership function (11).
3. Calculating the forecasted y-pattern from (10).
4. Decoding the forecasted y-pattern using transformed equations (5)-(8) to get the monthly electricity demand for consecutive months: i*+1, i*+2, ..., i*+m.

4 Experimental study

In this section, the proposed FNNR method was applied to model and monthly electricity load demand was forecasted. Then results were compared with results of several reported statistical and machine learning methods for load demand forecasting. Data used in this research were taken from the publicly available ENTSO-E repository (www.entsoe.eu). They included monthly electricity demand for four European countries: Poland (PL), Germany (DE), Spain (ES) and France (FR). The time range of data was 1998-2015 for PL, and 1991-2015 for other countries. We constructed the forecasting models for 2015, using data from previous years to model learning. Two variants of forecasting were considered:

- Variant A – a model generated forecasts for all 12 months of 2015 (i^* was an index of December 2014, $m = 12$),
- Variant B – for each month of 2015 a separate model was created which generated one step ahead forecast (12 models created for i^* corresponding to: December 2014, January 2015, ..., and November 2015, $m = 1$).

The model parameters, σ and n, were selected using grid search in leave-one-out cross-validation procedure. The searched values of n were 3, 4, ..., 24 and $\sigma = a \cdot d_{med}$, where d_{med} was a median of distances between x-patterns in the training set, and $a = 0.02, 0.04, ..., 1$.

Tables 1-8 present optimal values of parameters and Mean Absolute Percentage Errors (MAPE) obtained with these parameter values: validation errors ($MAPE_{val}$) and test errors ($MAPE_{tst}$ for 2015). Accordingly to the tables, the selection of the best way of pattern definition seems to be difficult. Results depend on the time series features, such as a trend and level of random, irregular influences. The optimal x-pattern lengths vary between 8 and 24 depending on time series and pattern definition. Note that the optimal lengths are rarely equal to the annual cycle length, which is characteristic for these time series.

Fig. 2 demonstrates test errors for individual months in both variants, A and B. Note that variant B, which generates one step ahead forecasts, does not always provide better results than variant A, in which the forecast horizon is 12 months. Errors for successive months are very varied. This is caused by the significant contribution of the random component in data.

Examples of the forecasted y-pattern construction are presented in Fig. 3. Grey lines in these figures are the x- and y-patterns from the training set. A darker shade of grey indicates x-patterns which are closest to the query pattern and y-patterns paired with them. These patterns have higher value of the membership function (11), and consequently greater impact on the forecast. The query pattern and the true y-pattern paired with it are drawn with thick solid lines. The forecasted y-pattern is drawn with dotted line. Moreover, the optimal input pattern lengths are different for different pattern definitions (see Tables 1-8).

In Tables 9 and 10 results of comparative models are shown: ARIMA, exponential smoothing (ES) and Nadaraya-Watson estimator (N-WE) [12]. The proposed

DOI: 10.1051/itmconf/20171502005

Table 1. Results for PL, variant A.

Patterns	n	a	$MAPE_{val}$	$MAPE_{tst}$
(1)-(5)	19	0.20	2.95	3.81
(2)-(6)	10	0.24	3.02	2.68
(3)-(7)	10	0.24	3.00	2.55
(4)-(8)	17	0.16	2.83	1.94

Table 2. Results for DE, variant A.

Patterns	n	a	$MAPE_{val}$	$MAPE_{tst}$
(1)-(5)	8	0.16	3.24	3.39
(2)-(6)	12	0.26	3.09	2.18
(3)-(7)	12	0.26	3.11	2.30
(4)-(8)	10	0.24	3.16	3.07

Table 3. Results for ES, variant A.

Patterns	n	a	$MAPE_{val}$	$MAPE_{tst}$
(1)-(5)	24	0.14	2.82	1.99
(2)-(6)	22	0.22	2.75	1.86
(3)-(7)	9	0.26	2.81	2.71
(4)-(8)	21	0.22	2.83	2.08

Table 4. Results for FR, variant A.

Patterns	n	a	$MAPE_{val}$	$MAPE_{tst}$
(1)-(5)	13	0.16	3.21	3.95
(2)-(6)	12	0.14	3.28	3.56
(3)-(7)	12	0.14	3.31	3.65
(4)-(8)	12	0.16	3.36	3.53

Table 5. Results for PL, variant B.

Patterns	n	a	$MAPE_{val}$	$MAPE_{tst}$
(1)-(5)	10	0.20	2.37	3.57
(2)-(6)	10	0.20	2.20	1.90
(3)-(7)	10	0.18	2.18	1.71
(4)-(8)	10	0.14	1.95	1.38

Table 6. Results for DE, variant B.

Patterns	n	a	$MAPE_{val}$	$MAPE_{tst}$
(1)-(5)	8	0.14	2.46	2.43
(2)-(6)	8	0.18	2.51	2.46
(3)-(7)	8	0.18	2.50	2.40
(4)-(8)	10	0.16	2.49	2.68

Table 7. Results for ES, variant B.

Patterns	n	a	$MAPE_{val}$	$MAPE_{tst}$
(1)-(5)	24	0.14	2.77	2.42
(2)-(6)	15	0.24	2.25	2.89
(3)-(7)	15	0.24	2.33	2.97
(4)-(8)	13	0.20	2.27	3.44

Table 8. Results for FR, variant B.

Patterns	n	a	$MAPE_{val}$	$MAPE_{tst}$
(1)-(5)	22	0.18	3.08	3.65
(2)-(6)	14	0.14	2.90	3.70
(3)-(7)	14	0.16	2.93	3.76
(4)-(8)	14	0.14	2.89	3.91

FNNR model belongs to the same group of nonparametric regression methods as N-WE; thus, results of both models are similar. When comparing errors of all models, it can be concluded that FNNR is competitive with other models, but it should be noted also that the classical ES model outperformed all other models in six of eight cases.

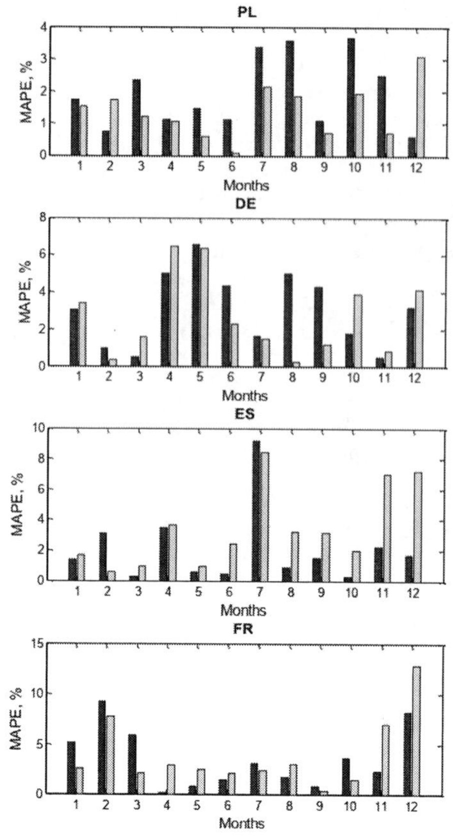

Fig. 2. Forecast errors for consecutive months of 2015. Variant A - dark bars, variant B - light bars.

Table 9. MAPE of the forecasting models, variant A.

Model	PL	DE	ES	FR
FNNR (4)-(8)	1.94	3.07	2.08	3.53
ARIMA	2.08	2.54	2.67	4.02
ES	1.92	2.32	2.17	3.02
N-WE	2.03	3.12	2.08	3.56

Table 10. MAPE of the forecasting models, variant B.

Model	PL	DE	ES	FR
FNNR (4)-(8)	1.38	2.68	3.44	3.91
ARIMA	2.02	2.56	2.18	3.91
ES	1.92	2.32	2.16	2.98
N-WE	1.35	2.72	3.42	3.99

5 Conclusion

This work proposes a practical methodology to forecast the monthly electric energy consumption using fuzzy nearest neighbour regression. This model is based on the assumption that the similarity of the input patterns implies the similarity of the output patterns paired with them. The patterns representing time series fragments are the key element of this approach. They unify data, reduce nonstationarity and filter out the trend.

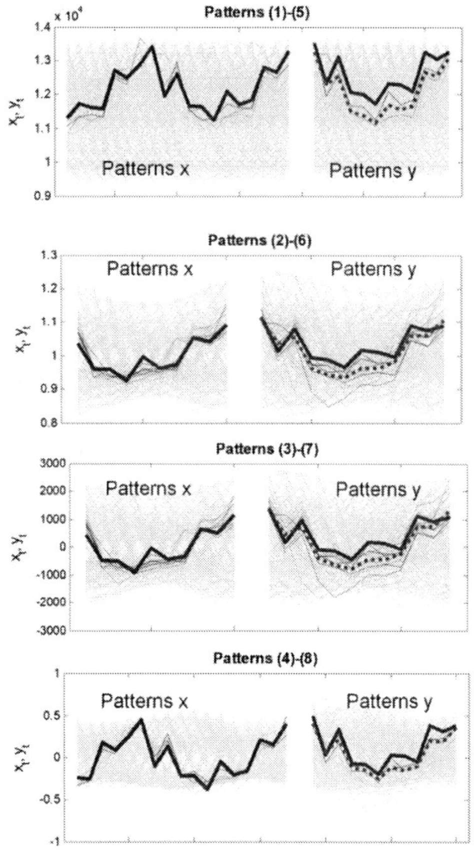

Fig. 3. Construction of the forecast pattern when using different pattern definitions for PL data, variant A.

The main advantages of the model are the simple and understandable principle of operation and only two parameters to estimate: the length of the input pattern and the width of the membership function. Models with fewer parameters have better generalisability and do not require complex learning procedures.

We demonstrate the effectiveness of our approach on real-world data. Comparing with commonly used methods, such as ARIMA and exponential smoothing, the proposed model results in similar errors on average. Better performance of the model is observed for more regular time series with lower noise component and stable relationship between input and output patterns. The factors which decrease this stability are the nonlinear trend and heteroscedasticity of time series.

References

1. J.-F. Chen, S.-K. Lo, Q.-H. Do, Information, vol. 8, 31, (2017)
2. P.C. Chang, C.Y. Fan, J.J. Lin, Int. J. Electr. Power Energy Syst., vol. 33, pp. 17–27, (2011)
3. M. Ghiassi, D.K. Zimbra, H. Saidane, Electric Power Systems Research, vol. 76, pp. 302-316, (2006)
4. M. Gavrilas, I. Ciutea, C. Tanasa, IEEE Conf. Electricity Distribution (2001)
5. C.L. Hor, S.Watson, S. Majithia, IEEE Transactions On Power Systems, vol. 20 , pp. 2078-2085, (2005)
6. E. González-Romera, M. Jaramillo-Morán, D. Carmona-Fernández, IEEE Transactions On Power Systems, vol. 21, pp. 1946-1953, (2006)
7. M.M. Elkateb, K. Solaiman, Y. Al-Turki, Neurocomputing, vol. 23, pp. 3-13, (1998)
8. Z. Hu, Y. Bao, R. Chiong, T. Xiong, Energy 84, pp. 419-431 (2015)
9. M. Jaramillo-Morán, E. González-Romera, D. Carmona-Fernández, Electrical Power and Energy Systems, vol. 49, pp. 253-263 (2013).
10. G. Dudek, Applied Soft Computing, vol. 37, pp. 277-287, (2015)
11. G. Dudek, P. Pełka, Przegląd Elektrotechniczny (Electrical Review), r. 93(4), pp. 62-65, (2017) (in Polish)
12. G. Dudek, P. Pełka, Proc. 18th Conf. Electric Power Engineering (EPE'2017), pp. 1-6, (2017).
13. G. Dudek, Applied Soft Computing, vol. 36, pp. 422-441, (2015)

Reliability and cost optimisation of complex electric power networks using ant colony algorithm

Łukasz Piątek[1,*]

[1]Częstochowa University of Technology, Institute of Information Technology, 42-200 Częstochowa, Poland

Abstract. The article presents a new approach towards reducing an overall cost of electric power network with maintaining its reliability. Goals are achieved by implementing an ant colony algorithm with a cut-set method as a method for reliability evaluation. The algorithm solves the problem of multi-objective optimisation, where both the network cost and network reliability index, known as unavailability, should be minimalised. The network cost is considered as a linear function of overall length of network's connections. For reliability evaluation in the cut-set method, real empiric data of hazard rate for overhead power lines are used. Parallel-series network structure, equivalent by means of reliability to analysed network, is generated through the cut-set method to compute unavailability of trial solutions. Sections of the structure are generated on the basis of minimum cut set, found by the algorithm for finding one- and two- minimum cuts. As used algorithm for finding minimum cuts has linear complexity, the evaluation of trial solutions is computationally effective. An example, presented in this article, provides figure of optimal network configurations found by the algorithm.

1 Introduction

Overhead power lines are exposed to various environmental conditions that threatens outages in continuity of electric power delivery to the end consumers. Various consumers have different endurance to power outages. As a typical household can bear up with several hours power outage, for industrial, business and utility customers, lack of power continuity could create damage and money loss. As some customers are willing to pay more for reliability of their power supply, it is up to electric grid operator to provide the best possible service.

To achieve better reliability, the grid operator has two choices. Either an improvement in reliability of key components of the network or an introduction of some redundancy into the network. Both solutions come with drawbacks because usually they require additional spending on building and maintaining the network structure. Improving reliability of such network components as overhead lines cannot be easily done, as they are vulnerable to weather conditions. Thus, fast response repair teams are used, that can repair a fault as soon as possible. The second solution, adding redundancy, provides alternative paths the can be used while some components of the network are damaged and under repair.

The designing of such power network that has the highest reliability with the lowest cost states a multi-objective optimisation problem. This paper addresses this problem by providing an approach based on adapted ant colony optimisation algorithm [1]. The modification introduced into the algorithm involves pheromone value updates based on the value of fitness function, heuristic search of solution space, as well as effective reliability estimation of trial solutions.

Similar problems of optimisation were considered in different studies. The approach in which there is a need to maintain connections between all network nodes was analysed in [2,3]. Networks where only k nodes must be connected where considered in [4,5,6,7]. Simulated annealing was considered in [8]. Heuristic algorithms were proposed in [4,5,6]. Genetic algorithms for network optimisation were proposed in [9,10,11]. In [11] a Monte-Carlo method it was used to evaluate the reliability of networks coded by the individuals in population. In [10,11] the reliability estimation it was done by computing the sum of maximum distance between nodes. The author of this paper also designed a genetic algorithm for this problem in [12].

2 Ant colony optimisation

Observation on biological processes and behaviours led to design of new algorithms used in problem solving. The ant colony optimisation algorithm is based on how insects search for food. First proposed in [1], it consists in allowing a population of ants searching for trial solutions by exploring the solution space through random walking. The crucial point is that each ant deposits an information about recent movements in the form of chemical component called in biology a pheromone. The levels of pheromones are then read by other ants and used by them as a guidance in their walk.

* Corresponding author: l_piatek@el.pcz.czest.pl

The more ants choose a path, the more likely it is for an ant to decide to follow that path. The probability of an ant choosing to move from state x to state y is given by:

$$p^{k}_{xy} = \frac{\tau^{\alpha}_{xy} \cdot \eta^{\beta}_{xy}}{\sum_{z \in allowed_y} \tau^{\alpha}_{xy} \cdot \eta^{\beta}_{xy}} \quad (1)$$

where τ_{xy} is the pheromone level assigned to the connection (x, y) and η_{xy} is some heuristic value designed to help ants achieve their goal. Parameters α and β allow to control balance between pheromone levels and the heuristic value η_{xy}. Just as in nature, the pheromone level is not constant but changes over algorithms iteration. First, as said before, every ant deposits some amount of pheromone while passing from state x to state y, and second the pheromone fades over time. The evanescence of pheromone is regulated by the pheromone evaporation coefficient ρ of the algorithm. For each connection the pheromone level τ_{xy} is updated as follows:

$$\tau_{xy} \leftarrow (1-\rho)\tau_{xy} + \sum_k (\Delta\tau^{k}_{xy}) \quad (2)$$

where $\Delta\tau^{k}_{xy}$ is the pheromone value deposited by k-th ant.

The classic application of ant colony optimisation algorithm is for the traveling salesman problem. Other applications are e.g. Job-shop scheduling problem, Quadratic assignment problem, Capacitated vehicle routing problem. This paper presents a new approach where ant colony algorithm has been adapted to deal with a multi-objective problem of optimisation of network structure. The adaptation involves a heuristic based on connections reliability and a new way of updating pheromone levels that takes into consideration the value of fitness function.

3 Network reliability estimation with cut-set algorithm

The evaluation of network reliability is necessary to correctly identify the best solution and to guide the algorithm presented in this paper. For this purpose, the cut-set method is used [13]. This method allows to compute reliability of connection between two distinct nodes in a graph. The nodes are called the source s and the sink t. The cut-set method consists in replacing a complex network by an equivalent, by means of reliability parallel-series structure. Each section of the equivalent structure is generated based on one minimal cut. The minimal cut is a set of connection that when damaged, cause disconnection of source and sink, but the repair of any single connection of this set can restore the link between source and sink. Finding all minimal cuts in a graph is a separate task. Any algorithm for finding these cuts can be used.

Once minimal cuts are found the computation depends on whether network components are renewable or not. As in this paper, the object of interest is a power grid, it is necessary to assume that connections can be repaired. Because power lines are put into an operation after fault in time that is usually short, compared to their operational time, we can assume that simultaneous outage of more than two connection is very unlikely. This allows to consider only 1 and 2 elements minimal cut in the cut set method. Of course, any catastrophic failure, that involves all area of the network, like flood or weather storms, can destroy many connections. But for such events network redundancy is not enough to prevent failure anyway. In this paper, we deal only with connections failures that are independent.

The reliability of network connections is usually measured empirically. Failures of the network elements occur with the mean time which is called Mean Time To Failure – MTTF. The repair rate is referred as Mean Time To Repair – MTTR. These are the two network component reliability indicants used as data for the optimisation algorithm. Additionally, an index named failure rate λ is frequently used for describing components reliability. Since a typical component of electric power network is characterised by a constant failure rate, the number of failures occurring over a period of time referred to original, total population has the exponential distribution. Then, the relationship between component's MTTF and failure rate can be defined as MTTF= λ^{-1}.

The study in [14] provides values of failure rate and MTTR for various components of electric power networks. For typical overhead distribution lines, the value of hazard rate per circuit kilometre λ_P is 0.0625 per year and the value of MTTR is 4.0 hours, regardless of line length. Since the failure rate value is given per circuit kilometre, we need to treat a line as a series of one-kilometre length lines. Then, the following formula is used to compute reliability indices of a connection between two network's nodes:

$$\lambda = d * \lambda_P \quad (3)$$

where d is the connection length in kilometres.

For renewable components the unavailability U parameter can also be used. The unavailability is the probability of the component being in the down state. Similarly, the unavailability of power network is the probability of the network to be in the down state, in which the network is unable to fulfil its designed task. As power networks are most of the time on stand-by, it is easier to operate with the unavailability parameter.

Once we have identified all minimal cuts and formed on that basis the equivalent parallel-series network, formulae provided in [13] can be used to compute the unavailability of the network. For one component section the unavailability is:

$$U_{1comp} = \lambda \cdot MTTR \quad (4)$$

where λ is the hazard rate of the component in this section and MTTR is the component's mean repair time. For a parallel connection of two component, with hazard rate respectively λ_1 and λ_2 and the repair time $MTTR_1$ and $MTTR_2$, the unavailability is:

$$U_{2comp} = \lambda_1 \cdot \lambda_2 \cdot MTTR_1 \cdot MTTR_2 \quad (5)$$

Formulae (4,5) are only valid if *MTTF»MTTR* for each network's component. This is true for most real electric power systems. Finally, the unavailability of the parallel-series network U_n is obtained by adding up the unavailability of each of the m – one- element sections and each of the n – two-elements parallel sections:

$$U_n = \sum_{i=0}^{m} U_{1comp}^{i} + \sum_{i=0}^{n} U_{2comp}^{i} \quad (6)$$

The special case is when a network has no minimal cut through less than three components. In such a case, because the probability of simultaneous failure of more than two components is very low in practical systems, the network is treated as perfectly reliable with the unavailability value of zero.

4 Algorithm for optimisation of reliability of complex network

The presented in this chapter algorithm solves the multi-objective optimisation problem of designing a network connecting two nodes with the maximum reliability and minimum cost. It is assumed, that the reliability and cost of network connections are provided as data. The designed network connects source node s with sink node t through network nodes defined as data for the algorithm. Algorithm can use arbitrary subset of nodes for constructing the network.

As mentioned in chapter 3, it is more convenient to operate with unavailability rather than reliability parameters. With unavailability, the fitness function for optimisation is defined as follows:

$$F(x) = w_C f_C(x) + w_U f_U(x) \quad (7)$$

and

$$w_C + w_U = 1 \quad (8)$$

where w_C is the network cost weight and w_U is the network unavailability weight. $f_U(x)$ is the network unavailability. The normalised network cost in $f_C(x)$ is defined as the sum of costs $C(i, j)$ of every connection from connections set E, compared to the cost C_{max} of full connected network:

$$f_C(x) = \frac{\sum_{(i,j)\in E} C_{(i,j)}}{C_{max}} \quad (9)$$

The argument of functions (7) and (9) is a vector of bits values representing the network's connections set. The vector length is $n(n-1)/2$, where n is the number of nodes. If there is a connection between two nodes, the appropriate bit is set to one.

The goal of the algorithm is to minimise the fitness function. The algorithm consists of seven steps. The first two are preparations steps, while the rest are in a loop that is an iteration of algorithm.

Step 1) start with positioning all N_{ant} ants on the source node.

Step 2) apply a start value τ_0 of pheromone for every connection.

Step 3) through successive iteration move ants from node x to node y with probability given by formula:

$$p_{xy}^{k} = \begin{cases} \dfrac{\tau_{xy}^{\alpha} \cdot \eta_{xy}^{\beta}}{\sum_{z \in allowed_y} \tau_{xy}^{\alpha} \cdot \eta_{xy}^{\beta}} \\ 0 \quad if\ d(y,t) > d(x,t) \end{cases} \quad (10)$$

where η_{xy} is the value of MTTF of a connection between nodes x, y. It is forbidden for any ant heading to the sink, to move to a node which is located further from sink node than the node where the ant is. If an ant reaches out the sink node, it starts moving back to the source node. For returning ant the probability of move is:

$$p_{xy}^{k} = \begin{cases} \dfrac{\tau_{xy}^{\alpha} \cdot \eta_{xy}^{\beta}}{\sum_{z \in allowed_y} \tau_{xy}^{\alpha} \cdot \eta_{xy}^{\beta}} \\ 0 \quad if\ d(y,s) > d(x,s) \end{cases} \quad (11)$$

Thus, we have two kinds of ants. Moving to the sink or moving to the source. In implementation of the algorithm two data structure are used for each kind of ants. For every iteration, ants that arrived at the sink node are moved to the data structure responsible for returning ants. Similarly, ants returning to the sink are moved to the data structure responsible for ant going to the sink node.

Eventually this creates the situation that is similar to one found in nature. Ants travel back and forth, transporting food from sink node to source node.

In every move each ant deposits pheromone $\Delta\tau_{xy}^{k}$. The value of additional pheromone is stored in the data structure used later when applying formulae (2) in Step 6.

Step 4) compute the value of fitness function (7). The graph for the evaluation is created on the basis of connection pheromone level. Every connection where the level of pheromone passes a threshold value τ_T is considered as a valid graph edge.

Step 5) compare computed fitness function to fitness function from previous iteration. If the fitness function is worse than that from previous iteration, apply special procedure. First, compute the average pheromone value for all connection from the graph from previous iteration. Then, set this average as a new pheromone value for every connection from previous iteration. Thus, connection with low pheromone level, that are important for improving fitness function are revived by increasing the level of pheromone.

Step 6) finally, for every iteration pheromone evaporation is applied (3) and new pheromone from ants' move is deposited.

Step 7) if there is no improvement of the value of fitness function during 10*n iterations, end the algorithm. Otherwise jump to *Step* 3.

Conducted tests of the performance of the algorithm showed that it can have reduced computational time compared to genetic algorithm. The genetic algorithm maintains a set of trial solutions and evaluates all these solutions for every iteration. The presented algorithm evaluates only one solution every iteration. Although the presented algorithm requires more iteration to find solution, the completion of every iteration is faster than in genetic algorithm. However, unlike the genetic algorithm, the algorithm finds only one solution. Sometimes it is desirable to see a set of solutions with the best value of fitness function. The network operator can then choose the desirable network configuration. The algorithm does not always find the best solution but a solution with fitness function very close to the best solution. Even then, that solutions are sufficient in practical application. It should be noted that the cut set method together with statistical relations (4-6) do not provide the exact value of unavailability. It only is an estimation.

5 Example

To illustrate the algorithm, we include an example of optimising seven nodes notional network. The problem is to find a set of connections that form a network which provide minimum unavailability at reasonable costs. The failure rate and MTTR for overhead electrical lines are given in [14]. For a typical overhead line, the failure rate has the value of 0,0625 fault per year per circuit kilometre. Mean time to repair has the value of 4.0 hours, regardless of line length. The coordinates of the nodes are given in table 1. It is assumed that the cost of a line is proportional to the line length and that the normalised cost is as (9).

Table 1. Node coordinates list.

Node	x [km]	y [km]
1 (source)	0	24
2	11	11
3	28	30
4	42	11
5	66	22
6	73	5
7 (sink)	106	15

In order to balance the values in function (7), the weights are used. The normalised network cost takes on values between 0 and 1. In the sample network, its unavailability takes on values around 10^{-6}. Thus, the adopted values of weights are w_C=0,00007 and w_U=0,99993. Algorithm users can freely set up these weights to fit their needs. The value of ant algorithm parameters is: N_{ant} = 400, τ_0=0.1, τ_T=0.1, α=1, β=0.5, ρ=0.1.

The optimal network found by the algorithm is shown in Figure 1. The value of unavailability for this network is $2.8*10^{-6}$ and the normalised network cost is 27% of the full connected network. By multiplying the computed network unavailability by the number of hours in a year, we can also express the network unavailability as 0.025 hours/year.

6 Conclusions

The presented algorithm provides an effective way to find reliable, low-cost networks. The adaptation of ant colony algorithm involves defining new heuristic suitable for the problem, new procedure to guide the algorithm by the value of fitness function, as well as using the cut-set method to evaluate reliability. For performance reasons, the algorithm uses the time-efficient, linear algorithm for finding one- and two-elements minimal cuts in the cut-set method.

The algorithm has better computational efficiency than the genetic algorithm applied to similar problems. Nevertheless, only one solution is found. It is not possible to find a set of solutions with best value of fitness function. The algorithm deals with renewable network components. The algorithm can be used in designing reliable electrical networks.

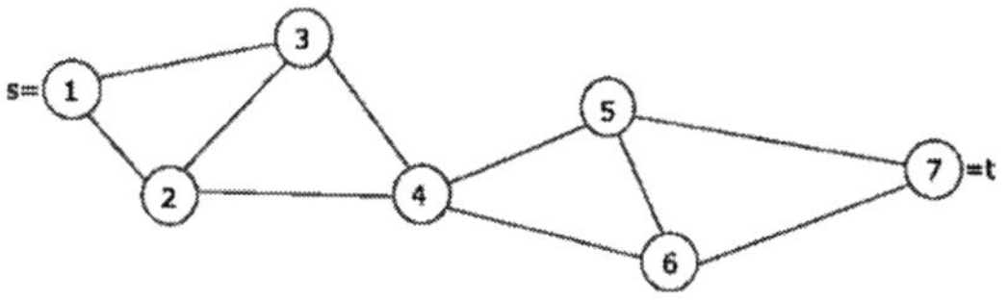

Fig. 1. The optimal network found by the algorithm.

References

1. A. Colorni, M. Dorigo et V. Maniezzo, actes de la première conférence européenne sur la vie artificielle, Paris, France, Elsevier Publishing, 134-142, (1991).
2. B. Dengiz, F. Altiparmak, A.E. Smith, IEEE Transactions on Reliability, **46**, Issue 1, pp. 18-26, (January 1997).
3. M. Gen, R. Cheng, S.S. Oren, Advances in Engineering Software, **32**, Issue 9, pp. 731-744, (September 2001).
4. K.K. Aggarwal, Y.C. Chopra, J.S. Bajwa, Microelectronics and Reliability, **22**, Issue 3, pp. 341-345, (May 1982).
5. K.K. Aggarwal, Y.C. Chopra, J.S. Bajwa, Microelectronics and Reliability, **22**, Issue 3, pp. 347-351, (May 1982).
6. Y.C. Chopra, B.S. Sohi, R.K. Tiwari, K.K. Aggarwal, Microelectronics and Reliability, **24**, pp. 911-913, (March 1984).
7. B. Liu, K. Iwamura, Computers and Mathematics with Applications, **39**, pp. 59-69, (January 2000).
8. V. Ravi, B.S.N. Murty, P.J.Reddy, IEEE Transactions on Reliability, Vol. **46**, pp. 233-239, (March 1997).
9. A. Kumar, R.M. Pathak, Y.P. Gupta, IEEE Transactions on Reliability, **44**, pp. 63-72, (January 1995).
10. A. Kumar, R.M. Pathak, Y.P. Gupta, H.R. Parsaei, Computers and Industrial Engineering, **28**, pp. 659-670, (May 1995).
11. B. Liu, K. Iwamura, Computers and Mathematics with Applications, **39**, pp. 59-69, (January 2000).

DOI: 10.1051/itmconf/20171502006

12. Ł. Piątek, M. Rudnicki , International Journal of Applied Electromagnetics and Mechanics, **43**, pp. 93-102, (2013)
13. R. Billinton, R. Allan: *Reliability Evaluation of Engineering Systems*, 2nd ed. (New York: Plenum Press 1992).
14. R. E. Brown: *Electric Power Distribution Reliability*, Boca Raton (CRC Press 2009)

ITM Web of Conferences **15**, 02007 (2017) DOI: 10.1051/itmconf/20171502007
CMES'17

Using methods of the reverse engineering to carry personalised preoperative stabilisers out on the example of vertebrae of human spine

Ewelina Kozłowska[1,*], and *Jarosław* Zubrzycki[1]

[1]Lublin University of Technology, Faculty of Mechanical Engineering, Institute of Technological Systems of Information, Nadbystrzycka 38A, 20-618 Lublin, Poland

Abstract. There is presented the possibility of using modern rapid prototyping methods in the form of reverse engineering on the example of lumbosacral spine of a human spine at the level of L3-L. Reverse engineering techniques were directly used in the generation and processing of point clouds from a real object. The point clouds were generated by using specialised Mimics software and basing on the results of computer tomography and magnetic resonance imaging using. In the next stage of processing, there was the export of the appropriately converted point clouds to STL format, compatible with CAD programs. The reverse engineering process took four steps: digitisation of the real object, using non-invasive imaging techniques (CT, MRI), processing the point clouds from the digitisation stage, construction of a surface model of the real object basing on the processed point clouds, realising of the real object's copy by using rapid prototyping techniques. In the final stage, the model of the complex spinal's part was tested to MES strength analysis.

1 Introduction

Nowadays, diseases and injures of a spine are included to illnesses from the group called: civilisation diseases. Thanks to the technological progress in the field of providing medical services, treating diseases and back injuries, there was a significant quality improvement in services dedicated to people from the risk group, as well as for people after different kind of accidents.

Long-term examinations contributed to the more effective development of non-invasive methods of the medical imaging, such as computed tomography (CT) and magnetic resonance (MRI), which acquiring a lot of more accurate data about anatomical structure of the entire human organism, particularly spinal's areas (entire area or partial).

Development of new technology in medicine, such as reverse engineering [4, 12,13] and artificial intelligence [1], enables to extend the chances in the field of the treatment of illnesses and damages of a spine. In the work, there was submitted to consideration directly concerning the most profitable choosing of treating definite diseases and damages of a spine on lumbar-cross-shaped sections at the level L3-L5.

Currently, thanks to the intensive developing techniques of implantology, the spondylosis is increasingly becoming redundant because of its invasive [13-15]. The spondylosis is a surgical procedure of anastomosis neighbouring vertebrae with the assistance of specialist implants, which are needed to accelerate the process of osseous concretion with the simultaneous stabilisation of vertebrae in a correct way in the area of damage.

However, it is not always possible to use the treatment method for this type of illnesses by implantation an imitating degenerated implant or the degraded vertebra called artificial disk.

There is the possibility of developing directly, minimally invasive but efficient solution for a patient, in medical and technical aspect, which enables to direct support in treating process and further rehabilitation of a patient.

Applying the reverse engineering in the imitative process of the vertebra requires gaining and analysis of clinic necessary data describing the real state of the vertebra and the constitution of the patient. A database of possible solutions is being created in the area of curing diseases and injuries of the spine. Applying methods of the rapid prototyping enables to design and to produce personalised endoprosthesis, in the form of the implant directly reflecting the injured vertebra [5, 10]. As a result, doctors receive above action data enabling them to make a rational decision in relation to the considered case, requiring the treatment at an outpatients' clinic on the neurosurgical and orthopaedic branch.

The direct source of ambulatory data is a process of analysis and aggregating of illnesses and traumas of backbone on the cross for small of back segment of patients, and results of non-invasive methods of diagnostics of medical (medical of imaging) image [9] presenting physiology and pathological changes that take place in the organism of patient.

* Corresponding author: ewelina.kozlowska16@wp.pl

Presently, the techniques of medical image used in medicine due to the widely developed possibilities which allow to execute researches considerably perfecting the process of medical diagnostics also help a simulation and preparation of operating procedures [10].

The main applied methods during non-invasive diagnostic of the medical imaging are examinations making with the help of the computed tomography (CT) and of magnetic resonance (MRI) [10]. In the article there is described process of the three-dimensional modelling situated in the lumbosacral part of spine at the level L3 - L5. In the next stage results coming directly from CT examinations and MRI made without applying the contrast will be used in the process of designing personalized endoprosthesis in the form of the implant reflecting the degenerate vertebra. The injured vertebra must be replaced to remove disease or the back injury from the patient at least invasive way. Afterwards there will be brought closer the process drawing the expert program up (advisory) basing directly on a clinic data included in the card of stories of the illness of the patient as well as on latest results of examinations coming from the computed tomography and the magnetic resonance.

2 Anatomy and generalised functions of human spine

The spine of the man is a column directly providing the support for the whole body. It directly consists of vertebrae and their connections. In its construction there are singled out the following elements:
1. 33-34 vertebrae,
2. 23 intervertebral disks (of so-called disks),
3. the spinal cord along with spinal nerves,
4. cords [8, 13].

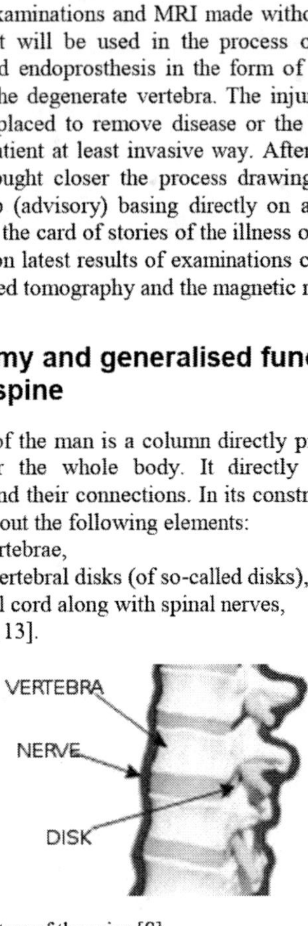

Fig. 1. Structure of the spine [8].

In the spinal construction five segments can be separated:
• cervical segment - 7 vertebrae keeping the head (C1 - C7),
• deep segment - 12 vertebrae holding ribs (Th1 - Th12),
• the lumbar region - 7 vertebrae building the back wall of the belly (L1 - L5),
• the sacrum - healing over of 5 vertebrae (connection with the pelvis) - S1,
• coccyx - 3-4 integrated vertebrae (Fig. 2) [8, 9].

Generalised functions of human spine:
• holding the weight of the body and ensuring the balance,
• ensuring the fastening of other bones and elements of the human body,
• amortisation of shocks,
• protection of the spinal cord and spinal nerves.

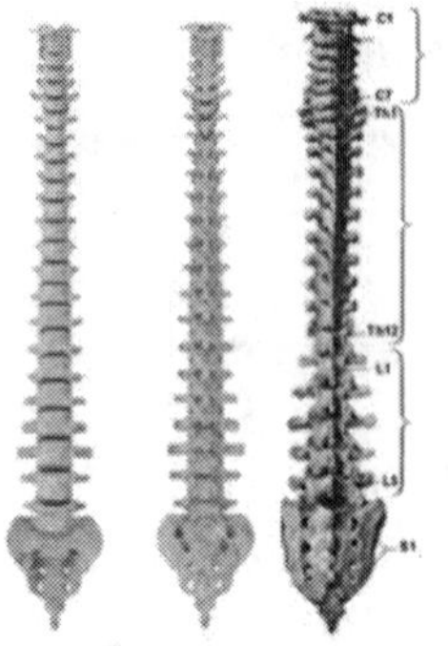

Fig. 2. Structure of the spine [17].

During the biological analysis of the structure of the lumbar vertebrae there is possibility to observe distinct flat spinous processes horizontally, placed towards the back, enabling to make moves of bending over towards the back of the lumbar region. Simultaneously, in the same part of the spine there are situated strong spinous crosswise, growing on the border of the hard core and the bow because of growing together residual lumbar ribs with the proper transverse process. The analysed lumbar region is deprived of all holes in transverse processes and the area joints for ribs.

Conducting anatomical analysis of the sacral vertebrae, it is possible to observe upper and bottom indents of the integrated sacral vertebrae, creating four pairs of sacral foramens joined directly to the vertebral canal. Sacral foramens are serving estuary to frontal branches, whereas towards the back branch of the dorsal cross nerves.

In the analysed lumbosacral part of the spine spinous processes (joint) are merging into the bone slat (median sacral crest), simultaneously integrated joint processes form two side slats (indirect cross backs). However crosswise processes are knitting into two lofty side sacral crests, and residual cross ribs form side parts of the sacrum. Owing to such a structure, vertebral canal is concealed directly under the centre comb and ends with the bottom hole, by which the filum terminale of the spinal cord is originating. Four appearing joint areas are creating the cross section: upper (merging with the last lumbar vertebrae) and two side (merging with bones of the pelvis at the lower ending) and caudal tangent with the coccygeal bone [6].

In the human spine, divided ranges are intervertebral disks between themselves, and the integrity of the structure is stabilised with cords of the spine constituting the depreciation simultaneously. Thanks to such a structure of the spine the outcome of curves is coinciding with the division, forming the straight silhouette at the man.

The intervertebral disk is built from the annulus fibrosus (outside part) and nucleus pulposus (inner part).

 DOI: 10.1051/itmconf/20171502007

Correctly formed nucleus pulposus is built in approx. 80% of water and is subject to the laws of hydrodynamics. Simultaneously, it plays crucial role during flexible moves in the spine through the migration into the side opposite to the made movement through the spine, however while bending the spine ahead, nucleus Atherosclerotic nucleus is moving in back direction. Mainly the intervertebral disk is serving as the shock absorber of the entire spine of the man (Fig. 3).

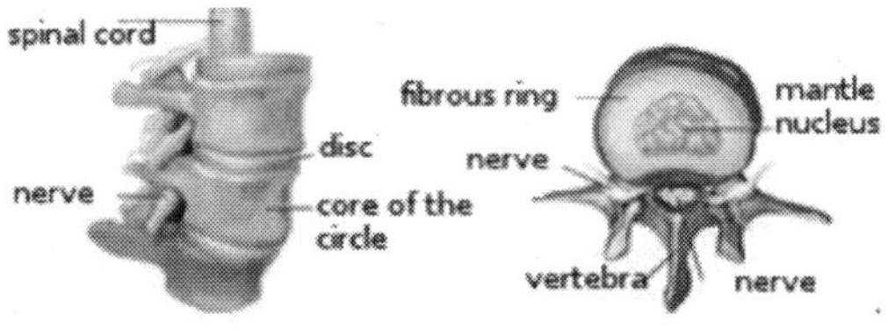

Fig. 3. Structure of the intervertebral disk [3, 16].

3 Computer support of medical diagnostics imaging

Today, the development of medical diagnostic techniques is directly based on computer-aided diagnosis (CADx) methods, which are widely used in medicine [11]. The usage of computer-aided diagnosis in medical imaging techniques (mammography, radiography, computer tomography, magnetic resonance imaging, ultrasonography) is the process of detection and evaluation of pathological changes. The wide spectrum of usability allows it to be used in clinical radiology, and in particular for the detection of spinal injuries and injuries in all its sections. In the concept of CADx, the radiologist attaches great importance to the prediction made by the computer, but the final decision belongs to the radiologist. By applying the CADx concept, the radiologist makes a good diagnosis, which, in case of doubt, can help with the suggestions of the computer [7, 15].

Computer image diagnostics support is intended to enhance the human ability to solve key problems of image interpretation. Upgrades and improvements to imaging techniques and alleviation detection and differential diagnosis algorithms are possible thanks to the ongoing research on CADx algorithms. Currently, CADx technologies are implemented directly into the PACS system, followed by the creation of collections of pathologies of similar nature.

Dynamic technological development and simultaneous improvement of the functionality of computer software and its use in digital imaging enable the use of CADx in most diagnostic methods [8]. However, the way to use CADx depends directly on the doctor's preference (radiologist), who can make a final diagnosis and help with the suggestion of the program. Unfortunately, the above procedure has a negative side, which is a waste of time devoted to conducting the study and issuing the opinion. In the above case, it is better to perform a preliminary analysis of the image by the computer and then subject it to a secondary evaluation by a doctor. Despite the extensive capabilities of the computer, even in this case there is a risk of missing some significant damage or deformation not captured by the computer [1, 6].

4 Design 3D cross for small of back segment at the level L3-L5

3D modelling process of the lumbosacral segment at the level L3 - L5 was made based on tomographic findings and after the stage of the process of the digitisation [12] (Fig. 4).

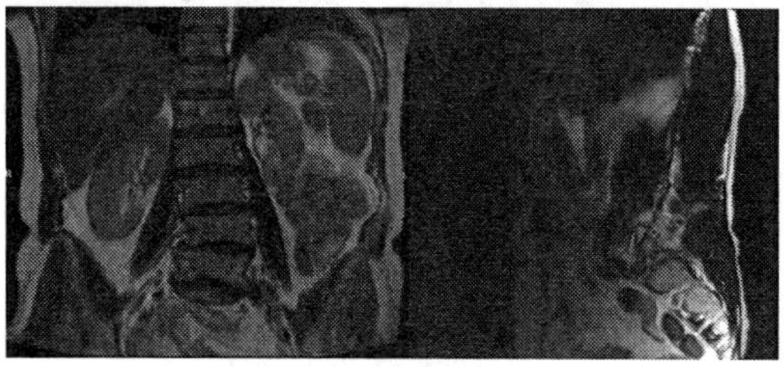

Fig. 4. DICOM image of the lumbar-sacral segment.

Basing on DICOM 2D obtained from CT and MRI using the Materialise Interactive Medical Image Control System images, there was generated the 3D model of the spine in a particular section. The above program is a specialised software for directly processing medical images and creating 3D models, as well as the "3Matic" module for numerical analysis.

After loading a series of DICOM images into Mimics, the program displays the 2D image in three dimensions (Fig. 4, 5): frontal, axial, sagittal and 3D view. The axial view comes directly from the import of a series of images, while the frontal and sagittal views are generated automatically by the program. In Mimics, all views are associative, i.e. each time you click on the image with the left mouse button, it automatically updates the other views [2, 11] (Fig. 5, 6).

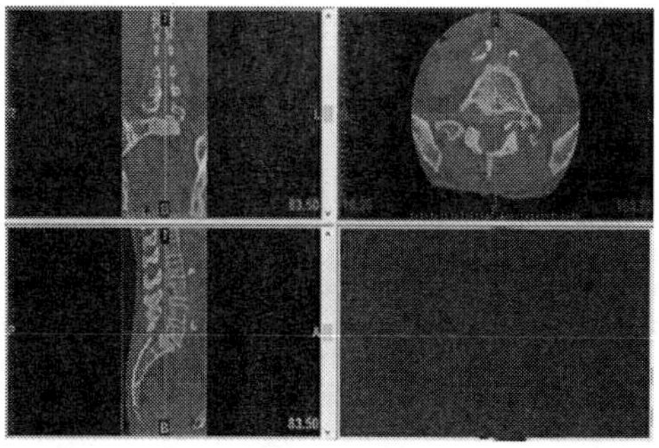

Fig. 5. View of the spine in the Mimics program after loading series of 2D photographs.

Then, after loading the DICOM files and transforming using the main tools like "Thresholding" (selection of tissue type using the HU range), "Growing Area" (selection and interpolation of areas that meet the criteria of the previous operation) and "Calculate 3D" The process of generating a 3D skeleton model of the spine as a digital model of a real object. The obtained 3D

DOI: 10.1051/itmconf/20171502007

model is a numerical reflection of the real dimensions of the spine of a given patient. Thanks to the above operations there is the possibility to generate a model which is adapted individually to the anatomical dimensions of a particular patient. The final effect of modelling is shown in Fig. 7.

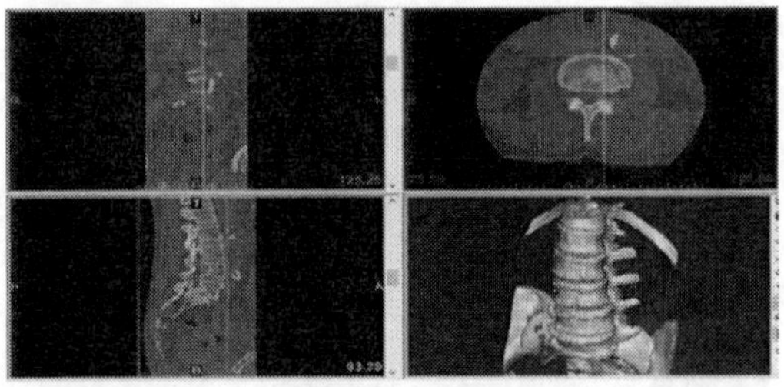

Fig. 6. View after loading series of photographs with the generated model 3D.

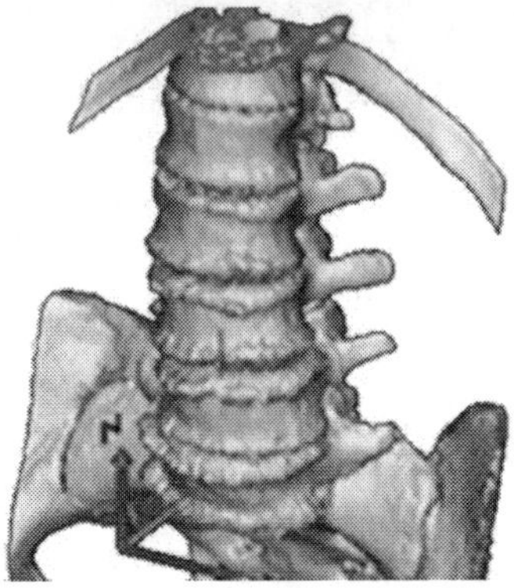

Fig. 7. Model 3D recreated section with the visible discontinuities of surface.

Visible 3D model of surface discontinuities area natural phenomenon occurring during model generation. They result are directly from the quality of data in DICOM format. However, with the "Edit Mask-Draw" function, it is possible to remove small deformation model imperfections, which the further stage could lead to incorrectly generating a correct STL file.

The next step in 3D lumbar-area modelling is the process of reconstructing its geometry by exporting the generated STL file to CAD programs for further processing and design of the surface model and CAE programs for numerical calculation and numerical simulation. Despite the many practical features of the Mimics program to further analysis the model which was generated in a above program there is a necessity of export to the Geomagic Design X. In the first stage, the Geomagic Design X program removes the generated discontinuities and distortions. Due to the presence of multiple discontinuities in 3D, it is not possible to generate a valid and complete STL file. In order to improve the quality of the file, there was used the feature for save the "Point Cloud" which enables to save data directly into point clouds which are not dependent on each other and described coordinates.

Further stages of L3-L5 lumbar-sacral spine modelling and reconstruction were generated in the Geomagic Design X program. The Geomagic Design X system provides a module for 3D design using digitized data. The "Digitized Shape Editor" is located in the shape design module "Shape". The above command is used directly to process data in the form of point clouds, triangular or quadrilateral networks.

Through the import of the STL file generated in the Mimics program next step is loading it as the cloud points which will be subjected to reconstruction is taking place. In a dialog box in part "Preview", after the keystroke "Update", a visualisation of the imported cloud points and activating are taking place of automatic creating the surface (Fig. 8).

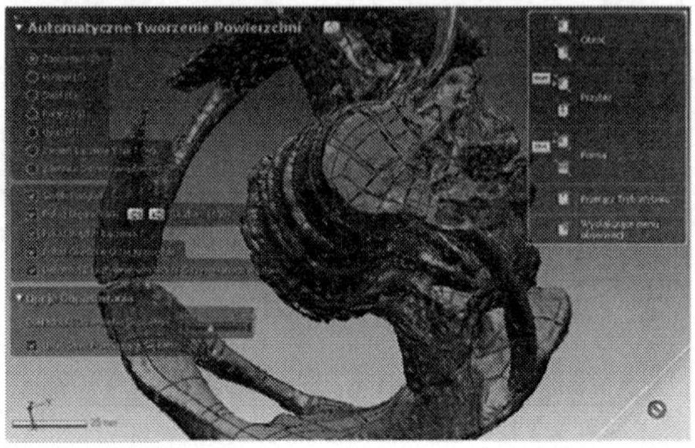

Fig. 8. Loading cloud points to the Geomagic Design Program X.

In Fig. 8 a model of the cloud points depicting the entity object, having elements in their outline incorrectly defined and classified creating the cloud of points and making it difficult for the area to create is visible on the model (red points).

In the further step generating the spot model in the form of networks of curves circumscribing established outlines of vertebrae is taking place what enables the visualisation of the shape of the skeleton of spine 2D (Fig. 9).

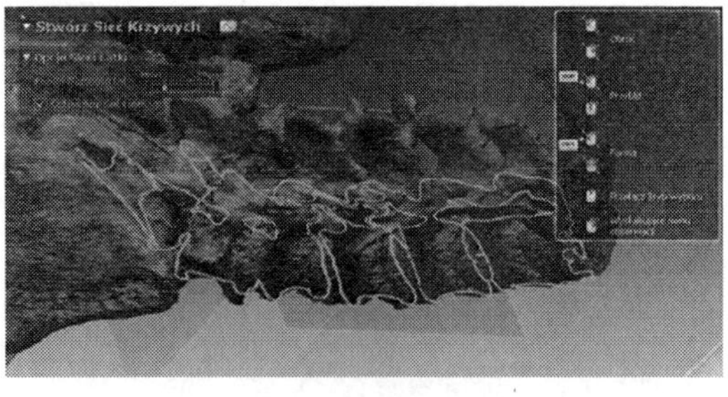

Fig. 9. Creating Curve Networks in Geomagic Design X.

After realising the process of generating 3D from the cloud points of 3D model containing only one L3 vertebra, a created surface model is essential to conduct the numerical analysis in CAE programs. The STL file generated correctly in the Geomagic Design Program X enables the NX 11.0 CAD/CAM to create the surface model in the program from the group, with STL which has got extended orders and modules enabling the export and direct processing of files (Fig. 10).

Fig. 10. 3D model of L3 vertebra with visible surface discontinuities.

In the next stage of the modelling of the cloud basing on correctly generated cloud points there is creation of the grid squares situated on points of the virtual surface allowing to draw the surface model (Fig. 11).

Fig. 11. 3D model of L3 vertebra with visible nodes on a grid of polygons.

The working out of a correct model of a geometric recreated object occurs through the processing of cloud points obtained directly from the process of digitalisation and because of the removal of all points of measuring unconnected with recreated the object (filtration of sounds, smoothing, filling losses in the structure of surface).

Therefore, the processed measuring cloud points grows into a spatial object that consists of quadrangles. The projected model is subjected to further treatment, through repairing multiparty nets (polishing, implementation of losses, cleating of different groups of polygons, treatment of border or sharpening of corners). Those operations aim to prepare appropriate data that allow realising the last phase of model creation of the real geometric as shreds of surface.

The designed surface model in the next stage was converted into the homogeneous solid. For that purpose, they used the function "Join" and "Close Surface" from the module "Reverse Engineering" (Fig. 12).

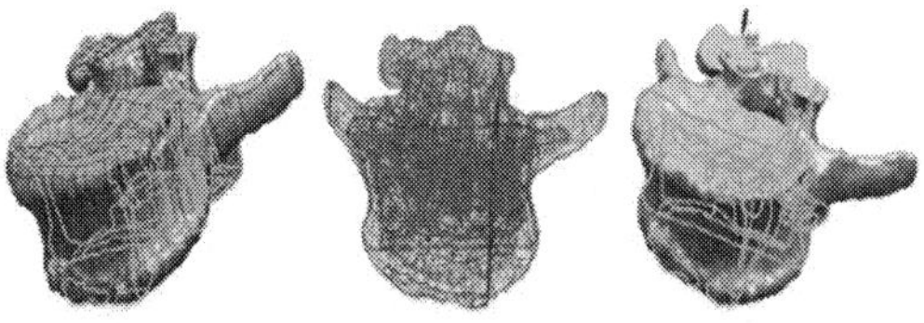

Fig. 12. Surface model of the L3 vertebra with visible cross-sections.

After the realisation of all stages of reconstructing of the real model and the rendering of virtual model 3D a solid model of the L3 vertebra came into existence (Fig. 13).

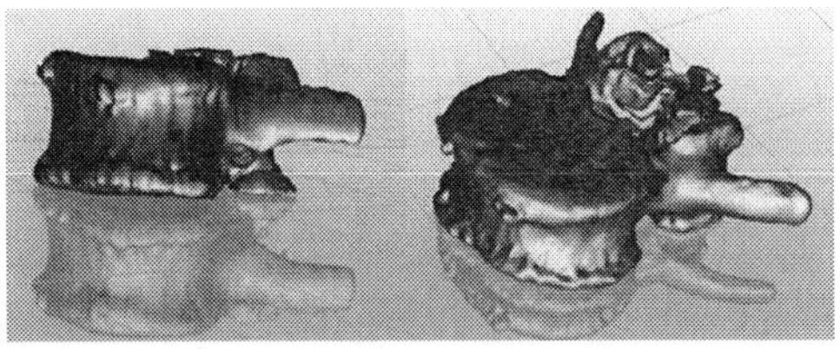

Fig. 13. The solid model of the L3 vertebra was formed from the transformation of the surface model.

The designed solid model allows analysis of the type and size of the deformation of the vertebra in the NX NASTRAN module in NX 11.0 (Fig. 14).

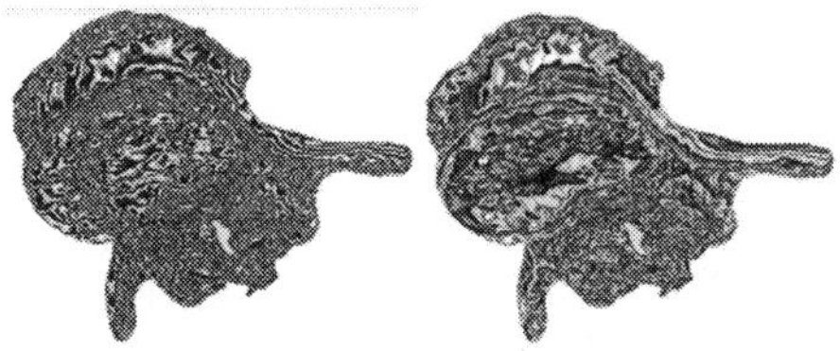

Fig. 14. 3D model of the L3 vertebra with visible deformations and structural defects.

Analysis of the type and size of the deformation shows that the greatest damage to the vertebrae of the vertebral column occurs at the site of the direct adhesion of the intervertebral disc to the vertebrae. The above damages result from both the errors generated during model generation from CT and MRI images and actual damage caused by a disease or spinal injury.

Visible significant deformations on the model in places of planting the spinal cord, of dorsal root of the spinal cord and on the body of vertebra surcharges of the checked length of the lumbosacral spine are effects at the level L3–L5 and strong emphases triggered by transferring the intervertebral disk surrounding the above disease section.

5 Numerical analysis of the L3 vertebra in the CAE program

A L3 vertebra model of the lumbar-sacral spine of the human spine was developed in Abaqus program. A strength analysis was realised in the case of a pressure load on the upper surface of the modelled L3 vertebra. This load allowed us to directly determine the state of stresses in each region of the disc. The isotropic material Ti-6A1-4V material was used for numerical analysis: Young's modulus: E=1.74e+04 MPa and Poisson's number: v = 0.44.

Static analysis was conducted directly on the designed L3 vertebra surface area model to obtain the regions which are exposed to damages - the largest concentrations of stresses according to the Huber-Misses hypothesis. According to the energy hypothesis of strain

DOI: 10.1051/itmconf/20171502007

deformation, the form of strain is the energy of the strain which must satisfy the strength condition [18]:

$$\sigma_{red} = \frac{\sqrt{2}}{2}\sqrt{(\sigma_1-\sigma_2)^2+(\sigma_2-\sigma_3)^2+(\sigma_3-\sigma_1)^2} \le k_r \quad (1)$$

Above regions were encumbered with the pressure p = 190 000 MPa, acting evenly on an upper surface of the L3 body of vertebra (Fig. 15).

Fig. 15. Model vertebra L3 with the visible load.

And so, on the lower area of the analysed vertebra the numerical model was placed getting back all degrees of freedom. (Fig. 16).

Fig. 16. Model vertebra L3 with visible placing.

The analysis allowed us to directly appoint the stress states of the Huber-Misses hypothesis and the value of the deformations caused by the pressure effect of the set value in the L3 region.

6 Results and discussion of the numerical analysis of the L3 vertebra

The numerical analysis of the state of overloading in the form of the pressure was performed for the physiological L3 vertebra of the lumbosacral spine segment. The upper surface of the L3 vertebra has been overloaded evenly.

The analysis of results, achieved from calculations done basing on assumptions of the hypothesis Hubera-Misesa, indicates on the biggest concentration of reduced stresses in regions adjoining directly to the isthmus of the L3 vertebral (Fig. 17).

Fig. 17. Model vertebra L3 with visible deformations.

In case of the functioning of vertebras in real conditions these places are the most overloaded and exposed to coming into existence of microfractures, caused by longer working hours in the appearance of damage of the vertebra and the intervertebral disk.

With the inflicted value of load by the pressure p = 190 000 MPa reduced strains were shaped on the level -1,560e-04 and 1,424e+04 MPa.

There were observed the minimal translocation of an upper surface causing dent of a central surface vertebra L3 giving -2,608e+04 and 8,434e+03, which occurred the point accumulations of the tensions in vertebra L3 (Fig. 18).

Fig. 18. Model vertebra L3 with visible displacement.

The maximum deformations caused by the effect of the pressure appeared directly in regions of spot masses of stresses at the level -1.008 e+ 04 and 1.513 e+ 04 MPa (Fig. 19).

Fig. 19. Model of vertebra L3 with visible deformations caused by pressure.

7 Conclusion

The paper presents the possibilities of lumbar spine reconstruction technique at the level L3-L5. The reconstruction process was conducted basing on serial photos taken by computer tomography and magnetic resonance imaging.

The basic element of modelling process of the model (spine vertebra) was 2D serial photos of the backbone at the level L3-L5 recorded during CT and magnetic resonance imaging, and directly recorded in DICOM format.

The computerised engineering support system enables to direct design of a virtual vertebral column model, showing the doctor spatial representation of the structures visible in CT images as cross-sections. The designed virtual model can be used in the surgical planning process and, at the same time, Intraoperative measurement is one of the methods of rapid prototyping.

Technique for the design and deformation of bone tissue directly forming the vertebral body is also used for planning and carrying out surgical procedures consisting in filling cavities in a vertebral structure with a specially prepared cement structure. Basing on DICOM images obtained by computer imaging methods, pathological areas of the vertebrae may undergo a reconstruction process in the CAD system. Thanks to the wide design capabilities of CAD systems a damaged vertebra design is created which is reconstructed in the next stage by direct removal of deformation and distortion of the vertebral body. After designing both models, they are overlapped and pointed where there are defects and discontinuities to inject the specially prepared agent. However, there are also cases in which the damage is already over-developed, and the usage of cement will not produce the desired effect. In this case, another implant treatment or stabilisation treatment is used.

The analysed 3D model of vertebra L3 lumbosacral section of human's spine was generated in the shape of cloud point at the beginning of the process by Mimics Materialise program, afterwards the model was exported to STL format and loaded into Geomagic Design X program, then in the next stage exported into CAD/CAE NX 11.0. In the paper, there was presented the scheme of actions to be done in sequence to obtain correct 3D model which can be used to conducting numerical analysis in program for strength calculation. In further parts of the work there were done calculations of strength and static deformations. Those calculations do not take into consideration vertebra cracking. Despite avoiding vertebra cracking, there is a possibility to predict damage level.

The intensive development of multidisciplinary techniques and medicine fields enables to decrease time for modelling, making an implant or stabiliser significantly and decreasing treating and rehabilitation time too.

The calculations above do not consider the fracture (fracture) of the vertebrae. Despite the absence of cracking, the extent of damage and disease progression can be determined directly based on the fracture.

The paper presents a direct influence on the physiological load on the lumbar-sacral spine segment on the functionality and viability of the L3 vertebrae. Numerical analyses of static MES enabled to definite interdependencies between the physical state of the tissues, the mechanical load and the potential for lesions in certain regions of vertebra L3.

In the later stages of the work, the model will be developed by modelling L3-L5 lumbar spine vertebrae by modelling the entire vertebrae of a defined angle of lordosis. The above modelling and numerical analyses will be aimed at broadening the knowledge of biomechanical phenomena occurring directly in the system (L3-L5 spine model) where the implant is implanted in place of the damaged vertebra as well as in the system with spine stabilisation applied in the lumbar-sacral stretch in the states of natural functioning as well as stress states.

References

1. E. Eilam, *Reversing: Secrets of reverse engineering* (John Wiley & Sons, 2011)
2. T. Várady, R. R. Martin, J. Cox, Comput. Des. **29**, 255–268 (1997)
3. J. Zubrzycki, M. Braniewska, Mechanik. **90**, 46–47 (2017)
4. J. Barrat, *Our final invention: Artificial intelligence and the end of the human era* (Macmillan, 2013)
5. J. Zubrzycki, N. Smidova, in Applied Mechanics and Materials **613**, pp. 172–181(2014)
6. T. Grimm, *User's guide to rapid prototyping* (Society of Manufacturing Engineers, 2004)
7. J. Tritt-Goc, Tomografia magnetyczno-rezonansowa, Foton **96**, 157-165 (2007)
8. R. Putz, R. Pabst., *Atlas anatomii człowieka-Sobotta* (Elsevier Urban & Partner, 2001)
9. M. Mańko, J. P. Zubrzycki, R. Karpiński, J. Technol. Exploit. Mech. Eng. **1**, 33–57 (2015)
10. R. Karpiński, Ł. Jaworski, J. Zubrzycki, others, Adv. Sci. Technol. Res. J. **10**, 240–246 (2016).
11. A. Bochenek A., M. Reicher, *Anatomia człowieka* (Wydawnictwo Lekarskie PZWL, 1998)
12. L. Tsochatzidis, K. Zagoris, N. Arikidis, A. Karahaliou, L. Costaridou, I. Pratikakis, Pattern Recognit. **71**, 106-117 (2017)
13. K. Kiszka, J. Haduch, M. Pasowicz, Przegląd lekarski **66**, 437-447 (2009)
14. S. B Bartkowski, *Chirurgia szczękowo-twarzowa* (Collegium Medicum UJ, 1996)
15. http://drkregoslup.pl/leczenie-chirurgiczne-stabilizacja-ledzwiowa/leczenie-chirurgiczne/stabilizacja-kregoslupa (08-2017)
16. https://www.twojkregoslup.pl/kregoslup/budowa-i-funkcje-ogolne-kregoslupa (08-2017)

ITM Web of Conferences 15, 02008 (2017)
CMES'17

DOI: 10.1051/itmconf/20171502008

Virtual road concept as a tool for road quality research

Marcin Badurowicz[1,*], and *Jerzy* Montusiewicz[1]

[1]Lublin University of Technology, Faculty of Electrical Engineering and Computer Science, Institute of Computer Science, Nadbystrzycka 36B, 20-618 Lublin, Poland

Abstract. Road quality assessment using crowdsourced data gathered by smartphone users, based on acceleration data, is an interesting subject on using modern technology for improvements of the infrastructure. The algorithms – for both road quality assessment and detection of different elements on the road – need to be tested, especially in the field. To facilitate building sets of different data and sharing them in a standardised way, the authors propose extraction of known road fragments with known types of surface degradation and construction of virtual streams of data, thus "virtual roads". The procedure for data extraction and building a database of segments, combining them into virtual road, as well as testing real-world algorithm using the constructed virtual road are presented in the paper.

1 Introduction

Road quality research, using crowdsourced data acquisition based on the smartphone accelerometer is an intriguing issue, especially in the rural areas, where professional grade assessment equipment cannot be used from the economic point of view. Crowdsourced methods are hugely discussed in the literature, and in previous works of the authors.

The huge and still rising set of data acquired from the crowdsourcing is a challenge in the fact of proper testing different algorithms, especially by different research groups where there is no such a possibility to drive repeatedly over a finite set of real roads to test different possibilities and situations. On the other hand, one research group may drive multiple hours in search for a specific road artefact. Multiple research in the terrain is time-consuming and not efficient. There is also a need for recording and storage of road artefact data in case of creation of new detection algorithms and techniques.

To cope with that, the authors are willing to propose the "virtual road concept". The virtual road is a digital-only set of data, carefully extracted and processed from the real-world data, to be used in building and testing different kinds of detection algorithms for potholes, and other kinds of road quality factors; as well as overall road quality assessment methods, all based on accelerometer data.

The goal of this paper is to present the concept of virtual road, its creation procedure as well as data acquisition which was needed before the actual construction, and performing the virtual road assessment using indicators already published by the authors.

1.1 Similar research

There is a set of established works on the topic of measuring road quality using crowdsourced mobile devices, especially smartphones and similar, as they are equipped with necessary environmental sensors, e.g. accelerometers, starting from the 2008 systems Nericell and TrafficSense [1], through the usage of different sensors, like microphones [2]. There is a research in detection mechanisms for finding different constructional and degenerative changes on the road surface. There are also works covering road quality calculation from accelerometer data [3, 4], especially in the crowdsourcing manner [5, 6] Finally, there are also discussed proposals of creation of road artefacts database for further reuse [7]. The authors were discussing usage of streaming analytics in the cloud computer pattern for such a research, as well as described road artefact detection algorithms [8, 9].

Taking into account results of all the previously performed experiments, which used different cars from different manufacturers and in a different overall state, there is always a direct connection between data recorded by the accelerometer and road profile [3]. The lacking factor now is no possibility of testing new road artefacts detection methods on the set of roads not existing in the real world, the need of maintaining a huge internal database of accelerometer readings and overall problems with sharing these data between scientists.

2 The virtual road concept

When driving on a road, drivers may encounter different types of road surface changes, impacting negatively their overall experience. Potholes, speed bumps, bad road

* Corresponding author: m.badurowicz@pollub.pl

segments, manhole covers and similar were dubbed as "road artefacts" by the authors in the previous research [8] and such term will be used in this paper.

The whole concept is based on the idea of the "building blocks", virtual road components, a digital representation of a signal when driving over the specific kind of the road artefact. These blocks may be set up into time-series, allowing to reduce the amount of non-significant data and simulating different kinds of roads; with different types of artefacts of different quantity. The building blocks database will be used to create the virtual road and will be published later for the other researchers to base similar research on.

To create the virtual road, there is a need to perform a series of operations on data acquired during regular experiments, until they are ready for the actual virtual road construction procedure.

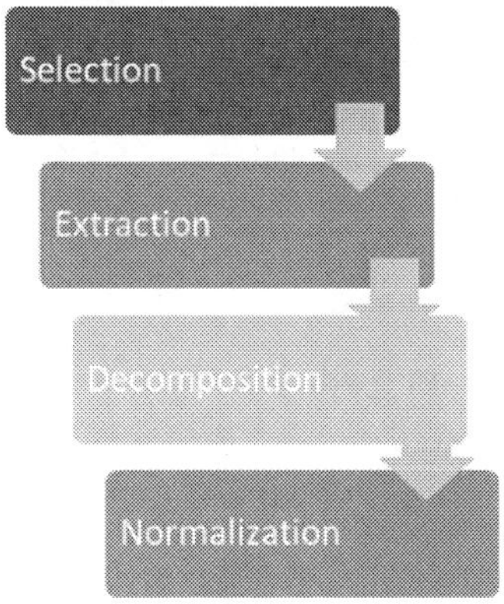

Fig. 1. Steps of building a virtual road components database.

Based on the procedure presented in the Figure 1, the authors are willing to describe it step-by-step, starting with the selection and classification.

2.1 Road artefact classification

First, there is a need of **selection** between types of road artefacts. To create a set of "building blocks", creation of simple classification of the road artefacts, based on their types is indispensable.

The authors have divided road artefacts to the 3 main classes: wide, narrow and point. Examples of the common road artefacts and their respective classes are presented in the Table 1.

The base of the classification is the length of the road artefact in comparison to the length of the car, or the car's wheelbase. The "Wide" class corresponds with the road artefacts longer drastically than the wheelbase – wide road artefacts are commonly relative to the overall surface type – for example paved or rubble road surface.

The "Narrow" class is a set of road artefacts which are about the size of the wheelbase – this includes railroad crossings, tramway crossings, some speed bumps or elevated pedestrian crossings. The "Point" type is smaller than wheelbase, typically the size of the only one wheel – many potholes, manhole covers (raised or lowered), smaller speed bumps.

The generic road artefact types, presented in the Table 1, are the base for building blocks for the virtual road. These 8 types are used and may be used to construct the virtual road with. There is also one more building block type – a baseline asphalt road, being a representation of signal for a regular, good quality road.

Table 1. Road artefact types and classes.

Road artefact	Class
Paved road	Wide
Rubble road	Wide
Railroad crossing	Narrow
Elevated pedestrian crossing	Narrow
Wide speed bump	Narrow
Pothole	Point
Manhole cover	Point
Speed bump	Point

2.2 Data capture and extraction

As mentioned in the introduction, the acquisition device for the whole concept was the smartphone device, mounted in the car in a stable way, so all vibrations from the road profile could be directly represented in the accelerometer data.

The acquisition device, smartphone, reads accelerometer data in 3 dimensions (X, Y and Z) with a frequency of 10 Hz, as well as current time (synchronised with the NTP protocol), speed and location (from the internal Global Positioning System receiver).

As the previous research in this topic suggests, there is a need for calibration to remove tiny accelerometer oscillations [8]. The coordinate-agnostic Global Coordinate System should be also used [10], so the device does not need to be orientated in any specific way. Global Coordinate System is a reorientation procedure, allowing the acquisition device to record acceleration values in axis N, E and Z2, which are the magnetic north, the magnetic east, and vertical (perpendicular to the Earth's surface), respectively.

To gather data about acceleration values from the different types of coordinates, the study was performed, using two cars, driving over the same route, covering multiple roads of different types, in the real-world environment.

One of the cars was an older Peugeot 1007 from 2006, a B-class car, while the second one was a 2014 Toyota Auris, C-class limousine of a higher standard. The differentiation was to achieve different results in overall road readings. The same acquisition device, procedure and methodology were used as in the previous research by the authors: two devices Lumia 820 with the same software were lying in the central tunnel of the car, with a calibration procedure started before entering the traffic [8].

The video camera mounted in cars was used for additional data acquisition step as well as verification. As software the authors were using allowed only to mark geographical position of the road artefact by the user, audio commentary of the artefacts was recorded by the

DOI: 10.1051/itmconf/20171502008

video camera, along with the overall view of the road. HD action camera (GoClever Action Silver) was mounted with a view on the road in the Peugeot 1007, with a microphone recording inside of the car, an example of the video camera frame is in the Figure 2.

That means that after road artefact encounter, the passenger in each car was tapping the device to mark the position and commented loudly type of the artefact using simple descriptions: "pothole, left", "bump" and similar.

Fig. 2. Example of the video camera data feed frame.

The video/audio tracks and the acceleration data streams from two devices were synchronised using the NTP protocol by time, allowing the authors to perform the data extraction by searching in the audio track names of desired artefacts, searching for the same position in the acceleration track, and final **extraction** of the known road artefact data. This step was performed manually using software developed by the authors and not automated. Software is presented in the Figure 3. The visual recording of the road was also checked with audio commentary to remove false identification from inside the car.

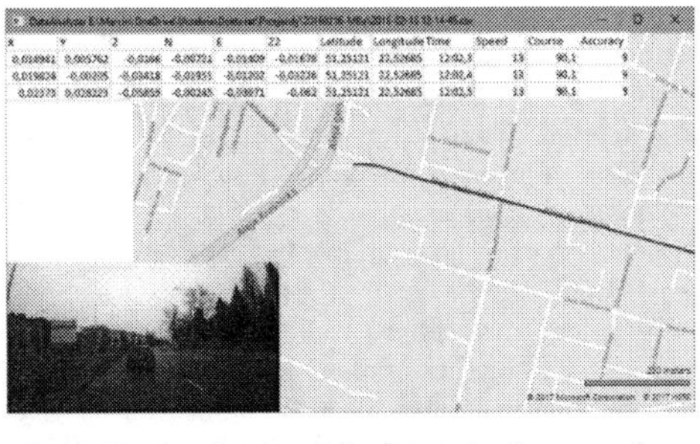

Fig. 3. Application developed for data extraction procedure.

Because driving over the road artefact is not instant, especially in the class of "wide" artefacts, every road artefact segment, and thus building block of the virtual road, must be composed of multiple data points.

Based on the previous research in this topic, the representative time was chosen as 50 data points, (5 seconds).

In case of road artefacts which are recognisable from the vertical acceleration stream (every "narrow" and "point" class), the procedure of extracting the signal was to use 25 data points (2.5 seconds) before the highest peak of vertical acceleration as well as 25 data points after this peak. This 50-data point wide data segment was a base for the proposed digital representation of the road artefact type.

2.3 Decomposition of the individual streams

Signals from driving over the road artefact were the series of tuples – or multi-dimensional vectors in other words, so they needed to be decomposed into multiple parallel streams. Some data which is acquired along with acceleration data needed to be discarded, as was not useable in the virtual road concept. Data streams acquired when driving:
- X, Y, Z axis acceleration;
- N, E, Z2 axis (reoriented) acceleration;
- Current speed;
- Current time;
- Current GPS coordinates.

In this set, both Time and GPS coordinates were not used in the virtual road concept, as they are linked to the real-world fragment.

For every road artefact data extracted in the previous step, these two were discarded. While removal of timestamp did not need additional description, based on the previous research, GPS readings accuracy was only sufficient only to mark the position of the road, but not to mark its beginning or ending position. This data is also needed only in the real-world context, so it was discarded completely.

2.4 Data normalisation procedure

Of course, different artefacts, even in the same class, may be very different one from another in the signal received from driving over them, so the "building blocks" representing the real-world road artefacts are going to be different one from another and there will be multiple recordings for the same road artefact type.

But this problem seems to be more complex as data obtained from two different cars are going to be vary significantly.

To cope with former, the authors are willing to normalise data, based on the known factor, ANF. Accelerometer Normalisation Factor is being calculated when the test car is driven over the known road segment with the same speed as the baseline car. The road segment is carefully selected as one of the A-class road segments, for example newly built highways [11]. First, every sample of dataset (X) to be normalised (x) are scaled into values from the set between minimum and maximum values for the baseline (Y), resulting in normalised x (x_n). This is presented in the equation 1:

$$x_n = x - \min(X) \cdot \frac{\max(Y) - \min(Y)}{\max(X) - \min(X)} + \min(Y) \quad (1)$$

Then, the final ANF should be calculated as an average of every normalised x_n divided by every baseline value (y_i), as presented in the equation 2.

$$ANF = \frac{\sum \frac{x_{n_i}}{y_i}}{n} \quad (2)$$

This procedure is applied to every element of an input vector, thus the final ANF is a vector consisting of

DOI: 10.1051/itmconf/20171502008

normalisation values for reoriented acceleration values, and is also a transformation of acceleration values from the regular units (g) into dimensionless values. Similarly, the authors are not using real time as the base for the calculations, but data points number, which is also dimensionless.

The normalised values for every known road artefact are grouped into classes of the road artefacts, building the database of different readings for the same road artefact class. The Figure 4, below, presents the example of normalised readings of an example road artefact, the speed bump.

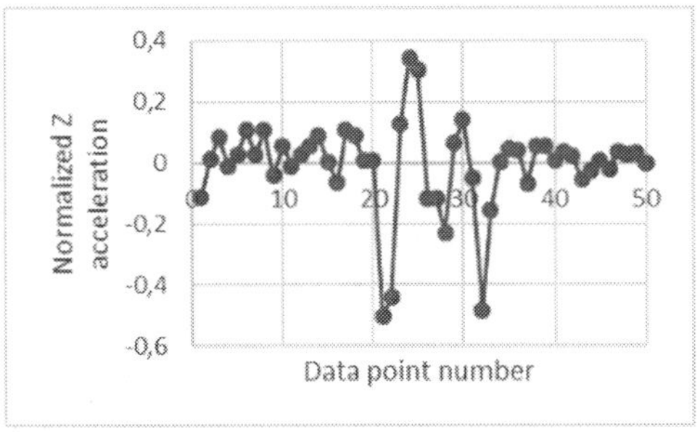

Fig. 4. Normalised readings when driving over a speed bump.

On the other hand, in the Figures 5 and 6 there are signals for road artefacts from different classes: pothole and rubble road, respectively.

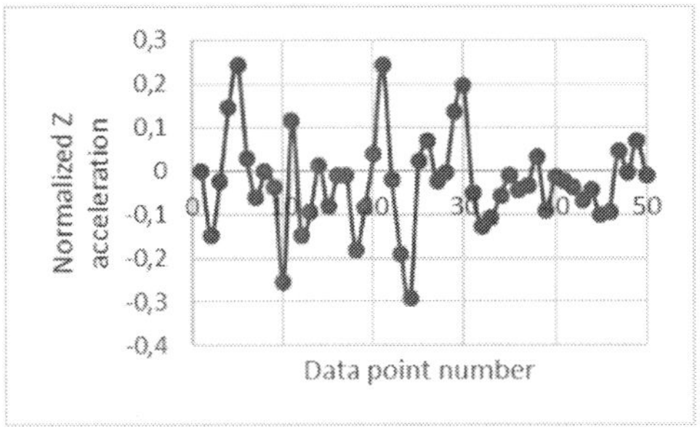

Fig. 5. Vertical acceleration readings (normalised) when driving over a pothole.

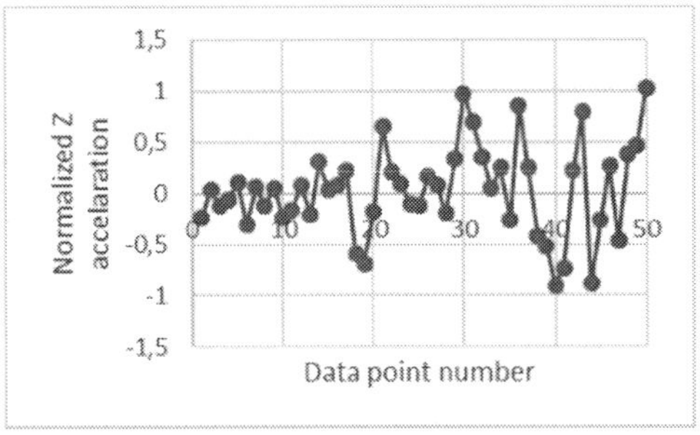

Fig. 6. Vertical acceleration (normalised) on a fragment of a rubble road.

The first problem mentioned – the variety of road artefacts in the same class – is not the problem itself in the case of this research. The authors are willing to include multiple road artefacts of the same class to achieve more diversity in the constructed virtual road, similar to the real world.

3 Virtual road construction

Providing the set of the virtual road artefacts classes, and the "building blocks" of the virtual road gives a possibility to prepare a virtual road, consisting of baseline asphalt fragments and including different types of road artefacts, to check different scenarios and test algorithms. The procedure, presented in the Figure 7, is based on three steps.

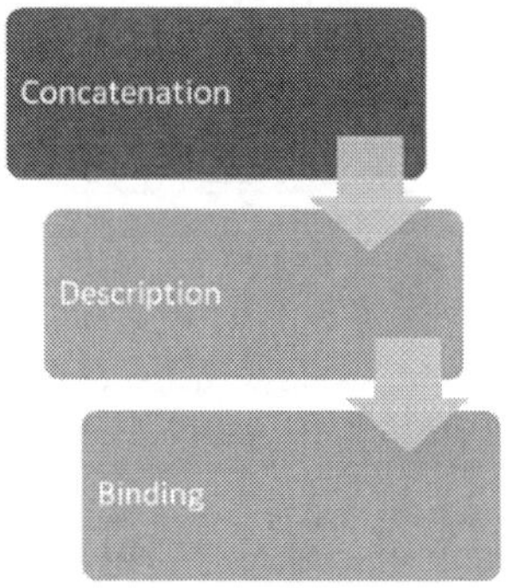

Fig. 7. Procedure for the actual virtual road construction.

The "building blocks" are simply needed to be settled one after another – to **concatenate** their signals. Because data streams for each road artefact type are not tuples anymore, as have been decomposed, there is also a possibility that virtual road concept will be a set of less than all original data streams, allowing testing of different techniques.

3.1 Concatenation of the building blocks

For the testing of the algorithms the authors already have worked with, the building blocks were used of only Z2 (global)-axis acceleration. The building blocks for the virtual road were chosen pseudo-randomly in the manner of:
- between 1 and 10 good road segments,
- between 10 and 20 rubble road segments,
- between every 5 road segments there may be a speed bump.

When selection of the given road artefact class (good road, rubble road, speed bump), the actual signal representing the artefact (variant) was selected randomly from the set of normalised signals for this road artefact class.

The built for the purpose of generation of virtual road application was however able to not only generate a virtual road data file for further research, but also had a possibility to save a virtual road and load it from a simple text file, with a strict format.

3.2 Text description of the virtual road

To achieve greater readability and reusability, the authors proposed a method for **description** of the virtual road timeseries in a text format.

 DOI: 10.1051/itmconf/20171502008

The format is based on the number of segments of the given type, then a capital letter describing the road artefact and optionally a lowercase letter describing the variant (the specific real-world equivalent) of the road artefact. The letters are corresponding to the road artefacts classes, as presented in the Table 2. Such format was pseudo-randomly generated, within rules described in the previous section.

Table 2. Road artefact types and corresponding letter code.

Road artefact	Code
"Good" road	A
Paved road	B
Rubble road	C
Railroad crossing	D
Elevated pedestrian crossing	E
Wide speed bump	F
Pothole	G
Manhole cover	H
Speed bump	I

Using the proposed system, for example 3Aa 1Ia 2Ab 3Ad 4Aa will be the equivalent of the virtual road consisting of 3 "good" road segments (variant a), then 1 speed bump (variant a), and then again 9 segments of the "good" road segments – but first 2 variants b, 3 variants d and 4 times variant a.

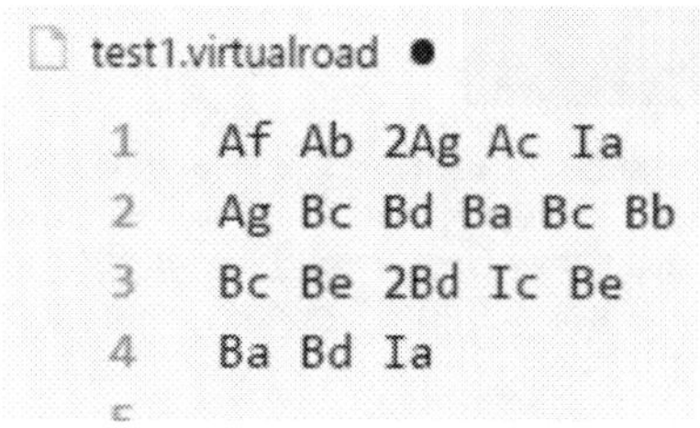

Fig. 8. Example description of the virtual road to be constructed.

Such text file, which another example is presented in the Figure 8 above, was read by the application and the actual values were placed in place by selection of the proper road artefacts from the internal database.

3.3 Virtual GPS coordinates binding

Sometimes, there is a need for using data that was discarded in the decomposition stage – in this case being this a GPS data stream, which the authors are willing to use, because requirements of road assessment method used to validate the virtual road concept.

To cope with that, the authors are changing the signal into a stream of tuples consisting a vertical acceleration and corresponding fake GPS coordinates. These coordinates will be used only for the testing procedure, which is based on it, and for easily visualisation.

In general, there may be not only need for GPS coordinates, but also for other values (i.e. speed), which is used in some road artefact detection methods.

4 Experimental results

To confront validity of the constructed virtual road against its real world counterpart, the authors used the same procedure for assessing the virtual road quality as for the real roads – the Road Relative Unevenness Index (RRUI) was used [11].

The virtual road was generated, which text representation has been presented in the previous example in the Figure 8. Such data was than bound with virtual GPS coordinates. In the presented solution, the authors decided to include fake GPS coordinates changing by the 20 meters per second, which was an equivalent of speed in urban areas. To calculate from the meters into GPS degrees, the WGS84 parameters for the Earth were used [12], resulting in usage of a value 111.32 km per decimal degree.

To achieve best results in presenting an output in the paper, only longitude was changed, meaning the virtual road goes straight from the south to the north.

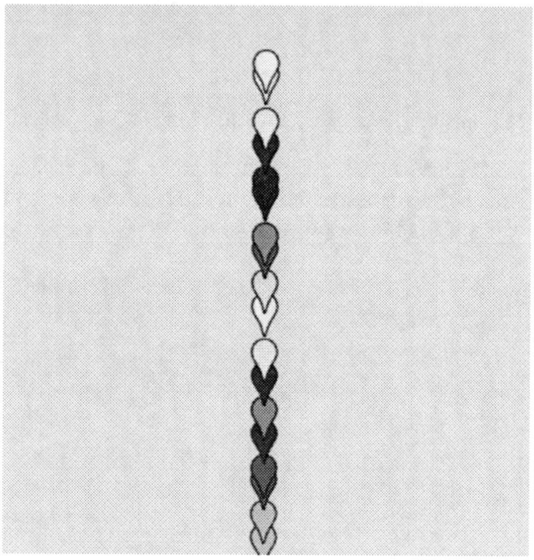

Fig. 9. RRUI pushpins of the virtual road.

As presented in the Figure 9, the RRUI assessment scale has been translated into a 7-point colour scale, marking start points of every new road segments assessed, which were 50 meters long, with a pushpin of different colour. The darker the colour, the higher the respective RRUI value was, marking greater relative unevenness, thus the lower overall road segment quality. Internally, the RRUI scale is a dimensionless value ranging in the case of the presented road from -2 to 51, where -2 is a road better than the baseline, and 51 is worse.

As the virtual road was constructed first from the "good" road segments, the initial RRUI values for the first road segments are low – from about 5 to 22, where high value of 22 is in the position of the road artefact, the speed bump, as is presented in the Table 3. The next virtual road segments, described in the Table 4, built upon rubble road fragments, should be much higher, which is correct again, with values ranging from 23 to even 50.

Table 3. RRUI values for virtual road fragments constructed from good roads.

Segment number	RRUI value
1	14.56
2	13.04
3	18.23
4	14.16
5	10.07
6	22.18
7	10.07
8	5.58

Table 4. RRUI values for virtual road fragments constructed from rubble roads.

Segment number	RRUI value
16	23.76
17	28.26
18	20.91
19	51.49
20	26.42

To check an another algorithm, the road artefact detection method using the modified Z-STDEV method [8] was performed on the same virtual road as before. Using these methods, road artefacts which were included in the virtual road (on position of 6th, 17th and 21st segment, respectively), were detected with one false positive in data point number 106, which is in a second good road segment (class A, variant b).

Similar tests were performed for the set of multiple generated virtual roads and in every case the RRUI values were correctly corresponding to the type of the road.

5 Conclusions and future work

The virtual road concept is an interesting way of constructing purely artificial signals for road artefacts algorithms detection training and testing. The authors also believe that building a normalised database of road artefacts and marking them in a simple text format would be useful for sharing virtual roads between researchers.

The streams, which were decomposed after the extraction phase, may be modified before virtual road construction. The authors are aware of the possibilities of this step, introducing changes to different data streams – for example scaling the Z-acceleration value to simulate deeper potholes. This step is not taken into consideration at this stage, but it will be one of the interesting points in the future for the presented concept.

The second shortcoming in the presented stage of research is no possibility for road artefacts of "point" class to be included into the middle of the "wide" class, being placed only sequentially. This is the next issue for the authors to solve into changing the presented concept to cover more realistic simulation of the real-world roads.

References

1. P. Mohan, V. N. Padmanabhan, R. Ramjee, Nericell: Rich Monitoring of Road and Traffic Conditions using Mobile Smartphones. Proc. 6th ACM Conf. Embed. Netw. Sens. Syst. - SenSys '08 (2008), p. 323
2. A. Mednis, G. Strazdins, M. Liepins, An. Gordjusins, L. Selavo, 136 (2011), doi:10.1007/978-3-642-22185-9
3. V. Astarita, R. Vaiana, T. Iueleand, V. C. Maria, V. P. Giofre, F. De Masi, Procedia - Soc. Behav. Sci. 111, 242–251 (2014)
4. M. Afrin, M. R. Mahmud, M. A. Razzaque, 2015 IEEE Int. WIE Conf. Electr. Comput. Eng. WIECON-ECE 2015, 495–498 (2016)
5. K. Laubis, V. Simko, A. Schuller, C. Weinhardt, Proc. 50th Hawaii Int. Conf. Syst. Sci., 1582–1591 (2017)
6. Lars Forslöf, Hans Jones, J. Civ. Eng. Archit. 9, 485–496 (2015)
7. L. C. Gonzalez, R. Moreno, H. J. Escalante, F. Martinez, M. R. Carlos, IEEE Trans. Intell. Transp. Syst., 1–13 (2017)
8. M. Badurowicz, J. Montusiewicz, in Information and Software Technologies, ICIST 2015 [WOS], R. Damasevicius, G. Dregvaite, Eds. (Springer, Berlin, Germany, 2015; http://link.springer.com/10.1007/978-3-319-24770-0_43), pp. 503–514
9. M. Badurowicz, T. Cieplak, J. T. Montusiewicz, Appl. Comput. Sci. 12, 18–27 (2016)
10. J. Smolka, M. Skublewska-Paszkowska, Hum. Syst. Interact. (HSI), 2016 9th Int. Conf., 126–132 (2016)
11. M. Badurowicz, T. Cieplak, J. Montusiewicz, in Computer Networks: 23rd International Conference, CN 2016, Brunów, Poland, June 14-17, 2016, Proceedings, P. Gaj, A. Kwiecień, P. Stera, Eds. (Springer International Publishing, Cham, 2016; http://dx.doi.org/10.1007/978-3-319-39207-3_31), pp. 360–369
12. World Geodetic System 1984, (available at http://spatialreference.org/ref/epsg/wgs-84/)

DOI: 10.1051/itmconf/20171502009

Web service for solving optimisation problems using swarm intelligence algorithms

Yuriy Tryus[1], *Andrii* Geiko[1], and *Grygoriy* Zaspa[1,*]

[1]Cherkasy State Technological University, Computer Science and Information Technology Department, 460 Shevchenko Blvd, 18006, Cherkasy, Ukraine

Abstract. In this research the web service which provides a user a possibility to use swarm optimisation algorithms for solving continuous and discrete extreme problems online was created. The web service includes facilities for data input and making numerical experiments related to optimisation problems research with swarm intelligence methods. For swarm intelligence algorithms numerical experiments, the user can tune the following parameters: problem dimension; swarm parts number; inertial, cognitive, and social coefficients. The user can also select algorithm stop criterion and set the number of algorithm iterations. During the optimisation problem solving, the iteration process graphical visualisation is done. Also, the protocol is being formed and it can be exported to .xls file. Checking the algorithms in the web service was made on well-known test functions and conditional and non-conditional optimisation problems; some of them are built into the service.

1 Introduction

Solving real optimisation problems existing in different areas of science, technology, or economics is complicated with existence the following properties of their mathematical models: search surface complex topology, multi-extrema, non-smoothness, multiple dimensions. In this case, using natural processes imitation based methods, which are borrowed from nature, and realise adaptive randomness of a search is expedient. The methods based on self-organised population systems collective behaviour modelling are among this kind of methods. For instance, an are ant colony optimisation (A. Colorni, M. Dorigo, V. Maniezzo (1991)), particle swarm optimisation (PSO) (J. Kennedy, R. Eberhart (1995)), grey wolf algorithms (Mirjalili S. (1995)), bee colony optimisation (D. Karaboga (2005)), glow-worm swarm optimisation (K. N. Krishnanand, D. Ghose (2005), monkey search algorithm (A. Mucherino, O. Seref (2007)), school of fish algorithm (Bastos-Filho C.J.A., de Lima Neto F.B., Lins A.J.C.C., Nascimento A.I.S., Lima M.P. (2008)), cuckoo search algorithm (X.-S. Yang, S. Deb (2009)), bats gang algorithm (Yang X. S. (2010)), bacterial optimisation (Niu Ben, Wang Hong (2012)) etc. The primary algorithm between them is particle swarm optimisation (PSO) offered by Kennedy J. i Eberhart R. in [1] and developed by many other researchers (see, for instance, [2]-[6]). It is popular, primarily because it can be used for effective solving the wide range of optimisation problems including continuous, discrete, binary, and multi-criteria optimisation.

2 Problem topicality

The swarm intelligence methods are diverse and can be used for solving problems in different domains: telecommunication, aviation, space research, robotics, medical care, biology etc. That is why swarm intelligence and its researching methods are a topical scientific direction. In many universities over the world, students of mathematical and computer faculties study PSO algorithms in courses related to the modern optimisation and operation research methods. Using these methods for solving real optimisation problems, computer mathematical software such as MatlabR2017a can be used. It has *particleswarm* function in Global Optimisation Toolbox [7]. It realises PSO method. But Matlab is a proprietary desktop application and is not available for everybody. So, creating a system which would provide the users the possibility to use PSO algorithms for solving training and practical optimisation problems online and include the theoretical information and facilities for making experiments for researching and developing PSO algorithms is a very topical problem.

The research objectives are a) reasoning of necessity of web service for solving optimisation problems of different type online with swarm intelligence methods development; b) requirements analysis to the main information technologies being used on the way of the web service creation; c) the web resource design and creation; d) making numeric experiment for testing the methods realised using test functions.

* Corresponding author: g.zaspa@chdtu.edu.ua

2.1 Problem definition

A web resource being called PSO Service should implement the main algorithms of swarm intelligence specifically canonical PSO algorithm, its adaptive and hybrid variants for solving continuous and discrete optimisation problems, one- and multi-criteria extremum problems. The web service user should have the possibility to a) enter objective function, constrain functions, direct constraint of variable values; b) set parameters for problems solving specifically: swarm dimension, search margins, calculation precision, relations type between swarm particles ("star", "ring", "random"); c) get problem solution in view understandable for the user. Also, the user should have the possibility to see the function value for every iteration. The service should visualise the PSO algorithms iterations for two-dimensional optimisation problems. For solving conditional optimisation problems, specifically mathematical programming the penalty function method with possible choosing among different penalties and setting their options will be used. For making numerical experiments on PSO algorithms ability to work and effectiveness the service should include a set of known test functions. Besides, the service should be multilingual, include general information on PSO algorithms, and their use for solving different kinds of optimisation problems.

3 PSO algorithms realised in PSO Service

Here, the non-conditional global optimisation problem should be solved, formulated as the problem objective function $f(X)$ in D search region minimisation:

$$f(x) \to \min,\ x \in D \subset R^d, \tag{1}$$

where D is hypercube of d dimension, x – f function vector argument; x^* – objective function $f(x)$ global minimum point.

Brief consideration of the PSO algorithms of solving problem of kind (1) which are realised in the PSO Service web service is presented.

3.1 Canonical PSO algorithm for continuous optimisation problems

In canonical PSO particles swarm is a set of decision points which transfer in space searching for global optimum. While moving the particles try to improve the value found before and exchange the information with their neighbours.

The set of swarm particle positions is defined as:

$$X = \{x_1, x_2, ..., x_s\}, \tag{2}$$

where s is particles number in a swarm.

The position of particle i is the set of its coordinates $(x_{i1}, x_{i2}, \dots, x_{id})$ in search region D of dimension d, $i = \overline{1,s}$, . For optimisation 20-50 particles are usually enough [4]. The swarm allows to find global optimum even it has particles number less than search region dimension d.

In the beginning of PSO algorithm work the particle swarm is initialised the random way. If there is no any prior information about the function being optimised the simplest way to generate the initial particles coordinates is to use the following formula:

$$x_{ij} = rand(x_{j\min}, x_{j\max}), \tag{3}$$

where x_{ij} is coordinate number j of the particle number i, $i = \overline{1,s}$, $j = \overline{1,d}$, ;

$rand(x_{jmin}, x_{jmax})$ is a random number with uniform distribution on interval defined with search region borders for coordinate j.

The set of particles speed vectors is also associated with the swarm:

$$V = \{v_1, v_2, ..., v_s\}, \tag{4}$$

All speeds can be considered as equal to 0 on the initial stage. But the practice shows that the following formula gives the best results [3]:

$$v_{ij} = [rand(x_{j\min}, x_{j\max}) - x_{ij}]/2, \tag{5}$$

where v_{ij} – speed component number j of particle i.

This way of defining the initial speeds guarantees that no particle gets out of search region in the next iteration.

On the following algorithm steps, the speed and particles positions components are changed according the formula (in coordinate form):

$$v'_{ij} = wv_{ij} + c_1r_1(p_{ij} - x_{ij}) + c_2r_2(g_j - x_{ij}), \tag{6}$$

$$x'_{ij} = x_{ij} + v'_{ij}, \tag{7}$$

where $i = \overline{1,s}$, $j = \overline{1,d}$, ; v'_i and x'_i are new speed and position respectively of particle i; p_i is the best previous position of the particle i; g is the best solution found with whole swarm; w is the *inertial coefficient*; c_1 is the *cognitive coefficient*; c_2 is the *social coefficient*; r_1, r_2 are random numbers uniformly generated on [0;1] interval; they are different for every coordinate.

If during optimisation a particle gets out of search region its respective speed component is set to zero and the particle returns to the nearest border.

The inertial coefficient w means the influence the previous speed of the particle on its new value. The cognitive coefficient c_1 describes the degree of the particle individual behaviour and its intention to get back to the best solution found with it. The social coefficient value c_2 sets the degree of collective behaviour and the intention to move to the particle neighbours' best solution.

Vectors $v_p = c_1r_1(p - x)$ та $v_g = c_2r_2(g - x)$ define the cognitive and social components of the particle new speed. The random numbers r_1, r_2 add randomness to the

 DOI: 10.1051/itmconf/20171502009

search. Fig.1 shows geometrical interpretation of swarm particles movement direction correction rule (6)-(7).

The coefficient values are selected within ranges [3]:

$$\begin{cases} 0 < w \le 1, \\ 1 < c_1, c_2 \le 2. \end{cases} \quad (8)$$

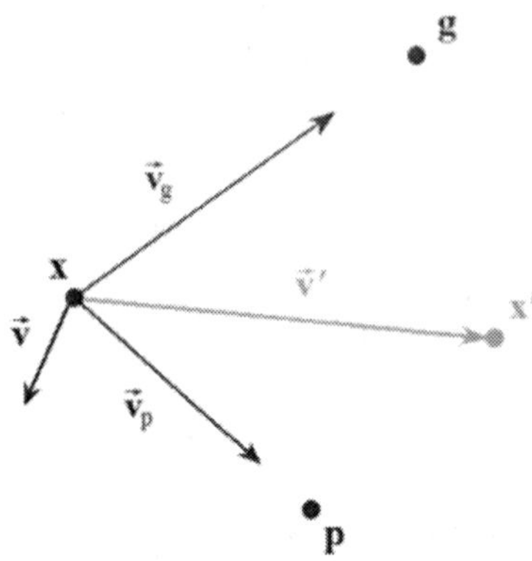

Fig. 1. Geometrical interpretation of rule (6)-(7).

For keeping balance between local and global search the coefficients c_1 and c_2 values are usually equal. Stagnation analysis in [4] has shown that coefficients values $w = 1/2\ln 2 \approx 0.719$ and $c_1 = c_2 = 0.5 + \ln 2 \approx 1.193$ provide in most cases the best results and search stability. Also, there are different ways of setting the swarm parameters dynamically, but adaptation needs some additional initial algorithm iterations: it can lead to objective function (necessary for optimum search) evaluation number increase.

The swarm remembers the best solutions found by particles in separate and by the whole swarm. On initialization the initial particles positions are considered as the best. On every following algorithm iteration after using formulas (6)-(7) the individual best position of every particle p_i and the whole swarm best solution g are corrected according to the rule:

$$\begin{cases} p_i = x_i, \text{if } f(x_i) < f(p_i), \\ g = p_i, \text{if } f(p_i) < f(g). \end{cases} \quad (9)$$

So, PSO algorithm scheme is the following:

1. Set algorithm parameters and initialize the swarm: $X = \{x_1, x_2, ..., x_s\}$.
2. Find the best individual value p_i for every swarm particle i, $i = \overline{1,s}$, and the best value for the whole swarm g.
3. Find new positions x_i', $i = \overline{1,s}$, for all swarm particles according to the formula (6)-(7).
4. Check the iteration loop condition. If the condition is *true*, go to item 5, else go to item 2.
5. Output the results.

The typical for optimisation population algorithms conditions are used to determine the iteration loop finish, specifically, reaching maximum iterations number, reaching acceptable solution precision, iteration process stagnation.

3.2 Adaptive particle swarm optimisation

In PSO algorithm there are some free parameters (w, c_1, c_2, s) which can have different optimal value for different problems. Choosing algorithm parameters free values vector is called its adaptation strategy. An adaptive PSO which decreases the probability of premature convergence is Inertia-adaptive PSO Algorithm (IA-PSO) offered in [8]. In this algorithm the inertial coefficient individual value is calculated for every swarm particle by formula:

$$w_i = w_0(1 - dist_i / \max\_dist) \quad (10)$$

where $w_i = rand(0.5, 1)$; $dist_i$ is current Euklid distance from particle i the best solution of swarm g found by the swarm in current iteration, max_dist is the biggest distance from a particle to the best solution found by the swarm in current iteration, e.g. $\max\_dist = \arg\max_{i=1,s}(dist_i)$.

Because of the inertial coefficient correction offered the gravity between swarm and particle increases on significant distance between particle and global swarm solution g. This is because in this case w parameter goes down and the particle stops go further from g. To avoid premature convergence, it is necessary to be sure that the particles have enough mobility on late optimisation stages. For reaching this goal the formula of the particles position correction (7) is modified according to dependency $x_{ij}' = (1-\rho)x_{ij} + v_{ij}'$, where ρ is a random value with uniform distribution on [-0.25, 0.25] interval.

3.3 Hybrid PSO algorithm with local search with Nelder-Mead method

Most of swarm intelligence algorithms allow to research search region D but are less effective on researching its small areas. In particle swarm optimisation the local search quality can be improved using lower values of w coefficient. But it decreases the probability of global solution localization and causes premature algorithm convergence. That is why it is better to make local search using special methods (gradient descent method, conjugated gradients method, Newton and quasi-Newton methods) [9]. One of the most popular and effective local search methods which doesn't demand objective function derivatives evaluation is Nelder-Mead algorithm [10]. It is also called deformed polyhedron method.

Hybrid particle swarm optimisation process can be divided into two stages.

1. Optimisation with particle swarm algorithm and global optimum approximate value locating.
2. Search for more exact solution with local search method. In this case local optimisation method gets values found during particle swarm optimisation and

continues locating optimum for the best result found with PSO after several iterations.

3.4 Particle swarm optimisation for discrete problems

In integer optimisation and combinatorial optimisation, the global solution which belongs to some discrete set of possible values should be found. First, the particle swarm optimisation algorithm was developed for solving continuous optimisation problems but using it for solving integer and binary optimisation problems is possible.

In the simplest case the integer programming problems can be solved with using continuous PSO with rounding the result to the nearest integer [11]. So, the solution is searched in D continuous region and before the evaluation of the corresponding objective function value the solution coordinates are recalculated by formula $x_i = [x_i + 0.5]$, where $[x]$ is x number integer part extraction operation. This variant of PSO is called *Discrete Particle Swarm Optimisation* (DPSO).

In [12] BinaryParticle Swarm Optimisation (BPSO) is offered.

3.5 Using PSO for solving mathematical programming problems

Swarm algorithms are intended for solving global optimisation problems on hypercube. For conditional linear and nonlinear optimisation in particular mathematical programming of kind:

$$f(x) \to \min,\ x \in X, \qquad (11)$$

where

$$X = \left\{x \in D \subset R^d \,\middle|\, g_i(x) \le 0, i = \overline{1,m},\ g_i(x) = 0, i = \overline{m+1,s}\right\} \qquad (12)$$

the external penalty functions method with *power* or *non-smooth penalty function* (on user choice) is used. Then the problem (11)-(12) in external penalty functions method comes down to solving the problems sequence of kind:

$$F_k(x, r_k) = f(x) + R(x, r_k) \to \min,\ x \in D \subset R^d, \qquad (13)$$

where $k = 1, 2, ...,$; r_k is penalty coefficient which corresponds the condition $r_k > 0,\ \lim_{k \to \infty} r_k = +\infty$, $R(x, r_k)$ is one of external penalty functions.

It is known (in, for instance, [9]) that if points $x^{(k)}$ are the points of global minimum of $F(x, r_k)$, $k = 1, 2, ...,$ then the sequence $\{x^{(k)}\}$ converges to global minimum point x^* of $f(x)$ function on X set i.e. (11)-(12) problem solution.

In practice the problem (13) is solved with fixed penalty parameter value r_k, which is a rather big positive value. The (11)-(12) problem solution will be obtained with some accuracy. In PSO Service parameter r_k has default value of 1000 but it can be changed in parameters input window for increasing problem (11), (12) solution accuracy.

4 PSO Service web resource development tools

PSO Service has web interface. The server side is developed using *Java* programming language and *Spring* framework, the client side is developed using *Thymeleaf* web patterns, *Bootstrap* for decoration. *Hibernate* object-relational mapping system is used for working with database. Table 1 includes information of the tools main characteristics.

Table 1. Overview of technologies for PSO Service web resource implementation.

Technology	Logo	Description
Java	Java	Object-oriented multi-platform programming language
Spring		Universal open-source framework for Java platform
Thymeleaf		XML, XHTML and HTML5 patterns engine for Java
Hibernate		Object-relational mapping framework for Java
Bootstrap	B	Tools set for setting web sites and web applications view

5 PSO Service main function

PSO Service web resource is bilingual (*English* and *Ukrainian*). It allows the user to learn the theory on swarm algorithms principles, solving optimisation problems with some of these algorithms.

On the main page (Fig. 2) user can read the information about swarm intelligence optimisation methods (*Algorithms*), use algorithms implemented in the service for solving optimisation problems (*Solve*), get information on working with the service (*Help*), get developers contacts (*Contact Us*), authorize (register) in the system (*Login*).

Having learnt about the algorithms user can move to solving the optimisation problem but first he needs to login or register if the user is not registered yet ("*Login*" function).

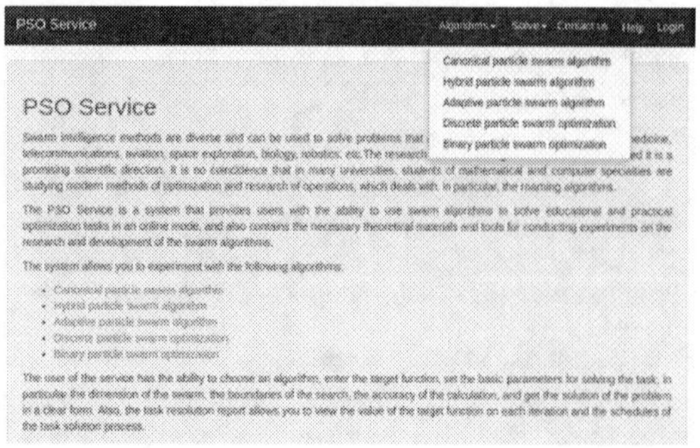

Fig. 2. PSO Service start page.

After login user can start solving optimisation problems with selected algorithm: *Canonical PSO*, *Hybrid PSO*, *Adaptiv PSO*, Discrete *PSO*, *Binary PSO*. If user wants to optimise objective function which is not among test functions he must fill the form on Fig. 3, specifically, the function field using calculator-like buttons and the fields "*Error*", "*Dimension of the task*", "*The size of the swarm*", "*Search space*".

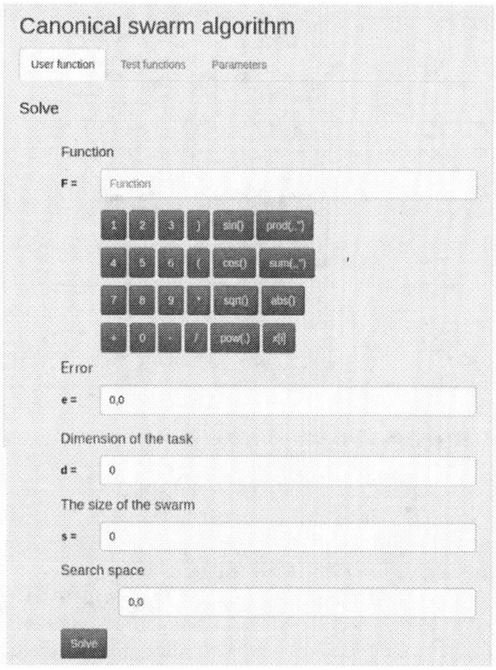

Fig. 3. User function input form.

Also, user can use some of predefined functions on "*Test functions*" tab (Fig. 4).

For making numeric experiments user is to set swarm optimisation parameters on "*Parameter*" tab (Fig. 5). User can change inertial, cognitive, social coefficients, set stopping criterion, number of repetitions etc. User can set, clear, or set default parameter values.

Having made the necessary settings and having sent the form to the server user gets the problem solution. The results page has three tabs: "*Calculation*", "*Input parameters*", and "*Graph/Protocol*".

The "Calculation" tab contains the algorithm execution main results (Fig. 6).

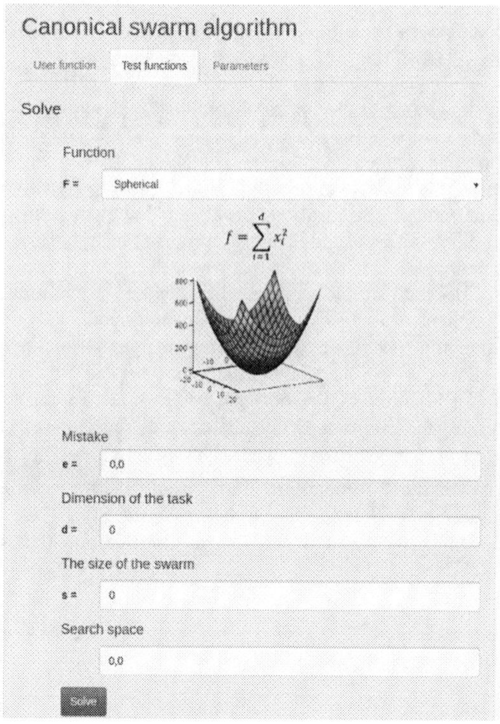

Fig. 4. Test function selection form

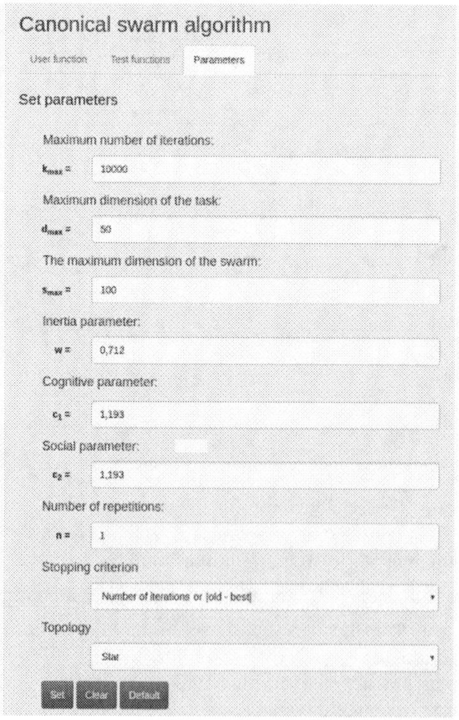

Fig. 5. Particle swarm optimisation parameters setting form.

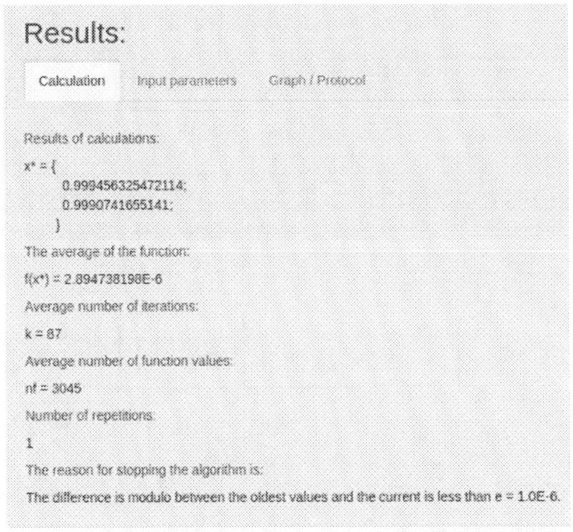

Fig. 6. "Calculation" tab for Rosenbrock function.

"Input parameters" tab includes main parameters of the problem and algorithm (Fig. 7).

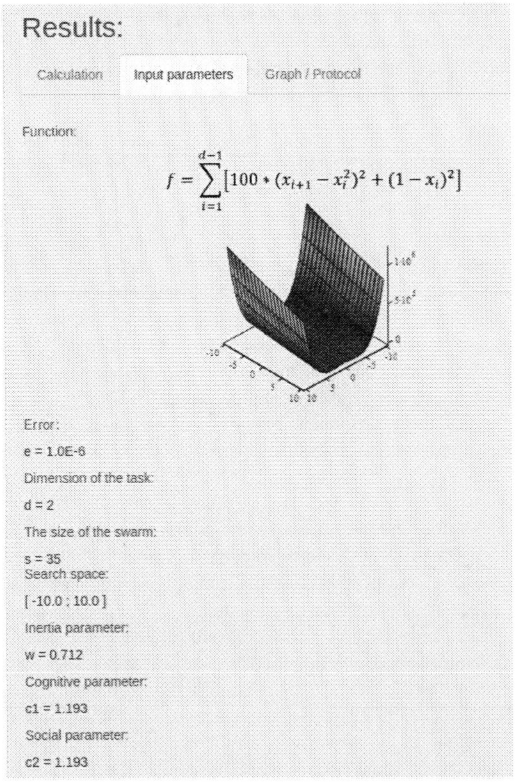

Fig. 7. "Input parameters" tab for Rosenbrock function.

"Graph/Protocol" tab (Fig. 8) contains the dynamic visualisation of optimisation process with the selected algorithm when problem dimension equals 2 and objective function value change on every iteration diagram. Also, problem solution protocol in .xls format is available.

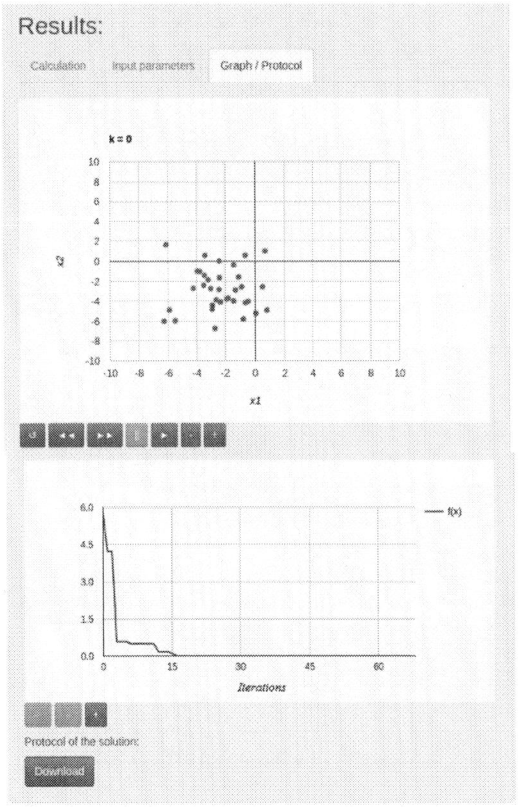

Fig. 8. "Graph/Protocol" tab.

6 Numeric experiment

The set of test functions (Table 2) is used for testing the algorithms implemented in web service accuracy. They are non-linear functions used for optimisation algorithms characteristics assessment, specifically accuracy, convergence etc.

All these functions except Schwefel and Rosenbrock have global minimum in point $x^* = (0, 0, \ldots, 0)$ with zero value in it.

Non-separable Rosenbrock function (f_4) is one of the most complex. Most of classical optimisation algorithms cannot find its global optimum $x^* = (1, 1, \ldots, 1)$.

Schwefel function (f_8) global optimum is geometrically remote from the best local minimum. That is why the search algorithm is inclined to go to wrong direction. The global solution for this function is in point $x_i^* \approx 420.9687$, $i = \overline{1, d}$, what is near search region border.

Griewank function (f_6) is a typical separable multi-modal objective function which has many local optimums. Because of topology like that the search algorithms incline to get stuck in one of local optimums.

For making the condition equivalent for every optimisation algorithm being tested the same parameters values were used on every algorithm run.

Table 2. Test functions.

Function name	Formula	*D*	$f(x^*)$
Spherical	$f_1 = \sum_{i=1}^{d} x_i^2$	[-20; 20]^d	0
Elliptic	$f_2 = \sum_{i=1}^{d} (10^6)^{\frac{i-1}{d-1}} x_i^2$	[-20; 20]^d	0
Schwefel 1.2	$f_3 = \sum_{i=1}^{d} \left(\sum_{j=1}^{i} x_j \right)^2$	[-10; 10]^d	0
Rosenbrock	$f_4 = \sum_{i=1}^{d-1} [100(x_{i+1} - x_i^2)^2 + (1 - x_i)^2$	[-10; 10]^d	0
Rastrigin	$f_5 = \sum_{i=1}^{d} [x_i^2 - 10\cos(2\pi x_i) + 10]$	[-5; 5]^d	0
Griewank	$f_6 = \frac{1}{4000} \sum_{i=1}^{d} x_i^2 - \prod_{i=1}^{d} \cos\left(\frac{x_i}{\sqrt{i}} \right) + 1$	[-100; 100]^d	0
Alpine	$f_7 = \sum_{i=1}^{d} \lvert x_i \sin(x_i) + 0.1 x_i \rvert$	[-10; 10]^d	0
Schwefel	$f_8 = -\frac{1}{d} \sum_{i=1}^{d} x_i \sin\left(\sqrt{\lvert x_i \rvert} \right) + 418.983$	[-500; 500]^d	0
Ackley	$f_9 = -20 \cdot \exp\left(-0.2 \sqrt{\frac{1}{d} \sum_{i=1}^{d} x_i^2} \right) - \exp\left(\frac{1}{d} \sum_{i=1}^{d} \cos(2\pi x_i) \right) + 20 + e$	[-30; 30]^d	0
Weierstrass	$f_{10} = \frac{1}{d} \sum_{i=1}^{d} \sum_{k=0}^{20} 0.5^k \cos(2\pi 3^k \times \times (x_i + 0.5)) - \sum_{k=0}^{20} 0.5^k \cos(\pi 3^k)$	[-0.5; 0.5]^d	0

In Table 3 the results of using PSO Service for Table 2 test functions solution using canonical PSO with "star" type connections between the swarm particles are given, $\bar{f}_p$ is the average of the function minimal values evaluated on every of n algorithm repetitions, $\bar{k}_p$ is iterations number average on every of n algorithm repetitions, $f_{\min}$ is function minimum value received after n algorithm repetitions, $x_{\min}$ is the vector for which the function has $f_{\min}$ value, $\| x_{\min} - x^* \|$ is Euklid distance between points $x_{\min}$ and x^*.

On testing the condition $| g_{old} - g_{best} | < \varepsilon$ was used as PSO algorithm stop criteria. In the formula g_{old} is the previous best value of objective function for the whole swarm, g_{best} is new best value of objective function for the whole swarm, ε is calculation error.

The following algorithm parameter values for test functions were used:

f_1, f_2, f_3: d=20, s=40, w=0.8, c_1=1.193, c_2=1.193, ε=1e-10, n=20;

f_4: d=20, s=40, w=0.8, c_1=1.193, c_2=1.193, ε=1e-8, n=10;

f_5: d=10, s=40, w=0.9, c_1=1.193, c_2=1.193, ε=1e-20, n=10;

f_6: d=5, s=40, w=0.8, c_1=2, c_2=2, ε=1e-10, n=10;

f_7: d=2, s=20, w=0.9, c_1=1.193, c_2=1.193, ε=1e-15, n=10;

f_8: d=5, s=30, w=0.8, c_1=2, c_2=2, ε=1e-15, n=10;

f_9: d=20, s=40, w=0.8, c_1=1.193, c_2=1.193, ε=1e-15, n=10;

f_{10}: d=10, s=40, w=0.8, c_1=1.193, c_2=1.193, ε=1e-10, n=10.

Table 3. Canonocal SPO testing results.

Function name	$\bar{f}_p$	$\bar{k}_p$	$f_{\min}$	$\| x_{\min} - x^* \|$
Spherical f_1	1.239993E-8	336	1.64199E-09	4.05215E-05
Elliptic f_2	1.8203008E-8	454	1.62316E-09	3.83073E-05
Schwefel 1.2 f_3	4.4579857E-7	1235	4,2467E-08	3.67142 E-4
Rosenbrock f_4	1.0211102335	3892	0.001712	0.05409512
Rastrigin f_5	0.0994959057	9093	2.66E-13	3.6716E-08
Griewank f_6	0.0187167926	9620	3.53E-09	0.000181
Alpine f_7	2.7E-14	1974	9.11E-16	6.44406E-15
Schwefel f_8	52.355709479	5774	0.000113	0.000105469
Ackley f_9	0.6810145400	759	3.24185E-14	3.67643E-14
Weierstrass f_{10}	0.1090308553	462	1.0353E-09	5.93228E-12

Numeric experiment on test functions result has shown that canonical PSO realized in PSO Service is efficient. Also, there was successful efficiency check for other algorithms realized in PSO Service.

7 Conclusion

1. Particle swarm optimisation is widely used in machine learning, specifically for neural networks learning and pattern recognition, parametrical and structural optimisation (shape, size, and topology) in design, biochemistry, biomechanics etc. Considering its effectiveness, it can compete with other global optimisation methods and low algorithmic complexity contribute to its implementation ease.

2. The most promising research directions in this field are theoretical research of particle swarm optimisation convergence reasons and related topics in swarm intelligence and chaos theory areas, combining different algorithm modifications for solving complex tasks, research of SPO in multi-agent computer systems, and research on possibilities of including more complex natural mechanism analogues in it.

3. In this research, the creation of web service for optimisation problems solving with swarm intelligence methods is based. Also, the main requirements for the system like that are set and information technology for its creation is described. The description of the main algorithms and methods implemented in the service and the results of numeric experiment for checking them on test functions are given.

4. In the future, new features are planned to be implemented in PSO Service software: other collective intelligence algorithms; more convenient conditional optimisation functional limitations input; increase methods number of moving it to global optimisation problem on hypercube; extend variants of PSO hybridisation with other local extremum search methods; add new embedded test functions.

DOI: 10.1051/itmconf/20171502009

5. Created PSO Service can be used: for teaching students, who study swarm intelligence methods; by scientists, who search for different optimisation problem types solving methods effectiveness; by users who use optimisation methods for decision making on real extreme problems solving.

References

1. J. Kennedy, R. Eberhart, *Proceedings of IEEE International conference on Neural Networks*, 1942 (1995)
2. J. Kennedy, R.C. Eberhart, *IEEE Conference on Systems*, 4104 (1997)
3. R. Eberhart, Y. Shi, *Proceedings of the 2000 International Congress on Evolutionary Computation*, 84 (2000)
4. M. Clerc, R. Eberhart, IEEE Trans. Evolut. Computation, **6**, 58 (2002).
5. M. Clerc, *Particle Swarm Optimisation* (USA: ISTE Ltd, 2006)
6. R. Poli, J. Kennedy, T. Blackwell, Swarm Intell., **1**, 33 (2007)
7. Particle Swarm Optimisation Algorithm, https://www.mathworks.com/help/gads/particle-swarm-optimisation-algorithm.html, date of relevance: July 12, 2017
8. M.I. Zhaldak, Y.V. Tryus, *The basics of optimisation theory and methods* (Cherkasy, Brama-Ukraine, 2005)
9. K. Suresh, S. Ghosh, D. Kundu, A. Sen [at el.], Intell. Syst. Design and Applicat., **2**, 253 (2008)
10. J.A. Nelder, R. Mead, Computer J., **7**, 308 (1965)
11. V.Y. Galchenko, A.N. Yakimov, *Particle swarm optimisation population methaeuristic algorithms* (Cherkasy, Tretyakov A.N., 2015)
12. J. Kennedy, R.C. Eberhart, *IEEE Conference on Systems, Man and Cybernetics*, **5**, 4104 (1997)

DOI: 10.1051/itmconf/20171503001

An investigation into multi-domain simulation for a pantograph-catenary system

Paweł Zdziebko[1,*], *Adam* Martowicz[1], and *Tadeusz* Uhl[1]

[1]AGH University of Science and Technology, Department of Robotics and Mechatronics, Al. A. Mickiewicza 30, 30-059 Krakow, Poland

Abstract. There is a need for modelling various phenomena present in a pantograph-catenary structure, (e.g. wave propagation and its reflections, friction, aerodynamic and electromagnetic forces), which allows for a more reliable study on the dynamic behaviours of a railway pantograph, particularly in the case of high-speed trains. Hence, the creation a complex pantograph-catenary multi-domain model should help to effectively meet the above-mentioned requirements. The work presents a co-simulation approach to investigate the pantograph-catenary dynamic interaction. The elaborated co-simulation algorithm assumes data exchange between multibody models of a pantograph and finite element model of a catenary. The presented approach explores multi-domain phenomena that have an influence on the pantograph-catenary interaction. The nonlinear finite element catenary model takes into account the slackening of droppers, relatively large displacements and contact with the pantograph's slider, while the multibody model of the pantograph considers friction forces and suspension springs. Additionally, aerodynamic forces caused by wind acting on the pantograph were computed using the fluid structure interaction method and implemented in the dynamic simulation. The influence on the pantograph-catenary interaction caused by electromagnetic force acting on the pantograph was investigated, along with the influence of the locomotive's vertical vibrations and tilt.

1 Introduction

Pantograph – catenary system is used commonly in electrified railways. Its effective performance is necessary for maintaining high speeds during train travel. The contact force, as well as its variation, is one of the parameters which describes the quality of the pantograph-catenary interaction. It is measured between the pantograph's head and contact wire (Fig. 1) and can significantly fluctuate during a ride due to vibrations in this complex system. Therefore, contact force fluctuations may cause unexpected drops of the current flow efficiency and further power decreases. Numerical simulations have become an essential tool in the analysis of a pantograph-catenary system. Exact modelling is crucial for conducting proper analysis of the performance in this structure, especially at the design stage of its components.

For many years, pantograph-catenary dynamic interaction has been the interest of many authors. The newest models for simulating catenary structure are mainly based on the Finite Element Method (FEM) and The Finite Difference Method (FDM) [1,2]. Finite element (FE) models allow one to consider the most important phenomena present in a catenary: wave propagation in the structure and nonlinearity (contacts, large displacements, slackening of droppers). In the case of the pantograph model, the newest publications [3], employ Multi Body Dynamics models (MB) because they allow them to consider a desired number of degrees of freedom (DOF), realistic kinematic behaviour and inertia properties of particular components.

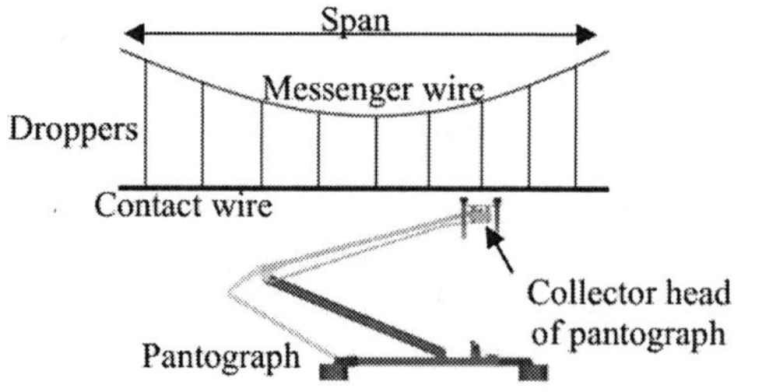

Fig. 1. Pantograph-catenary system.

* Corresponding author: zdziebko@agh.edu.pl

 DOI: 10.1051/itmconf/20171503001

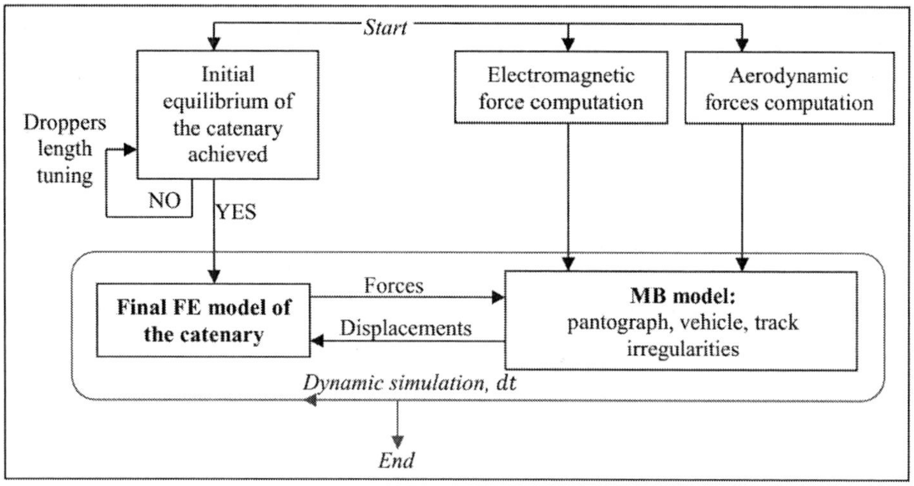

Fig. 2. Co-simulation algorithm.

The paper is organized as follows. After the present introductory in Section 1, Section 2 addresses the co-simulation algorithm proposed by the authors. The new approach considers multi-domain phenomena and it is an extension of the author's previous work [4]. Next, Section 3 describes the contributing models employed in the co-simulation. Subsequently, co-simulation results are presented in Section 4 and finally Section 5 concludes the paper.

2 Co-simulation algorithm

As mentioned before, the pantograph dynamic interaction with overhead lines (catenary) is disturbed by many different phenomena. The authors propose the co-simulation algorithm, which is presented in Fig 2. Firstly, three separate models are built:

- Initial FE model of a catenary – it is used to compute the final configuration of a catenary (with required pre-sag under gravity and tension of wires) by optimization of the droppers length.
- Model for electromagnetic force computation - this force acting on a collector strip is then passed into the MB model of a pantograph.
- Fluid structure interaction (FSI) model for aerodynamic forces computation - those forces acting on each component of a pantograph are passed into MB model.

Secondly, when final configuration of a catenary is achieved, the MB coupled track – rail vehicle – pantograph model is established. Finally, the dynamic solution of pantograph – catenary interaction is computed with certain time step dt in the co-simulation between FEM and MB codes. MB code is run first, and passes the displacement of the collector head (considering train speed and track excitations) into FEM code of a catenary. Subsequently, FEM code computes dynamic solution of a catenary under actual position of the collector head and passes the reaction forces acting on it into the MB code, where the MB model is calculated including forces delivered by FEM code. The procedure is repeated to cover the total simulation time [5].

3 Multi-domain models

3.1 FEM catenary model

The catenary structure employed in the analysis is similar to the one presented by the European Standard PN-EN 50318 [6]. It has a single contact wire and nine droppers per one span (which is 60 m long). The height of the traction catenary is 1.2 m and it has a stagger of ±0.2 m. For the pantograph-catenary dynamic interaction, ten identical spans were modelled (see Fig. 3.) Catenary wires are stretched by tensioning forces: 20 and 16 kN respectively for contact and messenger wires. The messenger wire is fixed to the poles while the contact wire is constrained by registration arms mounted to the poles by prismatic joints. Poles are not modelled (adequate constrains are between 'space' and particular node). A precise description of employed catenary system can be found in ref [6].

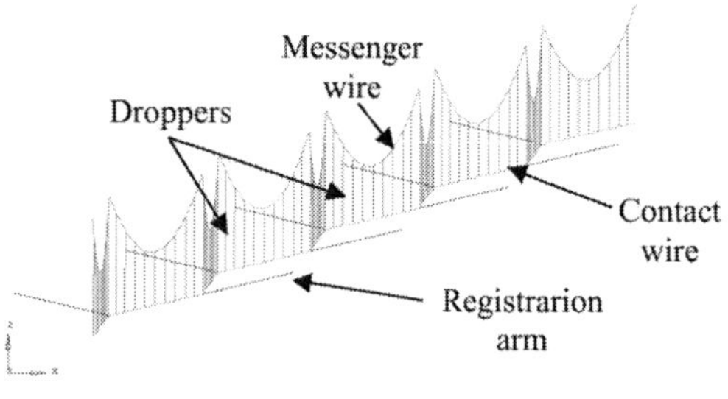

Fig. 3. Finite element model of the catenary.

FEM was employed for simulating the catenary system. The FE mesh was prepared in the Altair HyperMesh program, while the dynamic solution is computed in MSC.Marc solver (recommended for nonlinear models). The types of used FEs are listed in Table 1.

DOI: 10.1051/itmconf/20171503001

Table 1. Types of elements in the FE model.

Component	Element type
Contact wire	Beam, el. 98
Messenger wire	
Registration arms	
Droppers	Nonlinear spring

The employed beam elements (el. 98) include transverse shear effects. The detailed elements description can be found in [7]. In reality, droppers are made from steel cable and exhibit nonlinear characteristics. They can transfer tensioning load, but they are very compliant to compression (they do not transfer compressing loads). Therefore, the droppers are implemented in the model as nonlinear springs with the characteristics presented in Fig. 4.

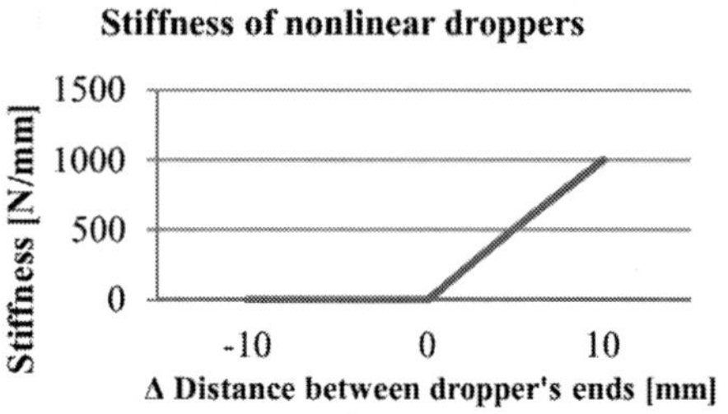

Fig. 4. Characteristics of nonlinear droppers.

Relatively large displacements are also considered in the FE model of the catenary. Dynamic transient solution of the proposed catenary model allows to consider wave propagation phenomena (and its reflections) in the structure. The wave propagation time was computed and it is 0.498s, therefore wave propagation velocity is equal to 433.7 km/h.

3.2 Multibody rail vehicle-pantograph model

The work concerns the pantograph mounted on the roof of a rail vehicle. The pantograph mechanism employed in this paper represents one which is commonly used in Europe - 160ECT (produced by EC Engineering). Basic kinematic diagram is presented in Fig. 5.

Dots in the image represent revolute joints (rotational axis is perpendicular to ZX plane), while triangle symbols represent montage points at the frame. Kinematic pairs in the pantograph mechanism are described in Table 2.

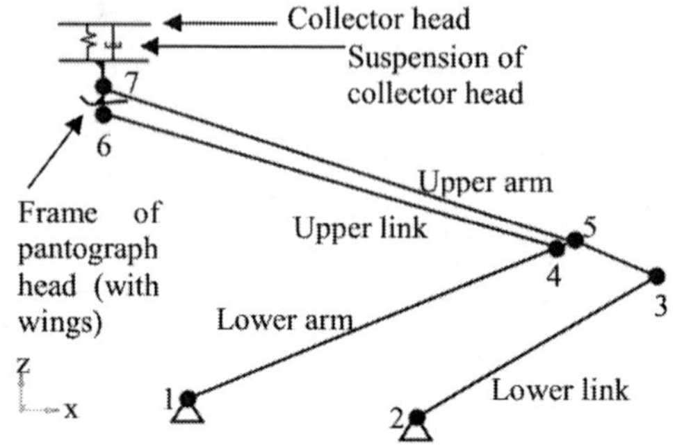

Fig. 5. Kinematic model of pantograph used in MB simulation.

Table 2. Revolute joints in pantograph.

No.	Body 1	Body 2
1	Frame	Lower arm
2	Frame	Lower link
3	Lower link	Upper arm
4	Lower arm	Upper link
5	Lower arm	Upper arm
6	Upper link	Frame of pantograph head
7	Upper arm	Frame of pantograph head

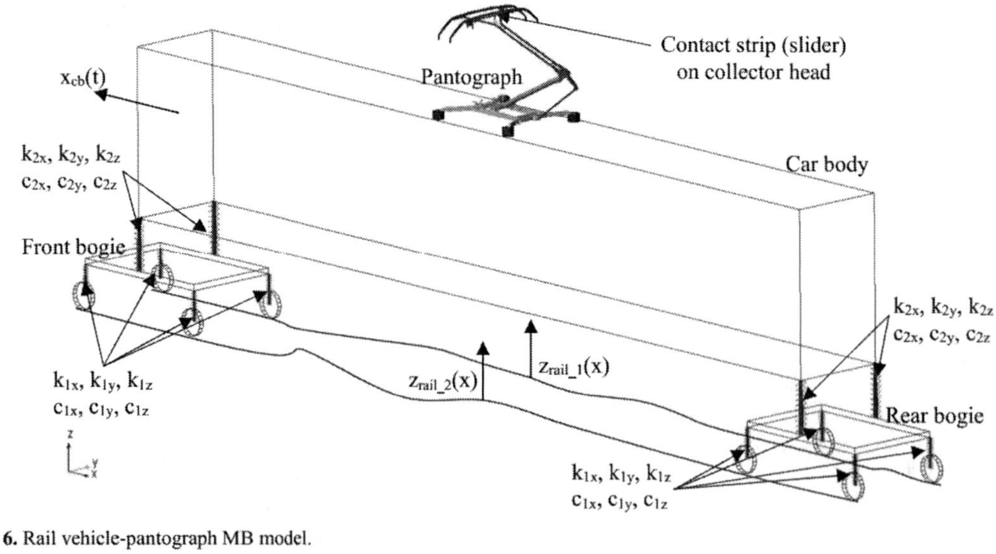

Fig. 6. Rail vehicle-pantograph MB model.

Friction coefficient was introduced into revolute joint no. 1, that total friction force in pantograph is compliant with the norm PN-EN 50206 [8].

The Authors propose to take into consideration vibrations and tilt of a car body on which the pantograph is placed. Those kinematic constrains can significantly disturb the pantograph- catenary dynamic interaction. Utilized model of a rail vehicle assumes 6 DOF of a car body, see Fig 6. Therefore, it allows us to take into account the most important directions of pantograph excitation from car body (vertical, lateral vibrations and tilt). The model of the rail vehicle was built according to [9]. The parameters of the two-level suspension system in the model are shown in Table 3.

Table 3. Parameters of rail vehicle suspension system.

Symbol	Description	Unit	Value
k_{1x}	Longitudinal stiffness of primary suspension	N/mm	2667
k_{1y}	Lateral stiffness of primary suspension	N/mm	3925
k_{1z}	Vertical stiffness of primary suspension	N/mm	880
c_{1x}	Longitudinal damping of primary suspension	Ns/mm	0
c_{1y}	Lateral damping of primary suspension	Ns/mm	0
c_{1z}	Vertical damping of primary suspension	Ns/mm	170
k_{2x}	Longitudinal stiffness of secondary suspension	N/mm	20
k_{2y}	Lateral stiffness of secondary suspension	N/mm	476
k_{2z}	Vertical stiffness of secondary suspension	N/mm	828
c_{2x}	Longitudinal damping of secondary suspension	Ns/mm	0.5
c_{2y}	Lateral damping of secondary suspension	Ns/mm	80
c_{2z}	Vertical damping of secondary suspension	Ns/mm	53

Other geometrical parameters as well as mass and inertial properties of the car body and bogies were taken according to the model from ref. [9]. Those values are consistent with the parameters of rail electric multiple units (EMU) utilized in Poland, therefore the authors consider that proposed rail vehicle model represents realistic constrains of a pantograph on high-speed trains.

The proposed model allows us to set the required course of displacement on the longitudinal direction of a train. However, in this paper, constant speed of 160km/h was set in all dynamic simulations.

Railway track irregularities are commonly considered to be an important source of vibration excitation in high speed train systems. Therefore rail irregularities on the vertical direction were taken into consideration in numerical analysis (independently for right and left rails, denoted as $z_{rail\_2}(x)$ and $z_{rail\_1}(x)$ in Fig 6, respectively). Moreover, each wheel is excited with regard to appropriate phase delay (resulting from the distance between them). Irregularities of a track are often considered as random variables and are described by power spectral density, which includes information of the amplitudes and wavelengths of a particular irregularity. In this paper, the authors used PSD formula describing track irregularities (eq. 1) according to ERRI B176 [10, 11].

$$S_A(\lambda) = 2\pi \frac{A_A \cdot \Omega_c^2}{\left(\left(\frac{2\pi}{\lambda}\right)^2 + \Omega_r^2\right)\cdot\left(\left(\frac{2\pi}{\lambda}\right)^2 + \Omega_c^2\right)} \tag{1}$$

where $\Omega_c = 0.82$ rad/m, $\Omega_r = 0.02$ rad/m are cut-off frequencies and λ is the wavelength in [m]. A_A in [m·rad] is the amplitude of track irregularities and value of $2.1 \cdot 10^{-7}$ is a guideline for good quality track, while the value of $6.1 \cdot 10^{-7}$ corresponds to low quality track. In this paper $A_A = 4.1 \cdot 10^{-7}$ m·rad was used and is interpreted as for a middle quality track.

On the basis of selected power spectral density, waveforms of track irregularities were computed using an inverse discrete Fourier transform with random phases. Used waveforms for right ($z_{rail\_1}(x)$) and left ($z_{rail\_2}(x)$) rails are presented in Fig. 7.

In normal operational conditions, the tilt of a car body is not present (it is on a negligible level), because track irregularities have too small amplitudes to significantly cause undesirable tilt of a car body. In reality, abnormal situations also show up, e.g. track cross level error (the situation when one rail is not at the exact same level like the second one) or uneven load distribution inside a coach can cause additional tilting of a car body which can additionally disturb pantograph-catenary dynamic interaction. Vibrations and tilt of pantograph's frame excited from a car body are presented in Fig. 8 and Fig. 9.

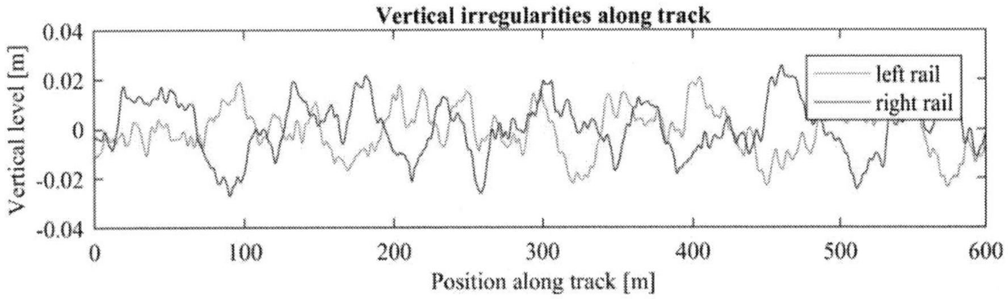

Fig. 7. Track irregularities for left and right rail.

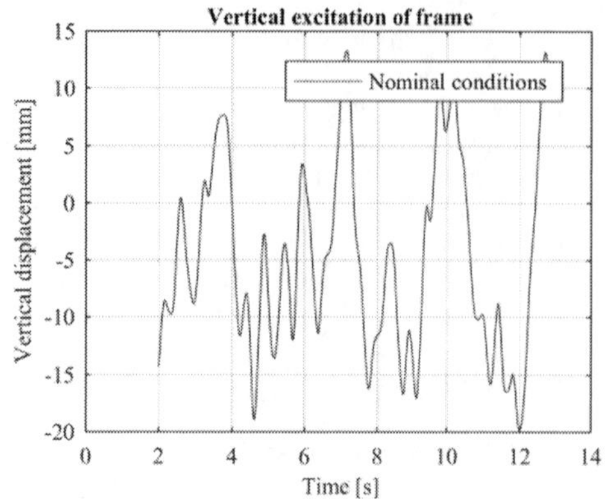

Fig. 8. Vertical displacement of pantograph's frame originating from car body vibrations.

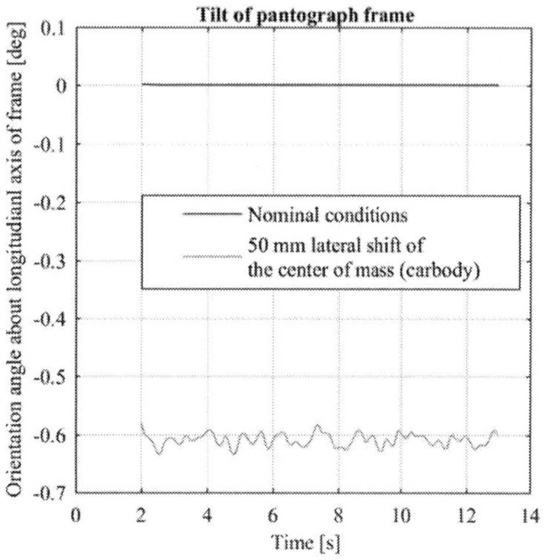

Fig. 9. Tilt of pantograph's frame originating from car body vibrations.

3.3 Aerodynamics

Wind flow is a source of large external forces acting on a pantograph mechanism in high-speed trains. Therefore, aerodynamic forces highly affect the nominal contact force (90-110N for DC systems according to [12]) between contact strips and contact wire. The static pressure exerted by a pantograph on a catenary wire should be kept within specific limits defined as functions of a train speed. On one hand, an increase of the pantograph static pressure makes the lifetime of contact strips shorter because of the high friction forces in that contact pair. On the other hand, at high-speed travel, the contact force shall be greater than for the normal operational condition to provide better contact at high speeds. The above-mentioned arguments confirm the need to take aerodynamic forces into account in pantograph-catenary dynamic simulations. In the paper, the authors prepared a numerical model of the pantograph for FSI analysis. Geometry of the pantograph was slightly simplified (small parts like screws and nuts were removed), but all important parts (frame, arms, links, pantograph head and wings) were modelled accurately.

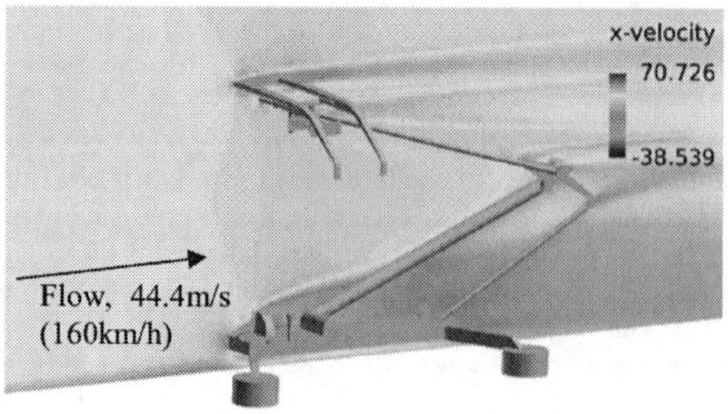

Fig. 10. FSI results: horizontal velocity in the middle plane.

The virtual wind tunnel has the following dimensions: 5x5x15m. The total number of elements in the model is 20300456. The model was computed in Altair AcuSolve solver, where a steady simulation strategy assumes Spalart-Allmaras turbulence model. In Fig. 10, velocity magnitude in the horizontal (longitudinal) direction is presented in the middle cross-section plane (legend is in m/s). Directly from the FSI analysis, drag, lift, and cross forces acting on each part of the pantograph were computed. Afterwards, those forces were introduced in the MB pantograph model and total uplift force, considering aerodynamic forces, was calculated. Finally, to validate the proposed FSI model, the obtained results were compared to those ones taken from the experiment in a real wind tunnel. The results of validation are presented in Table 4.

Table 4. Comparison of results from FSI model and experiment (flow speed: 160km/h).

Uplift force	
Proposed model	135.9 N
Experiment	136.4 N
Error	0.15%

The obtained results proof high accuracy of the proposed FSI model. Computed drag, lift and cross forces acting on components of the pantograph are considered to be within acceptable bounds.

3.4 Electromagnetic force

In normal operational conditions, a carbon strip mounted on the pantograph's collector head is in continuous contact with the contact wire, which carries current. In accordance with the principle of electromagnetic induction, a magnetic field is induced around the contact cable. The contact strip (through which the current also flows) is placed inside that magnetic field. Therefore, electromagnetic force is acting on the contact strip. This analysis aims to estimate that force. In the analysis, the magnetic field induced by three spans of a catenary were taken into consideration. The contact slider – contact wire schematic configuration is shown in Fig. 11, and it presents dimensions and currents in the

 DOI: 10.1051/itmconf/20171503001

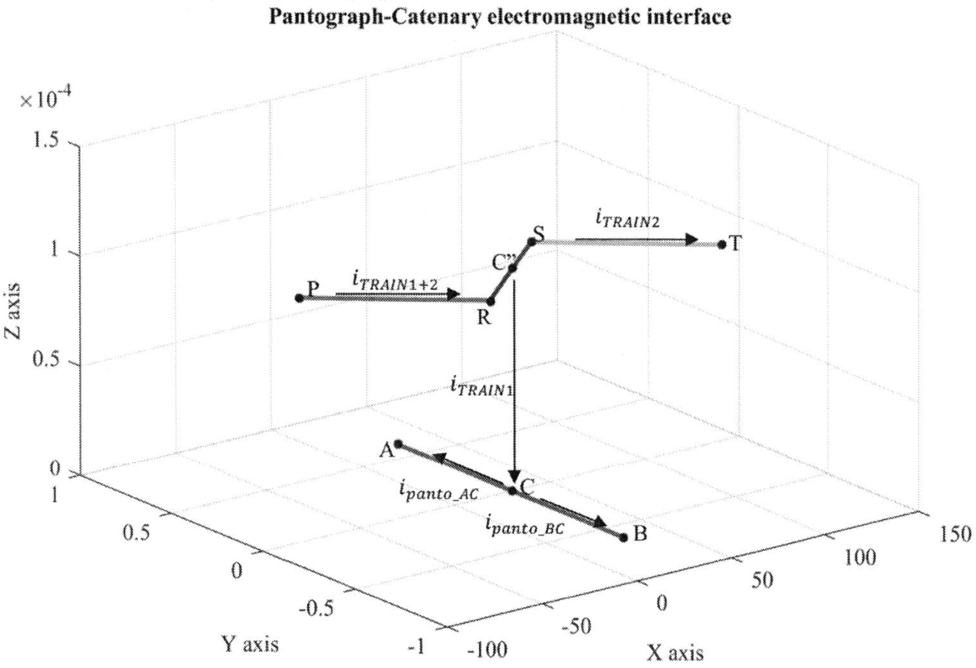

Fig. 11. Currents in electromagnetic pantograph-catenary interface. X axis: position along track [m], Y axis: dimensions of catenary stagger and width of contact strip [m], Z axis: distance between surfaces of contact wire collector strip [m] (global origin is located in the middle of contact strip). Different scales apply to the system coordinates.

employed structure. Section $|PRST|$ is the contact wire while $|AB|$ is the contact strip. The utilized scenario of energy supply assumes two trains travelling one by one which are powered from the same traction current converter plant. According to the European Standard [13] in high-speed lines (3 kV DC system) train currents may be as high as 3500A per one train. In the calculations the authors have taken following data:

$$i_{TRAIN1} = i_{TRAIN2} = 3500\ [A]$$
$$i_{TRAIN1+2} = 7000[A]$$
$$i_{PANTO\_AC} = \frac{i_{TRAIN1}}{2}[A]$$
$$i_{PANTO\_BC} = \frac{i_{TRAIN1}}{2}[A]$$

The distance between the slider strip and contact wire is to be $\overline{|CC"|} = 0.0001m$ and it represents irregularities in that interface [14]. The dimensions of the catenary spans correspond to the model presented in Section 3.1, while the length of the contact strip $\overline{|AB|} = 1.2m$. Magnetic induction at points on the contact strip is computed using the Biot-Savart's law. Segments of wires with current, which are taken into account in computing the magnetic field induction, are presented in Table 5.

The magnetic field in a particular point, e.g. B, can be computed using the Biot-Savart's rule. Let us first consider one finite length wire with current, e.g. $|RC"|$, see Fig. 12.

Table 5. Current in wires.

Wire with current	Current [A]
$\|PR\|$	$i_{TRAIN1+2}$
$\|RC"\|$	$i_{TRAIN1+2}$
$\|C"S\|$	i_{TRAIN2}
$\|ST\|$	i_{TRAIN2}

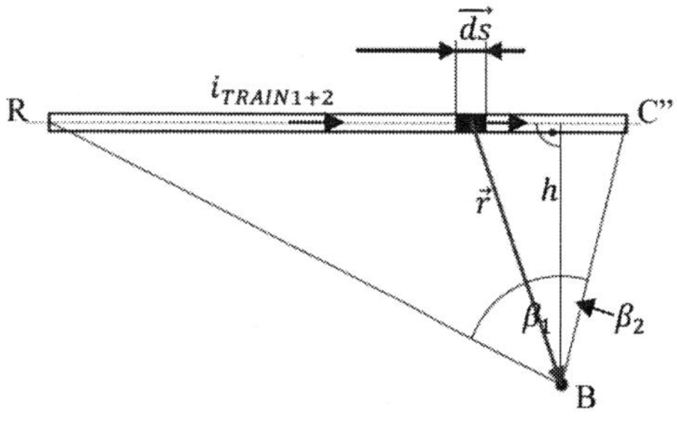

Fig. 12. Magnetic field from a finite length wire with current.

Magnetic induction at point B is (2):

$$\overrightarrow{B_{RC"}} = \frac{\mu_0\mu}{4\pi}\int \frac{i_{TRAIN1+2}\,\overrightarrow{ds}\times\vec{r}}{r^3} \qquad (2)$$

Where:

$\overrightarrow{B_{RC''}}$ – *magnetic field from wire* $|RC''|$

μ_0, μ – *permeability of vacuum and relative permeability of air, respectively*

$\overrightarrow{ds}$ – *current length element in dir. of current*

$\vec{r}$ – *the full displacement vector from the wire element to the point at which the field is being computed*

Solving the integral, we may find the magnitude of magnetic field at point B (h, β_1, β_2 are visible in Fig. 12):

$$\left|\overrightarrow{B_{RC''}}\right| = \frac{\mu_0 \mu\, i_{TRAIN1+2}}{4\pi h}(\sin\beta_1 + \sin\beta_2) \quad (3)$$

The direction of $\overrightarrow{B_{RC''}}$ is determined from the right hand rule. The magnetic field induced by other segments of wires is calculated in a similar way. The total magnetic field at point B is

$$\overrightarrow{B_{Total\_at\_B}} = \overrightarrow{B_{PR}} + \overrightarrow{B_{RC''}} + \overrightarrow{B_{C''S}} + \overrightarrow{B_{ST}} \quad (4)$$

The above procedure is repeated for all points at the contact strip ($|AB|$ is divided into n segments of the length dl - in the analysis $dl = 0.1mm$) and magnetic field at each segment $\overrightarrow{B_{seg}}$ is computed separately according to the procedure presented for point B.

The contact strip also carries current (i_{panto_AC} or i_{panto_BC} depending on which side the actual segment is located), therefore at each segment dl, electromagnetic forces acting on it can be computed from (5)

$$\overrightarrow{F_{seg}} = i_{seg}\overrightarrow{dl} \times \overrightarrow{B_{seg}} \quad (5)$$

The total electromagnetic force acting on the collector strip is

$$\overrightarrow{F_{TOTAL}} = \sum_{i=0}^{n} F_{i^{th}seg} \quad (6)$$

Components in the global coordinate system of electromagnetic force which are computed for, and presented in the above scenario, equal

$$F_{TOTAL\_X} = 31.6N, \quad F_{TOTAL\_Y} = 0N$$
$$F_{TOTAL\_Z} = -0.013N$$

The obtained results were implemented into the MB model. Electromagnetic forces with the largest longitudinal component slightly reduce the contact force between the collector head and contact wire. A graph representing its influence on the contact force for different working heights of the pantograph is presented in Fig 13. Vertical lines represent the lowest and highest working range of the pantograph (for Polish operational conditions). The highest influence of the electromagnetic field on the contact force is observed for working heights of between 1.5 to 1.8 m – in this range, electromagnetic force causes a reduction of the nominal pressure up to 2N.

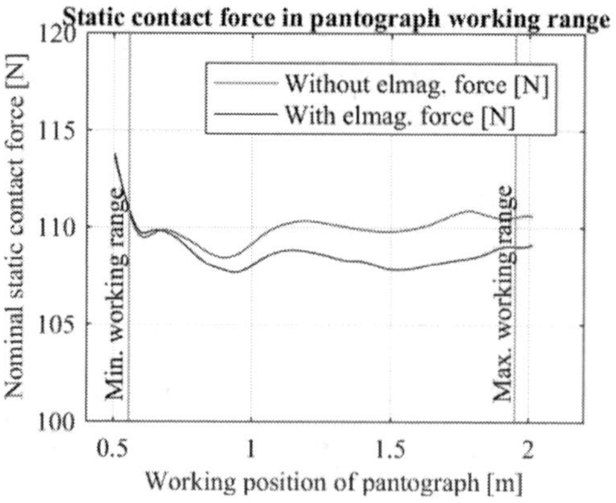

Fig. 13. Influence of electromagnetic force on actual static contact force in the pantograph working range.

4 Co-simulation results

The analysis of pantograph – catenary dynamic interaction was performed according to the proposed multi-domain co-simulation algorithm (Section 2). All important phenomena concerning this system were taken into account (precisely presented in Section 3). In the analysis, the static contact pressure exerted by the collector head on the contact wire was set to 100N and train travel velocity was equal to 160km/h. In order to assess the quality of the pantograph-catenary interaction, it is necessary to refer the results to reference values. According to Ambrosio et al. [15], the European Standard PN-EN 50367 describes important factors of the pantograph-catenary interaction quality (and their ranges of acceptance) by means of:

a. mean contact force: F_M
b. standard deviation of contact force: $\sigma < 0.3F_M$
c. max. contact force $F_{MAX} < 350N$
d. max. contact wire uplift at steady-arm $CW_{up} \leq 120$ mm
e. max. pantograph vertical amplitude $\Delta_z \leq 80$ mm
f. percentage of arcing NQ $\leq 0.2\%$

The results received in proposed co-simulation procedure are presented in Fig. 14 and 15.

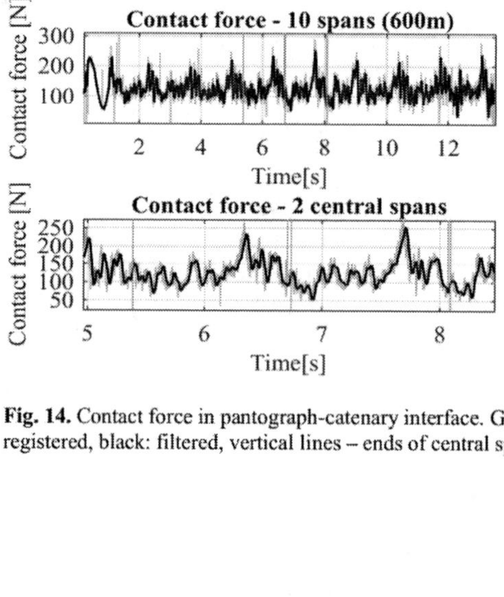

Fig. 14. Contact force in pantograph-catenary interface. Grey: registered, black: filtered, vertical lines – ends of central spans.

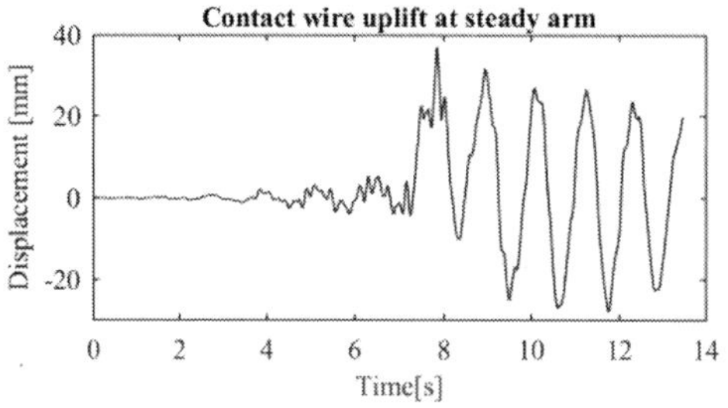

Fig.15. Contact wire uplift at steady arm between span no. 6 and 7.

According to [6] statistical parameters from the simulation were computed for the segment of two central spans after filtration (low-pass filtering, cut-off frequency 20Hz). The results are presented in Table 6. All parameters are within acceptance ranges.

Table 6. Results for pantograph – catenary interaction.

Parameter	Simulation	Allowed ranges [15]
Mean contact force, F_M	126.4N	108.4 – 134.8N (for v=160km/h)
Standard deviation of contact force, σ	34.8N	$\sigma < 0.3F_M$ $\sigma < 37.92N$ $(for\ achieved\ F_M)$
Max. contact force, F_{MAX}	253.9N	$F_{MAX} < 350N$
Max. contact wire uplift at steady-arm, CW_{up}	37mm	$CW_{up} \leq 120\text{mm}$
Max. pantograph vertical amplitude, Δ_Z	50mm	$\Delta_Z \leq 80\text{mm}$
Percentage of arcing NQ*	0	NQ ≤ 0.2%

*NQ is assumed to be a proportion of driving time, when the contact force is zero.

5 Conclusions

The main conclusions resulting from this paper are:

- In a pantograph-catenary interface, there are a number of different sources of disruption (vibrations of a contact wire and rail vehicle, aerodynamic and electromagnetic forces acting on a pantograph).
- Aerodynamic effects generate the highest external forces acting on a pantograph, which disturb its interaction with the catenary. FSI model was fully validated with the results from the wind tunnel experiment.
- 160ECT pantograph properly interacts with the catenary when the speed of train travel is equal to 160km/h. The pantograph is expected to have performance accurately at lower speeds because aerodynamic effects are less significant.
- The proposed co-simulation algorithm allows us to consider the following important phenomena that have a non-negligible influence on a pantograph-catenary interaction. These are: wave propagation in cables, nonlinear droppers, large displacements, real kinematic chain of a pantograph, aerodynamic and electromagnetic forces acting on a pantograph and excitation from a rail vehicle.
- The presented algorithm might be used in the optimization of a pantograph-catenary interaction in virtual prototyping approach and that will be the subject of future work.

The work was supported by the AGH Grant no. 15.11.130.627 and carried out employing the infrastructure of the Centre of Energy AGH (Czarnowiejska 36, 30-054 Kraków, Poland).

References

1. S. Bruni, J. Ambrosio, A. Carnicero, Y.H. Cho, L. Finner, M. Ikeda, S.Y. Kwon, J.P. Massat, S. Stichel, M. Tur, W. Zhang, Veh. Sys. Dyn. **53**, 412-435 (2015).
2. P. Zdziebko, T. Uhl, Arch. of Trans. **39**, 77-85 (2016).
3. Y. Song, H. Ouyang, Z. Liu, G. Mei, H. Wang, X. Lu, Mech. and Mach. Theory **115**, 35-59 (2017).
4. P. Zdziebko, A. Martowicz, T. Uhl, in: *Projektowanie Mechatroniczne: zagadnienia wybrane* (AGH, Kraków, 2017) (to be published) – in Polish.
5. MSC.Adams Co-simulation Interface (2017).
6. PN-EN 50318 (2002).
7. MSC. Marc Element Library (2016).
8. PN-EN 50206:2010.
9. K. Zboinski, M. Dusza, Non. Dyn. **89**, 863-885 (2017).
10. K. Karttunen, E. Kabo, A. Ekberg, Wear **314**, 78-85 (2014).
11. P. Kurowski, A. Martowicz, T. Uhl, G. Lasko, Arch. of Trans. **23**, 23-35 (2011).
12. PN-EN 50367:2012.
13. PN-EN 50388:2012.
14. Y. Zhao, G. Wu, G. Gao, W. Wang, L. Zhou, Jour. of the China Rail. Soc. **36**, 28-32 (2014) – in Chinese.
15. J. Ambrosio, J. Pombo, M. Pereira, Theo.& App. Mech. Lett. **3** (2013).

 DOI: 10.1051/itmconf/20171503002

Charging process analysis of an opposed-piston two-stroke aircraft Diesel engine

Łukasz Grabowski[1,*], *Konrad* Pietrykowski[1], and *Paweł* Karpiński[1]

[1]Lublin University of Technology, Faculty of Mechanical Engineering, Department of Thermodynamics, Fluid Mechanics and Aviation Propulsion Systems, Nadbystrzycka 36, 20-618 Lublin, Poland

Abstract. This paper presents the research results on a 1D model of an opposed-piston two-stroke aircraft Diesel engine. The research aimed at creating a model of the engine in question to investigate how engine performance is affected by the compressor gear ratio. The power was constant at all the operating points. The research results are presented as graphs of power consumed by the compressor, compressor efficiency and brake specific fuel consumption. The optimal range of compressor gear ratio in terms of engine efficiency was defined from the research results.

1 Introduction

Civil aviation widely uses four-stroke piston spark ignition engines to power light aircrafts such as: planes, helicopters, gyroplanes, unmanned aircrafts or UAVs. This paper shows the model of a supercharged two-stroke compression-ignition aircraft engine. This type of engine is discussed in few works only so we have decided to investigate it more broadly in view of the fact that the two-stroke compression-ignition opposed-piston Diesel engine shows many advantages, such as a simple design, a good balance, high thermal efficiency, high specific power and lower wear rate due to friction forces compared to the four-stroke row engine [1].

Its drawback is increased fuel consumption and higher emission of toxic substances compared to the four-stroke engine. It was therefore resolved that the two-stroke Diesel engine carbon dioxide emission ought to be reduced [2].

A supercharged system is a key component of the model of the entire engine because a correctly defined process boosts the results in a correctly mapped working process. A correctly mapped forced induction is fundamental due to the changes in pressure in the combustion chamber and engine performance. Forced induction can be carried out with a turbocharger or a mechanical compressor. The simulation and the experimental research into a turbocharger for car engine applications using a 1D modelling of the transient and intransient flow is discussed in [3]. The research showed that time lag in the model of a turbocharger enables us to more efficiently predict the change in pressure and oscillations in mass air flow because of more precisely calculated mass and energy stored in the device, which is obtained by a better described propagation of pressure waves.

In two-stroke engines, apart from parameters and the design of the forced induced system, cylinder scavenging is important. This process was investigated using a simulation model and in experiments [4]. The paper [5] discusses the numerical optimization of basic operational parameters of the two-stroke aircraft Diesel engine using a patented valve that improves scavenging.

Cylinder filling efficiency in two-stroke engines is inversely proportional to scavenging efficiency. Keeping balance between these two processes is fundamental for engine performance and brake specific fuel consumption. The optimization method involving improved performance and reduced fuel consumption for a forced induced compression-ignition two-stroke aircraft engine is described in [6].

The paper [7] presents different methods for force inducing the two-stroke aircraft Diesel engine. A mechanical compressor increases brake specific fuel consumption more than a turbocharger. However, this method of forced induction has a lower mass and dimensions, which is particularly important while designing aircraft structures.

According to [8], power absorbed by the compressor changes with the required amount of air and EGR valve and results from engine mapping. Two different gear ratios between the compressor and the drive shaft for different speeds and load conditions can improve engine performance.

The resulting engine power is affected not only by forced induction but also the nature of combustion. The researchers [9] investigated combustion in the opposed-piston two-stroke Diesel engine for heat release rate and combustion time.

This work presents a model of an engine and a forced induction system for the designed 100 kW opposed-piston two-stroke Diesel engine aircraft system and gives an account of preliminary simulation. The research aimed at using the created engine model to investigate the effect of compressor gear ratio on the engine performance.

*Corresponding author: l.grabowski@pollub.pl

The software used to develop the model was AVL BOOST, which is a tool dedicated for 1D modelling of piston internal combustion engines. This tool allowed us to map processes that occur in the combustion engine and parameters of the operating medium in the cylinder for the defined operating conditions and offered faster calculations than in 3D models. However, we could combine 1- and 3D models to obtain more information and more predictable simulation results [10]. Carlucci, Ficarella and Trullo [11] followed this approach to study the process of charge exchange in a two-stroke six-cylinder aircraft Diesel engine. The AVL BOOST also allows studying the correlation between the engine configuration and engine performance, *e.g.* power, torque, fuel consumption, *etc*.

2 Model of the engine

The model of the engine is created in the AVL BOOST. It was a zero-dimensional physical modelling based on the principle of conservation of mass and energy. The assumption is as follows: the 1D flow in the ducts means that the pressure, temperature and speed distribution calculated from the dynamic equations of gas correspond to average cross-sectional values in the duct. Flow losses due to phenomena that occur in the 3D space at individual points in the engine are considered as specific flow rates. This kind of model enables transient-state research that successfully reflects varied flight conditions, *e.g.* pressure, temperature. This model calculates the flows of air and fuel in the intake system and the cylinder including force inducing devices as well as combustion and the flow of exhaust gas to the environment.

The principles of the model of the engine are as follows:

1) a 1D model shows the length and diameter of engine elements for the flow only;

2) a model of the two-stroke engine;

3) identical geometries of all three cylinders;

4) twice as large surfaces of piston bottoms than in reality to reflect the nature of the operation of the opposed-piston engine;

5) geometric orientation of cylinders is neutral to their performance;

6) temperature of the component walls is constant;

7) the system for charge exchange is symmetric;

8) fuel is supplied directly to the combustion chamber forming a mixture of a given composition.

The AVL Boost model of the engine required some simplifications due to the nature of the 1D mathematical model. These are as follows:

1) the flow of medium through the intake system and exhaust passes mainly round-section ducts;

2) the components of a large volume and a small distance between the outlet and inlet sections to be modelled by means of the Plenum (constant-volume element);

3) the fuel flow depends on mass flow rates and is automatically regulated to maintain a desired mixture composition.

Our model shows the entire route of the charge flow between the air inlet and outlet. The model reflects the process of fuel injection into the cylinders and combustion that occurs there. The calculation uses the module for the two-stroke engine which consists of (Fig. 1): an air inlet, a compressor connector, a compressor, an air cooler, intake and outlet ducts, cylinders (changes in cylinder volume), and Plenum-created elements (of fixed volume). Preliminary research of 3D models of charge exchange enabled us to introduce similar-to-a-target-solution dimensions of the parts of the intake and outlet system including channels in the engine block. To obtain maximum filling, other elements of these systems were optimized.

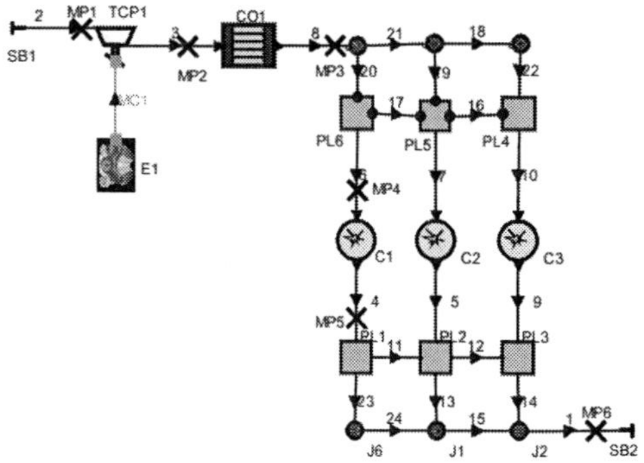

Fig. 1. Model of the engine.

The engine model is based on the mass flow in the intake and outlet ducts, and refers to the equations for the isentropic circular cross-section flow. Mapping the geometry of the intake ducts and other key elements is performed with a view to obtaining as favourable as possible conditions for engine operation. The model also is based on the assumption of mechanical radial compressors. To correctly model charge exchange, there are introduced the characteristics of window opening. The model includes the mechanical centrifugal compressor (radial), whose key function is to increase the degree of engine filling (volumetric efficiency). Supplying more air into the engine (compared to a naturally aspirated engine) contributes to obtaining higher indicated mean efficient pressure. The use of compressors in the two-stroke engine is necessary to provide for charge exchange.

The model of an air cooler is between the compressor and the Plenary element 1 placed before the cylinders. There are defined the dimensions, mass flow rate and temperatures of the air inlet, coolant and air outlet.

The compression ratio was defined by the correlation of the cylinder volume and the combustion chamber and the combustion chamber volume, which means that theoretical coefficient of compression is specified so as to ensure highly efficient scavenging.

The parameters of heat transfer into the cylinder walls were defined in line with the data from the literature and the Woschni heat transfer model, entered into the applied software (1978).

 DOI: 10.1051/itmconf/20171503002

The model of combustion is based on the Vibe model. The model of heat release is calculated from the semi-empirical formula with the given parameters. The following values of the parameters are applied in the model: combustion start -3 deg, combustion time 35 deg, shape m = 1, parameter a = 6.9. These parameters were specified with reference to the data from the literature [12].

The friction resistance was also defined in the engine by specifying the mean friction pressure.

3 Model of the supercharger system

Modelling circulation parameters of the forced induced engine requires creating and implementing mathematical models for individual elements of the system, including a centrifugal compressor. The mapping of a compressor with its real characteristics is affected much by the quality of mapping its performance under varied load conditions, which can be successful if precise calculations and relevant physical characteristics of the analyzed phenomena are included in the mathematical description.

Forced induction is also important due to pressure changes resulting from the use of aircraft engine (in line with the ISA). The compressor can increase pressure when filling the engine so the required power at the nominal altitude is maintained.

If analyzing the requirements and principles behind the model of thermodynamic process in terms of forced induction, the first aspect is that a compressor drive in this model is a mechanical connection with a crankshaft. Due to a direct connection of a compressor and a crankshaft, a mechanically forced induced engine reacts faster to considerable changes in load due to rapid changes in power equipment settings than the turbocharged engine. However, a compressor drive consumes a significant part of the engine-generated energy (up to 10%) and reduces overall circulation efficiency. Therefore, the transfer of energy to compressor drives should be included in the model.

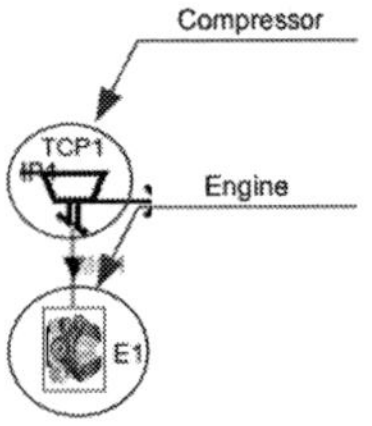

Fig. 2. Model of a compressor connected with the model of an engine.

The first stage of 1D modelling was the application of a simplified model of forced induction. Here, compression, isentropic efficiency and mechanical efficiency were defined. These values were, however, constant throughout the operating cycle, independent of the instantaneous changes in the mass flow rate. This model also disregards the impact of engine speed on the performance of a charging compressor. These values were as follows: compression π = 1.28; isentropic efficiency η_v = 0.8; mechanical efficiency η_v = 0.97. These values were as preliminary in the calculations and were specified from the literature data [13] and verified in the models of engines.

The model of a boost system uses Rotrex C30-94 compressor of a planetary transmission gear, capable of increasing engine speed several times, the required compression ratio (approx. 1.3) and the mass flow rate (0.145 kg/s) for the maximum power (100 kW) fall within the range of compressor efficiency above 65%.

To obtain the required engine power (100 kW), the gear ratio of the compressor is 1:13.9. A real compressor has a built-in reducer of a gear ratio of 1:9.5 so an additional gear belt ratio of 1:1.46 should be applied. For such a gear ratio and crankshaft speed of 4000 rpm, the rotor speed of the compressor is about 55500 rpm.

Fig. 3 shows the changes in compression and mass flow of a boosting compressor in a single operating cycle. The compression ranges from 1.26 to 1.42 and mass flow rate from 0.085 to 0.2 kg/s. These fluctuations result from wave phenomena in the intake system.

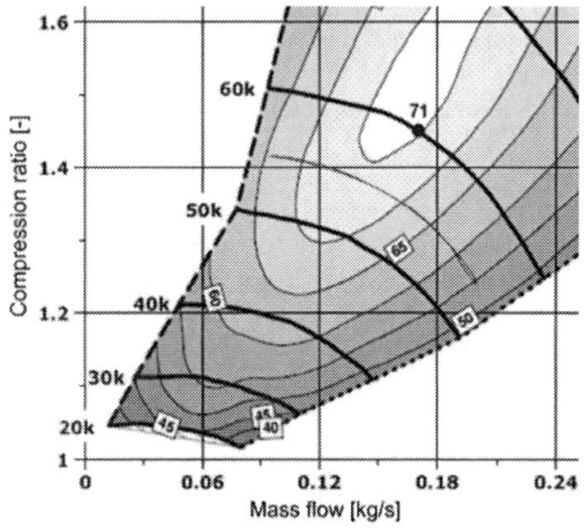

Fig. 3. Selected characteristics of the C30-94 compressor [14].

Pressure pulsations were introduced intentionally by appropriately selected dimensions of the intake system. This solution enables a dynamic forced induction and a reduced power to drive the turbocharging compressor.

When this characteristic of the compressor was applied the power to drive it increased from 4.2 kW to 6.7 kW so specific fuel consumption also rose from 203 to 207 g/kWh. This is due to the assumed in the earlier calculation high isentropic efficiency and pressure pulsation. Pressure pulsation triggers temporary changes in compressor efficiency, which reduces its mean efficiency and increases the demand for the power to drive the compressor.

4 Studying the results

This section discusses the research results on performance of our engine versus the charging compressor gear ratio.

For a fixed geometry of the intake system, calculations were conducted for different values of forced induced

 DOI: 10.1051/itmconf/20171503002

pressure changed by modifying the ratio of the gear driving the compressor. The AVL BOOST has no option to set the ratio of an additional gear between a crankshaft and a compressor. Therefore, the gear ratio of the compressor was regarded as the total gear ratio if mechanical efficiency is 98%. To obtain the required cruise power N = 73 kW, the dose of fuel was changed in each case to obtain higher AFR (Fig. 5). The calculations were made for 7 variants of the supercharger pressure (Table 1) ranging from pd_1 = 2.21 bar to pd_7 = 1.35 bar. The environmental conditions were defined by pressure p_o= 1 bar and temperature t_o = 15 C. Due to the value of ambient pressure, the forced induced pressure and compression in the subsequent cases are identical.

Table 1. Summary of the configuration of the forced induced pressure.

No.	Symbol	Total gear ratio [-]	Forced induced pressure [bar]	AFR [-]
1	pd_1	1:25	2.21	32.70
2	pd_2	1:22.2	1.92	32.28
3	pd_3	1:20	1.73	30.92
4	pd_4	1:18.2	1.59	29.54
5	pd_5	1:16.7	1.49	28.70
6	pd_6	1:15.4	1.41	26.27
7	pd_7	1:14.3	1.35	24.43

Figure 4 shows the operating points for the research variants of forced supercharger pressure. In all the cases, similar ranges of variation of supercharger pressure were recorded and we could therefore conclude that this phenomenon is correlated with the dimensions of intake ducts.

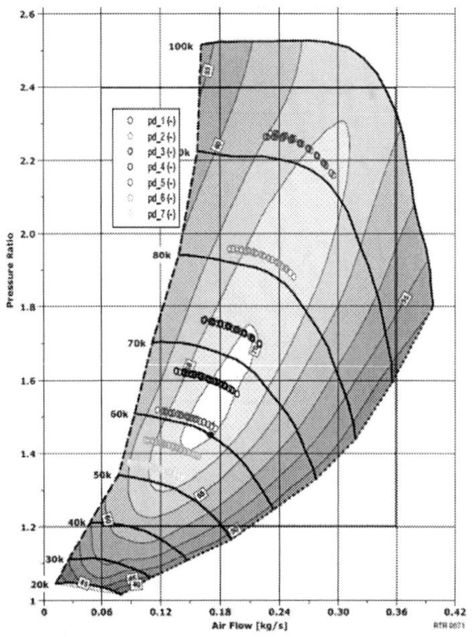

Fig. 4. Operating points for different variants of the forced induced pressure in the characteristics of the Rotrex C30-94 compressor [14].

The highest mean compressor efficiency was obtained for variants pd_3, pd_4, pd_5 (Figure 8) which were more than 69%. The lowest efficiency was obtained for the maximum forced induction (version pd_1) of about 64%. The efficiency in the last two variants, *i.e.* pd_6 and pd_7 is slightly lower and amounts to approximately 1%. Fuel mass given in mg/operating cycle (Fig. 7) increases from 71 g at 1.35 bar to 90 g at 2.21 bar, which results from the fact that the compressor needs more mechanical power.

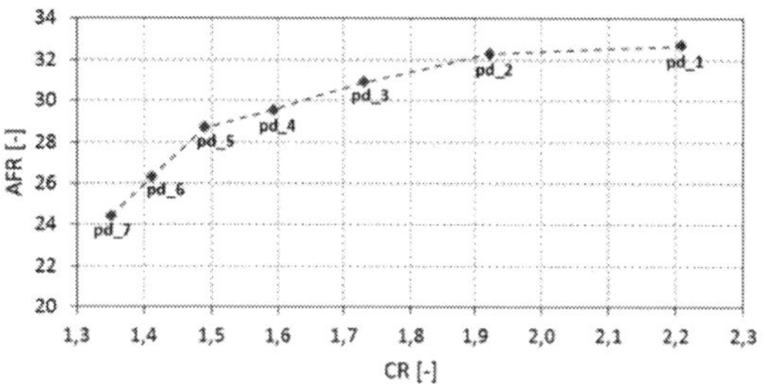

Fig. 5. AFR vs. compression ratio.

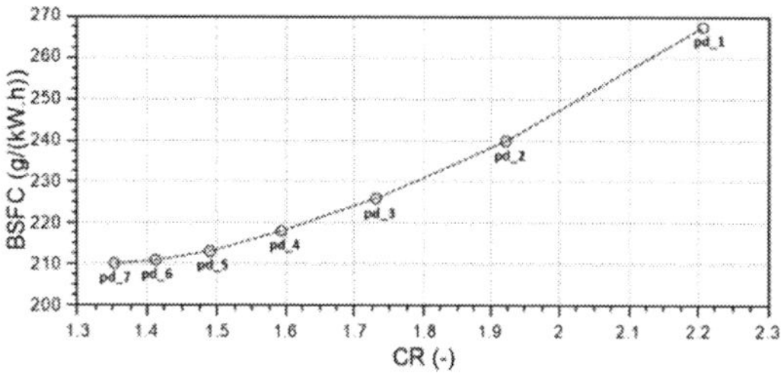

Fig. 6. Brake specific fuel consumption vs. compression ratio.

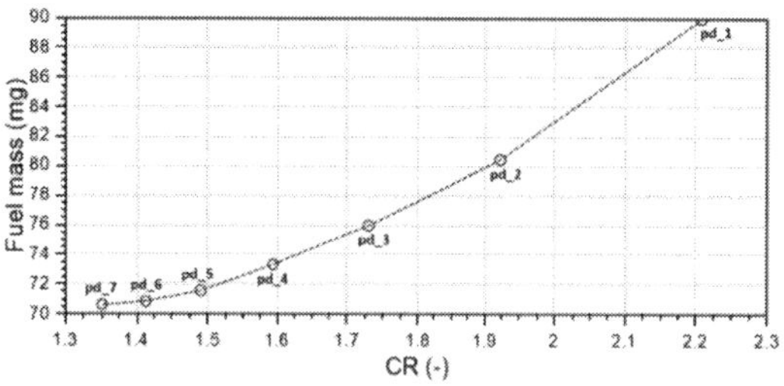

Fig.7. Supplied fuel mass per operating cycle vs. compression ratio.

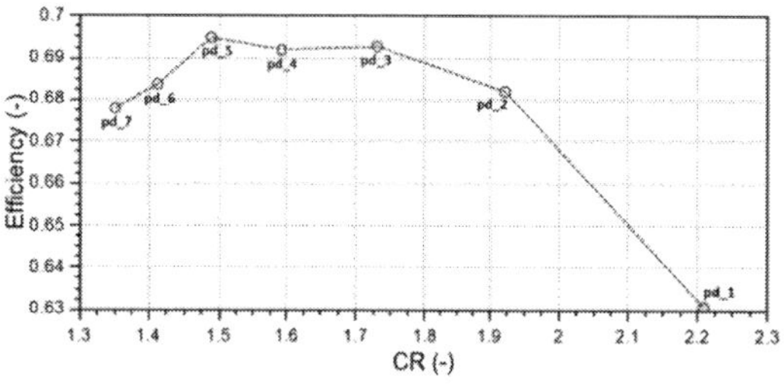

Fig. 8. Compressor efficiency vs. compression ratio.

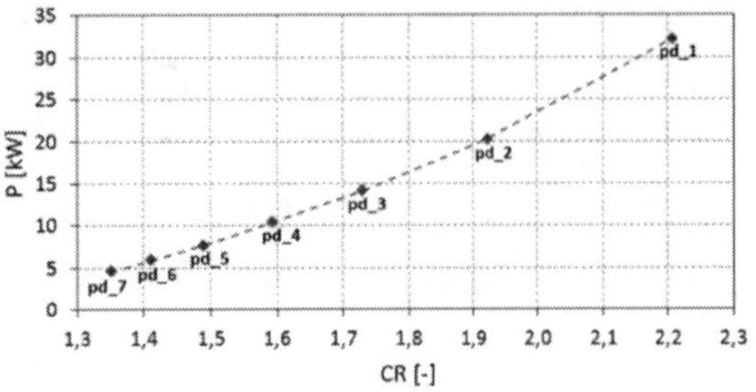

Fig. 9. Mechanical power consumed by the compressor vs. compression ratio.

5 Conclusions

Regarding the mechanical power needed by the compressor (Fig. 8), high compression deteriorates the overall efficiency of the engine (see Figure 3). Figure 6 shows the change in the specific unit fuel consumption from 210 g/kWh to nearly 270 g/kWh, which reduces the overall efficiency from 41% to 32% compared to the unchanged indicated efficiency.

Any further analysis and stand research should refer to the last four variants of forced induction, *i.e.* pd_4, pd_5, pd_6, and pd_7 because for higher pressure, the engine needs too much power so specific fuel consumption increases.

This work has been realized in the cooperation with The Construction Office of WSK "PZL-KALISZ" S.A." and is part of Grant Agreement No. POIR.01.02.00-00-0002/15financed by the Polish National Centre for Research and Development.

References

1. J. P. Pirault, M. Flint, *Opposed Piston Engines: Evolution, Use and Future Applications* (SAE International, 2010)
2. P. Tribotte, F. Ravet, V. Dugue, P. Obernesser, N. Quechon, J. Benajes, R. Novella, D. De Lima, Procedia Soc. Behav. Sci. **48**, 2295–2314 (2012)
3. V. DeBellis, S. Marelli, F. Bozza, M. Capobianco, Enrgy Proc. **45**, 909–918 (2014)
4. Y. Liu, F. Zhang, Z. Zhao, Y. Dong, F. Ma, S. Zhang, Appl. Therm. Eng. **104**, 184–192 (2016)
5. G. Cantore, E. Mattarelli, C. A. Rinaldini, Enrgy Proc. **45**, 739–748 (2014)
6. P. Carlucci, A. Ficarella, D. Laforgia, G. Trullo, Enrgy Proc. **82**, 31–37 (2015)
7. P. Carlucci, A. Ficarella, D. Laforgia, A. Renna, Energy Convers. Manag. **101**, 470–480 (2015)
8. S. Naik, D. Johnson, J. Koszewnik, L. Fromm, F. Redon, G. Regner, K. Fuqua, Practical Applications of Opposed-Piston Engine Technology to Reduce Fuel Consumption and Emissions, SAE Technical Paper 2013-01-2754 (2013)
9. F. Ma, Ch. Zhao, F. Zhang, Z. Zhao, Z. Zhang, Z. Xie, H. Wang, Energies **8**, 6365-6381 (2015)
10. J. Bohbot, M. Miche, P. Pacaud, A. Benkenida, Oil Gas Sci. Technol. – Rev. IFP **64(3)**, 337-359 (2009)
11. P. Carlucci, A. Ficarella, G. Trullo, Energy Convers. Manag. **122**, 279–289 (2016)
12. G. P. Blair, *Design and Simulation of Two-Stroke Engines* (Society of Automotive Engineers, 1996)
13. J. B. Jones, R. E. Dugan, *Engineering Thermodynamics* (Prentice Hall, 1995)
14. www.rotrex.com/media/7df7470f-60fb-4413-837e-ebe527259be4/PDF/Rotrex%20Technical%20Datasheet%20C30%20Range_pdf

Comparison of macroscopic descriptions of magnetization curves

Radosław Jastrzębski[1,*] , and *Krzysztof* Chwastek [1]

[1]Częstochowa University of Technology, Faculty of Electrical Engineering, 42-201 Częstochowa, Poland

Abstract. The paper considers a qualitatively different behaviour of two phenomenological hysteresis models. The first one is the widespread Jiles-Atherton description, which is based on the "effective field" concept. The other model is the proposal by the Brazilian research team GRUCAD. First order reversal curves simulated with the latter formalism do not exhibit negative slopes. This feature is in accordance with the experiment.

1 Introduction

Among different descriptions of magnetization curves [1] the phenomenological Jiles-Atherton (JA) model [2,3] has attracted a lot of attention of the scientific and engineering community due to a relatively simple mathematical form and physical interpretation attributed to its few parameters. However, soon after the JA model gained popularity, many researchers have come to the conclusion that this description has a number of drawbacks. The original iterative estimation procedure proposed by Jiles *et al.* [4] sometimes gave non-consistent sets of model parameters [5, 6]. Therefore most of the research focused on alternative methods for estimation of model parameters, and was usually based on artificial intelligence approaches [5-16]. At the same time several authors realized that in order to obtain a reliable representation of more complex magnetization cycles, including minor loops, reversal curves *etc.* it would be necessary to update the values of model parameters in accordance with certain functional dependencies [6, 10, 17-20]. The abundance of different model versions and the focus set mostly on the advantages of one estimation method over another have blurred serious problems with the model equations themselves.

The model developed by researchers from a Brazilian group GRUCAD [21-23] was originally meant as a „patch'' to the JA description. In our paper we shall show that the introduced modifications introduce a new quality.

2 JA model equations

The fundamental ideas behind the JA model are the concept of anhysteretic curve and the effective field. Anhysteretic curve is the purely reversible $M-H$ dependence. Jiles and Atherton suggested for its description the use of the so-called modified Langevin curve, given with Eq. (1)

$$M_{an} = M_s\left(\coth\frac{H_{eff}}{a} - \frac{a}{H_{eff}}\right), \qquad (1)$$

where a (A/m) and M_s (A/m) are model parameters, whereas H_{eff} is the so-called effective field interpreted as the true field with the magnetic material, different from the applied field. The specification "modified" in the name of the curve is to focus the attention of the user on this fact (ordinary Langevin function uses the applied field H as its argument). The effective field is given in the first approximation with the relationship

$$H_{eff} = H + \alpha M, \qquad (2)$$

where dimensionless α is yet another model parameter, describing the mutual interactions between magnetic moments within the material. It should be recalled that the concept of effective field enables incorporating in the description the cooperative contributions from different physical phenomena affecting the shape of hysteresis loop *i.e.* $H_{eff} = H + \alpha M + \sum_i H_i$, thus a generalized model in the self-consistent form $M = M(H_{eff})$ [24] may easily be created. For example Sablik and Jiles [25] suggested introducing an additional field term related to magneto-mechanical coupling; this model extension has attracted a lot of attention of researchers concerned with non-destructive testing methods [26, 27]. The idea of effective field may also be helpful when explaining the peculiarities of magnetization process in novel soft magnetic materials like Fe-based composites [28In general, the concept allows for qualitative and quantitative description of the effects from complicated factors and phenomena, which would otherwise be very difficult to analyze.

* Corresponding author: radoslaw@jastrzebski.info

DOI: 10.1051/itmconf/20171503003

The fundamental equation of the JA model relates the derivative of total magnetization with respect to the effective field to the weighted difference between anhysteretic and total magnetizations, which may be written as an ordinary differential equation (ODE)

$$\frac{\mathrm{d}M}{\mathrm{d}H_{eff}} = \frac{M_{an} - M}{k\delta}, \quad (3)$$

where $\delta = \pm 1$ is introduced to distinguish the ascending and the descending hysteresis loop branches, whereas k is a model parameter (if expressed in (A/m) taking values comparable to quasi-static coercive field strength). In certain versions of the model instead of total magnetization M, its component related to irreversible magnetization processes is used. These modifications do not alter significantly the logical derivation of the final formula or the qualitative model behaviour. A step-by-step derivation is provided *e.g.* in [29-30], whereas the differences between subsequent versions and their consequences are discussed in [31].

As pointed out in the previous paragraph, the model developers attempted to decompose total magnetization into its reversible and irreversible components. The original paper [2] did not include this idea, however most cited papers on the JA model, [3] and [4], offered such a possibility. The formulas in both papers showed slight differences, yet it can be stated that the decomposition was performed by weighting the appropriate contributions with a new dimensionless parameter c. The interpretation of this parameter was the ratio of differential susceptibility $\mathrm{d}M/\mathrm{d}H$ to differential anhysteretic susceptibility $\mathrm{d}M_{an}/\mathrm{d}H$ at the origin of the coordinate system. At this point it should be remarked that the Langevin function has a singularity at zero, which makes the parameter imprecisely defined. The Maclaurin expansion only partially resolves the problem. The value of the c parameter is subject to unexpected, sudden variations if a stochastic optimization routine applied for solving the estimation problem is repeated several times to ensure convergence [32].

Another paper [33] criticized the approach based on decomposition of total magnetization, considering it artificial and borrowed from the Preisach model [34]. In order to improve the representation of minor loops including asymmetric ones, Chwastek introduced a decomposition of total susceptibility into two terms, but both were modulated with an instant magnetization value, thus the modified JA description [20] resembled to a certain extent the product of the Preisach model [34]. It was, however, necessary to update the values of two model parameters depending on the magnetization level.

Yet another modification had to be introduced, namely the introduction of additional pseudo-variable δ_M in front of the difference $M_{an} - M_{irr}$, which appears in an equation similar to Eq. (3). The definition of δ_M is given below:

$$\delta_M = \begin{cases} 0: \mathrm{d}H/\mathrm{d}t < 0 \text{ and } M_{an} - M_{irr} > 0 \\ 0: \mathrm{d}H/\mathrm{d}t > 0 \text{ and } M_{an} - M_{irr} < 0 \\ 1: \text{in all other cases} \end{cases}, \quad (4)$$

or in a more brief notation:

$$\delta_M = 0.5\left[1 + \mathrm{sign}\left((M_{an} - M_{irr})\mathrm{d}H/\mathrm{d}t\right)\right]. \quad (5)$$

In the relationships (4) and (5) there might be slight modifications relative to the model version, namely the difference might be $M_{an} - M$ and $\mathrm{d}H/\mathrm{d}t$ may be replaced with $\mathrm{d}B/\mathrm{d}t$ for the inverse model [20, 35]. The meaning of δ_M pseudo-variable may be inferred from Fig. 1. Its role is to suppress the irreversible magnetization component after a sudden field reversal. Such a modification may be partially justified with the words by E. Della Torre *et al*: "In general, the slope of the $M - H$ curve is different before and after a turning point. The reason for the difference in slope is that before the turning point, the change in magnetization is due to both the reversible and the irreversible processes. After the turning point, only the reversible process contributes. This is due to the fact that the irreversible process does not return any energy to the applied field; hence, $\mathrm{d}M/\mathrm{d}H$ is zero for this process" [36].

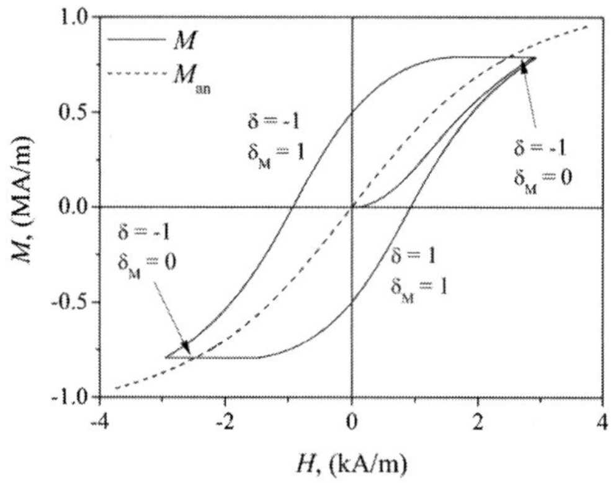

Fig. 1. A hysteresis loop and an anhysteretic curve simulated with the JA model that includes δ_M term.

The necessity to patch JA model equations with the δ_M pseudo-variable described above was mentioned for the first time in the paper on estimation [3] and explicitly appeared in Ref. [37]. However, it cannot be traced back to any ideas inherent in the model and thus should be considered as an artificial means to patch the JA equations. Without this pseudo-variable, after a sudden change of the sign of the magnetic field a loop fragment with a negative differential susceptibility would be observed. This could lead to the instability of FEM codes when calculating the local field distribution taking into account the hysteresis of the medium [38, 39].

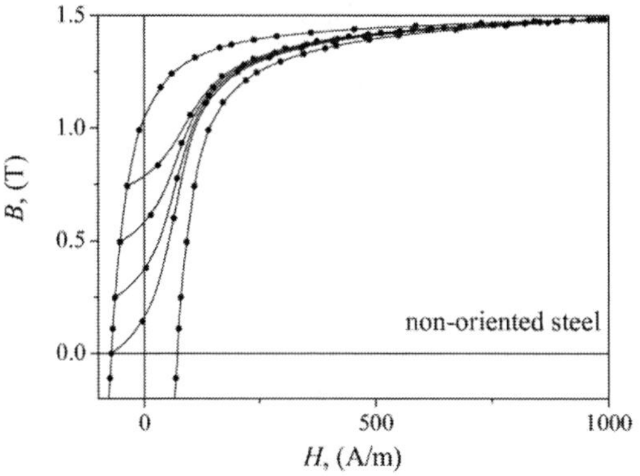

Fig. 2. Experimental first order reversal curves for a non-oriented steel measured using Epstein frame [40].

Experiments were carried out on custom- designed laboratory set-ups, enabled introducing bias signal to the excitation, thus allowing us to obtain asymmetric minor loops and reversal curves indicate unambiguously that after sudden field reversal the slope of the $M-H$ curve is non-zero, cf. Fig. 2. Therefore, the introduction of δ_M pseudo-variable is not fully justified. The term has been introduced *post hoc* to the quasi-static model equations to remove the undesirable model behaviour and its existence is in a strict connection to the form of the right hand side expression in (3), whose interpretation may be that the instant value of magnetization $M(t)$ on the hysteresis loop is obtained by introducing an appropriate offset from the instant value of anhysteretic magnetization $M_{an}(t)$.

Yet another interesting problem concerns the following point: both quantities are implicitly related by time (they should be calculated for the same time instants). We are therefore faced with two possibilities: either the anhysteretic (hysteresis-free) curve in reality does exhibit hysteresis if we follow rigorously all definitions, or that it should be calculated from a self-consistent implicit relationship similar to (1), as suggested *e.g.* in Refs. [41, 42]

$$M_{an} = M_s\left(\coth\frac{H+\alpha M_{an}}{a} - \frac{a}{H+\alpha M_{an}}\right). \quad (6)$$

It should be noted that in the latter case the „effective field'' has been implicitly redefined, which may have had consequences for the derivation of the final formula for differential susceptibility. Moreover, for certain sets of model parameters M_s, α and a the curve (6) may become S-slanted and go into the second and fourth quadrant of the $M-H$ plane, which in turn implies the non-physical shape of the hysteresis loop or failure of the estimation procedure [33, 39]. More information on the issue may be found in the Appendix.

Considering the first of the aforementioned cases, *i.e.* all definitions of respective quantities are rigorously followed, it is instructive to follow the logical scheme of calculations for the case of the so-called inverse model, *i.e.* when in the calculations the input quantity is magnetic flux density B [30, 35, 43]. The toned for using the inverse model stems from the fact that in magnetic measurements, carried out in accordance with international standards IEC 60404, the shape of magnetic flux density waveform should be kept sinusoidal. Three quantities B, M and H are related to each other with the constitutive relationship $B=\mu_0(H+M)$, where μ_0 is free space permeability. Now treating $B(t)$ as the input quantity to the inverse model, the calculation sequence is as follows:

1. define B_{max}
2. set up initial values $M(t=0)\cong 0, H(t=0)\cong 0$
3. calculate $H_{eff}(t)=H(t)+\alpha M(t)$
4. calculate $M_{an}(t)=M_s(\coth H_{eff}(t)/a - a/H_{eff}(t))$
5. calculate instant value of dM/dB and for the prescribed dB/dt waveform calculate dM/dt
6. integrate dM/dt, thus determine the instant value of $M(t+\Delta t)$
7. calculate the value of field strength for the successive time instant $H(t+\Delta t)$ from the transformer constitutive relationship $H(t+\Delta t)=B(t+\Delta t)/\mu_0 - M(t+\Delta t)$
8. assume next time step and go to step 3.

Integration of model equations may be carried out using standard library Runge-Kutta routines. After several time steps the vectors $B(t), M(t)$ and $H(t)$ are created in computer memory.

Next the vectors $H_{eff}(t)=H(t)+\alpha M(t)$ and $M_{an}(t)=M_s[\coth H_{eff}(t)/a - a/H_{eff}(t)]$ are set up. The anhysteretic magnetization $M_{an}(t)$ may be represented on the $M-H$ plane as in Fig. 1, where it took the sigmoid shape, as in fact the dependence $M_{an}=M_{an}(H_{eff})$ was depicted with a broken line, or as in Fig. 3, where the dependence $M_{an}=M_{an}(H)$ is drawn. For creating both figures the same set of model parameters was used.

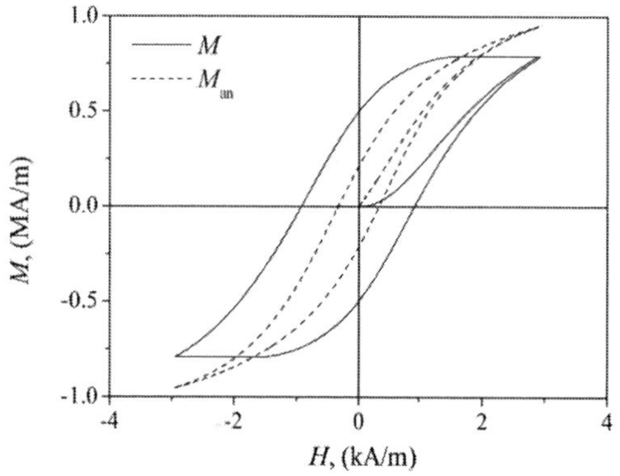

Fig. 3. The „anhysteretic'' curve given with the modified Langevin function in fact exhibits hysteresis [42]

Summing up to this point, the following remarks on the original JA description may be made:

- it is based on the concept of the effective field, which appears explicitly in the fundamental model equation (3) and the modified Langevin function (1) as the independent variable;
- the loop branches are constructed in the description by the introduction of an offset from the „anhysteretic'' curve (1);
- the direct introduction of the effective field as the argument to the „anhysteretic'' curve may lead to serious interpretation problems;
- it is necessary to „patch'' quasi-static model equations using the δ_{M} pseudo-variable in order to get rid of the regions with negative differential susceptibilities; this introduces characteristic flat fragments of the simulated curves;
- the proposed modification only partially resolves the problem; in fact this is an artificial solution. For a hysteresis description devoid of the global memory mechanism, such as the JA model [44], it is always possible to perform a coordinate transformation by moving the reversal point to the origin. After such transformation the slope of the $M-H$ curve (the differential susceptibility) should be positive and its value should correspond to the Rayleigh relationship.

3 GRUCAD model equations

In Refs. [21-23] the researchers from the Brazilian team GRUCAD introduced an alternative description. A distinct feature of the model is that the reversible and irreversible processes are separated from each other. Moreover their summing is carried out along the H axis. In the present paper we consider the set of equations, as outlined in Refs. [45, 46].

The reversible (anhysteretic) part is described with two equations:

$$H_{\mathrm{an}} = B/\mu_0 - M_{\mathrm{s}}(\coth\lambda - 1/\lambda), \tag{7}$$

$$\lambda = \frac{1}{a}\left[(1-\alpha)H_{\mathrm{an}} + \alpha\frac{B}{\mu_0}\right]. \tag{8}$$

The parameter α plays the role of a weighting coefficient between the flux density B and the reversible field strength H_{an}. The parameter a adjusts the level of λ for the Langevin function in (7).

$$\frac{\mathrm{d}H_{\mathrm{h}}}{\mathrm{d}B} = \frac{H_{\mathrm{HS}}(\coth\lambda_{\mathrm{H}} - 1/\lambda_{\mathrm{H}}) + H_{\mathrm{h}}}{\gamma\delta} \tag{9}$$

with λ_{H} defined as

$$\lambda_{\mathrm{H}} = \frac{H_{\mathrm{h}} + \delta H_{\mathrm{HS}}}{a} \tag{10}$$

where $\delta = \pm 1$ is the sign of $\mathrm{d}B/\mathrm{d}t$.

Equations (9) and (10) constitute the description of irreversible (hysteretic) part. It is given in the form of an ordinary differential equation whose independent variable is the flux density B. Thus the GRUCAD proposal is inherently an inverse (B - input) model. Such models are useful in two dimensional FEM analyses of magnetic circuits, where the magnetic vector potential $\mathbf{A}$ may be introduced ($\mathbf{B} = \mathrm{curl}\mathbf{A}$)[a].

Finally the reversible and irreversible contributions are described by

$$H = H_{\mathrm{an}} + H_{\mathrm{h}} \tag{11}$$

The model parameters are $\alpha, a, \gamma, H_{\mathrm{HS}}$ and M_{s}. The effects of their variations on the shape of hysteresis loop is discussed in [45]. The physical meaning of parameters α and a is altered when compared to the JA model. The first one no longer represents the cooperative effects resulting from the presence of „effective field''– the concept of effective field is simply abandoned. Finally the relationship (11) deserves attention. The branches of hysteresis loop are obtained by introducing appropriate offsets from the truly anhysteretic curve along the H axis and not along the M axis, as in the original JA model. Therefore, there is no need to patch the model equations using the inconvenient δ_{M} pseudo-variable. The model is somewhat similar in spirit to other descriptions consistent with the laws of irreversible thermodynamics, like those discussed in Refs. [47-49].

Figure 4 depicts a saturating hysteresis loop and some first order reversal curves (FORCs) simulated with the GRUCAD model. The model parameters were specified after [45], namely: $\alpha = 9.48\cdot 10^{-5}$ (-), $a = 40$ (A/m), $M_{\mathrm{s}} = 1.2\cdot 10^{6}$ (A/m), $\gamma = 0.12$ (A/m) and $H_{\mathrm{HS}} = 50$ (A/m). These parameters may describe well a sample of non-oriented electrical steel. The FORCs were chosen to start on the descending branch of the major loop. A precise choice of their beginning may be controlled in Matlab using the *events* feature available for the ODE solvers. In the inset a zoomed fragment of the $B-H$ plane is shown. It is evident that the FORCs do not exhibit negative slopes. Moreover, it can be stated that the first FORC almost overlaps the virgin curve, which is depicted with a broken line in the inset (the initial conditions for ODE were set to illustrate the resemblance between these two curves). The similarity of the pattern exhibited by magnetization curves was noticed more than a century ago [50]. It was practically implemented in a simple phenomenological hysteresis model [51,52].

---

[a] A standard quasi-static H - input hysteresis model like the original Jiles-Atherton model may always be transformed to the inverse form using the identity $\frac{\mathrm{d}M}{\mathrm{d}H} = \frac{\mathrm{d}M}{\mathrm{d}B}\frac{\mathrm{d}B}{\mathrm{d}H} = \frac{\mathrm{d}M}{\mathrm{d}B}\mu_0\left(1+\frac{\mathrm{d}M}{\mathrm{d}H}\right)$

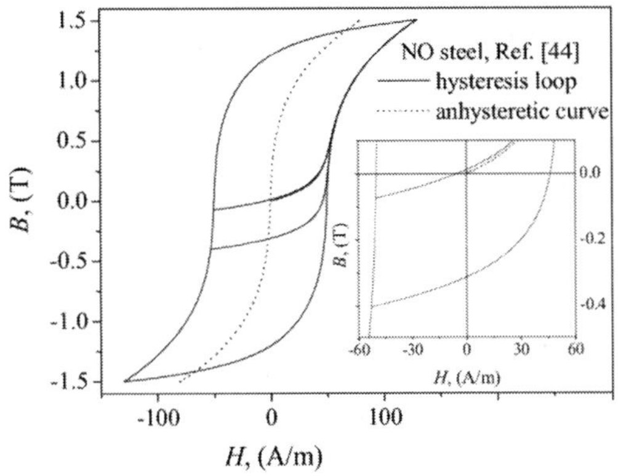

Fig. 4. The hysteresis loop at B_m =1.5 T and two FORCs simulated with the GRUCAD model.

4 Conclusions

In the paper an analysis of two phenomenological hysteresis models was carried out. First of them was the widespread Jiles-Atherton proposal, and the other one – a description advanced by the Brazilian research team GRUCAD. Attention was focused on qualitatively different behaviour of the models after sudden field reversal. First order reversal curves simulated with the GRUCAD description do not exhibit negative slopes, which makes this model useful for implementation in FEM codes.

Appendix

Recently Padilla *et al.* published a paper [39], which focused on the problem of negative slopes of hysteresis loops modelled with the JA model for specified sets of parameters. In the Appendix we apply the analytical approximation to the inverse Langevin function to illustrate the issue for the anhysteretic curve given with expression (6).

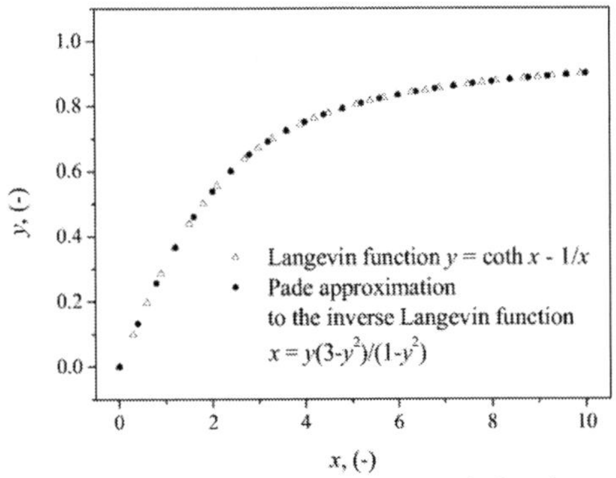

Fig. 5. The Padé approximation to the Langevin function.

The Padé approximation [53, 54] yields a simple formula for the inverse of the Langevin function

$$\text{if} \quad y = \coth x - 1/x \quad \text{then} \quad x = y\frac{3-y^2}{1-y^2} \tag{A1}$$

As it can be inferred from Fig. 5, the approximation offers results acceptable for engineering purposes.

Applying the considered approximation to (6) we obtain a relationship that may be solved for H :

$$H + \alpha M_{\text{an}} = \frac{aM_{\text{an}}}{M_{\text{s}}}\left(\frac{3-\left(\frac{M_{\text{an}}}{M_{\text{s}}}\right)^2}{1-\left(\frac{M_{\text{an}}}{M_{\text{s}}}\right)^2}\right) \tag{A2}$$

The critical set of model parameter shall occur when the derivative $\mathrm{d}H/\mathrm{d}M_{\text{an}}\big|_{M_{\text{an}}\to 0}$ takes the zero value. It means the following relationship between α, a and M_{s} :

$$\alpha_{\text{crit}} = \frac{3a_{\text{crit}}}{M_{\text{s}}} \tag{A3}$$

A Similar derivation may be carried out for the GRUCAD model parameters.

References

1. A. Iványi, *Hysteresis models in electromagnetic computation* (Akadémiai Kiadó, Budapest 1997)
2. D. C. Jiles, D. L. Atherton, IEEE Trans. Magn. **19**, 2183 (1983)
3. D. C. Jiles, D. L. Atherton, J. Magn. Magn. Mater. **61**, 48 (1986)
4. D. C. Jiles, J. B. Thoelke, M. K. Devine, IEEE Trans. Magn. **28**, 27 (1992)
5. D. A. Philips, L. R. Dupré, J. A. Melkebeek, IEEE Trans. Magn. **31**, 3551 (1995)
6. D. Lederer, H. Igarashi, A. Kost, T. Honma, IEEE Trans. Magn. **35**, 1211 (1999)
7. P. R. Wilson, J. Neil Ross, A. D. Brown, IEEE Trans. Magn. **38**, 1049 (2001)
8. A. Salvini, F. Riganti Fulginei, IEEE Trans. Magn. **41**, 1100 (2005)
9. K. Chwastek, J. Szczygłowski, J. Magn. Magn. Mater. **314**, 47 (2007)
10. R. Szewczyk, J. Phys. D : Appl. Phys. **40**, 4109 (2007)
11. R. Marion, R. Scoretti, N. Siauve, M.-A. Raulet, IEEE Trans. Magn. **44**, 894 (2008)
12. M. Toman, G. Štumberger, D. Dolinar, IEEE Trans. Magn. **44**, 1098 (2008)
13. A. P. S. Baghel, S. V. Kulkarni, IET Electr. Power Appl. **6**, 689 (2012)
14. R.-A. Naghizadeh, B. Vahidi, S.H. Hosseinian, COMPEL **31**, 1293 (2012)
15. Lu Hai-Liang, Wen Xi-Shan, Lan Lei, An Yun-Zhu, Li Xiao-Ping, J. Magn. Magn. Mater. **374**, 502 (2015)

16. M. A. Zaman, P. C. Hansen, L. T. Neustock, P. Padhy, L. Hesselink, J. Appl. Phys. **120**, 093903 (2016)
17. Z. Włodarski, IEEE Trans. Magn. **39**, 1990 (2003)
18. A. Benabou, J. V. Leite, S. Clénet, C. Simão, N. Sadowski, J. Magn. Magn. Mater. **320**, e1034 (2008)
19. K. Chwastek, J. Phys. D : Appl. Phys. **42**, 165002 (2009)
20. Zb. Gmyrek, Int. J. Numer. Model. **27**, 199 (2014)
21. P. I. Koltermann, L. A. Righi, J. P. A. Bastos, R. Carlson, N. Sadowski, N. J. Batistela, Physica B **275**, 233 (2000).
22. L. A. Righi, N. Sadowski, R. Carlson, J. P. A. Bastos, N. J. Batistela, IEEE Trans. Magn. **37**, 3353 (2001)
23. P. I. Koltermann, J. P. A. Bastos, N. Sadowski, N. J. Batistela, A. Kost, L. Jänicke, K. Miethner, IEEE Trans. Magn. **38**, 897 (2002)
24. P. Andrei, A. Stancu, H. Hauser, P. Fulmek, J. Optoelectr. Adv. Mater. **9**, 1137 (2007)
25. M. Sablik, D. C. Jiles, IEEE Trans. Magn. **29**, 2113 (1993)
26. K. J. Stevens, NDT&E **33**, 111 (2000)
27. Jianwei Li, Minqiang Xu, J. Appl. Phys. **110**, 063918 (2011)
28. A. Jakubas, P. Gębara, S. Seme, A. Gnatowski, K. Chwastek, Acta Phys. Pol. A **131**, 1289 (2017)
29. R. Venkataraman, P. S. Krishnaprasad, Proc. 37th IEEE Conf. Decision and Control, Tampa, USA, 1998, doi:10.1109/CDC.1998.757777
30. N. Sadowski, N. J. Batistela, J. P. A. Bastos, M. Lajoie-Mazenc, IEEE Trans. Magn. **38**, 797 (2002)
31. K. Chwastek, J. Szczygłowski, Przegl. Elektrotechn. **12**, 145 (2008)
32. R. Marion, N. Siauve, M.-A. Raulet, L. Krähenbühl, K. Chwastek, J. Szczygłowski, W. Wilczyński, A comparison of identification techniques for the Jiles-Atherton model of hysteresis, presented at XX EPNC 2-4.07.2008, Lille, France
33. S. E. Zirka, Yu. I. Moroz, R. G. Harrison, K. Chwastek, J. Appl. Phys. **112**, 043916 (2012)
34. I. D. Mayergoyz, Mathematical models of hysteresis, Springer, New York, 1991
35. B. Vaseghi, D. Mathekga, S. Rahman, A. M. Knight, IEEE Trans. Magn. **49**, 1637 (2013)
36. E. Della Torre, J. Oti, G. Kadár, IEEE Trans. Magn. **26**, 3052 (1990)
37. J. H. B. Deane, IEEE Trans. Magn. **30**, 2795 (1994)
38. H. Toms, R. Colclaser, M. Krefta, IEEE Trans. Magn. **2**, 982 (2008)
39. J. B. Padilha, P. Kuo-Peng, N. Sadowski, J. V. Leite, N. J. Batistela, J. Magn. Magn. Mater. **442**, 8 (2017)
40. http://www.zirka.dp.ua
41. D. C. Jiles, J. Appl. Phys. **76**, 5849 (1994)
42. Liu Qingyou, Luo Xu, Zhu Haiyan, Liu Jianxun, Han Yiwei, Chinese Phys. B **26 (7)**, 077502 (2017)
43. K. Chwastek, Math. Comput. Model. Dyn. Syst. **15**, 95 (2009)
44. T. Kleineberg, J. Magn. Magn. Mater. **166**, 315 (1997)
45. S. Steentjes, K. Chwastek, M. Petrun, D. Dolinar, K. Hameyer, IEEE Trans. Magn. **50**, 7300804 (2014), doi: 10.1109/TMAG.2014.2323250
46. R. Jastrzębski, K. Chwastek, I. Biondić, K. Miličević, Acta Phys. Pol. A **131**, 1228 (2017)
47. A. J. Bergqvist, Physica B **233**, 342 (1997)
48. F. Henrotte, A. Nicolet, K. Hameyer, COMPEL **25**, 71 (2006)
49. V. François-Lavet, F. Henrotte, L. Stainer, L. Noels, C. Geuzaine, J. Comput. Appl. Math. **246**, 243 (2013)
50. E. R. Madelung, Ann. Phys. **17**, 861 (1905)
51. S. E. Zirka, Yu. I. Moroz, IEEE Trans. Magn. **31**, 3509 (1995)
52. S. E. Zirka, Yu. I. Moroz, IEEE Trans. Magn. **35**, 2090 (1999)
53. A. Cohen, Rheol. Acta **30**, 270 (1991)
54. R. Jedynak, Rheol. Acta **54**, 29 (2015)

 DOI: 10.1051/itmconf/20171503004

Modelling of the temperature field that accompanies friction stir welding

Przemysław Nosal[1,*], and *Artur* Ganczarski[1]

[1]Cracow University of Technology, Institute of Applied Mechanics, Jana Pawła II 37, 31-864 Cracow, Poland

Abstract. The thermal modelling of the Friction Stir Welding process allows for better recognition and understanding of phenomena occurring during the joining process of different materials. It is of particular importance considering the possibilities of process technology parameters, optimization and the mechanical properties of the joint. This work demonstrates the numerical modelling of temperature distribution accompanying the process of friction stir welding. The axisymmetric problem described by Fourier's type equation with internal heat source is considered. In order to solve the diffusive initial value problem a fully implicit scheme of the finite difference method is applied. The example under consideration deals with the friction stir welding of a plate (0.7 cm thick) made of Al 6082-T6 by use of a tool made of tungsten alloy, whereas the material subjected to welding was TiC powder. Obtained results confirm both quantitatively and qualitatively experimental observations that the superior temperature corresponds to the zone where the pin joints the shoulder.

1 Introduction

At the end of the XX century Wayne Thomas and his team in The Welding Institute in Cambridge UK (TWI) developed and patented new technology of the materials joined by the friction with the mixing of plastic material in the welding zone. This technology is known as Friction Stir Welding (FSW) [1, 2]. FSW is a steady state process, which means that the material does not reach melting point and that it brings some advantages in the form of a defect free joint, which is known from traditional welding techniques [2, 3]. Frictional heat is generated by the friction between the tool and the base material to soften the material in the joining zone. Moreover, the plastic work of the material supplies additional portions of heat energy [4]. The mechanical influence of the tool leads to grain fragmentation inside the weld that has a positive effect on the mechanical properties of the joint. That is the reason why joints obtained by the use of FSW technology have better parameters than joints produced by more traditional methods e.g. MIG welding method [3]. FSW technology allows the joining of materials which are regarded as difficult to weld and materials which are dissimilar e.g. aluminium with copper [3]. At present, few countries around the world are developing this technology and are implementing it in various branches of industry like automotive, railway, aeronautic, ship-building and space industry [4]. Friction stir welding, besides its undeniable benefits in joining areas of different kinds of materials, has also other interesting applications. One of them is the possibility to manufacture composites via mixing one material phase, e.g. in the form of powder, in the other one which is a matrix (Fig.1). The received material manifests certain attributes of the Functionally Graded Materials (FGM) material.

Friction Stir Processing (FSP) is a variant of friction stir welding technology that includes all unique and beneficial features of this joining method [5-9]. FSP process allows us to change the properties of modified material through the creation of fine grained structures [5], surface composites [6-9] and modifications of cast alloys microstructure. One of the very few examples is research by Gandra [8], who observed that composite of AA5083 alloy and introduced SiC particles exhibit orthogonal gradients of composition and mechanical behaviour.

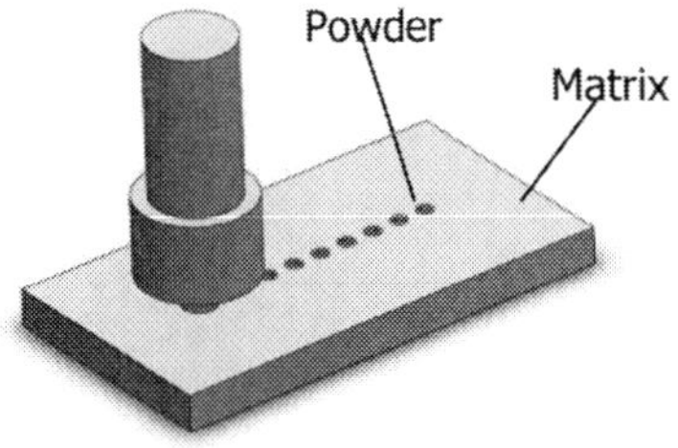

Fig. 1. TiC powder implementation approach onto the aluminium plate in the FSP process.

Liu et al. [9] observed that the best properties of reinforced aluminium matrix composites fabricated by FSP process are received after a few passes of the welding tool. The FSW process is associated with significant growth of the joining materials enthalpy,

*Corresponding author: nosal.przemyslaw@student.pk.edu.pl

 DOI: 10.1051/itmconf/20171503004

which allows their plastic strain and the adhesion phenomenon process. Moreover, the high temperature initiates the diffusion processes in materials that have different physical and chemical properties.

2 Mathematical model

Material flow around the mixing pin is a very complex phenomenon and it is determined by the specific geometry of the tool. Additionally, during the exact welding process compound of the direction of the rotary motion with linear motion of the FSW tool leads to asymmetric material flow in welding the zone, which causes asymmetry in the joint construction (Fig.2).

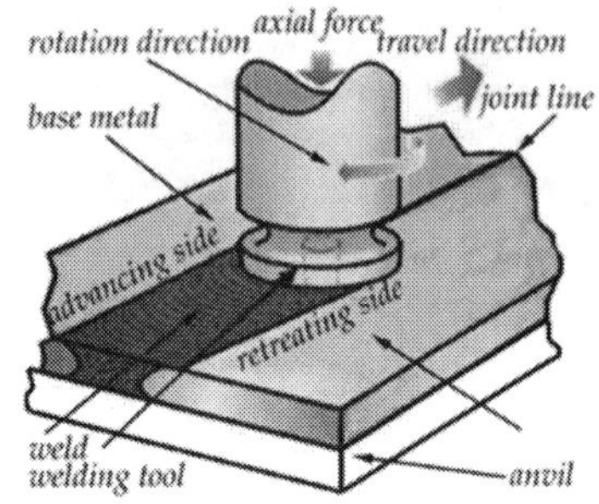

Fig. 2. Scheme of the FSW process [10].

With respect to this fact, we can distinguish two sides in the FSW joint: advancing side and retreating side. They depend on the relations between rotary and linear motion of the welding tool.

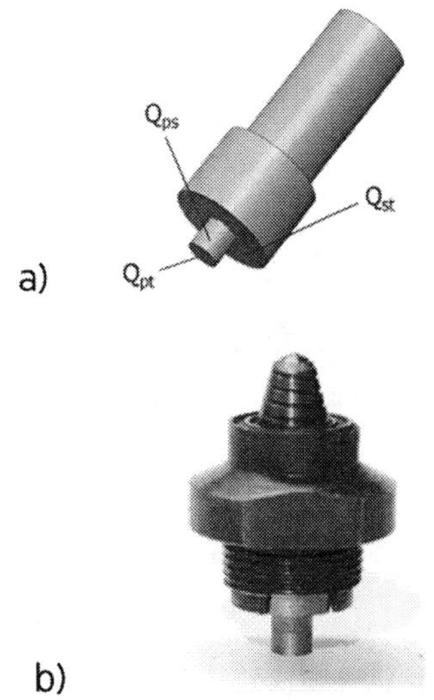

Fig. 3. FSW tool: a) respective tool areas that determine amount of generated heat energy, b) real welding tool with conical pin and cut spirals on the pin and the shoulder.

This article focuses only on the pure thermal issue of the first stage of the FSW process, namely the stage when the welding tool starts penetrating the work piece. Thermal modelling is the first step to predict the possible occurrence of voids or defects in the joint as well as microstructure evolution and finally properties of the joint [6]. All the processes are dependent on the heat that is generated during welding; therefore, thermal history has a large impact on the quality of the joint. Two basic components of the welding tool are the shoulder and the pin (Fig.3). Their basic function is generating heat energy by the friction between materials of the tool and the work piece. The second function is mixing of the plastic material inside the welding zone. During the FSW process, mechanical energy is transformed into heat energy. The fundamental assumption made by the whole energy that is generated during welding process is its division into the parts that are described respectively by active areas of the welding tool with the cylindrical pin. This energy can be described by the equation [10, 11]

$$Q_T = Q_{ps} + Q_{pt} + Q_{st} \tag{1}$$

Where the individual members respond to the heat which is generated on the pin sideQ_{ps}, pin tipQ_{pt}, shoulder tip Q_{st} and these components are described by the following integrals

$$\left.\begin{aligned} Q_{ps} &= \int_0^{2\pi}\int_0^{h} \omega\tau_{cont} r^2 d\varphi dz \\ Q_{pt} &= \int_0^{\frac{d}{2}}\int_0^{2\pi} \omega\tau_{cont} r^2 d\varphi dr \\ Q_{st} &= \int_0^{2\pi}\int_{\frac{d}{2}}^{\frac{D}{2}} \omega\tau_{cont} r^2 d\varphi dr \end{aligned}\right\} \tag{2}$$

Most of the welding tools have threads on the probe, which increase the effective friction coefficient and the frictional heat, which is generated by the pin [12]. Friction phenomenon occurring during FSW process is very difficult to model because welded material can stick to the material of the tool or can slip. Furthermore, it is non-linear and the value of friction coefficient is not constant. Between these two materials places where the material was melted may occur [6]. The basic equation for the heat generation at the contact surface is [2]

$$\dot{q}_V = \tau_{cont}\omega r \tag{3}$$

Where $\tau_{cont}$is the contact shear stress of the interface located between shear layer of workpiece and the tool, ω is a radial frequency and r is a radius of the tool. The next simplification deals with the spatial dispersion of the volume intensity heat source assumed in form of (compare [13])

$$\dot{q}_V = \frac{3Q_T r}{2\pi R^3 w_{sl}} \tag{4}$$

This is the extended version of Eq. (3) and manifests itself as excitation in the rotational symmetry and non-stationary Fourier equation

$$\lambda\left(\frac{\partial^2\theta}{\partial r^2} + \frac{1}{r}\frac{\partial\theta}{\partial r} + \frac{\partial^2\theta}{\partial z^2}\right) + \dot{q}_V = \rho c_V \frac{\partial\theta}{\partial t} \tag{5}$$

Variable w_{sl} means width of shear layer (Fig.4). In reality, the heat source is heterogeneous and it is closely connected with respective active surface, which nonlinearly affects the work piece, because the tool during the first stage of welding is penetrating material and the heat energy is lower than the heat energy at the end of this stage. In general, heat energy generated by the pin surface is negligible when compare to the heat energy generated by the shoulder so the energy Q_T in (4) can be substituted byQ_{st}, in thick plate $Q_{pin} \cong 10\%Q_T$ [6]. Mijajlovic and Milcic [10] presented analytical model for estimating the amount of heat generated during FSW, which is a process of mechanical energy transformation and considers many parameters such as: tool rotation speed, angular velocity, travel rate, tilt angle, axial force, torque, etc. Additionally, they estimated values of parameters which are mutually dependent and which influence the heat generation process.

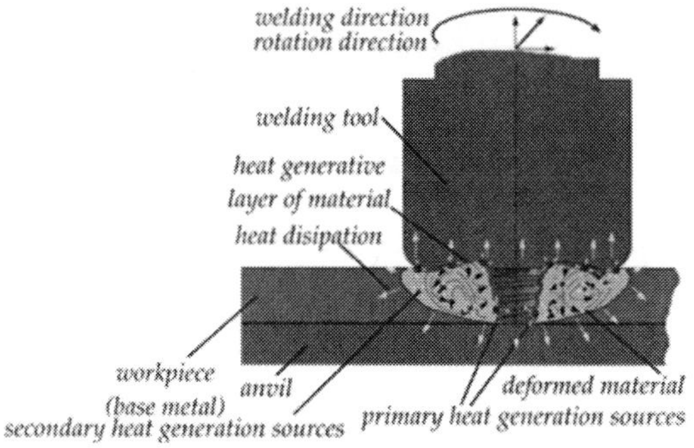

Fig. 4. Scheme of heat generation during FSW process [10].

In order to reach the unique solution of the Eq. (5) the following boundary conditions can be applied: θ_∞=20°C, adiabatic condition on the symmetric axis and the third order boundary conditions, which are modelling free convection on the extant boundaries of the showed area on the Fig.5.

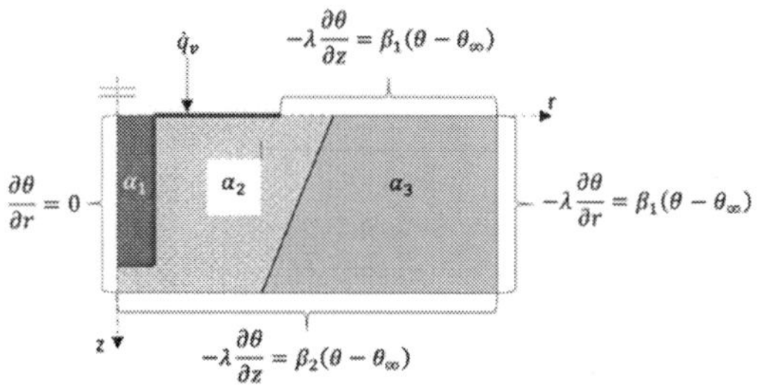

Fig. 5. Geometrical scheme and the boundary conditions of the analysed problem (von Neumann's stability coefficients for: tool-α_1, added phase - α_2 and matrix - α_3, respectively)

2D model is subject to validation and experimental verification. As the next stage, this model can be developed to a 3D model with heat generated by friction like in [14] and plastic work of material. Shi et al. [15] presented the model of heat transfer and material flow in Reverse Dual-Rotation FSW process by using CFD simulations. Similarly, Micallef et al. [16] used a thermo- fluid-flow model in computational fluid dynamics simulation to predict material flow and heat gradients. The coupled thermal flow model was also developed by Hamilton [17] to simulate FSP process by use of commercial software. Two different methods/models for rotating tools are presented by Kang et al. [18]. They assumed heat generation only by the plastic deformation not by friction. Finite volume method with Eulerian's description approach was adopted. First, the model assumes that only the contact area rotates and the plastic zone grows as heat spreads. Secondly, the model provides a plastic deformation zone named the rotation affected zone (RAZ), as a place where heat energy is generated. In work [19] Song and Kovacevic submitted equation for heat energy which is generated by the threaded surface of the pin.

$$Q_{pin} = 2\pi r_p hk\bar{Y}\frac{V_m}{\sqrt{3}} + \frac{2\mu k\bar{Y}\pi r_p hV_{rp}}{\sqrt{3(1+\mu^2)}} + \frac{4F_p\mu V_m\cos\varphi}{\pi} \quad (6)$$

where $\varphi = 90^\circ - \psi - \tan^{-1}(\mu)$,$V_m = \frac{\sin\psi}{\sin(180^\circ-\varphi-\psi)}v_p$, $V_{rp} = \frac{\sin\varphi}{\sin(180^\circ-\varphi-\psi)}v_p$, $v_p = r_p\omega$.r_p- radius of the pin, h - the thickness of the work piece, ψ- an angle of the threads helix, F_p- the translation force during the FSW process and $\bar{Y}$- the average shear stress of the welded material. Attallah [11] presented alternating direction implicit (ADI) scheme to solve unsteady heat transfer equation by finite difference method. Stability of ADI scheme is only restricted to being non-oscillating and finite. Time increment was described by equation

$$\Delta t < \frac{(\Delta x)^2+(\Delta y)^2+(\Delta z)^2}{8\alpha_{max}} \quad (7)$$

Appropriate value of time step allows accuracy and convergence of simulation.

3 Example

The simulation of the first stage of FSP process was made experimentally. The rotating welding tool penetrated the work piece with constant feed-rate and after achieving the set depth it was dwelling for a few seconds (Fig. 6). The workpiece and the matrix were made from aluminium plate and the second phase was titanium carbide powder.

Table 1. Physical properties of the tool material, TiC powder and the aluminium plate.

Material	ρ[g/cm²]	λ[W/cm·K]	c_v[J/g·K]
Tool –Tungsten alloy	19.25	175	0.135
Added phase –TiC powder	4.92	110	0.520
Matrix -Al 6082-T6	2.71	172	0.894

DOI: 10.1051/itmconf/20171503004

Comparison of the basic physical properties of materials used is shown in Tab. 1. Solution of the boundary problem (5) is achieved by using finite difference method based on the fully implicit scheme [20]

$$\alpha\left(1-\frac{\Delta r}{2r_i}\right)\theta_{i-1,j}^{new}+\alpha\theta_{i,j-1}^{new}-(1+4\alpha)\theta_{i,j}^{new}+\alpha\theta_{i,j+1}^{new}+\alpha\left(1+\frac{\Delta r}{2r_i}\right)\theta_{i+1,j}^{new}=-\theta_{i,j}^{old}-\frac{\dot{q}_V\Delta t}{\rho c_V} \qquad (8)$$

Where the symbol $\alpha=\frac{\lambda}{\rho c_V}\frac{\Delta t}{(\Delta r)^2}$ denotes von Neumann's stability coefficient, the magnitude of each differs for subsequent three materials. The differential scheme (8) is applied to each point of mesh at a size 35×90 that discretizes rectangle domain. Magnitudes of unknown sat points located outside the domain are expressed by combinations of unknown points by use of appropriate boundary conditions (see Fig. 5). Finally, the set linear algebraic equations, which exhibit a structure typical for sparse systems, are solved by use of the conjugate gradient method and the row-indexed compact storage [20].

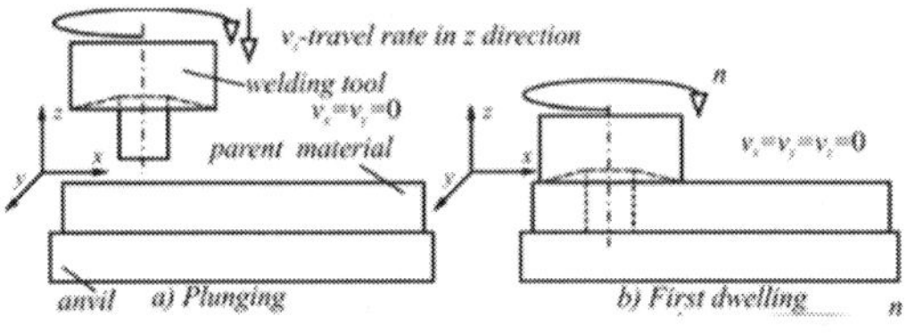

Fig. 6. Scheme of FSW/FSP first stage [10].

Obtained results involving the evolution of the temperature field shown in Fig. 7, confirm quantitatively and qualitatively temperature distribution field as well as theoretical results published by others scientists and empirical results reach by experimental tests. The presented model predicts achievement of the maximal temperature near to 310°C in location where the tool pin and shoulder meets. The temperature field is homogeneous in the thermo-mechanical affected zone (TMAZ). The temperature drops from 300°C to 270°C outside this zone of what is typical for heat-affected zone (HAZ). Individual zones in produced composite respond to the zones that are known from FSW joint (Fig. 8). Their heterogeneous areas are correlated with the obtained results from numeric simulation.

a)

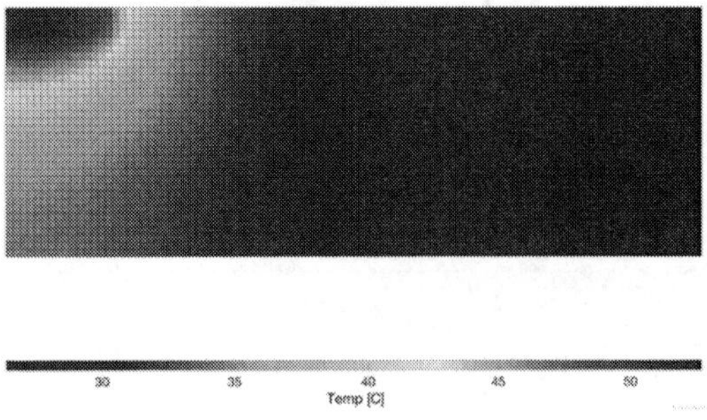

b)

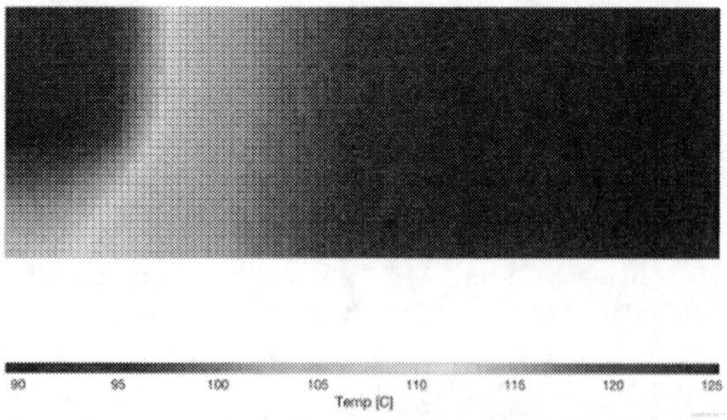

c)

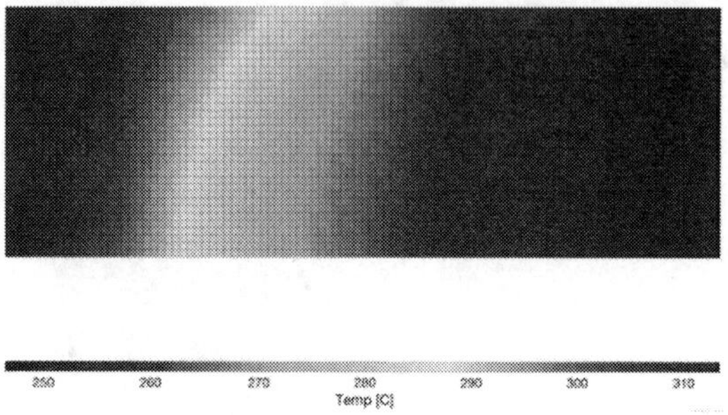

Fig. 7. Temperature fields in different period of the tool penetration in the base material: a) time 4 s – depth of the pin 0.8 mm, b) time 18 s – depth 3.6 mm, c) time 33 s – depth 6 mm.

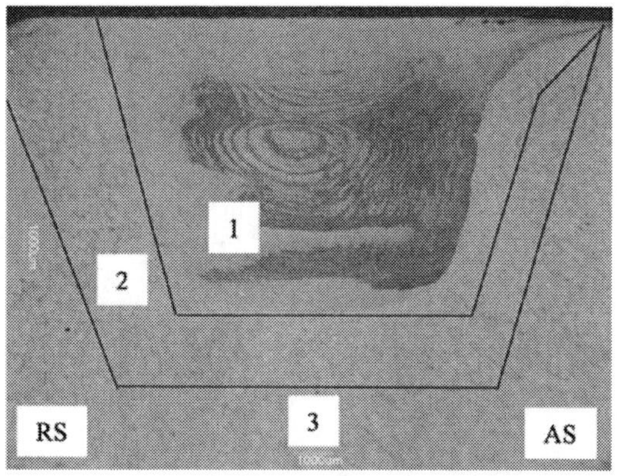

Fig. 8. Macrostructure image of sample Al-TiC composite produced with one pass of welding tool: AS – advancing side, RS – retreating side, 1 – weld nugget, 2 – thermo-mechanically affected zone, 3 – heat affected zone.

Observed black curves in the weld are introduced by TiC powder. An incomplete mix of TiC powder in to the aluminium matrix is visible as a result of one passing of the welding tool.

4 Conclusions

The following concluding remarks may be formulated for obtained results:

- the maximal temperature is lower than the melting temperature of aluminium which corresponds to FSP process property – solid state joining process,

- the highest temperature occurs in place where the pin and shoulder meet, which corresponds with the place where generated heat has no possibility of dissipation,
- the occurrence of the highest temperature was observed around 33 seconds at the end of the first stage of FSW process,
- isotherm dispersion outside the TMAZ area corresponds to the HAZ area observed in the photos from the metallurgical test.

Further research will focus on a mathematical model that would include mechanical influences on heat energy generation. Additionally, there will be an attempt to analytically predict some of the parameters (e.g. axial force, torque) which have large impact on the amount of generated energy but cannot be measured without specialized equipment.

The author is indebted to Dr Marek Hebda for help in metallography research and valuable clues. Presented sample of FSP composite was produced in Electriscompany, ELECTRIS, Mickiewicza 6, 32-650 Kęty, Poland.

References

1. W.M. Thomas, E.D. Nicholas, J.C. Needham, M.G. Murch, P. Temple-Smith, C.J. Dawes,*Improvements relating to friction stir welding*, US Patent No 5460317, 1999
2. D. Lohwasser, Z. Chen, *Friction Stir Welding. From basics to applications* (CRC, Cambridge, 2010)
3. P. Nosal, *Mechanical properties of friction stir welded copper plates joints*(AGH, Kraków, 2016)
4. A. Ambroziak, *Friction welding of materials with various properties* (OW PolitechnikiWrocławskiej, Wrocław, 2011)
5. R.S. Mishra, M.W. Mahoney, *Friction Stir Welding and Processing* (ASM International, Ohio, 2008)
6. M.S. Węglowski, S. Dymek, C.B. Hamilton, Bulletin Of The Polish Academy of Sciences Technical Sciences **61**, 4 (2013)
7. Y.X. Gan, D. Solomon, M. Reinbolt, Materials**3**,329-350 (2010)
8. J. Gandra, R. Miranda, P. Vilaca, A. Velhinho, J. Pamies Teixeira, Jour. of Mat. Proce. Tech.**211**, 1659-1668 (2011)
9. Q. Liu, L. Ke, F. Liu, C. Huang, L. Xing, Mate. and Design**45**, 343-348 (2013)
10. M. Mijajlowic, D. Milcic, Chap. 11, InTech, 2012
11. M.M. Attallah, *Friction stir welding of aluminium alloys. Microstructure-property development* (LAP, Birmingham, 2007)
12. M. Schwartz, *Innovations in Materials Manufacturing, Fabrication and Environmental Safety* (CRC, New York, 2011)
13. Y.J. Chao, X. Qi, Trans. of the ASME **125**, 138-145 (2003)
14. G.S. Reese, *Analytical Thermal Model of Friction Stir Welding with Spatially Distributed Heat Source*(BYU Scholar Archive, Brigham, 2012)
15. L. Shi, C.S. Wu, H.J. Liu, JMEPEG**23**, 2918-2929 (2014)
16. D. Micallef, D. Camilleri, A. Toumpis, A. Galloway, L. Arbaoui,SAGEJou. **230**, 1-26 (2015)
17. C. Hamilton, M.S. Węglowski, S. Dymek, MMMS and ASM Inter. **46B**, 1409-1418(2015)
18. S.W. Kang, B.S. Jang, J.W. Kim, *A study on heat-flow analysis of friction stir welding on a rotation affected zone*, JMST **28**, 3873-3883 (2014)
19. M. Song, R. Kovacevic, IJMTM**43**,605–615 (2003)
20. W.H. Press, S.A.Teukolsky, W.T. Vetterling, B.P. Flannery, *Numerical recipes in fortran*(Cambridge University Press, Cambridge, 1993)

Stochastic higher order finite element elasto-plastic analysis of the necking phenomenon

Michał Strąkowski[1,*], and *Marcin* Kamiński[1]

[1]Łódź University of Technology, Faculty of Civil Engineering, Architecture and Environmental Engineering, Department of Structural Mechanics, Politechniki Avenue 6, 90-924 Łódź, Poland

Abstract. The principal goal of this work is to investigate an application of the stochastic perturbation technique of the 10th order in coupled thermo-elasto-plastic analysis of tension of the steel elastic bar exposed to fire with thermally dependent material characteristics. An ambient temperature, calculated from the fire curve after ISO 834-1, equivalent to the fire exposure of the steel structure is treated here as the input Gaussian random variable. It is uniquely defined by the constant mean value at outer surfaces of this element, where material parameters of the steel as Young modulus, yield strength, heat conductivity, capacity and thermal elongation are considered all as highly temperature-dependent. Computational implementation known as the Stochastic Finite Element Method is carried out with the use of the FEM system ABAQUS and computer algebra system MAPLE. It uses both polynomial and non-polynomial local response functions of stresses and displacements. The basic probabilistic characteristics of time-dependent structural response are determined (expectations, coefficients of variation, skewness and kurtosis) and verified with classical Monte-Carlo simulation scheme and semi-analytical technique for input coefficient of variation not larger than 0.20. Finally, probabilistic convergence of all three methods versus increasing input uncertainty level is investigated.

1 Introduction

The influence of a fire on the behaviour of steel structures is always very interesting. It is clear that the total time from a fire ignition up to its failure strongly depends on the basic structural materials as well as on the protection cover systems. The determination of this influence is very challenging. There are a number of simulation techniques available like fuzzy sets theory, analytical methods, and others based on Taylor expansions methods known as perturbation techniques. In this paper, we use the well-known Monte Carlo method, semi-analytical method and 10th order perturbation method implemented with the FEM originating from the Second Order Second Moment method [1]. We compute four probabilistic coefficients to gain more information about probability density function itself. The basic probabilistic moments, as well as coefficients like skewness and kurtosis, are calculated by full symbolic expansion of integral definitions and partial differentiation of the nodal response functions with respect of gas temperature, which is a random input variable. In this paper, we consider strength test of elasto-plastic steel cylindrical specimens subjected to fire known from literature. In this case, all physical and mechanical material properties like Young modulus, yield strength, conductivity, heat capacity and linear, thermal elongation are temperature-dependent. Gas temperatures are taken from ISO-834-1 [2]. It is the main load in this paper.

2 Theoretical background

We use the random variable b and its probability density function $p_b(x)$. The first probabilistic moment of this variable is defined as [3]

$$E[b] \equiv b^0 = \int_{-\infty}^{+\infty} b p_b(x) dx, \tag{1}$$

where b^0 means the average value of the b. Based on Eq. (1) we determine variance of b as follows

$$Var[b] = \int_{-\infty}^{+\infty} [b - E(b)]^2 p_b(x) dx. \tag{2}$$

The third and fourth central probabilistic moments and related coefficients may be defined based on the classical definitions from the probability theory. The basic idea of the stochastic perturbation approach is to expand all the resulting state functions, and all input random variables of the given boundary initial problems via Taylor series about their expectations using the perturbation parameter marked ε. The random state

* Corresponding author: michal.strakowski@p.lodz.pl

DOI: 10.1051/itmconf/20171503005

function f referred to its parameter b around its mean value is given as follows

$$f = f^0 + \sum_{n=10}^{1} \frac{\varepsilon^n}{n!} \frac{\partial^n f}{\partial b^n} \Delta b^n , \tag{3}$$

where

$$\varepsilon \Delta b = \varepsilon \left(b - b^0 \right), \tag{4}$$

is the first variation of variable b about its expected value. Further, we analyze the expected values of displacements $u(b)$ defined analogously to the (3) by its expansion via Tylor series as follows

$$\begin{aligned} E[u(b)] &= \int_{-\infty}^{+\infty} u(b) p_b(x) dx = \\ &= \int_{-\infty}^{+\infty} \left(u^0\left(b^0\right) + \sum_{n=10}^{1} \frac{\varepsilon^n}{n!} \frac{\partial^n u(b)}{\partial b^n} \Delta b^n \right) p_b(x) dx \end{aligned} . \tag{5}$$

Notice that this power expansion is valid only if the state function is analytic and the all series converge. Therefore, any criteria of convergence should include the magnitude of the perturbation parameter which is taken as equal $\varepsilon = 1$ in several practical computations and here also [4-5]. It brings for the input random variable with symmetric probability density function in the second order perturbation approach

$$\begin{aligned} E[u(b)] &= u^0\left(b^0\right) + \frac{1}{2!} \frac{\partial^2 u(b)}{\partial b^2} \mu_2(b) + \cdots + \\ &+ \frac{1}{10!} \frac{\partial^{10} u(b)}{\partial b^{10}} \mu_{10}(b) \end{aligned} , \tag{6}$$

where $\mu_n(b)$ denotes nth order the central probabilistic moment of the variable b. We use 10th order approach so the orders higher than the 10th are neglected. Variance of the displacements state function approach as

$$Var[u(b)] = \mu_2(u(b)) = \int_{-\infty}^{+\infty} (u(b) - E[u(b)])^2 \, p_b(x) dx \tag{7}$$

using the 10th order approach we define the 3rd central probabilistic moment as

$$\mu_3(u(b)) = \int_{-\infty}^{+\infty} (u(b) - E[u(b)])^3 \, p_b(x) dx , \tag{8}$$

and the 4th central probabilistic moment may be given as

$$\mu_4(u(b)) = \int_{-\infty}^{+\infty} (u(b) - E[u(b)])^4 \, p_b(x) dx . \tag{9}$$

Based on classical definition of the variance we can calculate the coefficient of variation as follows

$$\alpha(u(b)) = \frac{\sqrt{Var[u(b)]}}{E[u(b)]} . \tag{10}$$

We compute skewness and kurtosis to recognize the type of probabilistic distribution of the state function. Skewness is a ratio of the third central probabilistic moment and the third power of standard deviation as

$$S(u(b)) = \frac{\mu_3(u(b))}{\left[\sqrt{Var[u(b)]}\right]^3} , \tag{11}$$

and kurtosis is given as

$$\kappa(u(b)) = \frac{\mu_4(u(b))}{\left[\sqrt{Var[u(b)]}\right]^4} - 3 . \tag{12}$$

Let us mention that, at this stage, the proposed procedure is independent from a choice of the initial probability distribution function. However, a satisfactory probabilistic convergence of the final results may demand various lengths of the expansions for random variables with different distributions. In these computations, we assumed Gaussian probability distribution for this function.

The approach is based on the iterative fully-coupled deterministic FEM [6] equations as follows

$$\begin{cases} \mathbf{C}^{(\alpha)} \dot{\boldsymbol{\theta}}^{(\alpha)} + \hat{\mathbf{K}}^{(\alpha)} \boldsymbol{\theta}^{(\alpha)} = \hat{\mathbf{R}}^{(\alpha)} \\ \mathbf{K}^{(\alpha)}(\boldsymbol{\theta}) \mathbf{q}^{(\alpha)}(\boldsymbol{\theta}) = \mathbf{R}^{(\alpha)} \end{cases} , \tag{13}$$

where $\mathbf{C}$ is heat capacity matrix, $\boldsymbol{\theta}$ is nodal temperatures vector, $\hat{\mathbf{K}}$ is heat conductivity matrix, $\mathbf{K}$ denotes stiffness matrix, $\mathbf{R}$ is matrix of nodal loads, $\hat{\mathbf{R}}$ is a matrix of thermal loads and $\mathbf{q}$ is the solution displacement vector and α indices the test number relevant to Response Function Method [7]. We solve Eq. (13) in ABAQUS system eleven times because we assume dispersion of the random variable in range $b = [b^0 - \Delta b; b^0 + \Delta b]$, where $2\Delta b = 0{,}05b$. Further determination of the probabilistic moments from Eqs. (5÷12) thanks to the symbolic derivation of all partial derivatives with respect to the given random variable b. We carry out all these calculations in computer algebra system MAPLE.

3 Numerical simulation

The numerical example consists of the necking of an axisymmetric steel specimen phenomenon [8]. The initial length of the bar is $2L_0$ and the initial radius is R_0. We assume that $L_0 / R_0 = 4$. We take a small cut of its right bottom corner $\Delta R = 0.005 R_0 \text{x} 0.05$ mm to ensure necking in the middle of it (Fig. 1).

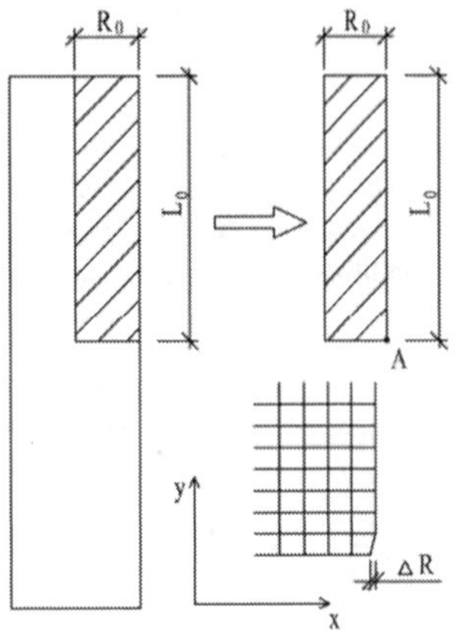

Fig. 1. Geometry and imperfection of the specimen.

4-noded FEM elements [9] are used for discretization and divide the specimen into 1600 pieces signed in ABAQUS system - CAX4RT. It means that we use 4-noded transient heat transfer elements with reduced integration. Each node has 7 degrees of freedom (6 mechanicals and 1 thermal which is temperature). Mechanical boundary conditions are u_x=0 on the left side, u_y=0 on the bottom side and forced displacement u_y=0.125L_0 applied to the top side (Fig. 2). In the beginning temperatures are set equal to T_0=20°C for all nodal points, while fire simulation is performed by setting surface temperatures equal to these calculated from the fire curve after ISO 834-1 [2] statement. We apply these temperatures to the top and the right side of the sample (Fig. 2).

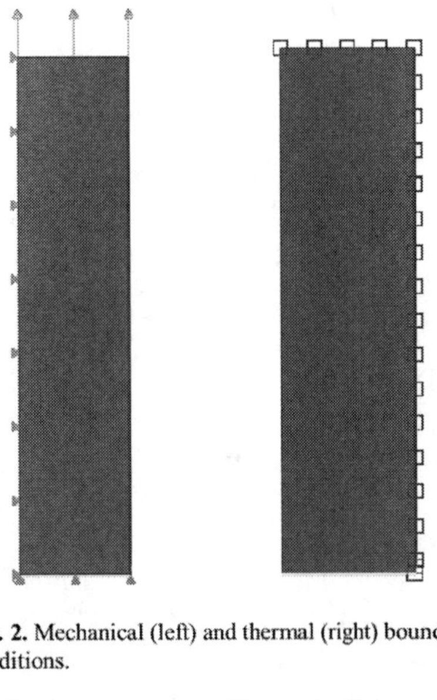

Fig. 2. Mechanical (left) and thermal (right) boundary conditions.

Steel properties as Young modulus, conductivity, yield strength, specific heat and coefficient of elongation are highly temperature-dependent based on EN 1993-1-2 [10]. We use fully coupled thermal-stress analysis because we look at the phenomenon where element temperatures affect stress and displacement values. 11 tests were carried out with the time period equal to 900 seconds for 11 different values of the gas temperatures. It is a random variable $b \equiv T_s$ in this case.

Figure 3 shows the resulting displacement distribution on deformed specimens at the moment t=300, 900 s. We can see the difference between the length of it at the beginning and the end of the experiment. Necking is visible as well.

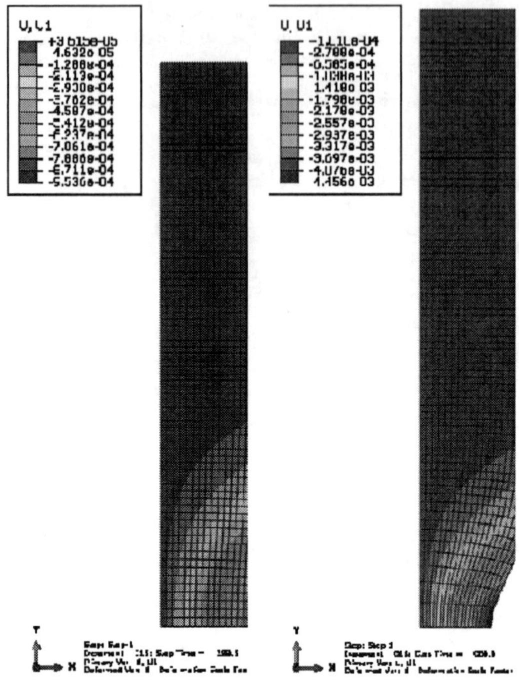

Fig. 3. Displacement at t=300, 900 s.

To describe the response of the structure we determined 18 different polynomial functions based on results from the FEM tests (Fig. 4).

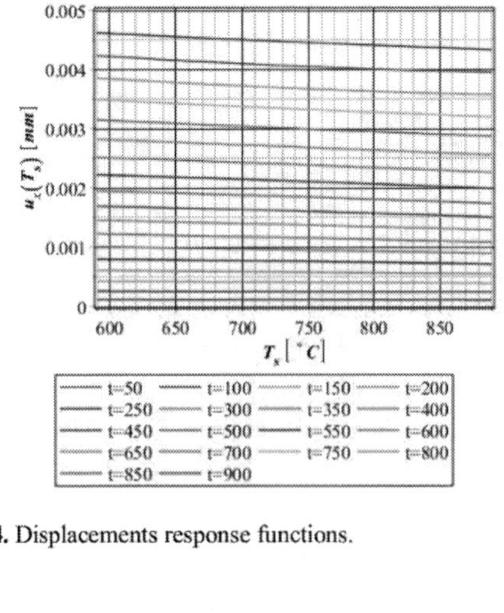

Fig. 4. Displacements response functions.

DOI: 10.1051/itmconf/20171503005

We are reminded that gas temperature is a random variable in this case with its input coefficient of variation $\alpha(T_s)$ in the range of [0.0,0.2]. Figures 5÷8 show expectations, coefficients of variation, skewness and kurtosis for extreme horizontal displacements in point *A* (Fig. 1) of the specimen as a time-dependent function. The expectations of displacements are concentrated around its mean values. We can observe very good convergence of the results obtained from semi-analytical (AM), Monte Carlo Simulation (MCS) and perturbation technique (PM). Expectations (Fig. 5) are independent of the input coefficient of variation $\alpha(T_s)$.

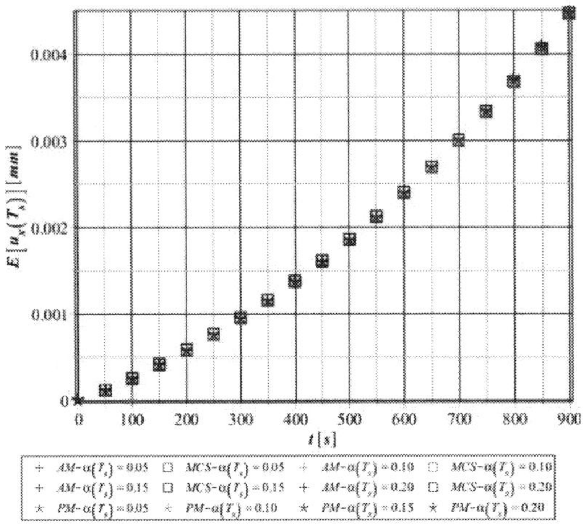

Fig. 5. Expectations of displacements.

Coefficients of variation (Fig. 6) are close to zero (small dispersion) for Monte Carlo Simulation (MCS) and it means that values of displacements are concentrated round its means. Using perturbation method (PM) and semi-analytical method (AM) for these calculations we obtain smaller values of the output than the input coefficients of variation in entire range $\alpha(T_s) \in [0;0.20]$.

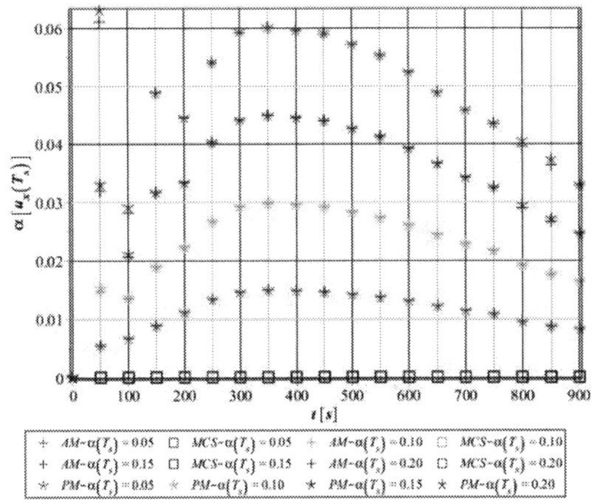

Fig. 6. Coefficient of variation of displacements.

Studying skewness distributions (Fig. 7) we can see fluctuations at the beginning and at the end of the analysis. According to this plot we can conclude that horizontal displacements u_x distribution is Gaussian when $\alpha(T_s) \leq 0.05$ because in this range skewness is close to zero based on all three approaches. Using Monte Carlo Simulation (MCS) we obtain smaller values in comparison with other methods. During all the time, skewness has positive and negative values, which brings problems with recognition of its distribution.

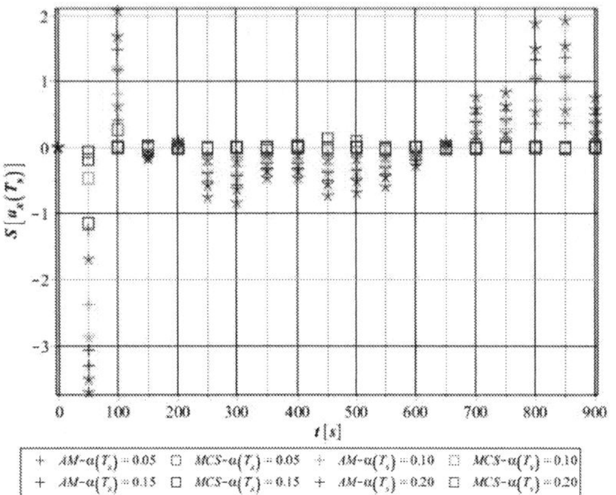

Fig. 7. Skewness of displacements.

Kurtosis of displacements (Fig. 8) takes only positive values. It means that most values of displacement are concentrated round its means. Using the Monte Carlo Simulation (MCS) and semi-analytical method (AM) we gain values close to zero. Kurtosis takes extreme twice about 50 and 800-850 second. A larger value of input coefficient of variation provides a greater value of kurtosis based on perturbation technique (PM).

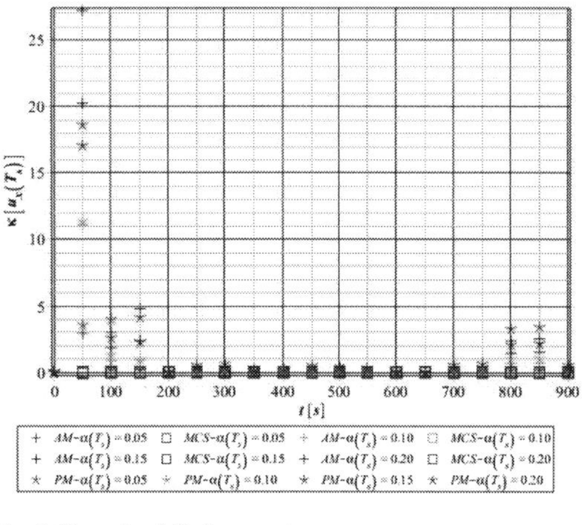

Fig. 8. Kurtosis of displacements.

4 Conclusions

Based on this computation we can say that the probability distribution of horizontal displacements u_x

 DOI: 10.1051/itmconf/20171503005

under fire conditions is Gaussian if we narrow the range for the input coefficient of variation to $\alpha(T_s) < 0.05$. This rule refers to all three methods. Moreover, we have shown the common usage of FEM software ABAQUS and computer algebra system MAPLE, which is crucial to use Stochastic Finite Element Method (SFEM) [11]. We focused on the influence of different methods on the results of stochastic analysis. In such non-linear problems, we have to use different functions to fit the response of the structure (Fig. 4). The information about displacement distribution obtained in this study can be used in the stochastic reliability analysis of the steel structures under fire according to both First and Second Reliability Methods [3]. We show that in fire conditions methods based on the Taylor series can be an alternative for the well-known Monte Carlo Simulation (MCS).

References

1. M. Kleiber, T. D. Hien, *The Stochastic Perturbation Method for Computational Mechanics* (Wiley, 1992)
2. ISO 834-1, Fire-resistance tests. Elements of building construction - Part 1: General requirements
3. M. Kamiński, *The Stochastic Perturbation Method for Computational Mechanics* (Wiley, 2013)
4. M. Kamiński, M. Strąkowski, Arch. Civ. Mech. Eng. **13**, 248-253 (2013)
5. M. Kamiński, P. Świta, Comput. Struct. **89**, 1241-1252 (2011)
6. R. Putanowicz, J. Jaśkowiec, P. Pluciński, A. Stankiewicz, Mech. Ctrl. **32**, 152-163 (2013)
7. M. Kamiński, J. Szafran, CMES: Comput. Model. Eng. Sci. **83**, 143-168 (2012)
8. N. Aravas, Int. J. Numer. Meth. Eng. **24**, 1395-1416 (1987)
9. O. C. Zienkiewicz, R. C. Taylor, *The finite element method, Vol. I, 4th Edition* (McGraw Hill, 1989)
10. EN 1993-1-2: Eurocode 3: Design of steel structures - Part 1-2: General rules - Structural fire design
11. G. Stefanou, M. Papadrakakis, Comput. Methods Appl. Mech. Eng. **193**, 139-160 (2004)

DOI: 10.1051/itmconf/20171503006

The assessment of water loss from a damaged distribution pipe using the FEFLOW software

Małgorzata Iwanek[1,*], and *Paweł* Suchorab[1]

[1]Lublin University of Technology, Faculty of Environmental Engineering, Nadbystrzycka 40 B, 20-618 Lublin, Poland

Abstract. Common reasons of real water loss in distribution systems are leakages caused by the failures or pipe breakages. Depending on the intensity of leakage from a damaged buried pipe, water can flow to the soil surface just after the failure occurs, much later or never at all. The localization of the place where the pipe breakage occurs is relatively easy when water outflow occurs on the soil surface. The volume of lost water strongly depends on the time it takes to localize the place of a pipe breakage. The aim of this paper was to predict the volume of water lost between the moment of a failure occurring and the moment of water outflow on the soil surface, during a prospective failure in a distribution system. The basis of the analysis was a numerical simulation of a water pipe failure using the FEFLOW v. 5.3 software (Finite Element subsurface FLOW systems) for a real middle-sized distribution system. Simulations were conducted for variants depending on pipes' diameter (80÷200 mm) for minimal and maximal hydraulic pressure head in the system (20.14 and 60.41 m H_2O, respectively). FEFLOW software application enabled to select places in the water system where possible failures would be difficult to detect.

1. Introduction

Water pipe breakages are an inherent part of the exploitation process of every water distribution system (WDS). Except the main consequence of pipe failure being the limitation of a water delivery to customers, pipe breakages are also common reasons of real water losses. According to World Bank data, 45 million m^3 of water is lost each day only in developing countries, due to WDS failures [1]. Water can be lost through burst leakages with a significant pressure loss or through background leakages. [2]. Burst leakages are relatively easy to detect as their effects are usually easy to notice (e.g. water outflow to the soil surface, suffusion holes, a lack of water delivery). Depending on the intensity of leakage from a damaged buried pipe, water can flow to the soil surface just after the failure occurrence or much later. It is also possible that water will never outflow to the soil surface, especially from hard-to-detect background leakages [3]. As the volume of lost water strongly depends on the pressure head and the time taken to localize the place of a pipe breakage, over the past decades numerous leakage detection methods and tools have been developed.

One of the most considerable ways to localize the place of pipe breakage is the usage of noise correlators, geophones, ground-penetrating radars, etc. The application of hardware tools is usually highly effective but also highly time-consuming and not applicable in real-time WDS diagnostics [4]. The other way to detect water leakage is the use of software methods. The most popular indirect leakage diagnosis methods are: Negative Pressure Wave method (NPW), Minimum Night Flow method (MNF) [5] and hydraulic modelling in combination with Supervisory Control And Data Acquisition system (SCADA). Moreover, the best results in water leakage detection are achieved by combining hardware and software methods – the area of possible water leakage is restricted thanks to software applications and then the leakage is localised by hardware tools.

Quick and precise detection and location of a pipe breakage results in relatively small water losses and therefore in small economic loss. However, water companies endeavour to limit the amount of lost water to an economically reasonable level. The optimal water loss level is defined as the amount of lost water of which financial value is smaller than the costs of water leakage minimising efforts [6]. That is why water companies decide to only repair pipe breakages that cause burst leakages with visible effects, often ignoring background leakages.

The aim of this paper was to select places in the analysed water system where possible pipe failures would be difficult to detect, and to predict the volume of water lost between the moment a failure occurs and the moment of water outflow to the soil surface. It was expected that the results could support the decisive process made by the water company to whether the breakage in a selected area is of a high or low repair priority. The basis of the analysis was a numerical simulation of a water pipe failure using the FEFLOW software (Finite Element subsurface FLOW systems) for a real middle-sized distribution system. Simulations were conducted for variants depending on the

* Corresponding author: M.Iwanek@wis.pol.lublin.pl

DOI: 10.1051/itmconf/20171503006

pipes' diameter for minimal and maximal hydraulic pressure head in the system.

2. Methodology

The analysis of water losses from a damaged water pipe was performed for the real WDS in the Lublin Region. It is a middle-size water supply system of an urban character, providing water to approx. 10 000 customers. The branched-looped water network is made of both traditional materials (steel, cast iron, asbestos-cement) – 55.32%, as well as from plastic pipes (PVC, PE) – 44.68%. Pipe diameters vary from DN80 up to DN200. The total length of the water network is approx. 54 km. The geometrical structure of an analysed WDS with diameter differentiation is presented in figure 1.

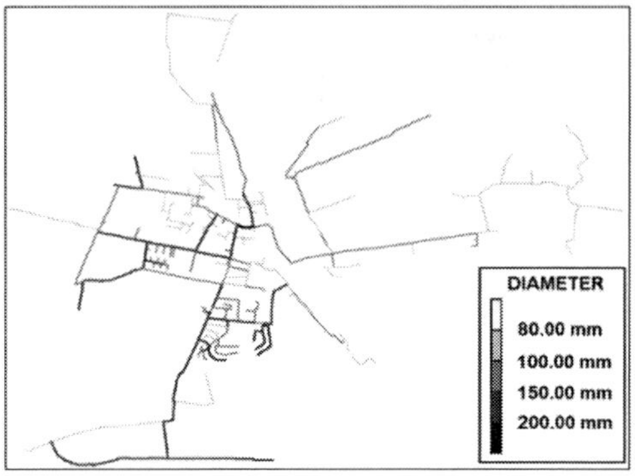

Fig. 1. Geometrical structure of the analysed WDS.

The analysed WDS is operated in non-zone configuration, but with relatively small daily pressure head variations in particular water supply nodes (e.g. pressure head amplitude in reference nodes equals to 3÷4 m H_2O maximum). However, the pressure head in the WDS in general varies from minimal acceptable levels (20 m H_2O) up to maximal acceptable levels (60 m H_2O). The contour plot of a hydraulic pressure head is presented in figure 2. "A" mark stands for the node with the lowest pressure head (20.14 m H_2O), and "B" mark identifies the node with the highest pressure head (60.41 m H_2O).

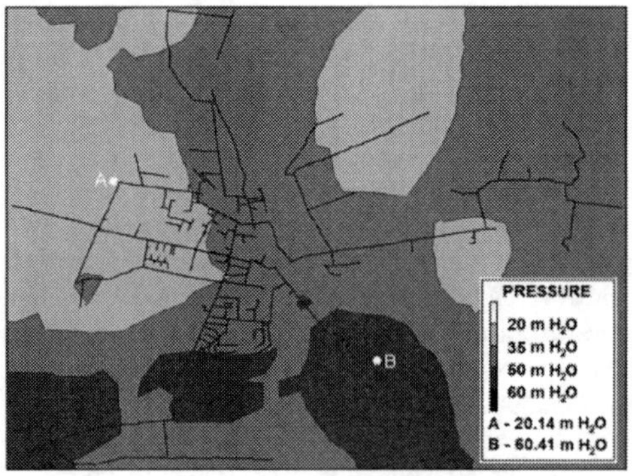

Fig. 2. Contour plot of a hydraulic pressure head in the analysed WDS.

Simulations of water losses were conducted for 12 different variants depending on the pipes' diameter and hydraulic pressure head in the system. The water supply network is composed of 4 different-sized traditional pipes (DN/ID 80, 100, 150, 200) and 2 different-sized plastic pipes (DN/OD 110, 160). Conducted variants of simulations included pipes of all size diameters in the analysed WDS and two extreme values of hydraulic pressure head inside pipes. Numerical simulations of a water pipe failure were performed in FEFLOW v. 5.3 software. It is a widely applied tool in simulations of water, pollutants, mass and heat transport in porous media and fractured media in two (2D) and three dimensional (3D) modelling. FEFLOW is based on Finite Elements/Volumes Methods (FEM/FVM) used to solve groundwater flow equation of both saturated and unsaturated conditions. The software was developed in ANSI C/C++ which allowed its application in Windows and Linux environments [7].

To simulate water pipe breakage in FEFLOW software, 2-dimentional soil profile was used, reflecting the cross section of a real water pipe environment in an excavation. Dimensions of a soil profile: 20 m x 5 m. The cross section of a soil profile with water pipe is presented in figure 3.

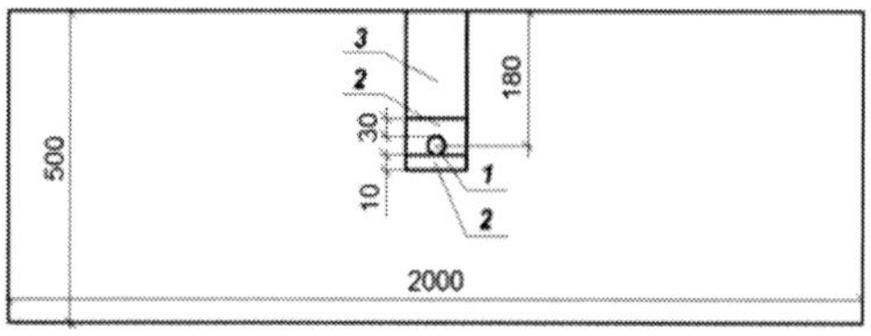

Fig. 3. Scheme of cross section of soil profile with the water pipe (1 – inner pipe diameter, 2 – sand, 3 – natural ground in and around the excavation; dimensions in cm).

Properties of unsaturated soils were characterized by the van Genuchten (1980) relationship given by formula 1. Hydraulic parameters of simulated soils are presented in table 1.

$$K = K_S S^l \left[1 - \left(1 - S^{\frac{1}{m}}\right)^m\right]^2 \qquad (1)$$

$$S = \frac{\theta - \theta_r}{\theta_s - \theta_r} = \frac{1}{[1 + (\alpha H)^n]^m} \qquad (2)$$

$$m = 1 - \frac{1}{n} \qquad (3)$$

where:
K – hydraulic conductivity, m/s
K_S – saturated hydraulic conductivity coefficient, m/s
S – degree of saturation given by formula (2)
l – fitting parameter [9]
m – fitting parameter given by formula (3)
θ – volumetric water content, m^3/m^3
θ_r – residual volumetric water content, m^3/m^3
θ_s – saturated volumetric water content, m^3/m^3
α, n – fitting parameters, 1/m, -
H – pressure head, m

 DOI: 10.1051/itmconf/20171503006

Table 1. Hydraulic parameters of soils used in simulations.

Soil (according to Fig. 3)	*S*	*Ks*	*P*	*α*	*n*
	-	10^{-4} m/s	% (vol.)	1/m	-
No 3 in and around the excavation	0.442	0.488	0.217	2.968	1.652
No 2 in the excavation	0.323	0.626	0.366	2.800	2.239

P – porosity

Two different types of boundary conditions were used in the model: the first-type (Dirichlet) boundary condition on the leak ($H = 20.14$ m H_2O or $H = 60.41$ m H_2O) and on the bottom of a soil profile ($H = -1$ m H_2O – specific value in a water retention curve), as well as the second-type (Neumann) as a boundary condition on the top of a soil profile. A scheme of soil profile with an exemplary finite element mesh (12843 elements for DN200) and boundary conditions are presented in figure 4.

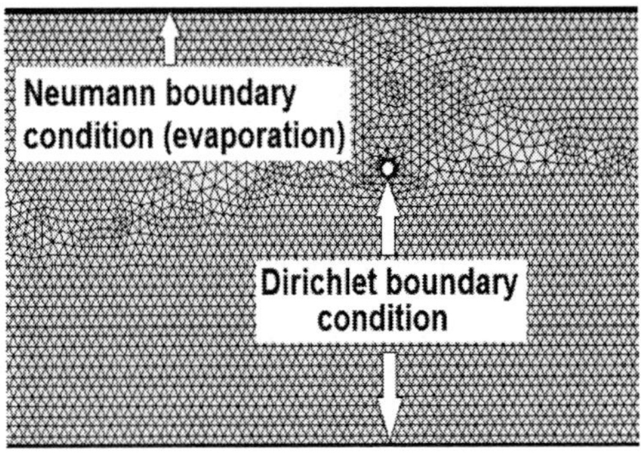

Fig. 4. Boundary conditions and finite element mesh of the exemplary analysed soil profile (for pipe diameter DN200).

The breakage of a pipe was shown as a circular 5 mm unsealing of the pipe. That assumption was made on the basis of results of numerical investigations [8,9] and laboratory investigations [10] of water outflow to the soil surface after the water pipe breakage. Pipes' diameters (nominal (DN) and inner (ID)) and areas of a leak are presented in table 2.

Table 2. Pipes parameters and leakage areas.

DN (mm)	ID (mm)	Wall thickness (mm)	Unsealing width (mm)	Unsealing area (m^2)
80	80	n/a	5	0.0025
110	97.1	6.45		0.0030
100	100	n/a		0.0031
160	141.2	9.4		0.0044
150	150	n/a		0.0047
200	200	n/a		0.0063

The result of each FEFLOW simulation variant was the time of water outflow to the soil surface. Based on obtained results, the total volume of lost water during the moment of a pipe breakage and water surface outflow was calculated using formula (4) [11].

$$V = Q \cdot t = A \cdot C_d \cdot \sqrt{2 \cdot g \cdot H} \cdot t \qquad (4)$$

where:
V – volume of lost water, m^3
Q – unitary water outflow, m^3/s
t – time of water outflow to the soil surface, s
A – leak area, m^2
C_d – discharge coefficient, -
g – gravitational acceleration, m/s^2
H – pressure head, m

The C_d coefficient, given by formula 5 – symbols as in formula 4, is defined as the ratio between the actual flow discharge (Q_{act}) and theoretical discharge resulting from the Torricelli's theorem.

$$C_d = \frac{Q_{act}}{A \cdot \sqrt{2 \cdot g \cdot H}} \qquad (5)$$

Q_{act} was estimated on a basis of the numerical simulation in the FEFLOW software, assuming outflow of water through an orifice to sand medium (parameters as for soil No 2 in the excavation – tab.1.). Details about the profile are given in figure 5.

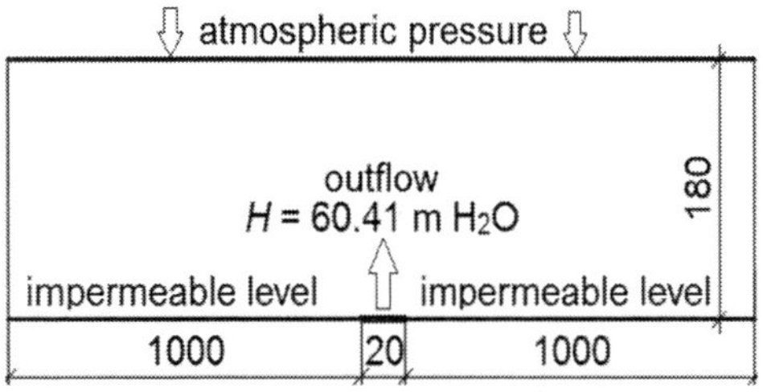

Fig. 5. Scheme of cross-section of soil profile to estimate C_d coefficient (dimensions in cm).

The final result of the study was the priority of the estimation of areas in which the water supply system had the highest possibility of hard-to-detect background leakages. Priorities were accepted in reference to the time between water pipe breakage and water outflow to the soil surface. Three types of water losses zones priorities were estimated:

- high priority where potential background leakages will be difficult to observe and where it is highly recommended to systematically implement software and hardware leakage detection methods,
- medium priority – implementation of hardware leakage detection methods is recommended occasionally,
- low priority – only software leakage detection methods are recommended due to the high possibility of water outflow to the soil surface together with significant pressure loss.

Investigations were carried out in the following order: estimation of a C_d coefficient, simulation of time of water outflow to the soil surface in FEFLOW software, the calculation of the amount of lost water during the pipe breakage identification of water supply priority zones.

3. Results

In accordance to presented methodology, the calculated value of discharge coefficient C_d equals 0.166. In water outflow to the soil medium, the efficiency of C_d is clearly lower when compared to the outflow to the atmosphere, because of higher flow resistance and contraction.

Results of numerical simulations of time between the moment of water leakage from a damaged pipe and the water outflow occurrence on the soil surface are presented in table 3. As expected, the smaller leak area (pipe diameter) and pressure head, the higher value of time. The extreme result was observed for the case of diameter DN80 and pressure head $H = 20.14$ m H_2O – time of water outflow equalled nearly 14 minutes. On the contrary, the shortest time was noted in the case of diameter DN200 and pressure head 60.41 m H_2O – time equalled 2.38 minutes. Exemplary saturation of soil profiles in moments of water outflows for extreme mentioned variants are presented in figures 6 and 7.

The amount of lost water due to pipe breakage is also presented in table 3. The case of the highest time value of water outflow on the soil surface does not result in the greatest amount of lost water as the volume of lost water depends also on leak area and the pressure head. The extreme case of the amount of lost water is DN200 and $H = 2.14$ m H_2O in which during 7.65 min of water leakage, over 9.5 m^3 of water is lost into the ground. Considering all the factors influencing the volume of lost water, the general observed trend is that the bigger the leak area for the same value of pressure head, the bigger the amount of volume of lost water. However, as the bigger pressure head results in shorter water outflow time on the soil surface, volumes of lost water are smaller for the pressure head equalled to 60.41 m H_2O than for 20.14 m H_2O.

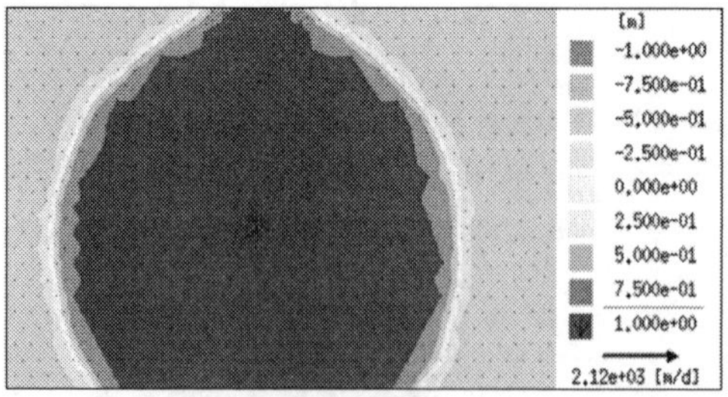

Fig. 6. Saturation of soil profile in the moment of water outflow (DN80, $H = 20.14$ m H_2O).

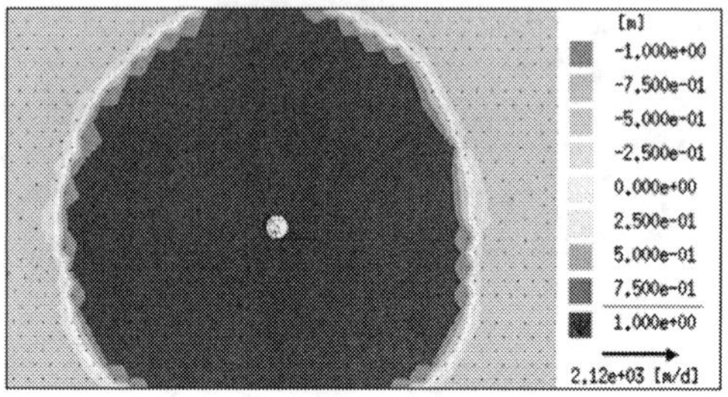

Fig. 7. Saturation of soil profile in the moment of water outflow (DN200, $H = 60.41$ m H_2O).

Graphical illustrations of dependence between the leak area, pressure head and the time of water outflow and volume of lost water are presented in figure 8. General observed dependence is that the lower the pressure head and smaller the leak area, the higher the water time outflow and therefore the bigger water losses.

Table 3. Simulation results of water time outflow and volume of lost water.

Pressure head (m H_2O)		**20.14**		**60.41**		C_d	**20.14**	**60.41**	**20.14**	**60.41**
Parameter		t		t			Q		V	
DN (mm)	**ID (mm)**	**s**	**min**	**s**	**min**	**-**	**m^3/s**	**m^3/s**	**m^3**	**m^3**
80	80	839.14	13.99	198.55	3.31	0.166	0.0083	0.0144	6.96	2.85
110	97.1	662.46	11.04	183.92	3.07		0.0101	0.0174	6.67	3.21
100	100	608.24	10.14	160.28	2.67		0.0137	0.0180	6.31	2.88
160	141.2	536.75	8.95	153.02	2.55		0.0146	0.0238	7.86	3.88
150	150	520.49	8.67	151.14	2.52		0.0156	0.0269	8.10	4.07
200	200	459.17	7.65	142.90	2.38		0.0207	0.0359	9.53	5.13

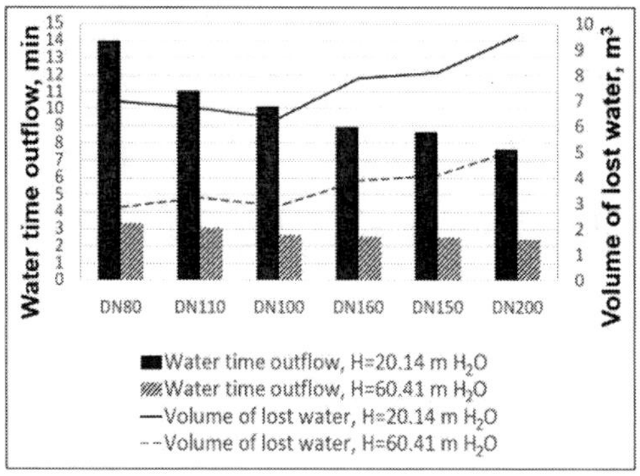

Fig. 8. Time of water outflow and the volume of lost water for different pipe diameters and pressure heads.

On the base of obtained results, in the analysed WDS three different types of priority zones were marked, which are presented in figure 9. The zone of high priority was estimated mainly in areas of low pressure head with small pipes diameters. In high priority zones, it is strongly recommended to systematically implement software and hardware water loss detection methods. In medium priority zones the occasional application of hardware tools should be sufficient, while in low priority zones software methods should be efficient enough.

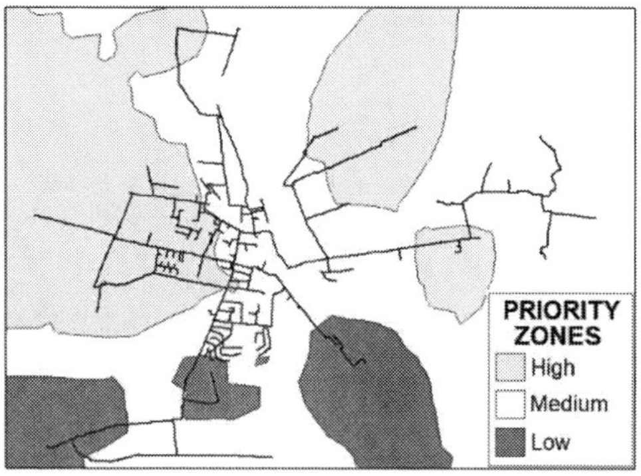

Fig. 9. Water losses priority zones of the analysed WDS.

4. Conclusion

The obtained results confirmed the expected conclusions that time of water outflow increases simultaneously with diminishing the leak area and pressure head. In the article, the most unfavourable case of possible leak area was taken into account (water leaking on the whole pipe circuit) which resulted in relatively short water time outflow. In real conditions, pipe holes which occur during pipe breakages are often of a smaller size – which will probably result in higher water outflow time and bigger water losses. Moreover, in these investigations it was assumed that water will always outflow to the soil surface, but it does not always happen in real soil conditions.

Dividing WDS into zones of different priorities of applying software and hardware methods of water losses detection may be helpful for the water company in developing the most efficient water losses strategy reduction. FEFLOW software is a very useful tool to estimate the water time outflow without laboratory examinations. In the case of analysed WDS, it is highly recommended that zones of high priority should be regularly monitored because of a high possibility of significant water losses during the occurrence of background leakages. Areas of a low pressure should be monitored the most carefully, because the potential pipe failure may be invisible on the soil surface and therefore may result in long-term water leakage.

In general, water companies should individually develop a water losses reduction strategy together with establishing the acceptable economical level of water losses. It is worth remembering, that a small level of real losses results not only in economic benefits of the company, but also positively influences on the company's social perception. Customers are not obligated to pay for the water losses and the company may demonstrate that the environment protection and realization of sustainable development principles are essential points of the company's managing policy.

References

1. World Bank Press Release: www.worldbank.org/en/news/press-release/2016/09/01/the-world-bank-and-the-international-water-association-to-establish-a-partnership-to-reduce-water-losses [access 8.08.2017]
2. L. Berardi, D. Laucelli, R. Ugarelli, O. Orazio Giustolisi, Procedia Eng. **119(1)** (2015)
3. M. Iwanek, P. Suchorab, M. Karpińska-Kiełbasa, PERIOD. POLYTECH-CIV. **61(4)** (2017) DOI: 10.3311/PPci.9728
4. J. Stańczyk, E. Bursztа-Adamiak, D. Poor, GWiTS **5(91)** (2017) DOI: 10.15199/117.2015.5.4 (in Polish)
5. Z.Y. Wu, M. Farley, D. Turtle, Z. Kapelan, J. Boxall, S. Mounce, S. Dahasahasra, M. Mulay, Y. Kleiner, *Water Loss Reduction* (Bentley Institute Press, 2011)
6. S. Speruda, Radecki R. *Ekonomiczny poziom wycieków. Modelowanie strat w sieciach wodociągowych* (Warszawa: Translator S. C., 2007) (in Polish)
7. M.K. Widomski, D. Kowalski, M. Iwanek, G. Łagód, *Modeling of Water Flow and Pollutants Transport in Porous Media With Exemplary Calculations in FEFLOW* (Lublin University of Technology Monography, 2013)
8. P. Suchorab, B. Kowalska, D. Kowalski, Annu. Set. Env. Protection **18(2)** (2016)
9. M. Iwanek, D. Kowalski, M. Kwietniewski, Ochrona Środowiska **37(4)** (2015) (in Polish, with English abstract)
10. M. Iwanek, B. Kowalska, E. Hawryluk, K. Kondraciuk, Eksploat. Niezawodn. **18(2)** (2016) DOI: 10.17531/ein.2016.2.16
11. J. E. van Zyl. Procedia Eng. **89** (2014)

DOI: 10.1051/itmconf/20171503007

The modelling of condensation in horizontal tubes and the comparison with experimental data

Rafał Bryk[1,*], *Holger* Schmidt[2], *Thomas* Mull[2], *Thomas* Wagner[2], *Oliver* Herbst[2], and *Ingo* Ganzmann[2]

[1] Warsaw University of Technology, Institute of Heat Engineering, 00-661 Warsaw, Poland
[2] AREVA GmbH, Thermal Hydraulics and Components Testing, 91052 Erlangen, Germany

Abstract. The condensation in horizontal tubes plays an important role in determining the operation mode of passive safety systems of modern nuclear power plants. In this paper, two different approaches for modelling of this phenomenon are compared and verified against experimental data. The first approach is based on the flow regime map developed by Tandon. Depending on the regime, the heat transfer coefficient is calculated according to corresponding semi-empirical correlation. The second approach uses a general, fully empirical correlation proposed by Shah. Both models are developed with utilization of the object-oriented, equation-based Modelica language and the open-source Open-Modelica environment. The results are compared with data obtained during a large scale integral test, simulating a Loss of Coolant Accident scenario performed at the dedicated Integral Test Facility Karlstein (INKA) which was built at the Components Testing Department of AREVA in Karlstein, Germany. The INKA facility was designed to test the performance of the passive safety systems of KERENA, the new AREVA boiling water reactor design. INKA represents the KERENA containment with a volume scaling of 1:24. Components heights and levels over the ground are in the full scale. The comparison of simulations results shows a good agreement.

1 Introduction

KERENA is a medium-capacity boiling water reactor developed by AREVA GmbH [1]. It utilizes several innovative passive safety systems that ensure safe depressurization and heat removal in case of an emergency or accidents, including a loss of coolant accident (LOCA). The main components of KERENA passive safety features are a Containment Cooling Condenser (CCC), Emergency Condenser (EC), Passive Pressure Pulse Transmitters (PPPT) and a Passive Core Flooding System. These systems utilize basic physical laws such as gravity and free convection. Therefore, in order to operate them, there is no need for an external power supply and additional devices such as pumps. This in turn leads to a significant reduction of costs during construction of the plant.

In order to examine the performance of the passive systems, a dedicated test facility was built at the Components Testing Department of AREVA in Karlstein, Germany. The Integral Test Facility Karlstein (INKA) represents the KERENA containment with a volume scaling of 1:24. Components heights and levels over the ground are kept in the full scale. The test rig setup allows us to investigate each component individually as well as the entire system during integral tests. For the purposes of testing the integral system, INKA is equipped with several large vessels corresponding to individual pools and vessels of KERENA. The steam accumulator of the Karlstein large valve test facility GAP represents the Reactor Pressure Vessel (RPV). The vessel has a storage capacity of 1/6 of the KERENA RPV. In order to simulate the decay heat of the core, the vessel is fed with steam by a Benson boiler with maximum power output of 22 MW. The vessel is designed for operational pressure up to 160 bars so experiments may be performed starting from the operation point of boiling water reactor of 70 bars.

Compartments of the KERENA RPV are represented by three separate vessels: Flooding Pool Vessel, Pressure Suppression Pool Vessel and Drywell Vessel. An additional tank is used as a representation of Shielding/Storage Pool Vessel.

Due to the appropriate instrumentation concept and the density of sensors, data obtained during experiments performed at the facility may be utilized for validation of thermal-hydraulic codes. Therefore, measurements collected during one of the tests performed in the framework of the Integral Experimental and Analytical Safety Cases for Design-Basis Accidents Featuring Passive Systems were used to validate condensation models developed in open source OpenModelica Connection Editor environment utilizing Modelica language.

Modelica is a non-proprietary, object-oriented, equation-based language to conveniently model complex physical systems [2]. Several environments utilize the language. Models described in this paper were developed with OpenModelica Connection Editor – an open-source

* Corresponding author: test-labs@areva.com

environment developed by Open Source Modelica Consortium for industrial and academic usage [3]. Modelica is equipped with a large multi-domain, open-source library. Components encapsulated in the library may be developed; enhanced and other objects may inherit their features.

2 Operation of KERENA Emergency Condenser

The two main purposes of the Emergency Condenser are heat removal and depressurization of the RPV in abnormal conditions. The principle of the EC operation is presented in Fig. 1.

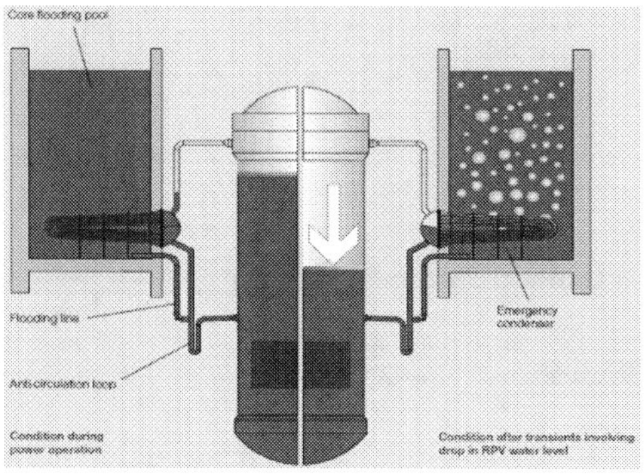

Fig. 1. Operation scenarios of the Emergency Condenser.

The EC is composed of a large number of slightly inclined tubes. The tubes are submerged in the Core Flooding Pool and they are filled with water during regular operation of the plant. Should the water level in the RPV decrease, the water drains from the tubes and the steam from the RPV enters the EC tubes. Due to the contact of the tube walls with the cold water at the secondary side in the Core Flooding Pool heat transfer occurs, leading to condensation of the steam in the tubes. Since the tubes of the Emergency Condenser are inclined slightly, the condensate flows back to the RPV due to gravity. The Emergency Condenser is connected to the RPV without any isolating elements and it comes into action automatically without any need for electrical power. Thus, the device is operated according to the natural law of communicating vessels.

3 Models development

3.1 System description and governing equations

The analysed thermal-hydraulic system was a straight pipe slightly inclined to horizontal orientation. The flow is governed by the mass, energy and momentum conservation equations, as 1-dimensional, partial differential equations, to which the following assumptions are applied:

- uniform velocity is assumed on cross sections,
- longitudinal heat diffusion term is neglected,
- accumulation is considered in each volume,
- influence of the angulation of the EC pipes is neglected,
- thermal-hydraulic properties are calculated on the basis of pressure and enthalpy at each node,
- boundary conditions are assumed according to the measurements,
- temperature at the external wall of the tube is taken from measurements,
- tube wall is assumed to have constant thermal conductivity,
- heat transfer coefficient at the secondary side is neglected,
- tube length and wall thickness are modelled according to EC specification
- tube bundle influence is not taken into account.

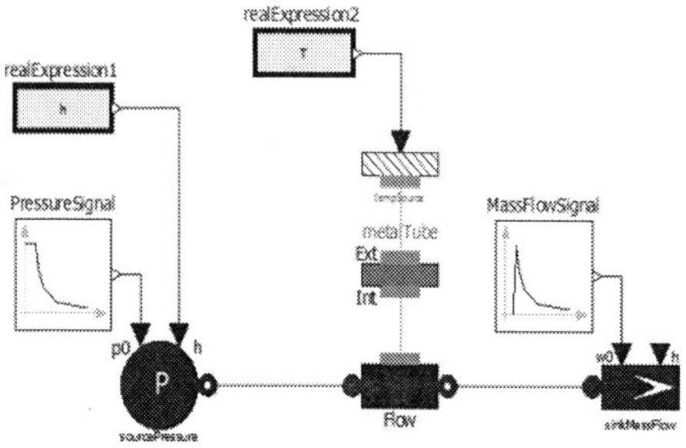

Fig. 2. Object-oriented configuration of the model.

The model was set up by adopting an extended version of the Flow1DFV2ph component of the Modelica ThermoPower library. The component, in which the balance equations are discretized by means of the finite volume approach, was modified by formulation of momentum balance equation for each volume instead of the whole component. This leads to a more detailed description by enabling the calculation of fluid properties and saturation conditions at each node, taking into account pressure in the particular node. Before modification, fluid properties at each node were evaluated either with respect to inlet or outlet pressure.

The mass balance equation in each volume is given by:

$$\dot{m}_j - \dot{m}_{j+1} = Al\frac{d\rho_k}{dt} \quad (1)$$

Taking pressure and specific enthalpy as state variables:

$$\dot{m}_j - \dot{m}_{j+1} = Al\left[\left(\frac{\partial\rho_k}{\partial h_k}\right)_p * \frac{dh_k}{dt} + \left(\frac{\partial\rho_k}{\partial p_j}\right)_h * \frac{dp_j}{dt}\right] \quad (2)$$

The energy balance equation in each volume is given by:

$$\dot{m}_j * h_j - \dot{m}_{j+1} * h_{j+1} + Q_k = Al(\rho_k\frac{dh_k}{dt} - \frac{dp_k}{dt}) \quad (3)$$

The momentum balance equation for each volume is given by:

$$p_j - p_{j+1} = \frac{l}{A} * \frac{dw_k}{dt} + \rho_k g l sin\theta + \Delta p_{k,f} + \Delta p_{k,a} \quad (4)$$

with acceleration and friction pressure losses for each volume given by:

 DOI: 10.1051/itmconf/20171503007

$$\Delta p_{k,a} = \frac{1}{A^2} * w_k{}^2 * \left(\frac{1}{\rho_{j+1}} - \frac{1}{\rho_j}\right) \quad (5)$$

$$\Delta p_{k,f} = \zeta \frac{2\Lambda_k l}{D A^2 \rho_k} * w_k^2 \quad (6)$$

Where Λ_k is calculated with Colebrook correlation:

$$\Lambda_k = \frac{0.332}{\ln\left(\frac{e}{3.5}+\frac{5.47}{Re_k^{0.9}}\right)^2} \quad (7)$$

In equations, 1-7 A and l stand respectively for tube cross section length of a single volume. Subscript K denotes volume number while J stands for the number of the nodes. Subscripts p and h mean derivatives with constant pressure and enthalpy respectively.

In the flow model, a replaceable module for heat transfer calculation was introduced. This allowed calling different models of heat transfer developed as separate objects.

Regarding the modelling of the heat exchange, new heat transfer components were introduced in order to implement correlations expressing two approaches described in subsections 3.2 and 3.3. These models were then called in the flow model by redeclare model clause in the text view of the whole system model presented as diagrams in fig. 2.

3.2 Flow regime map based model

The flow regime map based model was developed utilizing the map proposed by Tandon [8]. The model calculates the actual heat transfer coefficient by switching between several flow regimes dependent on semi-empirical correlations. Fig. 3 illustrates different flow patterns in horizontal tubes. The flow regime map, together with correlations applied in specific regime is shown in Fig. 4.

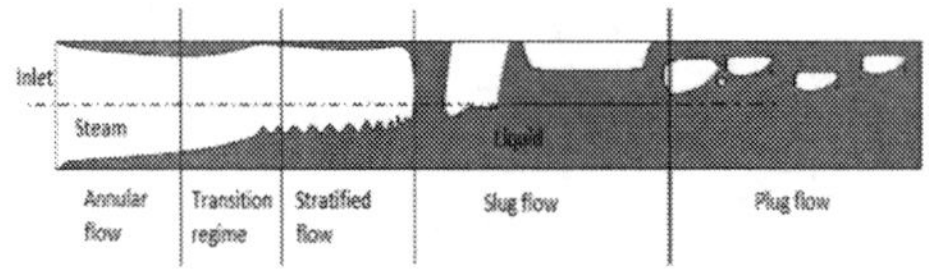

Fig. 3. Flow patterns during condensation in horizontal tube [4].

Tandon's flow regime map is based on the volume ratio of liquid and gas in a cross section area $(1 - \varepsilon)/\varepsilon$, where ε stands for the void fraction, and on the dimensionless steam velocity:

$$j_D^* = \frac{xG}{[gD\rho_V(\rho_L-\rho_V)]^{0.5}} \quad (8)$$

Since the EC pipes are initially filled with liquid water and due to possible single-phase condensate flow at the lower sections of the tube, an additional correlation for single phase liquid water flow was implemented into the model. Table 1 indicates all correlations and application ranges.

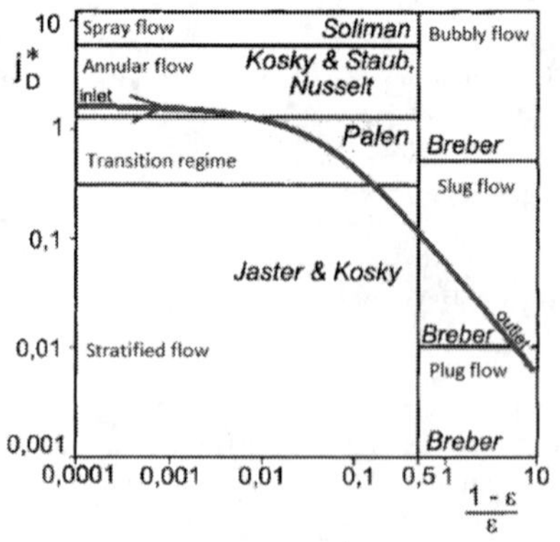

Fig. 4. Tandon's flow regime map [4].

In the table subscripts, Fi and V indicate whether a variable refers to condensate film or vapour. Further details of different pattern formation development of the correlations can be found in [8-14].

3.3 General correlation model

The implemented general correlation for condensation in horizontal tubes was developed by Shah [16]:

$$\alpha = \alpha_{SPh}\left[(1-x)^{0.8} + \frac{3.8 * x^{0.76} * (1-x)^{0.04}}{p_r^{0.38}}\right] \quad (9)$$

$$\alpha_{SPh} = 0.023 Re_L^{0.8} Pr_L^{0.4} \frac{\lambda}{D} \quad (10)$$

The correlation was developed by the analysis with a wide variety of experimental data. This includes different fluids and pipe orientation. Details of development of the correlation may be found in [16].

4 Results

Experimental data for model validation was obtained during the simulation of the Main Steam Line Break (MSLB), taking into account heat decay. The test was performed at the INKA test facility in the framework of the *EASY* Project. Since the MSLB experiment was conducted in order to investigate performance of the integral system rather than a single component, it should be emphasised that other components like Reactor Pressure Vessel, Drywell Vessel, Pressure Suppression Pool, Shielding/Storage Pool and Standpipe strongly influenced the behaviour of the Emergency Condenser. Furthermore, the test course assumed certain activities at certain points, which lead to strong transients and backflows. This can be seen in Fig. 5, where measured flows are plotted together with mass flow applied to the models. The red line, representing the inlet mass flow, is clearly delayed with respect to outflow measurement. Paradoxically, thermocouples inside EC tubes indicate an increase in water temperature (fig. 7) before steam enters the tubes according to Fig. 5.

DOI: 10.1051/itmconf/20171503007

Table 1. Heat transfer correlations with respect to flow regime.

Flow regime	Correlation	Range of application	Further details of calculations
Annular laminar [7]	$\alpha_{Ann\_lam} = 1.1025\,\lambda_{Fi} Re_{Fi}^{-0.33}\left[\frac{\rho_{Fi}(\rho_{Fi}-\rho_V)g}{\mu_{Fi}^2}\right]^{0.33}$	$1 \le j_D^* \le 6$ and $\frac{1-\varepsilon}{\varepsilon} \le 0.5$; $Re_{Fi} < Re_{wave}$	$Re_{Fi} = \frac{(1-x)GD}{\mu_{Fi}}$; $Re_{wave} = 0.392\left[\left(\frac{\sigma}{\rho_{Fi} g}\right)^{0.5}\left(\frac{g}{v_{Fi}^2}\right)^{0.33}\right]^{0.55}$
Annular wavy [11]	$\alpha_{Ann\_wave} = 0.8 Re_{Fi}^{0.11} * \alpha_{Nu}$	$1 \le j_D^* \le 6$ and $\frac{1-\varepsilon}{\varepsilon} \le 0.5$; $Re_{wave} < Re_{Fi} < Re_{crit}$	$Re_{crit} = \frac{100}{Pr_{Fi}}$
Annular turbulent [12]	$\alpha_{Ann\_tur} = \frac{\rho_{Fi} v^* c_{p,Fi}}{T^+(\delta^+)}$	$1 \le j_D^* \le 6$ and $\frac{1-\varepsilon}{\varepsilon} \le 0.5$; $Re_{crit} < Re_{Fi}$	$\delta^+ = 0.0504(0.5\,Re_{Fi})^{0.5}$ for $Re_{Fi} < 1000$; $\delta^+ = 0.0504(Re_{Fi})^{\frac{7}{8}}$ for $Re_{Fi} \ge 1000$; $T^+(\delta^+) = \delta^+ Pr_{Fi}$ for $\delta^+ \le 5$; $T^+(\delta^+) = 5\left[Pr_{Fi} + \ln\left(1 + Pr_{Fi}(0.2\,\delta^+ - 1)\right)\right]$ for $5 < \delta^+ < 30$; $T^+(\delta^+) = 5\left[Pr_{Fi} + \ln(1+5Pr_{Fi}) + 0.495\ln(\frac{1}{30}\delta^+)\right]$ for $\delta^+ > 30$; $v^* = \sqrt{\frac{D}{4\rho_{Fi}}\left(\frac{dp}{dx}\right)_{fric,TP}}$
Stratified [7]	$\alpha_{Strat} = \varepsilon^{0.55} * \alpha_{Ann\_lam}$	$j_D^* < 1$ and $\frac{1-\varepsilon}{\varepsilon} \le 0.5$	
Bubble, slug, plug laminar [14, 15]	$\alpha_{BSP\_lam} = 1.86\frac{\lambda_{Fi}}{D}\left(Re_{Fi} Pr_{Fi}\frac{D}{l}\right)^{0.33}\left(\frac{\mu_{Fi}}{\mu_w}\right)^{0.14}$	$\frac{1-\varepsilon}{\varepsilon} > 0.5$; $Re < 2000$	
Bubble, slug, plug turbulent [14, 15]	$\alpha_{BSP\_tur} = 0.024\frac{\lambda_{Fi}}{D}\left(Re_{Fi}^{0.8} Pr_{Fi}^{0.33}\right)\left(\frac{\mu_{Fi}}{\mu_w}\right)^{0.14}$	$\frac{1-\varepsilon}{\varepsilon} > 0.5$; $Re \ge 2000$	
Spray [9]	$\alpha_{spray} = 0.00345\,Re_{av}^{0.9} Pr^{0.33}\left(\frac{r}{c_{p,V}(T_S - T_W)}\right)^{-0.33}$	$j_D^* > 6$ and $\frac{1-\varepsilon}{\varepsilon} \le 0.5$	$Re_{av} = \frac{GD}{\mu_{mix}}$; $\mu_{mix} = \frac{x}{\mu_V} + \frac{1-x}{\mu_L}$
Single phase	$\alpha_{SPh} = 0.023\frac{\lambda}{D} Re^{0.8} Pr^{0.4}$	$\varepsilon = 0$	

This means that steam enters the EC tubes earlier than indicated in Fig. 4, but pressure transients resulted in negative mass flow at the inlet, which is presented as zero flow by the red line between 350 and 425 seconds of the test.

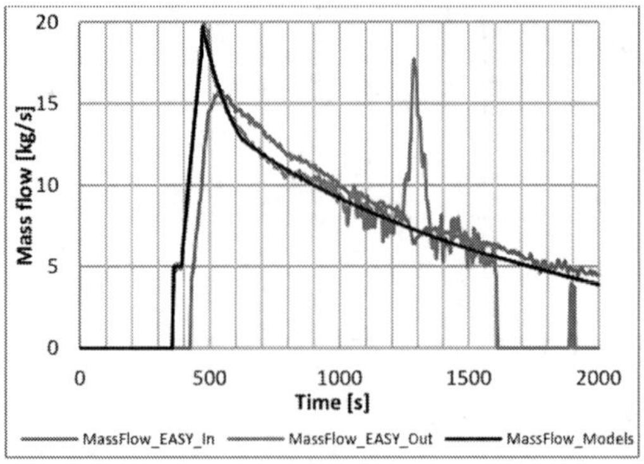

Fig. 5. Measured and modelled mass flows.

Similarly, the measured outflow peak after 1200 seconds and zero-flow after 1500 are caused by the performance of other devices. These cannot be taken into account during validation of condensation model. Therefore, only a part of the test was considered for further analysis. The time taken into account was between 0 and 1200 seconds of the test. This covers the initial operation of the system and the response of the EC to the fast transient caused by steam line break and corresponding RPV water level decrease. The break was initiated 350 seconds after the beginning of the test. At this time, steam from the RPV was released into the DWV, leading to a decrease of pressure in the RPV. This in turn resulted in RPV water level decreasing and forced the flow from the standpipe and EC into the RPV.

The mass flow between EC and RPV enabled the steam to enter the EC tubes (as presented on the right side of fig.1). The contact of saturated steam with cold tube walls leads to heat exchange and condensation of steam. The power of the EC calculated on the basis of measurements is presented in the figure below together with these computed by the models. The dashed lines indicate +/- 15% deviations from the measurement curve.

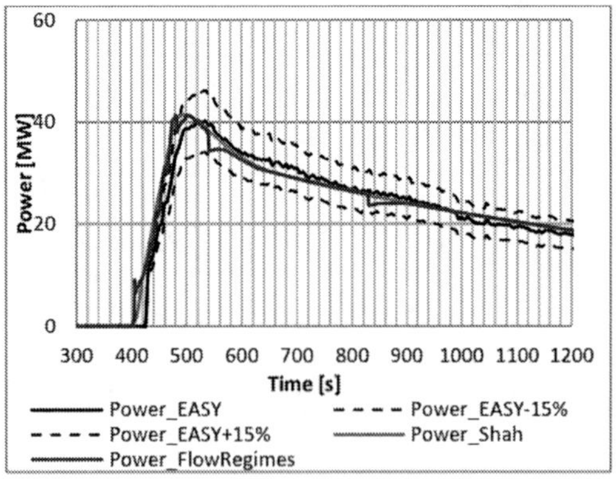

Fig. 6. Heat flow rates according to measurements and models.

Fig. 6 shows a good agreement between heat flow rate obtained from measurements and these computed by the models. Although both curves plotted according to simulations results have similar shapes, analysis of them clearly indicates their different origins. The blue line has a smooth shape, since power is calculated based on heat transfer coefficient correlation. Peaks of the red line result from switching to different correlations depending on transient flow regime. These peaks are even more apparent during analysis of single volumes. Nevertheless, both models provide satisfactory agreement.

Regarding the fluid temperature, two points were chosen to compare measurements against calculations. The first measurement point is situated before the angulation of the EC pipes i.e. at the x=4.423 m, where x is the longitudinal coordinate of the pipe. Corresponding calculation point was x=4.32 m. The discrepancy of around 10 cm is caused by the location of thermocouples during the test and nodalization of the models. The second measurement point is located at x=5.756 m i.e. at the bottom part of angulation of the pipe. The picture below shows temperatures measured at both locations and calculated values at corresponding positions. Solid lines represent the first point while dashed lines demonstrate fluid temperature at second point.

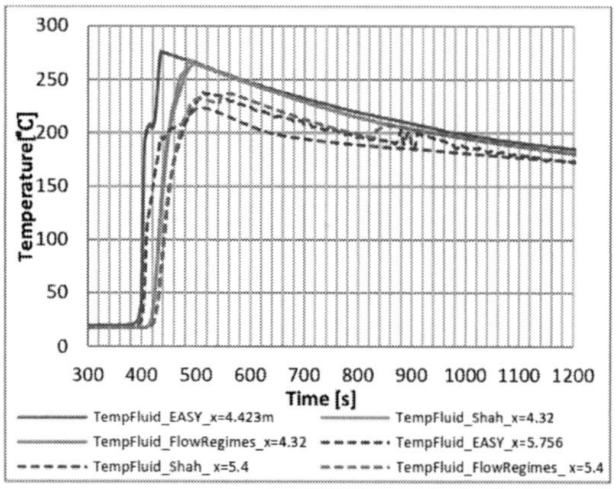

Fig. 7. Fluid temperatures.

Solid lines in Fig. 7 indicate that around 4.3 meters from the tubes inlet the flow is still a two-phase mixture at the saturation temperature. Fast transients of pressure and mass flows cause the delay of the temperature increase after initialization of the break. The dashed lines demonstrate that at the second position, the flow is already subcooled. The time delays of the transient in both models are still present, but it can be noticed that the flow regimes based model evaluates the temperature more accurately.

Considering the wall temperature, due to the falling water film and stratification, significantly different values were measured at the upper and bottom wall of the tube. Therefore, two measurements are presented in the chart below. Both come from the part of EC before the angulation, one represents the upper part of the wall, and the second the bottom part. Since the water film at the bottom part of the tube is thicker, the heat transfer at this part of the tube is lower, and so is the temperature. The figure below shows measured and calculated values.

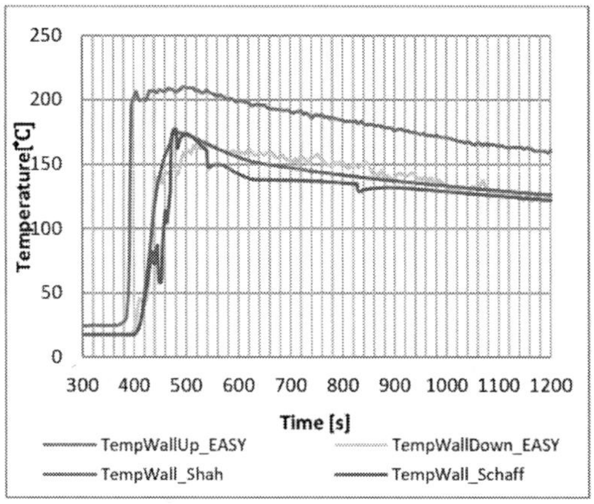

Fig. 8. Wall temperatures.

Since the wall temperature in both models was calculated without distinction of the top and bottom side, there is a significant discrepancy between measured and calculated values. Wall temperature calculated by the models is much better to evaluate than the one measured at the bottom part, while the measured one at the top part is around 40°C higher. This significant discrepancy will be the subject of future work, which will focus on modelling of condensation with taking into account falling film angle around the top of the tube.

5 Conclusions and outlook

Two different approaches for modelling of condensation in horizontal tubes have been investigated and implemented, with utilization of object-oriented Modelica language and OpenModelica environment. The aim of this work was investigation, verification and comparison of the performance of models applied with usage of open-source OpenModelica environment in order to draw conclusions considering their reliability in modelling of passive safety systems. The main conclusions are:

- both models predict heat flow rate with satisfactory accuracy,
- flow regime based model evaluates fluid temperature more accurately (fig. 7),
- since the flow regime based model switches between different heat transfer coefficient correlations, significant peaks of power and temperature were observed,
- these peaks may lead to discrepancies of parameters (wall temperature, fig. 8) and instabilities of simulations,
- due to thinner film layer at the upper wall part of the tube, the wall temperature is significantly higher which is not taken into account in the models,
- both models indicate correctly the point in which single-phase liquid water flow occurs.

References

1. R. Drescher, T. Wagner, S. Leyer, VGB PowerTech, **5**, 33-37 (2014)
2. https://modelica.org/
3. https://openmodelica.org/
4. A. Schaffrath, E.F. Hicken, H. Jaegers, H.-M. Prasser, Nuc. Eng. and Des. **188**, 303-318 (1999)
5. A. Schaffrath, A. Krüssenberg, A. Fjodorow, U. Gocht, W. Lischke, Nuc. Eng. and Des. **204**, 251-265 (2001)
6. Verein Deutscher Ingenieure, *VDI Heat Atlas*, Second Edition, Springer-Verlag Berlin Heidelberg (2010)
7. W. Nusselt, Zeitschr. Vereins Deutscher Ingenieure **27**, 541-546 and 569-575 (1961)
8. T.N. Tandon, H.K. Varma, C.P. Grupta, J. Heat Transfer **104**, 763-768 (1982)
9. E.N. Soliman, Int. J. Multi-Phase Flow **12**, 277-288 (1986)
10. S.S. Grimley, Trans. Inst. Chem. Eng. **23**, 229-235 (1945)
11. S.S. Kutateladse, *Fundamentals of Heat Transfer*, Edward Arnold (1963)
12. P.G. Kosky, W.F. Staub, AICHE J. **17**, 1037-1043 (1971)
13. H. Jaster, P.G. Kosky, Int. J. Heat Mass Transfer **19**, 95-99 (1976)
14. G.W. Breber, J.W. Palen, J. Taborek, Trans. ASME J. Heat Transfer **102**, 471-476 (1980)
15. E.N. Sieder, G.E. Tate, Ind. Eng. Chem. **28**, 1429-1435 (1936)
16. M.M. Shah, Int. J. Heat Mass Transfer **22**, 547-556 (1979)

 DOI: 10.1051/itmconf/20171503008

The modelling of two DOF joints controlled by elastic inner ties

Dominik Ożóg[1,*]

[1]Rzeszów University of Technology, Department of Computer and Control Engineering, ul. W. Pola 2, 35-959 Rzeszów, Poland

Abstract. This paper considers the mathematical model of two types of joints that can be used to connect the arms of two robots. The first of them is a simple revolute joint with one degree of freedom and the second is a universal joint with two degrees of freedom. Each of them is controlled using elastic ties that run in the joints and are connected to the inside of the arm joints. This paper describes a study of kinematics, dynamics properties and Extended Denavit-Hartenberg Notation parameters mentioned joints.

1 Introduction

In recent years, a rapid development of the techniques related to miniaturization has been observed. This phenomenon occurs especially in the areas of electrical engineering and mechanics. The use of new ideas enables the creation of devices that are increasingly available and the improvement of current solutions, which primarily contributes to their mobility [1, 2]. However, current technological possibilities do not allow for the production of devices with the desired degree of miniaturization in many areas. Examples of such areas are medicine and structures of minimally invasive surgical robots [3]. Currently surgical robots have arms under the surface of the skin, whose diameter is 10 millimetres [4, 5]. In the case of such redundant manipulators with many degrees of freedom, the creation of smaller structures becomes problematic. This is especially noticeable when trying to create smaller drive elements and gearboxes that drive additional joints. In addition, the power supply and control signals remain within the already installed mechanical systems in such a small space. It should be noted, that these devices must be resistant to long-term load. This is because minimally invasive surgical operations performed with such devices last much longer than classic open-patient operations. There are already electric motors for the construction of 5 mm diameter arms, but research at the Rzeszow University of Technology and the University of Rzeszow on gear sizes have shown a problem with the strength of the components used. The solution to these barriers can be a design that allows the installation of drive systems beyond the robot arms. By contrast, the forces generated by them are transferred to the mechanical system via flexible ties. This article presents the description of kinematics and dynamics for one joint from such a design. In this case, a universal joint with two degrees of freedom is used

2 Universal joint with elastic inner ties

A universal joint is a known construction commonly used as a clutch or transmission. It can also be used for manipulators, where, instead of the drive shaft, the cross connector joins two arms. In both cases, this structure has 3 degrees of freedom [6]. There are known applications of universal joints in the construction of robots. In medicine, it has been used in the construction of minimally invasive surgical robots [7]. Achieving three degrees of freedom in one joint is a great advantage that extends the work area and facilitates manoeuvring of the manipulator's arm by its user. This is very important in areas with a large number of obstacles in the work area. It is also very easy to achieve redundancy of the effector position. All it needs to do is to use 3 jointed universal joints or 2 such joints, and one with one degree of freedom.

In the present case, the tie rod is used to mutually position the manipulator arms connected by a universal joint. One of their ends is attached to the inner wall of the arm, which is further away from the base point. Through the holes in the cross connector these ties pass into the inner part of the arm closer to the base point. If the numbering of the manipulator's arms starts from the base point, the point of fitting the ties will be in the i+1 arm. On the other hand, after passing through the holes in the cross connector, they will be in the arm i. In this arm, the ties may be attached to the sliding mechanical constructions, which are the source of the forces acting on the arms when they are passed through the chain. Another solution is to transfer the strands outside the i-th arm to apply the drive mechanisms beyond the manipulator's arms.

The second of the described solutions is the purpose of this article. The kinematics of the described solution will be approximated without specifying the drive mechanism. After further consideration, one of the ends of the tie is always fixed to the inner wall i+1 of the arm, while the other end of the tie is inside the i-arm. The calculation of the coordinates of the end point of the

* Corresponding author: dominikozog@wp.pl

DOI: 10.1051/itmconf/20171503008

local reference system is taken into account. In this way, the anchor attachment point data can be substituted for the propulsion system regardless of whether it is within or outside of the arm.

2.1 Kinematics for two-dimensional system

When analyzing the universal joint in two-dimensional space, it is assumed that it is possible to rotate the arms relative to each other about an axis passing through the centre of one of the cross-member bushes. Two cases of such rotation should be considered. Each case refers to another crosspiece sleeve, depending on which sleeve has rotated the axis of rotation.

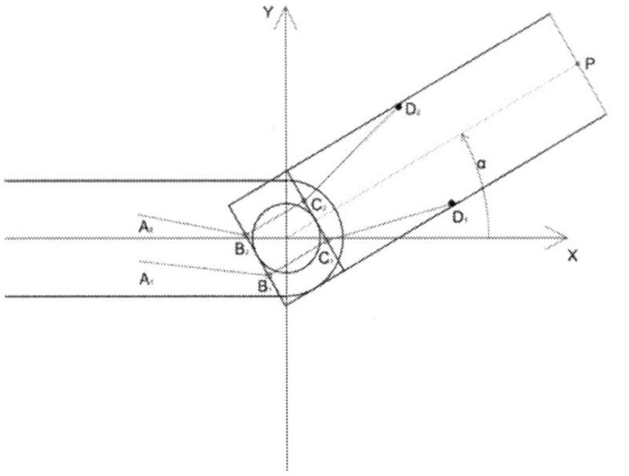

Fig. 1. Rotation around the sleeve of the cross connector connected by the i-th arm.

Fig. 1 illustrates the rotation situation around the coupling sleeve and the manipulator's arm. In this case, the sleeve rotates about the same angle as the arm i+1.

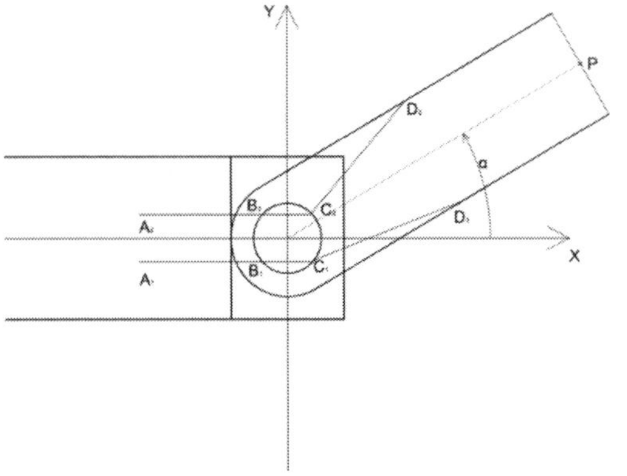

Fig. 2. Rotation around the sleeve of the cross connector connected by the i+1 arm.

This results in a change in the length of the cable ties, between the assembly point and the drive and the cross connector sleeve. Depending on the sign of rotation angle, one of the ties will be stretched, and the other will be shortened.

In the second case (shown in Fig. 2), the axis of rotation passes through the centre of the cross-member sleeve, where the sleeve is mounted to the inner wall of the i+1 arm. Rotation of arm i+1 does not affect the sleeve. This results in a change in the length of the section between the point of attachment of the tie to the inner wall of the arm and the hole in the cross-member sleeve on the side of the arm i+1.

If assuming that the length of the arm i+1 is L (from the centre of the joint), the position of the point P represents the vector:

$$\boldsymbol{P} = \begin{bmatrix} x_P \\ y_P \\ z_P \\ 1 \end{bmatrix} \quad (1)$$

,where

$$\begin{aligned} x_P &= L \\ y_P &= 0 \quad (2) \\ z_P &= 0 \end{aligned}$$

Using the Denavit-Hartenberg principle and performing actions that take into account the rotation of the Z-axis by the angle α we obtain the trivial task of forward and inverse kinematics [8-10]. The position of the point P after the transformation is given by the formula:

$$\boldsymbol{P'} = \begin{bmatrix} L\cos(\propto) \\ L\sin(\propto) \\ 0 \\ 1 \end{bmatrix} \quad (3)$$

In turn, the equation for inverse kinematics is:

$$\propto = tan^{-1}(\frac{y_{P'}}{x_{P'}}) \quad (4)$$

Equations 3 and 4 are satisfied by both articulation angles (Fig. 1 and Fig. 2).

2.2 The length of the ties as joint variables

Equations 3 and 4 would be forward and inverse kinematics patterns for a universal joint limited to one rotation axis. However, in the case of rotation angle control by the change of the length of the ties, this length should be regarded as a pivotal variable.

The scenarios that are shown in Fig. 1 and Fig. 2 are simplified models of the system described later in this article. To simplify the equations describing them, it is assumed that the points for the position of the two ties (A_i, B_i, C_i, D_i) are symmetric in relation to the axis OX in the initial - parallel - positioning of the arms. Point A_i refers to the assembly point of the i-th tie to the drive system and is defined by $A_i(x_{Ai}, y_{Ai}, 0)$ coordinates. Point B_i is the hole in the cross connector on the side of the i-th arm. It is located at a distance r from the axis OY, which is the value of the radius of the sleeve of the cross connector. This point is also at a distance j from the axis OX such that B_i(-

DOI: 10.1051/itmconf/20171503008

$r, \pm j, 0$). Similarly at the same distance from the OY axis, there is a point C_i and is located in the cross-member sleeve on the side of the i+1 arm. Point C_i is defined as $C_i(r, \pm j, 0)$. Point D_i is the place where the tie is attached to the inner wall of the i+1 arm. The radius of this arm is determined by R, and the thickness of its wall as e. The location of the tie is located h from the OY axis in the initial position of the arms. If t we define as the difference $t = R\text{-}e$, $D_i(h, \pm t, 0)$.

For a defined point, we can determine the length of the cable. Note that for the rotation around the sleeve of the cross connector connected by the i-th arm, two partial distances do not change.

$$\begin{aligned} |B_i, C_i| - const. \\ |C_i, D_i| - const. \end{aligned} \quad (5)$$

Similarly, in the rotation around the sleeve of the cross connector connected by the i+1 arm, however, other distances have a constant value.

$$\begin{aligned} |A_i, B_i| - const. \\ |B_i, C_i| - const. \end{aligned} \quad (6)$$

Let us define the sum of fixed values as l_{pi}:

$$\begin{aligned} l_{p1} &= |B_i, C_i| + |C_i, D_i| \\ l_{p2} &= |A_i, B_i| + |B_i, C_i| \end{aligned} \quad (7)$$

For the values defined above, we know that the lengths of the ties will be denoted by the formulas:

$$\begin{aligned} l_1 &= |A_i, B_i| + l_{p1} \\ l_2 &= |C_i, D_i| + l_{p2} \end{aligned} \quad (8)$$

To determine the value of the inverse kinematics, determine the value of the angle α knowing the values of equation (8).

$$\propto_i = tan^{-1}\left(\frac{n_i}{m_i}\right) + tan^{-1}\left(\frac{\pm\sqrt{m_i^2 + n_i^2 - k_i^2}}{k_i}\right) \quad (9)$$

where

$$\begin{aligned} k_i &= \frac{1}{2}\left(\left(l_i - l_{pi}\right)^2 - (z_{Ai} - z_{Bi})^2 - x_{Ai}^2 - y_{Ai}^2 - j^2 - r^2\right) \\ m_1 &= rx_{A1} - jy_{A1} \\ n_1 &= jx_{A1} + ry_{A1} \\ m_2 &= rx_{A2} + jy_{A2} \\ n_2 &= -jx_{A2} + ry_{A2} \end{aligned} \quad (10)$$

(9) shows that α_i is a set of two angles. This results from the possibility of obtaining the same length of the tie for two different values of rotation angles with opposite characters. Since the equations for two ties are solved, the common part of the sets obtained from (9) is unambiguous:

$$\propto = \{\propto \in R : \propto \in \propto_1 \wedge \propto \in \propto_2\} \quad (11)$$

Equations (2), (3), (8), (9) and (11) allow the calculation of the relationship of forward and inverse kinematics for the simplified case shown in Fig. 1. An analogous situation exists for the situation of Fig. 2 and the rotation for beta.

$$\beta_i = tan^{-1}\left(\frac{u_i}{v_i}\right) + tan^{-1}\left(\frac{\pm\sqrt{u_i^2 + v_i^2 - w_i^2}}{w_i}\right) \quad (12)$$

$$\begin{aligned} w_i &= \frac{1}{2}\left(\left(l_i - l_{pi}\right)^2 - (z_{Ci} - z_{Di})^2 - x_{Ci}^2 - y_{Ci}^2 - h^2 - t^2\right) \\ u_1 &= -hx_{C1} - ty_{C1} \\ v_1 &= tx_{C1} - hy_{C1} \\ u_2 &= -hx_{C2} + ty_{C2} \\ v_2 &= -tx_{C2} - hy_{C2} \end{aligned} \quad (13)$$

As with the Rotation around the sleeve of the cross connector connected by the i-th arm, (12) shows the sets of two solutions. For this reason, a formula analogous to Equation (11):

$$\beta = \{\beta \in R : \beta \in \beta_1 \wedge \beta \in \beta_2\} \quad (14)$$

2.3 The length of the ties for universal joints

In the case of a universal joint having 2 degrees of freedom, 4 ties are used. Two of them (l_1 and l_3) are responsible for the rotation around the OY axis by the angle β. The next two ties (l_2 and l_4) allow rotation around the axis OZ by an angle α. The mechanism of the universal joint and the 4 ties is shown in Fig. 3

.

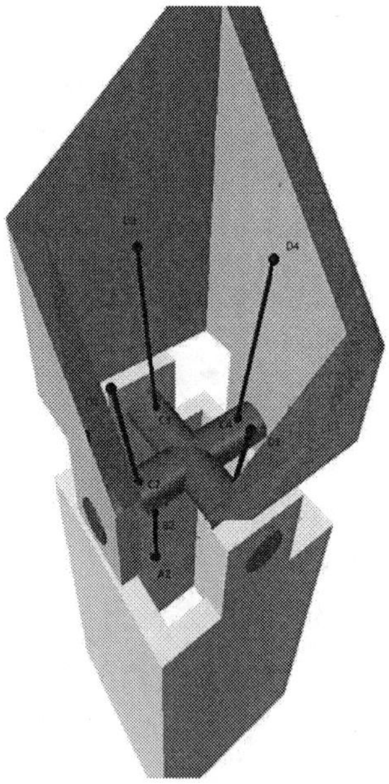

Fig. 3. Model of universal joint with 4 ties.

In Fig. 3, the points characteristic of the tie break are indicated in a manner analogous to Fig. 1 and Fig. 2. In further analysis of this paper, there is no simplification consisting in the symmetrical distribution of the puncture points. Each of these points can be placed anywhere in the joint and their coordinates are defined as A_i, B_i, C_i, D_i, where $i \in \{1, 2, 3, 4\}$.

To determine the kinematics equations for a universal joint with 2 degrees of freedom, Extended Denavit-Hartenberg Notation was used [11].

Table 1 lists the parameters of the universal joint controlled by the 4 ties. It included α and β rotation and the length of arm i+1. Using the Extended Denavit-Hartenberg Notation formula, we obtain the transformation matrix (15).

$$T=\begin{bmatrix} \cos(\beta)\cos(\alpha) & -\sin(\alpha) & \sin(\beta)\cos(\alpha) & L\cos(\beta)\cos(\alpha) \\ \cos(\beta)\sin(\alpha) & \cos(\alpha) & \sin(\beta)\sin(\alpha) & Lcos(\beta)\sin(\alpha) \\ -\sin(\beta) & 0 & \cos(\beta) & -Lsin(\beta) \\ 0 & 0 & 0 & 1 \end{bmatrix} \quad (15)$$

From the **T** matrix (15) the forward kinematics equation can be read. After their transformation and the use of trigonometric identities [12], we obtain inverse kinematics patterns for articulated variables that are values of Extended Denavit-Hartenberg Notation parameters.

$$\propto = tan^{-1}\left(\frac{y_P}{x_P}\right)$$
$$\beta = tan^{-1}\left(\frac{-z_P}{L\sqrt{1-\left(\frac{z_P}{L}\right)^2}}\right) \quad (16)$$

Determining the length of the ties for the universal joint, the two cases shown in Fig. 1 and Fig. 2 should be taken into consideration. For the system shown in Fig. 3, where there are 4 ties, A_i points are not subject to Denavit-Hartenberg conversion.

Table 1. Parameters of Extended DH-Notation.

Joint	Θ	d	β	a	α
i	0	0	var	L	var

The points B_i and C_i are only rotated around the axis OZ, while points D_i are subjected to two rotations. To determine the length of ties, define f(X, Y) such that:

$$f(X,Y) = \sqrt{\sum_{i=1}(X_i - Y_i)^2} \quad (17)$$

By using the function (17) you can define the length of the ties as:

$$l_i = f(A_i, R_Z B_i) + f(R_Z B_i, R_Z C_i) + f(R_Z C_i, R_Y R_Z D_i) \quad (18)$$

,where

$$R_Z = \begin{bmatrix} \cos(\propto) & -\sin(\propto) & 0 & 0 \\ \sin(\propto) & \cos() & 0 & 0 \\ 0 & 0 & 1 & 0 \\ 0 & 0 & 0 & 1 \end{bmatrix} \quad (19)$$

$$R_Z = \begin{bmatrix} \cos(\propto) & -\sin(\propto) & 0 & 0 \\ \sin(\propto) & \cos() & 0 & 0 \\ 0 & 0 & 1 & 0 \\ 0 & 0 & 0 & 1 \end{bmatrix} \quad (20)$$

Arrays A_i, B_i, C_i, D_i are vector-coordinated points of the corresponding points as in (1) and (3).

Constructing the inverse kinematics of the equation (18) is a very complex procedure. For this reason, it was decided to continue the work of transforming this equation by numerical methods.

3 Dynamics of universal joints with elastic inner ties

Determining the dynamic equations using the Lagrange method for situations of limiting the universal joint to a single rotation is a trivial task [10] and will be omitted in this document.

For further calculations, the arrangement as shown in Fig. 3 and for the axis OX directed parallel to the gravitational force, but with the opposite turn, is assumed. For these assumptions, the kinetic energy and the potential energy of the system were determined. The calculation assumed the distance from the centre of the joint to the centre of gravity of the arm i+1 equal to L_g and the mass of that arm was equal to *M*.

Around each axis there is another force moment affecting the arm and + 1. In relation to the axis OZ is the force T_α and around the axis OY T_β force. They are denominators (21) and (22).

$$T_\propto = \left(\left(\frac{d}{dt^2}\propto(t)\right)cos(\beta(t)) - \left(\frac{d}{dt}\propto(t)\right)cos(\beta(t))sin(\beta(t))\right)ML_g^2 - \quad (21)$$
$$-2MgL_g sin(\propto(t))cos(\beta(t))$$

$$T_\beta = \left(\frac{d}{dt^2}\beta(t)\right)ML_g^2 -$$
$$-\left(\left(\frac{d}{dt}\beta(t)\right)^2 - 2\left(\frac{d}{dt}\alpha(t)\right)^2\right)ML_g^2 cos(\beta(t))sin(\beta(t)) - \quad (22)$$
$$-2MgL_g L_g^2 cos(\alpha(t))sin(\beta(t))$$

4 Conclusion

This document shows the forward and inverse kinematics of the universal joint controlled by four elastic ties with force transmissions that can be mounted outside the robot arms. These forces affect the specific points of the manipulator's arm by the rods.

The high complexity of inverse kinematics has been demonstrated in the case of use as variable articulated lengths of ties. This problem will be the area of further research and may be solved by the use of numerical methods.

All calculations in the document concerned a single universal joint that is not associated with any further kinematic chain. The expansion of research into this area is planned in the near future.

References

1. A.B. Frazier ; R.O. Warrington ; C. Friedrich, IEEE Trans. Ind. Electron., **42**, 5 (1995)

 DOI: 10.1051/itmconf/20171503008

2. M. Rasouli, L. S. J. Phee, Expert Rev Med Devices, **7**, 5 (2010)
3. G. H. Ballantyne, F. Moll, Surg Clin North Am, **83**, 6 (2003)
4. I. K. McLeod, P. C. Melder, Ear Nose Throat J, March 2005, 170-173 (2005)
5. K. Ikuta, T. Hasegawa, S. Daifu, ICRA '03 (2003)
6. A. J. Mazzei Jr, A. Argento, R. A. Scott, JSV **222**, 1 (1999)
7. J. Shang, D. P. Noonan, C. Payne, J. Clark, M. H. Sodergren, A. Darzi, G.-Z. Yang, ICRA 2011 (2011)
8. P. I. Corke, IEEE Trans. Robot., **23**, 3 (2007)
9. H. Lipkin, IDETC/CIE ASME, **7**, 921-926 (2005)
10. R. Leniowski, *Podstawy Robotyki,* 25-84 (2005)
11. U. Thomas, F. M. Wahl, *Robotic Systems for Handling and Assembly*, **67**, 3-16 (2010)
12. L. Rade,B. Westergren, *Mathematics Handbook for Science and Engineering*, 118-133 (2004)

ITM Web of Conferences 15, 03009 (2017)
CMES'17
DOI: 10.1051/itmconf/20171503009

The optimisation of a water distribution system using Bentley WaterGEMS software

Karolina Świtnicka[1,*], *Paweł* Suchorab[1], and *Beata* Kowalska[1]

[1]Lublin University of Technology, Faculty of Environmental Engineering, Nadbystrzycka 40 B, 20-618 Lublin, Poland

Abstract. The proper maintenance of water distribution systems (WDSs) requires from operators multiple actions in order to ensure optimal functioning. Usually, all requirements should be adjusted simultaneously. Therefore, the decision-making process is often supported by multi-criteria optimisation methods. Significant improvements of exploitation conditions of WDSs functioning can be achieved by connecting small water supply networks into group systems. Among many potential tools supporting advanced maintenance and management of WDSs, significant improvements have tools that can find the optimal solution by the implemented mechanism of metaheuristic methods, such as the genetic algorithm. In this paper, an exemplary WDS functioning optimisation is presented, in relevance to a group water supply system. The action range of optimised parameters included: maximisation of water flow velocity, regulation of pressure head, minimisation of water retention time in a network (water age) and minimisation of pump energy consumption. All simulations were performed in Bentley WaterGEMS software.

1. Introduction

The main function of water distribution systems (WDSs) is to deliver water to consumers. That goal remains overriding to other purposes of the company, but nowadays water operators endeavour to achieve that goal in the most efficient way. To ensure optimal functioning of WDS, it is required from operators to take multiple actions, sometimes simultaneously. Therefore, the decision-making processes are often supported by multi-criteria optimisation methods. Significant improvements of exploitation conditions of WDSs functioning can be achieved by connecting small water supply networks into group systems. Among many potential tools supporting advanced maintenance and management of WDSs, a significant note is those tools that can find the optimal solution automatically, by the implemented mechanism of metaheuristic methods, such as the genetic algorithm [1].

In this paper, an exemplary WDS functioning optimisation is presented, in reference to a group water supply system. On the territory of the commune, 6 different WDSs are operated by one water company. In the near future, it is planned to create one group water system, supplying water to all inhabitants of the commune (over 20 000 people). As this process is of great complexity, it is divided into several steps, the first of all being the connection of urban WDS with the biggest rural WDS. In accordance to water company expectations and requirements, the aim of this paper was to define the required actions from the WDS operators in order to connect and optimise selected water distribution networks. The analysis of optimal connection and functioning solutions was based on computer simulations performed in Bentley WaterGEMS software.

2. Methodology

Analysed WDSs are located in the Lublin Region and supply water to two neighbouring settlement units: one of an urban (settlement A – approx. 10 000 consumers) and one of a rural character (settlement B – approx. 400 consumers). Branched-looped water distribution networks are made both of traditional materials (steel, cost iron, asbestos-cement), as well as from plastic pipes (PVC, PE). The total length of water network A is approx. 54 km and network B – approx. 66 km. Distribution pipe diameters in both systems varies from DN80 up to DN225. Presently, two analysed WDSs are operated individually. The geometrical structure of analysed WDSs with diameter differentiation and marked areas of settlements A and B is presented in figure 1. The territory of analysed WDSs varies significantly in geographical elevation (129.70÷181.05 m above sea level).

The required pressure head is provided by 4 pumps in 2 pump stations in each WDS (A1, A2 in WDS A and B1, B2 in WDS B). The functioning of all pumps stations is supervised by water company operators through on-line monitoring system. In each of the analysed systems, two specific moments in 24-hours of operation can be defined. The hour of maximum water demands (A – 19:00, B – 11:00) and the hour of the minimum water demands (A – 2:00, B – 5:00).

*Corresponding author: karolina2393@wp.pl

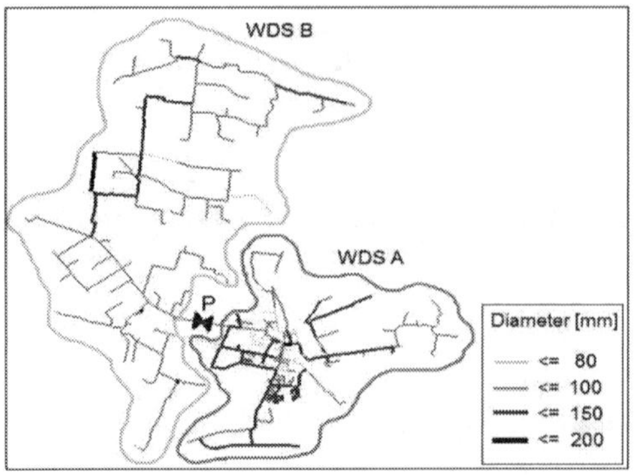

Fig. 1. Geometrical structure of analysed WDSs.

The analysis of optimal connection and operating solutions of analysed systems was based on computer simulations in Bentley WaterGEMS software. It is a widely applied [2-4] hydraulic modelling software for water distribution systems with advanced interoperability. Among many modules, WaterGEMS can be used to support the design process of new distribution systems, assess fire flow capacity, develop network flushing plans, identify water losses, manage and minimize energy use and prioritize pipe renewal [5].

Numerical model of analysed WDSs has a basic character – water supply networks were composed only of main distribution pipes, without households' connections. The model is composed of 548 junctions, 565 pipes, 2 pressure reduction valves (PRVs), 4 pumps and 3 water sources – reservoirs. Before the analysis, the model was validated and calibrated. The basis of calibration was the results of a pressure and flow measuring campaign. The percentage of measurement nodes fulfilled the required minimal amount of measurement points (2% of all nodes) for the designing and operation purposes [6]. The fitting of the model results to the real network conditions was estimated as satisfactory (determination coefficient R^2 of measured and simulated values equalled 97.7%). The total duration of the extended period simulation (EPS) time was 480 hours (20 days), with computation hydraulic time step 1 hour.

The process of connecting and optimising A and B WDSs was focused on fulfilling several requirements and expectations addressed by the water company, which the most important are as follows:

- minimisation of energy consumption for pumping and reducing energy costs,
- improvement of hydraulic operating conditions of WDS (increasing in-water flow velocity and prevention to secondary water contamination by decreasing of water retention time (water age)) by connecting network branches into looped structures,
- the assurance of proper water pressure head acquired by pump parameters modification and fittings installation.

Taking into consideration all expected results, the process of connecting and optimising analysed WDSs was divided into three succeeding steps:

1. Simulating the connection process by systems linked with additional pipes, simultaneously with pump parameters adjustment.
2. Optimisation of WDS with the use of the *Darwin Scheduler* module with implemented genetic algorithm – setting pump parameters in reference to expected pressure head values in nodes and required water flow velocity in pipes.
3. Additional optimisation – localizing new PRVs, new pipe connections and manual modification of pumps speed variable patterns.

The connection pipe (P) between A and B networks (length 1300 m, diameter DN110, material PE) is marked in figure 1. That connection pipe is already linking networks, but is not yet operated (closed gate valves at the beginning and at the end of the pipe). Together with the network connecting, the pumps parameters were expected to be adjusted accordingly. Additionally, after the WDSs connection, it was required to specify the hour of minimal and maximum water demands common for the whole system. In order to evaluate the effects of a systems' connection, it was necessary to examine basic hydraulic operating conditions: pressure head in nodes in the hour of minimal and maximal water demands, water flow velocity in pipes in the hour of maximal demands and water age in the network at the time step of the simulation. The results of hydraulic functioning conditions were used to further indicate the purposes of the systems' optimisation.

Darwin Scheduler module, based on the genetic algorithm, allows automatic optimisation of pumps operation by variable pump speed modifications, in order to fulfil requested hydraulic parameters. As a result, the lowering of the energy consumption is expected, and therefore the decreasing of pumping costs. In the analysed case, the automatized optimisation module was used for all 4 pumps which originally worked with 100% efficiency. Three general restrictive conditions for the whole WDS were assumed in the optimisation process: the minimum pressure head in water system (30 m H_2O in the urban area and 20 m H_2O in rural area), the maximum pressure head (60 m H_2O) and the maximum acceptable flow velocity for distribution pipes (3 m/s). The selected objective type of *Darwin Scheduler* module was to *Minimize Energy Cost* of pumping with simultaneous achieving of requested hydraulic parameters. The parameters of pump stations to optimize, such as decision type, speed properties and time specifications, are presented in the screen capture of a selected window of *Darwin Scheduler* module in figure 2. In order to calculate pumping costs, two industrial *Energy Pricing* tariff patterns were used: a) daily constant energy price (0.25 PLN/kWh) and b) price tariff with night price reduction (0.29 PLN/kWh during 6:00÷21:00, and 0.18 PLN/kWh in 21:00÷6:00). Additionally, during the optimisation process, the CO_2 emission was calculated in reference to energy use (the assumed CO_2 emission coefficient equalled 0.83 kg/kWh [7]).

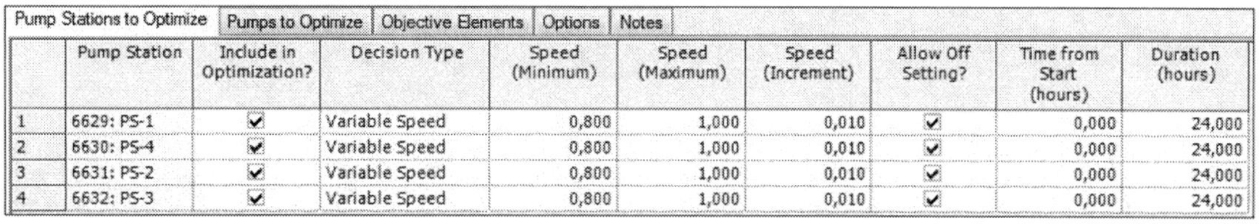

Pump Stations to Optimize	Pumps to Optimize	Objective Elements	Options	Notes					
	Pump Station	Include in Optimization?	Decision Type	Speed (Minimum)	Speed (Maximum)	Speed (Increment)	Allow Off Setting?	Time from Start (hours)	Duration (hours)
1	6629: PS-1	☑	Variable Speed	0,800	1,000	0,010	☑	0,000	24,000
2	6630: PS-4	☑	Variable Speed	0,800	1,000	0,010	☑	0,000	24,000
3	6631: PS-2	☑	Variable Speed	0,800	1,000	0,010	☑	0,000	24,000
4	6632: PS-3	☑	Variable Speed	0,800	1,000	0,010	☑	0,000	24,000

Fig. 2. Screen capture of a selected window "Pump stations to optimize" from *Darwin Scheduler* module.

Due to the fact that *Darwin Scheduler* module does not include the automatized possibility of new fitting installations or new pipe connections, it was decided to perform additional optimisation actions. The base for the additional optimisation was the WDS connection scenario from the optimisation module. The aim of the additional optimisation was to reduce pressure in the head in selected zones of increased pressure and limit the water age in the network. It was assumed by localizing 5 new PRVs and by designing 2 new pipe connections. Moreover, the pumps speed variable pattern of pump B1 was modified to further minimize pumping energy costs.

3. Results

In accordance to the presented methodology, the connection and optimisation process was divided into three steps. Connecting two separate WDSs into one system by linking two networks with a connection pipe resulted in dysregulation of proper system functioning. According to the simulation performed in Bentley WaterGEMS, in night hours (0:00÷3:00) the system operation was failing because several pumps could not deliver required water flow or pressure head. Therefore, it was necessary to manually modify pump properties: increase (pumps A1, A2) and decrease (pump B2) in maximal head. After several modifications, the simulated system was balanced and characterized with attributes presented in table 1 (parameters after step 1). After analysis of the WDS hydraulic parameters, several aspects of connection require improvement: an increase in minimum pressure head in the urban area, the reduction of maximum pressure head in the urban area, increasing flow velocity in pipes and lowering of water age. These goals were applied as the base for the automated optimisation.

The *Darwin Scheduler* module was used to automatically optimise the WDS in order to minimize the energy costs of pumping. The applied module modified pump speed variable patterns to adjust pump operation in reference to requested hydraulic restrictions. The main goal of the application of *Minimize Energy Cost* objective type of *Darwin Scheduler* resulted in a 1.4% reduction of daily pump energy consumption. It also resulted in smaller CO_2 emissions (reduction from 91.7 kg to 90.44 kg per day).Comparing hydraulic operation conditions of WDS before and after the optimisation in WaterGEMS, it is clearly visible that some of the requested parameters are achieved. The minimum pressure head in the urban area is set as the expected value (30 m H_2O). However, other pressure head values did not improve significantly (the decrease of maximum pressure head in urban area only by 0.5 m H_2O). Additionally water flow velocity and water age did not improve to an acceptable level. Detailed characteristics of the WDS after the optimisation in *Darwin Scheduler* module are given in table 1 (step 2).

Table 1. Results of WDS optimisation (step: 1 – connection, 2 – automatized optimization, 3 – additional optimisation).

Parameter	After step 1	After step 2	After step 3
Maximum pressure head (m H_2O) – urban area	69.70	69.20	60.00
Minimum pressure head (m H_2O) – urban area	28.39	30.00	30.00
Maximum pressure head (m H_2O) – rural area	54.19	54.19	50.10
Minimum pressure head (m H_2O) – rural area	20.17	20.17	21.77
Maximum flow velocity in pipe X (m/s)	0.00	0.00	0.13
Water age in node Yin 480 hour of simulation(hours)	287.14	287.14	8.47
Daily pump energy consumption (% to the original one)	100	98.6	97.5
Daily CO_2 emission (kg/d)	91.70	90.44	89.41

As the results of the automatic optimisation were not fully satisfactory, additional optimisation was focused on for further improvement of hydraulic conditions. In order to eliminate zones of higher-pressure head, 5 new PRVs were designed. The increase of water flow velocity in pipes was proposed by combining network branches into loops by 2 new connection pipes. The localisation of PRVs and connection pipes is presented in figure 3. Additionally, variable speed patterns of pump B1 was modified manually in order to obtain more satisfactory effects of energy consumption reduction. The modified variable speed pump B1 pattern is presented in figure 4.

Detailed effects of the additional optimisation are presented in table 1 (step 3). The installation of new PRVs resulted in proper water pressure head range in the system. Zones of increased pressure were eliminated, and the maximum pressure head value in the urban part of the water system equalled 60 m H_2O. Moreover, the pressure head in the rural area was lowered, which can positively influence the maintaining process of the WDS.

DOI: 10.1051/itmconf/20171503009

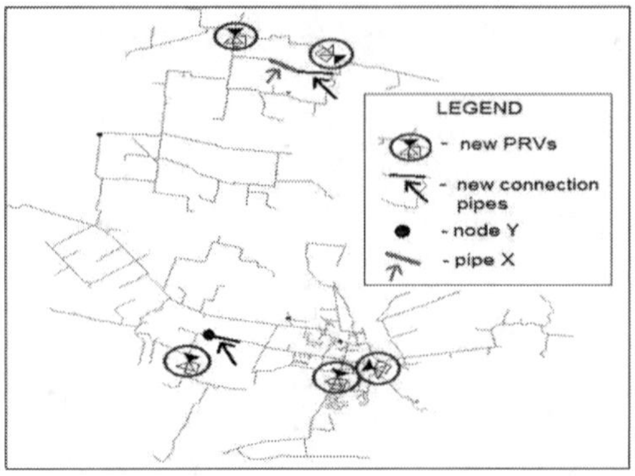

Fig. 3. Geometrical structure of the WDS with marked locations of new PRVs, new connection pipes and exemplary reference pipe and node.

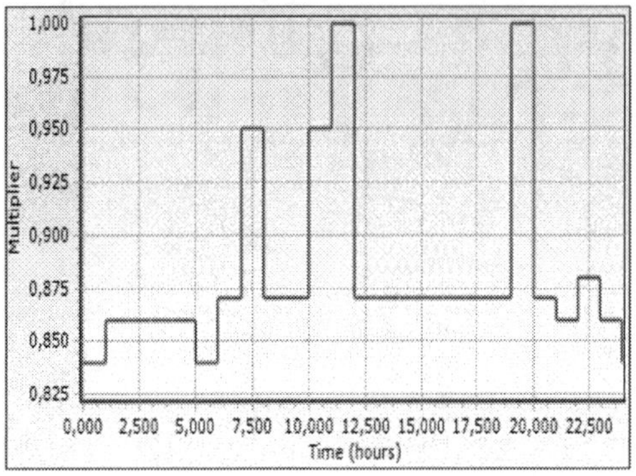

Fig. 4. Variable speed pattern for exemplary pump (B1) after modifications (other pumps are optimized in a similar way).

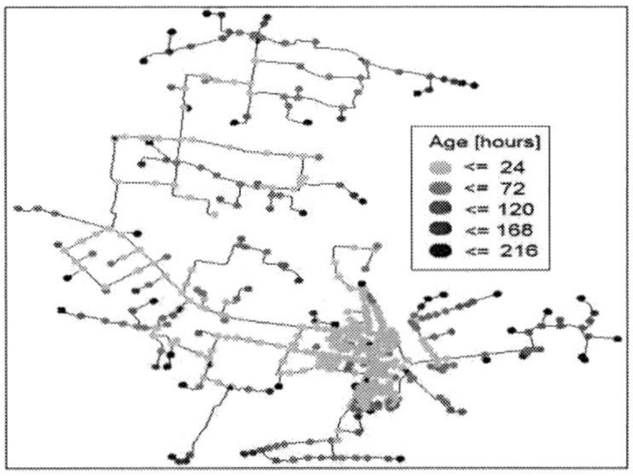

Fig. 5. Water age in the analysed WDS after additional optimisation in 480 hour of simulation.

New connections of branches into loops reflected in an increase of the flow velocity in these parts of the system. Locally, the flow velocity improved from 0.0 m/s up to 0.13 m/s. However, on the whole, WDS water flow velocities are generally still below the minimum recommended level (0.5 m/s). Additionally, new connections resulted in locally greater water age lowering (in reference node from 287.14 hour to 8.47 hour), but general water age in the system did not decrease significantly. The water age in nodes in 480 hours of the simulation is presented in figure 5. However, in the majority of nodes water flows to customers in less than 24 hours, which is shorter than permissible water age in a distribution network (1.3 day) [8]. By additional optimisation not only was an improvement of hydraulic conditions achieved, further energy costs were minimised (to the level of 97.5% of the original consumed energy).

4. Conclusion

The main result of this analysis is the action scenario for the water company, in which several steps are defined in order to connect and optimise maintained water distribution systems. In accordance to the obtained results, the application of proposed modifications in pump properties and the installation of new fittings would reflect in energy consumption minimisation and an improvement in hydraulic water conditions.

The performed analysis confirmed that the connecting of different water distribution systems into one group and optimising its operation is a process of great complexity. Therefore, it is always highly recommended to predict further actions in modelling software before its realization. The applied Bentley WaterGEMS software, with the genetic algorithm modules implemented, proved to be a very useful and universal tool, which can be used to support the decisive-making process of maintenance planning.

References

1. S. M. S.Abkenar, S. D. Stanley, C. J. Miller, D. V. Chase, S. P. McElmurry, SUSCOM **8** (2015)
2. J. Stańczyk, E. Burszta-Adamiak, D. Poor, GWiTS, **5** (2017) DOI: 10.15199/17.2017.5.4
3. C. Xiao, B. Li, G. He, J. Sun, J. Ping, R. Wang, Procedia Eng. **89** (2014) DOI: 10.1016/j.proeng.2014.11.203
4. P. Sage, Procedia Eng. **70** (2014) DOI: 10.1016/.j.proeng.2014.02.160
5. Bentley WaterGEMS Product Data Sheet www.bentley.com/en/resources/watergems [access 8.10.2017]
6. AWWA Engineering Computer Applications Committee, *Calibration Guidelines for water distribution system modelling* (1999)
7. KOBIZE, *Wskaźniki emisyjności CO_2 dla energii elektrycznej* (2014) (in Polish)
8. U.S. EPA, *Effects of Water Age on Distribution System Water Quality* (2002)

Application of neural networks in CRM systems

Agnieszka Bojanowska[1,*]

[1]Lublin University of Technology, Faculty of Management, Department of Marketing, Nadbystrzycka 38, 20-618 Lublin, Poland

Abstract. The central aim of this study is to investigate how to apply artificial neural networks in Customer Relationship Management (CRM). The paper presents several business applications of neural networks in software systems designed to aid CRM, *e.g.* in deciding on the profitability of building a relationship with a given customer. Furthermore, a framework for a neural-network based CRM software tool is developed. Building beneficial relationships with customers is generating considerable interest among various businesses, and is often mentioned as one of the crucial objectives of enterprises, next to their key aim: to bring satisfactory profit. There is a growing tendency among businesses to invest in CRM systems, which together with an organisational culture of a company aid managing customer relationships. It is the sheer amount of gathered data as well as the need for constant updating and analysis of this breadth of information that may imply the suitability of neural networks for the application in question. Neural networks exhibit considerably higher computational capabilities than sequential calculations because the solution to a problem is obtained without the need for developing a special algorithm. In the majority of presented CRM applications neural networks constitute and are presented as a managerial decision-taking optimisation tool.

1 Introduction

Customer Relationship Management (CRM) is a relatively recent concept developed on the border of management and marketing in the early 1980s. Major factors that spurred the developments in the field were:

- increasing the knowledge of customers on goods and services,
- discovering the profit of maintaining lasting relationship with best customers,
- the emergence of a global competition of goods and services,
- the growing need for managing large amounts of customer data,
- the rising interest in the profits from customer behaviour data processing.

Customer loyalty is widely recognised to be one of the key components of success in virtually any business. It can be increased through effective management of customer information and this will expand the business for the organizations [3]. Lin and Su [1] showed how CRM strategies if implemented effectively can increase customer value and provide satisfaction for customers.

It may be assumed that the concept of CRM is composed of two parts: the material component, *i.e.* IT systems of CRM including hardware, and the non-material, abstract element, which may be understood as the philosophy of CRM. The philosophy of CRM is the customer approach of a company rooted in its organisational culture. The system, on the other hand, consists of IT tools, such as databases and proper database processing tools. In order for customer management to generate profit for the company, the two foundations of the concept, the material and the abstract, must work in unison. The key problem with CRM is that the concept has yet to be conclusively defined in economics. The understanding of CRM is frequently situational, and depends on the current needs of a company deploying the system. Certain definitions highlight the systematic facet of CRM, *i.e.* the software and hardware, while others simultaneously treat the system as subordinate to the philosophy of building long-term relationships with customers. The approach presented in the following paper focuses on software tools dedicated for CRM systems; specifically, it is an attempt to apply artificial neural networks to a CRM system. Therefore, throughout when referred to in this paper CRM shall stand for: a system of tools and technologies aiding customer relationship management. In addition, it will be understood as a place where all data regarding customers, partners, suppliers or any stakeholders of any connection with the company is stored. This unique combination of technology and processes enables easy access to customer history, facilitates mailing management, preparation and issuing offers, focus on most profitable customers or management of salesforce.

2 CRM systems

At present there is an abundance of CRM systems in Poland and abroad. Software providers offer more or less complex solutions, which may be classified into several categories: dedicated for a particular enterprise (more expensive), universal, general-purpose solutions, which must be selected by a company considering its actual needs, and open source systems or systems based on

* Corresponding author: a.bojanowska@pollub.pl

popular software, *e.g.* MS Office. To select a proper CRM solution several aspects need to be taken into consideration, *e.g.*:

- What is the anticipated amount of customer data for the system to manage?
- What analyses is the software expected to carry out?
- Will the system be managed remotely or must it be centralised and company-based?
- Will the new software be compatible with systems already in use at the company (*e.g.* Enterprise Resource Planning, ERP)?
- Does the CRM solution provider offer employee training in both CRM system and philosophy?
- What is the budget of the company planning to implement CRM?

The type of CRM software and the shape of the whole system will depend on the way the company defines its needs; for instance, certain enterprises focus on increasing the number of loyal customers, whereas others seek to determine key customers in order to manage them.

An effective CRM system should:

- enable cloud storage of data;
- allow centralised administration of the programme;
- identify users entering and changing data, as well as all personnel with an access to the data, with regards to Personal Data Protection Act;
- enable multidimensional analysis of customer data;
- enable automation in contacts with customers;
- facilitate developing marketing campaigns, particularly the individualised ones;
- cooperate actively with social media.

CRM systems play an active role in aiding the work of the managerial board, and different departments in the company: marketing, sales, customer service, accounting and customer service; furthermore, it has an indirect impact on the work of other departments. It is therefore imperative that the system should work effectively and promote realisation of goals specified for the implemented concept of CRM. The subsequent sections of this paper show examples of implementations of neural networks contributing to the effectiveness of the system.

3 Neural networks in business applications

Similarly to CRM, artificial neural networks originate in 1980s, and the latter were developed side by side with the exploration of artificial intelligence. Initial studies which were critical to the developments in the field focused on building models of basic structures of the brain. A simplified operation of neural structures is shown in Fig. 1.

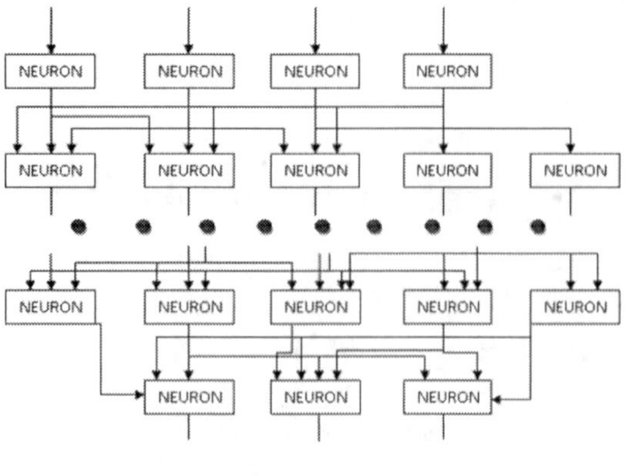

Fig. 1. Neural network [4].

Neural networks are found in a plethora of applications, such as:

- diagnostics of electronic systems,
- psychiatric examination,
- stock market forecasting,
- sales prediction,
- crude oil search,
- biological research interpretation,
- price forecasting,
- technological machine maintenance scheduling,
- scientific progress prediction,
- horse racing betting,
- production problems analysis,
- optimisation of commerce,
- raw material selection,
- employee assignment,
- industrial process management [4].

The range of applications for neural networks is notably wide, however, it mainly focuses on industrial and business processes. Such great has the effectiveness of neural networks become that the search for new potential implementations is unlikely to cease in the foreseeable future. Neural networks are frequently employed in banking to assess creditworthiness of customers, or in profitability assessment of investments. They may also be found in neuromarketing studies, where certain neurological patterns serve to build the image of future customer behaviour.

Predictive neural networks play a pivotal role in predicting events and patterns in economy. The unquestionable advantage of these networks is their capability to predict input signals based on mere observation of the training data set, without the knowledge of the nature of connections between the input and output. This feature may be successfully implemented in CRM systems, whose task is to predict future behaviour of customers based on behavioural data of usually unknown mechanisms.

What is more, CRM systems might benefit from another important feature of neural networks, the capability of networks to relate similar classes to each other. Data in software systems is often stored randomly, without a possibility of linking it. Neural networks are capable of

DOI: 10.1051/itmconf/20171504001

automated reasoning based on training data sets, which allows managers to learn about important links and aspects without overburdening them with unnecessarily detailed information, which could appear in CRM systems without neural networks.

Data analysis enabled by neural networks involves finding connections in the input data set. This sheds new light on economic analyses, and on the source of earlier failures and therefore aids preventing them in the future. This particular feature could prove decisive in prognosis of customer leaving the company and analysis of complaints.

4 Examples of neural network in CRM applications

Neural networks may be assigned to different tasks and used in different aspects of CRM systems. This section elaborates on four examples of such applications, which exemplify practical aspects of neural networks in the field of CRM: predicting customer expenses according to behavioural data, assessing profitability of building customer relationships, assessing probability of customer leaving the company based on complaint data, determining key customers. These aspects are of great value to managers, demanding that the CRM system implemented in their company generates substantial profit and minimises the risk of a failed investment.

4.1 Predicting customer expenses from behavioural data

In the overwhelming majority of cases when asked about their company objectives, managers respond that this is profit. It is for this reason that each company is predominantly focused on customer wallet share and the possibility of predicting customer future expenses. This knowledge allows businesses to develop reliable financial plans and opens investment opportunities. It is moreover important how often the purchases are made, and what the share of customer income the purchases constitute (wallet share). Predicting customer expenses was undertaken by R. Rada, B. Ruseti, N.Baçi and M.Karçanaj in "Artificial Neural Networks for Predicting Customer Expenses" [2]. In their study the data set consisted of approximately 620 records of customers respectively with input and output. Each customer was composed of different variables such as: age, marital status, social status, personal income, family income, number of persons in family and the output variable which were the expenses per month in mobile communication. The values of variables for the ones of the two-state variable domain were expressed by 0 or 1. After transforming them into numerical values, the next step was to normalize them. The authors used the min-max standardization method. ANN algorithm was implemented in Java programming language. After training the next process was to test the remaining 310 records of customers. The predictions were made using a numerous ANN models. After the predictions were generated, they were compared with the outcome in the testing dataset. There were several main conclusions from the study. First, main benefits of a company depend directly on the definition of the target consumer for a certain product. Seeing that traditional methods are not very effective, in this article they treated the possibility of integration of data mining techniques in business. They gave a general overview of the Artificial Neural Network in Mobile Communication. According to the received results they saw that this method was more efficient than traditional ones. The data examined were taken from a survey with about 930 records. They trained over 620 consumers and results of testing on the basis of training were very good. A comparative analysis of results from the study showed that the best network was based on the hyperbolic tangent function, which gave accurate results for 82.4% of test cases.

4.2 Profitability of building customer relationships

Profitability of building customer relationships is a broader issue than the previously presented one. The importance of the current problem is that both overestimation and underestimation of a customer may endanger the business and generate short- and long-term loss. It ought to be mentioned that it is not beneficial to build a relationship with every customer, as implementation of CRM in a company generates costs. In addition, companies bear costs of building relationships with each particular customer. These include both fixed and variable costs, and this is why correct estimation of benefits is critical. Such estimation frequently proves complicated owing to the fact that the idea of CRM refers to both countable/monetary benefits (wallet share) and uncountable ones (*e.g.* positive word of mouth about the company on the market); therefore, profitability assessment will depend on what sort of business is concerned. For the reasons presented above, neural networks may prove greatly helpful in deciding whether it is profitable to develop a relationship with a customer or not.

The example below builds on 10 factors which determine the profitability of customer relationship building in a CRM system (Table 1).

A - loyalty (weighted at 0.15 and value from 1 to 5, where 1 is the least sought value by a company, and 5 the most)

B - solvency (weighted at 0.1 and value from 1 to 5, where 1 is the least sought value by a company, and 5 the most),

C - purchasing regularity (weighted at 0.1 and value from 1 to 3, where 1 is the least sought value by a company, and 3 the most),

D - purchasing frequency (weighted at 0.1 and value from 1 to 3, where 1 is the least sought value by a company, and 3 the most),

E - positive word of mouth on the market (weighted at 0.15 and value from 1 to 7, where 1 is the least sought value by a company, and 7 the most),

 DOI: 10.1051/itmconf/20171504001

F - amount of a single transaction (weighted at 0.08 and value from 1 to 4, where 1 is the least sought value by a company, and 4 the most),
G - openness to communication with a company (weighted at 0.12 and value from 1 to 7, where 1 is the least sought value by a company, and 7 the most),
H - providing the company with information regarding the market (*e.g.* products of competition) (weighted at 0.09 and value from 1 to 3, where 1 is the least sought value by a company, and 3 the most),
I - unaffected relationship after recognised complaints (weighted at 0.05 and value from 1 to 3, where 1 is the least sought value by a company, and 3 the most),
J - openness to new products of a company (weighted at 0.06 and value from 1 to 4, where 1 is the least sought value by a company, and 4 the most).
The model of neural network was based on 25 records of customers. 10 inputs were applied into the network, which were defined in the preceding list. The output represents the answer to the question: "Is the investment in the relationship with this client profitable?." The answer is two-state. However, since the network generates a vector of weighted answers, this vector was converted by the criterion function to binary decisions. The form of the criterion function depends on the decision-maker and may take the following form:

$$Y_{dec} = 1 \text{ for } y >= y_{śr} \qquad (1)$$
$$Y_{dec} = 0 \text{ for } y < y_{śr}$$

where: Y_{dec} – a two-state decision variable;
y – a decision value;
$y_{śr}$ – a value generated by the network for a customer.
The model of such architecture was used to assess the profitability of building customer relationship in accordance with the concept of CRM. The structure of the network is presented in Fig. 2.

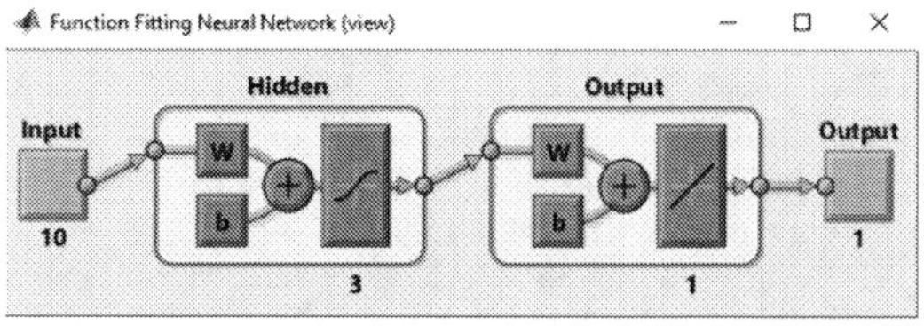

Fig. 2. Neural network used to model the process of assessing profitability of building customer relationship [Source: own work].

The model's adequacy was verified on the data obtained from 10 clients that were not included in the learning curve. The results of the simulation are shown in Table 1. Column K shows the result of the simulation and column L the calculated value based on the criteria. The simulation results show full compatibility, *i.e.* the network has learned to generate customer evaluation results in accordance with the adopted rating algorithm.

Table 1. Results of simulation of the cost-effectiveness analysis process of building relationships with customers using the neural network model.

	Customer features											
Customer ID	**A**	**B**	**C**	**D**	**E**	**F**	**G**	**H**	**I**	**J**	**K**	**L**
1	5	3	3	2	5	3	6	1	2	3	1	1
2	3	1	2	1	4	1	6	1	2	4	1	1
3	3	3	2	1	1	1	5	1	2	4	1	1
4	5	1	2	2	4	2	4	1	2	3	1	1
5	2	3	1	1	6	3	6	1	3	4	1	1
6	2	1	3	2	3	3	6	3	3	2	0	0
7	3	2	2	3	6	3	5	3	2	4	1	1
8	4	2	2	1	4	2	2	1	2	4	1	1
9	1	2	2	2	2	3	2	2	1	1	0	0
10	4	4	3	3	4	4	7	3	2	3	1	1

4.3 Assessing customer behaviour from complaint data

The lifecycle of a customer in a company is compared to the lifecycle of a product. Eventually a customer leaves the company, however, according to the principles of CRM, the company-customer relationship must be managed in such a way as to guarantee that the stage of the customer lifecycle which generates the highest profit to the company is maximally prolonged. Secondly, the company must strive to ensure that the customer leaving the company continues spreading positive word of mouth about the company on the market.
That is why it is of utmost importance to be able to predict when the customer will leave the company, and to prevent that occurrence. CRM systems implement certain information on customers, which enables making such predictions with a certain degree of probability. Algorithms for calculating the probability of such occurrences have been quite well developed, and are therefore capable of predicting whether and when the customer will leave the company. That is why, I attempted to analyse the problem in more detail, and included a conflict situation in the model. The conflict consisted in either a complaint about the service, which is not followed by any demands on the part of the customer, or a complaint about faulty goods, when the customer issues certain demands, *e.g.* compensation. Both types of complaint may be recognised or rejected, *i.e.* be resolved to the benefit of the company or to the customer's. CRM theory states that if the company-customer relationship did not suffer after the complaint was filed, the customer should maintain a high opinion of the company and stay with the company, regardless of the decision regarding the complaint. The status of the relationship following the conflict, however, is a separate issue. The presented example takes into account the following features regarding the behaviour and condition of a customer (Table 2):
A - incidental customer (0) loyal customer (5);

 DOI: 10.1051/itmconf/20171504001

B - customer satisfaction with claim (1 happy, 0 unhappy);
C - customer satisfaction with complaint (1 happy, 0 unhappy);
D - number of claims;
E - number of complaints.
The network model output for the customer described with these parameters may be expressed by three values of a variable defining such a customer:
1 – customer stayed with the company but reduced expenses;
2 – customer left the company;
3 – customer stayed with the company and raised expenses;
The model was developed on 50 records of customers. and verified based on data from the remaining 20 customer, which were not included in the training data set. The neural network consisted of 5 inputs and one output, as it may be seen in Fig. 3.

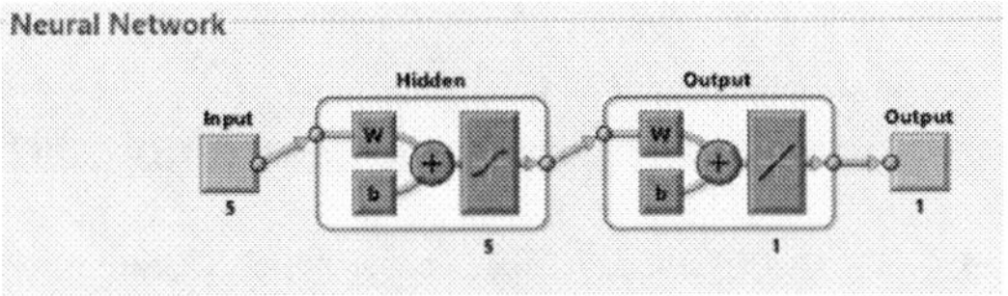

Fig. 3. Neural network used to model the process of predicting customer behaviour after complaint [Source: own work].

Table 2. Results of simulation of predicted customer behaviour based on data analysis of claims and complaints.

	Customer characteristics and behaviour					F	G
Customer ID	**A**	**B**	**C**	**D**	**E**		
1	0	0	0	0	5	1	3
2	1	0	3	1	10	3	3
3	1	1	1	0	5	3	3
4	1	0	10	1	7	2	2
5	1	1	2	1	0	2	2
6	0	1	8	0	2	1	1
7	0	0	4	1	3	3	3
8	1	0	7	1	5	3	3
9	1	0	4	1	7	3	3
10	0	0	10	1	10	2	1
11	1	1	9	0	4	3	3
12	0	1	7	1	0	3	3
13	1	1	0	0	6	3	3
14	0	0	0	1	6	3	3
15	1	1	7	0	5	2	2
16	0	1	7	1	2	1	1
17	1	1	4	0	6	3	3
18	0	1	7	0	10	2	1
19	0	1	0	1	5	2	2
20	1	0	1	1	10	1	1

Customer behaviour prediction by the trained network showed 85% agreement with actual customer response, not included in the training data set.
The results of the simulation are shown in Table 2. Column F shows the result of the simulation and column G the calculated value based on the criteria.

4.4 Determining key customers

Key customer is a consumer who will partake in the realisation of a long-term strategy of the enterprise [5]. To refer to largest purchasers as key customers is a gross simplification, and a pitfall which many modern companies are susceptible to. Maintaining a relationship with key customers is connected with significant investments, therefore in order not to expose the company to an unnecessary investment risk, key customers should be selected with due care. Different factors determine whether a customer becomes a key customer. Regardless of the type of business, certain frame must be marked for a customer to obtain the status of a key customer. In the analysed case the factors that were selected as relevant for the company were as follows (Table 3):
A- age – below 60;
B - marital status – married;
C - occupational status – regular employment or self-employed;
D - personal income – over 4000 PLN;
E - wallet share – over 25%;
F - purchase frequency – more than twice a month.
In order to build a training data set, the relationships of 50 customers with the company were analysed. The model based on the above criterion of division of customers into key (output value of model 1) and non-satisfying (logical value 0). The model may be employed to determine customers of great value to a business. The knowledge contained in the neural network depends on the expert's weight weights and the degree of their fulfilment. The resulting model was verified for 10 clients whose data were not in the learning curve. The network has learned to estimate the impact of the fulfilment of specified conditions on a key client and correctly identify such a client. Fig. 4 shows a schematic representation of the designed neural network.

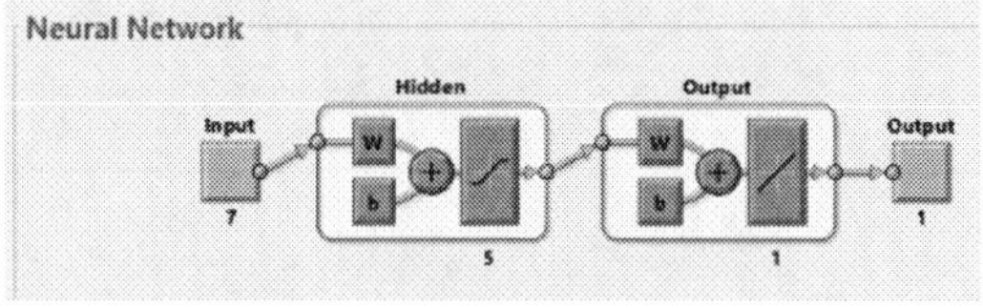

Fig. 4. Neural network used to model the process of determining key customer for a [Source: own study].

The input matrix consists of 6 attributes whose fulfilment determines the output value 1 (key customer) or 0 (customer not fulfilling the conditions). Thus created model proved to be accurate.

DOI: 10.1051/itmconf/20171504001

Table 3. Results of simulation of the process of selecting key customers for a company using a model in a neural network.

	Customer characteristics and behaviour						**G**	**H**
Customer ID	**A**	**B**	**C**	**D**	**E**	**F**		
1	50	1	1	4500	27%	1	1	1
2	73	1	1	7190	22%	4	0	0
3	18	2	1	1411	28%	2	1	1
4	27	2	0	2252	47%	2	0	0
5	64	1	1	6166	17%	2	0	0
6	25	1	0	4419	37%	4	1	1
7	38	2	0	3786	12%	1	0	0
8	21	1	1	9405	34%	2	1	1
9	44	2	0	2417	32%	3	1	1
10	79	2	0	3018	10%	4	0	0

As shown in Table 3 for all 10 simulated cases in the network, the simulation result (col. G) was consistent with the expert estimate (col. H). These examples show that the learned neuronal principles adopted by the expert can correctly identify the key customers of the company.

5 Summary

As stated in the Introduction, the study of four business cases was conducted in order to show how the application of neural networks may contribute to increasing the efficiency and productiveness of CRM systems. A brief introduction into the idea of CRM and neural networks was followed by presenting several applications of the method, according to [4]. The choice of the area of application of neural networks determined that economy and business solutions were in the focal point of the study.

The examples presented in the paper indicate that it could be possible to create a neural-network dedicated software tool for customer relationship management. This may be the subject of future research in the field of neural networks and CRM.

All model presented in the article were developed in MatLab software.

References

1. Y. Lin, H. Su, TQM and BI, **14**(6), 715 (2003)
2. R. Rada, B. Ruseti, N.Baçi, M.Karçanaj, J. of Edu. and Soc. Res., **6**(1), (2016)
3. J. Ranjan, V. Bhatnagar, The Learning Org., **18**(2), 135 (2011)
4. R. Tadeusiewicz, *Sieci neuronowe* (Akademicka Oficyna Wydawnicza, Warszawa, 1993)
5. S. Turek, *Najważniejsi klienci czyli Key Account Management w praktyce* (MT Biznes, Warszawa, 2006

DOI: 10.1051/itmconf/20171504002

Calculation of the number of branches of multi-valued decision trees in computer aided importance rank of parameters

Agnieszka Tiszbierek[1,*]

[1]Opole University of Technology, Faculty of Production Engineering and Logistics, Sosnkowskiego 31 Str., 45-272 Opole, Poland

Abstract. An elaborated digital computer programme supporting the time-consuming process of selecting the importance rank of construction and operation parameters by means of stating optimum sets is based on the Quine – McCluskey algorithm of minimizing individual partial multi-valued logic functions. The example with real time data, calculated by means of the programme, showed that among the obtained optimum sets there were such which had a different number of real branches after being presented on the multi-valued logic decision tree. That is why an idea of elaborating another functionality of the programme – a module calculating the number of branches of real, multi-valued logic decision trees presenting optimum sets chosen by the programme was pursued. This paper presents the idea and the method for developing a module calculating the number of branches, real for each of optimum sets indicated by the programme, as well as to the calculation process.

1 Introduction

Automation and control systems which are currently formed are complex, but many of their construction and operation parameters are interdependent, which in turn affects the quality and/or operating efficiency of these devices. This is why, it is not advisable to omit the above-mentioned dependencies during the machine design and construction process. It is also not possible to ignore them in the process of stating the importance rank of parameters of the analysed device, which should co-operate with testing and improving the device in creation [1,2]. The process of ascribing the importance rank was complex and complicated as far as calculations were concerned, and thus a computer programme based on an appropriate algorithm was created to aid the process. During works on the programme, the above-mentioned dependency of parameters was additionally taken into consideration, which led to creating appropriate modules (conditionality and a proxy variable) to protect dependent variables from division.

2 Description and operating principle of the programme

2.1. The algorithm of the programme

The basis of the programme is the Quine – McCluskey algorithm of the minimization of individual partial multi-valued logic functions [3,4,5]. Its basis is a mathematical equation composed of certain elements calculated on the basis of gathered and encoded data which, for the first stage of the algorithm, can be presented in the following form:

$$x_1 = i_k - i_z * w_z + i_z + p \qquad (1)$$

where:

x_1 – the number of step branches of the analysed decision tree,
i_k – the number of branches of the *k-th* level of the decision tree,
i_z – the number of simplifications of the *z-th* variable,
w_z – the valency of the *z-th* variable,
p – the number of branches, out of which the upper branches of the *k-th* level, which are not subject to simplification, were formed [6].

The main operation principle of the discussed algorithm is finding the **so-called primary implicants,** *i.e.* each combination which is not subject to further joining. The number of calculations which are made at the first stage depends on the number of analysed parameters and they are to be made this particular number of times. It is possible to proceed to the next step only after stating all possible minimums at a given stage. The next step consists in making compounds according

*Corresponding author: a.tiszbierek@po.opole.pl

to the parameter with the smallest value and then calculations for the other variables are repeated. If the minimum value was obtained at a given stage two or more times, the subsequent stages would have to be calculated collaterally, for each of the obtained minimums.

2.2 Computer programme

The window of the main programme consists of the main menu (*File* – the close the programme, *Data* – the download / save data, *Results* –save calculations and/or optimum sets, *Info* – information about the programme and the author), the area with loaded data (*KAPN*) and function buttons (*Add, Erase, Calculate and Check scopes*). Buttons *Calculate and Check scopes* are primarily off and they appear only when the data is loaded. The former causes automated calculation of each step and display of calculations together with interim results in a separate window run by clicking on the button *Results* (appearing after calculations are made); whereas the latter modifies values of these parameters, the maximum code value of which does not appear in any of the selected sets of parameters [7].

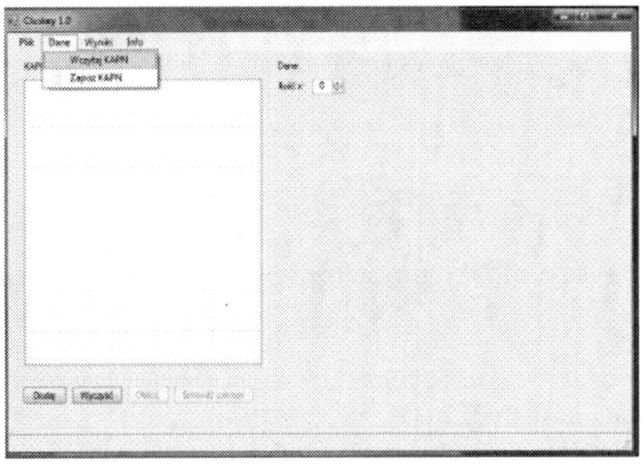

Fig. 1. Basic window of the programme with an option of downloading data.

Additional downloading of data causes the appearance of two subsequent buttons *Conditions* and *Proxy variable*, which start modules of new functionalities. The former imposes appropriate conditions of occurring parameters one after another. After it has been started, an additional window appears with a list of parameters and additional options, by means of which we may indicate which parameters and in which set should appear in subsequent optimum sets.

The algorithm of the programme, with the function of conditions, works similarly to the basic version of the programme. The only innovation is checking whether a newly calculated minimum is not included in the imposed condition (if it is at least the penultimate parameter). If this is the case, it has to be discarded and a new one has to be found. If, however, the minimum value is a parameter at the end of the imposed condition, the remaining parameters have to be set without additional calculations in an imposed sequence and then we should proceed to subsequent calculation stages. The second functionality of the programme is the possibility of coding an appropriately chosen group of parameters in one or a few proxy variables. An additional advantage of the programme is automation of coding of proxy variables. After choosing appropriate parameters, the programme sets a new value of the proxy variable by itself and assigns appropriately set verses with values of intrinsic parameters to a given code value.

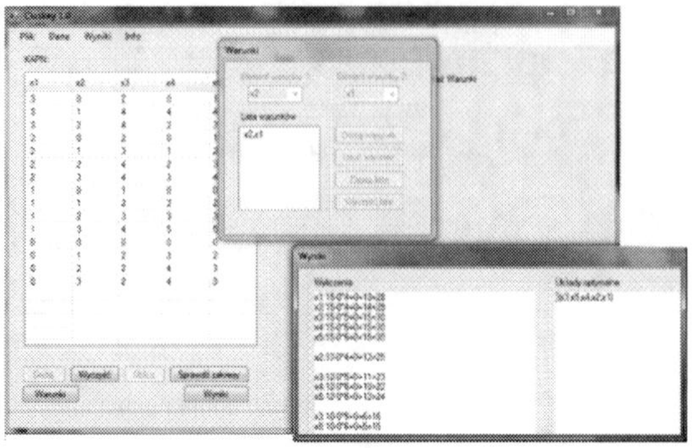

Fig. 2. The view of the programme with a condition and result window.

This process not only shortens coding time but also protects it from the mistakes resulting from the lack of due care or the plurality of variables and their significant multivalence. The process of calculation in the case of this functionality is based on the same algorithm of the programme as its basic version. After proxy variables have been coded, they are treated as "new" parameters (variables included in them are deleted from the table of parameters memory) and then a classic process of calculation with showing final results at the end takes place.

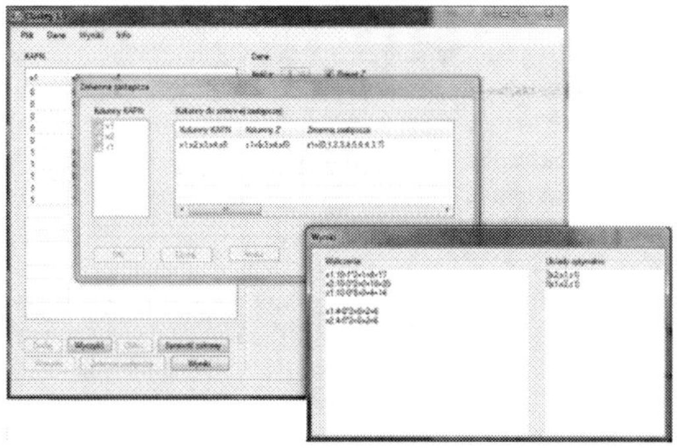

Fig. 3. The view of the programme with proxy variable and result windows.

3 Calculation complexity problems with dependent variables of similar importance

The problem described in [8] is a calculation complexity in the process of determining the importance rank of parameters of similar importance. Additionally, a dependence occurs between parameters and this aids to the difficulty of the process. An example presented in the elaboration is a set of symptoms in thyroid problems.

These symptoms were gathered during research and marked with appropriate x values. X parameters are bivalent (1 signifies the occurrence of the symptom, whereas 0 signifies its lack) and are set in eight variables x_1, x_2, x_3, x_4, x_5, x_6, x_7, x_8 written numerically in Table 1. Basic calculations were first made by means of rough sets, and then by means of bivalent decision logic trees [9]. Calculating an example by means of a computer programme was the next stage [10].

Table 1. A digital record of coding variables.

x_1	x_2	x_3	x_4	x_5	x_6	x_7	x_8
0	1	0	0	1	0	0	1
0	0	0	1	1	0	0	1
0	1	0	1	0	1	1	0
1	1	0	1	0	1	1	0
0	1	1	1	0	1	1	0

576 optimum sets we obtained, the analysis of which showed the lowest importance rank of parameters x_1 and x_3 as well as the dependency between parameters x_2 and x_4. This is why an idea of applying such functionalities as proxy variable and conditionality in case of subsequent calculations emerged. The realization of the idea of applying a proxy variable z, in which parameters x_1 and x_3 are included, with applying the condition of the sequence of parameters x_2 and x_4 at the same time, has is described in [8]. It has been forced in calculations that parameters x_2 and x_4 always occur in the following order x_4x_2, whilst parameters x_1 and x_3 become one variable z. Consequently the process calculation time is reduced along with the number of optimum sets to 120. Such a reduced number of optimum sets makes it possible to make further analysis in an easier way focusing only on remaining variables (x_5,x_6,x_7,x_8). A deeper analysis of obtained results would make it possible to assume that remaining parameters are of the same importance rank. It is thus possible to state that they are equivalent and independent.

4 Problem defining optimum sets of parameters on the basis of the number of real branches

4.1 The need and idea for the realisation of a new module of the computer programme

Although during calculations after drawing exemplary multi-valued logic trees for chosen optimum sets it turned out that they are different from each other in the number of real branches (highlighted on trees). For example, a tree representing a set $x_6x_5x_7x_2x_4x_8x_1x_3$ had 21 irreducible real branches, whereas with the set $x_7x_8x_4x_2x_6x_5x_1x_3$ it had 23 of them. A further analysis of the trees showed that up to a certain level, all trees are on the same level – possessing the same number of real branches, which is 13 (the fourth floor of the tree counting from the top).

Table 2. Exemplary optimum sets presented on shortened logic trees.

Exemplary trees with the placed set …x_1x_3 on upper branches

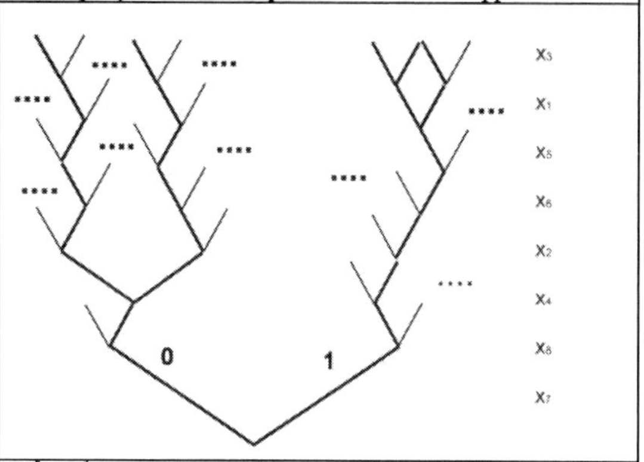

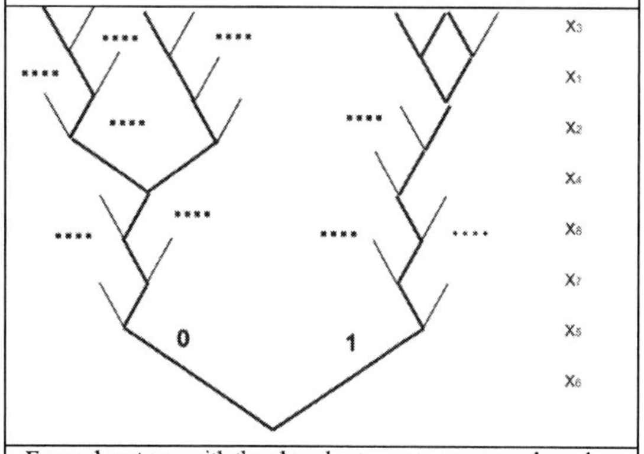

Exemplary trees with the placed set … x_3x_1 on upper branches

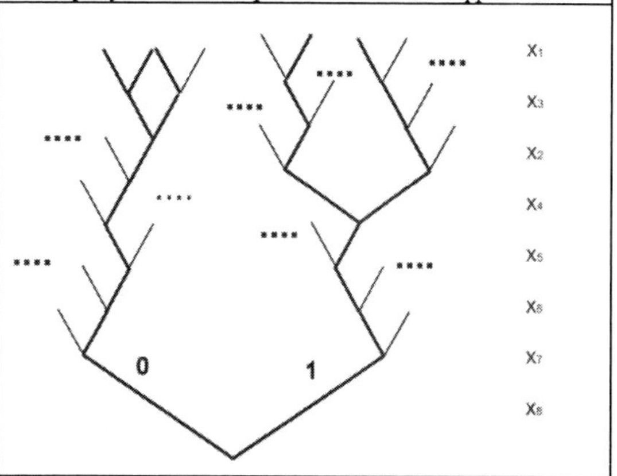

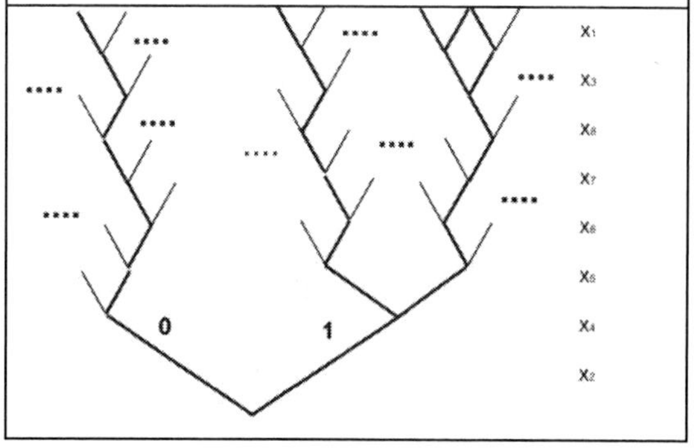

Therefore, the choice of the fifth variable was more important. For example, putting the variable x_7 after x_8 in the first set was beneficial (there are 15 real branches at this tree level), whereas choosing x_2 after x_6 was not (16 real branches). This problem showed how useful the method of checking irreducible real branches on multi-valued logic trees is. It enables finding the best sets of parameters among advantageous ones, even if the difference between them would be minimum.

The new functionality would be a module calculating the number of branches of real, multi-valued, logic decision trees on which optimum sets chosen by the programme will be located.

4.2 Modification of the basic algorithm

The creation of a new module requires modification of the basic algorithm and new software. The number of real branches has to be counted by summing their number on each level. In order to get to know the number of branches on each level, it is necessary to calculate the difference in the number of branches of *k-th* level of a decision tree and the number of simplifications of *z-th* variable multiplied by the valency of *z-th* variable for a given stage. Only the last stage should be calculated completely according to the basic algorithm. Sums obtained in this way have to be juxtaposed against each other. The smallest ones will indicate the most beneficial sets. Therefore, the pattern of the basic equation (1) has to be modified to the following form (for all stages, except the final one, it will be counted according to the basic pattern):

$$x_1 = i_k - i_z * w_z \qquad (2)$$

where:
x_1 – the number of step branches of the analysed decision tree,
i_k – the number of branches of the *k-th* level of the decision tree,
i_z – the number of simplifications of the *z-th* variable,
w – the valency of the *z-th* variable.

The value of each substage (each tree level) has to be counted according to the formula above. The value for the whole set will be used for comparisons. For example, for the set $\mathbf{x_8x_7x_6x_5x_4x_2x_3x_1}$ it is necessary to count values for particular x at a given stage as indicated below:

Step I (on x_1) : x_1:**5-1*2**+1+3=7
Step II (on x_3) : x_3:**4-0*2**+0+3=7
Step III (on x_2): x_2:**3-0*2**+0+3=6
Step IV (on x_4) : x_4:**3-0*2**+0+2=5
Step V (on x_5): x_5:**2-0*2**+0+2=4
Step VI(on x_6) : x_6:**2-0*2**+0+2=4
Step VII(on x_7) : x_7:**2-0*2+0+2=4**

Clearly, only the part of the equation which is included in the modified formula (the part in bold print) is to be taken into consideration in calculations. Finally, the following values will be obtained for each stage and their sum will give 21:

Step I (on x_1) : x_1:5-1*2=3
Step II (on x_3) : x_3:4-0*2=4
Step III (on x_2): x_2:3-0*2=3
Step IV (on x_4) : x_4:3-0*2=3
Step V (on x_5): x_5:2-0*2=2
Step VI(on x_6) : x_6:2-0*2=2
Step VII(on x_7) : x_7:2-0*2+0+2=4

The resulting value (21) is the number of all real branches for the calculated example. In this way, it is necessary to calculate values for each of optimum sets indicated by the programme and then compare these values. These sets which obtained the minimum value (or values) will be the most beneficial, even if their number is slightly different.

4.3 New functionality – data for a chosen device

An axial-flow vortex pump, which is a kind of a centrifugal rotary pump, was chosen for exemplary calculations. This pump is characterised by the following physical quantities (which influence its work efficiency):

- Q – describing the volume flow,
- H – describing the head,
- N – describing the power,
- η – describing the efficiency,
- n – describing the rotational speed,
- φ – describing the angle of setting the impeller blades.

Knowing the model construction we may connect the parameters describing its operation and elements responsible for its values. This, in turn, makes it possible to carry out a better analysis of the set of pump parameters from with a view to determine particular importance ranks and thus, an optimal set of construction and operation parameters.

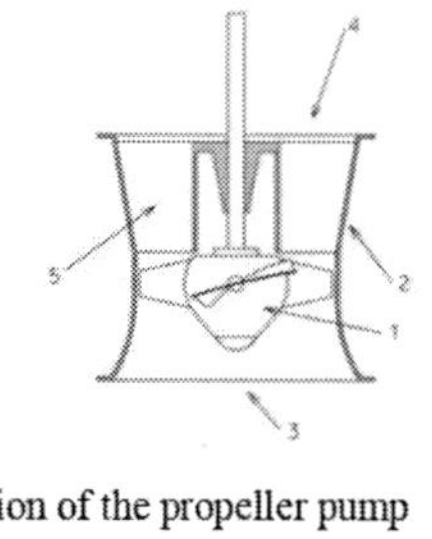

Fig. 4. Construction of the propeller pump

Pumps of this type are mainly used in industrial and energetic installations, particularly in power plants (most often in small water power plants). The research was carried out on the model turbine (designed and built with the use of construction elements of the serial axial-flow vortex pump 25P21-2 of the Warsaw Factory of Pumps and Valves - *Fabryka Pomp i Armatury WAFAPOMP SA*). The actual results had to be selected and coded in an appropriate way in order to make it possible to present it by means of a multi-valued logic function. Final values got an appropriate code number: φ,n=0,1,2,3; Q=0,1,2,3,4,5; H,N=0,1,2,3,4,5,6 [11].

Table 3. Coded results of parameter values

φ	n	Q	H	N
0	0	0	0	0
0	1	2	3	2
0	2	2	4	3
0	3	2	4	3
1	0	1	0	0
1	1	2	2	2
1	2	3	3	3
1	3	4	5	5
2	1	3	1	2
2	2	4	2	3
2	3	4	3	4
2	0	2	0	1
3	0	2	0	1
3	1	4	4	4
3	2	4	2	3

Computer-based calculations of entered values made it possible to obtain the following results :

step I:
x_1:15-0*4+0+13=28
x_2:15-0*4+0+14=29
x_3:15-0*5+0+15=30
x_4:15-0*6+0+15=30
x_5:15-0*6+0+15=30

step II (on x_1):
x_2:13-0*4+0+12=25
x_3:13-0*5+0+12=25
x_4:13-0*6+0+12=25
x_5:13-0*6+0+13=26

step IIIa (on x_2):	step IIIb (on x_3):	step IIIc (on x_4):
x_3:12-0*5+0+11=23	x_2:12-0*4+0+11=23	x_2:12-0*4+0+10=22
x_4:12-0*6+0+10=22	x_4:12-0*6+0+8=20	x_3:12-0*5+0+8=20
x_5:12-0*6+0+12=24	x_5:12-0*6+0+11=23	x_5:12-0*6+0+11=23
step IVa (on x_4):	step IVb (on x_4):	step IVc (on x_3):
x_3:10-0*5+0+6=16	x_2:8-0*4+0+6=14	x_2:8-0*4+0+6=14
x_5:10-0*6+0+5=15	x_5:8-0*6+0+4=12	x_5:8-0*6+0+4=12

Optimal systems:
f(x_3,x_5,x_4,x_2,x_1)
f(x_2,x_5,x_4,x_3,x_1)
f(x_2,x_5,x_3,x_4,x_1)
where: x_1-µ; x_2-n; x_3-Q; x_4-H; x_5-N

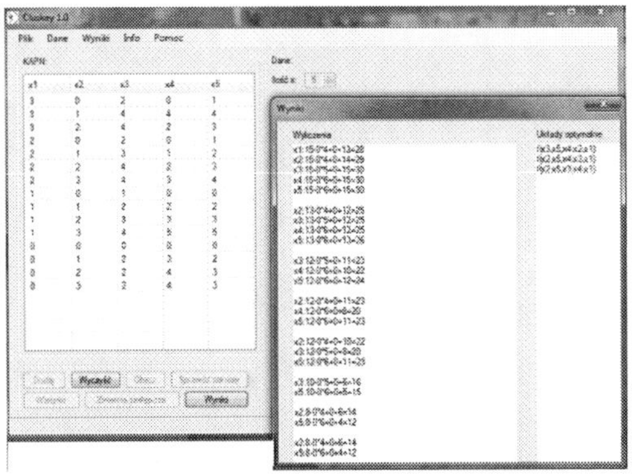

Fig. 5. Computer-based calculations for the analysed example.

As calculated. there should be three optimum (best) sets. However, further calculations (the calculation of the number of real branches for each set) show better and worse sets among them. Calculations for particular sets are presented below:

f(x_3,x_5,x_4,x_2,x_1)	f(x_2,x_5,x_4,x_3,x_1)	f(x_2,x_5,x_3,x_4,x_1)
x_1:15-0*4=15	x_1:15-0*4=15	x_1:15-0*4=15
x_2:13-0*4=13	x_3:13-0*5=13	x_4:13-0*6=13
x_4:12-0*6=12	x_4:12-0*6=12	x_3:12-0*5=12
x_5:10-0*6+0+5=15	x_5:8-0*6+0+4=12	x_5:8-0*6+0+4=12
sum: 55	sum:52	sum:52

It is thus possible to select two sets out of three, the number of real branches of which is the lowest value and these two sets (f(x_2,x_5, x_3,x_4,x_1), f(x_2,x_5,x_4,x_3,x_1)) can be acknowledged to be the most beneficial.

Certainly, knowing the construction and the process of pump operation, it is obvious that parameters φ and n should occur one after another because of their interactivity (requires applying an appropriate condition here). However, for the needs of the analysed problem it was resolved not to in order not to overshadow the calculation results.

5 Conclusion

An example described in this article shows how a computer programme can facilitate the process of calculating and how useful it can be in determining the importance rank of construction and operation parameters. A complex and time-consuming process lasts from a few to several dozens of seconds, it does not draw the user's attention and it eliminates many potential mistakes. Additional functionalities of the programme (such as the possibility of imposing conditionality of the set of parameters and creation of proxy variables) are equally important and useful. However, as shown, sometimes even these elements of the programme are insufficient, which is why the creation of the described module determining the number of real branches is an important solution recommended for proper functioning of the process of determining the importance rank of construction and operation parameters. A new module will facilitate the calculating process and confirm the credibility of obtained results. It will also solve the problem of obtaining a huge number of beneficial sets differing in the minimum number of real branches. This is why in subsequent stages, the described idea should be programmed and the obtained solution should be tested and described.

References

1. M. Giovanni, *Synthesis and optimization of digital circuits* (Wydawnictwa Naukowo - Techniczne, 1998)
2. C. D. Kozen, *Automata and Computability* (Springer, 1997)
3. E. J. McCluskey, Bell System Tech. Journal. **35**, 1417-1444 (1956)

 DOI: 10.1051/itmconf/20171504002

4. T. K. Jain, D. S. Kushwaha, A. K. Misra, *Optimization of the Quine- McCluskey Method for the Minimization of the Boolean Expressions. International Conference on Autonomic and Autonomous Systems* (ICAS, 2008)
5. R. Mohan Ranga Rao, IJAEST **3**(1), 12-14 (2011).
6. M. A. Partyka, A. Tiszbierek, Napędy i Sterowanie. **9/2015**, 132-139 (2015).
7. S. Ford, *Microsoft Visual Studio Tips* (Microsoft Press, 2008).
8. A. Tiszbierek, Napędy i Sterowanie, *Zastosowanie komputerowego wspomagania wyznaczania rangi ważności parametrów z funkcją zmienna zastępcza oraz warunkowość na rzeczywistym przykładzie ze zmiennymi zależnymi o podobnej ważności* (in review).
9. M. A. Partyka, *Podobieństwa i różnice przybliżonej klasyfikacji obiektów w ujęciu logiki i systemów informacyjnych dla CAD procesów decyzyjnych. XXIV Konferncja Zastosowań Matematyki* (Instytut Matematyki PAN, 1995).
10. A. Tiszbierek, M. A. Partyka, *XLV Konferencja Zastosowań Matematyki* (Instytut Matemtyki PAN, 73-75, 2016).
11. M. A. Partyka, R. Łuszczyna, M. Sojka, *Konferencja Komputerowe Zintegrowane Zarządzanie* (Oficyna Wydawnicza Polskiego Towarzystwa Zarządzania Produkcją, Opole, 292-299,2009).

 DOI: 10.1051/itmconf/20171504003

Cluster analysis to evaluate chemical compounds in the waters of piezometers from a drilling wastes landfill

Justyna Kujawska*[,1], and *Wojciech* Cel[1]

[1]Institute of Renewable Energy Engineering, Lublin University of Technology, 40B Nadbystrzycka Street, Lublin, Poland

Abstract. Storage of drilling wastes threatens the natural environment. Cluster analysis was employed for evaluation of the behaviour of chemical compounds in the water of piezometers from the drilling waste management facility in Luchów. Cluster analysis proved to be an efficient method, enabling the identification of groups of chemical compounds threatening the environment.

1 Introduction

The economic development of counties undergoes a rapid increase, the production of goods and services grows significantly, whereas the global population has doubled [1]. Hence, the consumption of energy and natural resources has increased as well. New sources of energy are constantly being sought. The prospecting of gas from both the conventional and unconventional sources gains increasing attention. Finding the balance between the economic development and the natural environment becomes a necessity. In order to improve the energy use efficiency, environmental protection and achieve the sustainable development, new legal regulations pertaining to efficient use of environment are being devised [2].

Poland is a country with high expectations regarding shale gas. Extensive works connected with shale gas exploration are being conducted [3]. These works may cause excessive pollution of the environment in the course of exploitation as well as other processes. Drilling yields substantial amounts of waste. New methods of their management are being sought. Additionally, new requirements related to their neutralization, storage, etc. are being devised. In Poland, since 2012, application of drilling waste in the reclamation of landfills and creation of waste treatment facilities was the commonly employed method of its management. The amendment to the Waste Management Act implemented in 2016, restricted filling the mining excavations with mining waste for reclamation and technological purposes. According to the updated regulations, mining wastes, depending on their type, may be stored:

- for up to 6 months – hazardous mining waste, the production of which could not be foreseen;
- for up to 1 year – mining waste other than hazardous or inert;
- for up to 3 years – non-polluted soil, mining waste other than hazardous produced in the course of exploratory works, waste produced during excavation, processing and storage of peat [4].

The properties of drilling waste depend on: the type of rock, drilling fluid, the employed technology. Drilling wastes contain the substances used for the preparation of drilling fluids, technological fluids, as well as the products of their degradation. The chemical compounds originating from drilling fluids found in the drilling wastes include: mineral oils, organic compounds such as: polymers (tannins, lignosulfonates, lignites, xanthan gum, guar gum; starch and cellulose derivatives, active surfactants, inorganic compounds (oxides and hydroxides of calcium, magnesium, sodium, potassium, as well as potassium or sodium chlorides) as well as barite, iron ores, galena, chalk, and limestone. Wastes also contain metal, not only the ones added to drilling fluids, but also originating from sedimentary rocks [5]. The compounds found in drilling wastes may migrate to surface and ground waters, as well as soils and plants [6].

In the case of waste landfills or waste treatment facilities, the main migration pathways include: leachates, surface run-offs as well as emission of gases and dusts. Leachates mainly originate from precipitation waters infiltrating through a landfill, leaching a whole range of pollutants.

Constant monitoring of a waste management facility and landfills is necessary, requiring devising the monitoring programmes. The presented research involved cluster analysis and multidimensional statistical analysis evaluating the behaviour of chemical compounds in the water from piezometers from the waste management facility in Luchów.

Cluster analysis aims at grouping objects into a common cluster. Clustering maximizes the homogeneity of objects belonging to the same groups and maximizes heterogeneity by assigning objects to different groups [7].

Cluster analysis is widely used in environmental engineering for studying the water quality [8,9]. Ashley

* Corresponding author: j.kujawska@pollub.pl

DOI: 10.1051/itmconf/20171504003

and Lloyd (1978) employed cluster analysis for the classification of ground waters depending on the content of organic compounds [10]. Frapporti et al. (1993) carried out cluster analysis of ground waters depending on the type of soils [11]. Cluster analysis was also employed for determination of the flow direction of ground waters on the basis of hydrochemical analyses of water [12,13].

2 Research methodology

2.1 Location

The studies were carried out at a drilling waste landfill situated in Luchów Górny, a village in Lubelskie Voivodeship, in Biłgorajski county, Poland. Samples of water were collected from the piezometers located in three spots around the landfill. In each spot, there were two piezometers collecting water from different depths (Table 1).

Table 1. Data pertaining to the ground water collection spots.

Name	Altitude [m]	Depth of a piezometer measured from the ground level [m]	Depth of a table of water measured from the ground level [m]
PC I Q	237	6.1	4.20
PC I T		17.0	4.40
PC II Q	229	6.2	5.10
PC II T		17.5	4.90
PC III Q	225	6.0	3.40
PC III T		16.5	3.10

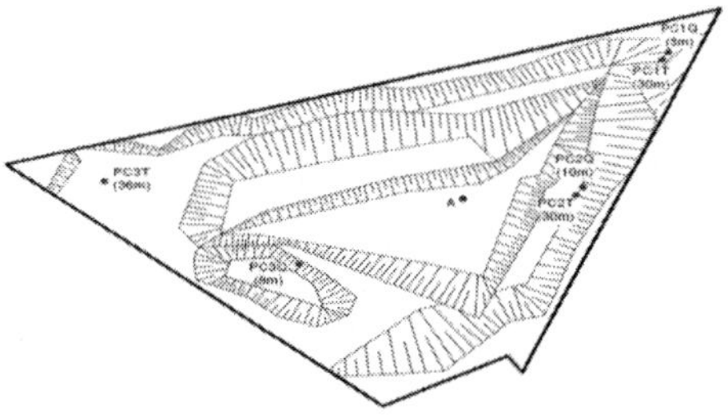

Fig. 1. Location of piezometers on the waste deposition landfill. The depth of piezometers is given in the brackets.

The piezometers marked as PC I are found in the northernmost and highest part of the landfill, which borders with agricultural land. The PC II piezometers are 230 m southwest from the highest landfill spot and lie in the area which is highly waterlogged. The PC III piezometers are 290 m west of the highest part of the landfill and are located on the sawmill area. The location of piezometers in relation to the landfill was presented in Fig. 1.

2.2 Samples collection and preparation

The samples were collected from March 2015 to December 2015. In total, 21 samples of water were collected.
The samples were taken according to the PN-76/C-04620/03 Standard.

2.3 Scope of research and analytic methods

The following parameters were determined in the collected samples:

- pH, by means of the potentiometric method, with a multifunctional VERSA STAR meter manufactured by ORION
- specific electrolytic conductivity, by means of a multifunctional VERSA STAR meter manufactured by ORION
- chloride content, with Mohr's method
- content of metals, including: Al, B, Ba, Cd, Cr, K, Na, Ni, Pb, by means of Inductively Coupled Plasma Optical Emission Spectrometry method, with JY 238 Ultrace (Jobin Yvon-Horriba, France)
- content of organic carbon, by means of TOC-5050A analyzer manufactured by Schimadzu (in line with PN-ISO 14335 standard)
- content of sulphate ions, determined by means of spectrophotometric method PBL/CH/28/06, issue 02 of 07.11.2011 on the basis of HACH 8051 method
- chemical oxygen demand was determined with spectrophotometric method, with HACH – LANGE PN – ISO 1575 2005 8051

2.4. Statistical analysis of clusters

Cluster analysis method was employed for grouping variables by means of Ward's method. Ward's method estimates the distance between clusters using the analysis of variance approach. This method aims at minimizing the sum of squares of deviation within clusters. The chemical properties of waters from piezometers from the waste management facility were used as variables. The analysis was carried out on the basis of Statistica software package (v. 13.1).

3 Results

The obtained results were presented in tables 2 and 3. The concentration of aluminium was below the detection limit; therefore, it was omitted in the statistical studies. The results of cluster analysis were presented in Figure 2. The cluster analysis enabled to identify three groups of chemical properties of waters. The first one, with the shortest average distance comprises pH, as well as the concentrations of B, Ba, Mn, and Ni. The second group includes the values of TOC, the concentration of Ca, K, Mg, Ng, chlorides bound with the group comprising the concentration of iron and COD. The third one, with the greatest distance between the identified groups, involves the values of conductivity, TDS and the concentration of sulphates. The obtained results of analyses were compared with the permissible values determined in the Regulation of the Minister of Environment of 21 December 2015 establishing the criteria for and the way of classifying the state of uniform parts of groundwater [14]. The concentrations of compounds in the examined waters indicate that none of them meet the criteria of border values for the waters of good chemical status. The lowest quality characterized the waters from PC IIIQ piezometer. The permissible values of as many as 8 parameters were exceeded, including: conductivity, TOC, concentrations of B, Ca, Mg, Mn, Ni (in all measurements) and Al (only in March). In the adjacent PC IIIT piezometer, 4 parameters were exceeded: TOC, B, Ca (in all measurements) and Al (in July). Both piezometers were located at a lower height than the others. In the remaining piezometers, only single parameters were exceeded: in PCIT and PCIQ piezometers – TOC, in the case of PCIIT – K, whereas in PCIIQ – Ca.

Table 2. Concentration of metals in the water taken from piezometers.

Piezometer	Sampling time	Al	B	Ba	Ca	Fe	K	Mg	Mn	Na	Ni
		[mg/dm^3]									
PC I T	III	<d.l	0.90	0.036	19.44	0.09	6.01	4.38	0.01	126.54	<d.l
	VII	<d.l	1.09	0.035	19.71	0.03	6.46	4.81	0.015	131.58	<d.l
	X	<d.l	0.98	0.03	24.45	<d.l	7.23	5.39	0.02	118.74	<d.l
PC I Q	III	<d.l	<d.l	0.03	141.47	0.11	2.61	11.88	0.03	24.2	<d.l
	VII	<d.l	<d.l	0.04	138.65	0.14	2.69	12.98	0.04	19.99	<d.l
	X	<d.l	<d.l	0.03	133.54	<d.l	2.99	12.87	0.01	16.57	<d.l
PC II T	III	<d.l	<d.l	0.099	115.71	<d.l	18.73	8.71	0.25	94.58	<d.l
	VII	<d.l	0.50	0.11	133.29	0.05	23.18	12.90	0.29	121.58	<d.l
	X	<d.l	0.44	0.12	161.83	<d.l	27.58	13.83	<d.l	130.11	0.003
PC II Q	III	<d.l	<d.l	0.05	328.62	0.24	3.70	45.46	0.09	90.57	<d.l
	VII	<d.l	0.14	0.062	324.06	0.23	4.07	50.01	0.16	92.53	<d.l
	X	0.99	0.18	0.07	236.96	0.33	3.09	45.26	0.18	99.01	0.004
PC III T	III	<d.l	2.05	0.02	467.33	0	3.63	99.33	<d.l	35.44	<d.l
	VII	4.32	2.33	0.013	466.99	0.15	4.84	99.87	0.01	33.66	<d.l
	X	0.02	2.28	0.01	445.28	0.26	4.23	88,0	<d.l	32.14	<d.l
PC III Q	III	5.02	1.91	0.02	469.28	0	11.26	120,09	1.74	24.58	0.09
	VII	<d.l.	2.05	0.013	500.08	0.01	13.18	116,75	1.35	22.36	0.08
	X	<d.l.	1.21	0.05	414.31	0	14.20	139,22	1.62	24.58	0.07
Border values[1)]		0.2	1	0.7	200	5	15	100	1	200	0.02

(light shading) parameters with the values within the standards of Class IV waters (waters of unsatisfactory quality)
(dark shading) parameters with the values within the standards of Class V waters (waters of poor quality)

Table 3. Characteristics of water collected from piezometers.

Piezometer	Sampling time	pH	Conductivity [µS/cm]	TDS	TOC [mgC/dm^3]	COD	Chlorides	Sulfate
				[mg/dm^3]		[mg/dm^3]	[mgCl/dm^3]	[mgSO$_4^{2-}$/dm^3]
PC I T	III	7.9	662	399	17.04	20	8.65	13.1
	VII	7.6	629	383.4	15.99	19.7	5.21	17.9
	X	7.5	634	388.21	16.32	25.1	8.79	13.1
PC I Q	III	7.3	794	468.65	18.26	32	54.02	59.5
	VII	7.2	723	439.80	16.25	31.3	52.06	62.3
	X	7.6	789	476.43	15.49	35.8	58.10	56.1
PC II T	III	7.9	669	399.05	8.66	35.6	70.56	63.9
	VII	8.1	689	416.41	8.57	34.7	80.35	198
	X	7.7	658	392.45	8.36	39.2	77.54	98.4
PC II Q	III	6.5	1058	667.27	6.06	39.5	129.03	68.8
	VII	6.3	1122	639.58	5.99	38.4	104.56	140
	X	6.4	1010	664.87	6.02	42.9	119.30	264
PC III T	III	6.3	2030	1230	20.12	20.2	49.50	1222
	VII	6.2	2045	1239	22.03	18.1	55.60	1100
	X	6.9	2033	1230	21.77	22.7	53.11	1278
PC III Q	III	6.9	2628	1626	15.48	22.58	10.89	1929
	VII	6.0	2731	1682.44	17.23	23.1	11.02	1128
	X	6.1	2654	1601.41	17.03	27.5	7.02	1152
Border values[1)]		6.5-9.5	2500	Not defined	10	Not defined	250	250

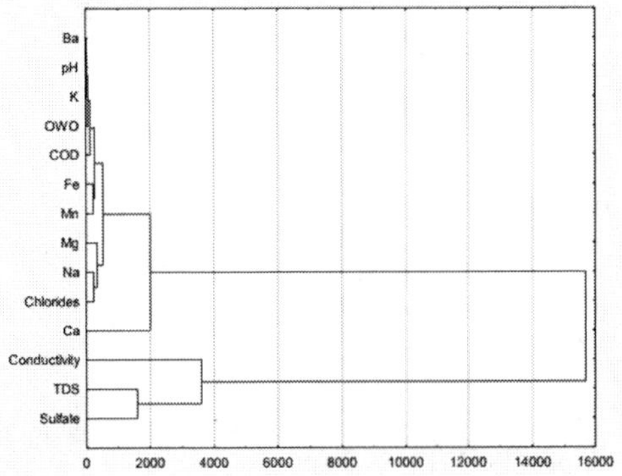

Fig. 2. Cluster analysis of selected parameters of water from piezometers from the waste management facility.

The obtained research results of water samples from piezometers indicate the elevated content of the compounds from the groups characterized by medium and greatest distance in the conducted cluster analysis. This confirms the applicability of cluster analysis for grouping the pollutants present in the ground waters.

4 Conclusion

The conducted cluster analysis involved classification of 21 water samples into three groups (low, medium and high pollution) on the basis of the chemical compounds in their content.

The results of cluster analysis may be used for optimizing the water quality monitoring programme, and enable to decrease the frequency of water sample

collection. In other words, in the further studies, the number of spots for monitoring the water quality may be reduced by selecting only the compounds which threaten the natural environment.
The results of statistical analyses allow to select the high-priority areas, thus aiding in the management of ground water quality.

References

1. A. Duda, J. Gołębiowska, A. Żelazna, Probl Ekorozw **11**, 1 (2016)
2. P. Rydzewski , Probl Ekorozw **11**, 1 (2016)
3. J. Macuda., M. Bogacki, J. Siemek., Probl Ekorozw 12, 1 (2016)
4. PolishWaste Management Act Dz.U. 2008 nr 138 poz. 865
5. J.M. Neff, Integr Environ Asses **4**, 2 (2014)
6. Y.Y. Fan, Wang G.C., Fu J.H., Zheng X.-H., Pet Sci Technol **32** (2014)
7. S. Javadi, S.M. Hashemy, K. Mohammadi, K.W.F. Howard, J Hydrol **549** (2017)
8. M. Hajigholizadeh, A. M. Melesse, Catena **151** (2017)
9. W. de Souza Pereira, G. Ferrari de Morais, D. de Azevedo Py Júnior, International Nuclear Atlantic Conference (INAC 2011 Belo Horizonte, MG, Brazil)
10. R.P. Ashley, J.W. Lloyd, J. Hydrol. **39** (1978)
11. G. Frapporti, P. Vriend, P.F.M. Van Gaans, Water Resour. Res. 29, 9 (1993)
12. J.A. Riley, R.K. Steinhorst, G.V. Winter, R.E. Williams, J. Hydrol. **119** (1990).
13. K.M. Ochsenkuhn,, J. Kontoyannakos, M. Ochsenku, J. Hydrol. **194**, 1, (1997).
14. Regulation of the Polish Minister of Environment of 21 December 2015

DOI: 10.1051/itmconf/20171504004

Computer modelling of the influences of a subsystems' interaction on energetic efficiency of biofuel production systems

Andrzej Wasiak[1,*], and *Olga* Orynycz[1]

[1]Bialystok University of Technology, Dept. of Production Management, Bialystok, Poland

Abstract. Energetic efficiency of biofuel production systems, as well as that of other fuels production systems, can be evaluated on the basis of modified EROEI indicator. In earlier papers, a new definition of the EROEI indicator was introduced. This approach enables the determination of this indicator separately for individual subsystems of a chosen production system, and therefore enables the studies of the influence of every subsystem on the energetic efficiency of the system as a whole. The method has been applied to the analysis of interactions between agricultural, internal transport subsystems, as well as preliminary studies of the effect of industrial subsystem.

1 Introduction

The problem of the efficiency of energy production has been raised relatively a long time ago mainly in terms of qualitative, philosophical or economic considerations [1]. Much later the quantitative approach was introduced [2]. To the measure of energetic efficiency the name EROEI (or EROI) "Energy Return on Energy Invested" was given. It was defined as a ratio of energy delivered by some production systems to the energy that has to be consumed in order to obtain, and deliver, the above mentioned amount of energy from a given resource. The EROEI is different than thermodynamic efficiency (η) of energy converter, which is established as the ratio of energy delivered by the converter to the energy input to that converter. Since some energy is lost (dissipated) during conversion, the values of thermodynamic efficiency are contained within the limits $0 \leq \eta \leq 1$. The EROEI, in contrast, does not consider the input of energy that has to be converted (in fact the input is rather a material resource). What is taken into account is the output of energy from the system, and the subsidiary fluxes of energy being necessary to assure performance of the system. Therefore, this index may assume values from zero to infinity. One may consider the lack of precisely defined calculations procedure as a disadvantage of EROEI. In fact, as indicated in [3], several procedures, and several definitions are used, and some resulting ambiguities have been discussed [4, 5]. Recently, the definition discussed in [6] was used by present the Authors, in a series of papers [7-10] devoted to the modelling of energy efficiency of agricultural subsystems in the rapeseed biodiesel production system.

It seems clear that the EROEI index can be equally applied to any energy production systems, including fossil fuels, biofuels, solar or wind energy. In recent years EROEI become an important characteristic for alternative energy production systems, because it is possible that in order to gain energy from alternative resources (intended to replace exploitation of fossil ones) one has to consume energy derived from fossil resources. In extreme cases, when the amount of consumed energy would exceed the amount of delivered, such processes would be lacking in sense.

It is also remarkable that fuel production systems are frequently composed of separated subsystems, having different characteristics, but connected to each other by fluxes of masses or energy. Therefore, establishing a way of separating the estimation of energetic efficiency of such subsystems, and the investigation of their interactions determining the total energetic efficiency of the production system become important.

2 Structure and efficiency of fuel production systems

The structure of fuel production systems can be classified as serial or parallel composition of components (subsystems) or a combination containing partly serial, partly parallel. Simple examples of such composed systems are shown in Fig. 1 and Fig. 2. The flux of material resources and semi-products are marked with black arrows going down. Fluxes of energy provided to facilitate conversion are plotted as grey arrows.

The EROEI indicator, in this paper is represented by the character (ε) because certain specific assumptions are introduced for the sake of computations.

* Corresponding author: a.wasiak@pb.edu.pl

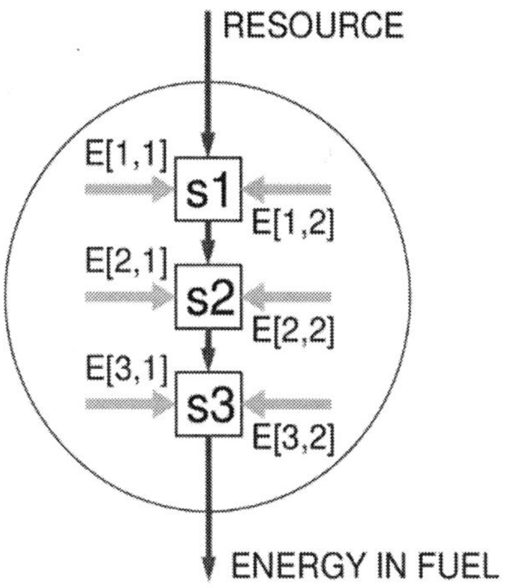

Fig. 1. Fuel production system with serial alignment of subsystems.

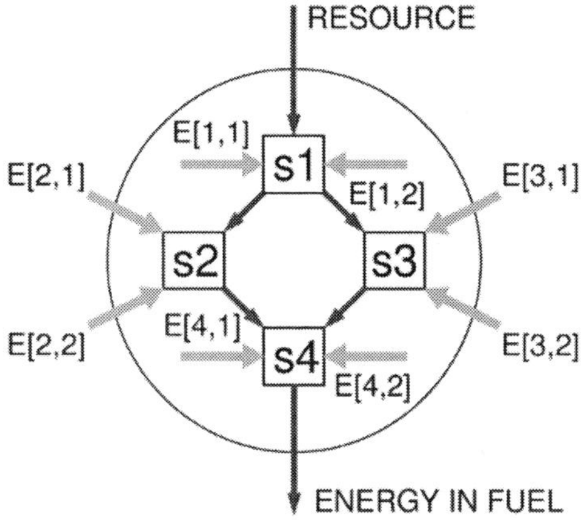

Fig. 2. Fuel production system with combination of serial and parallel alignment of subsystems.

The general definition of energetic efficiency of the production system is accepted as:

$$\varepsilon_{tot} = E_{fuel} / \Sigma E_{i,j} \qquad (1)$$

where index j numbers the energy fluxes going into a given i-th subsystem.

The partial energetic efficiency of the particular subsystem

$$\varepsilon_j = E_{fuel} / \Sigma\, E_i \qquad (2)$$

In both cases described by Eq. 1, and Eq. 2 the numerator of the expression is the same, and it is the final amount of energy obtained from the system (delivered out of the system). It has to be taken into account that determination of values of partial efficiency (according to Eq. 2) must be related to exactly the same quantity of resource entering the system, although during conversion processes both form and the quantity of related semi-products might be different than the amount of initial resource.

Accepting the above conditions one may easily derive the relation between total, ε_{tot}, and partial efficiency. Namely:

$$\varepsilon_{tot}^{\ -1} = \Sigma \varepsilon_j^{-1} \qquad (3)$$

It can be seen that for very particular cases, when all the values of partial efficiency in the system are equal, $\varepsilon_1 = \varepsilon_2 = \dots\dots = \varepsilon_I$, the total efficiency will be represented by:

$$\varepsilon_{tot} = \varepsilon_I / I \qquad (4)$$

where I is the total number of subsystems.

In the case when individual values of partial energetic efficiency differ from each other; such simple procedures described by Eq. 4 are not valid.

The same formulas can be correspondingly applied to the set of technological operations or processes occurring in the production system. It is important, however, to compute all partial, as well as the final, total efficiency with the same value of total energy obtained from production system from a well-defined, particular amount of resource. In the cases of biofuel production systems, the procedure might refer to particular mass of crops or to particular plantation area.

3 Numerical computations

Eq. 4 demonstrates that the addition of subsequent steps of production technology decreases the energetic efficiency of the whole production system. In the case of equal values of partial energetic efficiency of individual subsystems or processes, the simple form of Eq. 4 represents this decrease. When partial contributions are not equal, the general solution cannot be obtained. Consequently, the estimation of the effect for a particular case can be achieved by numerical computations. For the purpose of the present paper, examples of computations have been performed using the Microcal ORIGIN computer program.

For the first case, a system composed of two subsystems is considered. Fig. 3 presents results of computations of energetic efficiency of the whole system, ε_{tot}, as function of partial efficiency ε_i of both subsystems treated as independent variables. It is seen that the global efficiency of the system is represented by convex surface having local maxima corresponding to equal values of each contributing partial efficiency. The higher are values of both contributing characteristics, the higher is the resulting global energetic efficiency of the system. The maximum observed value of the global efficiency is reduced to half of the values of contributing ones.

DOI: 10.1051/itmconf/20171504004

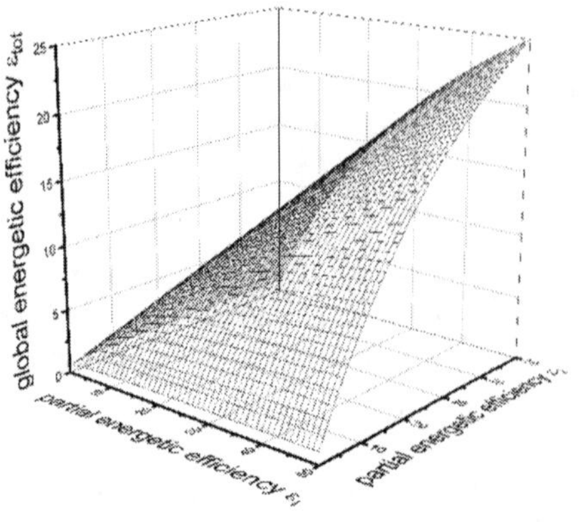

Fig. 3. Energetic efficiency of the production system composed of two subsystems.

Fig. 4 in turn shows the dependence of global energetic efficiency upon the first component efficiency (ε_1) being considered as an independent variable, for only several chosen values of the second component energetic efficiency, ε_2.

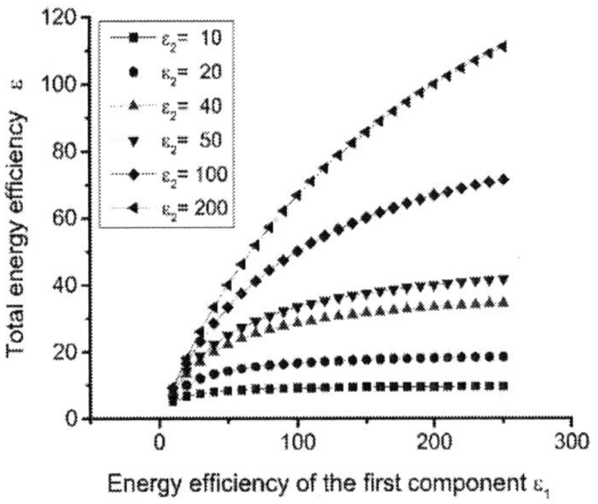

Fig. 4. Energetic efficiency of the production system composed of two subsystems computed for several combinations of subsystem's efficiency.

The dependence of total energetic efficiency of the system appears to be nonlinear function of the efficiency of the first component. The values of energetic efficiency of the system composed within a subsystem of low partial efficiency equal to ten are drastically reduced. Practically for all values of the ε_1 being between 50 and 250, the resulting global efficiency is always smaller than ten. In the other extreme case, when the value for ε_2, is taken as 200, the resulting global efficiency is also reduced with respect to the both components, and is always smaller than half of the value for larger component.

The results of calculations for a system composed of three subsystems are shown in Fig. 5. It was assumed that two of the subsystems have the same efficiency, while several values were chosen for the third one.

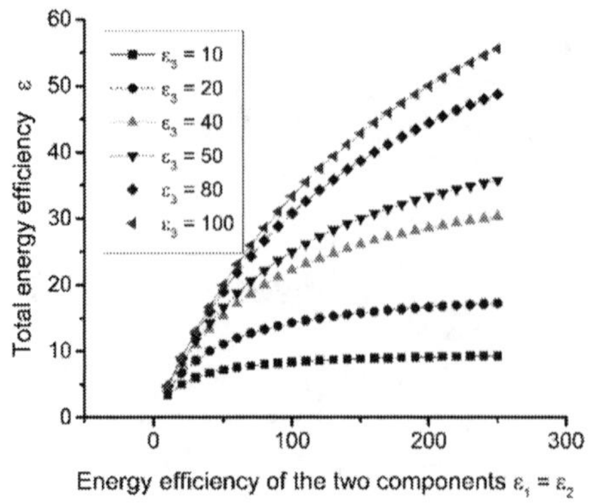

Fig. 5. Energetic efficiency of the production system composed of three subsystems.

This case also shows that global energetic efficiency of the whole production system is greatly reduced with respect to the efficiency of contributing subsystems. The highest reduction is observed when large differences between efficiencies of contributing subsystems occur. It is evident that even both of the two contributing subsystems exhibit high efficiency, the smallest contribution dominates over the remaining, reducing global value to the level even lower than that of the smallest contribution.

Taking into account that any production system is composed of several subsystems, in which several processes are realized, one should expect rather lower global energetic efficiency for the whole production system. Especially dangerous are those subsystems, or processes, which exhibit low energetic efficiency, since they strongly reduce the efficiency of the whole system. In addition, the number of subsystems is the factor strongly reducing the overall efficiency. It is also remarkable, that the highest efficiency is achieved when efficiencies of individual subsystems are close to each other.

4 Conclusions

The results of computations presented above demonstrate that computer modelling may serve as an effective tool for analysis of various technical phenomena giving deep insight into relationships between many factors.

The presented results of computations aim to investigate the effects of energetic efficiency of subsystems on the energetic efficiency of a whole system formed from those subsystems, and dedicated to fuel production allows one to formulate several concluding remarks.

First of all, it is evident that combination of several subsystems, each of which consumes energy, strongly reduces the global energetic effectiveness of the whole system. Moreover, the most favourable case is when the values of efficiency of each subsystem are equal or close to each other. In the case of large differences, the component of the lowest efficiency dominates reducing total efficiency.

Additionally, reducing the number of subsequent operations (the number of subsystems) leads to an increase in total energetic effectiveness.

Therefore, in designing fuel production systems one should carefully choose composing parts and processes in such a way that the efficiency of each contributing process is as high as possible, and combined processes show similar values of efficiency.

For processes of fuel production it seems necessary to achieve possibly high values of global efficiency of the system, otherwise harvesting energy from particular resources may consume more energy than the process delivers.

Considering alternative energy sources, including alternative fuels, the real replacement of fossil fuels might occur only when the total energetic efficiency of the production system is higher than one. Systems with smaller efficiencies should be avoided. Due to the existence of some rules of economic subventions given to alternative energy producers, one might expect the existence of a discrepancy between the efficient production of energy and economic profits as a result of that production. This point requires separate studies.

The proposed definition and the method of calculation of energetic efficiency of energy production systems offers the possibility of studying how individual components affect the global energetic efficiency of the whole system.

The elaborated rule of additivity for reciprocals of partial efficiencies of subsystems that results in assumed methodology opens new functionality for energetic efficiency indicators.

The research has been performed under the financial support from the Bialystok University of Technology. The Statutory Research Project S/WZ/1/2015 was financed from the scientific grant received from the Ministry of Science and Higher Education (MNiSW).

References

1. T. Martinez-Allier, R. Muradian, *Handbook of Ecological Economics* (Edward Elgar Publ., (2015)
2. C.J. Cleveland, R. Costanza, C.A. Hall, R. Kaufmann, Science, 225 (1984)
3. D.J. Murphy, C.A. Hall, Ann. N.Y. Acad. Sci., 1185 (2010)
4. Y. Zhang, L.M. Colosi, Energy Policy, 57 (2013)
5. D.J. Murphy, C.A.S. Hall, M. Dale, C. Cleveland, Sustainability, 3 (2011)
6. W.F. Pickard, Proceedings of the IEEE, **102**, 8 (2014)
7. A. Wasiak, O. Orynycz, IEEE International Energy Conference: ENERGYCON'2014, Dubrovnik, Croatia, (2014)
8. A. Wasiak, O.Orynycz, Agriculture and Agricultural Science Procedia, 7 (2015)
9. O. Orynycz, A. Wasiak, Appl. Computer Sci., **10**, 2 (2014).
10. A. Wasiak, O. Orynycz, Appl. Computer Sci., **11**, 1 (2015).

Estimation of uncertainty of laser interferometer measurement in industrial robot accuracy tests

Jerzy Józwik[1,*], *Elżbieta* Jacniacka[1] and *Dawid* Ostrowski[2]

[1]Lublin University of Technology, Mechanical Engineering Faculty, Department of Production Engineering, 36 Nadbystrzycka Street, 20-618 Lublin, Poland
[2]The State School of Higher Education, The Institute of Technical Sciences and Aviation, 54 Pocztowa Street, 22-100 Chełm, Poland

Abstract. The subject of this article is the assessment of measurement uncertainty of the Renishaw XL80 laser interferometer in MOTOMAN HP20 industrial robot inaccuracy test. The paper presents the methodology for estimating the measurement uncertainty of the system used in tests. Estimates of standard and extended uncertainty were calculated for the given research method. These uncertainties are based on the information included in the device calibration certificate (method B) but also on the basis of measurements and statistics (method A). The authors proposed their own research methodology, taking into account measurement capabilities of the applied system and the specific character of the robot work. Calculations employed universal computing systems based on standard algorithms. The results obtained from the research and calculations precisely defined key uncertainties allowing for objective evaluation of industrial robot errors identified by the Renishaw XL80 system.

1 Introduction

Uncertainty of laser interferometer measurement includes a series of uncertainties that take into account the impact of major errors affecting the measurement result. Uncertainty is defined as a certain "doubt" about the value of the measurement result. It is presented as an interval around the measurement result, within which, with a certain probability, lies the true value of the measured value [1-2]. Estimation of measurement uncertainty of an industrial robot presents a number of different difficulties, which are discussed in the paper [3]. Literature [4-9] indicates that although laser interferometry constitutes an effective method of measuring deviation of positioning accuracy the measurement result should be presented as a range, taking into account the uncertainty of its acquisition. In work [10, 11] ABB IRB 1600 robot positioning repeatability was determined using a laser interferometer. Measurements were carried out along three orthogonal linear tracks of 1000 mm (horizontal tracks) and 800 mm (vertical tracks). Each of the tracks had 5 measuring positions (standby time was 5s), specified in 30 repetitions. Positioning repeatability in horizontal axes was in the range of 6 μm to 14 μm while in the vertical axis 25 μm. Positioning deviation values do not exceed 650 μm. The authors of [10-11] as well as others [12-23], however, do not express the uncertainty of the obtained results. Similarly, in the work [13] n industrial robots errors (Cloos Romat 310, The ABB IRB 6400S and KUKA KR125) determined with analogous methods, do not have a specific measurement uncertainty. The largest recorded deviation of positioning accuracy was in the range of -1600 μm to +300 μm. However, these results may not be considered useful because of the lack of estimation of measurement uncertainty. According to the cited residual results [4, 5, 10-23], the inaccuracy of the robot is determined by a number of factors (geometric, dynamic, kinematic, thermal, systemic, structural) and each of them introduces a specific uncertainty to the measurement result. This assessment requires an accurately assessed measurement uncertainty [1-13]. Only proper estimation of uncertainty will allow for objective estimation of robot inaccuracies and selection of optimum measurement method [1, 2, 4]. Today the theory of measurement uncertainty has replaced the theory of measurement accuracy (measurement error). Uncertainty theory is commonly recognized, universal and accepted by all metrology organizations. So it is important to assume that the accuracy of any measurement is described by its uncertainty [1, 2].

2 Research methodology

2.1 Research set-up

In experimental work it is necessary to estimate statistical components of laser interferometer measurement uncertainty, the following positions and computerized measurement and calculation systems were used:

* Corresponding author: j.jozwik@pollub.pl

- MOTOMAN HP20 industrial robot, as a research object,
- Renishaw XL80 interferometer laser system with XC-80 compensator and temperature, pressure and humidity sensors,
- Excel computing environment,
- Statistica computing environment,
- Specialized software for data recording and analysis LaserXL,
- XCal-View 2.2 software.

2.1.1 Object of research

The object of accuracy and repeatability testing of positioning was a universal, six-axis industrial robot MOTOMAN HP20 from Yaskawa (Fig. 1a). Maximum lifting capacity of the robot is 20 kg. The robot is controlled by a YASNAC FS 100 controller. Fig. 1b shows the positioning of the robot on research station.

a) b)

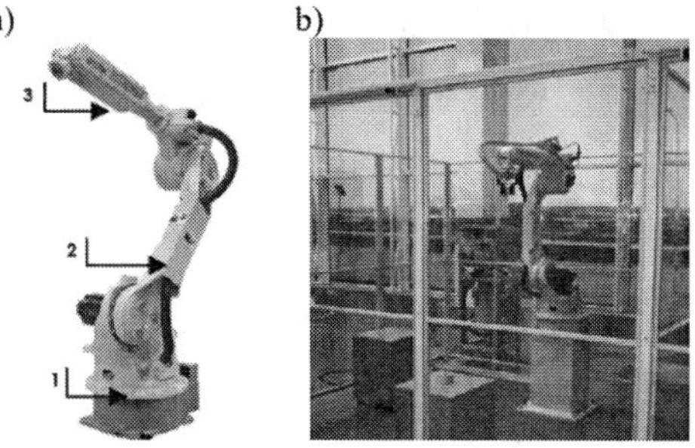

Fig. 1. Industrial robot HP20, a) general view, b) robot's positioning on research station, 1, 2, 3 – robot connectors.

All MOTOMAN HP20 work axes are rotary axes. The range of movement of individual axes of the robot are: S axis (±180 °), L axis (+155 °/-110 °), U axis (+255 °/-165 °), R axis (±200 °) +230 °/-50 °) and T axis (± 360 °). The axis of rotation of joint 1 is vertical, while the axes of joints 2 and 3 are horizontal axes. The maximum distance in the axis perpendicular to the base plane is 3063mm, and in the axis parallel to the base plane 1717 mm. Repeatability of positioning based on robot calibration certificate and the manufacturer's assessment is 0.06 mm.

2.1.2 Measurement set-up

Motoman HP20 industrial robot precision tests were performed with non-contact laser interferometry. The accuracy and repeatability of positioning was tested with the Renishaw XL80 system, composed of: a laser generating a stable radiation beam (Fig. 2a), an XC-80 compensator with temperature, pressure and humidity sensors (Fig. 2b), a set of optical measurement systems (Fig. 2c), a stand with a mounting table and specialized software for data logging and analysis LaserXL. The XL80 system enables measurement of linear and angular (rotational) deviations, flatness, straightness and perpendicular deviation. The accuracy of linear measurement declared by manufacturer is ±0.5 ppm and, thanks to the use of a compensator, is maintained throughout the range of changes in environmental conditions (temperature, pressure, humidity).

a) b)

c)

Fig. 2. Basic elements of the XL80 laser interferometer system, a) laser, b) XC-80 compensator with temperature, pressure and humidity sensors, c) optical measurement systems.

The tested robot was mounted on a stand guarded by special safety barriers, whose task is to ensure safety and stable working conditions of the robot and of the performed measurements.

2.2 Test settings

In the paper, the authors proposed their own research methodology, taking into account the measurement capabilities of the applied system and the specificity of the robot's work. This methodology is optimized for the duration of the test and the possibility of estimating the inaccuracy of the robot in industrial conditions. The diagram of the measurement method is shown in Fig. 3. The Cartesian coordinate system X', Y', Z' has been adopted for technical measurement of deviation accuracy with a laser interferometer. Measurement consisted of alternating (3 runs), bidirectional linear test. The target velocity of motion was vf = 50 mm/s. In each of the three selected planes (along the X', Y' and Z' axis), five parallel measuring paths (S1-S5) were defined at a distance of 250 mm (Fig. 3).

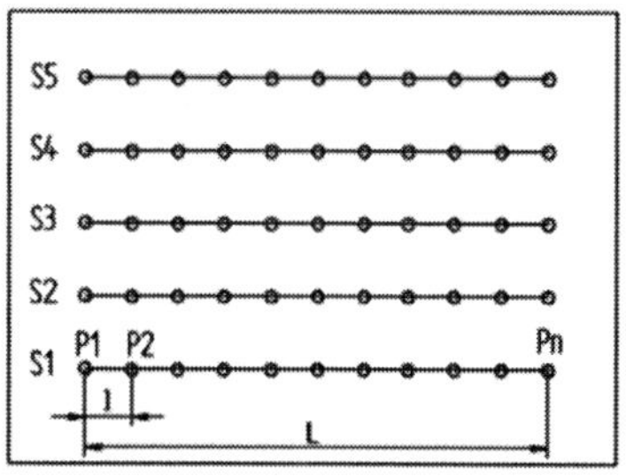

Fig. 3. Layout of paths and points of measurement of deviation of positioning accuracy, using a laser interferometer

On each path, measuring points (P1-Pn) spaced at l = 100 mm apart were specified. The lengths of the paths

(L) and the number of measurement points (np) were equal to:

along X' axis: LX' = 1100 mm, np = 12
along Y' axis: LY' = 1500 mm, np = 16
along Z' axis: LZ' = 900 mm, np = 10

The orientation of the accepted Cartesian coordinate system was presented in Fig. 4.

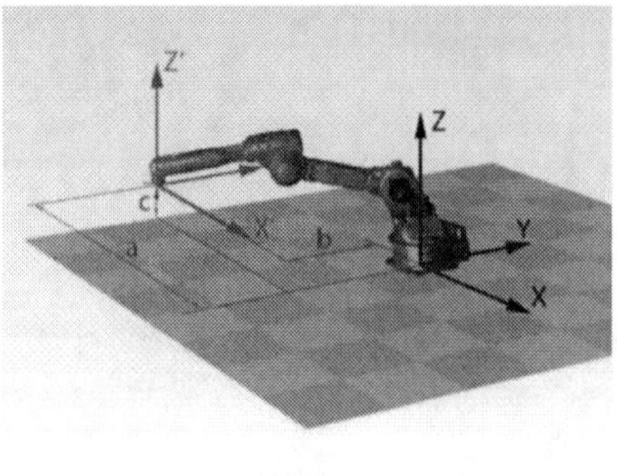

Fig. 4. Definition of the location of the beginning of the measurement coordinate system (X', Y', Z') in the base coordinate system of the robot (X, Y, Z) for linear measurements along the X and Y axes, where: a, b, c – displacement values in individual axes X ,Y, Z.

Such assumption makes it much easier to analyze the measured results. The definition of the starting position of the coordinate system (X', Y', Z') in the robot's base coordinate system (X, Y, Z) was dictated by the spatial accessibility of elements of the measuring system and the studied work space availability . The displacement in the Cartesian coordinate system were respectively a = 550 mm in the X axis, b = 1500 mm in the Y axis and c = 450 mm in the Z axis. The beginning of the coordinate system was defined in analogically to the Z axis. The Cartesian system displacement was b' = 1500 mm in the Y axis and c' = 200 mm in the Z axis. The arrangement of individual measuring paths along the selected axes X, Y and Z is shown in Fig. 5.

Measurement conditions, *i.e.* temperature, humidity and pressure were monitored by a laser interferometer meteorometer. Motoman HP20 was programmed offline. Before the first and after the last measuring point, a trip was programmed to compensate for the recurrence error. The test was carried out without load and with a load of 8.6 kg. The load selection was dictated by the most commonly used displacement mass by the robot effector. The measurements were made using a linear measurement algorithm, according to the ISO standard. XCal-View 2.2 software was used to record measurement data, including graph generation and extensive statistical analysis of results based on the applicable ISO standard.

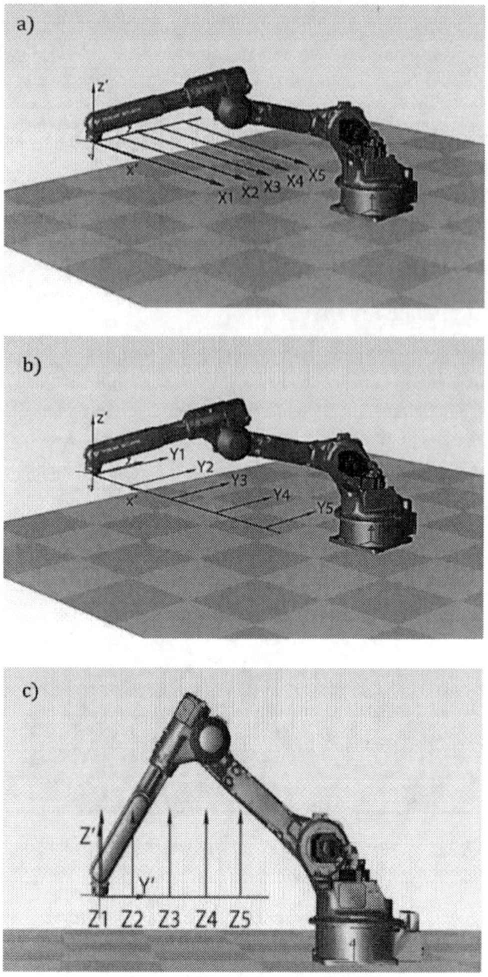

Fig. 5. Location of measurement paths in space while measuring a) along X axis, b) along Y axis, c) along Z axis.

3 Test results

The conducted experimental research allowed us to collect the data necessary for determining statistical uncertainty components. These results were used for the estimation with the A method. The example of experimental data in the direction of the Y axis is presented in Fig. 6.

Based on the conducted measurements, using specialized software XCal-View 2.2 we were able to determine the values of accuracy and repeatability of the positioning of the robot, such as: unidirectional positioning accuracy A ↑, A ↓, bidirectional positioning accuracy A, unidirectional positioning repeatability R ↑, R ↓, bidirectional R positioning repeatability, axial B feedback. For these values, a complex standard uncertainty and extended uncertainty were defined.

 DOI: 10.1051/itmconf/20171504005

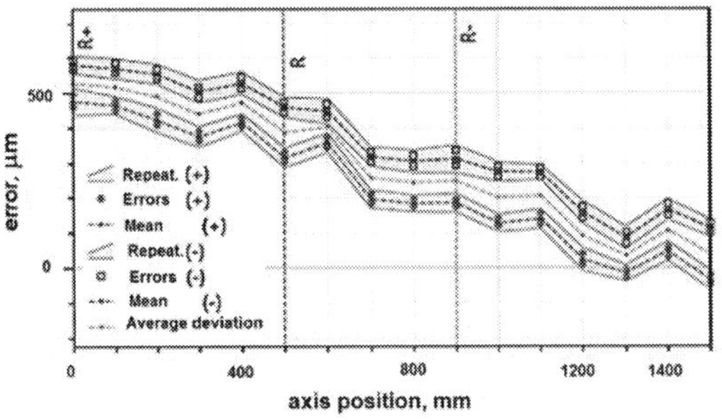

Fig. 6. Positional displacements along the selected measurement path Y1, in case of test without load.

4 Estimation of uncertainty

Uncertainty in laser interferometer measurement includes a series of uncertainties that take into account the impact of major errors affecting the measurement result (uncertainty budget). The analysis includes the following budget components: cosine error, Abbe error, dead path error, thermal error, maximum interferometer error, temperature sensor limit error, and thermal expansion factor error. Uncertainty resulting from the above errors was taken into account when estimating the uncertainty of measurement accuracy and repeatability of positioning. The information included in the calibration certificate of the device was used, along with the measurement results and statistical data of the results obtained in the measurements. Individual statistical components of measurement uncertainty were estimated with method A and non-parametric components were estimated with method B. Expanded uncertainty U was determined on the basis of equation (1)

$$U = \pm u_c k \qquad (1)$$

where: u_c – the adopted standard uncertainty, k – expansion factor.

Complex standard uncertaintyu_c, determined from the equation (2)

$$u_c = \pm\sqrt{u_A^2 + \sum_{i=1}^{n} u_{Bi}^2} \qquad (2)$$

where: u_A – standard uncertainty estimated with method A (statistical), u_{Bi} – standard uncertainties beyond the statistical method B.

Complex standard uncertainty includes both the uncertainty component u_A as well as all non-statistical components u_{Bi} estimated with B method. Standard uncertainty u_A estimated with method A (statistical) was determined on the basis of dependency (3)

$$u_A = \pm\sqrt{\frac{\sum(x_i - \bar{x})^2}{n(n-1)}} \qquad (3)$$

Where: x_i - single measurement result, $\bar{x}$ – measurement results average, n – number of measurements and the sum of squares of component non-statistical uncertainties determined with method B on the basis of the formula (4)

$$\sum_{i=1}^{n} u_{Bi}^2 = u_{Bcos}^2 + u_{BAbb}^2 + u_{Bdrm}^2 + u_{Bprz}^2 + u_{BT}^2 + u_{B\alpha}^2 \quad (4)$$

where: u_{Bi}^2 – component uncertainties resulting from device errors and measurement conditions, estimated by the method B, u_{Bcos}^2 – component taking into account the cosine error, u_{BAbb}^2 – component taking into account the Abbe error, u_{Bdrm}^2 – component taking into account the dead path error, u_{Bprz}^2 – component taking into account the influence of maximum limit error of interferometer, u_{BT}^2 – component taking into account the influence of the limit error of the sensor during temperature measurement, $u_{B\alpha}^2$ – component taking into account the error of estimation of thermal expansion coefficient.

Considering the specific character of measurements and setting of individual components of the interferometer during measurements, the cosine error values and the Abbe error were assumed to be insignificant and of negligible effect on the measurement result. Additionally, it was pointed out that the phenomena that are the source of these errors are controlled by the system software, and the measurement itself is only possible if all sources of their origin are removed. Another reason for not taking this error into account is that its value is so small that, relative to the expected measurement accuracy, the resulting uncertainty can be treated as a small second order.

4.1 Cosine error

A cosine error occurs when the laser beam is not parallel to the axis of motion of the robot arm (Fig. 7).

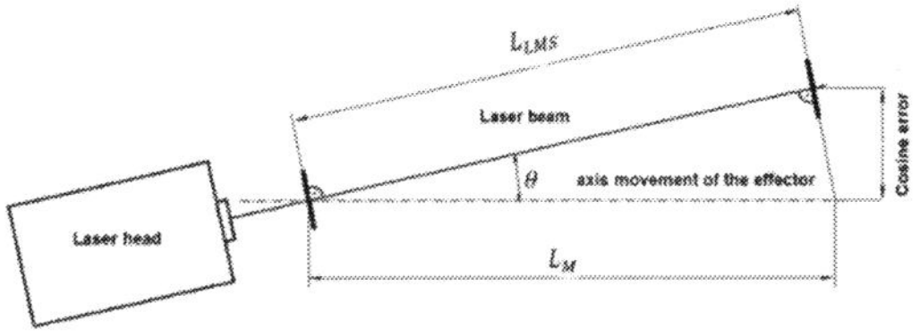

Fig. 7. Schematic representation of origin of cosine error resulting from the lack of parallelism of the laser beam to the axis of motion of the examined object.

Therefore, measurement distance L_{LMS} on the path from point A to point B, will be smaller than the actual path travelled by the L_M robot arm, according to the relationship (5).

$$L_{LMS} = L_M \cdot \cos\theta \qquad (5)$$

Minimizing this error should be done by aligning the laser beam with respect to the direction of travel of the robot shift assembly. It was therefore assumed that the uncertainty taking into account the lack of parallelism of the laser path to the axis of motion of the examined object is close to zero u_Bcos≈0. The basis for this assumption is the condition stating that the realization of the measurement in question is possible only when the error is small enough to allow the measurement to be fully carried out. In the analyzed case, the condition was fulfilled during the measurements. Adjustments of the

 DOI: 10.1051/itmconf/20171504005

interferometer used were controlled from the measurement system software, and the cosine error was small enough that the uncertainty of the source could be omitted.

4.2 Abbe error

Due to the fact that estimating the Abbe error is very difficult, it is assumed that the robot is included as part of the positioning deviation and is not considered as a source of measurement uncertainty. The Abbe error results from the angular movement of the robot arm with the reflector in a given direction when the measurement is made at a distance offset by a certain displacement (*i.e.* the distance called the Abbe offset) from the actual displacement point. Thus, the actual distance is different from the measured distance (Fig. 8).

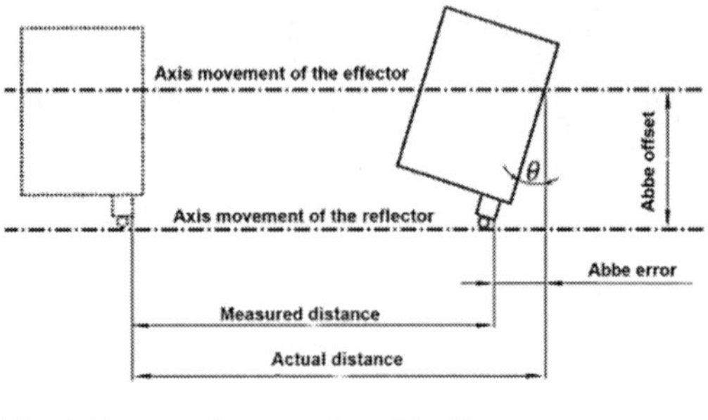

Fig. 8. Geometric interpretation of the Abbe error occurrence.

In the test case Abbe's offset was zero, because the headlight was the same as the robot's frame. It can therefore be assumed that the value of uncertainty also tends to zero, and therefore $u_{BAbb} \approx 0$.

4.3 Dead path error

Another factor affecting the result of the interferometric measurement is the dead path error (Fig. 9). Dead path error occurs when there is no compensation of the length of the laser beam path. Changing the length of a laser path occurs when ambient conditions affecting the wavelength of the measurement are drastically changing, for example the temperature of the material on which both the interferometer and the reflector are mounted changes, changing inter alia their distance.

Due to the negligible changes in temperature during measurements in a relatively long time, not exceeding 1.23 °C and the type of substrate on which these elements were placed and no change of dimensions (stable concrete slab), these factors were considered irrelevant from the point of view of the conducted measurement.

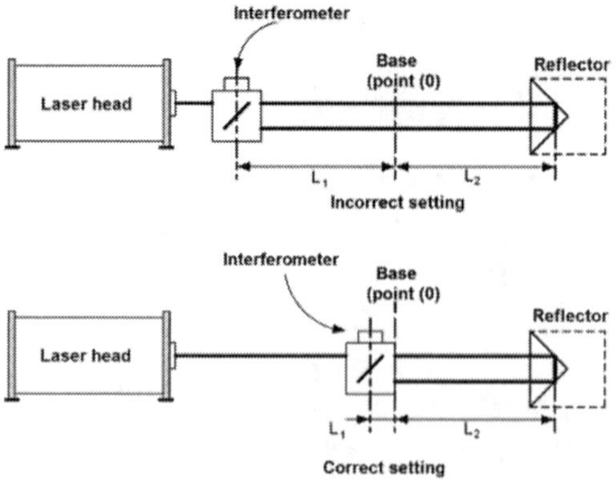

Fig. 9. Setting of the system optics, minimizing the dead path error value.

Literature analysis also shows [1, 2, 4] that the dead path error after normalization and proper placement of stationary and movable optics relative to a representative zero point (the so-called base) is negligible and there is no need to include it in budgeting of measurement uncertainty. This error is smaller, the closer the interferometer is to the "base" (Fig. 9). Correct alignment of the system optics, minimizing the value of the dead path error is shown in Fig. 9. Therefore, the standard uncertainty taking into account the dead path error is close to zero $u_{Bdrm} \approx 0$.

4.4 Uncertainty taking into account the influence of the maximum limit error of the interferometer

The uncertainty analysis included the effect of the maximum interferometer error, defined as *MPE*. The value of *MPE* error was derived from the qualification certificate of the device. *MPE* does not include corrections related to standardization of measurement results to the temperature of 20 °C. The resulting uncertainty of the instrument (laser)u_{Bprz} with adoption of normal probability distribution is based on the following relationship (6)

$$u_{Bprz} = \pm \frac{MPE}{2} x_{ij} \tag{6}$$

where: MPE – the maximum limit error estimated by the manufacturer with a known expansion factor k (in our case k= 2) when measured using environmental compensation stations - temperature, humidity, pressure (MPE = ±0.5 ppm - in the first year of use, MPE = ±1ppm - in the next three years of use) and x_{ij} – current preset position.

4.5 Uncertainty taking into account the influence of the temperature sensor limit error

With respect to the conducted robot tests, the impact of the sensor error on temperature was taken into account in estimating uncertainty. The general form of uncertainty of standard temperature measurement u_{BT} with many sensors used during the experiment and the adoption of a rectangular probability distribution can be determined from the formula (7):

$$u_{BT} = \pm \frac{\sum_{i=1}^{n} MPE_{T_i}}{n\sqrt{3}} \quad (7)$$

where: MPE_{T_i} – maximum limit error of the temperature sensor (according to calibration data provided by the manufacturer $MPE_{T_i} = \pm 0.1°C$, n – number of temperature sensors during the test).

For the case under consideration, the change in temperature values during the measurements did not exceed ΔT = 1.23 °C, and the uncertainty of standard temperature measurement u_{BT}, assuming rectangular probability distribution, was determined from (8):

$$u_{BT} = \pm \frac{MPE_T}{\sqrt{3}} \quad (8)$$

Limit error value MPE_T was adopted according to the calibration certificate on the level of ±0.1°C.

4.6. Uncertainty of determination of thermal expansion coefficient

Taking into account the possibility of extension of the standard, the standard uncertainty of the coefficient of thermal expansion was also determined $u_{B\alpha}$. The standard uncertainty of determining the coefficient of thermal expansion with rectangular probability distribution is described by the equation (9):

$$u_{B\alpha} = \pm \frac{\Delta\alpha}{\sqrt{3}} \quad (9)$$

where: $\Delta\alpha$ – limit deviation of the coefficient of thermal expansion.

On the basis of literature [1, 2, 4] it was assumed that the limit deviation of the coefficient of thermal expansion amounts to $\Delta\alpha = 1.5 \cdot 10^{-6}\ \frac{1}{°C}$, and the value$\alpha = 11.6 \cdot 10^{-6}\ \frac{1}{°C}$. Taking the above into account, the standard uncertainty of determining the coefficient of thermal expansion u_α was defined (10):

$$u_{B\alpha} = \pm \left(\frac{1.5}{\sqrt{3}} \cdot 10^{-6}\right) \frac{1}{°C} \quad (10)$$

Therefore finally, the sum of the uncertainty squares estimated with method B is (11):

$$\sum_{i=1}^{n} u_{Bi}^2 = u_{Bprz}^2 + \left(\frac{\partial x_{ij}}{\partial(\Delta T)} \cdot u_T\right)^2 + \left(\frac{\partial x_{ij}}{\partial(\alpha)} \cdot u_\alpha\right)^2 \quad (11)$$

and complex standard uncertainty u_C, may be determined from the equation (12):

$$u_c = \pm \sqrt{u_A^2 + u_{Bprz}^2 + \left(\frac{\partial x_{ij}}{\partial(\Delta T)} \cdot u_T\right)^2 + \left(\frac{\partial x_{ij}}{\partial(\alpha)} \cdot u_\alpha\right)^2} \quad (12)$$

Based on the adopted assumptions, computational formulas were developed, and the values of individual uncertainty components were calculated using the STATISTICA computerized computing system. An assessment of uncertainty of the load test was also conducted. Uncertainty and its components for measuring the positional displacements in the X, Y, Z axes at the maximum displacements during testing are given in Table 1.

Table 1. Uncertainty and its components during laser interferometer measurement of robot arm positioning in axes X, Y, Z, with and without static load

Uncertainty	X		Y		Z	
	1100mm		1500mm		900mm	
	m=0 [kg]	m=8.6 [kg]	m=0 [kg]	m=8.6 [kg]	m=0 [kg]	m=8.6 [kg]
	range of uncertainty values [μm]		range of uncertainty values [μm]		range of uncertainty values [μm]	
u_A	±7.137	±2.225	±10.262	±4.121	±0.843	±16.752
u_{Bcos}	±0.000	±0.000	±0.000	±0.000	±0.000	±0.000
u_{BAbb}	±0.000	±0.000	±0.000	±0.000	±0.000	±0.000
u_{Bdrm}	±0.000	±0.000	±0.000	±0.000	±0.000	±0.000
u_{Bprz}	±0.175	±0.175	±0.375	±0.375	±0.175	±0.375
u_{BT}	±0.058	±0.058	±0.058	±0.058	±0.058	±0.058
$u_{B\alpha}$	±0.001	±0.001	±0.001	±0.001	±0.001	±0.001
u_c	±7.140	±2.232	±10.268	±4.139	±0.861	±16.76
U	±14.28	±4.464	±20.536	±8.278	±1.722	±33.51

Table 1 seems to indicate that the predominant component of the uncertainty budget for laser interferometer measurement is the standard uncertainty u_A (the statistical component representing the difference in measurement results). The consequence of the significant value of standard uncertainty is the value of complex standard uncertainty u_c and extended uncertainty U. Other components of uncertainty resulting from the properties and characteristics of the laser interferometer ($u_{Bcos}, u_{BAbb}, u_{Bdrm}, u_{Bprz}$) and estimation of the impact of temperature on measurement accuracy u_{BT} and estimation of the coefficient of thermal expansion $u_{B\alpha}$ were shown to only slightly affect the value of complex standard uncertainty and extended uncertainty. From the results presented in Table 1 it can be stated that the application of load to the effector of the tested robot constitutes an important factor influencing the value of uncertainty. It can be stated that loading the effector with a mass of 8.6 kg gives a stabilizing effect on the examined axes X, Y, indicating that the conducted research works were of static character. The reverse effect of the uncertainty estimation was obtained, however, for the results from the tests in the direction of the Z axis (vertical axis). This can be explained by the significant elastic deformations of loaded robot arms.

5 Conclusion

Accuracy and repeatability of positioning are key parameters of industrial robots, which determine their usability in support of processes of different manufacturing technologies. However, the determination of these quantities must be related to the measurement uncertainty estimate used in the measurement method. Although modern robots have high reproducibility and relatively high accuracy to meet most applications, it should be borne in mind that these parameters may deteriorate over the machine operation time. This affects

DOI: 10.1051/itmconf/20171504005

the dispersion measure and the standard uncertainty of measurement for the measurement method used.

The analysis of uncertainty presented in the paper indicates that the components of the uncertainty budget take relatively low values, which proves that the method chosen for the given research task is correct. It should be noted that interferometer measurements are performed as static measurements, i.e. in certain positions and usually without load on the robot arm by the mass of the displaced elements. In addition, it is only possible to perform measurements at linear displacement in a certain plane. Any change in the measuring plane or travel direction of the robot arm, however, requires the optical system to be repositioned. The authors see some research opportunities using a laser interferometer to evaluate the effect of the weight ratio of the effector and robot arms on the different positions of the work space on the repetition of positioning. This was proved by the results of research and analysis of uncertainties during the static load of effector with mass of 8.6 kg. Of course, the robot arm might be loaded with more mass (max. load of the tested robot 20 kg), but from the experience of the authors the result is that the length of the measuring path is corrected by the system. This is due to the change of the trajectory with large overhangs and long paths (lack of parallelism resulting from increased bending moment and elastic deformation of the arms). It can therefore be concluded that the effect of weight of the effector and robot arms is crucial when assessing the uncertainty of the laser interferometer measurement, especially in the vertical axis (Z axis). Another problem is the inertia displaced by the mass arm assembly and the maintenance of the preset position (problem with position stabilization and position overshoot) which further limits testing under load.

Despite the limitations, the laser interferometer allows measurements to be performed with extremely low measurement uncertainty, which is very small with respect to the accuracy and repeatability of the positioning of industrial robots.

Although the method employed in this work is dedicated to CNC machine tools, it can also, to a certain extent and with some restrictions, be used to test the accuracy and repeatability of the positioning of industrial robots.

References

1. P. Majda, D. Rozkrut, Adv. Manuf. Sci. Technol. **39**, 3 (2015)
2. P. Majda, M. Pajor, Mechanik, 11 (2016)
3. J. Honczarenko, Roboty przemysłowe. Budowa i zastosowanie (WNT, Warszawa, 2010)
4. J. Józwik, D. Ostrowski, Wybrane problemy badawcze robotów przemysłowych (Wydawca Politechnika Lubelska, 2016)
5. J. Józwik, I. Kuric, L. Semotiuk, Communications, **3A**, (2014)
6. J. Józwik, M. Czwarnowski, Adv. Sci. Technol. Res. J., **9**, 28 (2015)
7. J. Jozwik, L. Semotiuk, I. Kuric, Adv. Sci. Technol. Res. J. **9**, 28 (2015)
8. G.M. Krolczyk, J.B. Krolczyk, R.W. Maruda, S. Legutko, M. Tomaszewski, Measurement **88** (2016)
9. A. Glowacz, Z. Glowacz, Infrared. Phys. Technol. **81** (2017)
10. M. Slamani, A. Joubair, I. Bonev, Ind. Rob. **42**, 6 (2015)
11. M. Slamani, A. Nubiola, I. Bonev, Ind. Rob. **39**, 1 (2012)
12. E. Feldshtein, J. Józwik, S. Legutko, Adv. Sci. Technol. Res. J., **10**, 30 (2016)
13. G.M. Krolczyk, P. Nieslony, J.B. Krolczyk, I. Samardzic, S. Legutko, S. Hloch, S. Barrans, R.W. Maruda, Measurement, **70** (2015)
14. P. Nieslony, G.M. Krolczyk, K. Zak, R.W. Maruda, S. Legutko, Precis. Eng. **47** (2017).
15. G.M. Krolczyk., R.W. Maruda, P. Nieslony, M. Wieczorowski, Measurement, **94** (2016).
16. K. Young, C.G. Pickin, Ind. Rob. **39**, 6 (2000)
17. G. Alici, B. Shirinzadeh, Mech. Mac.h Theory, **40**, 8(2005)
18. A. Angelidis, G.Ch.Vosniakos, Int. J. Precis. Eng. Man. **15**, 1 (2014)
19. C. Gong, J. Yuan, J. Ni, Int. J. Mach. Tool Manu. **40**, 14 (2012)
20. P.S. Shiakolas, K.L. Conrad, T.C. Yih, Int. J. Model. Simul. **22**, 3 (2002)
21. B. Shirinzadeh, Ind. Rob. 25, 1 (1998)
22. J. Swevers, W. Verdonck, J.DeSchutter, IEEE Contr. Syst. Mag. **27**, 5 (2007)
23. W. Zhenhua, X. Hui, Ch. Guodong, S. Rongchuan, L. Sun, Ind. Rob. **41**, 5 (2014)

 DOI: 10.1051/itmconf/20171504006

Experimental investigation of the degree of weakening in structural notch area of 7075-T6 aluminum alloy sheet welded with the RFSSW method

Andrzej Kubit[1,*], *Magdalena* Bucior[1], *Dawid* Wydrzyński[1],and *Łukasz* Bąk[2]

[1] Faculty of Mechanical Engineering and Aeronautics, Department of Manufacturing Processes and Production Engineering, Rzeszow University of Technology, Powstańców Warszawy 8, 35-959 Rzeszów, Poland
[2] Faculty of Mechanical Engineering and Aeronautics, Department of Materials Forming and Processing , Rzeszow University of Technology, Powstańców Warszawy 8, 35-959 Rzeszów, Poland

Abstract. The paper presents the methodology of the research determining the degree of weakening of the welded sheet obtained by the refill friction stir spot welding (RFSSW) method. The considered weakness is the effect of a structural notch resulting from penetration by the tool. RFSSW technology is a relatively new method of joining metals, which can successfully provide an alternative to resistance welding or riveting - traditionally used methods of joining thin-walled structures in the aerospace and automotive industries. The study presented in the paper focuses on the overlapping of sheet metal with 7075-T6 aluminum alloy combined in the configuration: 1.6 mm top sheet and 0.8 mm bottom sheet. Joints were assembled following the following process parameters: Welding time 1.5 s, the tool plunge depth in the range of 1.5 ÷ 1.9 mm, and the spindle speed of 2600 rpm. The analysis of the microstructure of joints revealed that along the edge of the tool path a structural notch is formed, the size and shape of which depend on the parameters applied. The paper describes the study consisting in punching the welded area along the formed notch in the upper sheet. The punching process was performed on a universal testing machine and the punching force was measured during the test. Based on the force value, the degree of sheet weakening in the notched area was determined. The smallest weakening was observed in joints made with the smallest tool depth, i.e. 1.5 mm, whereas the biggest weakening was obtained for tool depth of 1.9 mm. The load applied to the joints was equal to 5290N and 7585N respectively.

1 Introduction

The most commonly used materials in aircraft are aluminum alloys of the 2XXX [1] and 7XXX series [2, 3]. Main advantages of these alloys are their high strength and relatively low weight [4]. These properties have a significant role in the aviation industry because they reduce the weight of the aircraft and at the same time lower the operation costs. Aluminum alloys can be joined using various methods such as riveting, resistance welding [5] and adhesive bonding [6, 7]. Selecting the appropriate joining method depends on multiple factors, which include: the properties of joined alloys, geometric dimensions of the part (thickness of the joint elements) and the operational requirements.

Aircraft designs traditionally feature riveting as the main joining technique. The number of rivets used to join aircraft parts significantly increases the weight of the aircraft. An alternative joining method could be Refill Friction Stir Spot Welding (RFSSW) [8].

Considering the order of sleeve and pin movement, there are two different RFSSW methods. The article presents the variant in which the sleeve plunges into the joined material while the pin is responsible for pressing the material that was previously pushed out. Normally, the RFSSW process consists of four stages. The first stage consists in positioning the pin and sleeve on the top plane of the sheet metal joint. The clamping ring exerts constant pressure on the joined materials, which protects it from uncontrolled translation. Next the sleeve performs a rotational move and plunges into the material making it plastic and pushes it into the void formed by the receding pin. In the next stage, the pin plunges into the joint and the sleeve recedes upwards. The final stage consists of a quick positioning of the tool on the top surface of the metal sheet joint.

This is a relatively new technology, and therefore there are aspects of the process that are still under research [2, 9, 10]. The advantages of this method include: good mechanical properties, low energy consumption, the possibility of combining materials of different properties [11, 12]. However, determining the correct welding parameters causes multiple problems. The quality of RFSSW spot welds depends on such

*Corresponding author: akubit@prz.edu.pl

 DOI: 10.1051/itmconf/20171504006

welding parameters as" plunge depth [13, 14], tool rotational speed [9, 14, 15] and welding times [9, 14], which are a subject of analysis of numerous researches. Unfortunately, these articles offer no information regarding the degree of weakening of the welded sheet obtained by RFSSW method. It seems therefore important to carry out tests aimed at determining the relationship between process parameters and weakening of the weld joint.

The paper describes the study of punching the welded area along the notch formed in the upper sheet. The punching process was performed on a universal testing machine and the punching force was measured in the tests. Based on the force value, the degree of sheet weakening in the notched area was determined.

2 Method

This section presents the methodology for studying the weakening of a metal sheet joint that results from a structural notch and Thermo-mechanically Affected Zone (TMAZ) created during the welding process. The subject of the study was the RFSSW weld made in 7075-T6 Alclad aluminum alloy sheet. The chemical composition of the alloy is the following: Si - 0.4, Fe - 0.5, Cu - 1.2 ÷ 2, Mn - 0.3, Mg - 2.1 - 2.9, Cr - 0.18 ÷ 0.35, Zn - 5.1 ÷ 6.1, Ti - 0.2, other impurities: single 0.05, total 0.15, Al - rest. The joint was assembled on a HARMS WENDE machine using the tool presented in Fig. 1a.

a)

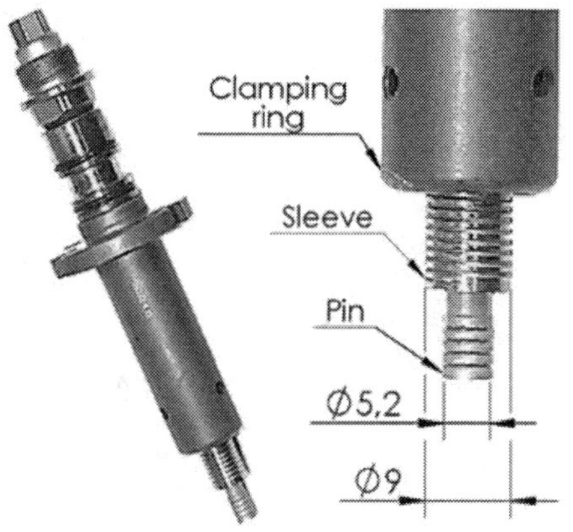

b)

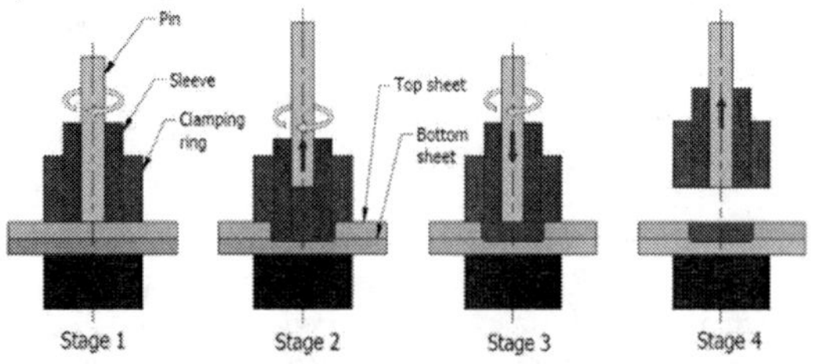

Fig. 1. Schematic illustration of the RFSSW processes (a) and a diagram of the constituent parts of the tools and their diameters (b).

The main parameters of the welding process are: tool rotational speed, tool plunge depth and plunging time.

The parameters in this study were selected on the basis of previous tests of the authors [14, 16]. A constant tool rotational speed of 2600 rpm and plunging time of 1.5 s were adopted in the study.

Because the size of the structural notch depends largely on the tool plunge depth, hence the tests comparing the joints made with the following depths: 1.5, 1.7 and 1.9 mm.

Fig.2 presents a lap joint used to determine the load carrying capacity of the joint and the weakening of the weld. The analyzed lap joint consisted of two pieces of metal sheet where the top one was 1.6 mm thick and the bottom one was 0.8 mm thick.

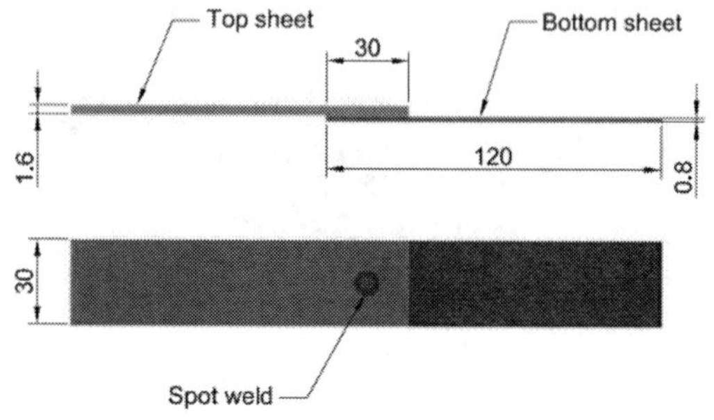

Fig. 2. The shape of the lap joint specimens used in the study.

The methodology of the study predicts weakening of only the top sheet in pure shear strength test. The lap joint in the presented configuration undergoes a tensile test, where the complex loading condition causes bending in the metal sheet. The requirement of maintaining an unbent state of the metal sheet joint after the strength test lead to the use of a special fixture that prevented bending of the metal sheet during loading, which is presented as a diagram in Fig. 3a, while a photograph of the test bench is presented in Fig.3b. The strength tests were performed on a ZWICK Z100 strength test machine with a jaw translation speed of 5 mm/min.

a)

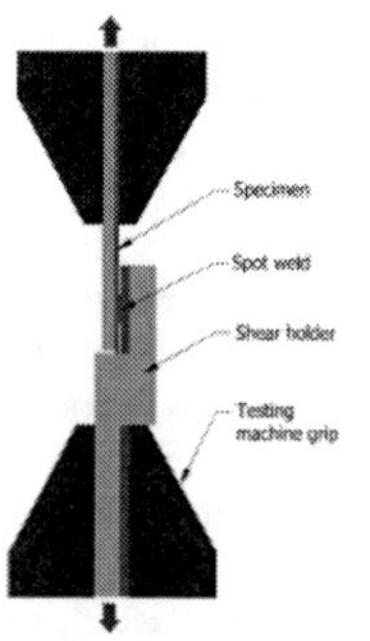

b)

Fig. 3. Diagram of the fixture used to apply pure shearing force in the static strength tests (a) and a photograph of the fixture in the test bench (b).

The primary assumption of the study was to determine the degree of weakening of weld along its circumference based on the punch force measurement around the stir area in a punch-matrix system for the top sheet, which is shown in Fig. 4a, while 4b presents the test bench and describes the parts of the set-up. The tests were conducted on a ZWICK Z50 strength test machine with a jaw translation speed of 10 mm/min. Besides measuring the ejection occurring in the weld area of the upper sheet, the force needed to punch through the solid sheet metal was also measured on the aforementioned test bench.

a)

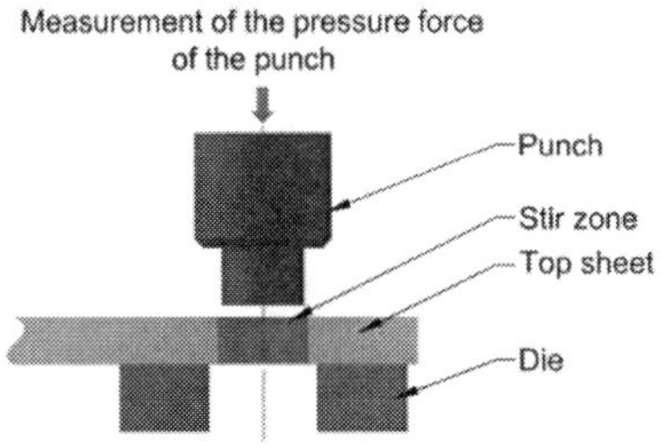

b)

Fig. 4. Diagram presenting the operation concept of the test bench used to determine the degree of weld weakening near the structural notch (a) view of the test bench (b).

Microstructural examination of the RFSSW specimens was carried out using Nikon Epiphot 300 light microscope with NIS-Elements V2.3 software. Macroscopic examination was also conducted using Opta-Tech X2000 stereoscopic microscope. Metallographic specimens were prepared according to ASTM E407 [17] following the given procedure: the samples were first cut using Struers Accutom-50 precision machine and then mounted in StruersPolyfast resin using Struers LaboPress-3. The micro-sections were prepared in accordance with the following procedure: grinding with abrasive papers down to 1200 grit size and polishing using diamond suspensions (9, 6, 3, 1μm). Micro-sections were etched with Keller's reagent, composed of 2 ml of HF, 3 ml of HCL, and 5 ml of HNO3 diluted in 190 ml of distilled water, and the etching time was about 10 s.

3 Results and discussion

After assembling the lap joint for each of the variants, selected samples were used as metallographic sections. The macrostructures of each tool plunge depth are presented in Fig. 5. The structure of the weld for every variant contains fine grains in comparison to the parent material. In some areas of the weld, plating can also be observed that results from the layered concentration near the base of the weld during process.

a)

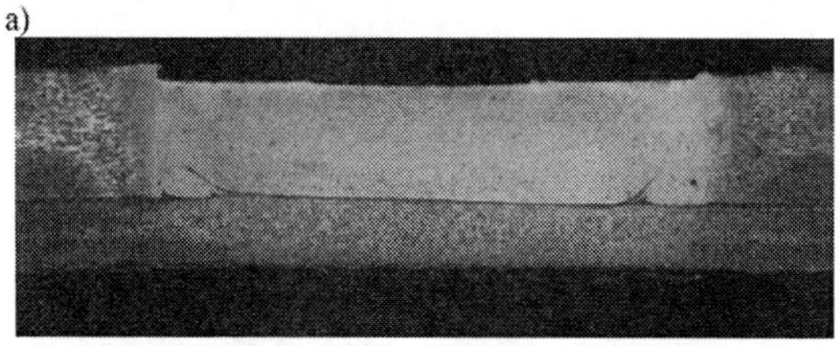

b)

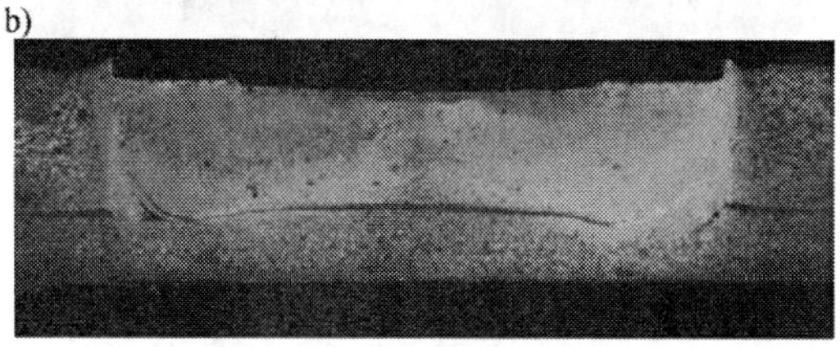

c)

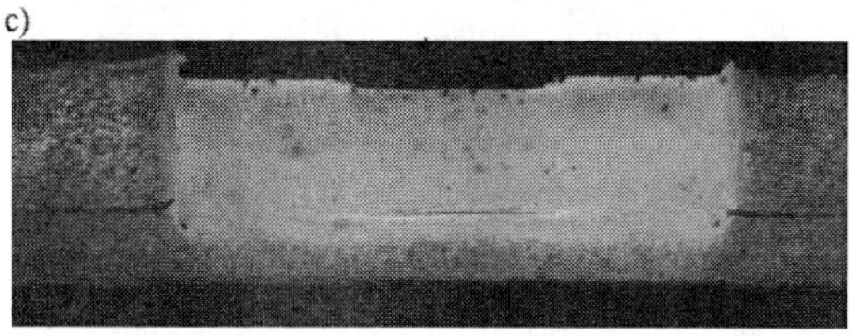

Fig. 5. The macrostructure of the welds created using different tool plunge depths: 1.5mm (a), 1.7mm (b) and 1.9mm (c).

The macrostructure views of the joint showed that every variant included a structural notch on the circumference of the weld in the penetration area of the external part of the tool, more precisely, a 9-mm-wide cylinder. Fig. 6 shows the microstructure of selected welds to illustrate the weakening phenomena. Fig. 6a presents the microstructure of the 1.5 mm deep tool plunge and highlights critical areas.

While conducting the static shear strength test that used a fixture ensuring pure shear, a unique observation was made for the samples with a 1.9 mm tool plunge depth, where disintegration of the weld area and the upper sheet occurred, as shown in Fig. 7.

a)

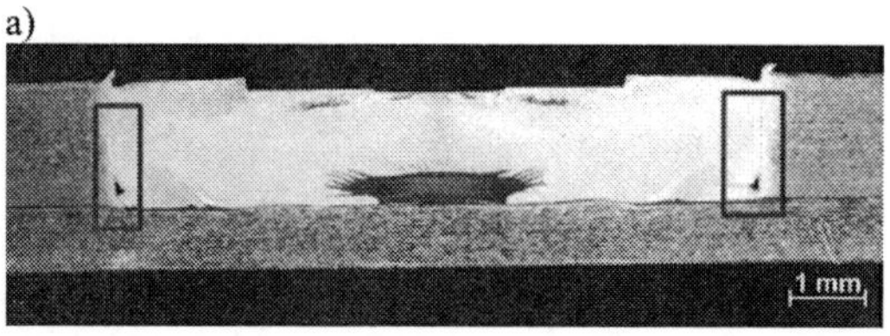

DOI: 10.1051/itmconf/20171504006

b)

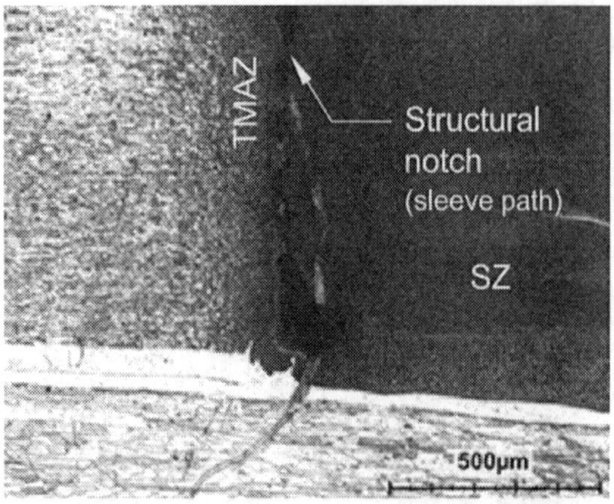

c)

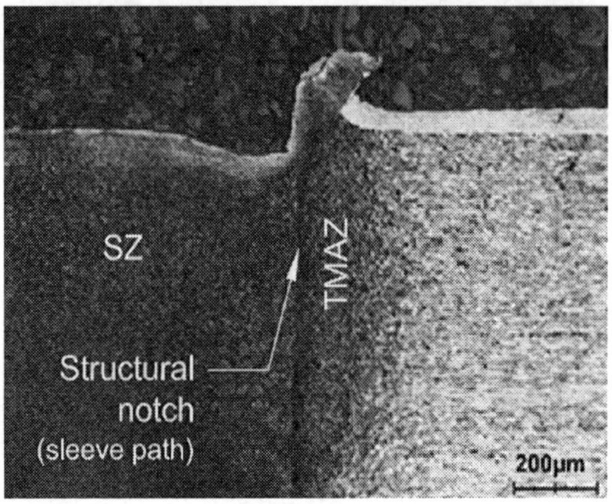

d)

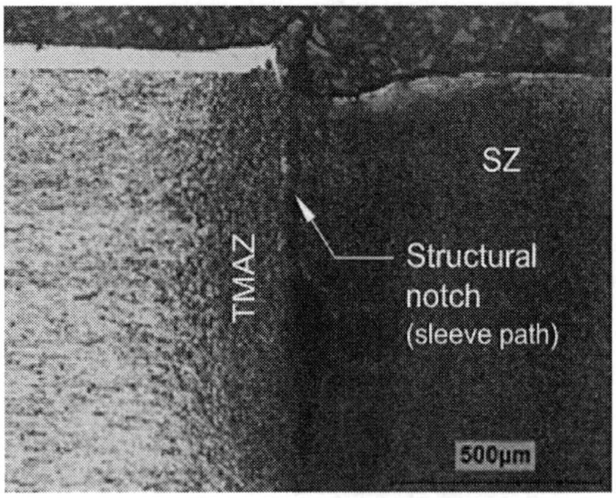

Fig. 6. The microstructure of the weld that highlights the structural notch area (a) and the microstructure showing the existence of structural notches for joints made with the following tool plunge depths: 1.5mm(b), 1.7mm (c) and 1.9mm (d).Marked areas: SZ – Stir Zone, TMAZ – Thermo-Mechanically Affected Zone.

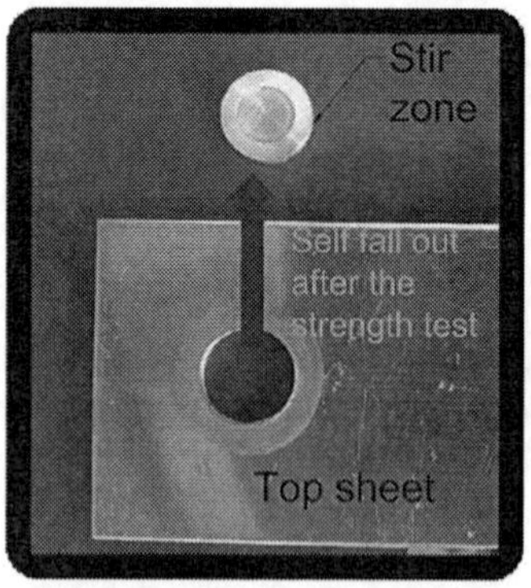

Fig. 7. The self fall out of the stir zone after the static shear strength test.

The ejection of the stir zone was conducted with three repetitions for each variant. The control for the analysis was the result of the punching test conducted on solid 1.6-mm-thick metal sheet on the aforementioned test bench. Fig. 8 presents sample curves for the stir zone ejection trials for each variant. Although the beginnings of the curves are similar, a tendency may be observed that the maximum force pushing out the stir zone decreases with an increase in the tool plunge depth during welding process.

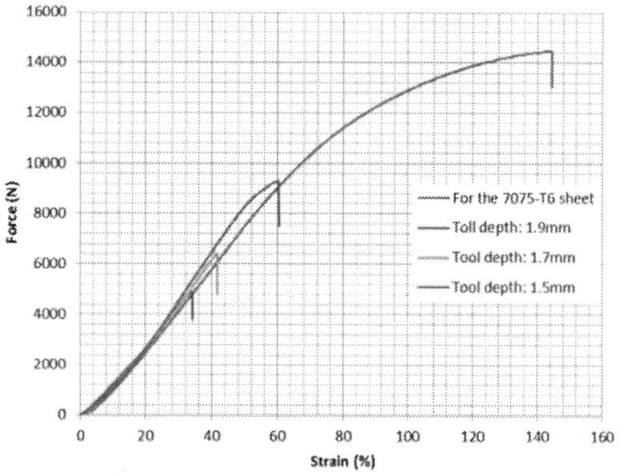

Fig. 8. Sample curves for the stir zone ejection tests for each variant and solid metal sheet.

Average values of the trails are presented in Table 1. The force values are used to determine the degree of weakening that is correlated to the strength of the lap joint subjected to pure shear.

Table 1. The average force values for ejecting the stir zone and the average load carrying capacity of the lap joints made using the RFSSW method depending on tool plunge depth.

Tool depth (mm)	Maximum force of the punch test (N)	Shear strength (N)
1.5	8970	7585
1.7	6125	6885
1.9	4824	5290

DOI: 10.1051/itmconf/20171504006

It was expected, that the greater tool plunge depth would result in a stronger joint. Meanwhile, the tendency is opposite: among the analyzed variants the most shallow tool plunge depth of 1.5 mm resulted in the highest load carrying capacity of the joint was 7585N. The maximum tool plunge depth of 1.9 mm resulted in a load carrying capacity of 5290N.

4 Conclusions

The main purpose of tests presented in this paper is presentation and verification of usefulness and reliability of the methodology used to determine the degree of weakening of a weld made using the RFSSW method on its circumference where the structural notches are.

This methodology is offers an original way of conducting experimental test for determining weakening of joints. This can be a complement to static strength testing and constitute a valuable tool in finding optimal process parameters. RFSSW welding, which is a relatively new technology, has a lot of potential and it appears capable of successively replacing the traditional methods of joining thin-walled metal structures in the future.

It is obvious that there is no single universal set of parameters for the RFSSW welding process, their selection always depends on factors such as the type of welded metal, the dimensions of the joints, the coatings used on the plates, *etc*. And the set of optimal parameters is not only guaranteed by the load capacity of the joint, but also provide a high-quality joint structure, devoid of structural defects, such as structural notches.

The presented research method is focused on fast, cheap and effective recognition of the existence of structural notches in the weld structure. The study shows a correlation between the degree of weakening of the joint and its load capacity, proving that lower the weakening shown by the method, which manifests itself in high value of force of the punch test, translates into higher loading capacity of the joint, this has occured for a 1.5 mm tool plunge depth.

The smallest weakening was demonstrated for the joint welded at the 1.5 mm tool plunge depth, while the greatest weakness was demonstrated for the weld joint at the tool plunge depth of 1.9 mm. These results coincide with the microstructural analysis of the joints. The larger tool plunge depth produces a larger structural notch due to a greater range of the sleeve path, which manifests itself in greater weakness.

The experiment revealed significant weakening in each variant of the joint due to the structural notch, as well as TMAZ (Thermo-mechanically Affected Zone) which is a natural phenomenon occurring in the welding process. As a reference to the results of the tests, an attempt was made to punch the hole in the intact 1.6 mm thick 7075-T6 Alclad aluminium alloy sheet using the described tooling. The value for maximum force at punching was 14500 N. For this value, the forces exhibiting weakening of the joint were 62% for the tool plunge depth 1.5 mm, 42% for the tool plunge depth 1.7 mm and 33% for the tool plunge depth of 1.9 mm.

Obviously, the method does contain a flaw, as the samples analyzed were used having undergone the static shear strength test of the lap joint, which could have weakened the weld area. The research shows that ensuring a pure shearing condition during the static shear strength tests results in the force required for punching through the sheet metal could be used to help determine the welding parameters for RFSSW in order to minimize the structural notch.

The aim of the paper was to present the original experimental method, but it requires further analyzes, conducting further studies for a greater range of welding parameters and correlating the results with microstructures of joints for the considered variants.

References

1. M. A. Sutton, B. Yang, A. P. Reynolds, R. Taylor, Mater. Sci. Eng. A **323**, (2002)
2. S. Zhikang, Y. Xinqi, Z. Zhaohua, C. Lei, L. Tielong, Mater. Des. **44**,(2013)
3. Y. Q. Zhao, H. J. Liu, Z. Lin, S. X. Chen, J. C.Hou, Sci. Technol. Weld. Joi. **19**, (2014)
4. T.Dursun, C.Soutis, Mater. Des. **56**, (2014)
5. P. Briskham, N. Blundell, L. Han, R. Hewitt, K. Young. SAE Technical Paper, (2006)
6. W. Zielecki, A. Kubit, T. Trzepieciński, U. Narkiewicz, Z. Czech. Int. J. Adhes. Adhes. **73**, (2017).
7. W. Zielecki, A. Kubit, R. Kluz, T. Trzepieciński. J. Adhes. Sci. Techn. **31**, (2016).
8. C. Shilling, J. D. Santos, Method and device for joining at least two adjoining work pieces by friction welding: Patent Application No. US0179682,(2002)
9. S. Venukumar, S. Yalagi, S. Muthukumaran, Trans. Nonferrous Met. Soc. **23**, (2013)
10. N. Pathak, K. Bandyopadhyay, M. Sarangi, S.K. Panda, J. Mater. Eng. Perform.**22**, (2013)
11. Y. Uematsu, K. Tokaji, Sci. Technol. Weld. Joi. **14**, (2009)
12. D. H. Choi, B. W. Ahn, C. Y. Lee, Y. M. Yeon, K. U. Song, S. B. Jung, Mater. Trans. **51**, (2010)
13. Y.Q. Zhao, H.J. Liu, S.X. Chen, Z. Lin, J.C. Hou. Mater. Des. **62**, (2014)
14. R. Kluz,A. Kubit, D. Wydrzyński,Technol. Assembly Automat. **2**, (2017)
15. L. Zhengwei, G. Shuangsheng, J. Shude, Y. Yumei, C. Peng Chai, J. Mater. Eng. Perform. **25**, (2016)
16. A. Kubit, R. Kluz,T. Trzepieciński, D. Wydrzyński, W.Bochnowski, Arch. Civil Mech. Eng.**18**, (2018)
17. American Society for Testing and Materials (ASTM). Standard practice for microetching metals and alloys. West Conshohocken: ASTM International (2007). Standard no. E407.

IT solutions supporting project management processes and the choice of software

Anna Kaczorowska[1,*], *Jolanta* Słoniec[2], and *Sabina* Motyka[3]

[1]University of Lodz, Faculty of Management, Department of Computer Science, Matejki 22/26, 90-237 Łodz, Poland
[2]Lublin University of Technology, Faculty of Management, Department of Enterprise Organization, Nadbystrzycka 38, 20-618 Lublin, Poland
[3]Cracow University of Technology, Faculty of Mechanical, Department of Manufacturing Processes, M6 Institute, Jana Pawła II 37 Avenue, 31-864 Cracow, Poland

Abstract. The article aims to present both the theoretical contents and proposed solutions to practical problems perceived in the support of project management. The process' approach was recommended as the basis of project management. The classification of instruments due to the implemented functions and structure of information systems, which support project management, was included, in conjunction with the type of projects handled. The maturity of project management information support was analysed. Project management areas that have the highest possibilities of information support were indicated, and a list of advantages of the use of information instruments was drawn up. Both the main criteria of software selection and the basic requirements presently formulated with reference to project management supporting systems were determined.

1 Introduction

The development of information systems for project management evolved from the supporting tools, in a rather narrow range, the work of just one person in a single project for solutions that assure the cooperation of many subjects in a multi-project environment [1, 2, 3, 4].

Presently, project managers and project teams mostly plan and control the accomplishment of undertakings as well as their reporting and documenting using information instruments. Information tools also assure a partial support of the processes of a projects initiation and definitions, project team organization, and administration of the results and collection of the project knowledge [5].

The software which supports the project management may increase the effectiveness of concrete project management, but it will never compensate for factual, methodical or organizational deficiencies occurring in the project implementing organization. Heerkens expressed this as follows: "The use of planning software without understanding of the essence of project management allows merely to create faster some lousy schedules" [6].

The use of an appropriate information infrastructure contributes to the enhanced quality of the management processes, including project management processes [2].

The information infrastructure of project management (IIZP) should be perceived as a platform that enables the functioning of project management and use of hard and soft instruments of management dedicated to that area.

IIZP consists of the following components: hardware, software, resources of information, knowledge[1], models[2] and procedures related to projects and networks. Owing to IIZP, new information systems for project management (ISfPM) arise. Contractors and people responsible for the project results and users who can use such instruments may significantly increase not only the capacity of their work, but also the reliability of undertaken decisions. They may also increase the chances of success in the project implementation.

However, the information system is only one part of the project management system. Its application depends on the project requirements related to IIZP and the scope in which the IIZP could support project management.

Despite the constantly growing level of advancement in an information system, they are not autonomic enough to independently specify more accurately the project goals and risks, and automatically organize the project team. However, the use of a properly chosen ISfPM may constitute an effective support of the decisive processes on all levels of the organization, by facilitated planning of the projects and controlling the correctness of their performance [7].

[1]Bases of knowledge enable the management of knowledge and decision-taking in extraordinary conditions.
[2]Models are gathered in the models base allowing for simulation of the consequences of undertaking various decisions.

* Corresponding author: annak@wzmail.uni.lodz.pl

DOI: 10.1051/itmconf/20171504007

2 The process model of project management

The increased interest in the process approach recently resulted from aspiring to meet the accomplishment of the buyers' expectations within very intensive competition. This approach pulls to the fore the dynamic aspects and process harmonization of activities and focuses on process effects [8, 9, 10, 11, 12].

Processes are the sets of consecutive activities in cause and effect relations to each other, carried out with the aim to reach the intended result (product, service), constituting in satisfaction of the client's important external or internal needs [5]. Consequently, projects are actually processes. Therefore, the projects management should be based on the process approach – its principles, models and methods [8].

Thus, an appropriate model for a projects presentation is the process model, because it enables a complete description of dynamic organizational phenomena, i.e. the projects.

The process model contains three main groups of project processes: executive processes, supportive processes and managerial processes.

Executive processes of the project (otherwise referred to as the operative or primary processes) have a direct share in the creation of the intended result – product and service. At the input of such processes, there are always concrete needs, whereas at the output – as an effect, the satisfaction of such needs should always occur. Executive processes are divided into two types: processes of designing the project's result (creative – consisting of a description, usually in the form of project documentation, intended result of the project) and project implementation processes (referring to the material implementation of the product/service of the project).

The project supporting processes – servicing or auxiliary – do not have any direct share in creating the project's outcome, but they create the conditions to achieve it. They comprise the processes of legal, staff-related, financial and accounting, informative, technological, and administrative support. The list of the project supporting processes is open and each time adjusted to the specific undertaking.

Managerial processes consist of harmonization of executive processes and supporting processes with the aim to reach the intended result of the project. Managerial processes include the processes of outlining the goals, organization and motivation, and control.

For every group of processes, i.e. executive, supporting and managerial processes we can determine a separate model structure (cycle) constituting a pattern of its course, repetitive in various projects. This allows us to formulate general principles and guidelines and use universal methods of project planning and implementation.

Recognition of repetitiveness of the projects model structure allows us to define those structures as cyclic ones. Therefore, in project management we deal with the projects lifecycle (cycle of projects), projects management cycle, cyclic processes of support (Figure 1).

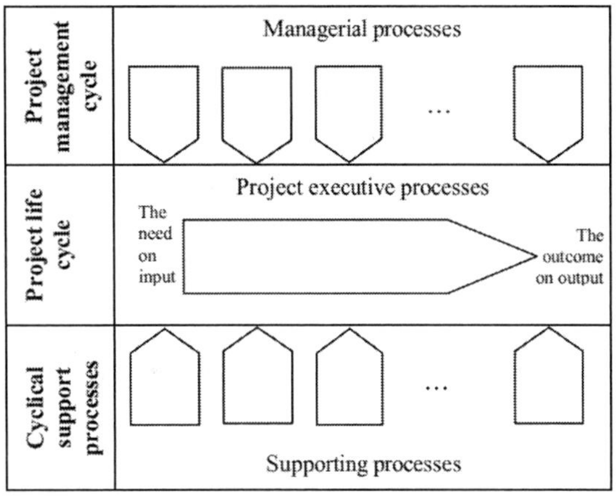

Fig. 1. General process model of project implementation (own preparation on the basis of [5, 13]).

In view of the diversity of projects, which is mostly manifested in their executory process, the search for a universal generalized model, which would describe the lifecycle of all projects, seems of little use. The effort is focussed on the development of the models of the respective types of projects, e.g. information projects, constructional projects, complex technological devices, and research & developmental projects [14].

The project management cycle is a generalized (model) presentation of the course of the project managerial process. This concept is based on the assumption that, despite a considerable differentiation of the goals and diversity of the projects' results, we can determine their common set of managerial processes. According to this concept [5, 13], the project management cycle consists of four phases (defining - DP, preparation - PP, workmanship - WP, finishing - FP) and eight managerial processes:

1) Initiation and definition of the project (DP),
2) Organizing a project team (PP),
3) Planning the project structure (PP),
4) Planning the project terms (PP),
5) Planning the project resources (PP),
6) Organization of the project performance (PP),
7) Guidance (control and coordination) of preparation of the performance of the project and actual performance of the project (WP),
8) Closure of the project (FP).

The structure of the project management cycle varies among different methodological standards [15].

Singling out three main groups of processes and their model structures, repeated periodically and forming a closed entirety, is of crucial importance for project management. It enables the description of the most important processes connected with project implementation. Therefore, the process' approach is recommended as the basis of improving the project management process, important for increasing the efficiency of the functioning of contemporary organizations [5, 9, 13].

DOI: 10.1051/itmconf/20171504007

3 Classifications of information systems for project management and their maturity

Information systems for project management (ISfPM) may be qualified to one of the following three classes [5]:

1) Project Management Support Systems – PMS,
2) Project Portfolio Management – PPM [16, 17],
3) dedicated tools and packages [18].

The functions, which should be accomplished by information systems of particular classes, are presented in Table 1.

Table 1. Functions implemented by particular classes of systems for project management (own work based on [1, 3, 5, 16, 19]).

ISfPM class	Functionality of the system
PMS	- Project planning support - Tasks management – preparation of the Work Breakdown Structure; determination of four types of relations between the current task and its preceding task: Finish to Start, Start to Start, Finish to Finish, Start to Finish; recognition of various types of tasks, such as: milestones, individual tasks, summary tasks, work packages; checking the status of single tasks and groups of tasks; generation of lists of particularly important activities, those with approaching accomplishment deadline, delayed activities or those threatened with a delay; implementation of the method of critical path; budgeting of tasks - Management of resources – records and assignment of resources for accomplishment of concrete tasks; analysis of loading the resources with work; signalling an excessive allocation of resources; calculation of resources work costs; - Views and reports – for various participants of the project with a possibility of their independent preparation and adjustment to concrete needs
PPM	- Coordination of tasks and allocation of resources at the level of the whole organization - Establishing the priorities for projects - Management of the projects portfolio risk - Management of users' rights - Processing larger data sets than those with the use of PMS - Group work using coordination packages of the group work - groupware - Multidimensional data analyses in OLAP technology
Dedicated tools and packages (DTaP)	- Advanced reporting in the language and currency of a given country, including the locally used measurement system, legal regulations of the activity of the project conducting organisation

From the point of view of the users working within one project, the PPM system functionality basically does not differ from the PMS system. Specific functions of PPM systems are used by the entities responsible for project coordination at a higher organizational level, e.g. strategic design offices or program managers or corporation projects portfolio.

Dedicated tools and packages are usually formed for the needs of concrete enterprises. Such programs should then meet all the needs connected with the project implementation. The use of specialist software may, among others, result from local conditions, specificity of the project or the area of the project which engages an organization from the branch in which the project manager is functioning. From the necessity to integrate the dedicated system with the information systems so far used in the organisation [18].

The use of various classes of systems for project management may be referred to the organization's project maturity levels [12, 17, 20, 21]. Classification suggested by Gasik [22] exhibits ascending nature in which each higher level contains the functionalities of lower levels and elements that are specific for itself.

The article omits Level 0, i.e. manual level, because the main information tool used on this level is the office software, such as: Ms Excel or Ms Word.

Level 1 – individual - is the first level on which information systems are used in project management. The software, which supports the project management, is used independently in respective projects, without communication inside the project portfolio or with other areas of organizational management. The project management support on this level comprises the creation of the Work Breakdown Structure (WBS), preparing the schedule, records of the progress of work, planning and records of costs, the analysis of the worked out value, and supervision of the project implementation. It is possible to transform the schedule into a plan through the allocation of resources, but software does not support the resources management because it does not contain any information about the use of resources in other projects. The way of using the information instruments depends on the project manager and, therefore, working time records are not kept in the system, especially when the resources work in many projects. The Ms Project Standard or Professional software in a one-stand version is most often used on this maturity level.

A Level 2 portfolio assumes the mutual management with the entire portfolio of all projects of the organization. ISfPM maintains one common repository, i.e. all dictionaries necessary to implement the projects. The most important dictionaries comprise dictionary of the types of projects, organizational structure, pool of the organization's resources, dictionary of the resource roles in projects, WBS templates, bookkeeping accounts structure, types of risks, and users' rights in the system. The most important functions implemented by ISfPM on this maturity level are working time records, supporting the common management of resources, mutual analyses of a projects performance, supporting of inter-project communication, supporting uniform reporting and unified standards of projects implementation. Among the

DOI: 10.1051/itmconf/20171504007

systems available on the market, the Primavera tools support the portfolio management best.

Level 3 – Integrated, requires the used ISfPM to be integrated with the organization's most important information systems which comprise the financial and accounting system - FK, human resources management system - HRMS, electronic information flow system - EIFS, customer relationship management system - CRMS, shopping support system - SSS, production support system - PSS. Information systems at this maturity level enable bilateral communication (ISfPM <—> concrete information system of the organization). They provide, through electronic mail or internet portals, information about the costs generated by the resources (for FK), type and evaluation of performed work (HRMS), contacts with project clients (CRM) and the status of conducted tasks (PSS). Presently, organizations planning to reach this maturity level must work out themselves the interfaces between individual systems. The existing ready solution – the closest to the requirements of the integrated level, is the SAP software (SAP Business One – Project Management).

Level 4 – Uniform, is characteristic of enterprises which have one application supporting all areas of the organization's management. All functions, which the organization needs for project management, are supported in one database. The organization's human resources and contractor in the project mean the same resource, described as one record in the base. There is no need to develop an interface and synchronisation with the FK system, because from the project level there is direct access to the accounts plan. Presently, there are no ready solutions that support project management at a uniform level.

Table 2. Maturity levels of ISfPM (own preparation on the basis of [5, 17, 22]).

ISfPM class	Maturity level of the system			
	1	2	3	4
PMS	+	–	–	–
PPM	–	+	+	partly
DTaP	+	+	+	–

Table 2 qualified ISfPM of a specific class to an appropriate level of maturity. Sign „+" stands for an achievement, whereas „–" is a failure to achieve a given maturity level by the system.

From the point of view of the organization that seeks the most appropriate ISfPM, the most important seems to be the classification involving technical issues of location of the programme, which distinguishes [23]:

1) Programs installed on the local server and users' computers,

2) Tools accessed through the Web interface,

3) Software as a Service (SaaS), i.e. the software provided by the network without the need of installation and maintenance of servers on the clients' part [24].

4 Advantages of IT support in project management

The share of respective phases of the project management cycle in the project's inputs and costs is differentiated. It will be mentioned after the results of the costing structure analysis with the participating 364 project management practitioners from the organizations which, within one year, carried out 16,110 projects for the total amount of $29.8b [25].

The defining phase lasts from several days to several weeks, its costs amounting to approx. 12%. The duration of the consecutive phase, i.e. preparatory phase is counted, (depending on the project complexity) in weeks, months and even years, its share in the project expenditures and costs amounting to about 20%. The performance phase is mainly the project implementation process, which is accompanied by the management processes of operation (control and coordination). This phase usually takes the majority of the time and the biggest cost of the project: performance – approx. 43%, operation – approx. 18%. The last phase – termination, lasts from several days to several months, its share in expenditures and costs amounts to approx. 7% [25].

Knowing the distribution of expenditures and costs between phases, it seems reasonable to seek software which would support the most difficult and expensive processes in the projects management cycle. Table 3 lists the advantages of information support of the processes occurring in the respective phases.

Table 3. Advantages of information support of individual processes of the project management cycle (own work based on [2, 5, 9, 19, 25]).

Supported processes within phases	Advantages of support
Initiation and definition of the project (defining phase)	- Taking the decision as to the date of starting or finishing on which the schedule of the project will be based and hence – what operations of the program will be undertaken - Preparing the definition or synthetic description of the project - Analysis of risk
Planning the project structure (preparatory phase)	- More precise distribution of works in the project, involving the most important factors affecting the result of the project
Planning the dates (preparatory phase)	- Reduction of the schedule development time - More effective arrangement of dates with the project participants - Assignment of a critical path
Planning the resources (preparatory phase)	- Clear determination of contractors' tasks and responsibilities - More optimum use of resources - Signalling an excessive allocation of resources and the so called „bottlenecks"

Supported processes within phases	Advantages of support
Planning the costs (preparatory phase)	- Reliable determination of costs for project-related works according to their original sources - Realistic calculation of planned costs and budgeting - Isolation of reserves (e.g. for risk management) in the project budget - Limitation of agreed penalties - Decreased freezing of capital
Organization of the project (preparatory phase)	- Faster communication between the project participants - On-going work of the team
Operation of the project (performance phase)	- Specification of the project evaluation indices - Recording the working time - Precise control of progress in the project at the time close to real time - Lower expenditures on the project operation - Faster response to deviations from the plan or changes of the plan - Increased transparency of the course of the project
Analysis of results and collection of project knowledge (completion phase)	- Fast access to the project data in submitted views and multidimensional reports - Higher transparency of the project due to improved reporting and documentation - Support in creation of the lessons learned documents - Shortening the time of preparing individual documents through developed templates

5 Criteria of the choice of an information system for the project management

The list of functions presented in Table 1 is not closed yet. It only shows which functions are offered by contemporary software supporting project management.

In some cases, the level of expansion of information tools enables, among others, management of the whole portfolio of projects. It allows communication among employees with the use of modern forms of transmission (electronic mail, chat, audioconferencing, videoconferencing, instant messaging, discussion forums, newsletters, project weblogs, enterprise information portals, workflow systems, desktop and real-time data conferencing, electronic meeting system, electronic whiteboard [26, 27]). Drawing up additional checklists or even the management of relations with clients is feasible.

The knowledge of the basic functionalities (Table 1) and advantages of ISfPM (Table 3) is the first step in the procedure of choosing appropriate software.

However, the knowledge of general functions itself should not be the basis for the selection of the system. This would cause a situation where to secure oneself for the future in view of possibly necessary functionalities of the system, excessively expanded programs would be chosen, i.e. the programs that are more complex in servicing and often contain unused functions.

Project managers or higher management of the organization should first, fully understand the specificity and scope of implemented projects and then determine the actual needs and requirements in relation to the software, and subsequently indicate the functionalities needed for the improvement of activities undertaken within the projects.

Subsequently, in order to facilitate the choice of appropriate software we could qualify the project, because of the level of its complexity, to one of the three groups:

1) Simple – short-term projects (up to three months), with simple tasks which require the basic support in management and involvement of capital up to 100 thousand PLN,

2) Moderately advanced – medium (up to one year) projects, characterized by the average size and scope of works, requiring a more comprehensive, though not complex yet, support in management, with the costs up to 10 million PLN,

3) Advanced – long-term (over one year) and complex projects which require a comprehensive support in management, generating costs above 10 million PLN.

The determination of the level of complexity of projects should result from the evaluation of the basic parameters of the project, i.e. time, scope, and cost.

In the case of proper interpretation of the level of project complexity, and qualifying it to one of the three mentioned groups, it becomes possible to better adjust the software to the needs of the project manager and team.

The least complicated projects generally do not require advanced information systems. Sometimes, there is no need to implement any additional solutions, apart from the tools already applied in organizations. In the case of such projects, the support in management may be reduced to checking if the assigned tasks accomplishment deadline is not exceeded. Other functions, such as reminders, and the possibility to share the information with other participants of the project should depend on, among others, the necessity of cooperation among the team members. For long-term projects, within this group of complexity, it may be necessary to draw up a schedule to support and facilitate the work and control.

Projects of the second level of complexity require more support, which is connected with the need to use additional functionalities, such as a possibility to draw up schedules and manage the tasks, the comprehensive management of resources, especially human resources, including, at times, communication through applied software and a possibility to analyse risks within the project. The schedules are presented, not only in the tabular form, but also in the Gantt chart. The software supporting the projects from the second complexity level differs from the one for simple projects in the way of task assignment, the level of determined responsibilities,

DOI: 10.1051/itmconf/20171504007

and methods of control of the project work implementation. The use of ISfPM should be the answer to the needs of coordination of numerous tasks with a clearly specified deadline, which requires the simultaneous work of several people e.g. on one document.

For projects of the highest level of complexity, the advanced ISfPM functions are useful. Apart from the possibility of time control, the management of people and tasks, there are the following indispensable functions: the analysis of the critical path, determination and control of the budget (the Earned Value analysis), risk analysis within the project, and the exchange of documents and drawing up reports. For such projects, the ISfPM facilitates both the project preparation and control of its course, and verification of the compliance of the basic plan with the results.

Whenever it is not possible to ascribe a specific project to one of the three levels of complexity (even after the analysis of its basic parameters), attention should be paid to the values of individual characteristics. It may appear that despite the simplicity of the performed tasks, one of the basic parameters of the project is conditioning the need to apply additional functionalities.

However, categorizing the project to an appropriate level of complexity is important due to the efficacy and efficiency of its implementation. When the project is qualified to an improper level of complexity and consequently the choice of an improper ISfPM, it may appear that a part of the work was redundant and additional activities for theoretical improvement of the project were undertaken unnecessarily.

Supplementary verification of the value of the basic parameters of the project will not always be sufficient. In such a situation, it is good practice to make an additional analysis of the factors presented in Table 4. The determination of additional factors consists of a detailed analysis of the basic parameters of a given project and results from classification of projects on account of the level of complexity.

The factors indicated in Table 4 are not the only factors that can be considered while selecting an appropriate ISfPM. Not infrequently, the specificity of the project affects the determination of additional variables which can also contribute to the choice of a concrete system.

To determine additional characteristics of ISfPM, it is worthwhile to check the detailed parameters of specific programs in the lists available in the network [28]. There are many available pages that enable a comparison and analysis of separate systems as well as simple training courses about recommended programs.

The studies conducted regularly in Poland by portal 4pm.pl, each time on a sample of over 1200 respondents, show a list of important traits that contribute to a positive or negative evaluation of applications used in project management. Particularly on portals in English (portal visual.ly), we can find a lot of information about the programs which support project management, about their possibilities, advantages and disadvantages.

Table 4. Additional factors helpful for the choice of ISfPM (own work based on [23, 24, 27, 29, 30, 31]).

Factor	Description of the factor
Time needed to prepare the project	- It may be: short (up to two weeks), average (up to three months), long (above three months) - More complex, longer projects require more time to prepare from the level of assumptions, through determination of necessary resources, drawing up a schedule and transforming it into a plan and pointing to a risk (such projects are included in the second group requiring more support or the third group requiring a complex support)
Type and amount of obtained results	- Object, product or service are the distinguished types of obtained results - The result may be either single or multiple (multiproduct) - The less complicated the results and the lower the input of work, the lower a given project should be qualified
Level of innovation	- There are non-innovative projects or those of a high, average or low innovation - Projects of a high degree of novelty, especially those orientated to products, require a higher information support and should be counted among advanced projects
Number of involved workers	- Increased number of workers increases communication difficulties
Need for communication	- It may be constant, occasional or rare - Requirements posed to ISfPM are higher with more employees involved in the project or with a higher on-going communication between members of the project team
Documentation	- It may be advanced (necessity of mutual development), medium-advanced (possibility of mutual development), simple (without cooperation) - Choice of programmes supporting the team work is pointless if new documents are not drawn up during the cooperation and the team does not work on such documents
Required level of control	- It may be complete, partial or may be confined only to supervision of time (simple projects) - Average-advanced projects require: verification of terms and costs, control of implementation of tasks and their correct performance - Advanced projects require: additional reports, supervision of direct work with documents, control of the accomplishment of tasks and budget

DOI: 10.1051/itmconf/20171504007

An important criterion of the choice of ISfPM is the system's architecture. The software installed on the local server and users' computers enable work without the need to be connected to the mains. Yet, it requires installation on each computer, which usually confines the possibility of group work. The tools available due to Web or SaaS solutions considerably facilitate the cooperation between the project participants. In certain cases, their drawback is the necessity to be on-line, because these are programs, which in other situations, will not be functioning.

The cost of the purchase, and maintenance of the system, is a subsequent determinant in the choice of support for project management information. Some available systems are completely or partly free of charge. In the latter case, we should pay special attention to users' opinions, level of data protection and limitation of the package, especially in the case of gratuitous programs to a specific functionality level or the number of users. These programs are often the best solution for the management of simple projects or those of a medium level of complexity, expressed in the number of tasks, resources and users. On the other hand, choosing paid software, apart from checking its functionality and installation mode, we should pay attention to the payment form and licensing rules. Some of the available ISfPM should be paid on a one-off basis during the purchase, whereas other ones use a subscription system, e.g. monthly fees for use. Otherwise, there are hybrid solutions, which use both payment forms at the same time. In order to carry out a complete and reliable cost analysis, apart from the licencing form and licence purchase cost, we should consider all other components of the implementation costs, i.e. the cost of actual implementation, cost of the purchase of additional software and hardware, cost of training courses, external services and connectors. The buyer should also be aware of some hidden costs depending on the licensing form, such as separate purchase of the carrier and licence to use some of the software, the additional lease of the licence for a specific time, fixed-time access to the www service etc. The final decision for the choice of the system should be guided by both its functions and total costs connected with its purchase, implementation and maintenance.

If no proper software is found, support through a dedicated solution (DTaP) is possible. Preparation of the question should then be preceded with a precise analysis of available solutions and an additional attempt to use several exemplary systems, both free of charge and payable. A dedicated solution usually generates much higher costs and a longer time of waiting for the delivery of the package, as compared to ready-made systems.

From the point of view of some organizations, an important aspect within the choice of an appropriate ISfPM is also the possibility of simultaneous management of many projects – a portfolio or a program of projects [32]. Only payable software exhibits such functionality, therefore before making the final decision, we should first consider if this function is important for the organization.

The four-level model of ISfPM maturity presented in section 3 is also practically very important for the implementation of information support for project management.

Firstly, it shows which requirements – gradually higher and higher – may be posed to ISfPM. Subsequently it shows the complexity of the process of implementing such a system. The individual level actually does not require any implementation efforts, and implementation of a fully functional system based on the Web interphase lasts exactly as long as putting the plug of the network cable of a special mini server supplied to the company into the socket of the local company network. Implementation of more advanced solutions usually does not exist without high work input. Preparation from scratch, testing and implementation of the dedicated software to the management of very complex projects may last for many months. The ISfPM maturity model may also be the basis for evaluation of the quality of the systems to be implemented.

To facilitate the choice of ISfPM, the rankings and comparisons of information solutions that support project management are drawn up. One of them, worthy of attention, is the regularly published report of the advisory company Gartner Inc., i.e. the Gartner magic square [33]. The report shows the position of individual systems and synthetically characterises each of the solutions available on the market.

6 Conclusions

The choice of the proper software for the organisation, or also sometimes for a specific project (DTaP), is one of the conditions of effective management of projects.

ISfPMs are used to support the preparation, implementation and completion of projects, and not their complete development.

The knowledge of the basic functions of ISfPM should be the first step in the process of choosing appropriate software.

It is very uncommon to have the possibility to assemble a fully functional system, especially for specific use (constructional projects, scientific projects, research and development projects) from one producer's offer. Therefore, the quality and reliability of connections among the modules gain greater importance.

For the same reason, the producer's market position drops and the prospects for the development of a specific product are decreasing. The customers will only expect from the suppliers some technical support and help in emergency situations. They will frequently have to apply for the dedicated solutions to niche companies supplying specialist modules.

Modular structure of the systems also allows an easier selection of software in view of its output. Specific gaps will be supplemented by the third companies' software.

In practice, the restrictions of generated reports disappeared, because it has become increasingly common to use standard relative databases. The professional SQL operator will be able, using the

DOI: 10.1051/itmconf/20171504007

commonly available software of the clients' databases, to search for any information and prepare any statements.

Implementation of methodologies in ISfPM is becoming less and less important. The ready systems usually offer a set of instruments used in project management, while more sophisticated methodological needs will satisfy the tools and dedicated packages.

The success of seemingly similar projects depends on the applied methods and specific tools as well as supporting, with information, systems for the management of undertaking work. However, no ISfPM will ever become a panacea for all the problems encountered by project managers and their contractors conducting the projects and

The current management occurring in the network process of decision-making is based not only on the fast acquisition of checked information, but also on knowledge. The most effective project approach is the one where people, using both the organization's information and knowledge resources, and implementing particular project procedures (using properly selected information instruments), construct gradually the intellectual capital and consequently acquire a new value of the organization functioning [23].

References

1. W. Chmielarz, *Zarządzanie projektami @ rozwój systemów informatycznych zarządzania* (Wydawnictwo Naukowe Wydziału Zarządzania Uniwersytetu Warszawskiego, Warszawa, 2013)
2. J. Kisielnicki, *Zarządzanie projektami. Ludzie – procedury – wyniki* (Oficyna a Wolters Kluwer business, Warszawa, 2011, 264-280)
3. E. Sońta-Drączkowska, *Zarządzanie wieloma projektami* (Polskie Wydawnictwo Ekonomiczne, Warszawa, 2012, 11-28)
4. W. Curlee, R. L. Gordon, *Successful Program Management: Complexity Theory, Communication and Leadership* (CRC Press, Boca Raton, 2014)
5. M. Trocki, *Nowoczesne zarządzanie projektami* (Polskie Wydawnictwo Ekonomiczne, Warszawa, 2012, 416-432)
6. G. R. Heerkens, *Jak zarządzać projektami* (RM, Warszawa, 2003, 127)
7. T. Schmidt, *Strategic Project Management Made simple. Practical Tools for Leaders and Teams* (John Wiley & Sons, Inc., Hoboken, 2009)
8. A. Bitkowska, *Zarządzanie procesami biznesowymi w przedsiębiorstwie* (Wizja Press & IT, Warszawa, 2009)
9. P. Cabała, *Metody doskonalenia procesów zarządzania projektami w organizacji* (Difin, Warszawa, 2016)
10. J. Charvat, *Project Management, Methodologies Selecting, Implementing, and Supporting Methodologies and Processes for Projects* (John Wiley & Sons, Inc., Hoboken, 2003)
11. V. LÉvÁrdy, T. R. Browning, IEEE Trans. on Eng. M. **56**, 600-620 (2009)
12. W. Chmielarz, *Information Technology Project Management* (Wydawnictwo Naukowe Wydziału Zarządzania Uniwersytetu Warszawskiego, Warszawa, 2015)
13. K. Schwalbe, *Information Technology Project Management 8e* (Cengage Learning, Inc., Boston, 2015)
14. M. Wirkus, A. Lis, *Zarządzanie projektami badawczo-rozwojowymi* (DiFin, Warszawa, 2012)
15. M. Trocki, *Metodyki zarządzania projektami* (Bizarre, Warszawa, 2011)
16. Peter W.G. Morris, J.K. Pinto, *The Wiley Guide to Project, Program & Portfolio Management* (John Wiley & Sons, Inc., Hoboken, 2007)
17. H. Kerzner, *Project Management: A Systems Approach to Planning, Scheduling, and Controlling Eleventh Edition* (John Wiley & Sons, Inc., Hoboken, 2013)
18. S. Y. Chadli, A. Idri, J. N. Ros, J. L. Fernández-Alemán, Juan M. Carrillo de Gea, A. Toval, *Software project management tools in global software development: a systematic mapping study* (SpringerPlus, https://www.ncbi.nlm.nih.gov/pmc/articles/PMC5121116/)
19. M. A. B. Ilyas, M. K. Hassan, M. U. Ilyas, *PMIS: boon or bane?* (Paper presented at PMI® Global Congress 2013—EMEA, Istanbul, https://www.pmi.org/learning/library/project-management-information-systems-overviews-5813)
20. M. Juchniewicz, *Dojrzałość projektowa organizacji* (Bizarre, Warszawa, 2009)
21. M. Trocki, *Zarządzanie wiedzą w projektach. Metodyki, modele kompetencji i modele dojrzałości* (Oficyna Wydawnicza Szkoły Głównej Handlowej, Warszawa, 2011)
22. S. Gasik, *Dojrzałość systemów zarządzania projektami* (Sybena Consulting, Warszawa, www.sybena.pl/dokumenty/MD-Model-dojrzalosci-systemow-ppm-V1.0.pdf)
23. A. Jaskanis, M. Marczewska, M. Darecki, *Zarządzanie projektami w administracji publicznej* (PRESSCOM, Wrocław, 2015, 235-266)
24. P. Poturalski, *Kluczowe aspekty wdrażania narzędzi do zarządzania projektami* (MDDP Business Consulting, Warszawa, http://www.mddp-bc.pl/know-how/narzędzia-it-wspierajace-zarzadzanie-projektami/art,91,kluczowe-aspekty-wdrazania-narzedzi-do-zarzadzania-projektami.html)
25. PM Network **12** (2006)
26. K. Kandefer-Winter, O. Nadskakuła, *Komunikacja w zarządzaniu projektami* (CedeWu, Warszawa, 2017)
27. K. Riemer, C. Steinfield, D. Vogel, Elec. Mark. **19**, 183 (2009)
28. D. B. Stang, M. Light, T. Jones, *Magic Quadrant for Project Portfolio Management, Worldwide* (www.gartner.com/doc/3728917/magic-quadrant-project-portfolio-management)

DOI: 10.1051/itmconf/20171504007

29. P. Simon, *Project management tips for a decision-support system implementation* (TechTarget Network, http://searchdatamanagement.techtarget.com/answer/Project-management-tips-for-a-decision-support-system-implementation)
30. P. Simon, *The Next Wave of Technologies. Opportunities in Chaos* (John Wiley & Sons, Inc., Hoboken, 2010)
31. A. C. Uzialko, *Choosing Project Management Software: A Buyer's Guide* (Business News Daily, http://www.businessnewsdaily.com/9976-project-management-software-buyers-guide.html)
32. Project Management Institute, *Project Portfolio Management. A View from the Management Trenches* (John Wiley & Sons, Inc., Hoboken, 2009)
33. P. Wyrozębski, *Zarządzanie wiedzą projektową* (Difin, Warszawa, 2014)

DOI: 10.1051/itmconf/20171504008

Milling gate vibrations analysis via Hilbert-Huang transform

Grzegorz Litak[1,*], *Marek* Iwaniec[1], and *Joanna* Iwaniec[2]

[1]AGH University of Science and Technology, Faculty of Mechanical Engineering and Robotics, Department of Process Control, Mickiewicz Alley 30, 30-059 Krakow, Poland
[2]AGH University of Science and Technology, Faculty of Mechanical Engineering and Robotics, Department of Robotics and Mechatronics, Mickiewicz Alley 30, 30-059 Krakow, Poland

Abstract. The study aimed at identification of the milling gate modal parameters: natural frequencies and mode shapes, in order to establish the working range free from resonant areas. The evaluation was performed under exploitation loads, generated in the course of the milling process. System responses, in the form of the acceleration time histories, were analysed with the application of the Hilbert-Huang method data analysis. Selection of this method was motivated by the fact that it does not require stationarity of the analysed and provides information about nonlinearities in vibration empirical modes. In contrast to classical nonlinear system identification methods, the Hilbert-Huang method does not require measurements of excitations acting on the system of interest nor linear system dynamic behaviour in a broad frequency range around any operating point.

1 Introduction

Vast majority of real mechanical systems is nonlinear to a certain degree. Nonlinear industrial systems can be classified as systems designed for work in nonlinear ranges of dynamic characteristics and systems of nonlinear properties resulting from fault appearance. Although the sources of nonlinear properties can vary [1, 2, 3, 4], all the nonlinear systems exhibit some common properties. In general, they do not follow the superposition principle and exhibit complex phenomena unusual for linear systems, such as jumps, self-induced and chaotic vibrations, changes in natural frequencies resulting from changes in excitation amplitudes and coexistence of many stable equilibrium positions. In view of these properties, classical identification methods cannot be used for the purposes of nonlinear system identification.

The main emphasis of the paper is put on the experimental identification of modal parameters (natural frequencies and mode shapes) of the milling gate subjected to operational loads resulting from the milling process. For the purposes of experimental data analysis, the Hilbert-Huang transform was used. The choice of this method was motivated by the fact, that it does not require stationarity of the analysed data and, therefore, various phases of milling process can be selected for analysis. Furthermore, it provides information about nonlinearities in vibration empirical modes. On the contrary to the classical nonlinear system identification methods, the Hilbert-Huang method requires neither the measurement of excitations acting on the system of interest nor linear system dynamic behaviour in a broad frequency range around any operating point [5, 6]. The possibility of system identification in the exploitation conditions is of key importance, since the measurement of operational exciting forces is difficult or impossible to carry out and diagnostics of many industrial systems, for the economical or technological reasons, should be carried out during the normal work [7]. Obtained results made it possible to establish the milling gate work range by avoiding resonant areas.

The structure of the paper is as follows: after short introduction, the measurement setup is presented. Then the consecutive steps of the Hilbert-Huang method application to the measured data are shown. Further part of the paper is devoted to discussion of the obtained results of analysis. Finally, the paper is concluded in the last section.

2 Experimental standing and measurement results

In the experiments, the typical milling process of different thickness of the cut layer realised on the milling machine was conducted. The excitation during the experiment was given by operational forces generated during the milling process. The aim of the investigation was identification of the modal frequencies as the function of the amplitude of local measured vibration. Assuming that in the set of the main attention was focused on the analysis of vibration of the gate of milling machines.

In the measurements the 28-channel dynamic analyser SCADAS III was used. The analyser was controlled by TestLab ImpactTesting software by Siemens-LMS. One channel was used for force measurements and 24 channels

* Corresponding author: g.litak@pollub.pl

 DOI: 10.1051/itmconf/20171504008

were used for measuring dynamical response in the chosen points of the gate. Additional 3 channels were used for registration the ambient vibrations as the referents points. Dynamical response was measured by the 3D piezoelectric accelerometers PCB 356A16 in the chosen points distributed on the gate.

The view of the experimental set and one of the typical experimental results are presented in Fig. 1.

(a)

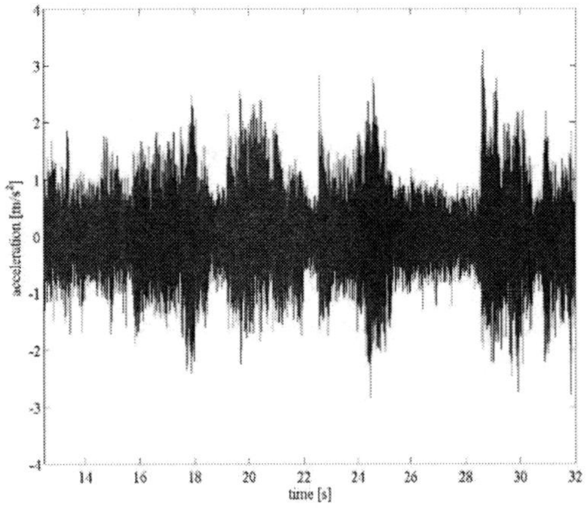

(b)

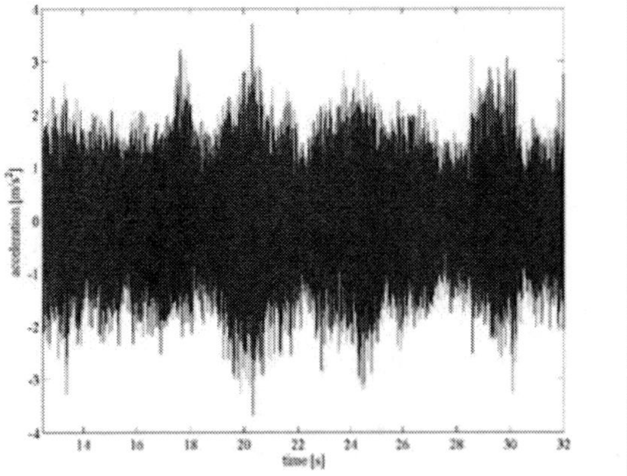

(c)

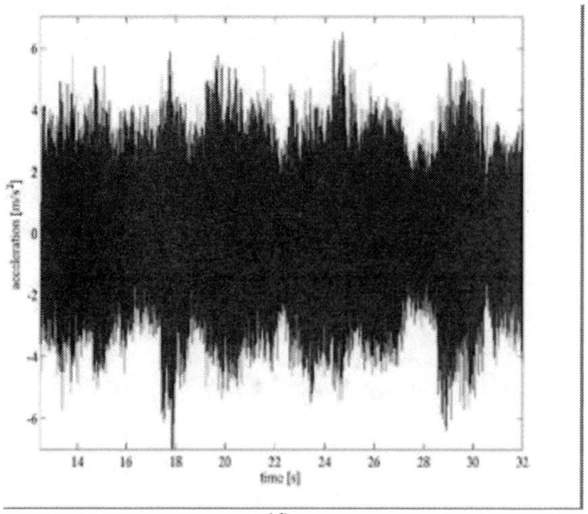

(d)

Fig. 1. Typical layout of a superelastic SMA oscillator. Milling gate (a) and (b) example of the time series measured during milling experiment. From the accelerometer located on milling gate, cases (b) – (d) denote the directions x, y, z.

3 Hilbert-Huang analysis

Performing the Hilbert-Huang analysis [8,9], the signal is firstly decomposed into the Huang empirical mode (EMD). In our case, the acceleration signal from the accelerometer (the second table point B4 modal analysis structure for fastening the spindle moulder) $A(t)$ (see Fig. 1). Operating frequency of milling was much higher than the sample rate 1024 Hz. In our analysis, we used even less frequency of 102.4 Hz (when every 10 value) to save the computation time. The main aim was to examine clamping structure vibrations of the milling machine. Obviously, $A(t)$ has three spatial components of A_i, where $i = x, y, z$, respectively. Figures 1b-d clearly indicate that the predominant direction of the vibration is determined by the z axis. In this direction, the amplitude of the vibration is the largest, and decreases subsequently in the y-direction and x-directions.

In addition to the above, it is worth paying attention to the phenomenon of intermittency in the presented series. This behaviour may indicate on a non-smooth milling or/and an emergence of the so-called vibration type chatter, with a considerable increase in the amplitude. Each of vector components of the acceleration vector $A(t)$ can be decomposed into the empirical modes $A_i^1(t)$, $A_i^2(t)$, ..., $A_i^m(t)$ [9-11] (Fig. 2a-c).

$$A_i(t) = \sum_{j=1}^{m} A_i^j(t) + r_m, \qquad (1)$$

where r_m stands for a truncation error on the m empirical mode. The subsequent j modal time series A_i^m is obtained after subtracting the corresponding means from the maximum and minimum values of the previous mode rm^{-1} (the envelope of the time series) interpolated by cubic polynomials along the time. From Fig. 2 it is clear, that only the first three have relatively large amplitudes. This is also shown in Tab. 1

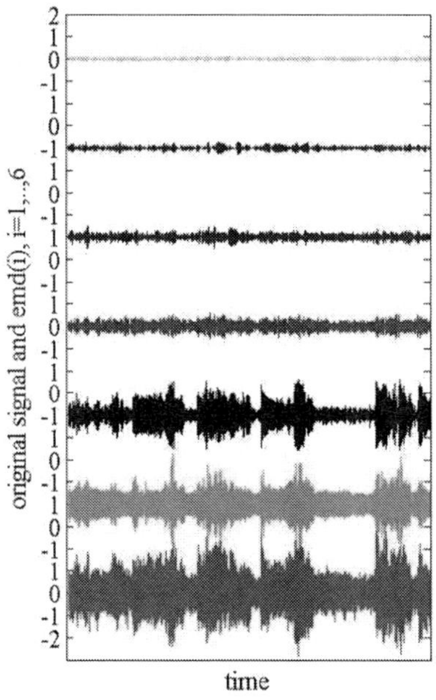

(a)

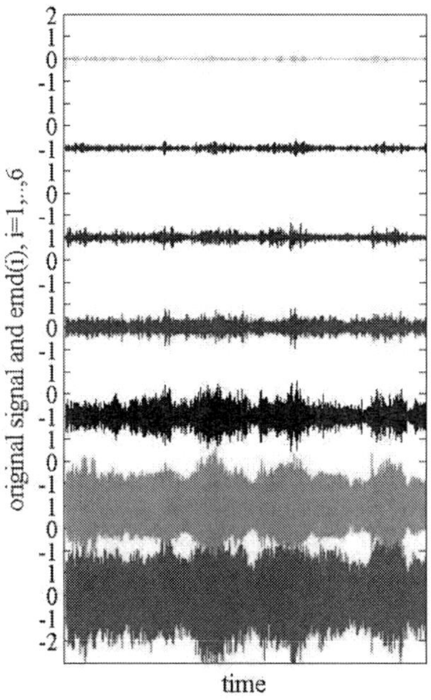

(b)

Fig. 2. Decomposition of the measured acceleration signals (acceleration in the x, y and z directions for (a)-(c), corresponding to Figs. 1(b)-(d), respectively into empirical modes. In the lower panel the original (measured) signal is placed and above in the subsequent higher panels the corresponding modes in the increasing order. It is worth to note that the first two modes have considerably larger amplitude values.

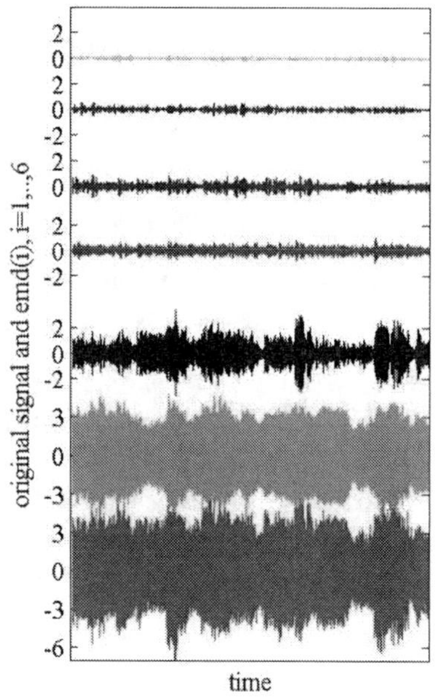

(c)

Fig. 2. Continuation.

In the next step, the Hilbert transform is performed in individual modalities, allowing for the separation of temporal amplitude, phase angle, and frequency. The Hilbert transform is defined from the original signal $g(t)= A_i^j(t)$ as follows:

$$\hat{g}(t) = \frac{1}{\pi}\int_{-\infty}^{\infty} \frac{g(\tau)}{t-\tau}\, d\tau. \tag{2}$$

Defining a new complex function z(t):

$$z(t) = g(t) + i\hat{g}(t) \tag{3}$$

We obtain information about the instantaneous amplitude, A, phase, θ, and frequency, f:

$$A(t) = \sqrt{(g(t))^2 + (\hat{g}(t))^2}\ ,$$
$$\theta(t) = \mathrm{arctg}\left(\frac{\hat{g}(t)}{g(t)}\right),\quad f = \frac{1}{2\pi}\frac{d\theta}{dt}. \tag{4}$$

Table. 1. Standard deviation σ of the original signal (measured) and corresponding standard deviations in successive oscillation modes.

	σ	Mod 1	Mod 2	Mod 3	Mod 4	Mod 5	Mod 6
x	0.653	0.748	0.646	0.235	0.176	0.108	0.051
	1.020	0.925	0.344	0.180	0.122	0.080	0.040
z	2.202	0.946	0.335	0.090	0.097	0.057	0.027

 DOI: 10.1051/itmconf/20171504008

The results of Hilbert-Huang transform (by considering first 6 modes presented in Fig. 2) are summarised in Fig. 3. They show high similarity in frequencies but differs by amplitudes. These tendencies were visible in Figs. 1 and 2.

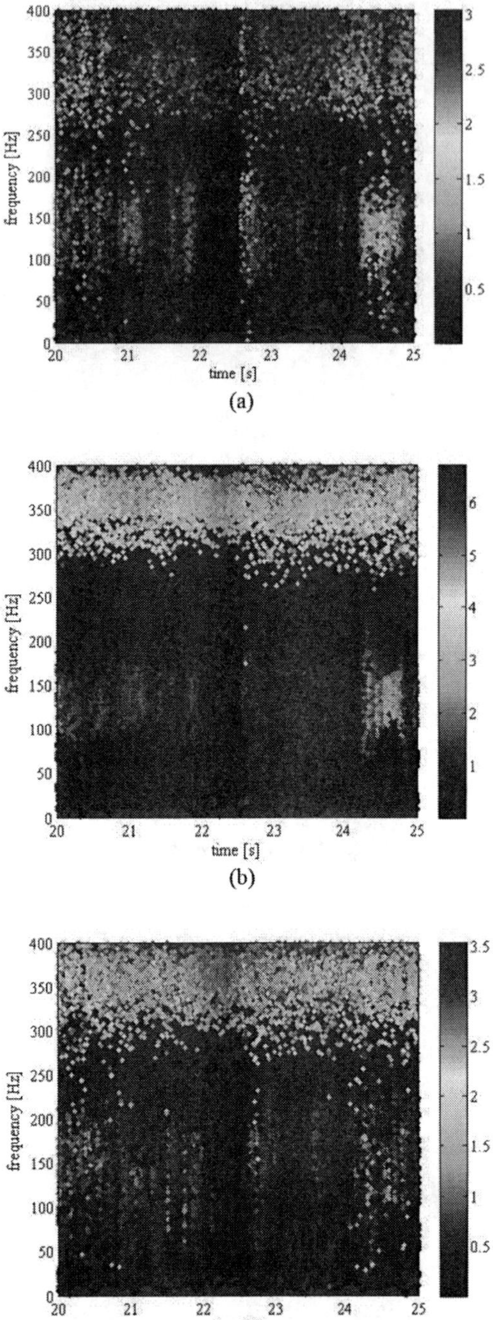

(a)

(b)

(c)

Fig. 3. Hilbert-Huang results for three considered cases. Colours denote to the amplitudes in a linear scale.

Furthermore, Figs. 4-6 show the distribution of instantaneous frequencies and amplitudes for the directions of x, y and z vibrations from the first four modes (Fig. 2). The resonance back bones matched to the results, indicating that the 2nd and 3rd fashion are non-linear. For these cases, the back-bone curve is clearly skewed and even changes the curvature with amplitude. It should be emphasised that linear systems are characterised by a straight back bone line.

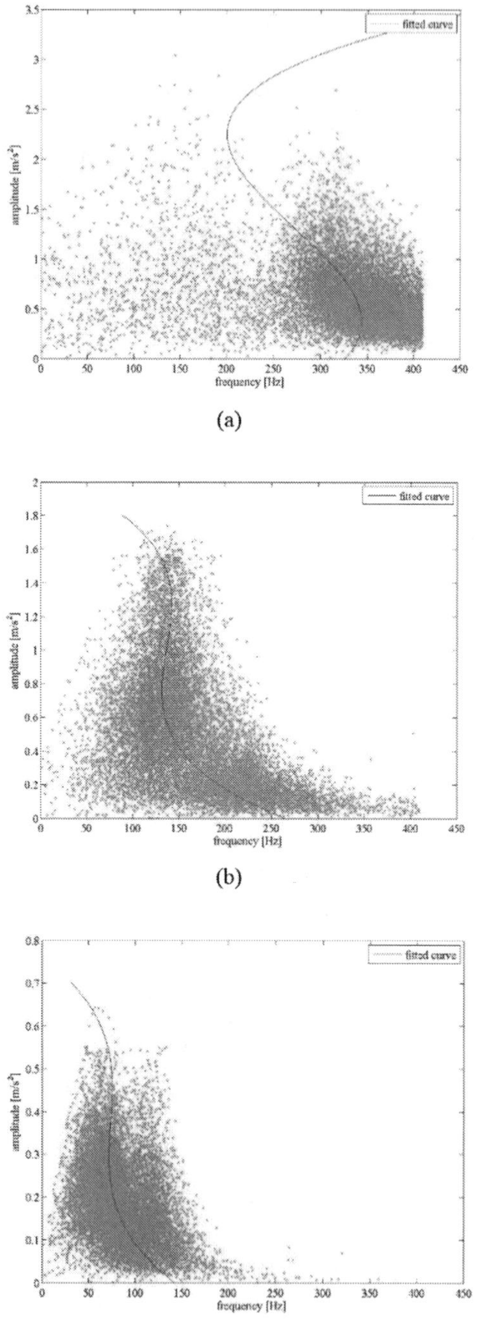

(a)

(b)

(c)

DOI: 10.1051/itmconf/20171504008

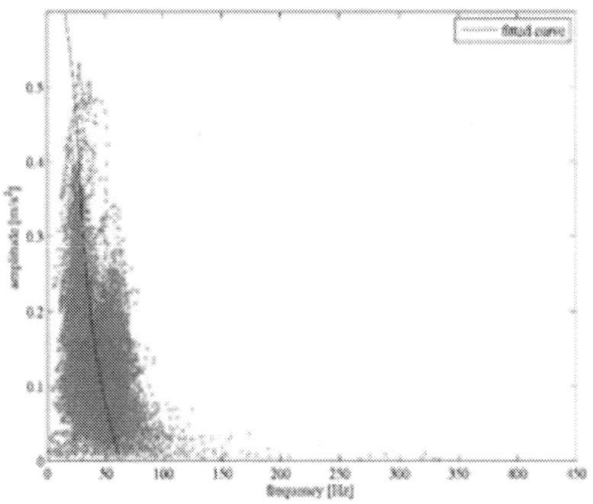

(d)

Fig. 4. Distribution of instantaneous frequencies and amplitudes for the first (a), second (b), third (c), and fourth (d) modes. The signal was measured in the x direction. The continuous line is a polynomial (third degree polynomial) matching the corresponding back bones of the resonance.

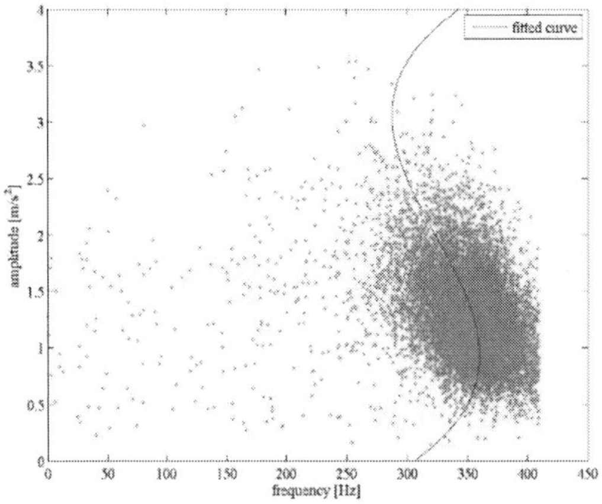

(a)

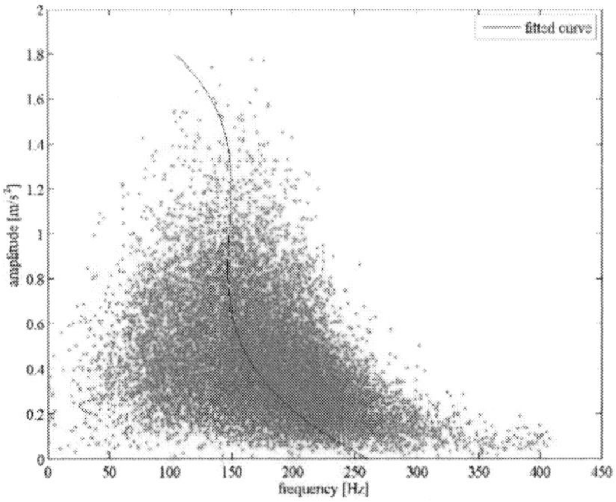

(b)

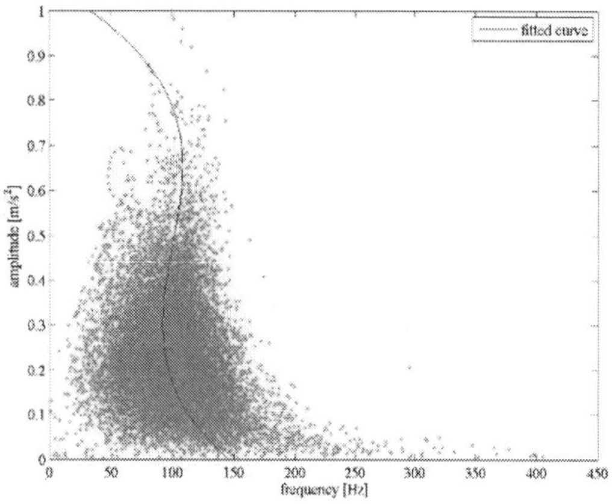

(c)

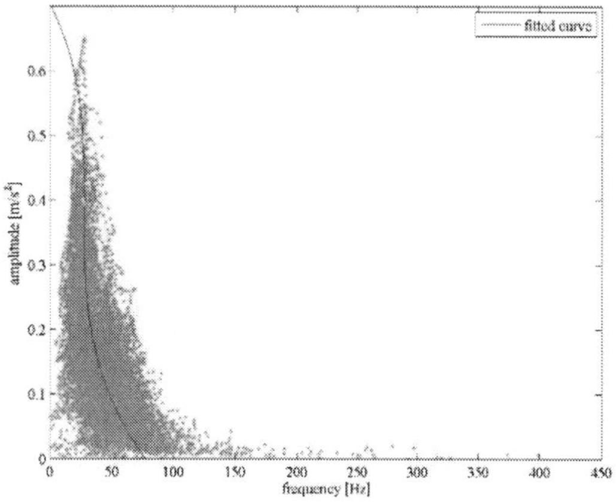

(d)

Fig. 5. Distribution of instantaneous frequencies and amplitudes for the first (a), second (b), third (c), and fourth (d) modes. The signal was measured in y direction. The continuous line is a polynomial (third degree polynomial) matching the corresponding back bones of the resonance.

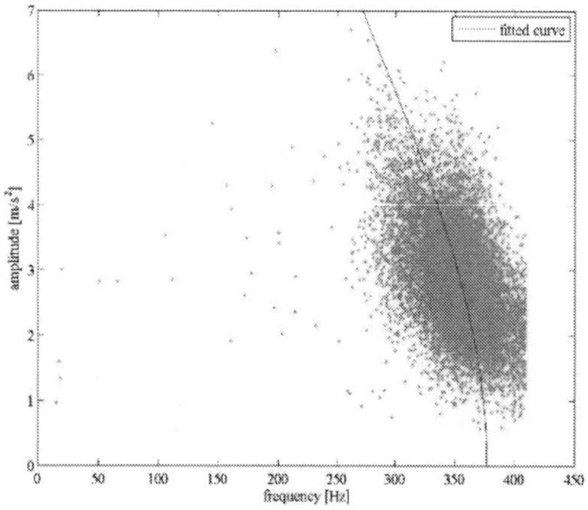

(a)

DOI: 10.1051/itmconf/20171504008

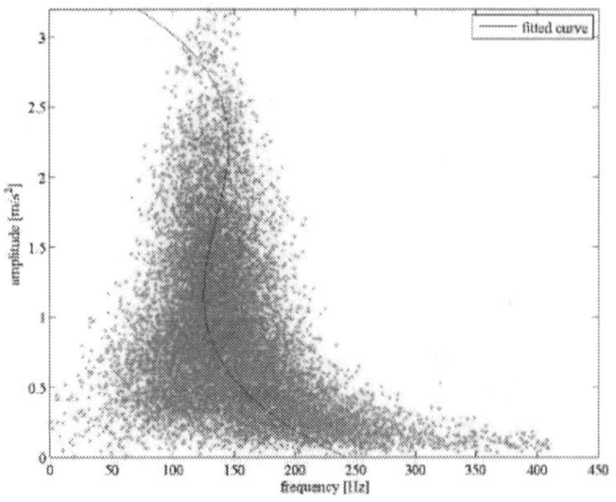

(b)

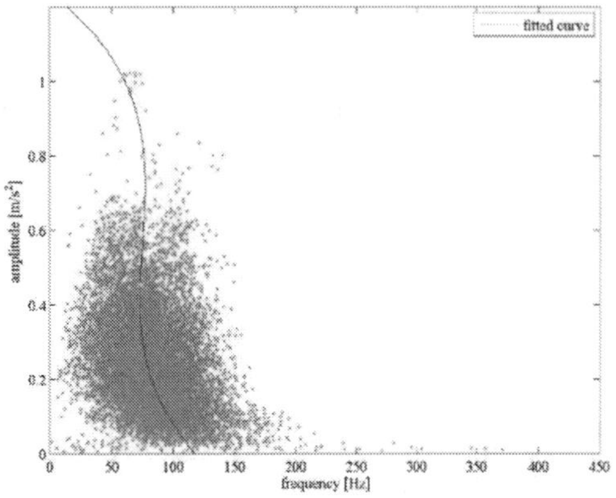

(c)

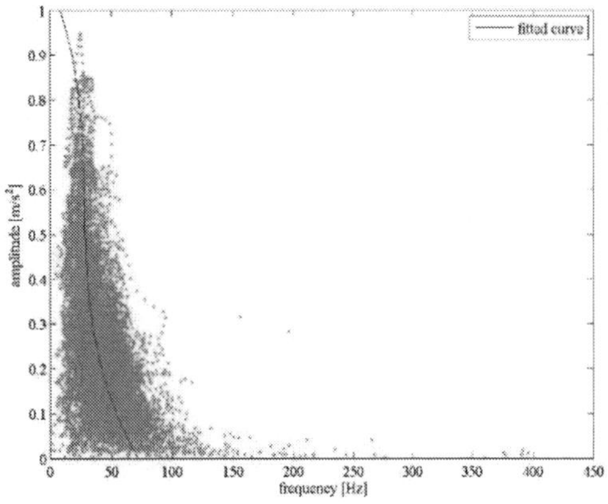

(d)

Fig. 6. Distribution of instantaneous frequencies and amplitudes for the first (a), second (b), third (c) and fourth (d) modes. The signal was measured in z. The continuous line is a polynomial (third degree polynomial) matching the corresponding back bones of resonance.

4 Conclusions

The presented analysis includes a number of positive features that can be used in modelling the frame and milling attachment of the milling machine. First, Huang's transformation does not require stationary time. For this purpose, various stages of the milling machine can be used. Secondly, this analysis provides information on nonlinear modal vibrations. Thirdly, the separation of empirical models has a broader meaning and the research technique can be enriched with other elements by creating the original hybrid technique dedicated to particular tasks.

References

1. F. Al-Bender, W. Symens, J. Swevers, H. Van Brussel, 2004, Int. J. Nonlin. Mech. **39**, 1721 (2004)
2. V. Babitsky, V.L Krupenin, *Vibrations of strong nonlinear discontinuous systems*, (Springer, Berlin, 2001)
3. G. Kerschen, K. Worden., A.F. Vakakis, J.C. Golinval, Mech. Systems and Sign. Process. **20**, 505 (2006).
4. A.H. Nayfeh L. Pai, *Linear and nonlinear structural Mechanics* (Wiley Interscience, New York, 2004)
5. M. Haroon, D.E. Adams., Y.W. Luk, ASME J. Vibr. Acoust., **127**, 483 (2005).
6. J. Iwaniec, Key Engineering Materials, **347**, 467 (2007)
7. J. Iwaniec, M. Iwaniec, Solid State Phenomena, **248**, 175 (2016).
8. D. Hilbert, *Grundzuege einer allgemeinen Theorie der linearen Integralgleichungen,* (Chelsea Pub. Co., New York, 1953)
9. N.E. Huang, Z. Shen, S.R. Long, M.L.C Wu., H.H. Shih, Q.N. Zheng, N.C. Yen, C.C. Tung, H.H. Liu, Proc. Roy. Soc. London A**454,** 903 (1998)
10. P. Wolszczak, K. Lygas, G. Litak, Advances in Science and Technology Research Journal **11**, 96 (2017)
11. G. Litak, K. Kecik, R. Rusinek, Latin Amer. J. Solids Struct. **10**, 133 (2013)

ITM Web of Conferences **15**, 04009 (2017) DOI: 10.1051/itmconf/20171504009
CMES'17

Scalability conditions of the measuring points network in a production system using the RFID selective gate technology

Monika Łobaziewicz[1,*]
[1]DataConsult Sp. z o. o., 30-614 Kraków, Poland

Abstract. The aim of paper is to present results of the research work in development conducted to define the scalability conditions for the measuring points network at production cells in the resistant to the electromagnetic disturbances system based on the patented selective RFID gate technology. This is a conceptual paper that includes research results of the selective RFID gate and its place in a production system, a communication model between two measuring points in a complex production system that was the input to do further analysis required to define its scalability conditions. The proposed approach provides the following results: order and speed of data reading from measurement points, data transfer between measurement points, access to database, communication between the database and measurement points, data archiving, procedures of changing the measurement point configuration in the system, including error handling and network re-configuration. It is believed that both practitioners and researchers will benefit from it.

1 Introduction

An infrastructure of modern production systems consists of technologically advanced machines, devices and production lines that automate manufacturing processes. The reliability of communication between elements of the infrastructure functioning in a production area is one of the important problems to ensure the quality, stability, security of production and the right identification of resources used in the production processes [1-4]. Nowadays, manufacturing enterprises expect such communication systems that provide fast and accurate data flow in real-time [5-7], are resistant to the disturbances of machines and devices placed in the production area and collaborate smoothly with other outside systems.

The above premises supported by an analysis of market trends and the identification of problems during the implementation of IT systems for production and warehouse logistics were a source of inspiration for a research conducted by DataConsult Sp. z o. o. in N=104 manufacturing companies. The research has shown that one of many technological barriers in the communication among elements of a manufacturing infrastructure is the electromagnetic disturbances caused by waves emitted by the machinery and production equipment, causing serious interruptions in the functioning of production lines. It turns out that the creation of a comprehensive measuring points system using a selective RFID gate technology meets the identified problems of enterprises because it allows to modify and develop production lines with new industrial automation components without the need to stop production or shorten the time required to make necessary changes to a minimum.

The aim of this article is to present the results of development work conducted to define the scalability conditions for the measuring points network at production cells in the resistant to electromagnetic disturbances system based on the selective RFID gate technology. An industrial research has been the input for the development work that included the communication model between two measuring points located at the production cells using the selective RFID gate. The model was based on the assumption of simultaneous independence of the communication from the configuration of industrial automation components, which enables a high level of security for collection, processing and sharing of data. This model is the result of a R&D project called "Resistant to electromagnetic disturbances prototype of the measuring points system at production cells based on the selective RFID gate technology" made by DataConsult Sp. z o. o. in 2011 - 2014.

The article consists of five parts. The first part is an introduction to the issues discussed in the article. The second part presents the concept of a selective RFID gate and its role in the communication system among measuring points located in production cells. In the third part the communication model between two measuring points using the selective RFID gate was presented, while the fourth part presents determinants of its scalability in a complex production system. The last part is a summary of the issues discussed in the article.

* Corresponding author: ml@un.pl

DOI: 10.1051/itmconf/20171504009

2 Selective RFID gate

2.1 Results of the research using EMC absorption materials to eliminate disturbances in radio communication systems

The state of the art [8-12], [17-19], shows that despite of its popularity RFID technology all over the world continues to develop and provides new opportunities for Automatic Data Capture (ADC) in various industries. The aim of using RFID is to share data encoded in tags and reading them in time and place required by adopted solutions of the system [13, 14]. The typical RFID activity is based on a set of antennas connected to a controller and embedded on a company specific structure called the RFID gate [15, 16]. It turns out that currently most of RFID gates used in enterprises operate in a non-selective way, meaning that not only the tags passing through the light of the gate, but also those next to, and even behind, can be read in the identification process. This gives an incorrect identification of objects moving along a production line.

Therefore, at the stage of industrial research carried out by DataConsult Sp. z o. o. in the cooperation with Institute of Logistics and Warehousing in Poznan, the approach as in the electromagnetic compatibility research was used, absorbing radio waves and eliminating unwanted phenomena such as collapse and reflection of waves. It has been verified whether and when it is possible to screen undesired back-light RFID antennas using EMC absorbers which is an important limitation for the correct objects' identification during production processes. The research was made by measuring the intensity of electromagnetic field in the fixed area behind the gate. For this purpose, a grid of measuring points was used to form a square of 1.2 m at the height of the RFID antenna (Fig. 1).

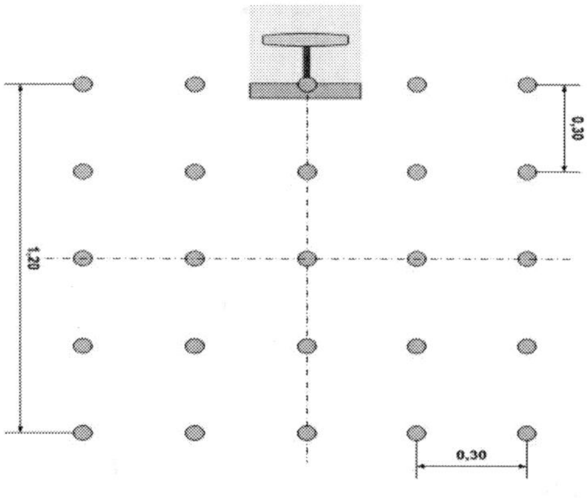

Fig. 1. A grid of measuring points for the measurement of field intensity situated out of the RFID gate.

The results have shown that the use of tapered absorbers significantly reduces the field intensity generated by the antenna out the RFID gate and thus eliminates unwanted tag readings [20].

2.2 Construction of the selective RFID gate

The research results presented above became the input to design the selective RFID gate (Fig. 2). It is the pioneering use of this device in the manufacturing industry and the second in the world [21]. In 2016 the Patent Office of the Republic of Poland has issued the patent PL 222069 B1 for the selective RFID gate [22].

Fig.2. Element of the selective RFID gate.

Actually, most of the RFID gates are based on metal constructions that can interact with radio waves used in RFID systems as information carriers. As a result, the RFID gate is not selective, meaning that it can read data from tags placed not only at its "light area" but also outside it. In addition, the typical RFID gates do not allow creating the area to be readable as a result of the characteristics of antennas used and often require the specific organization of the gate environment so that there are no undesirable markers in the range of the readings [23].

The selective RFID gate reads information from tags passing only through the gate's light that creates a space inside the device. As a result, the selective RFID gate is highly resistant to electromagnetic disturbances, allowing for densification of measuring points in production without increasing the risk of reading redundant tags. In the next part of the article its application in the communication among measuring points located on the production line is presented.

3 A communication model between two measuring points in a complex production system

This part presents two aspects. The first one concerns the ways of communication among measuring points that have been identified in industrial production systems. The results of the analysis were employed to develop a communication model between two measuring points in a complex production system.

DOI: 10.1051/itmconf/20171504009

3.1 Communication among measuring points in a production system

The industrial research concluded that in practice there are several variants of communication among measuring points in a production system.

3.1.1 Direct communication between RFID tags readers and Ethernet

One of the communication ways in a production area is the direct communication connecting RFID tag readers to Ethernet using standard LAN and WLAN ports. In consequence, it allows the configuration of communication among measuring points based on the IEEE LAN/ WLAN communication standard. Thus, it imposes a choice of communication interface based on TCP/ IP protocols. On the one hand, this solution makes it easy to implement it into the production system and on the other hand, a data collection for reporting and controlling requires additional devices. A disadvantage of this approach is the lack of certainty that the communication between the points is in the correct order according to the logic of a manufacturing process.

3.1.2 Measuring points communication based on database

The second approach is communication between measuring points and a database. The Ethernet and a built-in internal data communication protocol with RFID tags allow a reader to connect to an external data server. In this approach each of RFID readers may have an assigned IP address during a system configuration. It is also necessary to define the appropriate order of asking the measuring points about database server services. Reading RFID tag with a unique TID/EPC code by a specific reader places a resource at a specific location. Then, the information about the registration of transition of the marked resource is transferred to the database which can be saved for reporting purposes or for further processing. This solution gives good conditions for the scalability. Data collection by the database enables to monitor of all tagged assets, particularly "lost" and "abandoned" which can be identified by an operator having a portable RFID tag reader. The acquisition of data beyond RFID readers minimizes the effects of a device or software failure and the related consequences, e.g. the loss of already accumulated data.

3.1.3 Measuring points communication based on RFID tags with the limited access to Ethernet

The third approach of communication between measuring points is the use of RFID tags when the communication is limited by the access to Ethernet. The research concluded that RFID readers mostly work under Windows CE and have a flash memory. Each of measuring points equipped with the RFID reader independently reads a tag and directs it to a corresponding process on the basis of information contained in its memory. This is a secure solution but limits the ability to supervise measurement processes only to those measuring points that can be connected to Ethernet or where surveillance procedures have been defined. Although it is possible to include additional sums recorded in tags that inform about the quantity or a number of objects passing through the measuring points, however, due to the limited capacity of tags' memory, it can turn out to be inadequate. This solution makes each measuring point operate independently and requires the installation of up-to-date software that will work with all the other measuring points. This configuration requires manual and simultaneous software updating of the measuring points, which may cause the increase of system maintenance costs and require technical breaks.

The analysis leads to the conclusion that the optimal variant of communication among the measuring points is to use a database. There are new functionalities added to a communication system and additional measuring points are connected. It monitors the flow of objects and collects real time data. The communication based on the database enables a quick identification of technical problems that can appear during the work of measuring points.

3.2 The concept of a communication model between two measuring points in a complex production system

The analysis results of available communication approaches between measuring points have become the basis for creating a communication model between them based on the SQL database server (Fig.3).

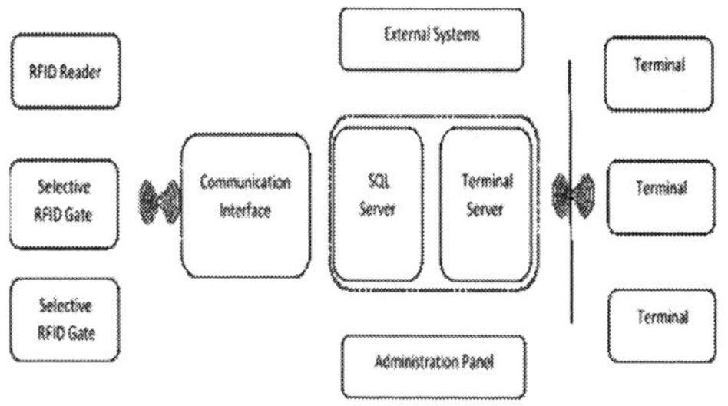

Fig. 3. The communication model between two measuring points with use of the selective RFID gate.

The model's architecture [24] requires a software installation in the measuring points linked to a database server by the dedicated service. The measuring points log to the terminal server that sends information generated from tags to the database. Simultaneously, parameters such as the time of tagging and the gate's IP are stored in the database.

In communication between two measuring points it is important to define what data will be transmitted between them. It enables the defining of measuring points behavior that is determined by the production line automation system, e.g. the objects flow with the defined

EPC code and the wrong weight may result the light or sound alarm on the line.

In a communication model between two measuring points many system components can be used, especially processes and technologies.

As far as processes are concerned, the recording of material in a production cell can be treated as an individual process or as a command generating new actions, the realization of which will be determined by a tagged asset passing through the next measuring point.

With reference to technology, a production order is when the input measuring point records objects through RFID tags before they come to a production hall and the output measuring point automatically registers and controls products leaving a production hall.

The research show that Universal Data Exchange Interface (UDEI) is the optimal standard for the communication between measuring points which is responsible for exchanging information with any external system, e.g. ERP. Using dedicated transition tables information is exchanged between the measuring points software and the system database tables. Data processing configurations are implemented in the database enabling the implementation of pre-defined measuring points actions.

To control and configure the system it has a dedicated communication interface that includes: Services, Cycle Tasks, Tasks, Templates, Document Types, Warehouses. The characteristics of configurations are available in ERP system. It has been assumed that the communication interface between two measuring points requires validation tests. The parameters of communication interface should be tested during the analysis and at the implementation stage. An important element required for the implementation of communication interface is to provide an adequate way of information presentation aggregated from the measuring points.

During the research work in development the communication model between the measuring points at the production line using the selective RFID gate was tested in conditions similar to the reality. The tests provided the following results: the correct identification of resources flowing through the RFID gate; the system has shown significantly improved resistance to electromagnetic disturbances; there are no limitations with the growing number of measuring points or distance shortening between production lines. The risk of reading an object from the next line does not exist; the system automatically records information in RFID tag; the easy integration with external IT systems and industrial automation devices; the high flexibility to adapt the system to the existing enterprise production system.

4 Scalability determinants of a communication model between measuring points in a complex production system

At the research work in development the factors that determine the scalability of the communication model between the two measurement points included: order and speed of data reading from each measurement points, data transfer between measurement points, access to database, communication between the database and measurement points, data archiving, procedures of changing the measurement point configuration in the system, including error handling and network re-configuration.

4.1 Measuring points synchronization

4.1.1 Order and speed of data reading from measuring points

The development work has shown that during a production process it is possible to define behaviour of measuring points depending on the recorded data. The order in which RFID tags are scanned is stored in system database tables, which is determined by the production process. The appearance of a non-registered element in the RFID tag at the measuring point triggers an action of industrial automation elements at the selective RFID gate. The use of triggers activates procedures for updating data in database tables when the problem occurs at a measuring point. Therefore, the chronology of RFID tags collected from the measuring points is controlled. The communication defined in this way causes the measuring points to have an active process of adding records in the transition tables. Because the RFID readers are equipped with Windows CE operating system, it is necessary to check if the process has started properly and works in all the measuring points. For this purpose, the system must register login and logout information of all the measuring points in the system database.

In a communication model between two measuring points the Universal Data Exchange Interface (UDEI) standard was used. As far as scalability of a system is concerned, it allows the setting of the time of cycle tasks and the defining of the time required for messaging between the database and each of the measuring point. In a production system this solution enables the communication speed between the measuring points to be adjusted to their configuration. It requires setting the maximum time necessary for communication between measuring points and the system database. These data should be established during the pre-implementation analysis and based on the identified production processes in the enterprise.

4.1.2 Data transmission principles between the measuring points

An indirect communication approach in which a database with indirect tables is used is also accepted in a communication model between two measuring points. In practice, this requires defining the appropriate behaviour of database triggers activating data transfer services. For this purpose, a detailed analysis of production processes creating the manufacturing system is necessary, taking into account the role of measuring

 DOI: 10.1051/itmconf/20171504009

points in the production line. Scaling of the communication system requires determining the level of significance of transmitted data among measuring points that can be divided into two groups:

- Critical data transmitted among measuring points with the highest priority and the shortest time available for the system. Such data may include, for example, events that threaten the safety of people working on the production line, resulting in unplanned downtime. For this type of data, it is assumed that the database triggers make active service that communicates with the measuring points to forward information about taking the action. Depending on the security level defined during the system implementation, it can be configured as a service whose aim is to perform tasks of the critical communication among measuring points;
- Non-critical data providing information. They can be communicated by services in cycle tasks that make work time flexible to ensure optimal company performance under the stable production conditions with the optimal network infrastructure load and database needed to ensure the critical communication.

4.2 Database management

The database management process consists of an access to database, the communication efficiency between the database and measuring points and data archiving.

4.2.1 Multiple access to the database

During communication between measuring points via database, the process is as follows: reading data from the measuring point → saving data in database tables → processing data → reading data from database tables → sending data to the measuring point. In this approach, reading and sending data from/to the measuring point is fully controlled. Problems can arise while saving, processing and reading data in database tables.

It is assumed that database tables are locked only during data saving while reading data from tables. The access to them for the other tasks is possible via the command "no lock". This significantly shortens the access time to data stored in the database and facilitates quick and repeated access to the data.

It turned out that it is a significant problem to save data in database tables. During this action, only one task "point-to-point" communication is available in the table when data is saved or updated. Database server with the appropriate parameters can significantly reduce the time of the table lock. During the research work in development, it was decided that the database server should provide a minimum 200 MB of RAM for each measuring point.

4.2.2 Communication efficiency between database and measuring points

When defining the scaling conditions of the designed system, it turned out that the communication efficiency between the databases and measuring points is time-dependent. In the basic configuration, the communication process is carried out as a single thread and tasks are queued. In order to improve the access time to the database, it is good to use multithreading. Then, the maximum benefit is in inverse proportion to the maximum number of configured threads available for the communication service. Two basic limitations have been identified in this case. First of all, when data is saved or updated the table in which the task is going on remains locked for others. As a result, the communication time between measuring points is reduced. Secondly, the use of multithreading means that tasks coming later through a communication service from a measuring point can be done in a new thread earlier. If there is no control over the order of executed tasks in every measuring point, introducing multithreading can result in the information about the lack of resources or generate extra actions by measuring points "seeking" production resources. The multithreaded service communication requires task modifications to control the communication time of preceding measuring points in the production hierarchy more accurately.

4.2.3 Data archiving

During the research work in development the following mechanisms to control the historical data collection have been implemented:

- Historical tables where data is transferred after a production process is finished. Data is not used by the database anymore for communication among the measuring points but can be used for accounting or reporting;
- Data archiving - the time after which the collected data is not used by users and should be exported to text files should be defined. Such a solution affects the security level of the system.

In communication among multiple measuring points the order of data flow is an important issue for the correct data exchange. Preceding measuring points in the production cycle should have access to the database in order to update data needed for the next points. It means saving all historical data collected by the measuring points identified as critical in the production process with the time of their reading at the measuring point in the database.

4.3 Procedures of configuration changes in the measuring points network

The next issue related to the scalability of communication among multiple measurement points, was failure handling and network reconfiguration. It turned out that as the number of measuring points increased, the configuration and communication complexity of the measurement and communication system exponentially increased. This can cause a significant increase of time necessary to remove any system failures or to plan longer and longer technical breaks to reconfigure measuring points. Thanks to Ethernet in the interface design, each measuring point

 DOI: 10.1051/itmconf/20171504009

has an assigned IP address, uniquely defining the location of the point in the production process. Assigning a unique MAC address for each measuring point makes data collection from the measuring equipment installed in it possible. This solution changes topology and reconfigures the measuring network by moving measuring points, if necessary.

4.3.1 Handling of failure in the measuring points

In order to handle failures in the communication between the measuring points correctly, the following parameters are required:

• Time to establish communication between measuring point and database, and

• Quantity and quality of data.

If during the communication between a measuring point and database any of the defined parameters in the system is not sent or is not in the right format, one of the following messages will be generated: critical (missing data required to continue a production process), or warning (information about the lack of data not directly affecting a production process).

Based on the defined failure information, the database trigger activates a failure procedure. It may contain different scenarios, such as devices restarting in a measuring point and attempting to re-establish communication, restarting the communication task, starting emergency procedures in other points, stopping the production process.

The research work in development concluded that in case of failure, the communication system gives an opportunity for analysis using historical data and data from measuring points located logically near to the failure point. It makes a remote pre-analysis of the event possible

4.3.2 Network reconfiguration

In the research work in development on the scalability of multiple measuring points, it was also required to define circumstances of system development and its modification. It has been pointed out that the measuring points system may be changed in three ways.

4.3.2.1 Changes of measuring points topography

The communication system is designed to independently identify measuring points, thus topography changes may refer to repositioning of the measuring points within the same topology. Each modification requires changing the corresponding IP or MAC address in the database and in the communication and production processes defined in the database. These works do not take much time because they rely on activating the script in the system database. The problem that can occur during a configuration change may be due to the limited ability to validate changes before they start.

4.3.2.2 Measuring points system development

The system topology development is accomplished by adding new IP/MAC address to the system database. They will remain inactive until it has configured the procedures and operations that will be used. The database development allows a new production system based on new measuring points to be tested.

4.3.2.3 Reduction of network of measuring points

The research work in development has shown that in order to reduce the number of measuring points the corresponding IP/ MAC address from the database must be removed and all processes and procedures using the deleted points are modified. The network reduction should be preceded by a detailed analysis of the network topology of measuring points in order to identify early production lines that could lead to unfinished final products. The problem that may occur during the network redundancy will be due to the limited ability to validate changes before they begin, which is similar to changing the topography of measuring points.

5 Conclusion

The input for the research work in development presented in the paper was the output of the previous research stages focusing on the design of selective RFID gate and then a communication model between two measuring points in a complex production system.

The main factors that determine the scalability of the communication between the measurement points network are as follows: order and speed of data reading from each measurement points, data transfer between measurement points, access to database and communication between the database and measurement points, procedures of changing the measurement point configuration in the system, including error handling and network re-configuration.

In case of order and speed of data reading from measurement points, the research shows that it is required to check at first if the process of reading RFID tags has started correctly and works in each of the points. Therefore, the system must register login and logout of all the measuring points in the database. The communication requires all the measuring points to have an active process of adding records in the transition tables. In addition, one should set the maximum time required for communication between measuring points and the system database.

The significance of transmitted data, critical or non-critical, between measuring points needs to be determined to make scaling of the communication system successful. It depends on the structure of manufacturing processes in a production system.

The next aspect is the access to database. The preceding measuring point in the production cycle should have access to the database in order to update data needed for the next point. The server should provide a minimum 200 MB of RAM for each measuring point.

 DOI: 10.1051/itmconf/20171504009

In order to improve the access time to the database, it would be good to use multithreading, even though it has some limitations. One of them is that tasks coming later through a communication service from a measuring point can be done in a new thread earlier. As a result, it can provide information about the lack of resources or generating additional actions by measuring points "seeking" production resources. It means that the multithreaded service requires task modifications to control the communication time of previous measuring points in the production hierarchy more accurately.

The research shows that as the number of measuring points grows up, the configuration and communication complexity of the system exponentially increases, and the probability of system errors is higher. Thus, the system is able to make analysis using historical data and data from measuring points located logically near to the failure point. It makes a remote pre-analysis of the event possible.

Any modifications of the measuring points network are related with the system topology. Adding or deleting measuring points requires changing the corresponding IP or MAC address in the database and in the communication and production processes defined in the database. This should be preceded by a detailed analysis of the network topology of measuring points in order to avoid serious problems at production lines.

Summing up, this work introduces the solution based on the theoretical research and some practical ideas that were designed by DataConsult in the collaboration with the Institute of Logistics and Warehouse. The research owes its success to the selective RFID gate that was patented in 2016. The validation of measuring points network with use of the selective RFID gate in a few production systems confirmed its effectiveness in real manufacturing conditions. Therefore, the results of the research work in development presented in the article can be useful for industry practitioners, IT experts, and researchers who might need to work together to build a smarter solution in the future.

References

1. D.C.Wimalachandra, B. Frank, T. Enkawa, Int J Ind Eng-Theory, **21**, 348-359 (2014)
2. D.A.Perry, B. Olson, P. Blessner, T.D. Blackburn, Systems Eng, **19**, 133-145 (2016)
3. C.K.H Lee, K.L. Choy, K.M.Y.Law, G.T.S Ho, J Manuf Syst, **3**, 412-422 (2014)
4. M. Ghobakhloo, S.H. Tang, S.H., Inform Technol Dev, **21**, 573-600 (2015)
5. J.Smart, A. Calinescu, L.H. Huatuco, Int J Prod Res, **51**, 362-379 (2013)
6. J. Santos, M.P.Garcia, M. Arcelus, E. Viles, J. Uranga, Int J Comput Integ M, **24**, 338-351 (2011)
7. C.Z. Zulkifli, H.N. Hassan, W. Ismail,S.N. Semunab, Acta Phys Pol A, **128**, B86-B89 (2015)
8. E. Dovere, S. Cavalieri, S., S.Ierace, Int J Prod Res, **55**, 1336-1349 (2017)
9. Z.X.Yang, W. Xu, P.K. Wong, X.B.Wang, Math Probl Eng, **575402** (2015)
10. M.Liukkonen, Int J Comput Integ M, **28**, 861-880 (2015)
11. K.M.Chen, J.C.Chen, R.A. Cox, Assembly Autom, **32**, 185-196 (2012)
12. H.H. Cheung, S.H. Choi, Comput Ind, **62**, 708-718 (2011)
13. N.Fescioglu-Unver, S.H. Choi, D. Sheen, S. Kumara, Inform Syst Front, **17**, 1369-1380 (2016)
14. H.A.Ringsberg, V. Mirzabeiki, Brit Food J, **116**, 104-124 (2014)
15. M. Keskilammi, L. Sydanheimo, M. Kivikoski, Int J Adv Manuf Tech, **21**, 769-774 (2003)
16. G.Huang, J.H. Li, X.D. Yuan, L.Gao, Y.Q. Rao, Int J Comput Integ M, **25**, 66-85 (2012)
17. E.W.T. Ngai, D.C.K. Chau, J.K.L. Poon, A.Y.M. Chan, J Eng Tech M, **29**, 112-130 (2012)
18. P.J.Zelbst, K.W.Green, V.E.Sower, P.M. Reyes, Int. J. Oper. & Prod. Manage., **32**, 329-350 (2012)
19. J.R. Hylsberg, Q. Zhang, T.S. Toftegaard, C. Turcu, (ed.)., InTech **18**, 319-334 (2011)
20. Maćkowiak, K. Sieczkarek, M. Logistyka, **4**, 4633-4639 (2014)
21. M. Łobaziewicz, A. Maćkowiak, K. Sieczkarek, Logistyka, **3**, 39-42 (2014)
22. Patent PL 222069 B1, Polish Patent Office, (2016)
23. A.Maćkowiak, K. Sieczkarek, Logistyka, **1**, 41 (2013)
24. DataConsult Sp. z o.o., Communication model between two measuring points in a production system, Research report, 1-16 (2014)

ITM Web of Conferences **15**, 04010 (2017) DOI: 10.1051/itmconf/20171504010
CMES'17

Selected aspects of the technological processes of the aircraft engine washing

Anna Rudawska[1], *Przemysław* Wrona[1], *Izabela* Miturska[1,*], and *Miroslav* Müller[2]

[1]Lublin University of Technology, Faculty of Mechanical Engineering, Department of Production Engineering, Nadbystrzycka 36, 20-618 Lublin, Poland

[2]Czech University of Life Sciences Prague, Faculty of Engineering, Agricultural Engineering, Kamýcká 129, 165 21 Prague, Czech Republic

Abstract. In production, treatment or operation of selected components it is not possible to avoid dirtiness. Given the fact that impurities significantly affect the quality, functionality and service life of a part, the surface preparation of such an element is a key aspect in the regeneration process. The choice of contaminant removal method will depend on the type of impurity. This operation is often treated as a separate technological process, during which the type and properties of the removed dirt are determined, along with assessing the degree of purity required to ensure the correctness of subsequent operation or operations. Rising expectations regarding specific effects of purification have led to massive improvement of traditional methods, and development of new processes that require the use of complex technologies. The present research was to prepare the technological process of washing selected parts of the aircraft turbine engine. The preliminary assumption was to use the LUA-1000 automatic washing processing line, which selecting and modifying particular parameters of the process in order to comply with the technological assumptions. The elaborated process includes the following stages: preparing the selected parts for washing, washing the engine parts and the quality control of washing. The primary objective of the study was to achieve the most beneficial parameters possible and, at the same time, to minimise the costs of the machine's work, as well as to preserve high quality and to comply with environmental standards.

1 Introduction

It is not possible to prevent the parts from becoming contaminated during the process of production and treatment. Due to the fact that the impurities affect a given part's quality, functionality and life, it is of high importance to prepare its surface during such processes as regeneration, bonding or protective and decorative coating [1-8]. Cleaning machines and devices is aimed at removing various types of impurities affecting the quality of the assembled products, both inside and outside. Due to some differences, it may be divided into four groups:

- in order to gain good-quality inseparable connections and joints, *e.g.* soldered connections and glue joints, welded joints made with the use of various methods, as well as in case of electric and frictional welding, it is necessary to clean the surface from the external impurities and the oxide film:
- high-precision components of control systems, such as mechanical, hydraulic or pneumatic, require proper cleaning of chemical substances and particles that adhere to their surface,
- less important system parts require cleaning the residues of the foreign matter on the surface because of the production process and that could lead to the corrosion process or accelerate the part's wear (the residues of surfactants, detergent solutions or cooling and lubricating liquids),
- the regeneration and renovation of certain parts may require the removal of damaged layers and coatings, *e.g.* decorative, protective and wear-resistant ones, as well as of the exploitation deposit in order to restore the original state [9, 10].

The selection of the method of removal depends on the properties of impurities, and the removal is often recognised as a separate technological process. During this process it is necessary to determine the kind and properties of the impurities that are to be removed, as well as the degree of purity that is required for the next operations or further exploitation to be correct. The method and process of cleaning, as well as the composition of detergent solutions and the method of quality control have to be determined. A proper choice and setting-up of various devices also plays a vital role in the process [11-17].

The following criteria have been shown to be helpful in the rational choice of the surface cleaning method:

- safety criterion,
- technical criterion,
- economical criterion,

* Corresponding author: i.miturska@pollub.pl

- maintainability criterion,
- ecological criterion [18, 19].

Massive improvement of traditional methods, as well as development of the new ones that require complex technologies has resulted from the need for specific effects of cleaning [20-24]. As the renovation of aircraft engines requires precision and accuracy, all the parts have to be cleaned properly. In order to meet these requirements, the newest cleaning methods are used. Their most vital advantages include optimal processes that generate low costs, as well as high quality of the cleaned parts. Moreover, verification as well as further stages of engine renovation may be carried out immediately after cleaning. Nowadays, automatic washing process lines are used in aircraft engine renovation process. Their work is always based on the technological process elaborated before. The process is adjusted to the purification method used by the machine or device, degree and kind of the impurities, type of the cleaned parts and the material they are made of.

The aim of this article was to elaborate a technological process of washing selected parts of the aircraft turbine engine with the use of the automatic washing process line. The process line in question belongs to the machine park of an aviation company operating in the Lubelskie region.

2 Research Assumptions

The article assumed the elaboration of the technological process of washing selected parts of the aircraft turbine engine in order to increase their durability and reliability. To this end, an automatic washing process line, which is specifically suitable in engine renovation process applications, was used. As the line works constantly, it is necessary to modify the process parameters in order to comply with the technological assumptions. Hence, substantial work has been undertaken in order to elaborate a detailed technological process.

2.1. Specification of the Used Parts

An aircraft turbine engine consists of the following basic parts: induction system, compressor, combustion chamber, exhaust system and co-working parts: gearbox, engine and airframe supplying systems (oil system, fuel system, air system). Due to the function of particular components, loads and working temperature, appropriate materials have to be used. The "cold" engine parts (induction system, gearbox and supplying systems) are made of materials of high specific strength (strength to density ratio), corrosion-resistance and mechanical properties' uniformity and stability. As far as the "hot" engine parts are concerned (combustion chamber, compressor and exhaust system), they have to comply with not only the aforementioned characteristic, but also have to be heat-resistant and show creep resistance in high-temperatures. Basic types of materials used for the modern aircraft turbine engines designs are: titanium alloys, steels and nickel-based superalloys. Light aluminium or magnesium alloys are used only in certain parts, e.g. fuselages, the first stages of compressors, gears or oil and fuel pumps. In addition, the composite materials are more and more often used in modern constructions.

An aircraft turbine engine scheme, as well as the alloys used for production of specific subassemblies are shown in Figure 1.

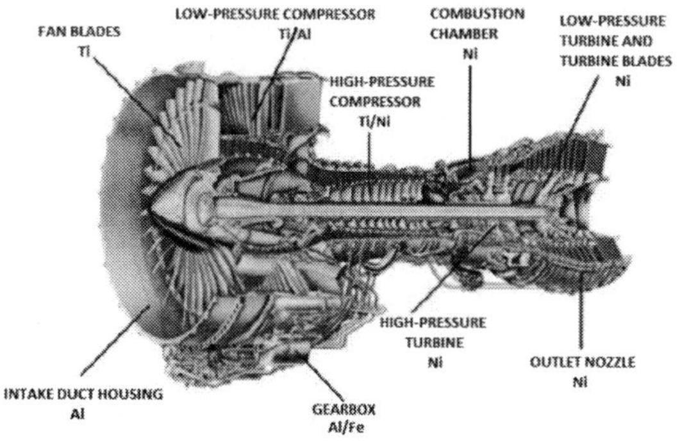

Fig. 1. An aircraft turbine engine's scheme and the alloys used for production of specific subassemblies [25].

The aforementioned factors contribute to the formation of various impurities. Depending on the material type and the degree of contamination, the engine parts designated for washing are divided into: steel alloys (slightly soiled steel, soiled steel, excessively soiled steel), titanium alloys (slightly soiled titanium, soiled titanium, and excessively soiled titanium), aluminium alloys and others (e.g. magnesium and copper alloys). Proper cleaning of aircraft engine parts during the renovation process leaves room for appropriate verification of the degree of the engine wear. It is of high importance to be careful so as not to damage the verified surfaces and not to break the original material of the washed part. Hence, it is necessary to qualify the soiled part properly so as not to cause its etching as a result of excessive washing. During the verification or regeneration processes, the renovated part may be soiled (e.g. with liquid or developer after having performed the liquid-penetration method, or with treatment liquid). In such cases, so called intermediate washing is being performed.

2.2 Cleaning Process

The cleaning process is fully automated, and employs two washing methods:
- ultrasonic cleaning-immersion,
- immersion.

Depending on the material of a given part is, as well as on the degree of contamination, four exemplary detergents by Chemtall company were used: Ardrox 6333A, Ardox A/185L, Ardox A/188, Ardox A/1873. The characteristics of all the detergents are shown in Table 1.

 DOI: 10.1051/itmconf/20171504010

Table 1. Characteristics of detergents used for washing [own study based on 26].

	Detergent			
	Ardrox 6333A	Ardox A/185L	Ardox A/188	Ardox A/1873
Reaction	alkaline	highly alkaline	highly alkaline	subacid
Washing Temperature	60 – 82 °C	82 – 88 °C	82 – 93 °C	80 – 85 °C
Concentration (volume)	5 – 30%	15 – 60%	10 – 25%	20 – 25%
Washing Method	ultrasonic cleaning-immersion	immersion	immersion	immersion
Type of the Removed Impurities	fats, oil films, lubricating grease, carbon deposit exhaust gases residues	fats, oil films, lubricating grease, carbon deposit enamel, corrosion	carbon deposit, scale formed in high temp.	carbon deposit, scale formed in high temperature
Type of Washed Parts	steel, titanium, aluminium, zinc, copper and magnesium alloys	steel and titanium alloys	heat-resisting and high-temp. creep resisting steel alloys	heat-resisting and high-temperature creep resisting steel alloys

After washing the parts with the use of the aforementioned detergents, they have to be rinsed in water in order to remove the chemical residues. Afterwards, prior to drying, the final rinse in demineralised water has to be performed: it will help to avoid streaks and seepage formation on the surface. In addition, the washed surface has the unloaded electrostatic charge, to which it owes higher soil-resistance.

2.3. Machines and Devices Used in the Washing Process

The cleaning process used in the research was fully automated and performed with the use of the LUA-1000 automatic washing process line produced by the ULTRON company. The technical specifications of the machine are shown in Table 2.

Table 2. Technical specification of the washer [own study].

Characteristic	Dimension	
Washer's external dimensions	Length	14200 mm
	Width	2150 mm
	Height	3150 mm
Basket's internal dimensions	Length	600 mm
	Width	600 mm
	Height	600 mm
Basket's maximum load	80 kg	
Manipulator's maximum load	130 kg	
Number of tanks (W)	11 + dryer	
Tanks' usable dimensions W1-W11	Length	850 mm
	Width	800 mm
	Depth	700 mm
Usable capacity of the particular tanks W1-W11	≈600 l	
Used materials	Steel 1.4301 Steel 1.4401 (tank W9)	
Power of heaters in the particular tanks	9 x 18 kW	
Power of drying heaters	18 kW	
Power of ultrasounds	2 x 3 kW	
Total installed power	189 kW	
Power supply	3x400 V/50 Hz + N + PE	
Water supply	1.5 – 8 bar	
Compressed-air supply	5 – 10 bar	
Washing and rinsing temperature range	10 – 90 °C	
Drying temperature range	20 – 110 °C	

The washer is fully automated and controlled by a PLC. The elements that are to be washed are transported in the placed on a trolley to the load area. W1-W11 tanks have to be filled with water, detergents and heated to the appropriate temperature. Depending on the types of washed parts, washing method and stages, different types of washing agents are used in each tank. The washing process with the use of an automated process line includes: preparation of the selected parts for washing, washing the engine parts and the quality control of the performed washing. An exemplary scheme of the washing process with the use of an automatic process line includes the following stages: inductive sensors in the load area detect the basket, after choosing the particular program and approving it by the operator, the basket is lifted by the manipulator and transported to the first tank in the process. After washing is done, the manipulator transports the basket to the next tank where the technological process is continued. Two tanks for washing are equipped with ultrasounds (W2 and W6). Tanks for preliminary and main washing (W1, W2, W6 and W8) are equipped with the off-line filtration system and the surface rinsing system. This system helps to remove grease from the liquid surface by enabling to take out the basket without letting the dissolved grease deposit on the washed part's surface. Every tank, apart from W2 and W6 (ultrasonic cleaning) is equipped with the liquid mixing system. In the case of the process tanks, the mixing pump starts to work 1 minute before the basket is put into the tank and continues to work until the basket is taken out of the tank. Thanks to this the bath is preliminarily mixed which improves washing efficiency. When it comes to the rinsing tanks, the mixing starts when the basket is lowered to the tank with water and continues until the basket is taken out of it. The whole process is fluent, and all the parts are transported from one tank to another subsequently. The final rinsing is performed in the tank with demineralised water, which is also equipped with the off-line filtration system, as well as the demineralisator that works in a closed system. Before the process is finished, the basked is transported to the dryer (W12), where the

DOI: 10.1051/itmconf/20171504010

washed elements are dried. Then, the manipulator transports the basket to the unload area and puts it on the trolley, which constitutes the final operation, following which the whole process is finished.

Due to certain technological reasons, the machine is equipped with two manipulators moving along the guides in two directions: along the washer and upwards-downwards. Each manipulator is driven by two engines with the use of motoreducers and a gear wheel – a toothed bar (lengthwise direction) and a screw drive (vertical direction). The maximum carrying power of the manipulators is determined in the technical specification. The baskets are made of wire net with large mesh, which enables removing the impurities from the washed parts. Water for tanks is taken from the water pipe network with the use of electro valves that control the filling of every tank separately. The water or detergent level is controlled by the level induction sensors installed at every tank, and is displayed on the PLC. The tanks are emptied manually with the use of ball valves. The tanks may be emptied separately or simultaneously, if necessary. The liquid may be removed by gravity or with the use of a pump. Every tank is equipped with a slideable cover that is closed when working. It prevents heat loss and liquid evaporation, as well as provides acoustic isolation. The covers are moved by pneumatic actuators.

2.4 Technological Assumptions

While working on the washing process, one has to aim for the most efficient parameters, the lowest costs connected with the machine's work (necessary utilities and time) and the highest quality possible. The main assumptions of the optimisation include:

- maximum use of the basket load space - taking into consideration the weight (80 kg) and volume (600 x 600 x 600 mm) limitations, the washing process should be performed with the lowest load possible. On the other hand, the washed elements should be arranged in such way as to enable the washing liquid and ultrasounds reach the whole washed surface. To this end, special baskets with barriers (Fig. 2) are used - they make the washing process efficient and safe.

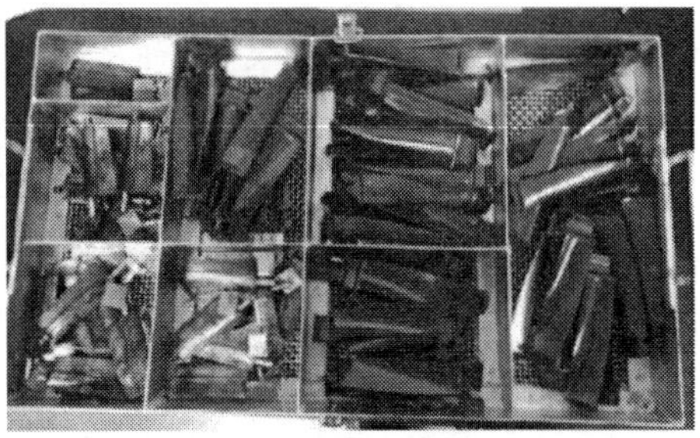

Fig. 2. The arrangement of the fan blades of the aircraft turbine engine in the basket before the washing process.

- temperature and bath time reduction - heating water and maintaining desired temperature in every tank with an average capacity of 600 l requires considerable resources of electric power. Temperature reduction by every single degree enables saving the electric power. However, bath time reduction leads to the increase in the number of charges during one change, which enables faster wear of the equipment.
- increase in the concentration - it reduces the bath time and temperature. However, it has to be underlined that too high a concentration may cause etching of the washed parts, and exceeding the specified value may not improve the washing quality.
- reduction of the bath change frequency - the washing liquid is being soiled during the washing process, which reduces its efficacy and affects the washing quality. As a result, the bath temperature has to be risen, the time of bath is prolonged, and the baths have to be changed more often. The cost of bath change (sewage treatment, removal of the current chemicals and replacing them with new ones) may amount even up to a few thousand zlotys. A limited number of bath changes may be reached through rinsing in water every time after the basket is taken out of the process tank. It prevents the chemicals from mixing thanks to the use of the off-line filtration system in every tank containing detergents.

3 Elaborated Technological Process

The technological process of washing selected parts of the aircraft turbine engine is one of the key stages in its renovation process. The use of the LUA-1000 automatic washing process line ensures the highest economic and environmental standards. The elaborated technological process presents the flow of washing of the most soiled steel elements of the aircraft engine, which is currently being renovated by the aviation company. It is the longest process during which the elements are washed in four tanks, rinsed in five tanks and then dried. The engine parts washed in the process include, inter alia, the parts of turbine - discs, discs protectors, blade-ring rotor blades and nozzles. The washing process consists of three stages: preparation of the selected parts for washing, washing the engine parts and the quality control of the performed washing. The flowchart washing process is shown in Figure 3.

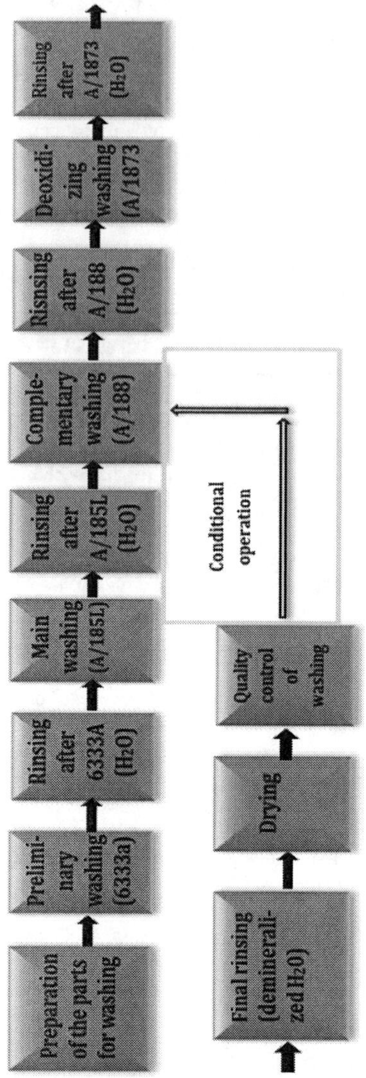

Fig. 3. Flowchart of excessively contaminated parts' washing.

The process also assumes performing a conditional operation, as its repeatability in terms of the detergents' effectiveness is high but the degree of contamination of renovated engines may differ depending on:
- the amount of overhaul life between renovations (the engine could still have some margin of hours, yet was sent for renovation due to the number of cycles),
- the environment where the aircraft was used,
- the technical culture standards according to which the engine was used,
- the reasons for being sent for renovation (*e.g.* a foreign body sucked into the engine).
The elaborated process was optimised according to the aforementioned technical assumptions. As a result, the highest quality and the lowest costs possible were achieved.

Table 3 shows the parameters for subsequent stages of the technological process.

Table 3. The parameters for particular stages of the technological process of the selected parts of the aircraft engine washing.

Stage	**Content of the operation/Process parameters**	
Preparation of the selected parts for washing	1. Put the basket into the trolley. 2. Arrange the parts in the basket so as to minimise the amount of liquid on the surface after the washing process. 3. Attach the parts to the basket using a stainless cord so as to prevent them from moving during washing. 4. Place the trolley with the basket on the loading station. 5. Select: "excessively soiled steel" process. 6. Turn the process on.	
Preliminary washing.	washing method	ultrasounds
	washing liquid	Ardrox 6333A
	concentration	15%
	Temperature T	65 °C
	bath time t_k	20 minutes
	draining time t_o	1 minute
	manipulator moving time t_m	1 minute
Rinsing after the preliminary washing	washing method	immersion
	washing liquid	H_2O
	concentration	-
	Temperature T	70 °C
	bath time t_k	1 second
	draining time t_o	1 minute
	manipulator moving time t_m	1 minute
Main washing	washing method	immersion
	washing liquid	Ardox A/185L
	concentration	50%
	Temperature T	90 °C
	bath time t_k	25 minutes
	draining time t_o	1 minute
	manipulator moving time t_m	1 minute
Rinsing after the main washing	washing method	immersion
	washing liquid	H_2O
	concentration	-
	Temperature T	70 °C
	bath time t_k	1 second
	draining time t_o	1 minute
	manipulator moving time t_m	1 minute
Complementary washing	washing method	immersion
	washing liquid	Ardox A/188

 DOI: 10.1051/itmconf/20171504010

	concentration	15%
	Temperature T	92 °C
	bath time t_k	10 minutes
	draining time t_o	1 minute
	manipulator moving time t_m	1 minute
Rinsing after the complementary washing	washing method	immersion
	washing liquid	H_2O
	concentration	-
	Ambient	temperature T
	bath time t_k	1 second
	draining time t_o	1 minute
	manipulator moving time t_m	1 minute
Deoxidizing washing	washing method	immersion
	washing liquid	Ardox A/1873
	concentration	25%
	Temperature T	85 °C
	bath time t_k	20 minutes
	draining time t_o	1 minute
	manipulator moving time t_m	1 minute
Rinsing after the deoxidizing washing	washing method	immersion
	washing liquid	H_2O
	concentration	-
	Ambient	temperature T
	bath time t_k	1 second
	draining time t_o	1 minute
	manipulator moving time t_m	1 minute
Final rinsing	washing method	immersion
	washing liquid	demineralised water
	concentration	-
	Ambient	temperature T
	bath time t_k	1 second
	draining time t_o	1 minute
	manipulator moving time t_m	1 minute
Drying	drying method	chamber drying
	washing agent	air
	concentration	-
	Temperature T	85 °C
	drying time t_s	4 minutes
	draining time t_o	-
	manipulator moving time t_m	1 minute
Grease residues level control	Apply the test liquid (45 mJ/m^2) on the surface. Check if the test liquid coat is even.	
Solid matter residues level control	1. Select a random washed element. 2. Rinse the element thoroughly with pure solvent. 3. Filter the solvent with the use of a test membrane.	
	4. Measure the membrane weight - the acceptable limit of the impurities' weight equals 100 mg per 1 kg of the controlled element If the acceptable limit of the impurities' weight is exceed, the conditional operation with the same process parameters should be undertaken. 5. Measure the amount and the size of the impurities - the acceptable amount of impurities equals 25 per 1 kg of the controlled element, provided that the biggest element is less than 0,015 mm. If the acceptable limit of the amount and the size of the impurities is exceeded, the conditional operation with the same process parameters should be undertaken.	

Figures 4 and 5 show the condition of the intake duct housing's internal surface before and after the washing process with the use of the automatic washing process line.

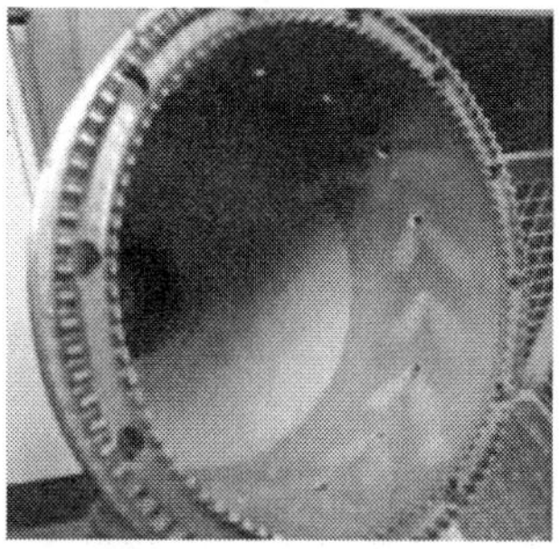

Fig. 4. The condition of the intake duct housing's internal surface before the washing process with the use of the automatic washing process line.

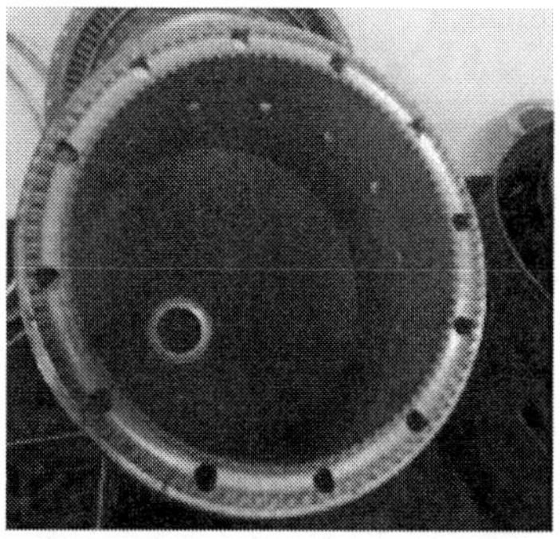

Fig. 5. The condition of the intake duct housing's internal surface after the washing process with the use of the automatic washing process line.

Based on the aforementioned figures, it may be stated that the elaborated technological process shows high effectiveness. However, it was only assessed visually.

 DOI: 10.1051/itmconf/20171504010

A more accurate result of the process is possible to acquire only after verification with the use of test membranes.

4 Summary and Conclusions

The following conclusions were drawn based on the presented analysis:
- various types of contaminants that deposit on the surface of elements during production and treatment process affect their quality, functionality and life;
- the components' surface cleaning methods during the renovation process play a vital role in restoring their usability and enabling for further technological process' stages;
- the already elaborated and used surface cleaning methods let their optimal use in specific cases, taking into consideration the assumed criteria in terms of reasonable costs and quality;
- the necessity of gaining the best effects possible has led to the improvement of traditional methods, as well as to the development of the new ones, which require the use of modern and advanced technologies;
- automation of the washing process has ensured the process' repeatability, high quality and environment protection requirements, as well as safety and hygiene at work;
- the degree of purity of the cleaned elements may be assessed both visually and with the use of various tools. The choice of a proper cleanliness degree assessment method depends on the type of element, designation and the cleaning method that was used;
- it is of high importance to optimise the technological process of washing when elaborating it, as it ensures obtaining the most efficient parameters, the highest quality and the lowest costs of the machine's work possible;
- the type of the washing process line, as well as washing methods, detergents and the order of subsequent stages are determined by the type of the washed elements, material they are made of and the degree of contamination.

The elaborated process of washing with the use of the automated process line improves the aircraft engine renovation process. The possibility of changing the parameters makes it easily adjustable to the constant technological development connected with using more and more modern construction materials and economical and environmental requirements.

References

1. R.D. Adams, Adhesive bonding. Science, technology and applications (Woodhead Publishing Limited, UK, 2010)
2. E.G. Babraj, D. Starikov, J. Evans, G.A. Shafeev, A. Bensaoula, Int J Adhes Adhes. **27**, 268–276 (2007)
3. G.W. Critchlow, D.M., Int J Adhes Adhes. **16**, 255–275 (1996)
4. G.W. Critchlow, K.A. Yendall, D. Bahrani, A. Quinn, F. Andrews, Int J Adhes Adhes. **26**, 419–453 (2006)
5. S. Legutko, Podstawy eksploatacji maszyn I urządzeń [Basics of Machines' and Devices' Exploitation], (Wydawnictwa Szkolne i Pedagogiczne, Warsaw, 2004)
6. W. Polini, L. Sorrentino, Int J Adhes Adhes. **27**, 1–8 (2007)
7. G. Rozovskis, J. Vinkevičius, J Adhes Sci Technol. **10**, 399–406 (1996)
8. Rudawska, Jo J Adhes Sci Technol. **26**, 1233–1247 (2012)
9. J. Andziak, Mechanik. 25-27 (1975)
10. T. Burakowski, Marczak J., Napadłek W., PIEl, Notebook no **228**, 126-132 (2006)
11. N. Brack, A.N. Rider, Int J Adhes Adhes. **44**, 20–27 (2014)
12. L. Domingues, J. C. S. Fernandes, M. Da Cunha Belo, M. G. S. Ferreira, L. Guerra-Rosa, Corros Sci. **45**, 149–160 (2003)
13. Jędrusik, J. Łunarski, TiAM. 1, 52-55 (2013)
14. B.B. Johnsen, F. Lapique, A. Bjørgum, Int J Adhes Adhes. **24**, 153–161 (2004)
15. O. Lunder, F. Lapique, B. Johnsen, K. Nisancioglu, Int J Adhes Adhes. **24**, 107–117 (2004)
16. S.G. Prolongo, A. Ureña, Int J Adhes Adhes. 29, 23–31 (2009)
17. Rudawska, J. Kuczmaszewski, Mater Sci-Poland. 24, 975–981 (2006)
18. T. Burakowski, T. Wierzchoń, Inżynieria powierzchni metali [Metal Surface Engineering], (WNT, Warsaw, 1995)
19. Koss, J. Marczak, Czas. Tech. 199-204 (2009)
20. W. Brockmann, P.L. Geiß, J. Klingen, B. Schröder, Adhesive bonding. Materials, Applications and Technology (Weinheim, Germany, Wiley-Vch Press, 2009)
21. R.E. Litchfield, G.W. Critchlow, S. Wilson, Int J Adhes Adhes. **26**, 295–303 (2006)
22. Rudawska, Int J Adhes Adhes. **50**, 235–243 (2014)
23. J.M. Sykes, Surface treatment for steel, in: Brewis DM. (Eds.), Surface Analysis and Pretreatment of Plastics and Metals (Applied Science Publishers, London, 1982)
24. P.R. Underhill, A.N. Rider, D.L. DuQuesnay, Int J Adhes Adhes. **26**, 199–205 (2006)
25. J. Konieczny, Armia. 4 **56**, 70 (2013)
26. https://www.hotwash.com.au/wp-content/uploads/2014/07/15-Product-Profile

ITM Web of Conferences 15, 04011 (2017) DOI: 10.1051/itmconf/20171504011
CMES'17

Supporting the climate of innovation in the SME sector - an application for SMEs

Sabina Motyka[1,*], *Jolanta* Słoniec[2], *Anna* Kaczorowska[3] and *Elżbieta* Łaciak[4]

[1]Cracow University of Technology, Faculty of Mechanical Engineering, Department of Manufacturing Processes, M6 Institute, Jana Pawła II 37 Avenue, 31-864 Cracow, Poland
[2]Lublin University of Technology, Faculty of Management, Department of Enterprise Organization, Nadbystrzycka 38 Str., 20-618 Lublin, Poland
[3]University of Lodz, Faculty of Management, Department of Computer Science, Matejki 22/26 Str., 90-237 Lodz, Poland
[4]Ericsson sp. z o.o., Czerwone Maki 87 A Str., 30-392 Cracow, Poland

Abstract. The ability of contemporary organizations to create innovative solutions in the areas of products, processes, organization and marketing requires proper technical and managerial skills and the skill of continuous learning. The pursuit and efforts of companies to upgrade their competitive position through innovation requires active involvement and improvement from all employees of an organization. Large organizations and international corporations usually have significant resources for developing and supporting management processes. SMEs are struggling with a range of economic, legal and psychosocial barriers, and difficulties in accessing innovative financing sources. Innovation management in smaller businesses, supported by recent data and analytical tools, has a potential for more effective and faster implementation. The aim of this article is to show the determinants and proposals of actions (organizational innovations) supporting the creation of a specific climate and culture of innovation in enterprises from the SME sector. In addition, the article presents an original solution in the form of a developed web application supporting innovation management in an enterprise. The presented utility tool is a proposal for organizational innovation dedicated to companies from the SME sector.

1 Introduction

Product innovation, process innovation, organizational innovation, and marketing innovation are fundamental factors in a company's position on the market. What is crucial is continuous development and improvement on the part of the owners, managers of businesses and the entire staff [1].

Proper business management fostering creation and development of innovative products requires continuous research and analysis of implemented innovative solutions [2]. Supporting the innovation management process with analytical tools aids correct and efficient implementation of innovation.

The main aim of the article is to present the current determinants of innovative activities carried out by enterprises from the SME sector, and describe the developed application, the task of which will be to build the innovation culture in these enterprises. With the tool for analyzing, evaluating and reporting activities to support the creation of an innovative organization, companies in the sector will be able to improve communication in the area of innovative activities, analyze the internal and external environment, including factors which generate and impede innovation, report ideas for improvement and formulate innovation projects in the area of products, processes, organization and marketing. The application is a proposal of an IT solution to be used in an enterprise by the management staff and in selected modules also by all employees.

2 Determinants of an innovative enterprise

In recent years, the subject of innovation management in organizations has been quite often taken in the literature [3-13]. The main researchers in this field of knowledge in Poland are [14-17]. However, studies of the determinants of innovation activities in the SME sector and studies showing support tools used in this respect are not very popular.

Awareness and understanding of actions taken in the field of building innovation and creating an environment which contributes to the development of new ideas are nowadays of great importance for the development of organizations [18].

The development of enterprises is mainly a qualitative phenomenon, which involves implementation of different changes and new solutions, including innovation of products, processes and technologies, as well as organization and management [19].

An innovative enterprise is an intelligent organization that constantly generates and implements innovation and is recognized by the consumer for its high level of novelty and competitiveness. The most

* Corresponding author: motyka@mech.pk.edu.pl

 DOI: 10.1051/itmconf/20171504011

commonly considered innovative features are the following [20]:
- wide range of R&D works,
- systemiatic implementation of new scientific and technical solutions,
- high share of innovation in technology, product and organizational of innovations,
- current introduction of innovations to market,
- the ability to anticipate the future,
- flexibility of action,
- creative teams of workers,
- the ability to use the innovation potential,
- continuous communiation with clients,
- continuous generation of innovations.

Another approach to defining an innovative organization and its components is the proposal of seven components listed in Table 1.

Table 1. Elements of an innovative enterprise [5].

No.	Element
1	Shared vision, leadership, the will to be innovative
2	Proper structure
3	Important people
4	Effective teamwork
5	High degree of commitment to innovation
6	Climate of creativity
7	Focus on the environment

The first element includes clearly articulated and shared objectives of an innovative enterprise – let everyone be actively involved in the achievement of strategic goal, by being as committed to innovation as the top management. In this perspective, the vision of the organization is defined as an idealized projection of the future, whose pillars are corporate values.

The organizational structure of an innovative enterprise should allow for creativity, learning and interaction. By stimulating and facilitating innovation, promoters, champions, gatekeepers and other employees play an important role in building of an innovative organization. Proper use of teams (in departments, between departments and between companies) to solve problems requires investment in the selection and building of teams. Active involvement of all employees in the continuous development of a company is another feature of an innovative organization. A company of this type has a positive approach to new ideas backed by the efficient incentive system. Modern innovative organization is oriented at both internal and external clients; it cooperates and has extensive contacts.

Innovation process management consists of the following stages [5]:
- searching – searching in the internal and external organization's environment in order to find signals of possible new innovations *e.g.* social changes, scientific discoveries, law changes;
- selecting – making decisions in accordance with the enterprise reaction;
- implementing – transformation of innovative ideas into product, service and/or process, followed by implementation in the internal and/or external market. At the stage of "implementing" it is necessary to take into consideration the following aspects:
- gaining necessary resources, *e.g.* finances, knowledge
- innovation project management,
- implementing innovations to the market,
- further development of innovation,
- capturing – undertaking operations to achieve profits from investments made.

Over the years, modern innovation management has been influenced by various management concepts, such as change management, technology management, R&D activity management, knowledge management [15].

Table 2. The impact of individual concepts on the development of innovation management [15].

Concept	Impact of the concept on innovation management
Change management	Continuing research on employee resistance to innovation regarded as an attempt to change the status quo
Technology management	Incorporation of technical innovation issues into innovation management
R&D management	Perception of R & D as one of the sources of innovation
Knowledge management	The role of knowledge and information as one of the sources of inspiration for the employees to undertake innovative activities, e.g. employee training, continuous education, databases for innovative projects

Innovation management in an enterprise is controlled by many factors of a political, social, economic and technological nature (Fig.1) [15].

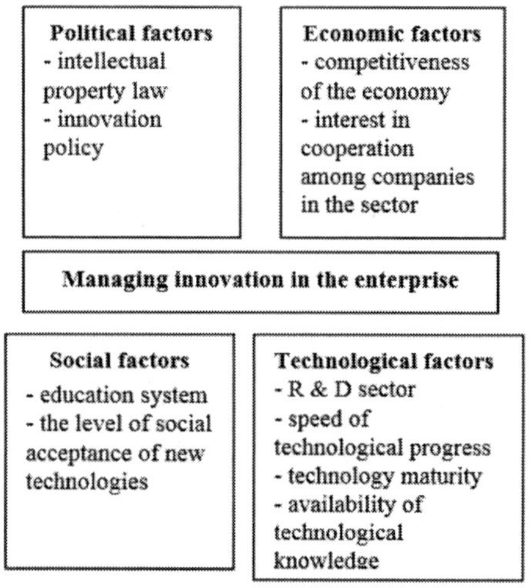

Fig. 1. External key factors influencing enterprise innovation management [15].

It is a skilful analysis of the organizational environment, its continuous assessment and planning new projects for implementation on the basis of the conducted analyses so as to enable maintaining the current position of the company and its further development.

The rules of innovation management that cover the key areas of the enterprise activity [21] can be formulated as:
- identification of the needs of customers and other stakeholders,
- strategic planning of product development and technology portfolio,
- organization of infrastructure that promotes innovation,
- maximization of innovative potential and resources that can be used,
- organization of teamwork, preparation of leaders and technical staff,
- creation of intellectual capital;
- shaping the culture of creativity and entrepreneurship,
- developing external networking.

The principles mentioned above supported by the developed organizational procedures in the area of innovation can significantly influence faster development of the company.

Large enterprises and foreign corporations can serve as an example of a stimulus for changes to organizations operating in the SME sector. They possess a great deal of knowledge acquired over many years of experience and innovative efforts, which, however, when copied directly, are unlikely to bring an intended effect. Each company should find its own solutions by developing specific routine procedures. Mastering innovation management procedures that guarantee success is not easy because they teach us what the company has learned over the long term by trial and error. A very important element in building the innovation climate in an enterprise is to eliminate the fear of the new. Innovation means doing something new that can have negative effects.

The skills described in Table 3 in most cases relate to actions and behaviors in the internal and external environment.

Table 3. Basic skills in innovation management [5].

Fundamental skill	Supportive behaviors
Recognition	Capturing from the environment the technical and economic hints that can trigger the change process
Proper gradation	Ensuring a good balance between the company's overall strategy and the proposed change (innovations are needed and as such should be introduced but not as a symptom of being "trendy", or because one does not want to lag behind)
Acquisition	Awareness of the constraints of the technical base of one's own company combined with the ability to connect to external resources of knowledge, information, technology, *etc.* Transfer of technology from various external sources and their integration in the right places in the company
Creation	The ability to create a part of technical solutions within own company through R&D department, specialist groups, *etc.*
Selection	Analyzing and selecting the optimum response to the signals coming from the environment; this will be in line with the company's own strategy and with its own abilities and resources or with the external technological resources.
Execution	Managing development projects for new products or processes from concept to implementation. Monitoring and controlling such projects
Implementation	Managing the process of implementing the innovation (technical or of another character) to ensure its acceptance and efficient use in the company
Learning	The ability to evaluate and analyze the innovation process and to identify unpleasant experiences for the improvement of management procedures
Business development	Setting routine procedures on the right track - in organizational structures, processes, corporate behaviors, *etc.*

Recognition, acquisition and selection are the key fundamental skills that allow collecting knowledge from the external environment.

3 Innovation culture in SMEs

It is widely recognized in a growing number of studies and research papers that the economic recovery largely depends on the improvement of the SME sector through research and development. Technology and innovation play a significant role in social and economic development. As a resource of economic competitiveness, the SME sector and innovation have become a priority issue. Innovation means activities that create value through knowledge and production growth. Innovation is driven by entrepreneurs who take risks, accept challenges and change things. In this respect, it is essential to promote policies that support innovation and technological transfer to the SME sector [22].

Business processes that combine strategy and operations are key drivers of innovation and the source of competitive advantage in the emerging global economy. However, continually tuning the strategy and business model requires from managers focusing attention on data analysis, which allows them to develop a new perspective on how the needs, interests and competencies of consumers, as well as the skills and credibility of suppliers are changing [23].

It can therefore be stated that the customer is increasingly involved in the implementation of innovation in the modern economy and that companies should expand their cooperation with other companies.

The latest report on the state of the sector of small and medium-sized enterprises prepared by PARP [24] indicates that in Poland there are 1.84 million companies, of which 96% are micro-enterprises. The report shows that the overall assessment of entrepreneurship in the country is quite good - most of the indicators analyzed are growing, particularly as regards the results obtained for micro and small businesses. Family businesses are becoming more and more important. The results of research conducted in the years 2009-2014 on machine tool enterprises belonging to the SME sector [25,26] show greater awareness and persistence in undertaking innovative actions by entrepreneurs of this sector, who in recent years have increasingly started to rely on their own R&D centers and on training their employees in the field of innovation. The high level of innovativeness in the surveyed companies is reflected in innovations introduced on a constant or frequent basis mainly in the area of products and processes. The prevailing problem among SMEs is the approach to marketing innovation. A large number of companies are collaborating with various external R&D centers and above all with universities. This type of cooperation enhances the possibilities of finding new solutions for companies. Furthermore, coupled with high awareness of the need for introducing changes and training own employees, it well reflects the state of innovativeness of Polish companies in the heavy machinery market and contributes to the increasing competitiveness with foreign companies [25,26].

The process of building a competitive advantage for SMEs on the domestic market and the related problems are often associated with a low level of innovation in companies. At the same time, the transfer of research results to the Polish economy producing development is often hindered by low interest among entrepreneurs in this type of activities, the commercialization of innovative research and the lack of competence in persons involved in this process [27].

To effectively mitigate these barriers, it is advisable to undertake educational and promotional activities aimed directly at employees of companies who have the greatest impact on the quality and development of the technology needed to compete in the market [28].

SMEs are becoming more and more committed to increasing innovation and competitiveness by spending a lot of money on this activity, but they rarely use measurable tools to gauge and evaluate the effects of such projects. Efficient management of an innovative company requires analytical tools that can be used in practice [29].

Innovation teams differ from typical design teams. They need other tools and other ways of thinking. They should be provided with the right dose of training and constant access to knowledge. At the same time, it is important to create a process that will allow them to introduce new ideas, teach their employees to observe the environment (both closer and further) in the search for new trends, technologies and changes in consumer behaviors, to implement the management system that will help capture ideas and encourage people to build and develop new ideas, to develop environments for creative work in enterprises and spaces for complete relaxation and regeneration [30].

The topic of SME support for innovation management tools is rarely taken up by researchers, as here for small and medium sized enterprises in the event of disastrous situations [31]. Tidd Bessant presents conditions and recommendations for building creative teams *etc* [5].

Excellent tools enhancing this education in the SME sector can be both ready-to-use applications as well as the supporting methods and models [1,29,32,33].

The analytical tools developed for SMEs include an application written in the MatLab language for measuring and evaluating enterprise innovation [29]. It is a simplified proposal to support and analyze innovation efforts of contemporary companies.

Another proposal is the DSM method and program for the evaluation of innovative solutions [32]. Both have been designed to assist decision makers in the implementation of these solutions and enabled systematizing the factors and criteria relevant for the selection of an innovative solution using the Analytic Hierarchy Process to address multi-criteria decision-making issues in order to determine the validity of factors relevant for evaluating innovative ventures. The method takes into account the following criteria: R&D resources, production resources, marketing, finance, management, market, product [32]. The tool can be used as a support for reported innovative solutions in the SZIP application described in this article.

Another approach is the model of behavior in creating the role of employees in the process of innovation [1]. The model identifies a specific set of actions to be implemented in companies in order to reduce barriers to innovation for the staff, which is a major problem in today's SMEs.

Creating culture and promoting innovation are long-term activities which require involvement of all employees in the organization. In the short term, focus should go to training employees and raising awareness of the principles and objectives of innovative activities. Building an innovative enterprise also requires the construction of an infrastructure to support innovation, an efficient flow of information between sections and communication on innovative projects [18].

4 Application supporting innovation management in an enterprise

Application supporting innovation management in an enterprise (SZIP) is a web application written in JEE. It has a client-server architecture. The client is implemented using JSP, HTML 5, CSS 3, JavaScript, and jQuery. Additionally, client-server communication uses AJAX techniques. The server part is written in Java using servlets. The application uses Apache Tomcat application server and Hibernate framework to communicate with MySQL database.

The application takes into account the conditions typical for developing organizations, outlined in the previous chapters, and business aspects that exist in the SME sector. The application is still being developed and its main features are outlined below.

DOI: 10.1051/itmconf/20171504011

Figure 2 shows the main window in the application.

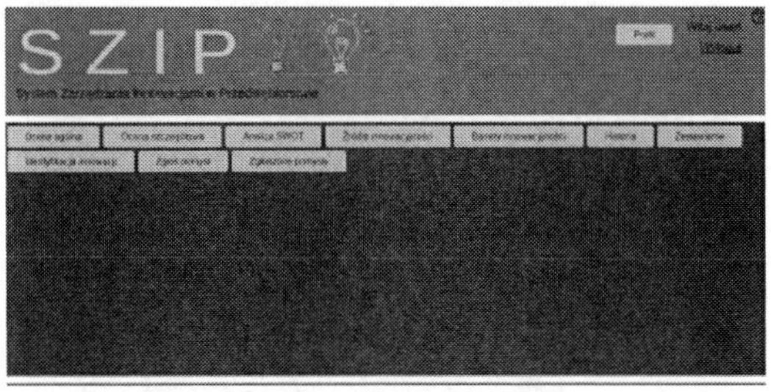

Fig. 2. The main window in SZIP application.

The following modules are available in the application:
- general evaluation of the company,
- detailed evaluation of the company,
- SWOT strategic analysis,
- analysis of innovation sources,
- analysis of innovation barriers,
- reporting innovative ideas (employee suggestion system) and their list,
- generating forms of reported innovative ideas,
- history of analysis,
- compilation of reports.

The application can be used by both the management staff and in part of the modules by other employees. Figure 3 shows the registration window in the application.

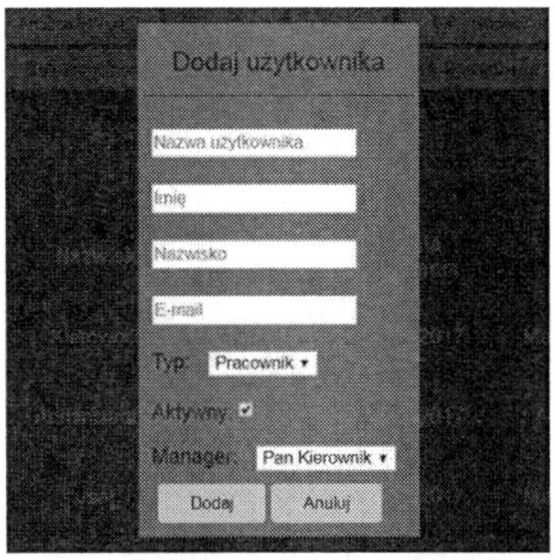

Fig. 3. User registration window in SZIP application.

The detailed evaluation and the general evaluation include the qualitative assessment of the surveyed enterprise and differ in the type of the analyzed elements that make up an innovative enterprise. Overall rating of the company (Figs. 4, 5 and 6) shows the results in the areas of strategy, processes, organization, relationships, and learning.

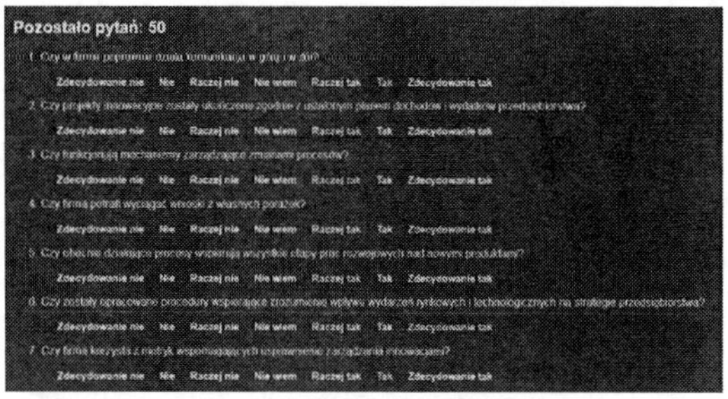

Fig. 4. Example of the general evaluation response in SZIP application.

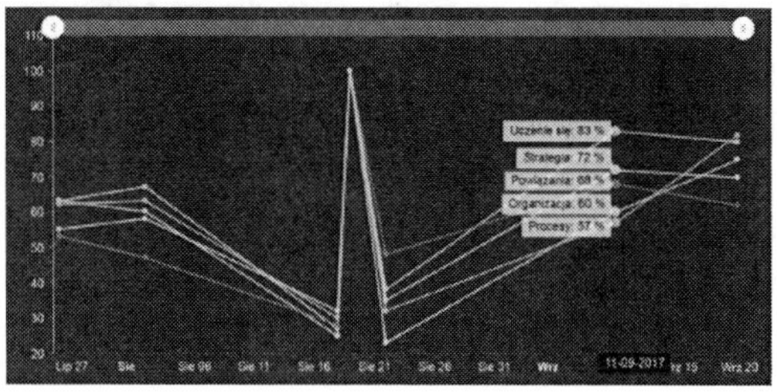

Fig. 5. Example of graphic results of the general evaluation in SZIP application.

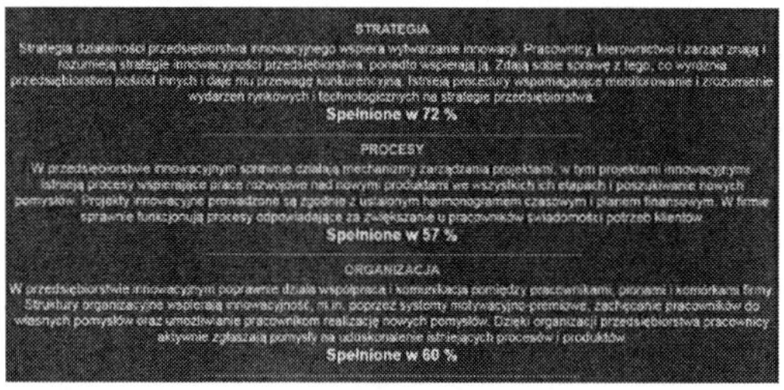

Fig. 6. Example of descriptive results of the general evaluation in SZIP application.

The detailed evaluation (Fig. 7) assesses the innovation climate of the company in the following areas:
- change of strategy,
- implementation of R&D projects,
- competence for innovation,
- searching for opportunities for innovation,
- security of innovative intellectual property,
- financing of innovative activities,
- supporting innovation.

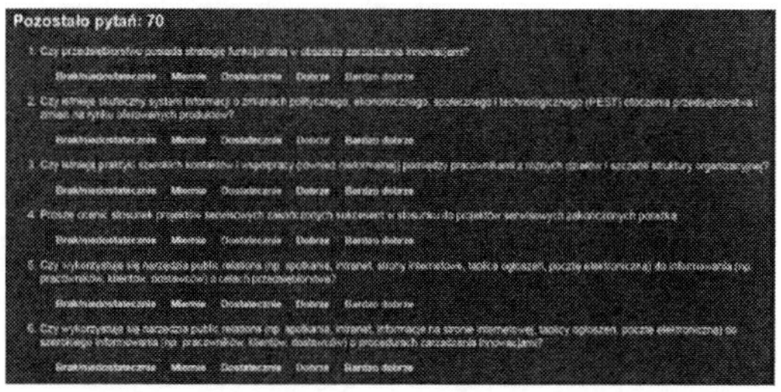

Fig. 7. Example of the detailed evaluation response in SZIP application.

The application enables conducting an analysis of the internal and external environment of the organization under examination, and to identify its strengths, weaknesses, opportunities and threats. With SWOT analysis it is possible to determine the strategic position of the company and discover which actions the company should take to make the best use of the situation existing in its internal and external environment. The conducted analysis identifies one of four strategic situations:
- Maxi-maxi strategy - The situation concerns an enterprise within which the strengths are prevailing and opportunities are dominating in the environment. This situation finds its reflection in a maxi-maxi strategy of strong expansion and diversified development.
- Mini-maxi strategy – In this case it is an enterprise in which the weaknesses prevail over the strengths, but the arrangement of external conditions is favorable for the company existence. The strategy of such a company should be to use the external opportunities while reducing or correcting the internal shortcomings.
- Maxi-mini strategy - in this situation, the difficulties in the company development are due to the unfavorable external conditions. The company can counter this situation with great internal potential and try to overcome threats by exploiting its many strengths to the maximum.
- Mini-mini strategy - in this situation, the company is deprived of any development opportunities. It operates in an unfavorable environment, and its potential for changes is small. There are no significant strengths that could be used to counter the threats and to improve the existing weaknesses.

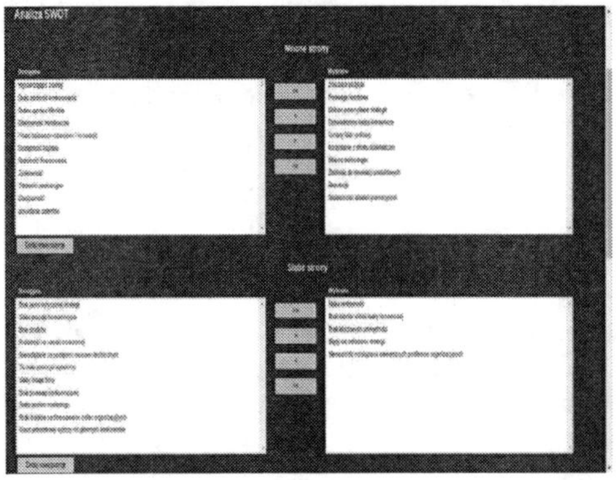

Fig.8. Example of the SWOT analysis in SZIP application.

A practical glossary of terms has been developed for the program, and it allows the user to learn the definitions of the terms used in the application. The glossary contains concepts used in the analysis which the user may find in the application, *e.g.* the definition of innovation types according to Oslo Manual, mission, vision, strategy, intellectual property protection, sources of and barriers to innovation, *etc.* (Fig. 9).

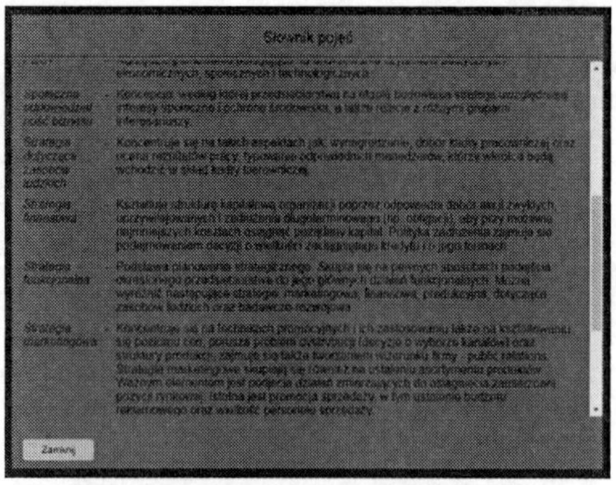

Fig. 9. Window showing the glossary of terms.

In order to avoid user errors, windows have been developed to provide information about entering invalid data or actions that are not permitted in a given data area.

Zgłoszone pomysły:

Tytuł pomysłu	Kategoria	Status	Data dodania	Autor	Data aktualizacji	Akcje
Test	Maszyny	Zamknięte	08-06-2017	Pan Pracownik	08-06-2017	
Stworzenie kanału tematycznego na forum branżowym	Relacja firma-klient	Otwarte	25-05-2017	Pan Kierownik	25-05-2017	
[illegible]						
Nowy projekt opakowania	Produkty	Otwarte	25-05-2017	Pan Kierownik	25-05-2017	
Zmiana reklamy zewnętrznej	Wizerunek firmy i promocja	Zamknięte	25-05-2017	Pan Kierownik	25-05-2017	

Fig. 10. List of innovative ideas submitted in SZIP application.

In the application after logging in, anyone can submit his own innovative idea in the area selected from the following nine areas:
- Company-customer relationships - ideas that improve customer service and streamline communication with partners, suppliers, and customers;
- Company image and promotion - activities related to the promotion of company brand, image and achievements;
- Organization of work and management, internal relations - activities to improve communication and information flow within the company, social issues, introduction of modern management methods, streamlining work organization in company departments and cooperation between these departments;
- Safety - ideas for improving the level of safety (people and data) and increasing the level of fire protection; protection of information security;
- Machines - suggestions for changes regarding technical improvements of machines, proposals for automation of production processes, *etc.*
- Organization of production - Ideas for organizational change in the area of production;
- Product - Ideas for structural and technological changes in the products offered by the company and ideas for new products.

Additionally, in the application it is also possible to generate forms for the reported innovative ideas in four areas: product, process, organization and marketing.

DOI: 10.1051/itmconf/20171504011

The form for the submitted innovation project includes the following data groups:
- author of the innovative solution and their personal data, description of innovation, legal status of innovation, market potential,
- title, abstract, keywords,
- description of the uniqueness of innovation in relation to current solutions,
- the period in which innovation was created,
- the level of innovation development,
- sources of funding,
- material inputs necessary for the implementation of innovations (material, personnel, apparatus, technology, financial, others);
- intellectual property rights (patent application, patent granted, trade secret),
- first commercial use or sale,
- anticipated number of potential recipients.

The analyses conducted by the application are reported and available in application history.

Fig. 11. Window showing the reports made in SZIP application.

The developed application significantly supports organizations in the SME sector in communication and actions promoting the culture of innovation. This enables every employee to make suggestions for improvement and submit an innovative solution. The reported ideas can be reviewed, evaluated and updated on a regular basis by managers. Employees have constant access to knowledge about ongoing projects and undertakings in the enterprise.

5 Conclusions

The versatility of the tool allows for its multiple uses. The application can be used not only in enterprises for a singular evaluation of innovation, but also to periodically monitor the level of improvements made and for current reporting of the ideas and solutions submitted by the company employees. The application can also be used in academic activities during classes with students.

The application helps entrepreneurs pay more attention to the role of innovation in achieving a higher level of competition, which should accelerate the development of small and medium-sized enterprises and increase their chances of competing against companies with higher financial resources. The proposed application will allow its users to identify the degree of modernity and innovation of companies operating in the SME sector and will support internal communication within the area of innovation management.

Observation of proposed solutions and using the application over the years will support the decision making process and provide a basis for capturing certain regularities in shaping the culture of innovation in SMEs.

References

1. S. Motyka, N. Łukasik, Maint. Probl., **99**, 31-42, (2015)
2. S. Motyka, *Ocena skuteczności wprowadzania innowacji technicznych w przedsiębiorstwach przemysłu maszynowego. Praca doktorska.* (Politechnika Krakowska, Kraków, 2007)
3. P.F. Drucker, *Innowacja i przedsiębiorczość. Praktyka i zasady* (PWE, Warszawa, 1992)
4. P. Kotler, F. Trias de Bes, *Winning At Innovation: The A-to-F Model* (Palgrave Macmillan UK, 2015)
5. J. Bessant, K. Pavitt, J. Tidd, *Managing Innovation: Integrating Technological, Market and Organizational Change* (Wiley and Sons, Hoboken, 2006)
6. C. A. Di Benedetto, C. M. Crawford, *New Products Management* (McGraw –Hill , New York, 1999)
7. L.V. Shavinina, *The International Handbook on Innovation* (Elsevier Science, London, 2002)
8. P. Trott, *Innovation Management and New Product Development* (Pearson Education Ldt, Essex, 2008)
9. T.H. Davenport, *Process Innovation: Reengineering Work Through Information Technology* (Harvard Business School Press, Harvard, 1993)
10. J. R. Baldwin, P. Hanel, *Innovation and Knowledge Creation in an Open Economy* (Cambridge University Press, Cambridge, 2003)
11. G. Hamel, *The Why, What, and How of Management Innovation* (Harvard Business Review, 2006)
12. J. Birkinshaw, G. Hamel, M.J. Mol, AMR **33**, 4, 825-845 (2008)
13. F. Morente, X. Ferràs, Intangible Capital **13**, 3 (2017)
14. P. Mielcarek, Procesy zarządzania innowacjami w przedsiębiorstwach przemysłowych (Difin, Warszawa, 2016)
15. M. Karlik, *Zarządzanie innowacjami w przedsiębiorstwie. Poszukiwanie i realizacja nowatorskich projektów* (Wydawnictwo Poltext, Warszawa, 2014)
16. A. Pomykalski, *Zarządzanie innowacjami* (PWN, Warszawa 2001)
17. K. Szatkowski, *Zarządzanie innowacjami i transferem technologii* (PWN, Warszawa, 2016)
18. S. Motyka, *Wewnętrzne uwarunkowania kreowania systemu innowacji w organizacji. Innowacje w zarządzaniu i inżynierii produkcji*, (R. Knosala, Ed.)

(Oficyna Wydawnicza Polskiego Towarzystwa Zarządzania Produkcją, Opole 2013, 209-219)

19. S. Motyka, *Pomiar innowacyjności przedsiębiorstwa. Komputerowo Zintegrowane Zarządzanie* (Oficyna Wydawnicza Polskiego Towarzystwa Zarządzania Produkcją, Opole, 2011, 160-168)
20. A. Sosnowska, S. Łobejko, P. Kłopotek, *Zarządzanie firmą innowacyjną (*Wydawnictwo Difin, Warszawa, 2001)
21. J. Łunarski, *System zarządzania innowacjami. Zarządzanie innowacjami – system zarządzania innowacjami* (Oficyna Wydawnicza Politechniki Rzeszowskiej, Rzeszów, 2007)
22. Z. K. Szabo, M. Soltes, E. Herman, E & M, **17**, 52-68, (2013)
23. PARP, *Raport o stanie sektora MŚP w Polsce* (Warszawa 2016), http://www.parp.gov.pl/publikacje/ebook/454
24. C.K. Prahalad, M.S. Krishnan, *Nowa era innowacji* (Wydawnictwo Profesjonalne PWN, Warszawa, 2010)
25. S. Motyka, A. Jachimowski, *Innowacyjność polskich przedsiębiorstw przemysłu maszynowego – bariery i szanse rozwoju, Komputerowo Zintegrowane Zarządzanie*, (Oficyna Wydawnicza Polskiego Towarzystwa Zarządzania Produkcją, Opole 2010, 256-264)
26. S. Motyka, J. Nowak, *Innowacyjność polskich przedsiębiorstw przemysłu maszynowego (stan aktualny i prognozy). Innowacje w zarządzaniu i inżynierii produkcji (R. Knosala, Ed.)* (Oficyna Wydawnicza Polskiego Towarzystwa Zarządzania Produkcją, Opole 2015, 128-140)
27. S. Łobejko, A. Sosnowska, *Komercjalizacja wyników badań naukowych. Praktyczny poradnik dla naukowców* (Urząd Marszałkowski Województwa Mazowieckiego, Warszawa, 2013)
28. K. Symela *Szkolenie specjalistów do spraw komercjalizacji innowacyjnych technologii.* (Edukacja Ustawiczna Dorosłych, **2**, 2014)
29. S. Motyka, P. Jarmuła, *Pomiar innowacyjności przedsiębiorstwa z wykorzystaniem środowiska Matlab. Innowacje w zarządzaniu i inżynierii produkcji* (R. Knosala, Ed.) (Oficyna Wydawnicza Polskiego Towarzystwa Zarządzania Produkcją, Opole 2016, 166-177)
30. S. Motyka, Zarządzanie Przedsiębiorstwem, **4**, 27-33 (2014)
31. M. Aba-Bulgu, M. Sardar, N. Islam, *Corporate Crisis and Risk Management. Modeling Strategies and SME Application* (Emerald Group Publishing, 2006)
32. J. Gawlik, S. Motyka, *Wspomagana komputerowo metoda oceny i wyboru innowacyjnego projektu - program DSM. Komputerowo Zintegrowane Zarządzanie* (Oficyna Wydawnicza Polskiego Towarzystwa Zarządzania Produkcją, Opole 2008, 328-337)
33. A. Kaczorowska, J. Słoniec, S. Motyka: *Comprehensive methods of evaluation and project efficiency account*, (Computer Science and Information Systems (FedCSIS), 2016 Federated Conference on, pp: 1159-1168, IEEE, ACSIS, **8**)

ITM Web of Conferences **15**, 04012 (2017) DOI: 10.1051/itmconf/20171504012
CMES'17

The effects of welding parameters on the tensile shear strength of refill friction stir spot welding of 7075-T6 aluminium alloy joints

Dawid Wydrzyński[1,*], *Magdalena* Bucior[1], and *Andrzej* Kubit[1]

[1]Faculty of Mechanical Engineering and Aeronautics, Department of Manufacturing Processes and Production Engineering, Rzeszow University of Technology, Powstańców Warszawy 8, 35-959 Rzeszów, Poland

Abstract. The paper presents the results of an analysis of the effect of welding parameters on the load capacity of joints obtained by the refill friction stir spot welding (RFSSW) method. This technology has a prospective application in aerospace and automotive industries, especially for aluminium alloys. The research was conducted for the overlapping joints made of two 7075-T6 aluminium alloy sheets with thicknesses 1.6 mm and 0.8 mm. Strength tests were conducted for two variants of loading the joint. The experiments were conducted according to statistical Hartley's plan PS/DS-P:Ha3. The welding times t was varied in the range of 1.5 ÷ 3.5 s, the tool plunge depth g in the range of 1.5 ÷ 1.9 mm, and the tool rotational speed n in the range of 2000 ÷ 2800 rpm. For these parameters the analysis of experiment reproducibility, impact significance and adequacy of equations were made. The results of analysis according to the design of the experiment (DOE) indicate that all analyzed parameters have significant influence on the load capacity of joints. The biggest load capacity of joints in the case of the first variant was gained at the maximum spindle speed of 2800 rpm, the minimum tool plunge depth of 1.5 mm and welding time of 1.5 s. For the second variant, the maximum load capacity of joints was gained at the spindle speed of 2400 rpm, tool plunge depth of 1.5 mm and welding time of 2.5 s.

1 Introduction

Refill friction stir spot welding and using RFSSW to fill craters is becoming one of the most commonly used methods of spot welding in the automotive [1, 2] and aviation [1, 3] industries. The technology, especially in the case of aluminium alloys, has a great amount of potential application. 7Xxx aluminium alloys [4, 5] are currently among the strongest and most commonly available materials used in the aviation and transport industries. In aircraft structures, the RFSSW technique [6, 7] can potentially be an alternative to resistance spot welding. An example of using this technology could be welding the fuselage covering of an aircraft. The attractiveness of this joining method results from technological-economical benefits like high production output and welding process stability that ensures high repeatability simplifying the automation of the process. Another benefit would be the ability of joining materials with different properties and better safety conditions in the workplace, when compared to traditional welding methods [4, 8, 9, 10]. This technique can be used to joint different types of materials like composites [11, 12], thermoplastics [13, 14] and non-weld-able alloys [4, 5].

Experimental design and optimization are tools that are systematically used to study various types of problems such as academic studies, development problems, and production problems [15]. Considering the high cost and the amount of time needed to conduct a study, the experimental design needs to be properly planned, the experiment needs to be properly conducted and the results need to be properly analyzed [16]. The literature on the topic has many descriptions of various methods and experimental design plans. Their methods are properly supported, as are their mathematical fundamentals and results analysis methodologies. Thus, it is possible to adapt an experimental design to any research problem that results in a rational study [17]. Patel et al. [18] use a DOE experimental design when determining the effect of welding parameters and tool geometry on the shear strength of joints made using FSW.

The RFSSW method requires the selection of appropriate welding parameters because they decide on

*Corresponding author: magdabucior@prz.edu.pl

 DOI: 10.1051/itmconf/20171504012

the quality of the joint. Incorrect process parameter selection causes the joints to have countless defects in the macro and microstructure, which will lower the strength of the joint [4, 19, and 20]. Thus, determining the parameters, which will affect the strength of a joint, is the most important. As a result, the Hartley static PS/DS-P:Ha$_3$ plan was selected to study the effect of process parameters of the RFSSW method on the shear strength of 7075-T6 aluminium alloy joints. The determined plan is characterized by its initial factors being selected in a strictly predetermined fashion and ordered according to certain rules that eliminate random factor selection. The plan is designed to minimize the number of experiments, and its greatest feature is the simplicity of the mathematical analysis [17].

2 Method

The effects of welding parameters on the tensile shear strength of refill friction stir spot welding were conducted according to statistical Hartley's plan PS/DS-P:Ha$_3$. The subject of the study was the overlapping joint of 7075-T6 aluminium alloy sheets with two different sheet thicknesses 1.6 mm and 0.8 mm (Fig. 1). The chemical composition of this alloy is Si - 0.4, Fe - 0.5, Cu - 1.2 ÷ 2, Mn - 0.3, Mg - 2.1 - 2.9, Cr - 0.18 ÷ 0.35, Zn - 5.1 ÷ 6.1, Ti - 0.2, other impurities: single0.05, total 0.15, Al - rest. The experimental investigations of the RFSSW process were conducted with an application of the HARMS WENDE machine. All experiments were realized with triple repetition. The significance level α = 0.05 was accepted for calculations.

The tool rotational speed n (rpm), tool plunge depth g (mm) and welding time t (s) were established as input factors. The output factor in the experiment was shear strength. Strength tests were conducted for two variants of loading the joint. In the first variant, with the usage of a special holder, pure shear strength was obtained (Fig. 2). For comparison, a tensile strength test was conducted in the second case. The single lap joint was subjected to tensile test on a ZWICK Z100 machine, where a grip translation speed was 5 mm/min.

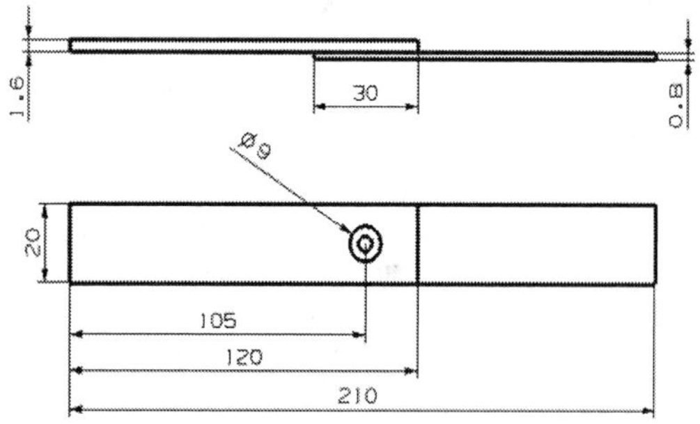

Fig. 1. Shape and dimension of the sample for RFSSW process.

The plan of the experiment implies the adoption of three levels of controlling factors: the minimum (-), the central (0) and the maximum (+). The welding times t was varied in the range of 1.5 ÷ 3.5 s, the tool plunge depth g in the range of 1.5 ÷ 1.9 mm, and the tool rotational speed n in the range of 2000 ÷ 2800 rpm. The matrix of the Hartley's plan with three input factors is presented in Table 1.

Table 1. Matrix of plan PS/DS-P: Ha$_3$.

Nr	x_1	x_2	x_3	x_1^2	x_2^2	x_3^2	x_1x_2	x_1x_3	x_2x_3
1	-	-	+	+	+	+	+	-	-
2	+	-	-	+	+	+	-	-	+
3	-	+	-	+	+	+	-	+	-
4	+	+	+	+	+	+	+	+	+
5	-	0	0	+	0	0	0	0	0
6	+	0	0	+	0	0	0	0	0
7	0	-	0	0	+	0	0	0	0
8	0	+	0	0	+	0	0	0	0
9	0	0	-	0	0	+	0	0	0
10	0	0	+	0	0	+	0	0	0
11	0	0	0	0	0	0	0	0	0

* x_1 - coded value of tool rotational speed n, x_2 - coded value of tool plunge depth g, x_3 - coded value of welding times t.

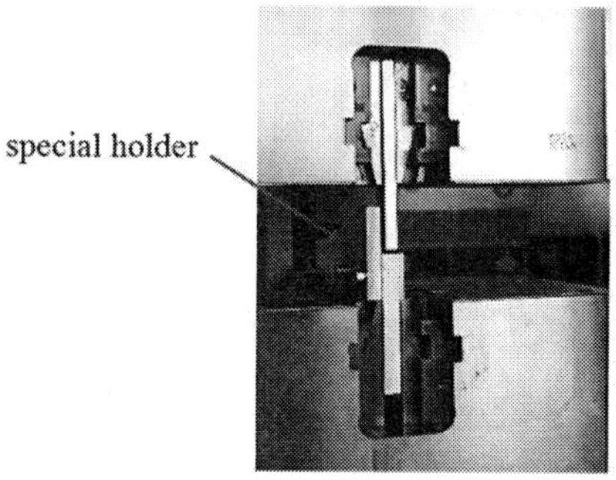

Fig. 2. Strength test example.

3 Results and discussion

The average load carrying capacity of joints resulting from the RFSSW method of various welding parameters is presented in Table 3.

Table 2. Average values of load capacity of joints obtained after welding process RFSSW method at differential parameters according to Hartley's plan.

No.	Input factors			Output factors	
	n rpm	g mm	t s	R_{tp} N	R_t N
1	2000	1.5	3.5	7000	6475
2	2800	1.5	1.5	7805	6480
3	2000	1.9	1.5	5510	4955
4	2800	1.9	3.5	5290	5265
5	2000	1.7	2.5	6735	5990
6	2800	1.7	2.5	6980	6260
7	2400	1.5	2.5	7350	6550
8	2400	1.9	2.5	6015	5535
9	2400	1.7	1.5	6720	6110
10	2400	1.7	3.5	6955	5910
11	2400	1.7	2.5	6400	5550

DOI: 10.1051/itmconf/20171504012

The joints that were tested for pure shear through the use of a special device (variant I) show a 8-20% higher load carrying capacity compared to samples that underwent a simple shear strength test (variant II) (Fig. 3). The slight difference between the two variants is presented in experiment No.4 where the maximum values of the experimental factors are tested, i.e. n = 2800 rpm, g = 1.9 mm and t = 3.5s.

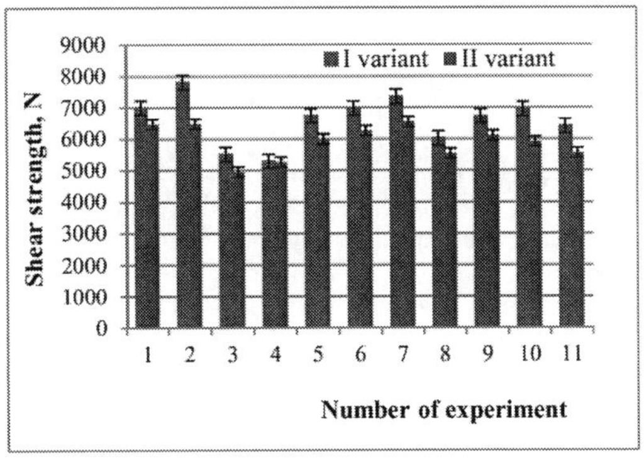

Fig. 3. Average values of shear strength of joints.

As a result of the regression, a functional relation between the process parameters (n, g, t) and the pure shear strength (R_{tp}), tensile shear strength (R_t), were determined. These relations are described by the following equations (Eqs. 1-2):

$$R_{tp} = 6763{,}29 - 890\,g - 1001{,}25nt \quad (1)$$

$$R_t = 5593{,}96 - 625\,g - 683{,}75nt \quad (2)$$

For the first variant, the biggest load capacity of joints, 7805 N, was gained at the maximum tool rotational speed of 2800 rpm. The minimum tool plunge depth of 1.5 mm and welding time of 1.5 s. A minimum value of load capacity of joints 5290 N, was obtained at experiment No. 4, for n = 2800 rpm, g = 1.9 mm and t = 3.5 s. For the second variant the maximum load capacity of joints 6550 N, was gained at the tool rotational speed of 2400 rpm, tool plunge depth of 1.5 mm and welding time of 2.5 s. At a tool rotational speed n = 2000 rpm, g = 1.9 mm and t = 1.5 s the smallest value load capacity of joints 4955 N, was observed. The tool plunge depth is very important parameter that affects the strength of joints. For the tool plunge depth of 1.5 mm the load capacity of joints is the best in turn for the g = 1.9 mm the worst.

When analyzing the results of the static shear strength tests (Figs. 4-14), 3 types of weld failure can be observed in RFSSW joints. The ones with the most shallow tool plunge depth n = 1.5 (Fig. 4a), (Fig. 5a) and the welding time of t = 2.5 s and rotational speed of n = 2400 rpm (Fig. 10b), (Fig. 11b) demonstrate weld shearing without damaging the welded material (type I) These also had the greatest load carrying capacity.

In the second case (type II) the welds were torn out of the thinner sheet metal, (Fig. 6a, 9a, 10a) and (Figs. 4b-9b, Figs. 12b-14b). This demonstrates the essential value of one the parameters, tool plunge depth g = 1.7 mm, which affect this type of joint. These joints have a lower load carrying capacity compared to other joints made with a lesser tool plunge depth.

a) b)

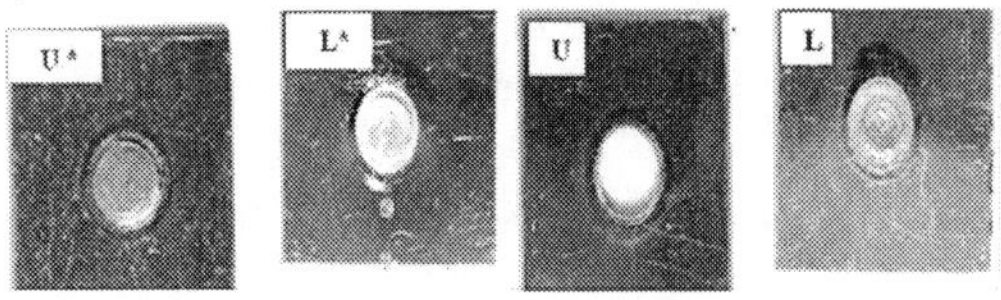

Fig. 4. Experiment No. 1. Illustration type of failure modes observed under the shearing test: a) I variant, b) II variant.
* (U) Bottom view of the upper sheet and (L) top view of the lower sheet.

a) b)

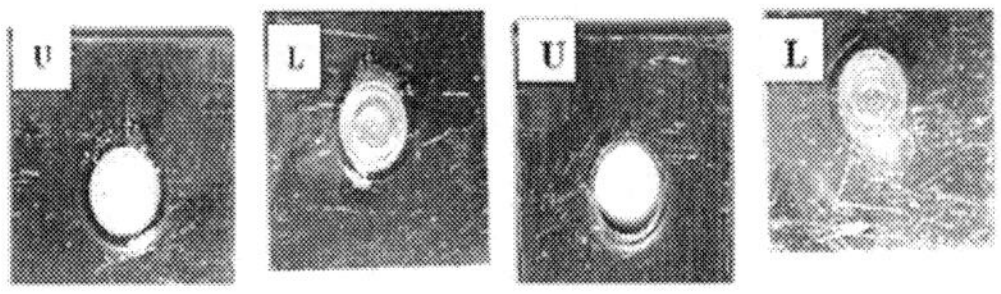

Fig. 5. Experiment No. 2. Illustration type of failure modes observed under the shearing test: a) I variant, b) II variant.

a) b)

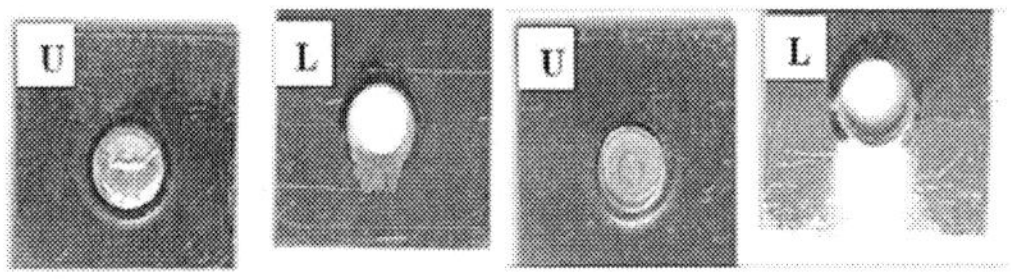

Fig. 6. Experiment No. 3. Illustration type of failure modes observed under the shearing test: a) I variant, b) II variant.

a) b)

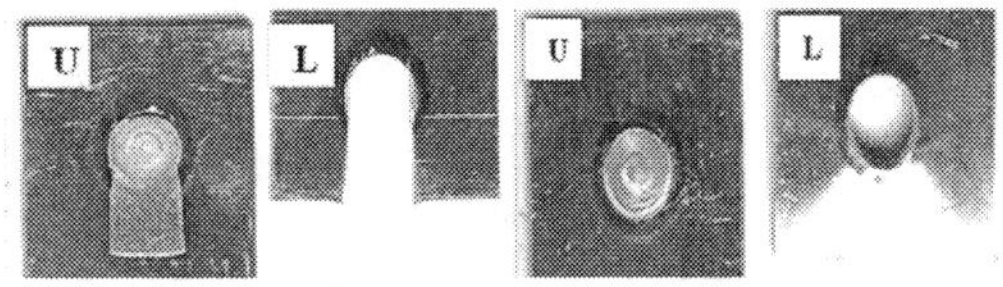

Fig. 7. Experiment No. 4. Illustration type of failure modes observed under the shearing test: a) I variant, b) II variant.

a) b)

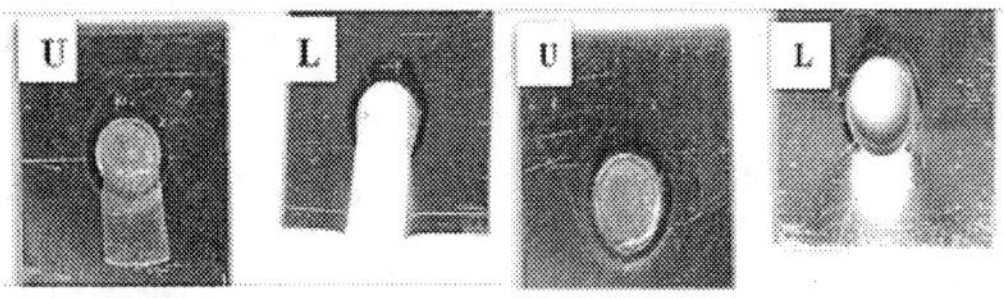

Fig. 8. Experiment No. 5. Illustration type of failure modes observed under the shearing test: a) I variant, b) II variant.

The type III failure occurred when there was a partial tear of the parent material and welded out of the bottom

 DOI: 10.1051/itmconf/20171504012

piece of sheet metal. This only occurred in the cases where the static shear strength test was conducted with the additional device (Fig. 7a, 8a, 11a-14a). The tool plunge depth of g = 1.7 mm and the rotational speed of n = 2400 rpm also had an effect on this type of failure.

a) b)

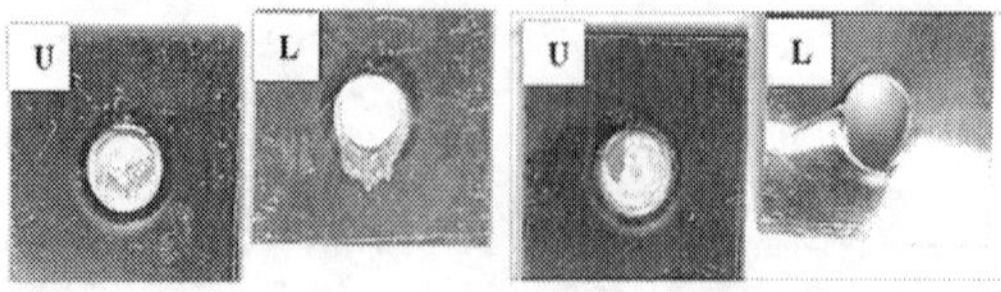

Fig. 9. Experiment No. 6. Illustration type of failure modes observed under the shearing test: a) I variant, b) II variant.

a) b)

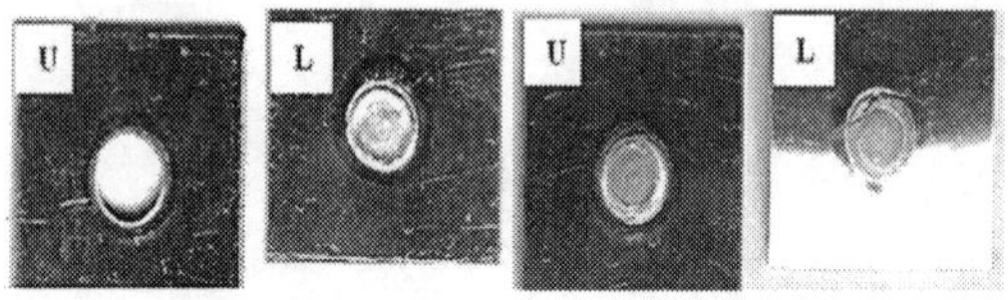

Fig. 10. Experiment No. 7. Illustration type of failure modes observed under the shearing test: a) I variant, b) II variant.

a) b)

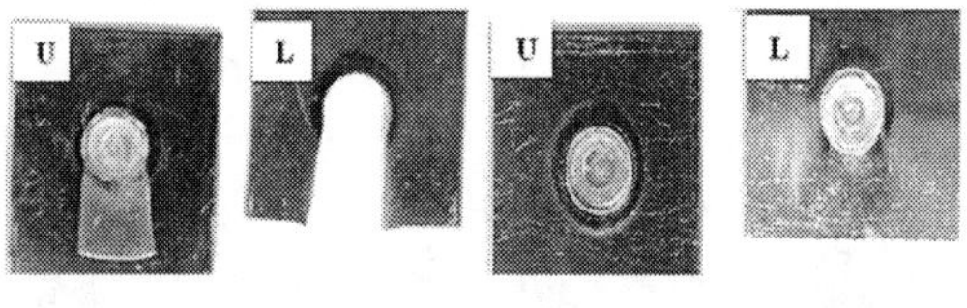

Fig. 11. Experiment No. 8. Illustration type of failure modes observed under the shearing test: a) I variant, b) II variant.

a) b)

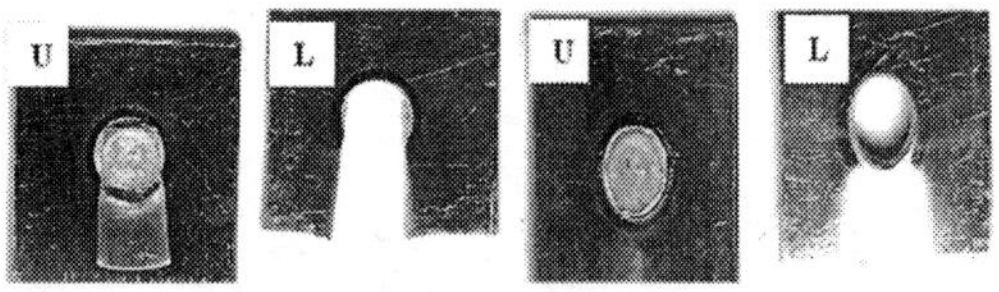

Fig. 12. Experiment No. 9. Illustration type of failure modes observed under the shearing test: a) I variant, b) II variant.

a) b)

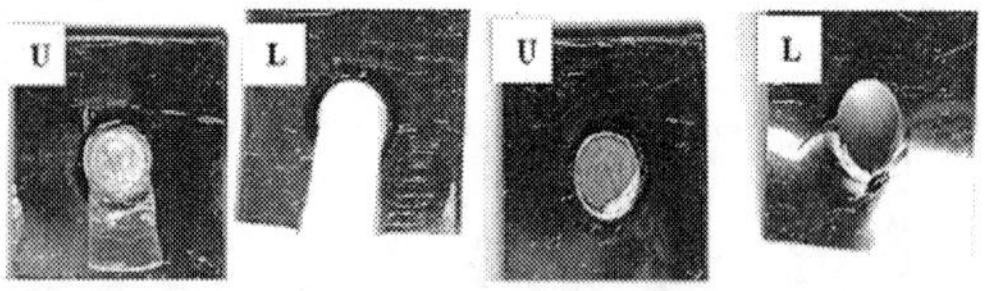

Fig. 13. Experiment No. 10. Illustration type of failure modes observed under the shearing test: a) I variant, b) II variant.

The high correlation coefficient of R = -0.91 shows that there is a strong relation between the tool plunge depth and the load carrying capacity of the joint. An increase in tool plunge depth reduces the load carrying capacity of the joint.

a) b)

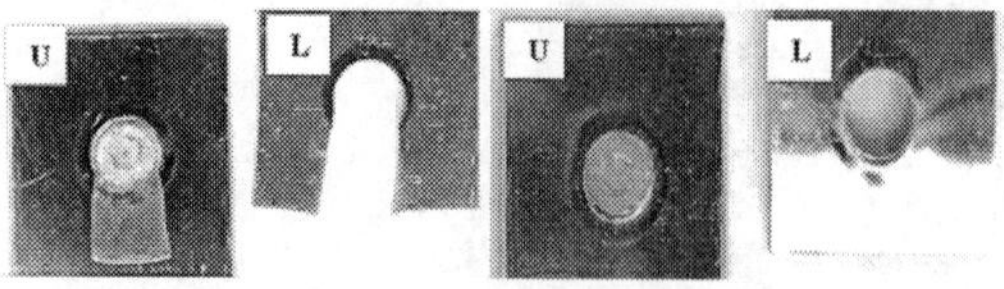

Fig. 14. Experiment No. 11. Illustration type of failure modes observed under the shearing test: a) I variant, b) II variant.

4 Conclusions

As the results of the investigation, the adequate equations were obtained describing the effect of welding parameters on tensile shear strength of the RFSSW process. These equations indicate that all analyzed parameters have a significant influence on the quality of welding joints. The load carrying capacity of the first variant is between 8 and 20% greater than the values obtained in variant II. The results of the static shear strength test demonstrate 3 types of mechanical failures for spot welds made with the RFSSW method. The tool plunge depth had a significant effect on the type of mechanical joint failure. An increase in tool plunge depth resulted in a decrease in the load carrying capacity. The greatest load carrying capacity was achieved for tests with an n = 1.5 mm tool plunge depth, while an n = 1.9 mm tool plunge depth resulted in the lowest load carrying capacity. This is confirmed by the high value of the correlation coefficient R = - 0.91. The confirmation of the test values and the calculated values weld load carrying capacity demonstrates that Hartley's plan is appropriate for this type of study.

References

1. M. Kenichiro, B. Niels, F. Livan, F. Micari, A. E. Tekkaya, CIRP Ann. Manuf. Technol. **62**, (2013)
2. D. Lohwasser, Proc. 2nd Int. Sym. FSW, **27-29**, (2000).
3. A. Paoletti, F. Lambiase, A. D. Ilio, Proc. CIRP **33**, (2015)
4. S. Zhikang, Y. Xinqi, Z. Zhaohua, C. Lei, L. Tielong, Mater. Des. **44**, (2013)
5. Y. Q. Zhao, H. J. Liu, S. X. Chen, Z. Lin, J. C. Hou. Mater. Des. **62**, (2014)
6. Y. Uematsu, K. Tokaji, Y. Tozaki, T. Kurita, S. Murata, Int. J. Fatigue **30**, (2008)
7. J. Eggers, HWI Weld Times **5**, (2012)
8. D. H. Choi, B. W. Ahn, C. Y. Lee, Y. M. Yeon, K. U. Song, S. B. Jung S.B. Mater. Trans. **51**, (2010)
9. C. B. Smith, J. F. Hinrichs, P. C. Ruehl P.C. Sheet Met. Weld. Conf. XI, **2-5**, (2004)
10. Z. H. Zhang, X. Q. Yang, J. L. Zhang, G. Zhou, X. D. Xu, B. L. Zou, Mater. Des. **32**, (2011)
11. E. Azarsa, A. Mostafapour, J. Manuf. Proc. **15**, (2013)

DOI: 10.1051/itmconf/20171504012

12. S. T. Amancio-Filho, Mater. Sci. Eng. **A 528**, (2011)
13. S. H. Dashatan, T. Azdast, S. Ahmadi, A. Bagheri, Mater. Des. **45**, (2013)
14. M. K. Bilici, A. I. Yükler, Mater. Des. **33,** (2012)
15. T. Lundstedt, E. Seifert, L. Abramo, B. Thelin, A. Nystrom, J. Pettersen, R. Bergman.Chemometrics Intelligent Lab. Syst. **42**, (1998)
16. A. Skowronek, Technical Transactions**, 104**, (2007)
17. M. Korzyński, *Metodyka eksperymentu* (WNT, Warszawa, 2013)
18. N. Patel, K. D. Bhatt, V. Mehta. Proc. Technol. **23**, (2016)
19. Y. Q. Zhao, H. J. Liu, Z. Lin, S. X. Chen, J. C. Hou, Sci. Technol. Weld. Join. **19**, (2014)
20. Y. Uematsu, K. Tokaji, Sci. Technol. Weld Joi. **14**, (2009)

Uncertainty measurement with the kinematic telescopic bar during industrial robot inaccuracy tests

Jerzy Józwik[1,*], *Elżbieta* Jacniacka[1] and, *Dawid* Ostrowski[2]

[1]Lublin University of Technology, Mechanical Engineering Faculty, Department of Production Engineering, 36 Nadbystrzycka Street, 20-618 Lublin, Poland
[2]The State School of Higher Education, The Institute of Technical Sciences and Aviation, 54 Pocztowa Street, 22-100 Chełm, Poland

Abstract. The subject of this article is the assessment of measurement uncertainty with the kinematic telescopic bar QC20-W in the accuracy test of MOTOMAN HP20 industrial robot. The measurements were performed to determine the statistical uncertainty of error measurement using the system applied. Analysed in tests was the robot's ability to recreate a circular outline through standard, complex and extended measurement of uncertainty measurement. The obtained results were served to perform rapid evaluation of robot inaccuracy. These uncertainties were based on the information included in the device calibration certificate (estimated with method B) but also on the basis of measurements and statistical data (estimated with method A). These components of the uncertainty budget take relatively small values ($u_c = 0.818 \div 4.130$; $U = 1.636 \div 8.260$ for $k = 2$, which proves that a properly selected method was applied to the research task. The method of research and calculation precisely identified key uncertainties allowing for an objective assessment of the industrial robot errors carried out with Renishaw the kinematic telescopic bar.

1 Introduction

Robot is an automatic device capable of handling operations in an industrial manufacturing process, equipped with a kinematic system consisting of at least three motion units and an autonomous programmable control system [1]. The cycle of manipulative and locomotive movements performed by the robot is most often reproducible, however, it may vary depending on the modification of the program, the information given or the state of the environment [1-3]. In industrial practice, robots perform movements with certain accuracy, which is variable during operation generating errors interfering of accuracy [4]. In the work [5] geometric and non-geometric errors are distinguished. Work [6] defines errors affecting the accuracy of a robot, dividing it into three basic groups: dynamic, structural, and kinematic. Another example of error classification is presented in [7], where four groups of industrial robot accuracy errors are distinguished: geometric, dynamic, thermal and system. There are few measuring systems that enable effective identification and evaluation of robot errors. The QC20-W Ballbar diagnostic system, dedicated to CNC machines, is used in this study. Although the use of the QC20-W Ballbar system for CNC machine tools is very wide [1, 4, 8-10], in robots it is applied on a much lower scale [2, 11-14]. Some authors [11,12] compare the accuracy of selected industrial robots with the use of a ball-shaped, telescopic kinematic bar. ABB IRB 1600 industrial robots were used in various tests (YZ and XZ) for variable radius values (100, 150 and 300 mm) and feed rates ranging from 20 to 700 mm/s. The measurement was taken at the selected point with defined coordinates (0, 65, 149 mm) in the robot base coordinate system. Based on the results, the authors [11, 12] determined the relationship between the programmed flow rate of the effector effect and the recorded deviations but did not determine measurement uncertainty. It is therefore possible to conclude that these results are not precise and are only subjective. Uncertainty of measurement, besides accuracy, repetition, resolution, or positioning errors, is one of the most important concepts in research procedures and in defining individual errors of industrial robots. [15] Measurement uncertainty is one of the most important tasks of utmost relevance to accurate measurement of a given quantity and includes a number of uncertainties regarding the impact of major factors influencing the measurement result [2, 15-19]. From literary analysis, it is clear that uncertainty is an interval that characterizes the scattering of values of a given quantity that may be reasonably attributed to the true value (measurand) calculated from the measured information. Uncertainty is defined as a certain "doubt" about the value of the measurement result. It is presented as the range around the measurement result, within which, with a certain probability, there lies the real value of the measured quantity (Fig. 1) [1, 2, 15-20].

* Corresponding author: j.jozwik@pollub.pl

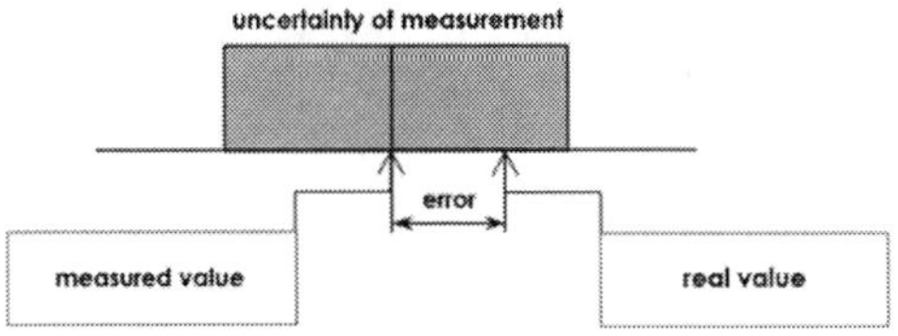

Fig. 1. Graphical representation of error and of measurement uncertainty [2].

Regarding the types of uncertainty, we distinguish standard uncertainty u or u(x), complex standard uncertainty Uc, and expanded uncertainty U or U (x) or Uc (x). [15] The estimation of measurement uncertainty of an industrial robot presents a number of different difficulties, discussed in the literature [2, 8], the evaluation of which requires an experimental study [2, 15, 19-26]. Only proper estimation of uncertainty will allow for objective estimation of robot inaccuracies and selection of the optimum measurement method [1, 2]. The theory of measurement uncertainty has now replaced the theory of measurement accuracy (measurement error). Uncertainty theory is commonly recognized, universal and accepted by all metrology organizations, and therefore it is important to remember that accuracy of any measurement is described by its uncertainty [20-26].

2 Research methodology

2.1 Research stands

During experimental works required for the estimation of statistical components of the kinematic telescopic bar, the following positions and computerized measurement and calculation systems were used:

- industrial robot MOTOMAN HP20, as research object,
- measurement system QC20 Ballbar,
- calculation environment Excel,
- calculation environment Statistica,
- specialized software for data recording and analysis Ballbar 20 v.5.

2.1.1 Object of research

The object of the conducted research was measurement uncertainty assessed with the kinematic telescopic bar of circular displacement of a universal of a six-axis industrial robot MOTOMAN HP20 by Yaskawa (Fig. 2a). All MOTOMAN HP20 work axes are rotary axes. The ranges of movement of the particular axes of the robot are: S axis (±180 °), L axis (+155 °/-110 °), U axis (+255 °/-165 °), R axis (±200 °) +230 °/-50 °) and T axis (±360°). The axis of rotation of joint 1 is vertical, while the axes of joints 2 and 3 are horizontal. The maximum distance in the axis perpendicular to the base plane is 3063 mm, and in the axis parallel to the base plane 1717 mm. Repeatability of the positioning based on the robot calibration certificate and the manufacturer's assessment is equal to 0.06 mm.

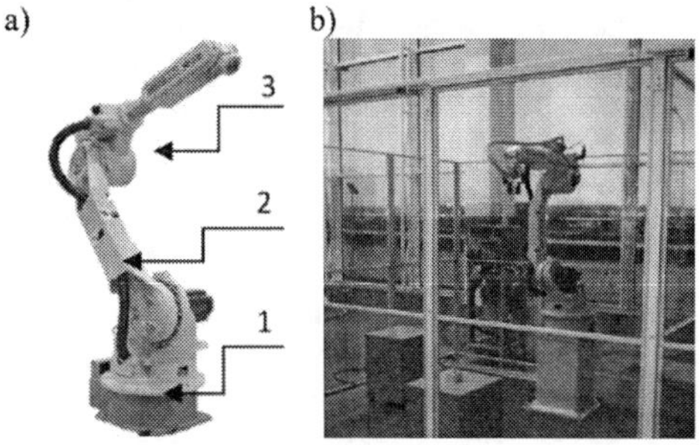

Fig. 2. Industrial robot HP20, a) general view, b) robot's positioning on research station, 1, 2, 3 – robot connectors.

The maximum lifting capacity of the robot is 20 kg. The robot is controlled by YASNAC FS 100 controller.

2.1.2 Measurement set-up

As part of the accuracy test of the Motoman HP20 industrial robot, Renishaw QC20-W Ballbar Telescopic Tester was used to identify non-measurable laser interferometer errors. The QC20-W Ballbar enables performing a circle test for different radius values in three planes. The detailed parameters of the measuring device are: resolution of the transducer (0.1 μm), measurement accuracy ± (0.7 + 0.003 μ L) μm in the temperature range of 15-25 °C, where L is the length at which the error is measured), maximal sampling rate (1000 Hz), data transmission range (10m, Bluetooth Class 2), operating temperature (0-40 °C). The basic elements of the system are:

- a 100 mm long kinematic ball bar assembly,
- a set of extension rods with lengths of 50, 150 and 300 mm,
- magnetic handles,
- a calibrator,
- software for recording and analysis of data.

The measurement consists in registering the deviation of the circle while performing the path by an effector on a circular path in the selected plane for a fixed location and speed of movement. The magnetic grip of the effector (1) is attached to the robot gripper and the magnetic base (4) is placed on the table prepared for this purpose. Between the elements (1) and (4), a displacement transducer (3) of a specified length is placed to measure the deviations of radius displacement over the whole range of circular paths.

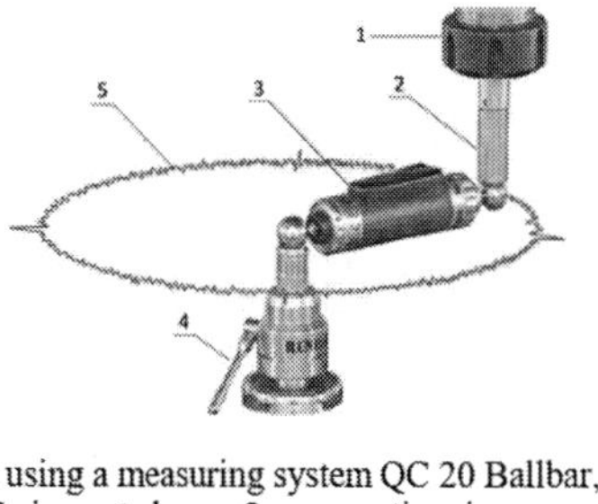

Fig. 3. Test using a measuring system QC 20 Ballbar,
1 – effector's impact sleeve, 2 – magnetic gripper,
3 – displacement transducer, 4 - magnetic base, 5 - circular measuring path.

DOI: 10.1051/itmconf/20171504013

Although the method is dedicated for CNC machine tools, to some extent and with some restrictions it can also apply to measuring accuracy of industrial robots, as demonstrated by the authors of these works [2, 5, 11, 12].

2.2 Test settings

The QC20-W Ballbar kinematic bar allows for identification of errors typical for numerically controlled machine tools such as: roundness deviation, slacks (reverse, transverse), lateral and parallel deviations, and many others; however, some of them may as well describe the properties of industrial robot. The authors point to the deviation of the circle as a representative measure of robot inaccuracies in a dynamic test during motion describing the ability of a path to follow on a circular outline. It should be pointed out that this type of path is often used when working in an industrial environment. Measurement using the QC20 Ballbar can be carried out in one of the three planes XY, XZ and YZ. As mentioned in section. 2.1.1, the first step prior to accession to the main works, was the establishment of research methodology. The measurement consisted in recording the radius deviation for a given radius while performing a circular path by an effector in a given plane for the selected path and feed value (Fig. 4). In order to ensure a constant linear velocity of the effector flow v_f, according to the recommended procedure, each programmed measurement motion was preceded by a runoff and a freefall. Measurements were carried out in the XY plane in the 0-360 ° range while moving clockwise CW and counter clockwise CCW (Fig. 4).

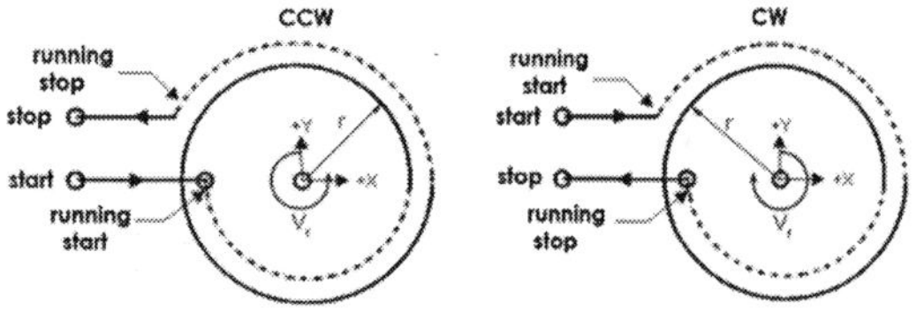

Fig 4. Method for accuracy measurement of HP20 robot with a telescopic kinematic bar QC20-W Ballbar by Renishaw [2].

The circle test was performed for three radius values r of the interpolated circle, equal to: 100 mm, 150 mm and 300 mm. The speed effector movement v_f was set at the levels of: 50 mm/s, 75 mm/s, 100 mm/s. For each test configuration, three repetitions were performed. Each change of length of the QC20 Ballbar telescopic bar was preceded by a calibration of the length of the bar arm. In the paper, the authors proposed their own research methodology, taking into account the measurement capabilities of the applied system and the specificity of the robot's operation. This methodology is optimized for the duration of the test and the possibility of estimating the inaccuracy of the robot in industrial conditions. Figure 5 shows the method for attaching a telescopic kinematic bar to a measuring stand. The magnetic centring base (2) is mounted in the robot axis on the metal element (1) placed on the table (in front of the robot) in the central part of the HP20 work area. This location of the centre of the interpolated circle in the base coordinate system of the robot is dictated by the most commonly used area of the robot. This does not mean, however, that this is the position in which the greatest value of the identified errors should be expected. The telescopic kinematic bar (3) was placed between the magnetic base (2) and the magnetic holder (4) attached to the robot's gripper (effector) (5). A computer with specialized Ballbar 20 V5 software was used to record measurement data, and was connected to a measuring device via Bluetooth.

Fig. 5. Method of attachment of the QC20-W Ballbar to the test position, where: 1 – metal element, 2 – centring magnetic base, 3 – telescopic kinematic bar, 4 – magnetic holder, 5 – robot's gripper (effector).

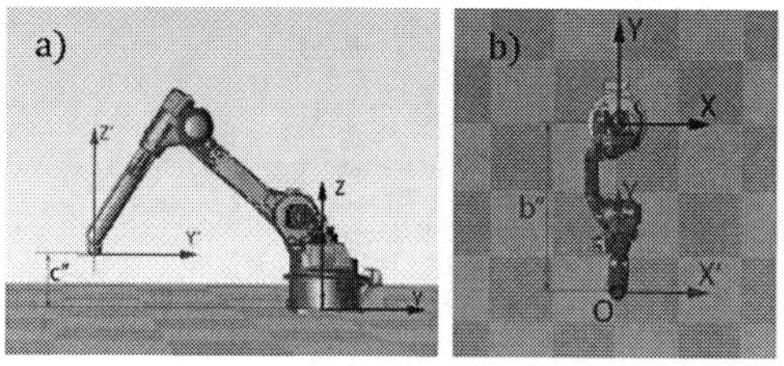

Fig. 6. Location of the centre of the circle in the base coordinate system of the robot: a) side view, b) top view.

In Fig. 6, the letter O denotes the centre point in the base coordinate system, with respect to which the position of the radius for realizing the circular outline path during the test is defined. The offset of the centre point in the Cartesian coordinate system relative to the beginning of the coordinate system of the studied robot were b" = 750 mm in the Y axis and c" = 200 mm in the Z axis, respectively. Measurement conditions, *i.e.* temperature, humidity and pressure were monitored and maintained on a constant level. The trials were conducted according to ISO 230-4 standard. Ballbar 20 V5 software was used to record measurement data, generate graphs and in extensive statistical analysis of results based on the ISO standard applied.

3 Test results

Considering the construction of the robot, the evaluation of the results of research with the kinematic telescopic bar was aimed at the analysis of universal inaccuracy measurements, the most important of which is the deviation of the circle and the path of the circular profile. The analysis of errors identified by the kinematic telescopic bar showed that most of them refer to the axis of the Cartesian coordinate system of CNC machine tools. Due to the fact that the errors identified by the kinematic telescopic bar may arise for reasons that do not occur in relation to robots (*e.g.* non-uniform stress of straight edge (no straight edge), mismatched parameters of the linear error compensation, overheating or damage to screws in one of the axes (lack of such screws) or misalignment of the guides), the authors focused on general measures of inaccuracy of industrial robots. In particular, they demonstrated the possibility of dynamic measurement of roundness deviation, which is a big advantage of the system and allows assessing the circular trajectory of a given radius in any angular position. For each measurement configuration, *i.e.* the radius r value, the path with a circular outline r and the linear velocity of the robot's effector v_f, experimental deviation of the circle was determined. Figure 7 presents an exemplary map of deviation of the circle in full angle defined for the measurement circle r = 100 mm and the constant linear velocity of the effector motion v_f = 50 mm/s.

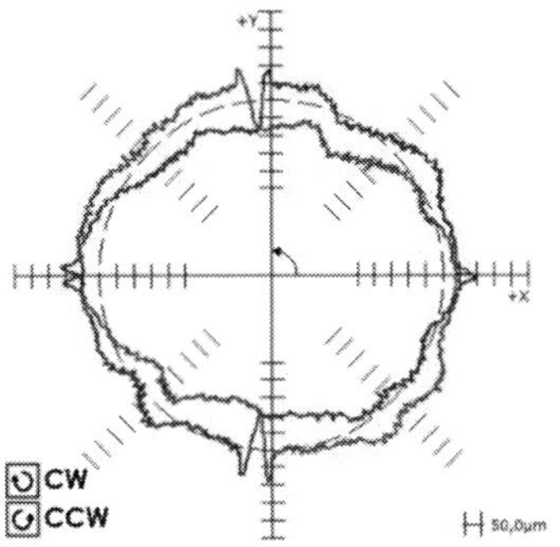

Fig. 7. Maps of the roundness deviation Δo for the interpolated circle with a radius r = 100 mm and vf = 50 mm/s.

The maps of the roundness deviation test for all conducted tests indicate "jumps" characteristic of backlash. The shape of the actual interpolated circle observed in test may indicate that there are loops in the kinematic arms of the robot related to their insufficient rigidity. The graphs also show the areas in which the deviation of the radius of amplitude is increased, especially at higher feed rates. This fact may be a sign of vibration during effector movement. It should be borne in mind, however, that while measurements using the QC20 Ballbar telescopic bar are local and refer to numerically controlled machine tools, the kinematic structure of industrial articulated robots is more complex. Therefore, full evaluation of accuracy requires examination of the whole robot work space. Also, the way of interpreting the different types of errors as well as their types should be adequate for the construction of the robot, its weight and working conditions. As with laser interferometer measurements, QC20 Ballbar testing should be performed when the robot is in a steady state. For this reason, only deviation of roundness Δ_o was assessed in the work.

The average value of roundness deviation Δ_o in the function of linear velocity of tool movement v_f and radius length r of the interpolated circle was presented in Figure 8.

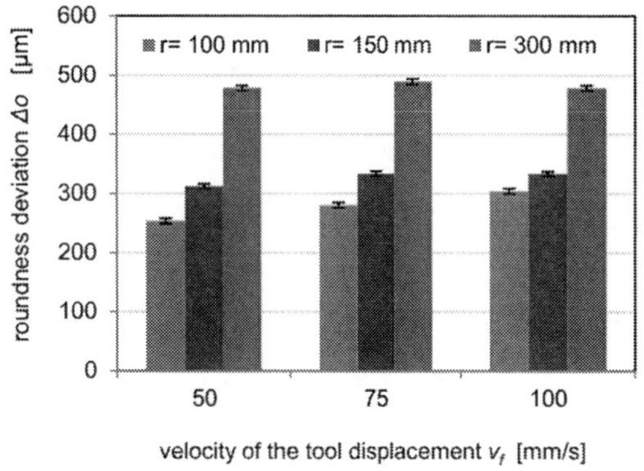

Fig. 8. Mean value of roundness deviation as a function of the linear velocity of the tool displacement v_f and the length of the r radius of the interpolated circle.

4. Estimation of uncertainty

The measurement uncertainty recorded with the QC20-W Ballbar the kinematic telescopic bar includes a number of component uncertainties, which take into account the effect of all errors affecting the measured result. As for laser interferometers, also for QC20-W Ballbar kinematics, these uncertainties are estimated based on the information contained in the calibration certificate of the instrument, but also on the basis of measurements and statistics. Table 1 presents the errors and uncertainties used in system testing. These data are based on QC20-W Ballbar calibration certificate. The other components of the kinematic measurement uncertainty of the telescopic bar were estimated on the basis of the results obtained with method A and non-statistical components were estimated with method B.

Table 1. The errors and inaccuracies of the kinematic telescopic bar used in the tests, determined on the basis of a calibration certificate (at an air temperature of 19.3 °C, 1029 mbar, relative humidity 42.3 %RH).

Positioning error	+/-0.5µm
Maximum error	+/-0.19µm
Uncertainty (k=2)	0.24µm
Uncertainty of pattern, for Zerodur during calibration of length: 50, 100, 150, 300 (k=2)	0.00085µm

Extended measurement uncertainty U using QC20 Ballbar system was determined from the equation (1).

$$U = \pm u_c k \quad (1)$$

where: u_c - complex standard uncertainty, k – expansion factor.

 DOI: 10.1051/itmconf/20171504013

Complex standard uncertainty u_c was derived from equation (2), including component u_A from the scattering of the measurement result estimated with method A (3), and the components of measurement uncertainty determined with method B, obtained from the non-statistic sources (4).

$$u_c = \pm\sqrt{u_A^2 + \sum_{i=1}^{n} u_{Bi}^2} \quad (2)$$

where: u_A – standard uncertainty estimated with the method A (statistical), u_{Bi} – standard uncertainties non-statistical, estimated with method B

$$u_A = \pm\sqrt{\frac{\sum(x_i-\bar{x})^2}{n(n-1)}} \quad (3)$$

where: x_i – single measurement result, $\bar{x}$ – average measurement results, n – number of measurements

Non-statistical components were designated as the sum of centring uncertainties u_{BP}, calibration uncertainties u_{BK} and calibration error u_{BE}. Equation (4) describes the sum of squares of uncertainties estimated by B method.

$$\sum_{i=1}^{n} u_{Bi}^2 = u_{BP}^2 + u_{BK}^2 + u_{BE}^2 \quad (4)$$

Taking into account the above, the complex standard measurement uncertainty with the kinematic telescopic bar was determined from the relationship (5):

$$u_c = \pm\sqrt{u_A^2 + u_{BP}^2 + u_{BK}^2 + u_{BE}^2} \quad (5)$$

Each component was determined with consideration of the data included in the system calibration certificate. Therefore, when accepting a rectangular distribution of probability, component u_{BP} was determined as (6), and component u_{BK} was determined from equation (7), with the assumption of normal distribution. Component u_{BE} was determined from equation (8) assuming a rectangular probability distribution.

$$u_{BP} = \pm\frac{0.5}{\sqrt{3}}\ [\mu m] \quad (6)$$

$$u_{BK} = \pm\frac{0.24}{2}\ [\mu m] \quad (7)$$

$$u_{BE} = \pm\frac{0.19}{\sqrt{3}}\ [\mu m] \quad (8)$$

In the analysed case, the uncertainty resulting from the temperature measurement and the coefficient of thermal expansion were omitted. The assumption of the negligence of the omitted uncertainties was taken into account because of the direct calibration of the QC20 Ballbar the kinematic telescopic bar prior to the measurement itself and a very short measurement time of several minutes. The authors of the study, performing the measurement in thermally stable conditions, where the ambient temperature did not change by more than 0.2 °C, assumed that during the measurement no significant changes in the length of the measuring arm during the test would occur. Thus, given the above, these uncertainties were considered insignificant. The uncertainty surrounding the pattern from the Zerodur was also omitted. The basis for not taking into account the uncertainty of extending the pattern is its extremely low (near zero) coefficient of thermal expansion. This coefficient is practically independent of the temperature, especially in the study range (Zerodur in the range 0 to 50 °C has an average thermal expansion at the level 0 ±0.007×10⁻⁶ K⁻¹).

Table 2. Uncertainty and its components when measuring deviation of the arm circumference of the test robot with the kinematic telescopic bar.

Uncertainty	R=100 mm		
	v_f=50 mm/min	v_f=75 mm/min	v_f=100 mm/min
	Range of uncertainty values [µm]		
u_A	±4.116	±3.648	±1.304
u_{BP}	±0.289	±0.289	±0.289
u_{BK}	±0.120	±0.120	±0.120
u_{BE}	±0.110	±0.110	±0.110
u_c	±4.130	±3.663	±1.345
U (k=2)	±8.260	±7.326	±2.690
	R=150mm		
u_A	±1.689	±1.348	±0.748
u_{BP}	±0.289	±0.289	±0.289
u_{BK}	±0.120	±0.120	±0.120
u_{BE}	±0.110	±0.110	±0.110
u_c	±1.721	±1.388	±0.818
U (k=2)	±3.442	±2.776	±1.636
	R=300mm		
u_A	±1.205	±2.317	±3.846
u_{BP}	±0.289	±0.289	±0.289
u_{BK}	±0.120	±0.120	±0.120
u_{BE}	±0.110	±0.110	±0.110
u_c	±1.250	±2.340	±3.860
U (k=2)	±2.500	±4.680	±7.720

Based on the assumptions adopted in the paper, computational formulas were developed and the values of individual components of uncertainty determined. Uncertainty and its components at k = 2 for kinematic deviation of the telescopic bar for the test arm length r = 100 mm, r = 150 mm and r = 300 mm, and the armature effect velocity equal to v_f (50, 75 and 100 mm/min) were included in Table 2. Table 2 clearly indicates that the dominant component of the uncertainty budget for measurement with the kinematic telescopic bar is standard uncertainty u_A (statistical component resulting from the scattering of measurement result) and the uncertainty of centring the QC20-W Ballbar on the station u_{BP}. The consequence of the high value of standard uncertainty is the value of complex standard uncertainty u_c and extended uncertainty U. Other components of uncertainty, *i.e.* uncertainty of calibration u_{BK} and calibration error u_{BE} only slightly affect the value of the complex standard uncertainty u_c and expanded uncertainty U. The results presented in Table 2 appear to indicate that for most of the studied cases (r = 100mm and r = 250mm) v_f of the robot during circular interpolation implies a decreasing tendency to change the uncertainty of measurement, which means that its accuracy increases. It should be added that any result of the deviation of the roundness in the measured operating conditions should be recorded with the appropriate (for the arm r and effector movement velocity v_f) uncertainty.

5. Conclusion

The analysis of uncertainty of kinematic measurement taken with the QC 20-W Ballbar telescopic bar show that components of the uncertainty budget take relatively small values ($u_c = 0.818 \div 4.130$; $U = 1.636 \div 8.260\ for\ k = 2$, which proves that a properly selected method was applied to the research task. Focusing on universal inaccuracies (roundness deviation) and the benefits of the QC 20-W Ballbar system for industrial robots enables evaluation of uncertainty of a circular trajectory of a given radius. The ability to identify circular deviation for different radius values of an interpolated circle allows an assessment of the impact of measuring length in the industrial robot space (from one attachment) to the uncertainty value of the measurement in all perpendicular planes, without changing both the program and the bar orientation. This significantly shortens the test run and the ability to assess standard uncertainty.

It should be noted that QC 20-W Ballbar the kinematic telescopic bar measurements are performed as dynamic measurements, i.e. in certain positions and typically without load on the robotic arm of the displaced elements.

Based on the results of the research and calculation, it can be concluded that the kinematic telescopic bar test results are subject to very low measurement uncertainty and therefore high accuracy.

References

1. J. Honczarenko, Roboty przemysłowe. Budowa i zastosowanie (WNT, Warszawa, 2010)
2. J. Józwik, D. Ostrowski, Wybrane problemy badawcze robotów przemysłowych (Wydawca Politechnika Lubelska, 2016)
3. W. Zhenhua, X. Hui, Ch. Guodong, S. Rongchuan, L. Sun, Ind. Rob. **41**, 5 (2014)
4. J. Józwik, P. Pieśko, G. Krajewski, Maintenance and Reliability, **47**, 3 (2010)
5. M. Abderrahim, A. Khamis, S. Garrido, L. Moreno,Accuracy and Calibration Issues of Industrial Manipulators (Industrial Robotics - Programming, Simulation and Applications, 2006)
6. K.L. Conrad, P.S. Shiakolas, T.C. Yih, Robotic calibration issues: accuracy repeatability and calibration (Proceedings of the 8th Mediterranean Conference on Control & Automation, Rio, Patras, Greece, 2000)
7. B. Greenway, Ind. Rob. **27**, 4 (2000)
8. J. Józwik, I. Kuric, L. Semotiuk, Communications **3A** (2014)
9. J. Józwik, M. Czwarnowski, Adv. Sci. Technol. Res. J. **9**, 28 (2015)
10. J. Józwik, L. Semotiuk, I. Kuric, Adv. Sci. Technol. Res. J. **9**, 28 (2015)
11. M. Slamani, A. Joubair, I. Bonev, Ind. Rob. **42**, 6 (2015)
12. M. Slamani, A. Nubiola, I. Bonev, Ind. Rob. **39**, 1 (2012)
13. K. Young, C.G. Pickin, Ind. Rob. **39**, 6 (2000)
14. B. Shirinzadeh, Ind. Rob. **25**, 1 (1998)
15. P. Majda, D. Rozkrut, Adv. Manuf. Sci. Technol. **39**, 3 (2015)
16. G.M. Krolczyk, J.B. Krolczyk, R.W. Maruda, S. Legutko, M. Tomaszewski, Measurement **88** (2016)
17. A. Glowacz, Z. Glowacz, Infrared. Phys. Technol. **81** (2017)
18. E. Feldshtein, J. Józwik, S. Legutko, Adv. Sci. Technol. Res. J. **10**, 30 (2016)
19. G. Alici, B. Shirinzadeh, Mech. Mach. Theory **40**, 8 (2005)
20. A. Angelidis, G.Ch.Vosniakos, Int. J. Precis. Eng. Man. **15**, 1 (2014)
21. C. Gong, J. Yuan, J.Ni, Int. J. Mach. Tool Manu. **40**, 14 (2012)
22. P.S. Shiakolas, K.L. Conrad, T.C. Yih, Int. J. Model. Simul. **22**, 3 (2002)
23. J. Swevers, W. Verdonck, J.DeSchutter, IEEE Contr. Syst. Mag. **27**, 5 (2007)
24. G.M. Krolczyk, P. Nieslony, J.B. Krolczyk, I. Samardzic, S. Legutko, S. Hloch, S. Barrans, R.W. Maruda, Measurement **70** (2015)
25. P. Nieslony, G.M. Krolczyk, K. Zak, R.W. Maruda, S. Legutko, Precis. Eng. **47** (2017)
26. G.M. Krolczyk., R.W. Maruda, P. Nieslony, M. Wieczorowski, Measurement **94** (2016)

ITM Web of Conferences 15, 05001 (2017)
CMES'17
DOI: 10.1051/itmconf/20171505001

An attempt to use FMEA method for an approximate reliability assessment of machinery

Krzysztof Przystupa[1,*]

[1]Lublin University of Technology, Mechanical Engineering Faculty, Department of Automation, Nadbystrzycka Street 36, 20-618 Lublin, Poland

Abstract. The paper presents a modified FMEA (Failure Mode and Effect Analysis) method to assess reliability of the components that make up a wrench type 2145: MAX Impactol TM Driver Ingersoll Rand Company. This case concerns the analysis of reliability in conditions, when full service data is not known. The aim of the study is to determine the weakest element in the design of the tool.

1 Introduction

Reliability is an ability of devices to fulfil specific functions. It is characterised by models of the functions of reliability distribution that are time-based. These models are associated with a function that determines the intensity of damage occurring during operation. These are statistical dependencies relating to a certain population of devices used under certain conditions. Reliability of devices is shaped by factors which can be attributed to groups related to:

• Construction,
• Use of materials and production technologies,
• Conditions of use.

When deciding on use of devices for specific tasks, conformity of their operating conditions with operating conditions intended by the equipment producer [1] should be determined. It is often necessary to use devices that are designed to operate under conditions not anticipated by the producer. There is also a scenario of determining suitability of the device for use in conditions that have not been tested to improve the construction of the device.

In practice, equipment acceptance tests are used. They involve adoption of certain number of failures and an assumption of significance level for which they are determined and then the probable time of use between consecutive failures. Unfortunately, these tests do not meet expectations regarding the probability of failure of individual parts of the device.

In order to determine reliability of the equipment, it is necessary to carry out extensive tests in a form of statistical dependence for specific conditions. It is a laborious and costly task.

In this work, an attempt was made to use the FMEA method to estimate reliability of an Ingersoll Rand Company 2145 MAX Impact™ wrench dedicated for the automotive industry and used in the mining industry.

2 Characteristics of the device and its conditions of use

The view of the device is presented in Fig. 1.

Fig. 1. A view of the wrench type 2145: MAX Impactol TM ¾" Driver produced by Ingersoll Rand Company (images taken from http://www.ingersollrandproducts.com).

Basic technical parameters of the impact wrench:

• Impact rate: 1 150 BPM.
• Max speed: 7 000 RPM.
• Average speed: 6 300 RPM.
• Noise level: 91.1 dB.
• Average air consumption: 241 L/min.
• Max air consumption: 906 L/min.

According to the producer, the impact wrench is designed for mounting and dismantling screw connections. This task is accomplished by generating a specific torque of impulsive force. Impact wrenches ensure that the maximum torsional moments are not exceeded during the screwing cycle, thus, ensuring repetitive tensile stresses in the bolts and preventing

* Corresponding author: k.przystupa@pollub.pl

 DOI: 10.1051/itmconf/20171505001

damage to the joints. Reliability and durability during assembly and disassembly of screw connections in service conditions of keys is a decisive feature of their usefulness [2]. This feature is supported by easy control, easy torque change and activation (switch on/off).

These types of pneumatic impact wrenches are designed for various types of assembly and repair works. In particular, they are widely used in the automotive industry, and less often in heavy works e.g. in mining. In hard coal mining, pneumatic wrenches are used during assembly and dismantling of mining enclosures, and therefore, used in significantly less favourable conditions. The operation of pneumatic impact wrenches in the mining industry is characterised by very difficult conditions, different from other applications. They include:

- High level of dustiness,
- High humidity, or even drowning a device in muddy waters,
- Haste in work,
- Difficult, frequently improper conditions of device storage (during brakes).

The minimum time required to use keys in their stand-by mode should ensure that all work, including the assembly of the mining enclosure, is performed in the full and exclude the risk of losses due to damage and loss of required functions.

An important problem is the serviceability of the device (wrench), understood as the ability to maintain or reconstruct a functional state after failure, to fulfil the intended tasks [3-4]. This problem is directly related to the construction of three main functional units: pneumatic motor, impactor and compressed air control unit. The pneumatic motor accelerates the inertia element, which periodically hits the projections of the wrench operating shaft. The outer mass, which is driven by the pneumatic motor, transmits through the resilient combination of energy colliding elements to the inertia element. By the stiffness of the seat-screw system (spring mechanism), the inertia element transfers energy to the screw, which in the model is treated as a substrate. During operation, the inertial element periodically collides with a very rigid screw-socket system and, as a result, a sudden (in-step) fall in inertial velocity occurs. The jumping drop and increase in inertial velocity, in turn, results in the generation of torque applied to a short impulse with a significant peak value. The double oscillation system used in impact wrenches enables very efficient transfer of energy from the impactor to the screw. The individual wrench solutions [8-11] differ in the construction of the inertial element and the coupling mechanism and in the inertia element.

In the present case, the system of pneumatic drive consists of 9 elements, the drive control system also consists of 9 elements, while the impactor consists of 8 elements.

3 FMEA method and its use

FMEA method (Failure Mode and Effect Analysis) does not use experimental data as standard, but it involves a methodical analysis of the future design. For this reason, the results of this analysis can be used to eliminate defects in future products or future processes [12]. Improvement in construction is obtained by identifying the causes of potential defects, analysing the possibility of failure, applying effective preventive measures and avoiding the occurrence of identified hazards in new products or processes. This is achieved with knowledge and experience from previous experiments and similar works.

The FMEA was developed in the 1960s for the needs of the American space program Apollo-Saturn. It is a current standard [12] in various industries, where the product is required to be particularly reliable for its safety. The method allows to verify the design, remove sources of product defects inherent in its design and build the production process. This saves a lot of money to cover the so-called "bad quality costs". As a result, cheaper and better-quality products can be produced; thus, a manufacturer gets a better competitive position.

The method works in a variety of conditions: in serial and unit production – wherever a defect in a product can expose the manufacturer to serious financial losses. Its use is consistent with the principle: it is better to prevent than repair.

Another important feature of the FMEA is that the project can undergo further analyses, introducing improvements that are increasingly effective in eliminating sources of product defects. Project analyses can provide new ideas to improve the properties of the product. In this way, the FMEA method is entered in the Deming Wheel Cycle: P-D-C-A (Plan-Do-Check-Act). It makes it very useful both in ISO 9001 quality assurance systems, as well as in TQM-compliant systems [13].

This method consists in assigning a numerical indicator to all components of the device, and then in ordering the part according to the likelihood of failure.

The FMEA does not specify, for example, the average time between failures or reliability function, but it can be used to describe a type, effect, cause and severity of damage to the functional units of a device that are globally defined by the so-called risk priority number. A high value of this number, referring to specific construction elements, indicates the appropriateness of using preventive measures to improve the reliability of the device.

Figure 2 shows the structure of the modified FMEA method. The introduced change consists in adapting this method to the analysis of an existing structure, rather than those being designed and used in the criteria determining the risk of failure in MTBF time. The method is limited to numerical determination of three characteristic elements of every construction:

- Probability of the event occurring (LPW),
- Severity of the event (LPZ),
- Detection of the event (LPO).

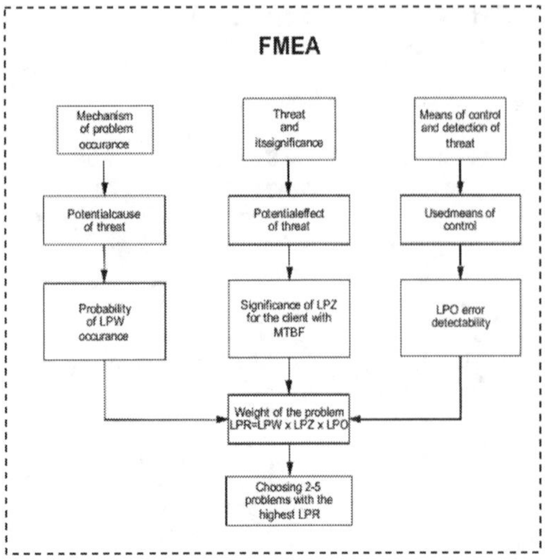

Fig. 2. The structure of a modified FMEA.

For each failure, one should calculate the LPR, which is an index of event priority (the weight of the problem), which is the product of the values of the indicators: LPW, LPZ, LPO. A high value of the LPR indicates a high importance of the part in the design of the device, with the high probability of failure of that part and the high severity of the failure for the user.

4 Practical example

A practical use of the FMEA method was demonstrated in the example of the impact wrench 2145: MAX Impactol. The basic element of FMEA analysis is the method of evaluating the value of three basic indicators (LPW, LPZ, LPO). They are necessary to prioritise the threat and are defined in Table 1. For each element of the structure, numerical weights of the problem are assigned. They were identified by answering the following questions:

1. What is the probability of a failure in each element (defining LPW)?
2. If the failure occurs, what is its severity for the user (defining LPZ)?
3. If the failure occurs, what is the probability of detecting this even by the user (defining LPO)?

For each of the questions listed, consider the values in Table 1 referenced for each element of the structure. The values were expressed by numbers of the answers to the individual questions. This value will be used to rank the "importance of events".

Additionally, to determine the LPW, LPZ and LPO values, MTBFs (Mean Time Between Failures) were estimated based on data provided by the service.

Mean time between failures of a given device is calculated from the formula [5-7]:

$$MTBF = \frac{t_e \cdot 100}{r} \qquad (1)$$

Where:
r – number of failure of a given element per 100 failures,
t_e – time of constant use (in this case c.a. 2480 hours).

The significance of MTBF for shaping the criteria LPW, LPZ AND LPO are presented in Table 1.

The LPW, LPO values (Table 1) were determined by estimation, based on work experience. On the other hand, LPZ (Table 1) was adopted using an additional criterion (MTBF) not part of FMEA analysis as defined in standard [12].

Table 1. Numerical evaluation of risk priority number (PRN).

<table>
<tr><th>Probability</th><th>LPW</th><th>Severity</th><th>LPZ</th><th>Detection</th><th>LPO</th></tr>
<tr><td rowspan="2">Failure occurrence within 2 years: 0<n<30.</td><td rowspan="2">3</td><td>MTBF >6000 hours.</td><td>1</td><td>Failure grows slowly and is easy to observe in exploitation.</td><td>2</td></tr>
<tr><td>MTBF 4000÷6000 hours.</td><td>2</td><td>Failure grows slowly and is difficult to observe.</td><td>5</td></tr>
<tr><td rowspan="2">Failure occurrence within 2 years: 30<n<120.</td><td rowspan="2">6</td><td>Failures grow in an observable manner and lead to small limitation of performing element's function MTBF is 2000÷4000 hours.</td><td>4</td><td rowspan="2">Failure grows fast and is observable.</td><td rowspan="2">7</td></tr>
<tr><td>For failures grow in an observable manner and lead to limitation of performing element's function MTBF is 1000÷2000 hours.</td><td>5-6</td></tr>
<tr><td rowspan="2">Failure occurrence within 2 years: n>120.</td><td rowspan="2">10</td><td>For failures grow in a step manner and lead to significant limitation of performing element's function MTBF is 500÷1000 hours.</td><td>7-8</td><td rowspan="2">Failure grows fast and is not possible to observe.</td><td rowspan="2">10</td></tr>
<tr><td>For failures that cause malfunctioning of the element and generate losses for the client MTBF<500 hours.</td><td>9-10</td></tr>
</table>

The numerical results of LPW, LPZ, LPO and numerical result of FMEA analysis in a form of LPR were calculated by multiplying LPW, LPZ and LPO values for individual parts of the wrench are presented in Table 2.

Table 2. Event priority number (PRN) was calculated according to FMEA methodology.

Name of system	**Element**	LPW	LPZ	LPO	**LPR**
Pneumatic drive	Rotor front bearing	10	10	1	100
	Rotor rear bearing	10	10	1	100
	Cylinder	10	3	10	**300**
	Casing	6	10	2	120
	Blade package	6	3	7	126
	Rear bearing distance of the rotor	6	3	2	36
	Engine washer	6	3	2	36
	Rotor	6	5	7	210
	Front plate	6	3	10	**180**
Drive control	**Buttons set**	10	10	10	**1000**
	Rotary valve assembly	10	5	5	250
	Trigger assembly	10	5	2	100
	Spacer inlet sleeve spacer	10	3	5	150
	Set of inlet assembly	6	3	5	90
	Filter of inlet sleeve	3	3	7	63
	Power control knob	3	5	2	30
	Rotary knob	3	5	2	30
	Inlet band sleeve	3	10	10	**300**
Impactor	Impactor chamber screw	10	3	2	60
	Impactor chamber sleeve	6	5	2	60
	Impactor chamber washer	6	3	2	36
	Hammer frame assembly	6	3	10	180
	Impactor hammer	6	5	2	60
	Impact chamber assembly	6	10	10	**600**
	Pin	6	10	2	120
	Hammer stroke	6	10	2	120

The results presented in table 2 are based on three functional assemblies consisting of the wrench construction. For each team, two parts with the highest LPR value were selected. These are important parts in the design of the device, with the high probability of failure of this part and high severity of the failure for the user. In the next step for these parts, a program of actions to improve the construction and the program of actions preventing the occurrence of the failure have been developed.

5 Conclusions

FMEA analysis can be used for a rapid, descriptive assessment of reliability of products construction for use in industry. The illustrated example of a practical use of the method shows that the FMEA method can be modified in a simple and fast way for specific needs, e.g. by considering other additional factors such as MTBF.

Because of the analysis for the Ingersoll Rand impact wrench Type 2145 MAX Impactol TM ¾ that the most likely damage was found in:

• Pneumatic motor assembly: cylinder and front plate,
• The control unit: the set of buttons and the sleeve of the intake assembly,
• Impact assembly: the impact chamber assembly.

The results obtained have been confirmed by the complex reliability analyses carried out by the wrench service company.

At this point I would like to thank prof. dr hab. eng. Stanisław Płaska (Lublin University of Technology) for providing figures and giving valuable comments and consultations during the preparation of this text.

References

1. M. Klank, Przegląd Górniczy. **12** (2013)
2. S. Szymaniec, Napędy i sterowanie. **18,** 66-72 (2016)
3. P. Fuchs, J. Zajicek, Maint. Rel. **15**. (2013)
4. W. Zhigang et al., SAE International Journal of Materials and Manufacturing. **9**, 303-314 (2016)
5. M. Morelli, Reliability and Maintainability Symposium (RAMS), 2016 Annual. IEEE (2016)
6. L. Xiaolei,. Revista de la Facultad de Ingeniería, **31.7** (2016)
7. R. Langlois-Berthelot, *Trwałość niezawodność funkcjonalność wyrobów przemysłowych* (WNT, Warszawa, 1972)
8. US 6.186 332 B1
9. US 6.889.778 B1
10. US 8.074.732
11. WO 2012/138721
12. IEC 812
13. PN-EN ISO 14121-1:2008

 DOI: 10.1051/itmconf/20171505002

Analysis of a planetary gear modelled with a contour graph considering the decision making complexity of game-tree structures

Adam Deptuła[1,*], *Józef* Drewniak[2], and *Marian A.* Partyka[1]

[1]Opole University of Technology, Faculty of Production Engineering and Logistics, Institute of Processes and Products Innovation, Ozimska 75, 45-370 Opole, Poland,
[2]University of Bielsko-Biala, Faculty of Machine Building and Computer Science, Department of Mechanical Engineering Fundamentals, Willowa 2, 43-309 Bielsko-Biała, Poland

Abstract. Analysis and synthesis of mechanisms is one of the fundamental tasks of engineering. Mechanisms can suffer from errors due to versatile reasons. Graph-based methods of analysis and synthesis of planetary gears constitute an alternative method for checking their correctness. Previous applications of the graph theory concerned modelling gears for dynamic analysis, kinematic analysis, synthesis, structural analysis, gearshift optimization and automatic design based on so-called graph grammars. Some tasks may be performed only with the methods resulting from the graph theory, *e.g.* enumeration of structural solutions. The contour plot method consists in distinguishing a series of consecutive rigid units of the analysed mechanism, forming a closed loop (so-called contour). At a later stage, it is possible to analyze the obtained contour graph as a directed graph of dependence. This work presents an example of the application of game-tree structures in describing the contour graph of a planetary gear. In addition, complex parametric tree structures are included.

1 Introduction

Engineering practice requires correct evaluation of the mathematical model describing a given system with certain variables. A proper mathematical model contains a group of functions joining different variables and describing connections between quantities in the system. Decision tables [1-3] and logical functions [3-4] can be applied in simulation of machine systems, for example described by ordinary or partial differential equations. It results from the fact that the occurring nonlinear elements can be divided into the finite number of linear elements (parts), and as a consequence we obtain a linear simulation system from the primarily single nonlinear system. Analysis and synthesis of mechanisms can be performed by means of versatile methods. These tasks can suffer from human errors. It is therefore reasonable to employ some alternative methods which will allow for comparison of results and for detection of almost unavoidable mistakes. The graph-based methods deliver such an alternative approach to modelling of a wide class of mechanical systems. There have also been other attempts to model planetary gears via diagrams *e.g.* Wolf's pictograms [5], but those methods did not further evolve due to lack of generalization and lack of connections with other branches of mathematics [6, 7]. Furthermore, a method based on the signal flow graph theory for modelling gears [8, 9] has not been too frequently used. On the contrary, the graph based methods have been independently and intensively developed in recent years all over the world (see *e.g.* [10]). The superiority of the graph-based modelling of mechanical systems consists in the fact that a graph is inseparably connected with other algebraic structures like *e.g.* sub-graphs (*e.g.* trees, cycles, cliques, paths), dual graph, matrices, vector spacer of cuts and cycles, structural numbers and matroids as well as algorithms connected with these structures. By interpreting the mechanical knowledge In terms of graphs, we solve mechanical problems via graph models [11]. Versatile linear graph models were considered by Tsai [12] but Hsu's graph models of gears [13] are utilized in the following considerations. In the case of analysis of car gearboxes, the graph-based methodology is also efficient when incorporating additional transformations of a basic graph according to a particular gear driver [5]. An introductory comparison of the graph and contour methods for analysis of gears was done by the authors in [5, 10].

The occurring constructional parameters influence courses of unknown functions dependent on time. A traditional analysis, such as Input → Output and Output → Input, of a given system with the dependence graph method enables obtaining the vertex groups of the following properties:

- elements inside the given group contain many information connections,

* Corresponding author: a.deptula@po.opole.pl

 DOI: 10.1051/itmconf/20171505002

- particular groups have few reciprocal information connections.

In this way we obtain constructional recommendations for obtaining multiple solutions from the dependence graph, so the selection of suitable sub-solutions can be done with use of tree classifiers from the neuron networks. Let us consider a different approach: the translation of the dependence digraph into the tree game structure [14, 15]. Constructional and service parameters may be joined in different mathematical ways, it is necessary to determine precise decomposition of the dependence graph in order to define decision variables. Such procedure may determine the area of feasible solutions allowing to select a proper optimization procedure.

2 The dependence graph for tree game structures

The equations of dynamics can be used for determination of mutual connections of all the functions dependent on time. As a result of notation and decomposition of the dependence graph of those functions we obtain the groups of distribution which describe properties of successive subsystems of the considered machine system and a set of suitable constructional and service parameters.

An oriented graph can be defined as an ordered pair of sets. The graph vertices are included in the first one, whereas graph edges, which is an ordered pair of vertices, are included in the other (Fig. 1) [14, 15].

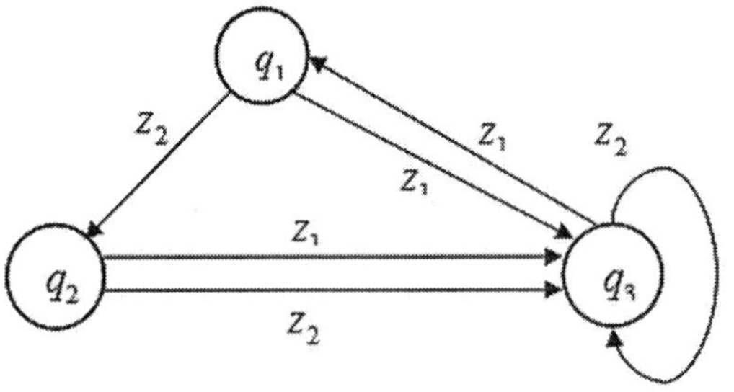

Fig. 1. An oriented game graph.

The oriented game graph shown in Fig. 1 is composed of a set of vertices Q:

$$Q=\{q_1,q_2,q_3\} \tag{1}$$

and of a set of edges Z, which is an ordered pair of vertices:

$$Z=\{z_1,z_2\}. \tag{2}$$

As a result of a graph distribution from the chosen vertex, a tree structure with cycles is obtained in the first step and then, a general game tree structure is obtained. Each of them has an appropriate analytical formulation G_i^+ and G_i^{++}.

A game tree structure is a part of a systematic research method. A start vertex q_i is chosen in the first step. Acting in accordance with the algorithm and assuming that the start vertex is q_1 it is possible to transform the oriented dependence graph presented in Figure 1 into an analytical formulation $G_{q_1}^+$, and then we obtain the following formulation as a result of the operation (3).

$$G_1^+=({}^0q_1({}^1z_1q_3({}^2z_1q_1,z_2q_3)^2,z_2q_2({}^2z_1q_3,z_2q_3)^2)^1)^0 \tag{3}$$

The next step is to obtain a tree structure that plays a parametrically defined expression [14, 15] (4):

$$G_1^{++}=({}^0q_1({}^1z_1q_3({}^2z_1q_1^1,z_2q_3^1)^2,z_2q_2({}^2z_1q_3({}^3z_1q_1^2,z_2q_3^1)^3,\; z_2q_3({}^3z_1q_1^2,z_2q_3^1)^3)^2)^1)^0 \tag{4}$$

It is possible to return to an earlier vertex and even to a start vertex from an appropriate end vertex; we obtain a game tree structure presented in Fig. 2.

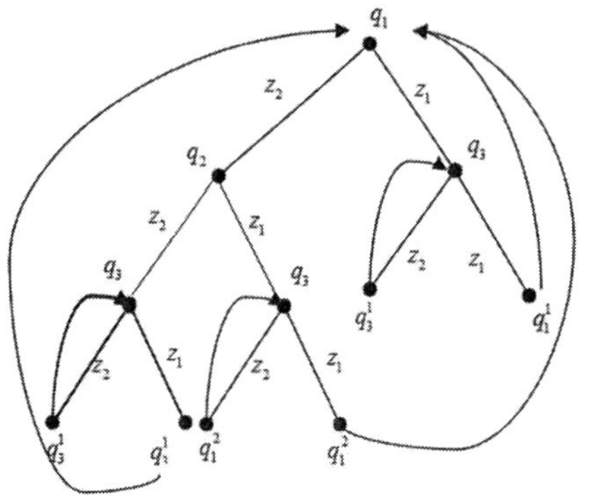

Fig. 2. Tree structure with cycles and the initial vertex q_1.

When we change cycles in the structure shown in Fig. 2 into decisions and/or expressions we obtain a game tree structure presented in Fig. 3.

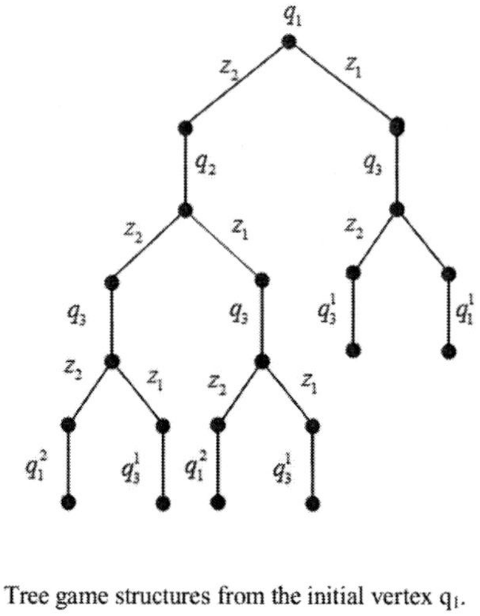

Fig. 3. Tree game structures from the initial vertex q_1.

The contour graph distribution from any vertex in the first stage leads to a tree structure with cycles, and next to a general tree game structure. Each structure has a proper analytic notation G_i^{++}- where *i*- is a vertex, from which the graph decomposition started, determining a way of transition from the dependence graph to the tree structure.

3 Graph based model of a planetary gear

The general idea of the graph-based modelling of mechanical systems consists in the following steps [16]:

- discretisation of a mechanical system. This means that appropriate simplifications have to be made. Some aspects are omitted. Some structural elements are considered as essential and they are interpreted as graph vertices. Some connections or relationships between these elements are abstracted. They are represented via graph edges, usually a system of weights can be assumed to the edges or vertices and edges,
- assignment of the graph to the mechanism (especially a planetary gear) based upon special rules. There are several different rules depending on the object of modelling and problems solved via the graph based method,
- derivation of special subgraphs, *e.g.* f-cycles or contours; these subgraphs can be singled out based upon the graph-theoretical algorithms, which causes that the approach is simple and algorithmic,
- listing the codes of these graph elements; the encoding rules are clearly defined what allows for avoidance of mistakes,
- generation of a system of equations in an algorithmic way using the codes; these codes allow for management (assignment) of indicators of variables existing in the considered equations in a straightforward manner,
- solution of the obtained system in a chosen algebraic manner to obtain desired angular velocities, ratios, forces, accelerations *etc.*

A similar routine may be formulated for the reverse order of activities, *i.e.* synthesis: going from the graph generation towards the creation of a gear functional structure [12].

The contour method of modelling of a mechanical system consists in creation of a special graph enclosing the contours, *i.e.* closed circles built of arrows connecting the vertices representing the elements of the system. Planetary gears are mechanisms consisting *inter alia* of geared wheels, in which at least the symmetry axis of one wheel encircles the main symmetry axis of the system. The wheel, which is going round, is called a planetary wheel or even just a planet or a satellite gear [16, 17, 18]. In case when all three basic gear parts are movable, then a planetary gear has two degrees of freedom and there is a demand for driving two different basic elements. In such a case, a system serves as a differential gear. The considered planetary gear is presented in Fig. 4.

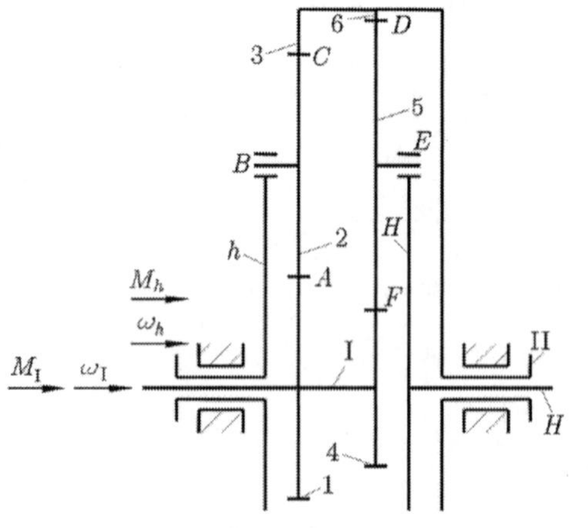

Fig. 4. Functional scheme of an exemplary planetary gear: *1, 2, ... ,6* sun wheels, wheels with internal toothings and planets; *h, H* – arms; *A, B, . . . , F* – characteristic points, *M* – angular velocities and torques.

A special planetary gear is considered. It encloses an internal closed loop formed by wheels and planets. The mobility W (DOF, degree of freedom) for the considered mechanism, *i.e.* planetary gear, can be calculated from the following formula [5, 6, 7, 19] (5):

$$W = 3n - 2c5 - c4 = 18 - 12 - 4 = 2 \tag{5}$$

where: n – number of links (movable elements), $c5$ – number of full joints (one degree of freedom), *e.g.* rotational type; $c4$ – number of half joints, *e.g.* meshing type; moreover $c5$ and $c4$ are equal to the total number of edges and diagonals of a polygon ($c5$ = 6) and the number of stripped edges ($c4$ = 4), respectively. We also have n = 6, *i.e.* the number of graph vertices – in the case of linear graph representation of a planetary gear.

For the considered gear, the following data for teeth numbers and the module are assumed: $z1$ = 15; $z2$ = 24; $z3$ = 63 (−63); $z4$ = 18; $z5$ = 21 and $z6$ = 60 (−60), m = 2 mm, where negative values of the teeth numbers are considered for the internal gearing in the case of the Willis method and one common module for all meshings was assumed.

3.1 Contour based model of the planetary gear

The contour method of modelling a mechanical system consists in creating a special graph enclosing the contours, *i.e.* closed circles built of arrows connecting the vertices representing the elements of the system [20, 21]. The contour graph of the gear presented in Fig. 4 is shown in Fig. 5.

 DOI: 10.1051/itmconf/20171505002

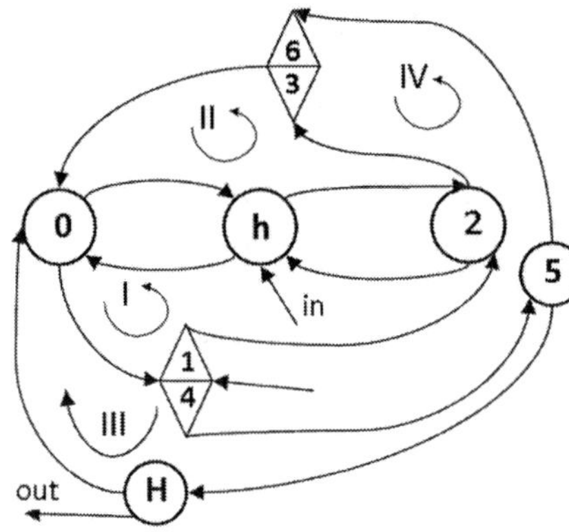

Fig. 5. Contour graph of the planetary gear.

The following contours can be distinguished for the contour graph (Fig. 5) of the considered planetary gear:
(I) $0 \to 1 \to 2 \to h \to 0$
(II) $0 \to h \to 2 \to 3 \to 0$
(III) $0 \to 4 \to 5 \to H \to 0$
(IV) $0 \to 4 \to 5 \to 6 \to 0$
Contour (I) generates the following pairs of indicators $\{(1,\ 0);\ (2,\ 1);\ (h,\ 2);\ (0,\ h)\}$. The contour can be interpreted as follows: we start it from 0, *i.e.* the support for the axis of wheel *1*, which is geared with planet wheel *2*, planet *2* rotates around its own axis mounted to *h* and rotates together with the arm *h* around the axis depicted by 0. This is the end of the contour:
$0 \to 1 \to 2 \to \mathrm{h} \to 0$.
Analogous explanations can be given for the tree remaining contours. The contour graph is presented in Fig. 5 where the special drawing rules by Marghitu [20] were applied.

Based upon the distinguished contours, the following system of equations may be developed [5, 6, 7, 19]:

$$\begin{cases} \vec{\omega}_{10} + \vec{\omega}_{21} + \vec{\omega}_{h2} + \vec{\omega}_{0h} = 0 \\ \vec{r}_A \times \vec{\omega}_{21} + \vec{r}_A \times \vec{\omega}_{h2} = 0 \\ \vec{\omega}_{h0} + \vec{\omega}_{2h} + \vec{\omega}_{32} + \vec{\omega}_{03} = 0 \\ \vec{r}_B \times \vec{\omega}_{2h} + \vec{r}_C \times \vec{\omega}_{32} = 0 \\ \vec{\omega}_{40} + \vec{\omega}_{54} + \vec{\omega}_{65} + \vec{\omega}_{06} = 0 \\ \vec{r}_F \times \vec{\omega}_{54} + \vec{r}_D \times \vec{\omega}_{65} = 0 \\ \vec{\omega}_{40} + \vec{\omega}_{54} + \vec{\omega}_{H5} + \vec{\omega}_{0H} = 0 \\ \vec{r}_F \times \vec{\omega}_{54} + \vec{r}_E \times \vec{\omega}_{H5} = 0 \end{cases} \tag{6}$$

where $\boldsymbol{r}_k$ are the position vectors of the points $k = A, B, \ldots, F$ shown in Fig. 4– means vectors of relative angular velocity of the *i*-th element in relation to the element *j*.

$$\begin{cases} -r_A\omega_1 - r_A\omega_{21} - r_A\omega_{h2} - r_A\omega_{0h} = 0 \\ r_A\omega_{21} + r_B\omega_{h2} = 0 \\ -r_C\omega_{h0} - r_C\omega_{2h} - r_C\omega_{32} - r_C\omega_{03} = 0 \\ r_B\omega_{2h} + r_C\omega_{32} = 0 \\ -r_E\omega_{40} - r_E\omega_{54} - r_E\omega_{H5} - r_E\omega_{0H} = 0 \\ r_F\omega_{54} + r_E\omega_{H5} = 0 \\ -r_D\omega_{40} - r_D\omega_{54} - r_D\omega_{H5} - r_E\omega_{0H} = 0 \\ r_F\omega_{54} + r_E\omega_{H5} = 0 \end{cases} \tag{7}$$

Every contour generates two equations connected with angular velocities and some other for forces and torques. The latter are not used in the present paper. The first equation for every pair is a sum of angular velocities designated by a contour, the second one is a sum of cross products of the respective arm multiplied by the angular velocity. In the second case, some summands are omitted when the arm length is equal to zero. The following relations are used for simplification of the system [5-7, 19].

To solve the system (6-7), the following actions were undertaken: we omit the arrows, *i.e.* we turn the vector equations into scalar ones. It can be done because the velocities act as vectors along the same direction. Moreover, the senses (*i.e.* orientations along the given direction) will be established via the solution of the system. The vector multiplication (cross product) can be simplified to the scalar one, because the angles between the arms and velocities in the case of gears with cylindrical wheels are always equal to 90°.

For other mechanisms enclosing cranks, pistons, cylinders, sliders, followers, *etc*, a detailed analysis of angles has to be carried out [5, 6, 7, 19, 20] to correctly simplify the system of equations. Assuming the cylindrical gear wheels, we have the relationships:

$$r_i = \frac{d_i}{2} = \frac{z_i \cdot m}{2} \tag{8}$$

where: m is the module of the i-th wheel, z_i – teeth number of the i-th wheel, $\mathrm{i} = 1, 2, \ldots, 6$.

Due to the fact that it is a universal design of multi-inputs and multi-outputs, all modules were assumed as equal. It causes that the module is not present in the formulas for the output angular velocities. In the case of different modules, these formulas would be a little more complicated but the system is also solvable. The solution to system of equations is as follows:

$$\begin{aligned} \omega_6 = \omega_3 &= \frac{\omega_H(2r_4 + 2r_5) - r_4 \cdot \omega_1}{2r_5 + r_4} = 66{,}65s^{-1} \\ \omega_h &= \frac{r_1\omega_1 + (r_1 + 2r_2)\omega_3}{2(r_1 + r_2)} = 84{,}03s^{-1} \end{aligned} \tag{9}$$

We obtained the same numerical results as in the case of the linear graph-based modelling – see formulas. It confirms the correctness of the performed modelling and analysis.

4 Application of dependence graphs in tree game structures in the analysis of planetary gear

The contour graph distribution from any vertex in the first stage leads to a tree structure with cycles, and next to a general tree game structure. Each structure has a proper analytic notation G_i^{++}- where *i*- is a vertex, from which the graph decomposition started determining a way of transition from the dependence graph to the tree structure. In work [19], a contour graph is transformed from each vertex to a tree structure.

For example, graph distribution from vertices *0, 2* leads to analytical expressions (Eq. 10-11).

$$\begin{aligned} G^{++}_{p0} = &({}^{0}0({}^{1}[II]\cdot h({}^{2}[II]\cdot 2({}^{3}[I]\cdot h^{1},[II\wedge IV]\cdot[6/3] \\ &({}^{4}[II]\cdot 0^{1})^{4})^{3},[I]\cdot 0^{2})^{2},[I\wedge III]\cdot[1/4]({}^{2}[I\wedge IV]\cdot 2({}^{3}[I]\cdot h^{2}, \\ &[II\wedge IV]\cdot[6/3]({}^{4}[II]\cdot 0^{3})^{4})^{3},[III\wedge IV]\cdot 5({}^{3}[III]\cdot H({}^{4}[III]\cdot 0^{4})^{4}, \\ &[IV]\cdot[6/3]({}^{4}[II]\cdot 0^{5})^{4})^{3})^{2})^{1})^{0} \end{aligned} \tag{10}$$

$$\begin{aligned} G^{++}_{p2} = &({}^{0}2({}^{1}[I]\cdot h({}^{2}[II]\cdot 2^{1},[I]\cdot 0({}^{3}[II]\cdot h^{1},[I\wedge III]\cdot \\ &[1/4({}^{4}[I\wedge IV]\cdot 2^{2},[III\wedge IV]\cdot 5({}^{5}[III]\cdot H({}^{6}[III]\cdot 0^{1}, \\ &[IV]\cdot[6/3]({}^{7}[II]\cdot 0^{2})^{7})^{6})^{5})^{4})^{3})^{2},[II\wedge IV]\cdot[6/3]({}^{2}[II]\cdot 0)^{2})^{1})^{0} \end{aligned} \tag{11}$$

Figures 6-10 show the structure from each of the vertices: *{1/4}, {6/3}, 0, 2, 5* of the graph (Fig. 5). In the next step, it is necessary to use complex structures.

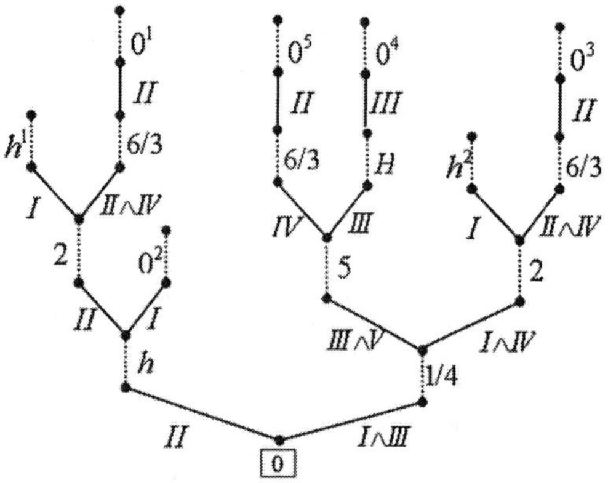

Fig. 6. Tree game structures from the initial vertex *0*.

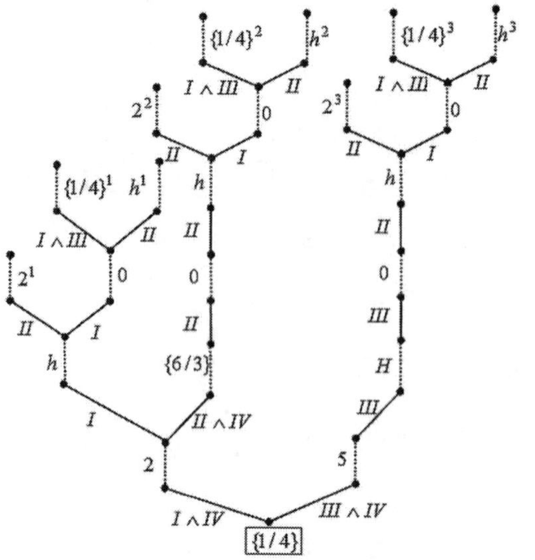

Fig. 7. Tree game structures from the initial vertex *{1/4}*.

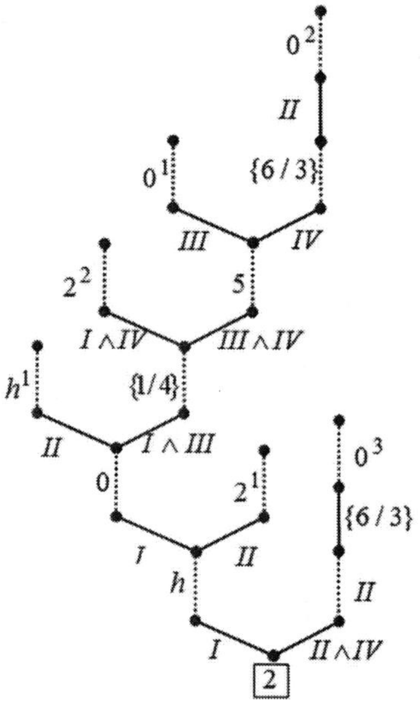

Fig. 8. Tree game structures from the initial vertex *2*.

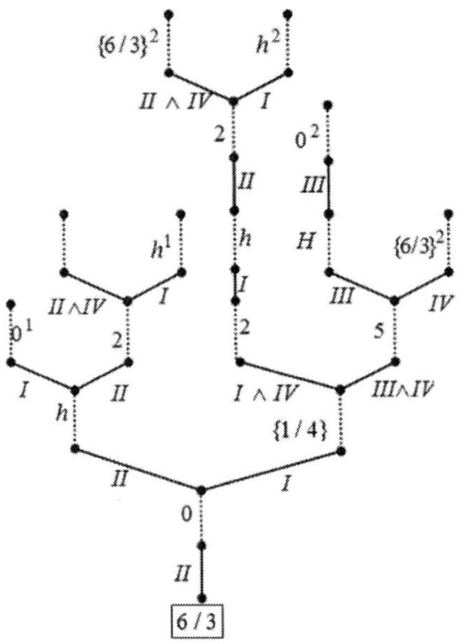

Fig. 9. Tree game structures from the initial vertex *{6/3}*.

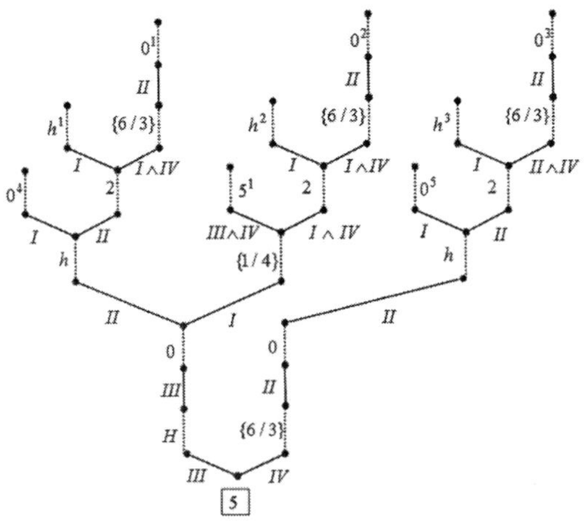

Fig. 10. Tree game structures from the initial vertex *5*.

4.1 Application of complex parametric structures

Complex decision trees join construction and/or operational parameters of similar properties as well as of the same adopted discretization of interval values. It makes it possible to decrease the calculation complexity in order to specify the most important subgroups of the whole set or system. In case of joining decision variables of separate properties and functions that they have in the system, a sub-analysis of a given set or system is obtained [14, 19, 22, 23].

Distribution graph of the dependency of Figure 5 gives a set D of trees:

$$D=\left\{G^{++}_{\{1/4\}},G^{++}_{\{6/3\}},G^{++}_{2},G^{++}_{5},G^{++}_{0}\right\}. \qquad (12)$$

In a complex parametric tree structure, by imposing all parametric game structures, from each vertex to the game structure from the predetermined vertex, we obtain a set S of total complex parametric structures:

$$S=\left\{S_{G\{1/4\}},S_{\{6/3\}},S_{G2},S_{G5},S_{G0}\right\}. \qquad (13)$$

Figures 11-12 show examples of complex structures: $S_{G\{5\}}$ and $S_{G\{3/6\}}$.

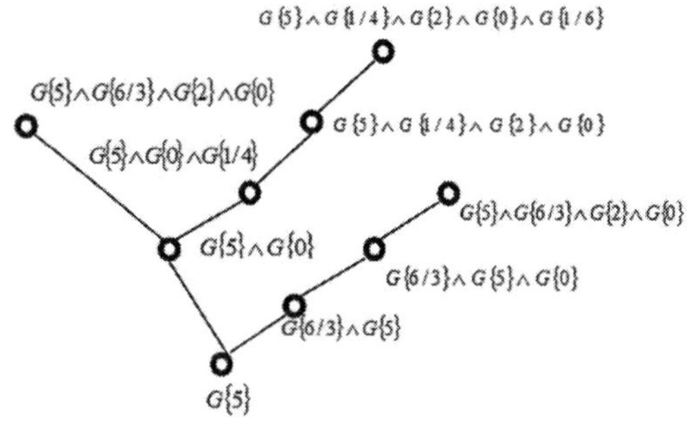

Fig. 11. Complex game structures from the initial vertex G *{5}*.

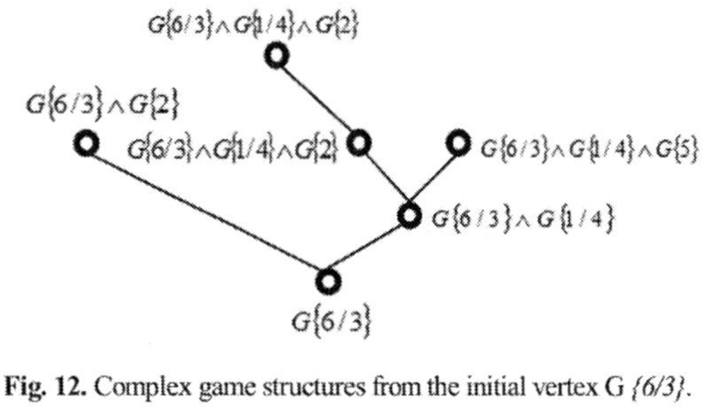

Fig. 12. Complex game structures from the initial vertex G *{6/3}*.

All transformations are suitable for the contour graph of the analyzed planetary gear [22,24].

Because complex structures can be n complete nodes γ_n, therefore, there exists a family B of sets of full nodes from all the sets of the set S:

$$B\subseteq\left\{\begin{array}{l}\{\gamma_1,\ldots,\gamma_n\}:\gamma_1,\ldots,\gamma_n\in G^{++}_{\{3/6\}}\wedge G^{++}_{\{1/4\}}\wedge G^{++}_{0}\wedge G^{++}_{2}\wedge G^{++}_{5},\\ \gamma_1\neq\gamma_2\ldots\neq\gamma_n\end{array}\right\}.$$

Nodes on complex tree structures describe decisions, analytical-algebraic transformations and states of a given parametric play structure.

5 Conclusions

Thanks to the construction of parametric structures from the primary contour graph, a better representation of the planetary gear can be obtained in two areas: structural analysis and the analysis of subordinate expressions.

- Structural analysis. Each parametric game structure describes an independently grouped relational preference system. Choosing the right parametric game structure requires defining a set of variable gears to determine strategies such as solving equation systems. For a given structure, it is possible to predict whether a given sequence of decisions in the structure leads to the determination of the complete path of the solution of the system of equations or to find the optimal value of a given parameter.

- Analysis of subordinate expressions. An expression describing the degree of subtraction of a given component graph, *i.e.* the parametric game structure, is parametrically marked with a pair of parentheses within which the expression being an analytical model is written. A string formed by symbols describing vertices and edges (structural parameters and / or functions) may be subject to factorization processes, *i.e.* to find the product form on the basis of the product of sums or decompositions – the distribution by a set of functions. By applying dependence graphs and game tree structures we are able to present a sequence of changes of arithmetic values of constructional and service parameters in order to obtain the required behaviour of the system for the planetary gear. Unlike traditional dependence graphs and tree classifiers, the dependence graph with game tree structures implement a connection of importance rank of vertices (states) and the height of the tree structure. Application of complex game trees does not change types and graphical shapes of such structures. They become more complex but they maintain the given structural properties resulting from the initial dependence graph. Thus, a local role of decomposition can be distinguished. The introduced decomposition decision eliminates interaction of constructional and service parameters because a designer can make a decision about only single changes and observations at the successive stages. The algorithmic method for forming game structures from the contour graph describes the optimization method of systematic search. The game structure describes a space of possible solutions in order to find optimum objective functions. There is a connection with other graphical structures which can be graphs in another sense, or even decision trees with node and/or branch coding. Such interpretation can lead to different types of logical trees.

References

1. A. Deptuła, Inter. J. of Appl. Mech. and Eng. **19(3),** 539-548 (2014)
2. M. A. Partyka, *Algorytm Quine'a- McCluskeya minimalizacji indywidualnych cząstkowych wielowartościowych funkcji logicznych* (Oficyna Wydawnicza Politechniki Opolskiej, Opole, 1999)
3. S. Berghofer, M. Reiter, LNCS **5674**, 147–163 (2009)
4. A. Deptuła, P. Osiński, M. A. Partyka, Arch. Civ. Mech. Eng. **13(4)**, 422-431 (2013)
5. S. Zawiślak, *The Graph-based Methodology as an Artificial Intelligence Aid for Mechanical Engineering Design* (Wydawnictwo Akademii Techniczno-Humanistycznej, Bielsko - Biała , 2010)
6. J. Drewniak, S. Zawiślak, Inter. J. of Appl. Mech. and Eng. **17**, 3, 791-798 (2012)
7. J. Drewniak, S. Zawiślak, PAMM **9**, 1, 547-548 (2009)
8. J. Wojnarowski, *Grafy i liczby strukturalne jako modele układów mechanicznych* (Wydawnictwo Politechniki Śląskiej, Gliwice, 1986)
9. S. Uematsu, Inter. J. of the Japan Soc. For Precis. Eng. **31**, 141-146 (1997)
10. J. Drewniak, S. Zawiślak, Mechanika **1**, 13-21 (2010)
11. J. Drewniak, S. Zawiślak, JTAM **48**, 2, 415-433 (2010)
12. L. W. Tsai, *Enumeration of kinematic structures according to function* (CRC Press, Boca Raton, USA, 2001)
13. C. H. Hsu, Inter. J. of Veh. Des. **13**, 3, 233-241 (1992)
14. A. Deptuła, M. A. Partyka, JAMRIS **5**, 3, 17-26 (2011)
15. A. Deptuła, Syst. in Manag. **2**, 2, 100- 112 (2014)
16. L. Muller, *Przekładnie obiegowe* (PWN, Warszawa, 1983)
17. P. Y. Huang, L. H. Lu, D. Z. Chen, CSME Conf. 13, 68-73 (1996)
18. G. Kron, AIEE Trans. **49**, 5, 683-685 (1930)
19. A. Deptuła, J. Drewniak, M.A. Partyka, Mechanik **7**, 640- 642 (2017)
20. D. Marghitu, *Kinematic chains and machine components design* (San Diego. Academic Press, 2005)
21. F. Freudenstein, ASME J. of Eng. for Ind. Ser. B **93**, 176-182 (1971)
22. W. Pijls, A. de B., Theo. Comp. Sci. **252**, 1-2, 197-215 (2001)
23. P. P. Shwnoy, Theor. and Dec. **44,** 2, 149-171 (1998)
24. Y. R. Shiue, R. S. Guh, Inter. J. of Adv. Manuf. Tech. **28**, 737-746 (2006)

 DOI: 10.1051/itmconf/20171505003

Assessing the damage importance rank in acoustic diagnostics of technical conditions of the internal combustion engine with multi-valued logical decision trees

Adam Deptuła[1,*], *Wojciech* Macek[2], and *Marian A.* Partyka[1]

[1]Opole University of Technology, Faculty of Production Engineering and Logistics, Institute of Processes and Products Innovation, Ozimska 75, 45-370 Opole, Poland,
[2]Opole University of Technology, Faculty of Production Engineering and Logistics, Department of Engineering and Work Safety, Generała Kazimierza Sosnkowskiego 31, 45-272 Opole, Poland

Abstract. This paper presents possible applications of acoustic diagnostics in inspecting the technical condition of an internal combustion engine with autoignition on the example of the Fiat drive unit with the common rail system. As a result of measuring the sound pressure level for specific faults and comparing the noise generated by the motor running smoothly, the detailed maps of changes in the acoustic spectrum may be generated. These results may be helpful in future diagnostics of internal combustion engines. In the paper, we present the results from the scientific works in the area of research, design and operation of internal combustion engines, conducted at the Department of Automotive Engineering, in cooperation with the Laboratory of Hydraulic Drives & Vibroacoustics of Machines at the Wroclaw University of Technology. The broader study has so far allowed us to develop an authoritative method of identifying the type of engine damage using game-tree structures. The present works assess the possibility of using multi-valued logic trees.

1 Introduction

The fault diagnosis technique is used to prevent early faults in a mechanical system. Acoustic emissions and vibration signals are well known for their suitability for monitoring conditions of rotating machineries. Most of the conventional methods for fault diagnosis using acoustic and vibration signals are primarily based on observing the amplitude differences in the time or frequency domain. The non-invasive diagnostics of internal-combustion engines are an exciting method for determination of occurring problems, because of the fact that they do not involve interference in the structure or disassembly of construction to identify a given failure [1]. Several methods of non-invasive engine testing have been already proposed, including the measurements of pressure and vibrations [1-2]. It is important to test engine operating parameters during the manufacturing process as well as in its subsequent operation [3]. This enables determining the optimal parameters of engine operation. The development of acoustic diagnostics is a promises quick and simple identification of engine damage. Nowadays this method is widely used in a range of applications, especially in the field of machine testing [4-5].

For designers of internal combustion engines, one of the major problems is to design a reliable system of OBD (On-Board Diagnostics), which is the system of auto-diagnostic of a vehicle. Unfortunately, such systems contribute to an increasing size and overall production and operation costs of internal combustion engines. However, the OBD system is currently the only widely used non-invasive diagnostic system. Other diagnostic methods rely on visual assessment of the engine, which forces interference in its structure in order to disassemble the damaged and other components to gain the access to the damaged part. If the acoustic wave spectrum measurement systems, eliminating the impact of background noise and other noise sources, which do not affect operation of the engine, were developed, it would be possible for each user to carry out in-house acoustic diagnostics. In recent years, we may observe a growing interest in the development of non-invasive diagnostic methods, particularly regarding internal combustion engines [6-7]. This article focuses on reliability assessment of acoustic diagnostics methods of internal combustion engine and the preparation of accurate diagnostic maps for Fiat 1.3 JTD engine 70 horsepower model. The measurements consisted in comparing the sound pressure level of specific frequencies for motor operation with induced defects with the smooth operation of the engine. The paper includes the results from the scientific work in the area of research, design and operation of internal combustion engines, conducted at the Department of Automotive Engineering, in cooperation with the Laboratory of Hydraulic Drives & Vibroacoustics of Machines at the Wroclaw University of Technology. A detailed description of the acoustic diagnostic method and the results of measurements were described in

* Corresponding author: a.deptula@po.opole.pl

DOI: 10.1051/itmconf/20171505003

previous articles [8-9]. In addition, the work [8] developed a decision system based on graphs and parametric game structures that identify the damage based on the spectral characteristics of sound emission. The game graph [9], taking into account the search algorithm, finds characteristic features, points and values (individual for each engine damage) by which the acoustical spectrum may be assigned to the specified type of damage. The present article takes into account the possibility of including multi-valued logic trees in an integrated decision system. An integrated decision system for logic machine learning was developed to test and identify the acoustic properties.

2 Measurements

The measurements were made on a Fiat diesel engine with the common-rail system. The tested model was Fiat 1.3 JTD, series designation - 188A9000. In total, 13 measurements of the acoustic wave spectrum for different cases of induced engine faults were conducted. The average L_{am} [dB], the maximum L_{Amax} [dB] and the minimum sound level L_{Amin} [dB] were measured by means of the following instrumentation: a modular sound pressure level meter with the record time history and analysis of the frequency (sound level meter, accuracy class I) from B&K type 2250 serial No. 2506429 with preamplifier type ZC 0032 serial No. 4112 and microphone type 4189 serial No. 2519832. The sound level meter meets the requirements set out in Regulation of Polish Tripartite Commission for Social and Economic Affairs from 28 May 2007 (Journal of Laws from, No. 105, item 717) and is confirmed by the current calibration certificate of the Regional Office of Weights and Measures in Wroclaw [8-9].

The test rig consists of eddy-current brake comprising the cooling system, the coil and the magnetic plate mounted on a shaft which is connected to the crankshaft of the engine. At the exit of the cooling channels two temperature sensors are provided. The rotational speed is measured by the rotation sensor and the torque is determined by the strain gauge force transducer attached to the arm of a known length to the brake housing, which is located on the bearings to obtain free rotation in the axis of the torque. The sound level meter was placed at a distance d equal to 1 m from the cuboid surrounding the motor housing with accordance to the standard (Fig. 1).

Fig. 1. Test rig for acoustic measurements [8-9].

The first study was to measure the background noise on the test bench, which is mainly caused by the flow of the coolant system of eddy current brake. Another 12 measurements were carried out for 6 different cases of the engine diagnostic condition during the load of 80Nm for two engine speeds - 1,000 rpm and 2,000 rpm. The time of each measurement was t = 10 s. These included the following cases [8-9]:

- engine in the perfect diagnostic condition, not warmed-up,
- engine in the perfect diagnostic condition after reaching the proper operating temperature,
- disconnected boost pressure sensor,
- disconnect the camshaft position sensor,
- disconnected injector No. 2,
- disconnected fuel pressure sensor.

2.1 Results of measurements

The results are presented in the form of spectral characteristics of acoustic pressure. The sound pressure level difference compared to the reference sound pressure level is depicted in the bar charts (Fig. 2, Fig. 3). An important parameter is the actual difference in sound pressure level, which specifies the real difference in the energy of the acoustic wave (Fig. 4, Fig. 5) [8-9].

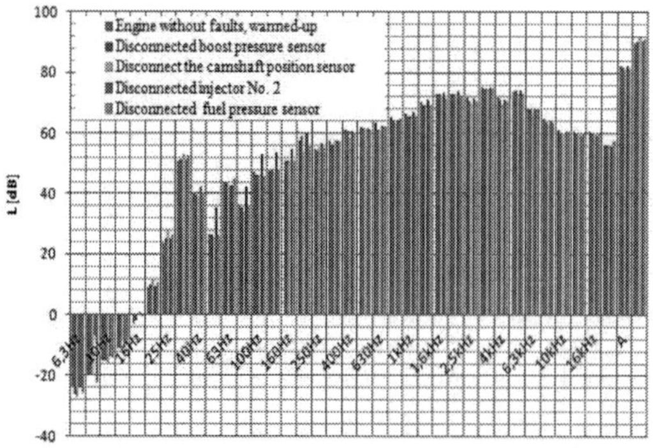

Fig. 2. Graph of the average sound pressure level L_{mA} for the speed of 1,000rpm.

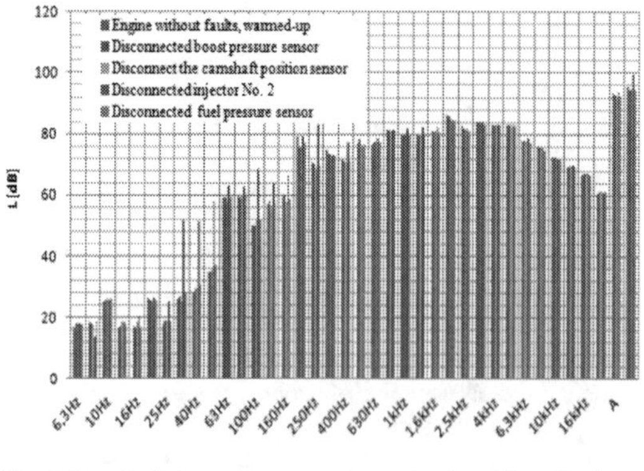

Fig. 3. Graph of the average sound pressure level L_{mA} for the speed of 2,000rpm.

DOI: 10.1051/itmconf/20171505003

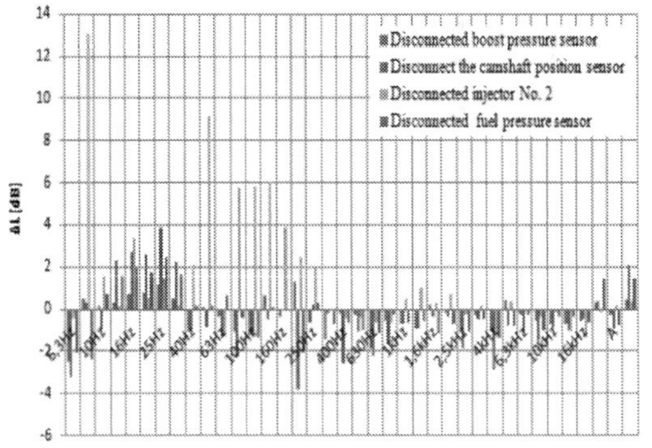

Fig. 4. Graph of differences in sound pressure level *ΔL* at the speed of 1,000rpm.

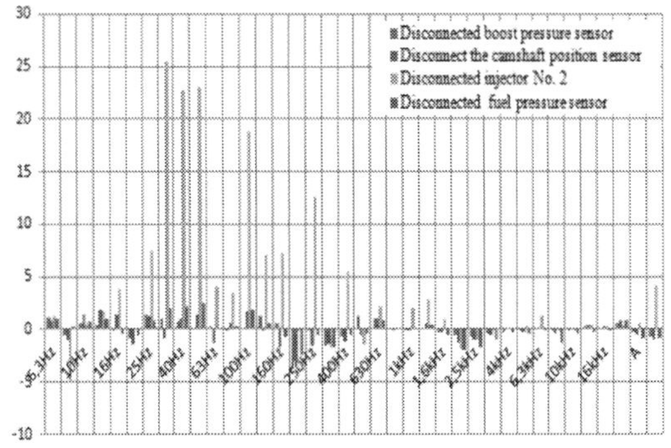

Fig. 5. Graph of differences in sound pressure level *ΔL* at the speed of 2,000rpm.

2.1.1 Results analysis

On the basis of Figs (2-5) failures occurring in the test engine can be clearly identified without the need of OBD system application. As the comparison of the sound level at all frequencies for each fault is time-consuming it is worth paying attention only for the characteristic changes in the acoustic spectrum for specific faults that are described below [8-9].

Boost pressure sensor disconnected - the main changes concern the frequency of 315Hz and 2.5 kHz, because there occurs a clear decline in the sound pressure level at the both the speed of 1,000 rpm and 2,000 rpm. These differences are about -1.5dB for 315Hz for both speeds, -1.01dB for 1000rpm and -0.66dB for 2,000rpm for a wave frequency of 2.5 kHz.

Disconnected camshaft position sensor - the best picture of the spectrum changes is produced during the test at a rotational speed of 1000 rpm. For higher speed, there were no significant differences.

Disconnected injector No. 2 - the only type of fault, which can be determined with an unaided ear. The strong increase in the noise level at low and medium frequency. It is worth pointing out here a correlation of enhancement for 1,000rpm and 2,000rpm, where increase of the pressure level of the first speed occurs at the doubled frequency values of higher speed.

Disconnected fuel pressure sensor - a situation similar to the one with disconnected camshaft position sensor, however, both strengthened and damped waves are smaller.

3 Discrete decision model

In order to create the method of decision trees it is possible to use heuristic search methods [10-11]. Actions undertaken in finding the solution can be classified as searching of objects of specified characteristics. For the identification of the type of engine damage, rules of procedure by the generate-and-test method focused on testing of a given property of an acoustic signal having been used.

It can be assumed that elements of the system form the set Q, relations – the set Ω, properties of elements – the set Γ, the objective function- F. It can be assumed that states of the engine are elements of the set:

$$Q=\{X_o,X_1,X_2,...,X_i...,X_m\} \tag{1}$$

where: X_i- i^{th} engine damage, if Q is a finite set.

For a given object $X_i \in Q$ it is possible to differentiate a certain level of sound pressure $\tilde{A}_i$. Then, sound pressure values, as possible states of the analysed damage X_i will be a finite set in this domain.

$$^{+}\tilde{A}_i=\{a_{i,1},a_{i,2},...,a_{i,j},...,a_{i,w}\} \tag{2}$$

where: elements $a_{i,\ j,}$ for i=*const* and j= 1, 2, mean highlighted values in this domain (frequency) $\tilde{A}_i$. Values of differences in the level of sound pressure ΔL for the frequency f [Hz]: {6.3Hz, 8Hz, 10 Hz, ..., 20kHz, A, Z} will be included in the analysis.

It is necessary to define the relation of the measured values of sound pressure levels values of given damages X_i in order to differentiate damages. The set Π_s composed of the ordered pairs (X_i, X_j) may be called the relation on the set Q. The set Π_s is a subset of the Cartesian product QxQ and takes the following form [8, 9, 12].

$$\Pi_S=\{(X_{i1},X_{j1}),(X_{i2},X_{j2}),...,(X_{is},X_{js})\} \tag{3}$$

where i_s, j_s are specific values of the sound pressure level of a given damage.

When searching a game graph, there will be a possibility of determining the kind of damage X_i in the scope of the relation of the Π_s type by means of defining highlighted values. Such a state can be presented in the following form of a vector:

$$\Pi_s^i=\left[\delta_s^i(\mathrm{j}_1),\delta_s^i(\mathrm{j}_2),...,\delta_s^i(\mathrm{j}_n)\right] \tag{4}$$

In the case of our analysis, there is the following combination table: X_1 - disconnected boost pressure sensor, X_2- disconnected shaft position sensor, X_3- disconnected injector No. 2, X_4 - disconnected fuel

DOI: 10.1051/itmconf/20171505003

pressure sensor. For example, Figures 6 and 7 show acoustic sound levels of specified failures: disconnected boost pressure sensor (failure I), disconnected camshaft position sensor (failure II), disconnected injector No. 2 (failure III), disconnected fuel pressure sensor (failure IV) in comparison to proper operation of engine.

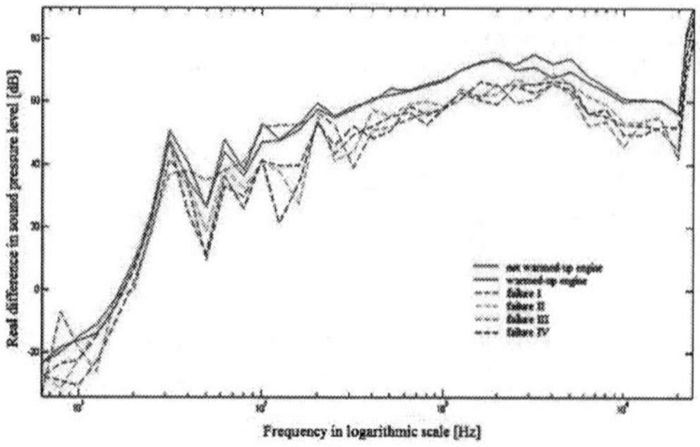

Fig. 6. Graph of sound pressure level for the speed of 1,000 rpm – comparison of all failures.

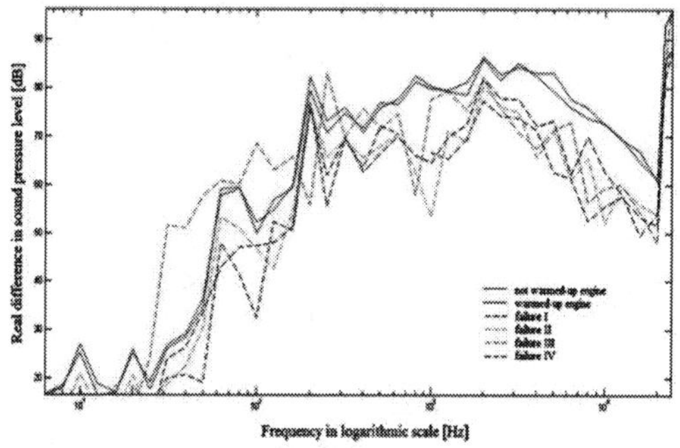

Fig. 7. Graph of sound pressure level for the speed of 2,000 rpm – comparison of all failures.

4 Multi-valued logic decision trees

Complex, multiple-valued logic functions determine the degree of importance of logic variables by changing the levels of the logic tree, from the most important (near the root) to the least important (on the top) because there is a generalisation of a Boolean quality index into a multiple-valued one: $(C_k - k_i m_i) + (k_i + K_i)$, where C_k – the number of branches of k- level, k_i – multiplicity of simplification on *k*-level, m_i – value of *i*- variable, K_i – the number of branches (k-1) – level, out of which branches of k-level were formed which cannot be simplified. In this way it is possible to obtain the minimum complex alternative normal form. All transformations are described by the so-called Quine–Mc Cluskey algorithm based on the minimization of individual partial multiple-valued logic functions.

4.1 Quine-McCluskey algorithm for the minimization of multi- valued logic functions

The Quine-McCluskey algorithm is capable of finding all prime implicants of a given logic function, *i.e.* there is a shortened alternative, normal form SAPN [13, 14]. The terms of incomplete gluing and elementary absorption play the main role in the search of prime implicants and are used for the APN of a given logic function.

The following transformation is called the consensus operation:

$$Aj_o(x_r) + \ldots + Aj_{m_r-1}(x_r) = A \tag{5}$$

where: $r = 1, \ldots, n$ and A- a partial elementary product, the literals of which possess variables belonging to the set: $\{x_1, \ldots, x_1, x_1, \ldots, x_n\}$.

$$\begin{aligned} & w_o Aj_o(x_r) + \ldots + w_{m_r-1} Aj_{m_r-1}(x_r) = \\ & = \left(min\{w_o, \ldots, w_{m_r-1}\}\right) \cdot A + \sum_{s=i_o,\ldots,i_{m_r-2}} w_s \cdot A \cdot j_s(x_r) \end{aligned} \tag{6}$$

where: w_i - polyvalent weighting factor.

For example using the formula:

$$Aj_0(x_r) + \ldots + Aj_{m-1}(x_r) = A, \quad Aj_u(x_r) + A = A, \tag{7}$$

where: $A = A(x_1, \ldots, x_{r-1}, x_{r+1}, \ldots, x_n)$,

$$j_u(x_r) = \begin{cases} m-1 \;, & u = x_r \\ 0 \;, & u \neq x_r \end{cases} \qquad 0 \leq u \leq m-1 \tag{8}$$

For example multiple-valued logical function $f(x_1, x_2, x_3)$, where $x_1, x_2, x_3 = 0, 1, 2$, written by means of numbers KAPN (Canonical Alternative Normal Form): 100, 010, 002, 020, 101, 110, 021, 102, 210, 111, 201, 120, 022, 112, 211, 121, 212, 221, 122, there is one MZAPN (Minimal Complex Alternative Normal Form) after the application of the Quine– Mc Cluskey algorithm based on the minimization of individual partial multi-valued logical functions having 13 literals (Figure 8) [13, 14].

$$\begin{aligned} f(x_1, x_2, x_3) = & \, j_o(x_1)\left(j_o(x_2) j_2(x_3) + j_1(x_2) j_o(x_3) + j_2(x_2)\right) + \\ & + j_1(x_1) + j_2(x_1)\left(j_o(x_2) j_1(x_3) + j_1(x_2) + j_2(x_2) j_1(x_3)\right). \end{aligned} \tag{9}$$

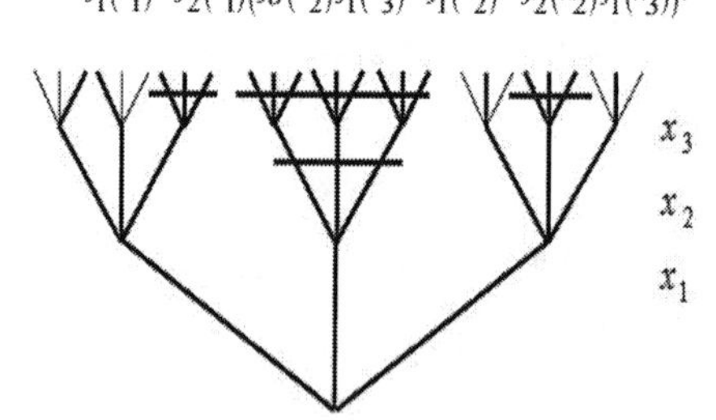

Fig. 8. A multi-valued decision tree for the parameters x_1, x_2, x_3 with an appropriate layout of levels.

4.2 Application of multi-valued logic decision trees

Based on the obtained measurements results (Figures 2-5) the morphological table for this specific case was prepared, as shown in Tables 1-2.

 DOI: 10.1051/itmconf/20171505003

Table 1. Morphological table of measurement results and logic values for n = 1000 [rpm].

Frequency [Hz]	X_1	X_2	X_3	X_4	Logic values
	n= 1000 rpm				
	0	0	0	0	
6.3Hz	-2.53	-3.27	-0.42	-2.06	0
8Hz	0.51	0.28	13.07	-2.35	
10Hz	0.12	-0.95	1.56	0.72	
12.5Hz	0.28	2.33	0.14	1.58	
16Hz	0.72	2.75	3.41	1.96	
20Hz	0.78	2.6	0.47	1.7	
25Hz	1.13	3.88	1.43	2.46	1
31.5Hz	0.5	2.24	0.17	1.66	
40Hz	-0.91	-1.08	2.07	0.11	
50Hz	0.09	-0.86	9.11	0.14	
63Hz	-0.35	-1.35	-1.28	0.67	
80Hz	-1.05	-2.21	5.77	-0.4	
100Hz	-1.2	-1.2	5.81	-1.37	2
125Hz	0.64	-0.47	6.06	0.01	
160Hz	-0.34	-0.02	3.9	-0.1	
200Hz	1.3	-3.84	2.44	-2.07	
250Hz	-0.62	0.19	1.93	0.29	
315Hz	-1.41	-0.19	-0.06	-0.73	

Table 2. Morphological table of measurement results and logic values for n = 2000 [rpm].

Frequency [Hz]	X_1	X_2	X_3	X_4	Logic values
	2000 rpm				
	1	1	1	1	
6.3Hz	1.09	0.87	1.31	0.96	0
8Hz	-0.58	-1.05	-4.99	0.25	
10Hz	0.51	1.4	0.42	0.67	
12.5Hz	0.44	1.85	1.75	1	
16Hz	-0.14	1.47	3.93	-0.43	
20Hz	-0.84	-1.39	0.2	-0.56	
25Hz	1.45	1.35	7.4	0.78	1
31.5Hz	0.96	-0.89	25.4	2.07	
40Hz	0.7	1	22.65	2.16	
50Hz	0.12	1.4	23	2.49	
63Hz	0.32	-1.36	3.97	0.1	
80Hz	-0.07	0.56	3.38	0.24	
100Hz	0.08	1.7	18.88	1.91	2
125Hz	1.31	-0.16	7.03	0.52	
160Hz	0.52	-1.7	7.18	-0.73	
200Hz	-3.76	-3.1	-0.02	-2.48	
250Hz	-0.14	-1.63	12.65	-0.64	
315Hz	-1.54	-1.4	-1.67	-1.68	

In the first step decision trees were used for the frequency range: up to 315 Hz. Arithmetic values of analysed parameters have been chosen in order to make an analysis in accordance with a table. In case of our analysis, there is the following combination table:

X_1 - disconnected boost pressure sensor;
X_2 - disconnected shaft position sensor;
X_3 - disconnected injector No. 2;
X_4 - disconnected fuel pressure sensor.

These values had been coded by means of logic decision variables for the needs of logic decision trees.

$$\begin{cases} X_1, X_2, X_3, X_4 = 1000\,[rpm] \sim 0 \\ X_1, X_2, X_3, X_4 = 2000\,[rpm] \sim 1 \end{cases};$$

$$f \in \langle 6{,}3 - 20 \rangle\,[\mathrm{Hz}] \sim 0;\; f \in \langle 25 - 80 \rangle\,[\mathrm{Hz}] \sim 1;$$

$$f \in \langle 100 - 315 \rangle\,[\mathrm{Hz}] \sim 2.$$

Based on the acquired measurements results, the morphological table for this specific case (and logic values) was prepared as shown in Table1 (for n = 1000 [rpm]) and Table 2 (for n = 1000 [rpm]).

The sets of highlighted values presenting features of sound pressure values for a given damage have been obtained from the analysis. The above summaries allow us to create a measuring ladder, which will make it possible to determine the type of engine damage on the basis of sound pressure measurements on chosen frequencies. For values in Table 1, according to the Quine-Mc Cluskey algorithm we obtain the following solution:

For Table 1:

X_1	X_2	X_3	X_4				
N	N	N	N				
C	C	C	N	*			
C	N	C	C		*		
C	C	C	C	*	*		
C	C	C	C				*
N	N	C	C	*	*	*	
C	N	C	C	*			*
N	N	N	C		*		
N	N	C	N			*	
N	N	C	N	*			
C	N	C	C	*			
C	N	C	N	*			
N	C	C	C				
N	N	N	N	*			

$$\begin{matrix} 1_0 \\ 2_0 \\ 3_0 \end{matrix} \begin{cases} N\,N\,N\,N \\ C\,C\,C\,- \\ C\,-\,C\,C \end{cases} 0 \qquad \begin{matrix} 1_1 \\ 2_1 \\ 3_1 \\ 4_1 \end{matrix} \begin{cases} -\,N\,N\,N \\ N\,N\,-\,C \\ N\,N\,C\,- \\ C\,-\,C\,C \end{cases} 1$$

$$\begin{matrix} 1_2 \\ 2_2 \\ 3_2 \end{matrix} \begin{cases} N\,N\,-\,N \\ C\,N\,C\,- \\ N\,C\,C\,C \end{cases} 2$$

The final solution is:

$$3_0 \wedge 1_1 \wedge 2_2 = C\,N\,C\,C\,.$$

DOI: 10.1051/itmconf/20171505003

For Table 2:

X_1	X_2	X_3	X_4				
C	C	C	C				
N	N	N	C				
N	C	C	N	*			
N	N	C	N	*			
C	C	C	C	*	*		
C	N	C	C	*			
N	C	C	C		*		
C	C	C	C	*			
C	N	C	C	*	*		
C	N	C	N		*	*	
N	N	N	N				*
N	N	C	N			*	*

$$\begin{array}{l} 1_0 \\ 2_0 \\ 3_0 \end{array}\begin{cases} N \; - \; C \; N \\ C \; C \; C \; C \\ N \; N \; N \; C \end{cases} 0 \qquad \begin{array}{l} 1_1 \\ 2_1 \end{array}\begin{cases} C \; - \; C \; C \\ - \; C \; C \; C \end{cases} 1$$

$$\begin{array}{l} 1_2 \\ 2_2 \\ 3_2 \\ 4_2 \end{array}\begin{cases} C \; - \; C \; C \\ C \; N \; C \; - \\ - \; N \; C \; N \\ N \; N \; - \; N \end{cases} 2$$

The final solution is:

$$2_0 \wedge 1_1 \wedge 1_2 = C\,C\,C\,C$$

$$2_0 \wedge 2_1 \wedge 1_2 = C\,C\,C\,C \; .$$

It should be noted that a similar analysis should be made for the frequency range: $f \in 400$ [Hz] - 20 [kHz].

The sets of highlighted values presenting features of sound pressure values for a given damage were obtained from the result of an analysis. The summaries above may be employed to create a measuring ladder, which will make it possible to determine the type of engine damage on the basis of sound pressure measurements on chosen frequencies.

5 Conclusions

The most advanced form of a substantially integrated system can be an integrated system, resulting from the integration of complex system-aided management of complex systems supported decision-making. (Fig. 9).

All operations were made in the real time during normal use of the motor housing. On the basis of the results received from the inference mechanism, resulting from the correlation between acoustic signals measured in a continuous manner and model signals placed in the data base, there would be an identification of the influence of particular signals from microphone on the chosen parameters of motor housing integrated logic system may be used for the integrated decision system. Proposed in the paper decision method for identifying engine failure based on sound emission spectral characteristics allows fast identification of specific damage. The complete procedure of fault engine diagnostics proposed by authors includes several steps [15, 16]. Firstly, the precise sound pressure level characteristics of an operating engine need to be obtained for efficient and damaged engine in order to gain engine acoustic maps. The next step is to create a morphological table with logical coding. The proposed decision method of identifying engine failure based on sound emission spectral characteristics, will allow the quick identification of particular damage.

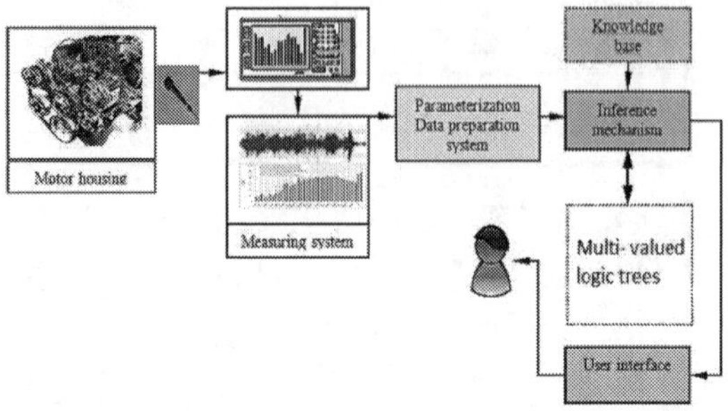

Fig. 9. Decision support system.

The method may be successfully used for subsequent measurements on the same engine and engines of the same kind. In the case, where a greater number of measurements are made, it may occur that a better classification system based on the tools with the reversed correlation matrix generated by the motor running smoothly may be proposed, which enables generating more detailed maps of changes in the acoustic spectrum. These results may be helpful in the future diagnostics of internal combustion engines. In the paper, the results of scientific work in the area of research, design and operation of internal combustion engines, conducted at the Department of Automotive Engineering, in cooperation with the Laboratory of Hydraulic Drives & Vibroacoustics of Machines at the Wroclaw University of Technology are included. The ongoing research works have already developed an authoritative method for identifying the type of engine damage using game-tree structures and induction machine [17].

References

1. S. Luft, *Podstawy budowy silników* (Wydawnictwo Komunikacji i Łączności, Warszawa, 2010)
2. L. Barelli, G. Bidini, C. Buratti, R. Mariani, Appl. Therm. Eng. **29**, 1707–1713 (2009)
3. S. Berghofer, M. Reiter, LNCS **5674**, 147–163 (2009)
4. P. Osiński, W. Kollek, Arch. Civ. Mech. Eng. **13**, 3, 313-321(2013)
5. M. Kirpluk, *Podstawy akustyki* (Wydawnictwo NTL-M. Kirpluk, Warszawa, 2012)
6. Z. Ranachowski, A. Bejger, Arch. of Acoust. **30**, 4, 465-472 (2005)
7. W. M. Adaileh, Appl. Mech. Mater. **295-298**, 2013-2020 (2013)

DOI: 10.1051/itmconf/20171505003

8. A. Deptuła, D. Kunderman, P. Osiński, U. Radziwanowska, R. Włostowski, Arch. of Acoust. **41**, 2, 345–350 (2016)
9. A. Deptuła, P. Osiński, U. Radziwanowska, Arch. of Acoust. **41**, 3, 449-460 (2016)
10. T. Mitchell, M-G. Hill, *Machine Learning* (McGraw-Hill Science/Engineering/Math, Boston, 1997)
11. S. M. Weiss, C. A. Kulikowski, *Computer systems that learn: classification and prediction methods from statistic, neural nets, machine learning and expert systems* (Morgan Kaufmann Publishers, 1991)
12. M. Merthold, D. Hand, *Intelligent data analysis. An introduction* (Springer-Verlag, Berlin, 1999)
13. A. Deptuła, Int. J. of Appl. Mecha. and Eng. **19(3)**, 539-548 (2014)
14. M. A. Partyka, *Algorytm Quine'a- McCluskeya minimalizacji indywidualnych cząstkowych wielowartościowych funkcji logicznych* (Oficyna Wydawnicza Politechniki Opolskiej, Opole, 1999)
15. D. Zhen, T. Wang, F. Gu, B. Tesfa, A. Bal, Meas. Sci. Technol. **24**, 13 (2013)
16. J. R. Quinlan, *Machine Learning* (Springer, 1986)
17. A. Deptuła, Inter. J. MTM **10** (7), 19-22 (2016)

 DOI: 10.1051/itmconf/20171505004

Calculating an electromechanical torque of a squirrel cage motor based on an axial flux obtained by the FEM

Janusz Petryna[1], *Jarosław* Tulicki[1], and *Maciej* Sułowicz[1,*]

[1]Cracow University of Technology, Faculty of Electrical and Computer Engineering, Institute of Electromechanical Energy Conversion, Warszawska Street 24, 31-155 Cracow, Poland

Abstract. This article presents a method of calculating an axial flux signal and how to use this signal to determine an electromagnetic torque for a squirrel cage induction motor in a steady state. Numerical analyses were performed using a two-dimensional field-circuit model for the squirrel cage induction motor. Most of analyses were performed for a symmetric motor and to estimate a magnitude of the electromagnetic torque, an algorithm was developed. The electromagnetic torque estimation method proposed in the article is easy to implement and does not require any interference into the drive system transmission. The analysis shows that it is possible for the induction motor to determine a load torque based on the axial flux measurement using a measuring coil applied to the motor, for example from the fan side.

1 Introduction

One of commonly used diagnostic signals in evaluating the condition of induction motors is the axial flux. This signal is easily measurable and allows to extract many useful information about the condition of the machine itself and about the efficiency of the electric power conversion by this machine [1-17].

One of ways to install a coil for the axial flux measurement is shown in Fig. 1.

Fig. 1. Method of axial flux measurement.

A coreless coil is a basic element used to capture changes in unipolar flux as a function of time. The winding direction of this coil is perpendicular to the main axis of the machine. The coil acts as a secondary winding, in which the voltage is induced by the motor alternating the flux. The flux instantaneous value is directly proportional to the induced voltage on the coil terminals. From the above, it results that the main metrological feature of the air coil is the lack of nonlinear signal distortion and the characteristic similar to the band-pass filter. In terms of mechanical construction, the most important parameter is the diameter of the carcass, which should not be less than the diameter of the rotor cage and greater than the internal diameter of the stator.

In laboratory and industrial practice, the measurement coil may be located on the non-drive side of the fan cover [16] or on the driving side inside of the bearing disk [14, 15], less frequently on the stator body. This axial flux signal can just be used in many ways, both for the diagnosis of the machine condition itself and for determining the rotational speed or torque of the motor shaft. In an ideal induction machine, axial flux should not be present. In fact, around the stator end windings and around the ring segments of the cage, flux of axial direction occurs.

Moreover, both in the stator and rotor circuits, various types of asymmetries due to geometrical inaccuracies in the distribution of windings or non-homogeneity of materials appear. As a result of this type of inaccuracy in the machine, small but noticeable axial flux is produced. The necessary conditions for the formation of axial flow through numerical analyzes are given in papers [2, 5, 7].

Accurate determination of axial flux time course is fully possible on the basis of the solution of the field-circuit equations of the three-dimensional model. An example of such a model with a measuring coil assembled is shown in Fig. 2.

* Corresponding author: msulowicz@pk.edu.pl

DOI: 10.1051/itmconf/20171505004

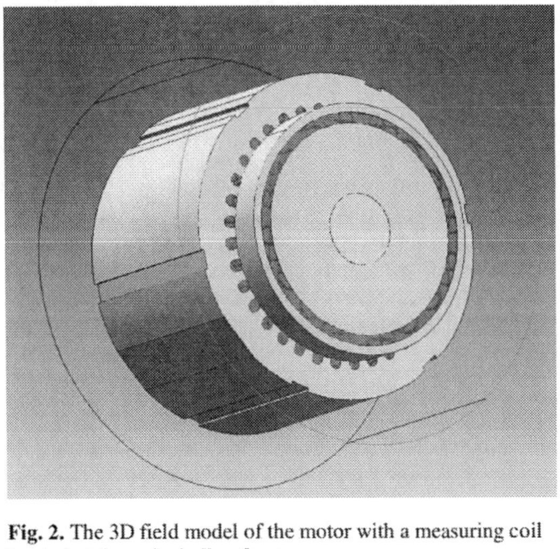

Fig. 2. The 3D field model of the motor with a measuring coil located at the end winding front.

This approach allows to define quantitative and qualitative character of the leakage flux waveforms in the shaft axis inside the motor.

With some simplifications [15], a 2D model can be used to determine the axial flux. In the following section, the application of the 2D model to determine the relationship between the axial flux signal and the load torque will be described. This relationship is extremely significant to develop a method for estimating the load torque of a machine based on the measurement of the axial flow only.

2 The 2D FEM model description

The field-and-circuit model was built on the basis of a deep-slotted Sg-112M low-power induction motor with two pairs of poles. The motor has a single-layer winding with a diametral winding pitch. Stator yoke geometry included an actual shape of the magnetic core with the assembly points of the machine body (the sheets do not have a circular shape as it is assumed in typical FEM models). Each slot of the stator yoke was divided between areas filled with winding copper wires and the surface covered by the wedge. The rotor cage bars are made as a cast of pure aluminum. Included in the geometric design of the rotor are also additional elements such as: an air-filled ventilation duct and a rotor shaft made of the machinery steel. Material of the stator and rotor cores were packed was the MU1000 type material of the linear characteristics. Due to the lack of possibility of including the stator and rotor end windings in the two-dimensional model, the stator end windings and cage end rings were included in the circuit part of the model. The above winding elements whose leakage fluxes are closed by the air, are defined as the passive RL two-terminal units. The accepted model of the motor does not take into account the rotor slots skew. For the purpose of calculations, it was set up that the maximum distance between the finite elements mesh points is 1.5 mm. The simulation time was 10 s at the integration step of 0.5 ms. The field-circuit 2D model of the induction motor was developed with use of the MagNet software. The following figures show the 2D model of the motor for which the calculations were performed (Fig. 3.), and the FEM grid for the entire rotor cross-section (Fig. 4.) and quarters of this model (Fig. 5.).

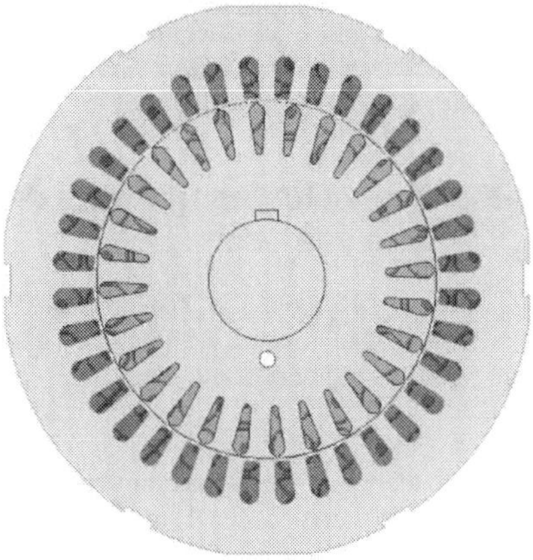

Fig. 3. The 2D model of the tested motor – main view.

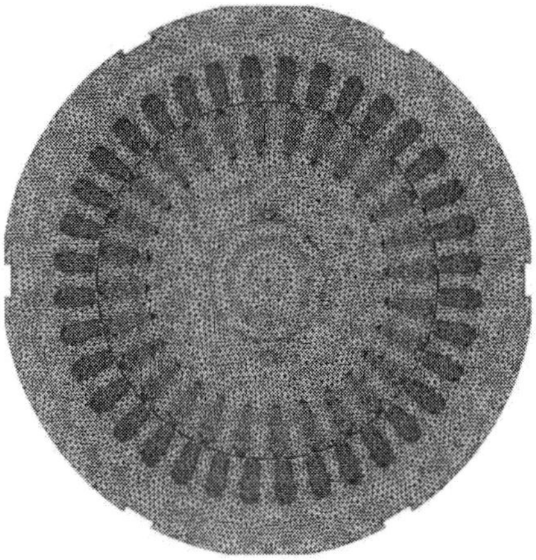

Fig. 4. The FEM grid for the complete 2D model.

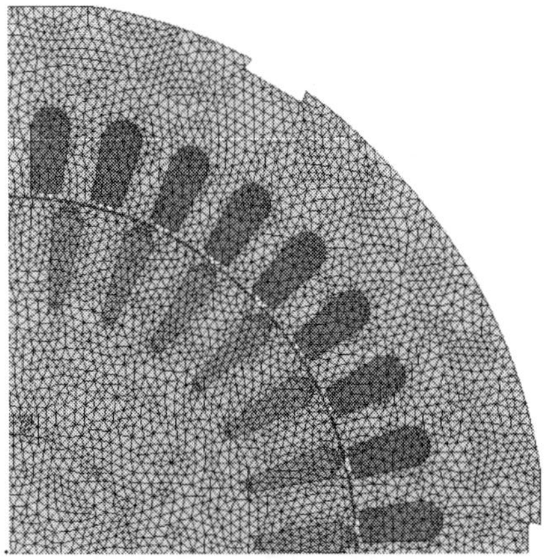

Fig. 5. The quarter of the FEM grid.

On the basis of the model described above, stator currents, rotor currents, currents in the joint rings segments, the voltage patterns on the stator windings and the electromagnetic torque are determined.

In the following part, the methodology for calculating the axial flux for the symmetrical motor is presented, with some imperfections in the rotor cage execution and negligible resistance asymmetry in the stator windings, being assumed.

3 Method of calculating the axial flux signal and its analysis

When calculating the axial flux signal, it was assumed that it was generated because of small asymmetries in the stator and rotor windings.

A two-dimensional FEM model was taken for the calculations, and it was assumed that the resulting waveform of the axial flux was a superposition of the stator and rotor fluxes. It can be described by the formula:

$$\Psi^0(t) = \sum_{n=1}^{m_s} \Psi_n^s(t) + \sum_{k=1}^{N_r} \Psi_k^r(t) \quad (1)$$

where: Ψ^0 – axial flux, Ψ^s – stator winding leakage flux, Ψ^r – leakage flux of the rotor joint ring segment, m_s – stator number of phases; N_r – rotor bars number.

It was assumed that the stator flux is produced by the end winding front connections. In turn, the flux of the rotor origin is associated with the joint rings segments.

In the axial flow calculation for the stator, 2.5% difference in the resistance of the individual phases was assumed.

The individual phase currents values were multiplied by the coefficients to obtain the asymmetry of the currents. The 1.025 coefficient was applied for L1phase, 0.975 for L2 phase and 1 for L3 phase.

The value of the leakage inductance was estimated. In further calculations, the value of L_{sr}=1.83*10^{-5} H was assumed.

The axial flux signal generated by the stator can be written as follows:

$$\Psi^s(t) = \sum_{n=1}^{m_s} I_n^s(t) \cdot L_{rs} \quad (2)$$

Similarly, the time course of the axial flux from the rotor can be calculated as:

$$\Psi^r(t) = \sum_{k=1}^{N_r} \Psi_k^r(t) \cdot L_{rr} \quad (3)$$

The leakage inductance value was determined. In further calculations, the value L_{rr}= 3.63*10^{-9} H was assumed.

The instantaneous value of the induced voltage in the measuring coil is expressed as a derivative of the axial flow over time:

$$e(t) = -N\frac{d}{dt}\phi^0 = -\frac{d}{dt}\Psi^0(t) \quad (4)$$

where: $e(t)$ – voltage induced in the measuring coil, N – proportional factor (number of turns of measurement coil).

All the following waveforms of stator windings, currents in the rotor cage bars and currents in the segments of the cage joint rings, the electromagnetic torque and the rotational speed of the rotor were calculated on the basis of the 2D model. The calculation included a symmetrical stator circuit, a symmetrical rotor cage, symmetrical 230 V power supply, a torque load of 7 Nm, and an average established speed of 1459 rpm.

Calculations were proceeded from the moment when the motor was switched on until the speed was established. Apart from the 7 Nm load, simulations were performed for 8, 9, 10, 11 Nm. An example of a current waveform in a segment of a cage joint ring is shown in Fig. 6.

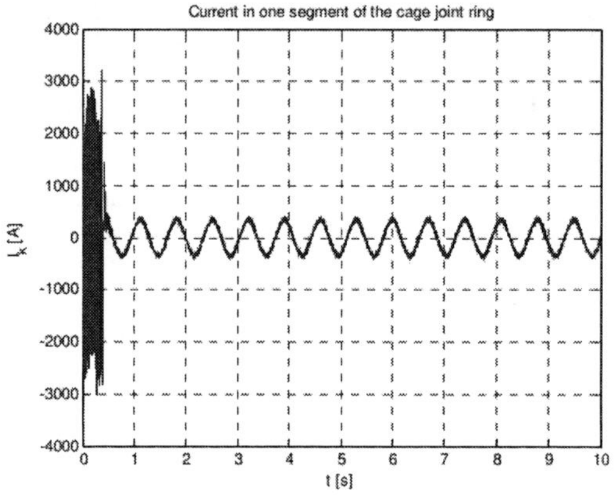

Fig. 6. Current waveform in 1 segment of the cage joint ring.

Fig. 7. shows the current waveform in the winding of the first stator phase and Fig. 8. – current waveform in 1 bar of the rotor cage.

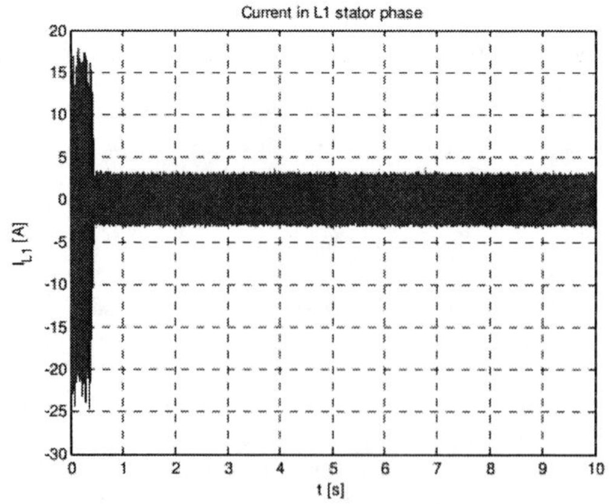

Fig. 7. Current waveform in the winding of the first stator phase.

DOI: 10.1051/itmconf/20171505004

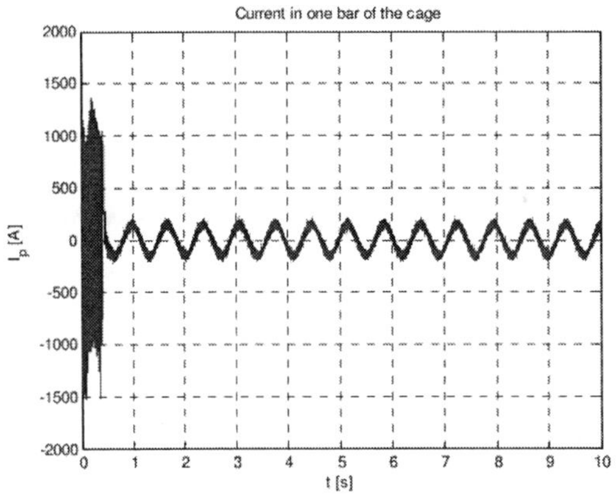

Fig. 8. Current waveform in 1 bar of the rotor cage.

The calculated axial flux according to formula (1) and the coil voltage based on formula (4) are shown in Fig. 9. to Fig. 11.

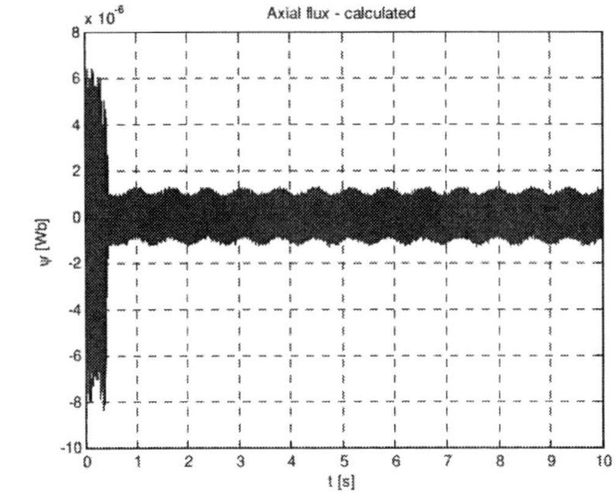

Fig. 9. The waveform of the calculated axial flux.

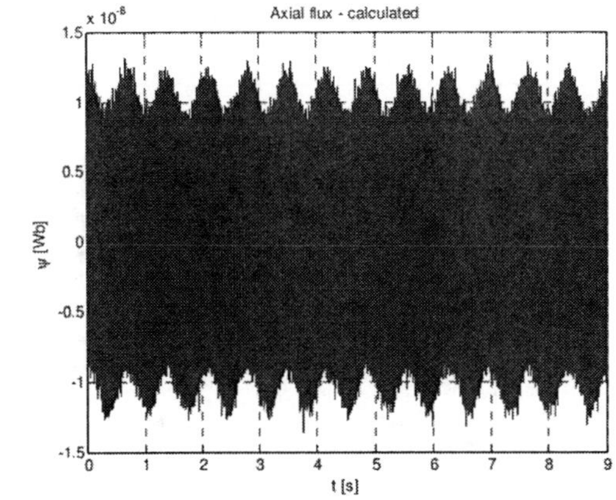

Fig. 10. The waveform of the calculated axial flux – steady state.

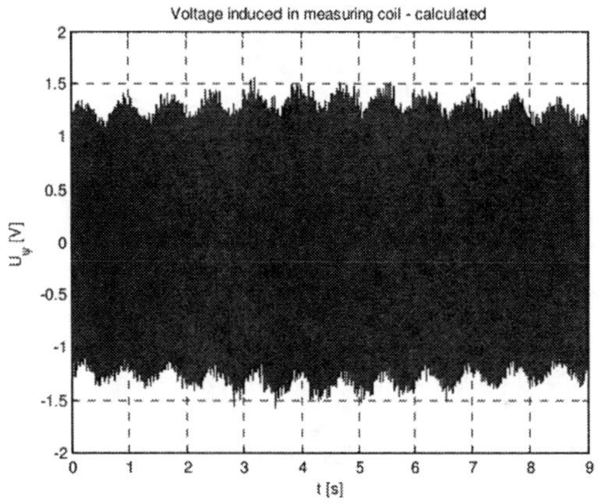

Fig. 11. The waveform of induced voltage in the coil – steady state.

Further analyses will consider the steady state signals of the machine operation. The amplitude spectra of the axial flux and its derivative are shown in Fig. 12. and Fig. 13. Fig. 14. and Fig. 15. present respectively the spectra of these both signals calibrated in dB.

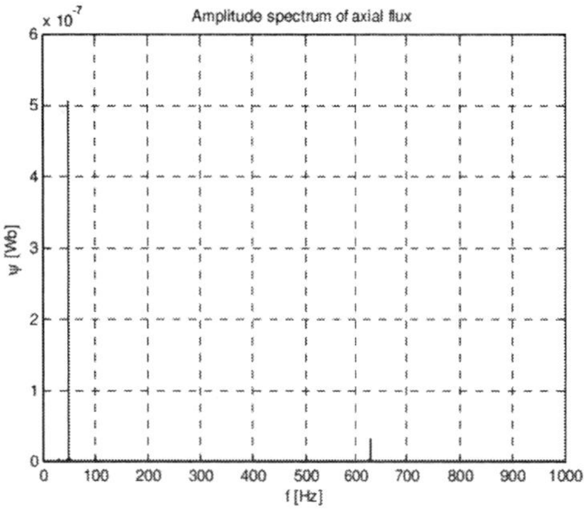

Fig. 12. Amplitude spectrum of the axial flux.

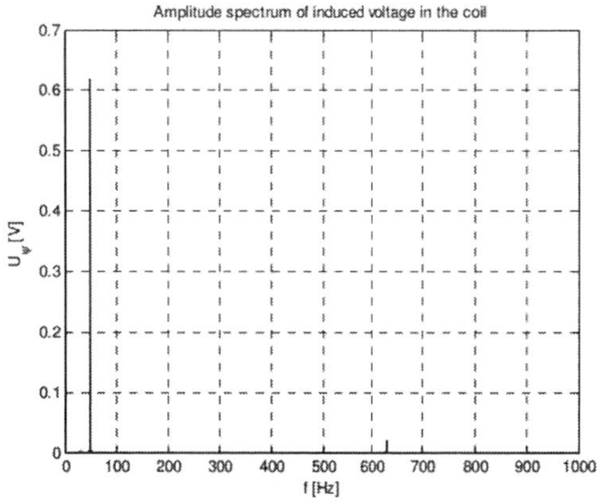

Fig. 13. Amplitude spectrum of the axial flux derivative.

DOI: 10.1051/itmconf/20171505004

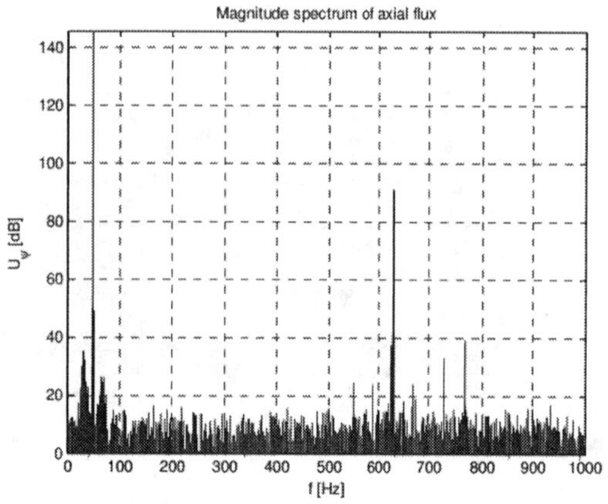

Fig. 14. Magnitude spectrum of the axial flux.

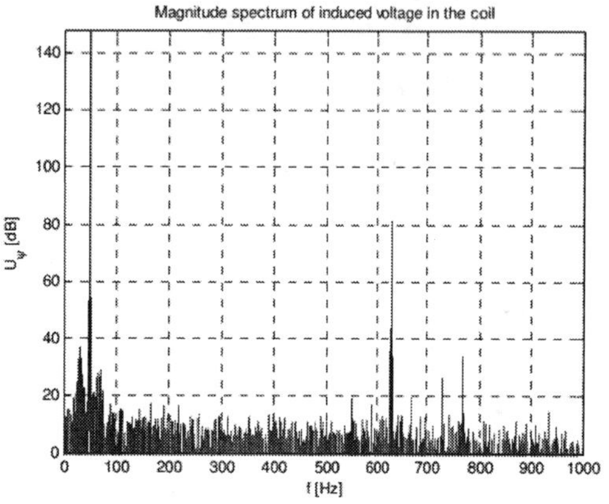

Fig. 15. Magnitude spectrum of the axial flux derivative.

Both in the signal of the axial flux and in its derivative some characteristic features can be noted, which can be uniquely identified as information related to a specific state of the machine.

In order to determine the magnitude of the torque on the basis of the axial flux, the dominant features of the signals from both the stator and rotor will be very important.

In Fig. 14. and Fig. 15, the dominant features as f0, sf0, principal slot harmonic and some additional characteristic frequencies used for diagnostics of the analysed motor, are seen.

4 Calculation of electromagnetic torque and its analysis

Directly from the field model, the time waveforms of the electromagnetic torque and the speed signals were calculated. The speed signal will also be useful in the load torque estimation algorithm. The electromagnetic torque and speed signals are shown in Fig. 16. and Fig. 17.

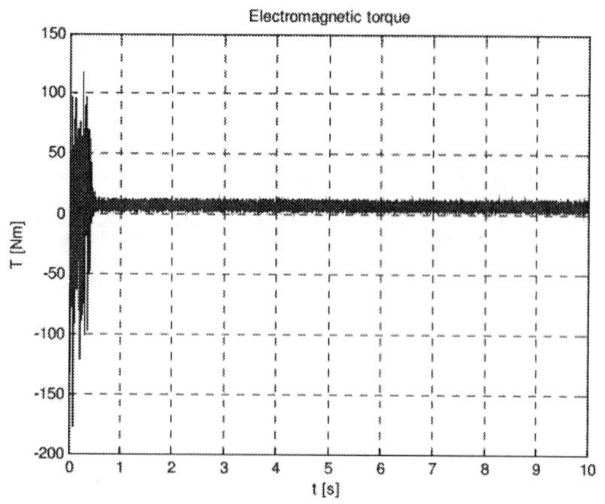

Fig. 16. Electromagnetic torque time course.

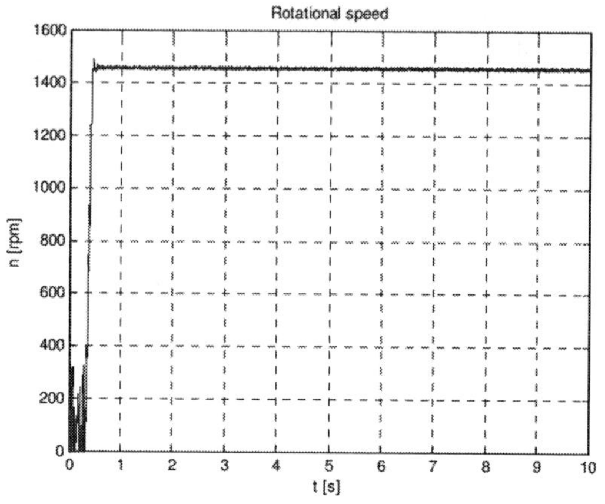

Fig. 17. Rotational speed time course.

As for the axial flux, the electromagnetic torque signal was analysed in steady state. An exemplary torque course in steady state is shown in Fig. 18.

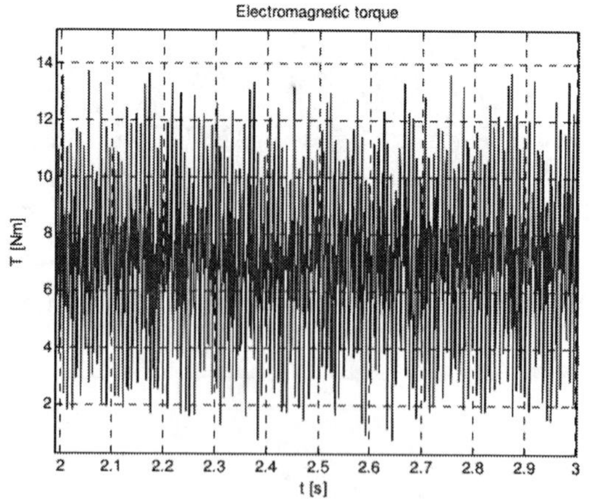

Fig. 18. Time course of the electromagnetic torque – enlargement in the interval of 1 second.

 DOI: 10.1051/itmconf/20171505004

Fig. 19. and Fig. 20. show the amplitude spectra of the electromagnetic torque calibrated respectively in linear and logarithmic scale.

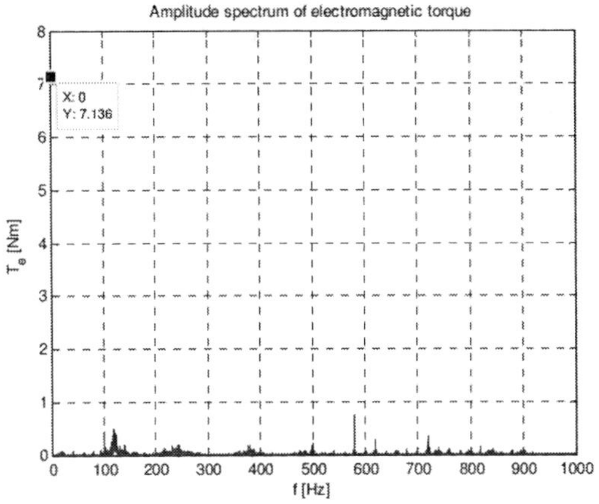

Fig. 19. Amplitude spectrum of electromagnetic torque.

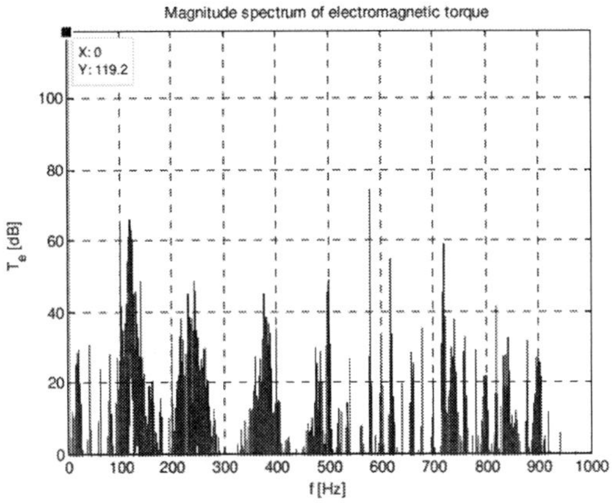

Fig. 20. Magnitude spectrum of electromagnetic torque.

The value of the amplitude of the zero-torque component determined from the time course of the calculated torque signal, will be an important feature in search of the relationship between the axial flow and the electromagnetic torque.

5 Torque estimation algorithm based on axial flux signal

The above analyses and developed method of the axial flux analysis allowed to observe and find the relationship between the characteristic features of the axial flux and the electromechanical load torque.

Fig. 21. to Fig. 23. present the characteristic features of the axial flux for the dominant components of the stator and rotor spectrum.

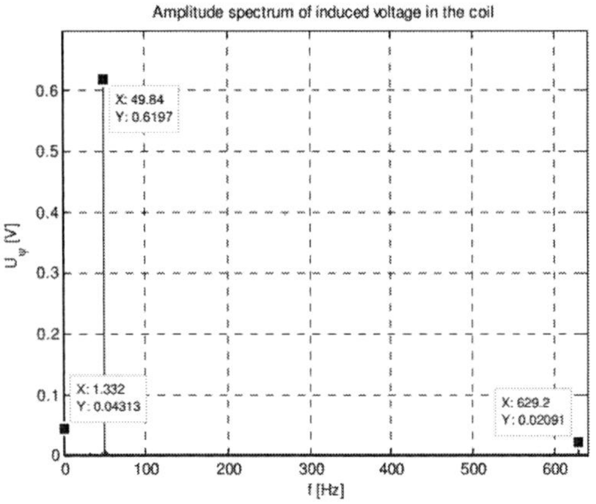

Fig. 21. Spectrum of axial flux – load torque 7 Nm, observation window from 0 to 640 Hz.

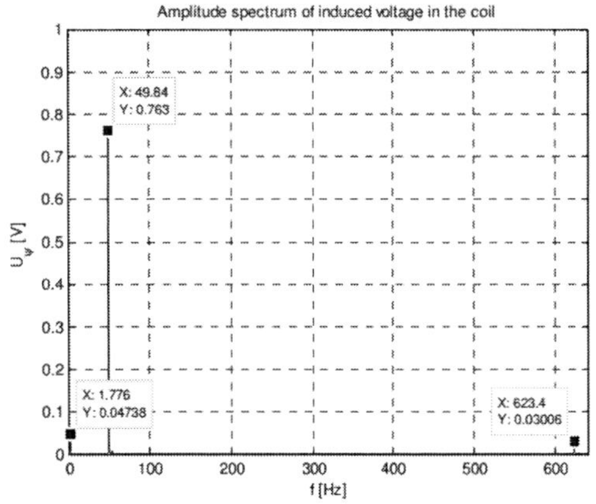

Fig. 22. Spectrum of axial flux – load torque 9 Nm, observation window from 0 to 640 Hz.

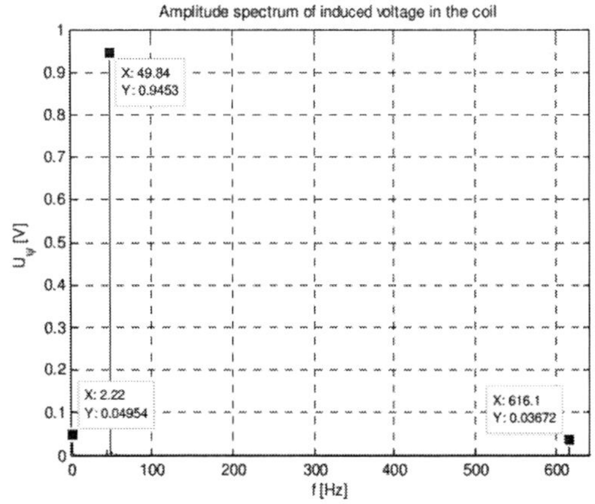

Fig. 23. Spectrum of axial flux – load torque 11 Nm, observation window from 0 to 640 Hz.

Table 1 summarizes the values of the characteristic features and torque values obtained on the basis of time courses of the axial flux and the electromagnetic torque.

Table. 1. List of the characteristic features of the axial flux and the electromagnetic torque.

T [Nm]	Amp s*f0 [V]	Amp f0 [V]	Amp fpsh [V]	RMS [V]	T / Amp f0 [-]
7.14	0.043	0.62	0.023	0.8826	11.52
7.96	0.032	0.68	0.029	0.9715	11.67
9.30	0.047	0.76	0.030	1.0866	12.19
10.77	0.041	0.87	0.020	1.2391	12.40
11.91	0.049	0.94	0.036	1.3465	12.60

In the table above, the letter signs are entered:
T – electromagnetic torque read out from the spectrum,
s – slip,
f_0 – frequency of the supply voltage,
f_{psh} – principal slot harmonic.

The characteristic calculation points shown in Table 1 are illustrated in Fig. 24.

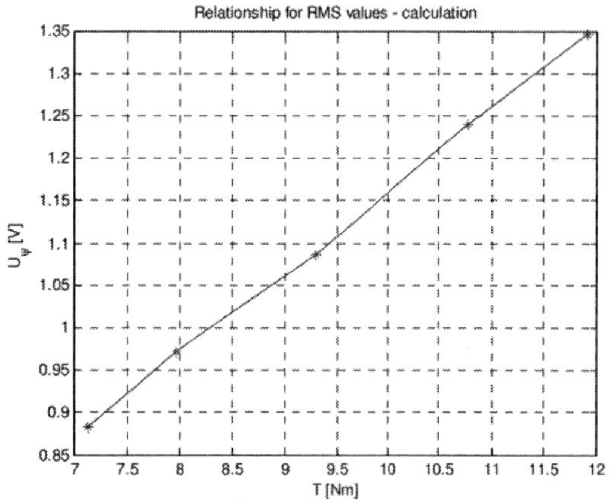

Fig. 24. Relationship between axial flux and torque.

Using the measurement coil shown in Fig. 1, a derivative of the axial flux for different motor loads was measured. About 5000 turns of the coil was wound on a resotex casing (frame) of about 150 mm diameter with an enamelled wire of 0.25 mm diameter.

The motor was loaded by the separate-excited generator. The motor and generator were connected by ROTEX flexible couplings. On the shaft, between the test motor and the load machine, a Data FLEX 22/50 torque meter was fitted, enabling a precise measurement of the load torque.

The characteristic amplitudes obtained from spectral analysis of the measured signal proportional to the derivative of the axial flux and the measured electromagnetic torque are summarized in Table 2 and graphically illustrated in Fig. 25.

In both Fig. 24. and Fig. 25, it can be observed a considerable similarity in the relationship between torque and the axial flux signal. These relationships show clearly that the load torque estimation algorithm based on the axial flow signal proved to be correct.

Table. 2. List of the characteristic features of the axial flux and the electromagnetic torque – measurements.

T [Nm]	Amp s*f0 [V]	Amp f0 [V]	Amp fpsh [V]	RMS [V]	T / Amp f0 [-]
0.16	0.001	0.371	0.003	0.431	0.43
4.22	0.007	0.518	0.004	0.550	8.15
9.50	0.029	0.864	0.013	0.902	11.00
13.81	0.042	1.197	0.017	1.226	11.54
17.40	0.073	1.445	0.019	1.511	12.04

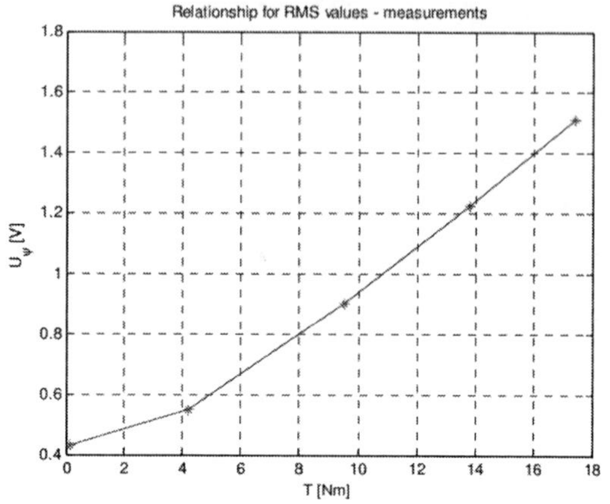

Fig. 25. Relationships between axial flux and torque – measurements.

6 Conclusions

The results of the calculations presented here point the utility of the axial flux signal for determining the shaft torque.

The direct determination of the torque from the flux for a specific machine requires the flux versus torque characteristic as in Fig. 26. for as many work points as possible. This is possible at the test station or on the

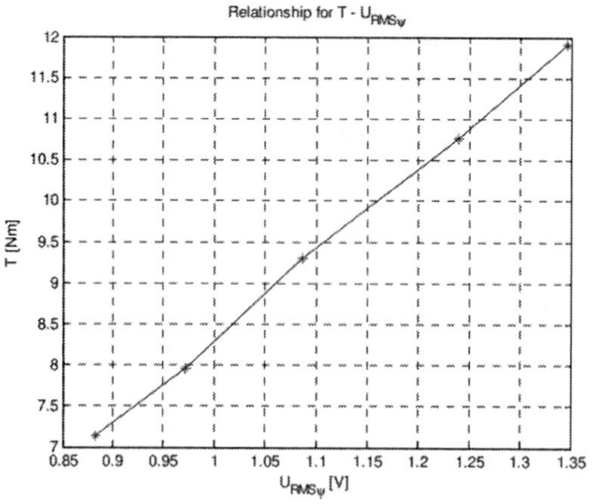

Fig. 26. Reverse relationship torque versus axial flux.

 DOI: 10.1051/itmconf/20171505004

machine workplace and makes sense when it is necessary to determine running load torque, machine efficiency or energy efficiency of the process.

Presented results of the research, which was carried out under the themes No. E-2/664/2017/DS, No. E-2/619/2016/DS, were founded by the subsidies on science granted by Polish Ministry of Science and Higher Education.

References

1. T. Assaf, H. Henao, G. A. Capolino, IEEE-ISIE, **2**, 815-819 (2004)
2. A. Bobon, Tran. on Elec. Eng., **1**, 3, 72-75 (2012)
3. A. Ceban, R. Pusca, R. Romary, IEEE Tran. on Ind. Elect., **59**, 5, 2082-2093 (2012)
4. P. Ewert, IEEE SME, 1-6 (2017)
5. L. Gołebiowski, M. Gołebiowski, M. Noga, J. Skwarczynski, Elektrotechnika i Elektronika, **25**, 2, 147-152 (2006)
6. H. Henao, C. Demian, G. A. Capolino, IEEE Tran. on Industry Applications, **39**, 2, 1272-1279 (2003)
7. W. Jarzyna, Seventh International Conference on Electrical Machines and Drives (Conf. Publ. No. 412), 141-146 (1995)
8. J. Penman, H. G. Sedding, B. A. Lloyd, W. T. Fink, IEEE Tran. on Ener. Con., **9**, 652-658 (1994)
9. W. Pietrowski, Przegl. Elektr., **87**, 6, 189-192 (2011)
10. W. Pietrowski, Przegl. Elektr., **88**, 7b, 20-23 (2012)
11. J. Siau, A. Graff, W. Soong, N. Ertugrul, Tran. on Ind. Elect., **51**, 107-121 (2004)
12. R. Pusca, R. Romary, V. Fireteanu, A. Ceban, Prog. in Electromag. Research, **50**, 37-59 (2013)
13. J. Tulicki, J. Petryna, M. Sułowicz, Tech. Tran. Electrical Eng., **3-E**, 99-113 (2016)
14. J. Tulicki, M. Sułowicz, J. Petryna, ISEF (2017)
15. V. Kokko, University of Oulu (2003)
16. D.G. Dorrell, W.T. Thomson, S. Roach, IEEE Tran. on Ind. App., **33**, 1, 24-34 (1997)
17. L. Frosini, A. Borin, L. Girometta, G. Venchi, IEEE Int. Symp. on SDEMPED, 356-363 (2011)

DOI: 10.1051/itmconf/20171505005

Chamber stapes prosthesis with an improved fastening of the membrane

Małgorzata Pudlik[1,*], *Henryk* Skarżyński[2], and *Monika* Kwacz[3]

[1]Institute of Radioelectronics, Warsaw University of Technology, Nowowiejska 15/17, 00-665 Warsaw, Poland
[2]Institute of Physiology and Pathology of Hearing, Mochnackiego 10, 02-042 Warsaw, Poland
[3]Institute of Micromechanics and Photonics, Warsaw University of Technology, św. Andrzeja Boboli 8, 02-525 Warsaw, Poland

Abstract. Stapes prostheses are dedicated to assisting patients with otosclerosis. Currently used stapes prostheses have the form of a piston and are called piston prostheses. However, the perilymph excitation by piston prosthesis is not perfect so a new chamber stapes prosthesis (ChSP) has been proposed. The ChSP allows for a more effective stimulation of the cochlea than the piston. A crucial element of the ChSP is a flexible membrane. The membrane transmits vibrations from the plate to the fluid that fills the chamber and then to the inner ear. In the first prototype, the membrane was glued to the chamber, which is an unacceptable solution in a real medical device because of a risk to patient's health. Therefore, there is a need to improve the membrane fastening. In this study, we present a new improved method for attaching the membrane to the chamber. A modified 3D model of the ChSP was build and an additional fastening element was designed. The design process of the fastening element was based on the analysis of deformation conducted for elements made of different materials. As a result the optimal geometry of the element was determined. The new fastening has been experimentally tested using the 3D printed model. Fastening the membrane to the prosthesis was a success and assessed as efficient. The results of the experimental tests confirmed that presented modification can be used in the ChSPs intended for clinical trials.

1 Introduction

As acoustic waves enter the human auditory system, they are transmitted through its consecutive parts: the outer ear, the middle ear, and the inner ear. In one of these, the middle ear, three ossicles: malleus, incus and stapes are located. The last one, stapes, is the smallest bone in the human body. The footplate of the stapes is attached to the oval window (OW) niche and transmits the vibrations thereof, stimulating the fluid (perilymph) which fills the cochlea (the inner ear). The middle ear is thus responsible for providing efficient transfer of sound from the air to the inner ear, which results in the excitation of the basilar membrane vibrations. Mechanical waves from the cochlea are finally converted to electric signals in neurones, which constitutes the hearing process, given that the auditory system functions properly. In certain cases pathologies may lead to hearing loss as a result of *e.g.* a disease called stapes otosclerosis. In otosclerosis, the mobility of the stapes is reduced or stopped completely. This is caused by non natural bone remodelling in the OW niche [1,2].

There are many factors contributing to otosclerosis, hence standard treatment cannot be applied. According to the scientific literature, the most effective treatment for otosclerosis is a surgical procedure [3-5]. The first otic surgery called stapedectomy was performed by Shea in 1956, by replacing the stapes with a prosthesis and covering the OW with a living membrane [6]. A modified operation, stapedotomy, is currently the most common technique for reconstructing the sound conducting mechanism of the middle ear and reducing hearing loss. During stapedotomy, a stapedial suprastructure is removed and then a small hole (0.4 - 0.7 mm) in the stapes footplate is made. Then a prosthesis is attached on one side to the long process of the incus and on the other the piston is inserted to the newly created hole. Currently used stapes prostheses, typically come in the form of a piston. They are called piston prostheses; an example of one of those is presented in Fig.1A.

For many years, scientists have conducted research concerning improvement of the stapes surgery performance. This research has resulted in the development of various types of piston prostheses. Currently used piston prostheses differ in the type of material (*e.g.* Teflon, titanium, gold), the piston diameter, or the attachment method to the incus [8-10]. However, only few researchers focus on improving efficacy, especially for frequencies above 2 kHz, which is the main problem with piston prosthesis. Therefore, in 2014, a new type of prosthesis was proposed [11]. It is called the chamber stapes prosthesis (ChSP) and its design is shown in Fig.1B.

* Corresponding author: mgpudlik@gmail.com

 DOI: 10.1051/itmconf/20171505005

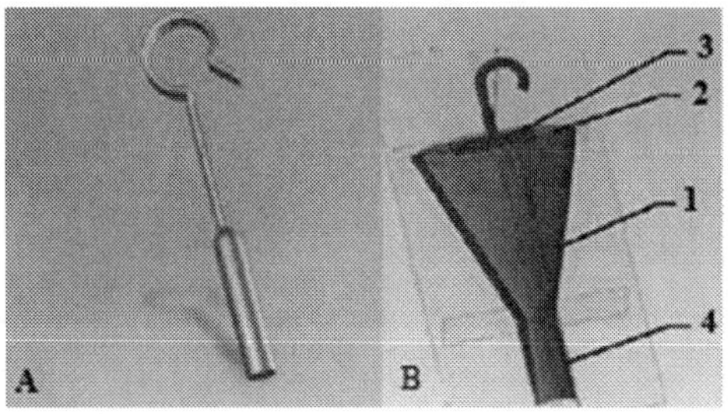

Fig.1. (A) Typical piston prosthesis [7]. (B) Design of chamber stapes prosthesis (ChSP) [12]. 1 – chamber filled with fluid, 2 –flexible membrane, 3 – plate with an attachment, 4 - capillary tube.

Computer simulations, carried out to preliminarily verify the functioning of the ChSP, have shown ChSP superiority compared to the standard piston prosthesis [12]. The tests consisted in studying the vibrations of the basilar membrane (BM) induced by the prostheses compared with the BM vibration in the physiological ear model. The results of the numerical simulation showed that the ChSP provides better excitement of the BM than the piston prosthesis and similar or slightly better than the physiological stapes footplate. The findings of the study suggested that the clinical use of the ChSP could give better postoperative hearing results compared with the standard piston prosthesis.

Then the experimental preclinical studies were conducted by measuring vibrations of the round window (RW) membrane in the human temporal bone before and after implantation of the ChSP prototype, using a scanning laser-Doppler vibrometer [13]. The measurement results confirmed the superiority of the ChSP over the piston prosthesis, especially for frequencies above 2 kHz.

The key element in the design of the ChSP is a flexible membrane (2 in Fig.1B). The membrane allows the vibration to be transmitted from the plate to the fluid filling the chamber. Except that, the membrane should ensure tight closure of the chamber. During the first experiments, the membrane, which was glued to the chamber, sometimes peeled off of the prosthesis. This is unacceptable in a real medical prosthesis because of a risk to patient's health. The loss of tightness of the chamber and resulting fluid outflow poses a risk of infection in the inner ear, which can lead to inflammation and, in the worst case, to a hearing loss. Therefore, there is a need to improve the fastening of the membrane.

1.1 Objective

The objective of this study is to find, determine and test the best solution for the detected problem associated with bonding the membrane to the chamber and holding it in the desired place in the prosthesis. To this end, (i) a conceptual design of a new membrane fastening should be created, (ii) a 3D-CAD model of the chamber should be modified, (iii) a new fastening element should be designed, (iv) the ChSP with the new membrane fastening should be manufactured and tested experimentally.

2 Methods

2.1 Conceptual design

The conceptual design (Fig.2) for an improved fastening of the membrane consists in creating a groove (2) in the prosthesis chamber (1) and subsequently placing an additional fastening element (3) in this groove. The additional element, resembling a ring, was designed to keep the membrane (4) and is applied on the open upper surface of the chamber.

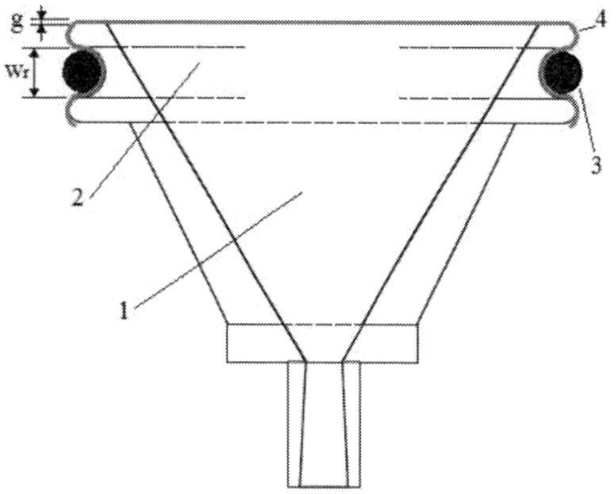

Fig. 2. Conceptual design for an improved fastening of the membrane. 1 – chamber, 2 – groove with height w_r, 3- additional fastening element, 4 – membrane with thickness g.

2.2 3D-CAD model and 3D print of the chamber

A modified 3D model (Fig.3A) of the chamber geometry was built using Autodesk Inventor 2015. The model was created according to the technical documentation according to all the design, technological and functional requirements, *e. g.* rounded edges that prevent damage to the membrane, and dimensions that ensure proper implantation of the ChSP in the middle ear. The 3D-CAD model was saved in the STL format.

The complete STL model of the chamber was checked using Autodesk Netfabb 2017. This software delivers all of the necessary tools to fix the geometry and prepare files for 3D printing.

To manufacture the modified ChSP, one of the additive manufacturing technologies, MJP (Multi Jet Printing), was. The MJP method allows printing complex geometries with high accuracy, even up to 16 microns. This feature is critical to manufacturing such small elements like a stapes prosthesis. The 3D model of the chamber was printed (Fig.3B) by *I.J. Paliga* (www.ijpaliga.pl) using a PolyJet3500 machine. The printer specifications and properties of the building material are collected in Table 1.

 DOI: 10.1051/itmconf/20171505005

Table 1. Printer specifications and properties of the material used in the manufacturing process of the modified ChSP.

Accuracy of 3D printing	layer thickness - 16 μm; resolution - 750 x 750 DPI
Work space	289 x 185 x 203 mm
Wares	minimum wall thickness - 0.2 mm; smooth surface, moving parts, functional and thin-walled models
Material	photopolymer (VisiJet M3 X) with properties: density - 1,1 g/cm^3; tensile strength - 49 MPa; bending strength - 65 MPa; modulus of elasticity E - 2168 MPa; heat resistance - 88° C

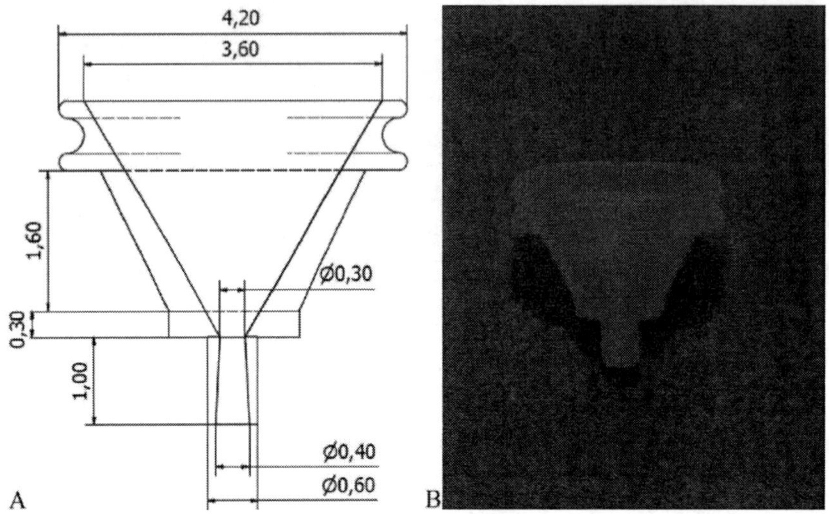

Fig. 3. Modified model of the chamber. (A) Front projection of model built using Autodesk Inventor 2015 (dimensions in mm). (B) 3D printed model.

2.3 Design of the fastening element

The fastening element in the shape of a ring was designed in two variants (Fig. 4): (1) an o-ring with a circular cross-section, and (2) a ring with a rectangular cross-section (a very short section of a thin-walled tube).

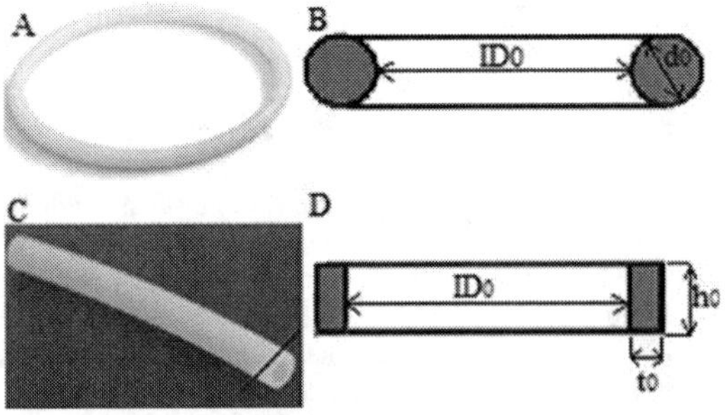

Fig. 4. Two variants of fastening elements. (A) a commercial o-ring (www.alliedmetrics.com). (B) the cross-section of the o-ring with ID_0 and d_0 dimensions. (C) a commercial thin-walled tube with marked cut-off location (www.printo3d.pl, self-modified), (D) the cross-section of the ring after cut off from the tube with ID_0, t_0 and h_0 dimensions.

Each of the rings was designed to be made from the following two materials that have a wide range of applications in medicine: (1) PTFE (polytetrafluoroethylene) and (2) PDMS (polydimethylsiloxane), the latter being a polymer from the group of siloxanes. Both these materials are biocompatible and non-toxic. During the designing stage the material properties such as the range of elasticity (ε), Poisson's ratio (ν) and Young's modulus (E) were taken into account (see Table 2). Stresses in the material, both during the application of the ring and during the whole period of use of the prosthesis should not exceed the elastic limit of the material. Moreover, it was necessary to adjust the ring dimensions to fit into the groove (see Fig.2).

Table 2. Material properties of PTFE and PDMS (data from: 1– [14], 2 – [15], 3 – [16], 4 – [17]).

	Range of elasticity at ambient temperature, ε (%)	Poisson's ratio, ν (-)	Young's modulus, E (MPa)
PTFE	10[1)]	0,46[3)]	20[1)]
PDMS	50[2)]	0,5[4)]	1,3[4)]

DOI: 10.1051/itmconf/20171505005

2.3.1 Round cross-section ring (o-ring)

For the round cross-section ring, the dimensions necessary to determine were: (1) the cross-sectional diameter (d_0), and (2) the inner diameter (ID_0) for the unstretched ring (Fig.4B).
Taking into account the thickness of the membrane in the groove (see Fig. 2), the cross-sectional diameter (d_1) after stretching was:

$$d_1 = w_r - 2g \tag{1}$$

where:
w_r – height of the groove (in this study w_r= 0.4 mm),
g – thickness of the membrane (in mm).

When calculating the cross-sectional diameter (d_0) of the ring before stretching, the relationship (2) between the transverse and longitudinal strains under the stretching force was used.
The manner in which the material deforms is determined by the dimensionless Poisson's ratio (ν); ν is defined, as the ratio of the transverse strain to the longitudinal strain :

$$\nu = -\frac{\varepsilon_p}{\varepsilon_w} \tag{2}$$

where:
ε_p – transverse strain (%),
ε_w – longitudinal strain (%).

In our modified ChSP, the longitudinal strain (ε_w) is defined as an increase in the ring circumference after stretching relative to the ring circumference before stretching:

$$\varepsilon_w = \frac{L_1 - L_0}{L_0} \cdot 100\ \% \tag{3}$$

where,
L_0 –circumference of the ring before stretching (in mm),
L_1 –circumference of the ring after stretching (in mm).

Similarly, the transverse strain (ε_p) of the round cross-section ring is defined as:

$$\varepsilon_p = \frac{d_1 - d_0}{d_0} \cdot 100\ \% \tag{4}$$

where,
d_0 – cross-sectional diameter of the ring before stretching (in mm),
d_1 – cross-sectional diameter of the ring after stretching (in mm).

Converting the equation (4) and using the relationship (2), the cross-sectional diameter (d_0) for the unstretched ring is:

$$d_0 = \frac{d_1}{1 - \nu \cdot \frac{\varepsilon_p}{100}} \tag{5}$$

Taking into account (1) and ν and ε values given in Table 2, the d_0 values were calculated from (5). The resulting d_0 values for different membrane thickness (g) are listed in Table 3.
To calculate ID_0, the circumference (L_1) of the groove that incorporates the membrane of thickness (g) was determined for different values of the g. Next, the L_0 values for rings before stretching and finally the ID_0 values were calculated using (2) and (3). The resulting ID_0 values for different membrane thickness (g) are listed in Table 3.

Table 3. The cross-sectional diameter (d_0) and the inner diameter (ID_0) of the unstretched o-ring for different thicknesses (g) of the membrane.

membrane material	g (mm)	d_0 (mm)		ID_0 (mm)	
		PTFE	PDMS	PTFE	PDMS
PTFE	0.05	0.31	0.40	2.93	2.15
PTFE	0.10	0.21	0.27	3.03	2.22
PDMS	0.10	0.21	0.27	3.03	2.22
PDMS	0.11	0.19	0.24	3.04	2.23
PDMS	0.12	0.17	0.21	3.06	2.25
PDMS	0.13	0.15	0.19	3.08	2.26
PDMS	0.14	0.13	0.16	3.10	2.27
PDMS	0.15	0.10	0.13	3.12	2.29

After determining the o-ring dimensions, the catalogues of different companies, both Polish and foreign, *i.e.* Apple Rubber (www.applerubber.com), the Precision Associates c(www.precisionassoc.com), were browsed to choose the o-rings with the d_0 and ID_0 dimensions close to the calculated values (Table 3). The chosen o-rings have been verified for the range of elasticity (10% for PTFE and 50% for PDMS, according to Table 2) and filling of the groove (less than 0.4 mm). After collecting the best fit of available o-rings, the strain values far exceeded the specified limits of elasticity. Finally, among analyzed rings, the PDMS o-ring with dimensions: ID_0= 1.5 mm, d_0= 0.25 mm (from Apple Rubber) was acceptable, with the greater allowed range of extensibility, *i.e.* around 120%.

2.3.2 Rectangular cross-section ring

For the rectangular cross-section ring (Fig. 4D), the dimensions necessary to determine are : (1) the inner diameter ID_0, (2) the wall thickness (t_0), and (3) the ring height (h_0) for the unstretched ring. The h_0 is likewise the length that should be cut of from the tube (Fig. 4C) to make the ring that properly fills the groove.
The ID_0 values are the same as the values calculated for the inner diameter of the o-ring (see Table 3).

The wall thickness (t_1) after stretching will be:

$$t_1 = g_r - g, \quad (6)$$

where:
g_r – depth of the groove (in this study g_r= 0.2 mm),
g – thickness of the membrane (in mm).

Similarly to equation (5), the wall thickness (t_0) before stretching is :

$$t_0 = \frac{t_1}{1 - \nu \cdot \frac{\varepsilon_p}{100}} \quad (7)$$

Taking into account (6) and the ν and ε values given in Table 2, the t_0 values were calculated from (7). The resulting t_0 values for different membrane thickness (g) are listed in Table 4.

Table 4. The wall thickness (t_0) and the ring height (h_0) of the unstretched ring for different thicknesses (g) of the membrane.

membrane material	g (mm)	t_0 (mm)		h_0 (mm)	
		PTFE	PDMS	PTFE	PDMS
PTFE	0.05	0.31	0.40	0.31	0.71
PTFE	0.10	0.21	0.27	0.21	0.51
PDMS	0.10	0.21	0.27	0.21	0.51
PDMS	0.11	0.19	0.24	0.19	0.47
PDMS	0.12	0.17	0.21	0.17	0.42
PDMS	0.13	0.15	0.19	0.15	0.38
PDMS	0.14	0.13	0.16	0.13	0.33
PDMS	0.15	0.10	0.13	0.11	0.28

Next, the companies producing thin-walled tubes of such small diameters were found: *i.e.* Adtech (www.adtech.co.uk), Bakai (www.well-tube.com), Nelco (www.nelcoproducts.com), and WITKO (www.witko.com.pl).
Finally, based on the strain and the groove height w_r=0.4 mm, the (h_0) values were calculated (Table 4).

3 Testing and results

Before experimental testing, a mounting plate was made to facilitate the assembly of the chamber, membrane and ring, which are very small elements. The plate model was created using Autodesk Inventor Professional 2015. After the simulation of fitting the plate with the chamber (function *collision analysis*), the plate model was sent to the *I.J. Paliga* company. The plate was manufactured with the same method as the chamber (see Section 2.2).
The PDMS membrane (www.spiggle-theis.com) with thickness of 0.15 mm, the rectangular cross-section rings cut off from the PTFE and PDMS tubes (www.witko.com.pl) and the printed chamber were used for testing (Fig. 5). Dimensions of the PTFE and PDMS rings were : ID_0 = 3.03 and 1.5 mm, t_0= 0.2 and 0.4 mm, $h_0 \approx$ 0.11 and 0.28 mm, respectively.

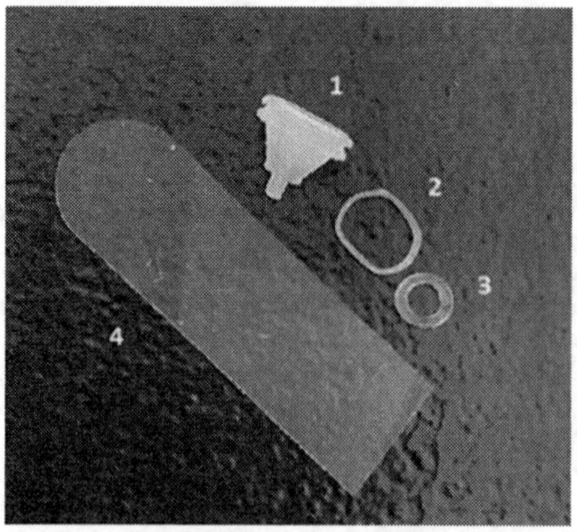

Fig. 5. Elements prepared for testing. 1 – chamber, 2 – PTFE ring, 3 – PDMS ring 4 – PDMS membrane.

At first, the rings were carefully and precisely cut off from the tubes. The length of the cut pieces was h_0. Subsequently, the chamber was placed onto the mounting plate and the membrane was placed on the upper surface of the chamber. Then the ring was manually stretched and inserted into the groove of the chamber using precision tweezers. Finally, an extra portion of the membrane was cut out. Fig. 6 shows the final result of the assembly procedure.
The stage consisting in stretching the PTFE ring in the elastic strain spectrum (0-10%, see Table 2) was extremely difficult. It was easy to enter into the flow range of the polymer (necking region), which ended with a rupture, even after the reduction in tensile strength.

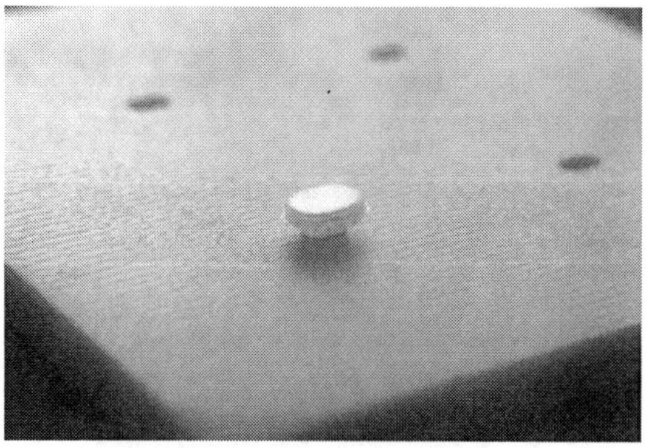

Fig. 6. Chamber with new fastening of the membrane located in the mounting plate.

In the case of a PDMS ring, significantly exceeding the 50% range of elastic strain, an attempt of stretching the ring was made. PDMS is a highly resilient biomaterial with properties similar to silicone rubber. The Young's modulus (E) of the PDMS is much lower than the one of the PTFE (see Table 2). Therefore, the more elastic PDMS proved a superior material for the designed fastening of the membrane. Unfortunately, purchase and testing of commercial PDMS micro o-rings (www.applerubber.com) was impossible due to their high price.

DOI: 10.1051/itmconf/20171505005

Long-term and safe use of the ChSP requires a ring that would never slip out of the groove. To meet this requirement, the friction coefficient between the ring and the PDMS membrane should be as large as possible. Among the pairs PDMS (membrane) - PTFE (ring) and PDMS (membrane) - PDMS (ring), the latter is definitely more favourable.

In summary, the new fastening of the membrane was successfully tested with both the PTFE and the PDMS ring, but the PDMS ring would prove a significantly better choice.

The described tests are preliminary, and were conducted to analyze possible ideas for an improved fastening, to check how it works during application in the chamber and to select the preferred ones. To fully verify the functionality of the presented solution, it will be used in a prototype designed for experimental research. This prototype will be implanted in a human temporal bone and its functioning will be assessed based on vibration measurements and observation of the stability of the membrane fixation.

4 Conclusions

This study presented modification of the ChSP design in order to improve the fastening of the PDMS membrane to the chamber. By drawing on the computational method we obtained the geometrical model and an additional element which composed a new membrane fastening. The dimensions of the prosthesis are adjusted to the dimensions of the middle ear. The additional element meets the material properties requirements and the dimensions fitting into the groove. Finally, the process of manufacturing the entire structure is relatively uncomplicated, which is important in the case of such a small element like a ChSP.

Our solution brings far-reaching benefits of its application in the prosthesis.

The improved fastening can ensure safe implantation and long-term use. The risk of loss of chamber tightness and fluid leak can be eliminated. Moreover, fastening the membrane with the ring will certainly enhance the stability of the entire ChSP after the vibrating plate is placed on the membrane. The ChSP with the improved fastening ensures reduced risk and helps to avoid some of the possible post-operative complications.

Further works on improving the ChSP are in progress and will be continued, as this prosthesis creates an opportunity for better postoperative hearing results in patients with stapes otosclerosis.

This work was partially supported by the Polish National Centre for Research and Development (Research Project No. 266457/NCBR/2015). No additional funding was received for this study.

References

1. F.H. Linthicum jr., Otolaryngol. Clin. North. Am. 26 (1993)
2. G. L. Davis, Am. J. Otolaryngol. 8 (1987)
3. E. Olszewska, S. Chodynicki, B. Łazarczyk, Otolaryngol. Pol. 56 (2001)
4. C.W. Cummings, Otolaryngology – Head and Neck Surgery (St Louis, Mosby, 1998)
5. T. Gierek, L. Klimczak-Gołąb, Otolaryngol. Pol. 6 (2007)
6. J. J. Shea, Ann. Otol. Rhinol. Laryngol. 67 (1958)
7. www.kurzmed.com/products/otologie/stapedioplasty
8. R. Tange, W. Grolman, W. Dreschler, Otol. Neurol. 25 (2004)
9. W. Grolman, R. Tange, Otol. Neurol. 4 (2005)
10. I. Wegner, J. Verhagen, I. Stegeman, R. Vincent, W. Grolman, Laryngoscope, 1 (2016)
11. W. Gambin, M. Kwacz, M. Mrówka, Komorowa protezka ucha środkowego (Patent PL 271562 B1, 2014)
12. M. Kwacz, P. Marek, P. Borkowski, W. Gambin, Hear Res. 310 (2014)
13. M. Kwacz, M. Sołyga, M. Mrówka, K. Kamieniecki, PloS one. 12 (2017)
14. Y. Kawano, Y. Wang, R. A. Palmer, S. R. Aubuchon, Polímeros, 12 (2002)
15. O. Akogwu, D. Kwabi, S. Midturi, M. Eleruja, B. Babatope, W. O. Soboyejo, Mater. Sci. and Eng: B. 17 (2010)
16. P. Smith, R. W. Zappe, Elsevier 21 (2004)
17. I. D. Johnston, D. K. McCluskey, C. K. L.Tan, M. C.Tracey, J. Micromech. and Microeng. 24 (2014)

Control and data acquisition system for rotary compressor

Marcin Buczaj[1], and *Andrzej* Sumorek[2,*]

[1]Lublin University of Technology, Inst. of Electrical Engineering and Electrotechnologies, Nadbystrzycka 38A, 20-618 Lublin, Poland
[2]Lublin University of Technology, Department of Structural Mechanics, Nadbystrzycka 40, 20-618 Lublin, Poland

Abstract. The rotary compressor (crimping machine) is a machine designed for making hollow forgings. The rotary compressor is a prototype device designed and built at the Technical University of Lublin. The compressor is dedicated to perform laboratory tests related to the hollow forgings of various shapes using different materials. Since the rotary compressor is an experimental device, there is no control and acquisition data system available. The article presents the concept and the capabilities of the computer control and data acquisition system supporting rotary compressing process. The main task of software system is acquisition of force and kinetic parameters related to the analysed process of the rotary forging compression. The software allows the user to declare the course of the forming forgings. This system allows current recording and analysis of four physical values: feed rate (speed of working head movement), hydraulic oil pressure at inlet and outlet of hydraulic cylinder and the torque of engine. Application functions can be divided into three groups: the configuration of the pressing process, the acquisition and analysis of data from the pressing process and the recording and presentation of stored results. The article contains a detailed description about hardware and software implementation of mentioned functions.

1 Introduction

Designing data acquisition systems for experimental devices is a complex task. The difficulties are introduced by an incomplete set of information about the functioning of the device, the course of the process under investigation, or an imprecise description of the expectations of end users of the device. For this reason, the practical implementation of the data acquisition and control system is a multiphase process. In the process of developing the system one can distinguish the following stages [1]:
- feasibility study (concept exploration),
- decomposition and definition (concept of operations, system requirements, high level design, detailed design),
- implementation process (hardware and software development, field installation),
- integration and definition (device testing, subsystem verification, system verification, system validation),
- operations and maintenance,
- changes and upgrades.

Depending on the complexity of the process under study, the number of measurement points and the financial means available, data acquisition and control systems vary widely. It is possible to build several channel information systems based on a single platform Raspberry Pi [2, 3, 4]. The advantages of such systems are low manufacturing costs and an open software platform. At the opposite extreme are highly complex systems requiring high-speed operation requiring programming of FPGA embedded systems [5, 6, 7]. A particular group of measurement systems are systems dedicated to the analysis of particle behaviour or cosmic radiation [8, 9]. In their case, it is often a problem to develop a sensor capable of recording the desired values. A wide range are universal systems adapted to the study of a particular phenomenon. In such cases, an often used environment for programming is LabVIEW (National Instruments) [10, 11, 12, 13]. This article presents the original measurement system dedicated to collecting information from an experimental rotary crimping machine. The system is based on the LabVIEW graphical programming environment.

2 Requirements

2.1 Rotary crimping machine

Hollow shafts and axles are used, inter alia, in drive systems of machinery, equipment and vehicles. The great popularity of hollow elements is primarily due to their comparable strength properties relative to full equivalents (mainly for bending and twisting loads), while their mass is significantly reduced. Currently the main recipients of the hollow elements are the automotive and aerospace industries [14]. For the purposes of the research, the construction of a forge assembly for rotary crimping, further called the rotary crimping machine, was developed. The designed and constructed machine had a segment structure and consisted of ten major sets [15]. Due to the originality of the study, there was no ready-made measurement system that could be used in such a device (Figure 1).

* Corresponding author: a.sumorek@pollub.pl

DOI: 10.1051/itmconf/20171505006

Fig. 1. The view of rotary compressor device with control panel

The device was experimental and subject to current modifications. Following an analysis of the way of its functioning, the physical quantities critical for the crimping process were determined. The design and construction of the measuring system included four parameters:
– feed (linear displacement of pistons in hydraulic cylinders);
– oil pressure in the hydraulic cylinder at the inlet and outlet of the hydraulic power supply;
– torque of the crimping machine's motor;
– critical value of the offset at which the cylinder feed should end.

2.2 System requirements

In the analysed case, the measurement system did not have to record all the possible process parameters. From the user's point of view, the behaviour of the electrical and hydraulic equipment is insignificant. It is not essential to measure such parameters as operating voltage, current value, the power of the device, or the time of switching contacts. The accumulated amount of information about the electrical parameters would rather describe the operation of the crimping machine's drive system rather than the crimping process itself. Redundant data on electrical parameters could slow down the acquisition of the remaining values and make it difficult to search the values needed for analysis.

The main parameters characterising the course of the crimping process are the displacement value of the actuator piston, the crankcase torque value and the pressure in the cylinders of the actuators. In terms of time constraints, two restrictions were specified. The measurement process series can reach a maximum of 20 seconds. All measuring channels at this time are to sample parameters at speeds of at least 100 measurements per second. At the same time, it should be noted that in the case of applicable sensors, each of them will have a different output. The requirements for the measuring system are shown in Table 1.

Table 1. Requirements for a data acquisition system.

Requirement	Description
Pressure measurement	Range: 0-25 MPa
	Accuracy: ±1%
	Number of measurement points: 2
Length measurement	Range: 0-50 mm
	Accuracy: ±0.05 mm
	Number of measurement points: 1
Torque measurement	Range: 0-10 kNm
	Accuracy: ±1%
	Number of measurement points: 1
Data recording	Cycle time: 20 s
	Recording frequency in measuring channel: 100 Hz
	Data file format: compatible with csv, tsv (for Notepad, Excel, Open/Libre Office)
Operating system	Windows XP – Windows 10
Communi-cation	USB 2.0 (Full-speed)

3 Hardware layer

The block diagram of the control and data acquisition circuits is shown in Figure 2. In the data collection part, the information flow is unidirectional (sensors, measurement cards, personal computer). The record is maintained for the duration of the crimping process by the declared cycle length (default 20 seconds). In the part responsible for automatic control of the device the PLC is programmed once. The program is implemented in the loop throughout the device. Control panel initialisation functions are as follows: main engine start-up, pump start-up, tool feed, tool feed rate control, tool off, system shutdown when tool position limit is exceeded, emergency shutdown of the entire system.

The selected Novotechnik TS 50 displacement transducer performs a direct displacement measurement. The transmitter housing is mounted to a fixed mechanical element. The transducer bar is attached to the moving piston of the pressure cylinder. Along the entire length of the sensor is mounted a resistance path. Thanks to this potentiometer it is possible to determine the position based on the ratio of the resistance corresponding to the position of the slider and the resistance of the whole path. Figure 3 shows the TS 50 after mounting on a test bench.

 DOI: 10.1051/itmconf/20171505006

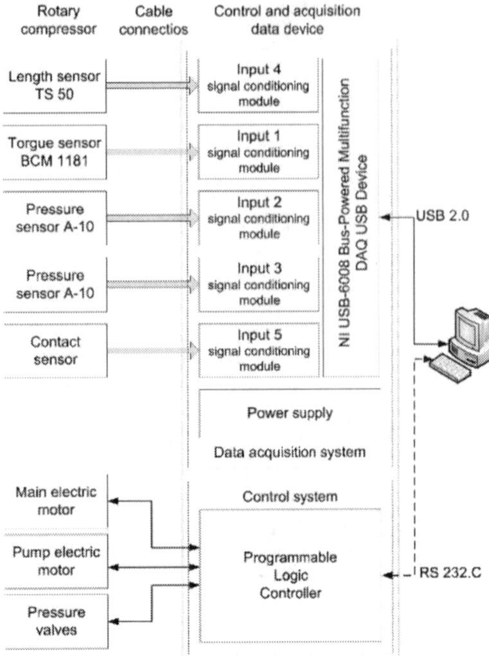

Fig. 2. A schematic diagram of the connection of the control and data acquisition system with the rotary crimping machine.

Fig. 3. Novotechnik TS 50 sensor after installation

Fig. 4. Wika A-10 sensor in laboratory conditions and after installation

The Wika A-10 pressure transmitter is dedicated to general industrial applications. It is compact and offers the accuracy required in this application. Because the user can select 0.25% or 0.5% nonlinearity, the one ensuring a higher accuracy of the measurement was selected. Typical applications include machines, machine tools, measuring and control technology, hydraulics and pneumatics, pumps and compressors. This range of applications perfectly matches the requirements of this project. Figure 4 shows the A 10 converter in laboratory conditions and after mounting on a test bench.

The BCM 1811 torque converter measures the torque transmitted by the rotating drive shaft. Throughout the BCM 1811 range, it is possible to measure torque from 5 to 150000 Nm at half-percent accuracy over the whole measuring scale. The output signal is of a frequency nature. There are transducer implementations coupled to a speed sensor. Figure 5 shows the BCM 1181 in laboratory conditions and after mounting on a test bench.

Fig. 5. BCM 1181 sensor in laboratory conditions and after installation

The National Instruments USB-6009 measurement card can measure and generate both analogue and digital signals. It is shown in Figure 6.

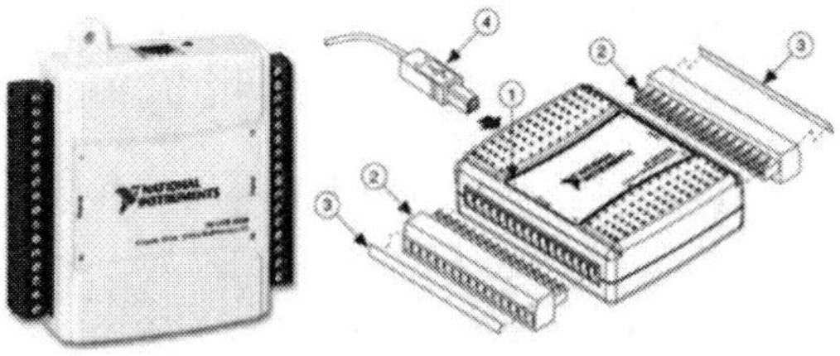

Fig. 6. National Instruments USB-6008 Data Acquisition Card (1 - overlay label with pin orientation guides, 2 - screw terminal connector plug, 3 - signal label, 4 - USB cable)

The card is equipped with 8 analogue inputs with a resolution of 11 bits to measure voltages ranging from –10V to 10V with respect to the ground terminal. It is possible to programmatically switch the voltage measurement method and connect the inputs in pairs to the 12 bit inputs. The card has two independent voltage sources with a resolution of 12 bits, generating variable voltages of any shape ranging from 0V to 5V. They were protected against overload by an internal 50Ω resistor connected in a series. The maximum sampling rate of analogue signals is 10 kS/s. The digital part of the card contains 12 digital connectors in two ports, both as input and output, two voltage sources of +2.5V and 5V, and ground output. The card is also equipped with a 12-bit counter.

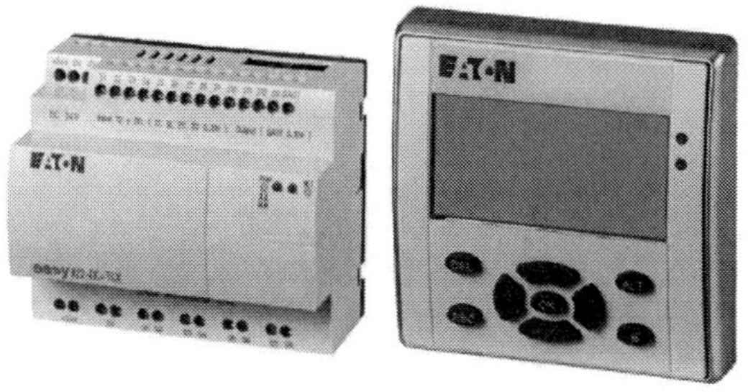

Fig. 7. Programmable Logic Controller EASY822-DC-TCX with MFD display unit

As previously mentioned, the controllable range of the control is governed by functions such as starting the motors, starting the pump drive, starting the tool feed, controlling the tool feed rate, moving the tool off, or shutting down the entire system. The control algorithm is performed physically using a Moeller EASY822-DC-TCX PLC. The EASY822-DC-TCX controller is capable of handling 12 digital inputs, 4 analogue configurations, 8 binary outputs. In order to enable the user to monitor the execution state of the program, the system is enriched with an MFD-80-B multifunction display (Figure 7).

4 Software layer

The data acquisition and control system is a hybrid. The EASY822-DC-TCX PLC controller is responsible for controlling the crimping machine. The NI-USB 6008 card is responsible for the measurement process.

4.1 Control system

The functions of the crimping machine control system with a PLC controller described in the previous section are combinatorial and sequential functions forced by the state of the input variables declared by the user in the panel of Figure 1.

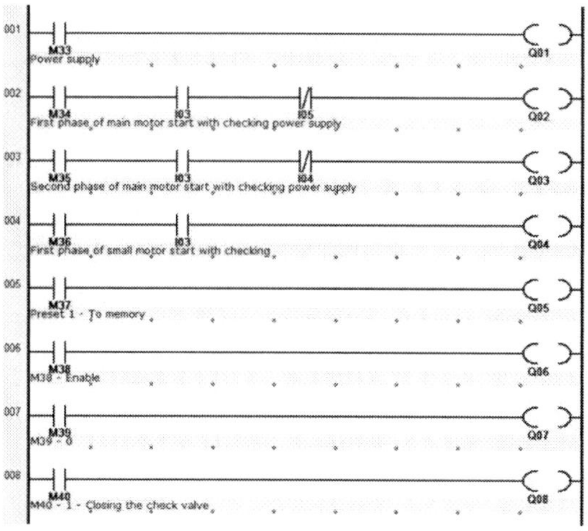

Fig. 8. A fragment of control system code (ladder diagram)

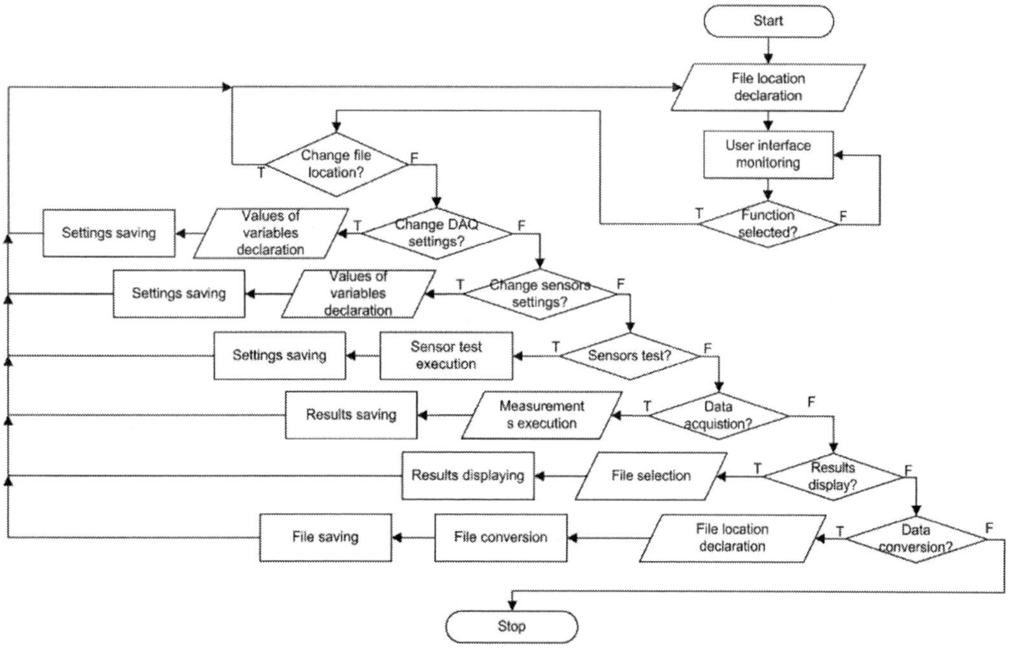

Fig. 9. Algorithm of program execution (user perspective)

 DOI: 10.1051/itmconf/20171505006

Four possible PLC programming methods have been selected for ladder diagram programming. This is a natural way of programming control systems in the relay contactor systems. The control of such components of the crimping machine as motors and valves is perfectly reproducible by means of a ladder diagram. The ladder language is so flexible that it can be used even for programming alarm systems [16, 17]. The control program was developed in the "easySoft-Pro" V6.92 (Eaton Industries GmBH) graphical environment.

4.2 Data acquisition system

The data acquisition system corresponds to the NI-USB 6008 measurement card system with sensors and software. The block structure of the components of the system is shown in Figure 2.

Data acquisition software has been implemented in a graphical development environment dedicated to the LabVIEW Professional Development System (2015, Service Pack 1, ver. 15.0.1, 64 bit, National Instruments). Figure 9 shows a scheme of a program algorithm that performs the functions listed in Chapter 2. The algorithm shows the interrelations of the program blocks for use by the end user. It describes possible action scenarios that can be initiated by the user.

The algorithm is based on the architecture of the state machine, so that from the user's point of view it is possible to go straight to the selected range of activities. The only exception to this rule is the mandatory declaration of the disk directory where the hardware settings and measurement results will be stored. Without this initial step, it is not possible to call other procedures. At any time, you can return to these settings and modify them.

All procedures supported by the measurement software can be divided into three groups:
- configuration operations,
- activities related to system component testing and measurement,
- presentation of results and operations on results files.

In the first, so-called configuration group of operations, there is a mandatory declaration of location on the disk space to save the configuration settings and the location of the measurement files. The configuration files store information about the correction factor of the card, data on where to save the files with the results, the duration of the measurement, or how to trigger the measurement series.

The second group of procedures refers to the operation of the measurement card. It has the ability to test card operation (physical operation of each channel). The key element is the recording of measurement data. It is possible to manually trigger the measurement procedure. It is more convenient, however, to initiate the measurement procedure after the declared travel displacement or torque is declared. Both of these stages should be preceded by a declaration of the maximum duration of the measurement series.

The third group of procedures supports the presentation and conversion operations. Immediately after a series of measurements, the results are gathered from the file in the place indicated by the user. The program gives you the opportunity to view them as a time chart. This means that displacement time graphs, two pressure graphs and torque are displayed. It is also possible to access results in the form of numbers corresponding to the values read from the sensors. Using converters built into the program, it is possible to convert a disk file into a spreadsheet format. Conversion to *.ods, *.xls, and *.xlsx formats was validated.

The measurement software works with the experimental crimping system. Its driving elements are high power electric motors, while the processed elements are metal parts, heated to temperatures above 1000°C. For this reason, after launching, the program presents a dialog box in which the user is informed about the need to read the instruction manual (Fig. 10).

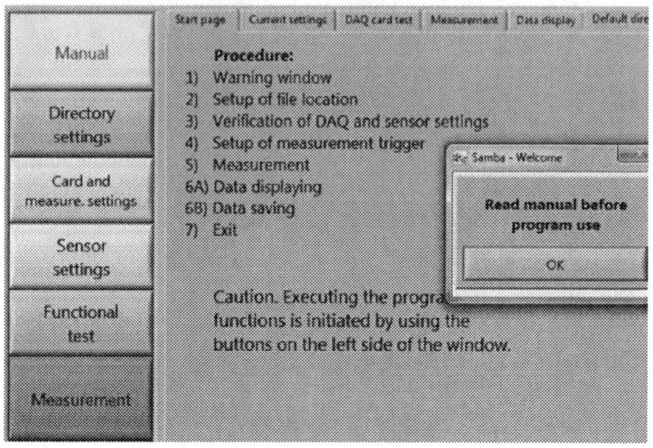

Fig. 10. Welcome window with a warning message

As soon as the program is started, the "Default Directory Settings" window (Figure 11, Figure 9) is displayed each time. You are asked to select two directories for the measurement files (*.lvm) and files with results converted to the spreadsheet format (*.xls).

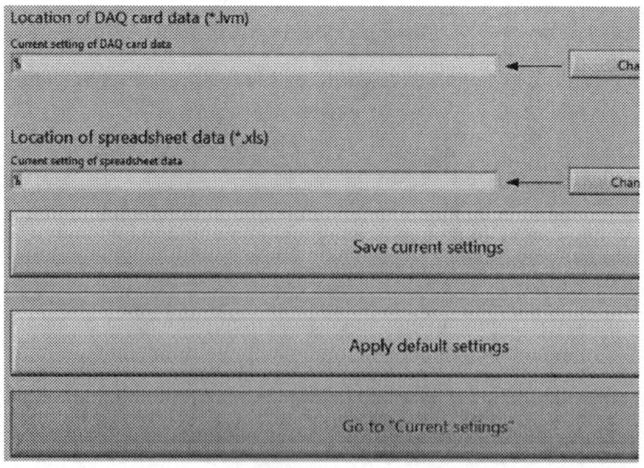

Fig. 11. Window for declaration of file directory location

Selecting the "Card and Measurement Settings" button (Figure 10) displays the "DAQ Setup" window (Figure 12, Figure 9). The left part of the panel is used to declare the settings, the right part is a control displaying the values that will be applied after selecting the "Save changes" button.

DOI: 10.1051/itmconf/20171505006

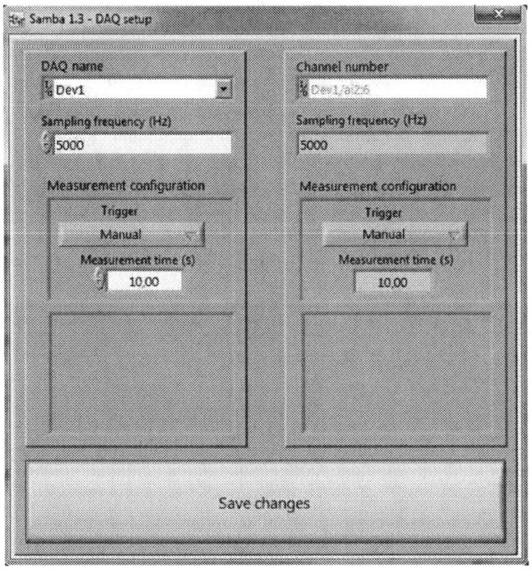

Fig. 12. DAQ setup window

The frequency of sampling of signals from individual channels is determined by the capabilities of the measurement card. In the current configuration it is possible to define a sampling frequency in the range of 1000 to 5000 samples per second (Fig. 12). Each channel is sampled at the same frequency.

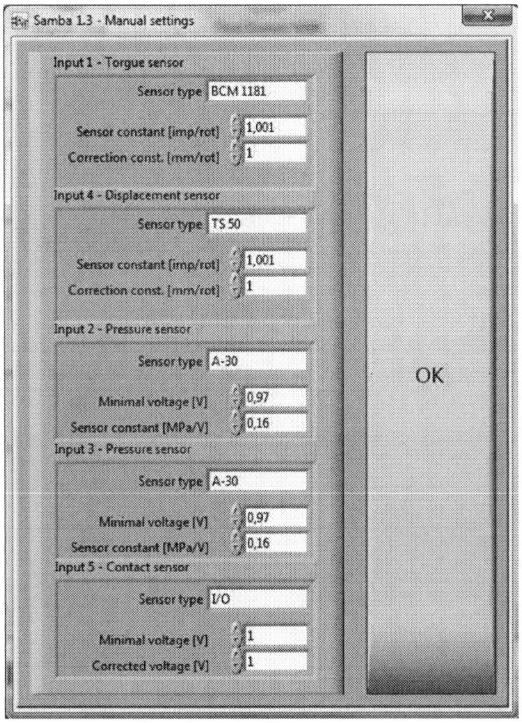

Fig. 13. Sensors setup window

By selecting the "Sensor settings" button (Figure 10), the "Settings" window will be displayed (Figure 13, Figure 9). The five data ranges concern five sensors that can be connected to inputs 1-5 of the measuring system.

Selecting the "Functional Test" button (Figure 10) initiates the "DAQ Card Test" window. The system performs two tests: communication with the measuring card and the presence of sensors.

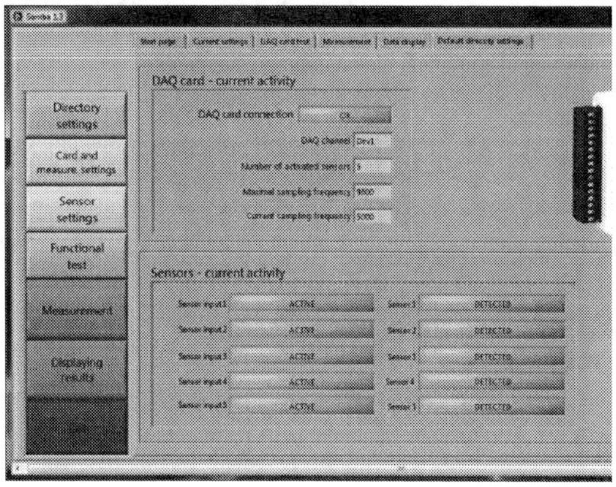

Fig. 14. Sensors and DAQ test window

The message "No connection to the data acquisition card" may suggest the following card issues:
- another device connected to the USB port (not necessarily a measuring board),
- another measurement card connected (a different Dev than the one selected),
- incorrectly declared sampling frequency,
- problems with the USB cable connection.

The lack of co-operation of the sensors with the system may result from:
- confusion of sensor inputs (position sensor connected instead of pressure sensor and vice versa);
- bad electrical connection of the sensors (unplugged plug, loose contacts, broken/fractured wire);
- connection of a device not intended for the system (another similar, but not hardware-compatible sensor),
- physical damage to the sensor (short circuit, open, malfunction).

The "Active" and "Detected" signals signal only communication with the expected data card channels in the expected voltage ranges. They do not guarantee a correct measurement because it depends on the sensor settings being run as described above.

The transition to conducting measurements is only useful after:
- declaring a place to record results,
- declaring how the card works,
- choosing how to start the measurement,
- entering sensor parameters.

The "Measurement" window (Figure 15) is displayed after selecting the "Measurement" button (Figure 10). If automatic triggering is declared, the "Measurement Settings" field indicates that the procedure is ready to run. If you choose to manually start the measurement, the system is waiting for the "Start" button to be pressed. Successful completion of the measurement procedure

 DOI: 10.1051/itmconf/20171505006

depends on the time elapsed since the initiation of the procedure and is signalled by the status bar.

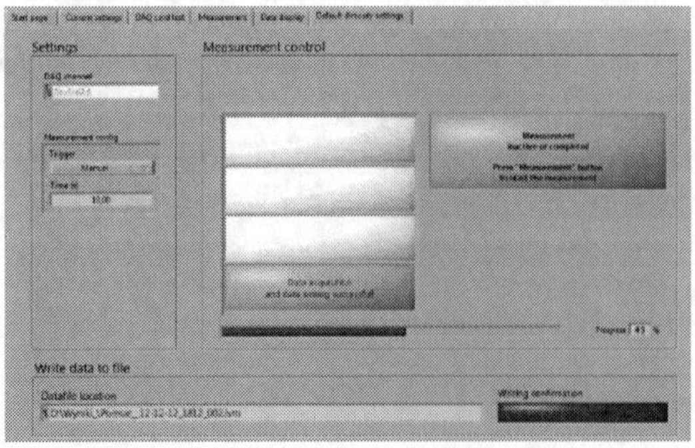

Fig. 15. Data acquisition window

In the data collection and presentation system there is no immediate relationship between the value recorded and the time presented. Multiple measurement series are possible and results may be reviewed at another time or place.
Loading of the presented data from the file is initiated by the button "Displaying results" (Figure 16). The result is a dialogue box that allows you to select a file. For a file to be correctly interpreted, it must be a result file of the measurement system, i.e. a file with an extension of *.lvm. It contains undistorted measurement data which, after rescaling, are presented in graphs and in a table (Figure 16). In the case shown in Figure 16, the upper timeline contains data from two pressure sensors, the lower passage – data from the displacement and torque sensors.

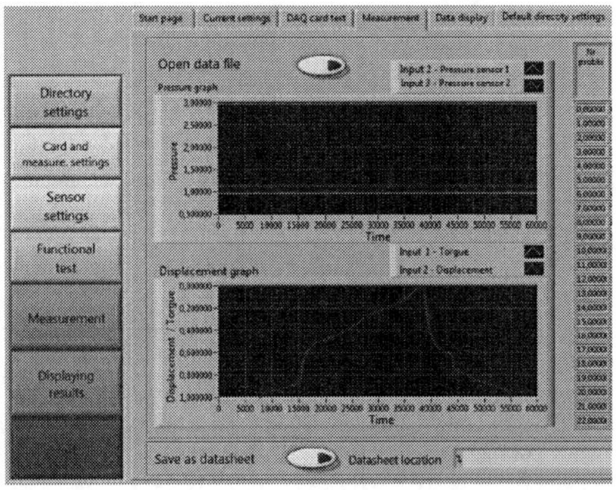

Fig. 16. Data presentation window

Conversion to a spreadsheet is possible by selecting the "Save as Datasheet" button located at the bottom of the window (Figure 16). As a result, a dialog box is displayed allowing you to choose where to save the file. The saved file is a text file of the *.tsv type. An effective record of the file is signalled by the filled "Datasheet location" field.

The LabVIEW Professional Development System programming environment provides tools for developing applications. One such tool is the ability to browse the structure of programs and subroutines (Figure 17). Overview of the hierarchy of two levels of data acquisition system subroutines shows that the previously described functions are implemented by one master program, 6 original subprograms, and 7 subprograms in the LabVIEW package. Further tracking of the structure shows the use of dozens of functions included in the package associated with the subroutines used.

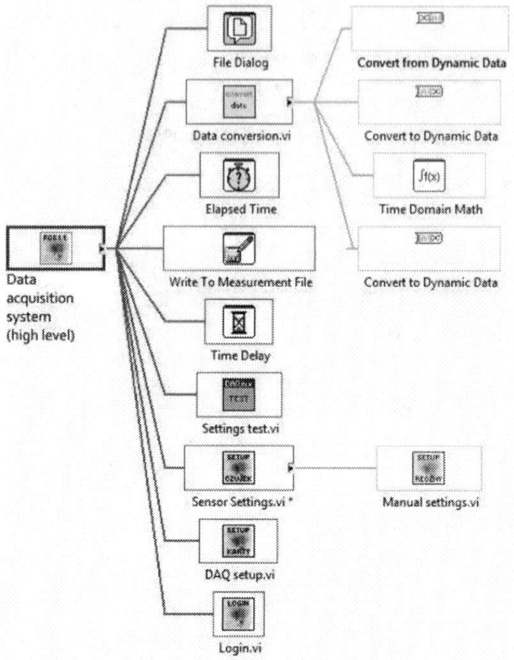

Fig. 17. The hierarchy of system components

5 Practical test results

The data acquisition and control system is used during the process of rotating hollow rotary hollows in laboratory conditions with the use of real materials. Practical research involved the development of hollow forgings from tubular blanks. An example data set obtained from the data acquisition system is shown in Figure 18 (force calculated on the basis of recorded pressure and torque).

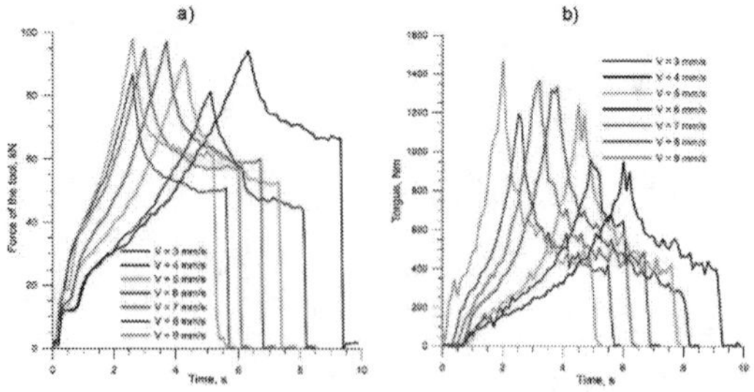

Fig. 18. The course of force (a) and torque (b) as a function of time (v - tool movement speed) [14]

DOI: 10.1051/itmconf/20171505006

As part of the research work on the crimping process outside of data recording, time and power characteristics based on FEM simulation were determined. On the basis of the recorded data and experimental research, differences in maximum forces and moments can be observed, as well as in their distribution over time. Experimenters find the reason for the discrepancy in terms of the difference between the actual thermal conditions and the ones taken into account [14].

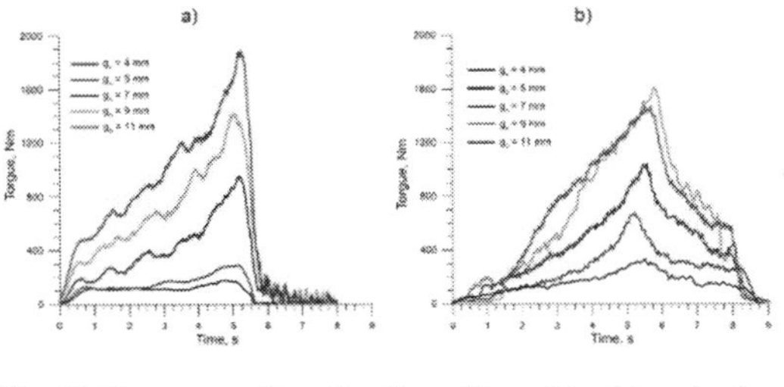

Fig. 19. The course of as a function of time: (a) - determined by the FEM method; (b) – experimental data [14]

6 Conclusions

The constructed and practically tested computer control and data acquisition system of the rotary crimping machine allows for the following conclusions to be drawn:

– it is possible to build a hybrid control and data collection system with the software included with the controllers and measurement cards;

– despite its simplicity, the system provides experimenters with all the data;

– during the development of the system, all stages of the professional design of data acquisition systems were involuntarily introduced [1];

– the ability to improve the functionality of the system was enhanced by introducing additional features, such as storing system settings (cards, sensors, triggers) based on the user's login name.

References

1. J.P.T. Mo, A. Sinha, *Engineering Systems Acquisition and Support*, Woodhead Publishing, 37-49 (2014)
2. A. Miha, Measurement **100**, 7-18 (2017)
3. J. Fletcher, W. Malalasekera, Energy **111**, 32–46 (2016)
4. A.J. Lewis, M. Campbell, P. Stavroulakis, Measurement **87**, 228–235 (2016)
5. J. Kryszyn, P. Wróblewski, M. Stosio, D. Wanta, W.T. Smolik, Measurement **101**, 28-39 (2017)
6. R. Rajpal, H. Mandaliya, J. Patel, P. Kumari, P. Gautam, V. Raulji, P. Edappala, H.D. Pujara, R. Jha, Fusion Engineering and Design **112**, 964-968 (2016)
7. J. Nieto, D. Sanz, P. Guillén, S. Esquembri, G.de Arcas, M. Ruiz, J. Vega, R. Castro, Fusion Engineering and Design **112**, 957-960 (2016)
8. J.A. Aguilar, A. Albert, F. Ameli, M. Anghinolfi, J. Zúñiga, Nuclear Instruments and Methods in Physics Research Section A: Accelerators, Spectrometers, Detectors and Associated Equipment **570**, 107-116 (2007)
9. Y. Sasai, Y. Matsubara, Y. Itow, T. Sako, L.X. González, Nuclear Instruments and Methods in Physics Research Section A: Accelerators, Spectrometers, Detectors and Associated Equipment **857**, 50-57 (2017)
10. H. Rezk, I. Tyukhov, M. Al-Dhaifallah, A. Tikhonov, Measurement **104**, 204-211 (2017)
11. J.B.W. Roeber, S.K. Pitla, M.F. Kocher, J.D. Luck, R.M. Hoy, Computers and Electronics in Agriculture **127**, 1-14 (2016)
12. F.J. Ferrero Martín, M. Valledor Llopis, J.C. Campo Rodríguez, J.R. Blanco González, J. Menéndez Blanco, Measurement **55**, 265-271 (2014)
13. H.Q. Liao, Z.R. Qiu, G.H. Feng, Procedia Environmental Sciences **10 part B**, 1188-1192 (2011)
14. J. Tomczak, *Studium procesów obciskania obrotowego odkuwek drążonych* (in Polish), Lublin (2016)
15. Z. Pater, A. Gontarz, J. Tomczak, T. Bulzak, Archives of Civil and Mechanical Engineering **15**, 917–924 (2015)
16. M. Wcislik, IFAC Proceedings Volumes **36**, 1, 37-40 (2003)
17. W. Bolton, *Programmable Logic Controllers*, Newnes, 111–146 (2009)

Description of anisotropy of magnetic properties for chosen grades of electrical steels

Piotr Borowik[1], and *Krzysztof* Chwastek[1,*]

[1]Częstochowa University of Technology, Faculty of Electrical Engineering, Aleja Armii Krajowej 17, 42-201 Częstochowa, Poland

Abstract. Magnetic properties of electrical steels, either grain-oriented or non-oriented ones, are significantly affected by anisotropy. The paper considers two descriptions for the dependence: power losses versus angle. The first one is based on the theoretical studies on texture from materials science. The other one is purely phenomenological, and based on an equation of ellipse in a rotated coordinate system.

1 Introduction

The designers of magnetic circuits in electrical machines need more and more efficient and accurate design tools supporting the CAD process, what implies an increasing interest of engineering community in modelling of nonlinear phenomena occurring at the core. Anisotropy is a phenomenon that complicates an analysis. Above all, this applies to grain-oriented steel, used as a core material in transformers, but the so-called non-oriented (alternator) steel used in rotating electric machines also exhibits substantial anisotropy [1-4]. Usually, magnetic properties and dependencies are provided by the manufacturers of electrical steels solely for the rolling direction (for the angle 0°). Sometimes, the dependencies are also given either for the transverse direction (for the angle 90°) or for the mixed samples (combination of strips cut for two characteristic angles, measured with the Epstein Frame). The worst magnetic properties are exhibited for angles cut at 54-60°. Therefore, a simple interpolation scheme based on two characteristic directions, such as the one proposed by Jesenik et al. [5], may not always be useful.
In order to describe anisotropic magnetic properties of electrical steels several approaches are available. One method is based on the so-called co-energy concept [6-8]. If a hysteresis phenomenon is being neglected, the co-energy density per volume module is given with an expression

$$w'(\mathbf{H}) = \int_0^{\mathbf{H}} \mathbf{B}(\mathbf{H})\mathrm{d}\mathbf{H} \quad (1)$$

The induction may be derived from (1) using (2).

$$\mathbf{B}(\mathbf{H}) = \mathrm{grad}_{\mathbf{H}}\, w'(\mathbf{H}) \quad (2)$$

An important feature of the co-energy approach is its consistence with the laws of irreversible thermodynamics. The possibility to describe magnetisation curves for chosen grades of grain-oriented and non-oriented electrical steels was verified in Refs. [7] and [3], respectively. At this point it should be recalled that the concept of co-energy is significant in the contemporary descriptions of electromagnetic energy conversion devices used for example in mechatronics, cf. e.g. [9-10] for details.

Some researchers with materials science background prefer to work with a description, which has its roots in the analysis of the so-called Orientation Distribution Functions (ODFs), commonly used in metallurgy for material characterisation [11-13]. The effective magnetic anisotropy of a polycrystalline sample should be obtained by averaging the grain anisotropy over the grain ODF, i.e. over the texture [14]. The magneto-crystalline energy, defined as an energy to rotate magnetic moments, averaged over the ODF and is given with the following expression

$$\begin{aligned} E(\alpha,\beta) &= \frac{1}{9n_4\sqrt{\pi}}\left[\frac{K_1}{5}+\frac{K_2}{55}\right]F_4(\alpha,\beta)+\ldots \\ &+\frac{1}{13n_6\sqrt{\pi}}\frac{K_2}{231}F_6(\alpha,\beta) \end{aligned} \quad (3)$$

where α,β define the specimen direction in the reference frame, n_4 and n_6 are constants, $n_4 = 0.64636$, $n_6 = 0.359601$ [12,13], K_1 and K_2 are anisotropy constants, whereas texture functions F_4 and F_6 are defined as in (4) and (5), respectively [15]:

$$F_4(\alpha,\beta) = \frac{1}{\sqrt{2}}C_4^{11}P_4^0(\cos\alpha)\cos 2\beta + C_4^{13}P_4^4(\cos\alpha)\cos 4\beta \quad (4)$$

$$\begin{aligned} F_6(\alpha,\beta) &= \frac{1}{\sqrt{2}}C_6^{11}P_6^0(\cos\alpha) + C_6^{12}P_6^2(\cos\alpha)\cos 2\beta + \ldots \\ &+ C_6^{13}P_6^4(\cos\alpha)\cos 4\beta + C_6^{14}P_6^4(\cos\alpha)\cos 6\beta \end{aligned} \quad (5)$$

C_4^{11} and C_4^{13} are parameters of texture, whereas expressions like P_4^q where $q = 0,2,4$ are Legendre's polynomials of the first kind, fourth order.

* Corresponding author: krzysztof.chwastek@gmail.com

For the angle $\alpha = 90°$ i.e. for the surface plane of the specimen, the relationship (3) yields an expression for a magnetic torque, which agrees considerably well with experimental data, as confirmed by several authors

$$M(\beta) = -\frac{dE(\beta)}{d\beta}. \qquad (6)$$

Further simplification of the above-given expressions follows if one accounts facts that the anisotropy constant K_2 is one order of magnitude lower than K_1. Then it is possible to neglect in the first approximation the second term in expression (3) (the one containing F_6). Thus magneto-elastic energy may be written as proportional to fourth order texture function $F_4(90°, \beta)$ only, as suggested by Birsan and Szpunar [15]. From (3) and (4) a simplified relationship may be written

$$E(\beta) \approx E_1 \cos 2\beta + E_2 \cos 4\beta. \qquad (7)$$

If the contribution of anisotropy constant K_0 is accounted, which is a free term independent from the direction of the applied field, then it follows that magneto-elastic energy, as well as any physical quantity of interest, which is directly related to it, may be written as the function of three first ODF coefficients, i.e.

$$A = A_0 + A_1 \cos 2\beta + A_2 \cos 4\beta \qquad (8)$$

In the formula (8) A may denote magneto-elastic energy, but also power losses, coercive field strength, permeability etc. The parameters A_i, $i = 0...2$ are given with the following auxiliary relationships obtained from measurements at the angles $\beta = 0°, 45°, 90°$

$$A_0 = 0.25\left[A(0°) + A(90°) + 2A(45°)\right], \qquad (9)$$

$$A_1 = 0.5\left[A(0°) - A(90°)\right], \qquad (10)$$

$$A_2 = 0.25\left[A(0°) + A(90°) - 2A(45°)\right]. \qquad (11)$$

From the above-given expressions it is evident that to predict angular variations for any magnetic quantity of interest, it is sufficient to carry out its values for three directions $\beta = 0°, 45°, 90°$. In this sense, this procedure is similar to the Lankford parameter used for anisotropy evaluation in metallurgical processes [16,17].

Apart from these two above-discussed methods, which have solid foundations rooted in the theories of electromagnetism and materials science, in the literature there are also some other semi-empirical approaches, for instance [18-20]. The model proposed in the next section may be considered as a member of this class.

2 Model proposal

The equation of an ellipse in Cartesian coordinates and the formulas for rotating coordinate system by the angle ϕ should be recalled. The ellipse equation is

$$\frac{x'^2}{a^2} + \frac{y'^2}{b^2} = 1, \qquad (12)$$

or in the parametric form

$$x' = a\cos t, \quad y' = b\sin t, \qquad (13)$$

where $t \in \langle 0; 2\pi \rangle$.

Rotation of the coordinate system is achieved by the transformation

$$\begin{cases} x = x'\cos\phi - y'\sin\phi \\ y = x'\sin\phi + y'\cos\phi \end{cases} \qquad (14)$$

where $\phi = 30°$ (this is the complementary angle to the angle 60 deg. shown in the Fig. 1). For simplicity of calculations it is assumed that the worst magnetic properties are at 60 deg.

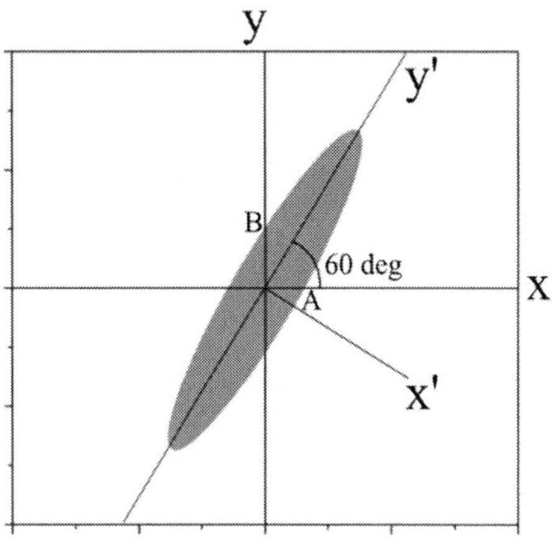

Fig. 1. The ellipse and the reference coordinate systems.

The required values for the prescribed direction may be easily found by means of interpolation. This approach should work for power losses at a constant induction level; for determination of field strength, the afore-described approach allows to determine just the magnitude of the vector $\mathbf{H}$; thus, it is necessary to supplement it with the auxiliary dependence obtained from the assumed material behaviour for the rolling direction [7]. The problem shall be the subject of further studies.

An important advantage of the presented approach is that it requires just two values of the considered quantity (for example power loss) for the orthogonal directions $0°$ and $90°$. These values determine the minor and the major axis of the ellipse. In this sense the approach is similar at the core to the co-energy based approach.

3 Modelling

3.1. Grain-oriented steel

In Ref. [15] Birsan and Szpunar provided a table concerning the measurements of power losses at 1.0 T for a fully processed (grain-oriented) SiFe steel, 0.22 mm thick. The values for the rolling and the transverse

 DOI: 10.1051/itmconf/20171505007

directions were a = 0.183 W/lb (0.404 W/kg) and b = 0.981W/lb (2.163 W/kg), respectively. As the measurements for the angle 45° were not available, it has been approximated as the average for 40° and 50°, $P_{45^\circ} = 0.5(0.766 + 0.846)$ W/lb = 0.806 W/lb, that is in SI units P_{45° = 1.777 W/kg. This value is necessary for the ODF-based method. Values for parameters A_i, $i = 0...2$ from (9) – (11) have been determined. Next, required values for intermediate angles from (8) were defined.

For the other considered description (the rotated ellipse in the Fig. 1), power losses for an arbitrary angle were set by determining the magnitude of an auxiliary vector quantity, which perpendicular components were equal to values for the points A and B in the Fig. 1.

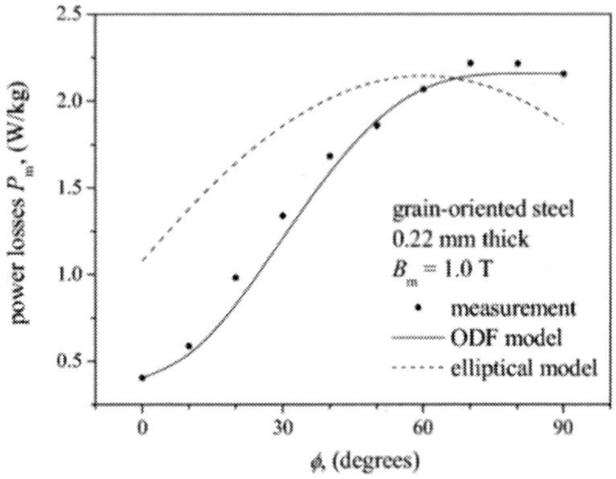

Fig. 2. The measured and modelled dependencies of losses vs. angle for grain-oriented steel.

Figure 2 depicts the modelling results for grain-oriented steel. It can be stated that the elliptical model describes the angular variation of losses only qualitatively for this grade of GO steel. The largest discrepancy between the measured and the modelled values occurs at $\phi = 0^\circ$; it is equal to 181.7%. The averaged error value is 49.8%, whereas for the ODF-based method – 5.4% [in the latter case the largest error value (observed for the angle $\phi = 20^\circ$) exceeded slightly 17.4%].

The paper by Shiozaki and Kurosaki [1] provided a discussion on the effect of processing technologies on the magnetic properties for typical non-oriented electrical steel sheets. The authors provided exemplary ODF figures and tried to establish the relationship between textures and typical figures of merit provided by the manufacturers: power losses at B = 1.5 T, f = 50 Hz and induction at H = 5000 A/m, f = 50 Hz. Technological processes considered by Shiozaki and Kurosaki are listed in the Table 1. Their detailed description is given in Ref. [1].

The processing conditions had a significant impact on the properties of the specimens, as could be inferred from the analysis of published measurement data. The problem of texture optimisation is of a great practical interest to metallurgists and materials science engineers, cf. for example [21-25]. Simultaneously, it should be remembered that the magnetic properties of ready-made parts of magnetic circuits may usually be worse than the properties of steels themselves due to stresses introduced during subsequent material processing [26-28].

Table 1. Technological processes considered by Shiozaki and Kurosaki [1].

Case	Process
1.	single reduction, cold rolling process
2.	conventional temper-rolling process
3.	special temper-rolling process
4.	conventional hot band annealing and single reduction cold rolling process
5.	special hot band annealing and single reduction cold rolling process

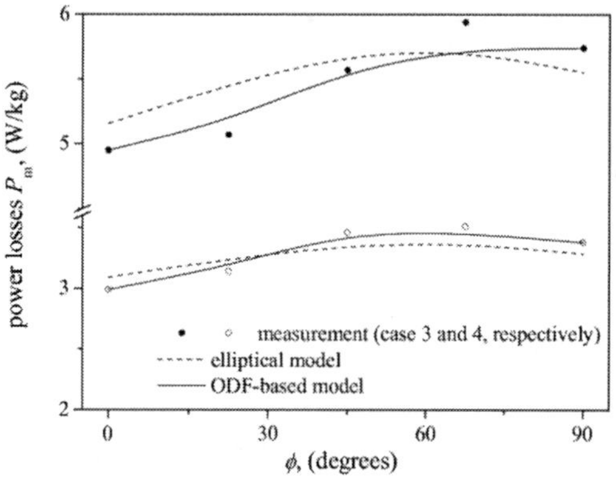

Fig. 3. Measured and modelled dependencies of losses vs. angle for chosen non-oriented steels.

The prediction capabilities of the proposed elliptical model, for the five aforementioned specimens subject to different processing routes, have been checked. The results are given in the Table 2 and in the Fig. 3 (just two cases for better visibility). In the Table "max. error" denotes the maximum value $\delta_{\%} = 100 \cdot |P_{\text{meas}} - P_{\text{model}}| / P_{\text{meas}}$, whereas "average error" means that averaging has been carried out for all possible angles from the range $\varphi \in \langle 0^\circ, 90^\circ \rangle$.

Table 2. Prediction errors for two considered models.

Process	ODF		elliptical	
	max. error	average error	max. error	average error
single reduction	1.1	0.3	2.6	1.5
conv. temper-rolling	1.2	0.3	3.7	1.7
spec. temper-rolling	3.4	1.1	7.7	4.2
conv. hot band annealing	1.4	0.6	3.9	3.3
spec. hot band annealing	0.3	0.1	6.0	2.9

It can be stated that the elliptical model yields slightly larger values of errors, yet it should be remembered that it requires less input information i.e. measurement data just for two directions. From data presented in the Table 1, it can be concluded that the ODF model might be particularly suited for engineering purposes. The elliptical model yields a relatively good qualitative and quantitative description of power losses vs. angle for the considered non-oriented steels.

4 Conclusions

In the paper two descriptions of anisotropic properties of electrical steels have been discussed in detail and their predicting accuracy was illustrated using some representative examples from the literature.

References

1. M. Shiozaki, Y. Kurosaki, Textures Microstruct. **11**, 159 (1989)
2. V. Paltanea, Gh. Paltanea, H. Gavrila, Electr. Electron. Eng. **2 (6)**, 383 (2012)
3. K. Chwastek, IET Electr. Power Appl. **7**, 575 (2013)
4. F. Martin, D. Singh, A. Belahcen, P. Raasilo, A. Haavisto, A. Arkkio, COMPEL **34**, 1475 (2015)
5. M. Jesenik, V. Goričan, M. Trlep, A. Hamler, B. Štumberger, J. Magn. Magn. Mater. **304** (2006) e820-e822
6. Th. Péra, Fl. Ossart, Th. Waeckerlé, J. Appl. Phys. **73**, 6784 (1993)
7. O. Biró, St. Außerhofer, K. Preis, Y. Chen, COMPEL **29**, 1482 (2010)
8. J. Krause, arXiv:1212.5163v1 [math-ph] 20.12.2012
9. H. H. Woodson, J. R. Melcher, *Electromechanical dynamics*, 3 vols. (Massachusetts Institute of Technology: MIT OpenCourseWare), available from: http://mit.ocw.edu.
10. M. Hadžiselimović, P. Virtič, G. Štumberger, T. Marčič, B. Štumberger, J. Magn. Magn. Mater. **320**, e556-e561(2010)
11. M. F. de Campos, Proc. XVIII IMEKO Congress, 17-22.09.2006, Rio de Janeiro, Brazil. Available from: http://www.imeko.org/publications/wc-2006/PWC-2006-TC4-037u.pdf
12. H.-J. Bunge, *Texture analysis in materials science - mathematical methods* (Butterworths, London, 1982)
13. O. Engler, V. Randle, *Introduction to texture analysis. Macrostructure, microstructure, and orientation mapping*, (CRC Press/Taylor & Francis Group, Boca Raton, 2010)
14. J. A. Szpunar, Textures Microstruct. **11**, 93 (1989)
15. M. Birsan, J. A. Szpunar, J. Appl. Phys. **80 (12)**, 6915 (1996)
16. W. T. Lankford, S. C. Snyder, J. A. Bausher, Trans. ASM **42**, 1197 (1950)
17. W. F. Hosford, R. M. Caddell, *Metal forming. Mechanics and metallurgy* (Cambridge University Press, Cambridge, 2007)
18. H. Vande Sande, F. Henrotte, K. Hameyer, COMPEL **23 (2)**, 579 (2004)
19. Hao Wang, Chang-sheng Li, Tao Zhu, Int. J. Minerals, Metall. Mater. **21**, 1077 (2014)
20. Th. Kefalas, A. G. Kladas, COMPEL **31 (2)**, 682 (2012)
21. M. Birsan, J. A. Szpunar, J. Appl. Phys. **81 (2)**, 821 (1997)
22. L. Kestens, S. Jacobs, Texture, Stress, Microstruct. (2008) doi:10.1155/2008/173083
23. D. Steiner Petrovic, B. Markoli, M. Ceh, J. Magn. Magn. Mater. **322**, 3041 (2010)
24. K. M. Lee, S. Y. Park, M. Y. Huh, J. S. Kim, O. Engler, J. Magn. Magn. Mater. **354**, 324 (2014)
25. P. Ghosh, R. R. Chromik, A. M. Knight, S. G. Wakade, J. Magn. Magn. Mater. **356**, 42 (2014)
26. A. Pulnikov, PhD Thesis, Ghent University, Ghent, Belgium, 2004
27. D. Gaworska-Koniarek, B. Szubzda, W. Wilczyński, J. Drosik, K. Karaś, J. Phys.: Conf. Ser. **303**, 012091 (2011)
28. S. Steentjes *et al.*, Proc. 4th Int. Conf. Electric Drives Production EDPC'2014, 30.09-1.10.2014, Nürenberg, Germany

DOI: 10.1051/itmconf/20171505008

Digital signal processing in ultrasonic based navigation system for mobile robots

Paweł Stączek[1,*]

[1]Lublin University of Technology, Mechanical Engineering Faculty, Department of Automation, Nadbystrzycka Street 36, 20-618 Lublin, Poland

Abstract. A system for estimating the coordinates of automated guided vehicles (AGV) was presented in this article. Ultrasonic waves for distance measurement were applied. Used hardware was characterised, as well as signal processing algorithms. The system was tested on wheeled mobile robot in model 2D environment. The results of working range and errors of position estimation were discussed.

1 Localisation systems for mobile robots

An important element of automated transport systems is the localisation system of vehicles (mobile robots) in their work environment. Many different techniques have been developed for automatic position measurement. GPS is fundamental for open-field applications, often supplemented by ground stations [1-4]. In an indoor environment, the following localisation methods are used:

• radio signals from the Wi-Fi networks [5-8],
• optical markers and automatic vision systems [9-13],
• distance measuring transducers (odometry) [4],
• onboard accelerometers, gyroscopes, compasses [14-17],
• noncontact distance measuring devices using: radar waves, infrared light, lasers (LIDARs) or ultrasonic waves [18-25].

Each of these localisation methods has limitations; ergo, hybrid systems are built using two or more of these techniques simultaneously.

2 Ultrasonic based localisation systems

One of the oldest methods of determining the distance between objects in space is the use of ultrasounds – the acoustic waves with frequencies above than 20 kHz. The simplest variant of this method is the measurement of a time lapse – since a "packet" of ultrasonic waves has been emitted, until it reaches a detector (a microphone). This time duration is simply called "time of flight" (TOF). Knowing the velocity of sound waves propagation in the medium separating the objects, the distance between them can be calculated [4,18].

While ultrasonic distance measurements are relatively easy to implement, precise determination of the relative position (direction) of the transmitter and receiver is difficult. The reason for this, is the problem of emitting a coherent and targeted ultrasound beam, and the low "directionality" of the microphone sensitivity. Hence, a number of transmitters and/or microphones distributed in space are used for precise positioning, along with accredited triangulation methods [18,21,23,24].

Advanced methods of modulating the emitted ultrasound packet and algorithms of processing of received waveforms have been developed [19,20,22,25]. They allow to precisely calculate the distance and direction of arrival of the wave. In the paper [22], the authors describe a method of modulation and demodulation of ultrasounds, which results in establishing a measurement error of 1mm (up to 3 m distance) and 3 degrees in direction (up to ±50 degrees) with one transmitter and two closely spaced microphones.

Despite the rapid development of other methods of locating objects (mainly vision systems), ultrasounds are still broadly used to detect objects (obstacles) and to navigate robots at distance up to a dozen meters. The advantages of ultrasound systems are: the simple operation principle, and the low hardware requirements for their implementation – hence, the low price of construction and implementation, in comparison to other systems.

3 Design requirements of the proposed localisation system

The goal of the author's work was to develop, build and test an experimental ultrasound-based locating system of mobile robot for indoor use. The basic requirements for the system were formulated as follows:

• estimating the coordinates of the mobile robot position on the plane (floor),
• operating range of a dozen or so meters, i.e. on the area of one room, corridor, small production hall,
• establishing an error of positioning of the order of several centimetres,
• installing ultrasound transmitters on walls/ceiling,
• installing a microphone on the mobile robot,
• setting all ultrasonic transmitters to operate on a single frequency; hence, they must transmit in the time sequence,

* Corresponding author: p.staczek@pollub.pl

• synchronising time of ultrasound emission/reception by radio frequency waves,
• establishing possible presence of minor obstacles in the way of ultrasounds (e.g. tables, chairs, people),
• implementing digitally signal processing algorithms from the microphone.

4 Structure of mobile robot localisation system

The block diagram of the developed location system for mobile robots is shown in Fig. 1. The system consists of:
• four piezoelectric ultrasonic transmitters of type UT-1640K-TT-2-R [26],
• microprocessor controller of ultrasonic transmitters including: 8-bit AVR microcontroller (Arduino UNO module), 433 MHz RF receiver and four H-bridge transistor drivers,
• wheeled mobile robot, on which following items have been installed (Fig. 2.):
  • microphone made in MEMS technology of type SPH0641LU4H-1 [27],
  • 32-bit Kinetis K64F-family microcontroller,
  • 433 MHz radio transmitter for digital data transmission,
  • ZigBee radio communication module,
• personal computer with ZigBee radio communication module.

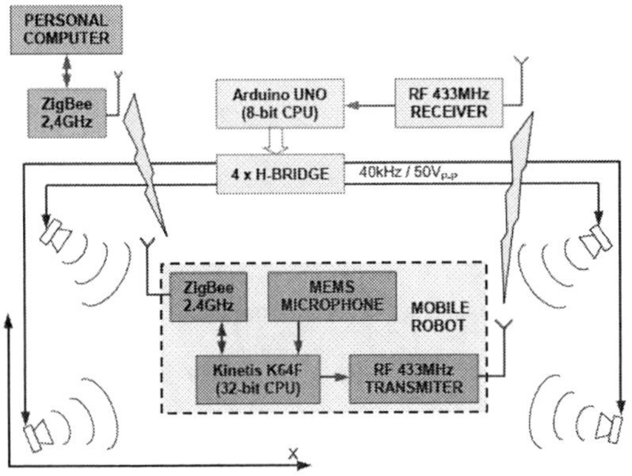

Fig. 1. Scheme of experimental ultrasonic-based system for mobile robot localisation

Four ultrasonic transmitters were mounted in the corners of the room of a rectangular floor (dimensions 10 m x 9 m), 2.5 m above the floor (Fig. 3). The axis of the transmitters were directed toward the central point of the flat floor, which was the working area of the mobile robot. In addition, four obstacles, i.e. large tables with a height of 0.8 m, were set on the floor. The microphone was installed on a mobile robot and the axis of mic. was upwardly vertical (Fig. 2.).

The estimation of the robot position on the plane was automatically made in one-second cycles, consisting of the following phases:

1. The microcontroller on the robot sends a data packet through 433 MHz radio link to the ultrasonic transmitters controller for synchronising the times of both microcontrollers.
2. The transmitters controller generates a sequence of 4 rectangular signals (Fig. 4.), one for each of the four ultrasound transmitters. Signals for successive transmitters are generated every 200 ms, driving only one transmitter at a time (during less than 0.7 ms).
3. During the transmission of ultrasounds, the robot's microcontroller records the signal from the microphone.
4. The recorded signal from the microphone is processed in the robot's microcontroller, and the time, needed for the sound wave to travel from each transmitter to the microphone, is calculated.
5. Results of time measurements (TOF) and "power" of the received signals are sent by ZigBee radio link to the desktop computer, where they are further processed (filtered) and the current position of the robot is estimated.
6. The computer sends calculated position to the microcontroller on the robot.

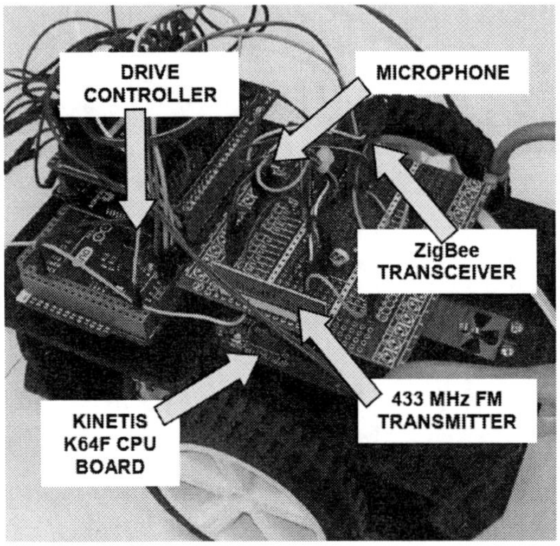

Fig. 2. Experimental mobile robot with components of ultrasonic-based localisation system.

At the current stage of the research, the algorithm for calculating the robot coordinates, based on measured time durations (TOFs), is performed on a desktop computer. The target algorithm will be implemented in the robot's microcontroller.

5 Signal processing algorithms

Distance measured using ultrasounds is particularly sensitive to disturbances caused by acoustic waves, generated in the vicinity of the microphone, and by interference with the ultrasonic waves reflected from other objects. The temperature of the surrounding air also has a significant impact on the distance measurement error (the velocity of sound in the air varies by nearly 2%/ 10°C). However, the influence of this factor is relatively easy to automatically compensate for.

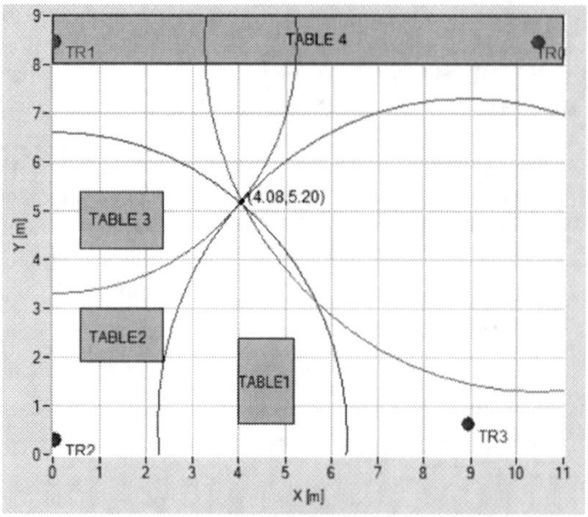

Fig. 3. Scheme of test-room floor (top view) and arrangement of localisation system elements; four ultrasonic transmitters marked TR0÷TR3 were installed at 2.5 m above the floor.

The significant effect on distance measurement accuracy has the way, the acoustic wave is generated, as well as filtering the signal received from the microphone. In typical applications, the ultrasonic transmitter periodically supplies the differential voltage signal in the form of several sinusoidal or rectangular cycles, with a frequency corresponding to the resonant frequency of the transmitter (Fig. 4.). In the turn, the time after which the "packet of waves" has been detected by a microphone is searched in received signal. Unfortunately, due to the inertia of the vibrating elements of the transmitter and the microphone, the recorded response is distorted (Fig. 4.). The amplitude of oscillations gradually increases, and then slowly disappears. Hence, this leads to difficulties with unambiguous detection of the beginning and the end of the packet, received by the microphone.

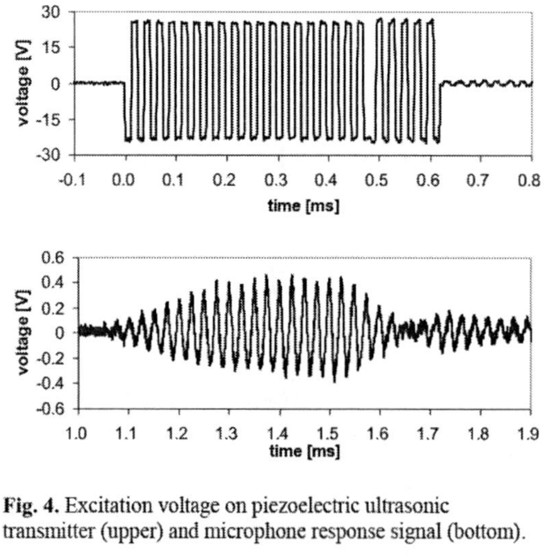

Fig. 4. Excitation voltage on piezoelectric ultrasonic transmitter (upper) and microphone response signal (bottom).

In order to decrease the time of fading of transmitter oscillations after emitting of the ultrasonic wave package, it was proposed to add active vibrations damping, by driving transmitter by the signal shifted in phase at 180 degrees. Fig. 4. shows voltage recorded at the terminals of the piezoelectric transmitter. There are 19 cycles of rectangular signal (basic wave packet), 5 cycles of the opposite phase, followed by fading oscillations of the voltage, generated by the oscillating piezoelectric (after emitting the whole package, the transmitter terminals are short-circuited by the H-bridge, in order to suppress further oscillations of piezoelectric).

The block diagram of the processing path of the microphone output signal is shown in Fig. 5. The pulse density modulation (PDM) output signal is applied to the RC low-pass filter input. The output voltage of the filter U_1 is digitised at 160,000 samples per second (4 times the frequency of the ultrasonic wave), and the results are stored in the memory buffer of the robot's microcontroller. The recorded signal is processed in the Butterworth digital pass band filter [28-32]. The transfer characteristics of the filter is shown in Fig. 5. Next, absolute values are calculated from the filter output A_F and then averaged in the moving average filter ($n = 32$). The time at which the A_{MAX} peak occurs in the A_{FMA} signal is assumed to be the moment of recording the centre of the waves packet sent by the transmitter, i.e. the TOF (after additional correction for the delay time introduced by the both digital filters).

The choice of a 40 kHz Butterworth filter was driven by its very high selectivity and a lack of ripples in the frequency characteristic, typical for other filters [28,31]. A second order filter with the general structure, shown in Fig. 6., was applied. Real numbers $a_1 \div a_6$ are filter parameters. The disadvantage of the selected filter is its relatively complex numerical implementation, requiring the use of multiplication and addition of floating point numbers. However, in the discussed solution, the filtering of the signal from the microphone took place after recording the samples in the memory buffer, so the efficiency of the numerical filter was a secondary issue.

Chosen signal waveforms are shown in Fig. 7. They were acquired for unfavourable position of the microphone relative to the transmitter (distance 9 m, the angle of incidence of the ultrasound waves to the axis of the microphone was about 60 degrees). The signal received by the microphone (results of sampling the voltage U_1 by ADC) is shown in upper chart. On the horizontal axis, consecutive samples were numbered (at rate 160 kSamples/s). Within the sample range of 130 to 230, the received packet of emitted waves by an ultrasonic transmitter is hardly visible. The middle chart shows the waveform of the A_F signal, i.e. the output from the pass band filter. The bottom graph shows the result of averaging the signal $|A_F|$ in the moving average filter – the A_{FMA} signal. The maximum value of A_{MAX} is around sample no. 210. Clearly, filters introduce a time delay, which should be included in calculating the final value of the arrival time of the ultrasonic packet from the transmitter to the robot's microphone.

DOI: 10.1051/itmconf/20171505008

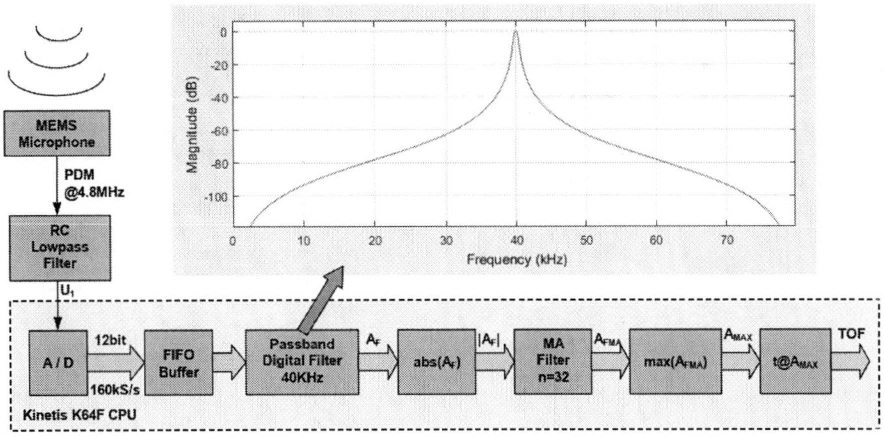

Fig. 5. Block diagram of digital processing of ultrasonic signal for distance measurement.

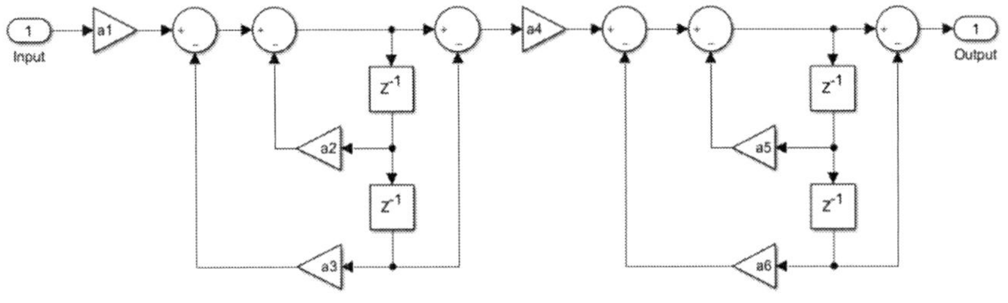

Fig. 6. Block diagram of 2nd order Butterworth pass band digital filter.

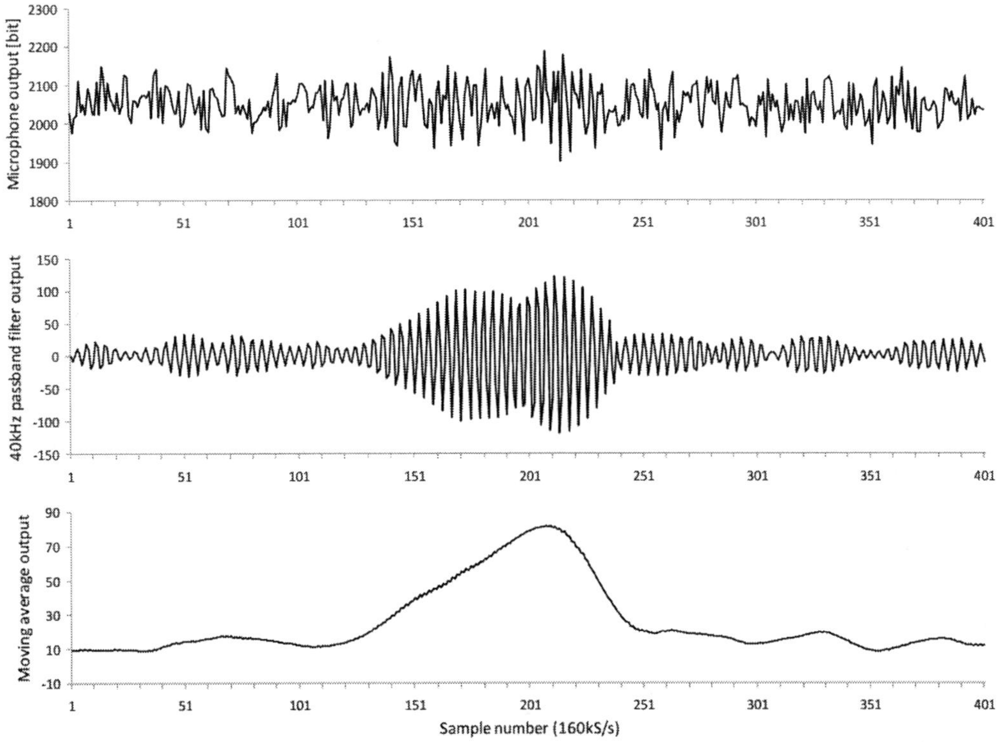

Fig. 7. Example of filtering of ultrasonic signal: microphone output (upper), 40 kHz pass band filter output A_F (middle), moving average filter output A_{FMA} (bottom); sampling frequency was 160 kHz.

DOI: 10.1051/itmconf/20171505008

In addition to the TOF value, the robot's microcontroller calculated the stipulated measure of "power" P of the received ultrasonic signal. This value was defined as the ratio of the value of the A_{MAX} to the average of all A_{FMA} values recorded throughout the whole acquisition window (corresponding to the time of flight for 13 m distance):

$$P = \frac{A_{MAX}}{A_{FMA}} \tag{1}$$

The factor P (1) contains information about the uncertainty of the result of TOF measurement (not in the strict statistical sense). In the conducted tests, the P value changed from nearly 1 (for pure noise from the robot environment) to about 30, for a microphone 3 m away from the ultrasonic transmitter (transmitter and microphone in one axis). Further tests have shown that the results of TOF measurements with a P-ratio of less than 3 are unreliable (there are severe distance measurement errors).

6 Calculating the coordinates of the robot

The coordinates of the robot on the plane were calculated based on four measured TOFs. The floor in the test room was flat and the robot's microphone was at a constant height (0.14 m). Ultrasound transmitters were stationary and known coordinates (Fig. 3.). The velocity of the wave propagation in the air for calculations was automatically corrected, according to the temperature of the robot's surroundings.

The TOF measurements, transmitted in one-second cycles to the desktop computer, were further filtered. Results with a P (1) factor less than 3.5 were ignored – as a result of high uncertainty. The remaining TOF results were processed by a median filter of 5 elements.

Calculation of robot coordinates on a plane, based on the distance from four points received by perpendicular projection of transmitter positions to the floor plane, is a non-complicated geometric task and does not require detailed discussion. For real distance measures (that include estimation error), up to 12 points of intersection of circles pairs with centres corresponding to the positions of four ultrasonic transmitters (Fig. 3.) are obtained. Some of the solutions are outside the floor area and should be ignored. Choosing the right solution – i.e. one of the two points of intersection of the circles around the opposing transmitters, requires an additional logical condition. The point is then chosen to be less distant from the solutions of the intersection of adjacent circles.

7 Testing the mobile robot localisation system

The built localisation system has been tested in the laboratory. The errors in estimating the position of the still robot at selected points of the room floor were calculated. The distance between the position of the robot's microphone, estimated by the system versus the actual position, is shown in Fig. 8. The positioning error does not exceed 0.15 m and its average value is 0.07 m. The error value increases slightly, as an object is moved away from the central point of the floor and near obstacles (up to 0.2 m next to one of the tables).

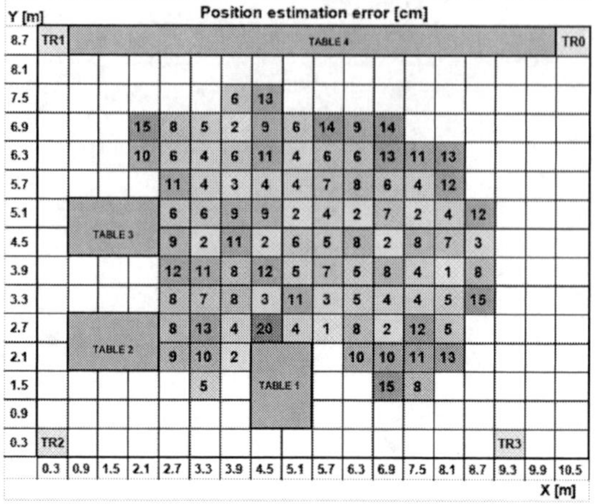

Fig. 8. A map of position estimation error on the room floor for experimental ultrasonic-based localisation system.

In the areas corresponding to the white fields in Fig. 8. at least three of the four measured distances from transmitters were unreliable (P<3.5). The reason for the low signal level received from distant transmitters is the large angle between the axis of the robot's microphone facing upwards and the direction at which the ultrasonic waves reach it (even up to 70 degrees). For this angle, value of microphone sensitivity is 10 times lower than the sensitivity along its axis. Likewise, the transmitter's sound pressure level decreases, as it moves away from its axis. Hence, complete amplification of the transmitter-microphone signal path can be several tens of times smaller at the extremities of the robot's working plane, than at its centre.

Improving the robot's localisation range can be achieved by increasing the voltage driving the ultrasound transmitters, but this will reduce the comfort of the occupants in the room (ultrasonic transmitters driven by rectangular voltage signal begin to be audible for humans). It may also affect the lifetime of piezoelectric transmitters.

8 Conclusions

An experimental system for locating a mobile robot on the plane, using ultrasonic waves, was developed and tested. The average value of the robot position estimation error was 0.07 m, while the maximum value did not exceed 0.2 m on a rectangular floor measuring 10 x 9 m.

On the periphery of the robot's work zone and near the major obstacles to ultrasounds, the localisation system was unable to determine the position of the robot.

However, thanks to the filtering algorithms used, the system ignores the uncertain results of the measurements; therefore, does not provide false information. In these areas, the robot control system should use other navigation mechanisms, such as odometry.

Further work is planned to increase the range of operation of the location system and to reduce the "dead" zones. The use of three microphones in a 120-degree horizontal spacing, and with an inclination of about 45 degrees from the vertical, should increase the sensitivity of the system. Additionally, it will be possible to determine the angular position of the robot axis relative to the reference coordinates system.

References

1. M. S. Kalaimagal and D. R. Sivaramakrishnan, in IJERT (2012)
2. Y. Zhang and D. P. Hong, Int. J. Precis. Eng. Manuf. **16**, 847 (2015)
3. R. Mautz, Geod. Ir Kartogr. **35**, 18 (2009)
4. B. Siciliano, O. Khatib, *Springer Handbook of Robotics* (2016)
5. C. W. Yuen, N. Balasubramaniam, and N. M. Din, in *2009 IEEE 9th Malays. Int. Conf. Commun. MICC*, 857–862 (2009)
6. A. M. Ladd, K. E. Bekris, A. P. Rudys, D. S. Wallach, L. E. Kavraki, IEEE Trans. Robot. Autom. **20**, 555 (2004)
7. P. Bahl and V. N. Padmanabhan, in *Proc. IEEE INFOCOM 2000 Conf. Comput. Commun. Ninet. Annu. Jt. Conf. IEEE Comput. Commun. Soc. Cat No00CH37064*. **2**, 775–784 (2000)
8. A. M. Ladd, K. E. Bekris, G. Marceau, A. Rudys, D. S. Wallach, L. E. Kavraki, in *IEEERSJ Int. Conf. Intell. Robots Syst.* **1**, 402–408 (2002)
9. J. Yu, P. Lou, X. Qian, and X. Wu, in *2008 ISECS Int. Colloq. Comput. Commun. Control Manag.*, 28–33 (2008)
10. L. Meier, P. Tanskanen, F. Fraundorfer, and M. Pollefeys, in *2011 IEEE Int. Conf. Robot. Autom.* 2992–2997 (2011)
11. M. Agrawal and K. Konolige, in *18th Int. Conf. Pattern Recognit. ICPR06,* 1063–1068 (2006)
12. R. Mautz and S. Tilch, in *2011 Int. Conf. Indoor Position. Indoor Navig.*, 1–7 (2011)
13. F. Dellaert, W. Burgard, D. Fox, and S. Thrun, in *Proc. 1999 IEEE Comput. Soc. Conf. Comput. Vis. Pattern Recognit. Cat No PR00149*, 594, **2** (1999)
14. J. M. Barrett, *Analyzing and Modeling Low-Cost MEMS IMUs for Use in an Inertial Navigation System*, (Worcester Polytechnic Institute 2014)
15. R. Cechowicz, Acta Mech. Autom. **11**, 104 (2017).
16. R. Zhang, A. Bannoura, F. Höflinger, L. M. Reindl, and C. Schindelhauer, in *2013 IEEE Sens. Appl. Symp. Proc.*, 38–42 (2013)
17. S.-Y. Lee and H.-W. Yang, Robot. Comput.-Integr. Manuf. **28**, 425 (2012)
18. F. Ijaz, H. K. Yang, A. W. Ahmad, and C. Lee, in *2013 15th Int. Conf. Adv. Commun. Technol. ICACT*, 1146–1150 (2013)
19. G. Kaniak and H. Schweinzer, in *2008 IEEE Sens.*, 1092–1095 (2008)
20. C. Walter and H. Schweinzer, in *2012 IEEE Int. Instrum. Meas. Technol. Conf. Proc.*, 1571–1576 (2012)
21. D. Bank, IEEE Sens. J. **2**, 597 (2002)
22. L. Kleeman, in *Proc. 1999 IEEERSJ Int. Conf. Intell. Robots Syst. Hum. Environ. Friendly Robots High Intell. Emot. Quotients Cat No99CH36289*, 1185–1190 vol.2. (1999)
23. H. Elmer and H. Schweinzer, J. Phys. Conf. Ser. **13**, 133 (2005)
24. J. D. Tardós, J. Neira, P. M. Newman, J. J. Leonard, Int. J. Robot. Res. **21**, 311 (2002)
25. S. Fazli, L. Kleeman, Robotica **24**, 433 (2006)
26. PUI Audio Inc., http://www.puiaudio.com/product-detail.aspx?categoryId=2&partnumber=UT-1640K-TT-2-R.
27. Knowles Corp., http://www.knowles.com/eng/Products/Microphones/SiSonic%E2%84%A2-surface-mount-MEMS.
28. A. V. Oppenheim, R. W. Schafer, J. R. Buck, *Discrete-Time Signal Processing* (1999)
29. I. W. Selesnick, C. S. Burrus, IEEE Trans. Signal Process. **46**, 1688 (1998)
30. S. N. Yao, T. Collins, P. Jančovič, Comput. Electr. Eng. **38**, 811 (2012)
31. S. J. Orfanidis, *Introduction to Signal Processing* (1996)
32. R. Sorrentino, G. Bianchi, *Microwave and RF Engineering* (2010)

ITM Web of Conferences 15, 05009 (2017) DOI: 10.1051/itmconf/20171505009
CMES'17

Implementation of PID autotuning procedure in PLC controller

Marcin Daniun[1], *Michał* Awtoniuk[1,*], and *Robert* Sałat[1]

[1]Warsaw University of Life Sciences, Faculty of Production Engineering, Nowoursynowska 164, 02-787 Warsaw, Poland

Abstract. In this paper, we present the automatic PID tuning procedure based on the Method of Moments and AMIGO tuning rules. The advantage of the Method of Moments is that the time constant and transport delay are estimated at the areas rather than on the individual points. This results in high resistance to the measurement noises. The sensitivity to measurement noises is a serious problem in other autotuning methods. The second advantage of this method is that it approximates plant during identification process to first order model with time delay. We combined the Method of Moments with the AMIGO tuning rules and implemented this combination as a stand-alone autotuning procedure in Siemens S7-1200 PLC controller. Next, we compared this method with two built-in PID autotuning procedures which were available in Siemens S7-1200 PLC controller. The procedure was tested for three types of plant models: with lag-dominated, balanced, and delay-dominated dynamics. We simulated the plants on a PC in Matlab R2013a. The connection between the PC and PLC was maintained through a National Instruments data acquisition board, NI PCI-6229. We conducted tests for step change in the set point, trajectory tracking, and load disturbances. To assess control quality, we used IAE index. We limited our research to PI algorithm. The results prove that proposed method was better than two built-in tuning methods provided by Siemens, oscillating between a few and even a dozen percent in most cases. The proposed method is universal and can be implemented in any PLC controller.

1 Introduction

A proportional-integral-derivative (PID) controller is still the most widely used algorithm in industrial applications [1]. Individual terms can be activated or deactivated to create P, PI, PD, and PID controller type but the most common amongst them is PI. The biggest problem with commissioning PID algorithm is a controller tuning process. Manual tuning can be time consuming. It also requires specialised knowledge and experience in the field of control engineering. The research conducted in the 1990's found that 80% of PID applications were badly tuned [2]. For this reason, more and more PID controllers are equipped with autotuning function. Many controller manufacturers as well as scientists are trying to develop new, more accurate and efficient autotuning procedures.

In this paper, we present the procedure for automatic tuning of the PID controller based on the Method of Moments, as a method of system identification, and the AMIGO tuning rules. The advantage of the Method of Moments is that the parameters of model are estimated at the areas rather than on the individual points. This results in high resistance to the measurement noises. We implemented Method of Moments + AMIGO as a stand-alone autotuning procedure in Siemens S7-1200 PLC controller. Next, we compared the method with two built-in PID autotuning procedures which were available in Siemens S7-1200 PLC controller.

2 PID autotuning

The tuning process of the PID controller is based on determining the value of the parameters of the controller. Selection of controller parameters is a difficult task; therefore, any manufacturers decide to implement special procedures to make tuning process automatically. This is called autotuning.

There are many different autotuning procedures but generally autotuning has two major phases: control object identification and controller parameters calculation. The most common approach in the first phase is to build a model of control object. Model should be as simple as it is possible, e.g. first order model with time delay (FOTD). Equations used to calculate the controller parameters, based on model parameters, are called tuning rules. The literature is abundant in different tuning rules, e.g. O'Dwyer in [3] shows 649 controller tuning rules dedicated for FOTD model only.

Autotuning methods can be divided into two groups. The first group uses step response, and the second is based on relay method. In the PLCs both methods are available. Siemens S7-1200 PLC controller can be an example with two autotuning procedures: Pretuning and Fine Tuning implemented in it.

The Pretuning procedure is based on the step response. Firstly, the controller is turned into manual mode and controller output u is forced to a value set by

* Corresponding author: michal_awtoniuk@sggw.pl

DOI: 10.1051/itmconf/20171505009

the user. Secondly, value of plant output y is observed until it reaches the point of inflection. Finally, the Chien-Hrones-Reswick tuning rules are used to calculate the parameters of the controller [4], [5].

The Fine Tuning procedure uses the relay method developed by Åström and Hägglund, which is based on the Ziegler-Nichols Ultimate Cycling method [6], [7]. The method involves a series of oscillations of the plant output y. Oscillations are forced by switching the controller output u between two values. The controller parameters are calculated with Ziegler-Nichols tuning rules. The relay method is still being developed. Currently, researchers are focused on asymmetric relay autotuning and adaptation the method to plants with more input and output signals such as TITO [8].

In this paper, we compare two build-in autotuning procedures with combination of Method of Moments and AMIGO tuning rules. The Method of Moments is a system identification technic based on the step response characteristic. In this method, the plant is modelled by FOTD model (1).

$$G(s)=\frac{K}{Ts+1}e^{-Ls} \tag{1}$$

where: K -- static gain, T -- time constant, L -- time delay.

FOTD model has tree parameters: static gain K, time constant T, and time delay L. In the Method of Moments parameters of model are estimated at the areas rather than on the individual points. In the fig. 1 graphical interpretation of the method are presented.

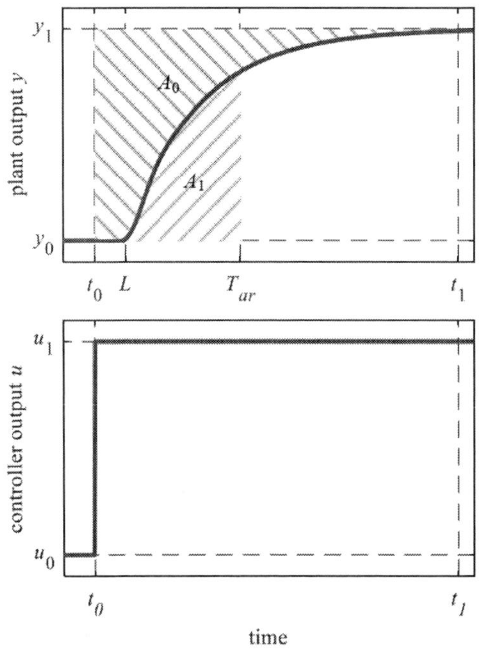

Fig. 1. Graphical interpretation of parameters calculated in the Method of Moments.

The static gain is calculated by equation

$$K=\frac{\Delta y}{\Delta u}=\frac{y_1-y_0}{u_1-u_0} \tag{2}$$

where: Δy -- increase of plant output, Δu -- increase of controller output.

Calculation of time constant T and time delay L is not as simple as static gain K and requires estimation of additional variables: area A_0 (integrated distance between steady-state of step response y_1 and plant output y), time T_{ar} (average residence time), and area A_1 (area under the step response up to time T_{ar}) according to equations (3-7)

$$A_0=\int_{t_0}^{t_1}\left[y_1(t)-y(t)\right]dt \tag{3}$$

where: y_1 -- steady-state of plant step response, y -- plant output.

$$T_{ar}=\frac{A_0}{K} \tag{4}$$

$$A_1=\int_{t_0}^{T_{ar}} y(t)dt \tag{5}$$

$$T=\frac{eA_1}{K} \tag{6}$$

where: e – Euler's number.

$$L=T_{ar}-T \tag{7}$$

After FODT model parameters estimation, calculation of controller parameters is performed with AMIGO tuning rules. These rules are well described in [7]. We used PI algorithm thus our controller had two parameters: proportional gain K_p and reset time T_i.

$$K_p=\frac{0.15}{K}+\left[0.35-\frac{LT}{(L+T)^2}\right]\frac{T}{KL} \tag{8}$$

where: K_p – proportional gain.

$$T_i=0.35L+\frac{13LT^2}{T^2+12LT+7L^2} \tag{9}$$

where: T_i – reset time.

3 Materials and methods

Fig. 2 presents scheme of data collection. Key elements used in our research were:

- PLC controller Siemens SIMATIC S7-1200,
- Data acquisition board National Instruments NI-6226,
- Matlab&Simulink software.

All autotuning procedures, i.e. Pretuning, Fine Tuning, and Method of Moments + AMIGO, were implemented in PLC controller. The first and the second procedure are available in S7-1200 PLC controller as a build-in functions. We implemented the third procedure by ourselves in the form of function block. We have programmed the PLC controller in ladder diagram language in TIA Portal V13 software. To calculate the integrals from equations (3) and (5) we used rectangle rule. The sampling period was 20 ms.

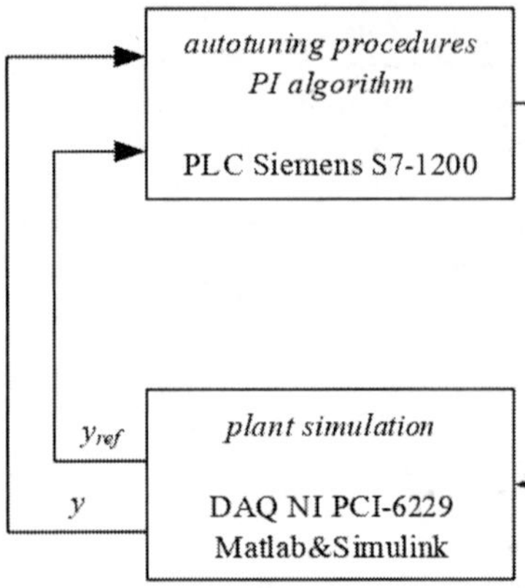

Fig. 2. Block diagram of data collection.

PID algorithm, a build-in function, is implemented in S7-1200 PLC controller as a technological function available in a standard library. The function block name is PID_Compact. According to manual PID algorithm is described as [4]:

$$u(s) = K_p\left[\left(by_0 - y\right) + \frac{1}{T_i s}\left(y_0 - y\right) + \frac{T_d s}{aT_d s + 1}\left(cy_0 - y\right)\right] \quad (10)$$

where: b -- proportional term weighting, y_{ref} -- set point, a -- derivative delay coefficient, T_d -- derivative time, c -- derivative term weighting.

The values of coefficients a, b, and c were estimated automatically during Pretuning and Fine Tuning process. In Method of Moments + AMIGO autotuning procedure we set them to a constant value equal 1. We limited our research to PI algorithm so T_d=0.

We simulated the plants in Matlab&Simulink. Communication between PLC and Matlab&Simulink was maintained throughout data acquisition board in the way depicted in the fig. 2. All signals, i.e. set point y_{ref}, plant output y, and controller output u, were voltage signals from the range 0-10V.

Autotuning procedures were tested with tree plants described by transfer functions (11-13). P_1 is a lag-dominated plant, P_2 is balanced, while P_3 is delay-dominated plant [9].

$$P_1(s) = \frac{1}{(s+1)(0.1s+1)(0.01s+1)(0.001s+1)} \quad (11)$$

$$P_2(s) = \frac{1}{(s+1)^4} \quad (12)$$

$$P_3(s) = \frac{e^{-s}}{(0.05s+1)^2} \quad (13)$$

To compare results of autotuning procedures three scenarios were prepared: step change in the set point, trajectory tracking, and load disturbances. Fig. 3 shows changes of set point signal in all scenarios. In the first case, step change of set point was applied in the 10th second to a value of 7V and the next step was applied in the 110th second to a value of 3V. Test lasted 210 s. In the second scenario, the set point was changed according to trajectory of response of the first order system $(5s+1)^{-1}$ to step change to value of 7V. In the 100th second another trajectory was applied. It was response of the first order system $(25s+1)^{-1}$ to step change to value of 3V. Test lasted 250 s. In the last scenario step change of set point was applied in the 5th second to a value of 5V, remaining constant to the end of experiment. Next, in the 155th second the load disturbance of value of 1V was applied to a plant input, i.e. controller output signal u. Test lasted 305 s.

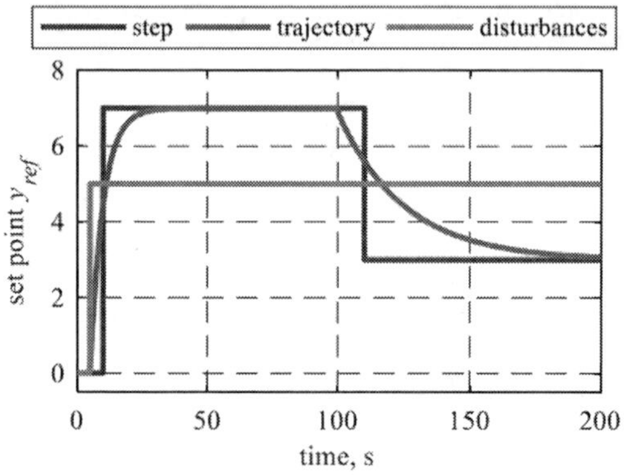

Fig. 3. Changes of set point signal in tree scenarios.

We compared all autotuning procedures from control performance point of view. To assess control quality, we used Integral Absolute Error (IAE) index:

$$IAE = \int \left| y_{ref}(t) - y(t) \right| dt \quad (14)$$

4 Results

Tables 1, 2, and 3 show results for lag-dominated plant P_1, balanced plant P_2, and delay-dominated plant P_3, respectively. The tables include controller parameters calculated by each autotuning procedures and values of control quality index IAE for step change, trajectory tracking, and load disturbances scenario.

After system identification process, by means of Method of Moments, plants P_1, P_2, and P_3 were modelled by M_1, M_2, and M_3 models described by following transfer functions

$$M_1(s) = \frac{e^{-0.12s}}{1.119s+1} \quad (15)$$

$$M_2(s) = \frac{e^{-1.962s}}{2.168s+1} \quad (16)$$

DOI: 10.1051/itmconf/20171505009

$$M_3(s) = \frac{e^{-1.052s}}{0.117s+1} \quad (17)$$

Fig. 4-6 show plant output and controller output signals during experiments in all three scenarios.

Table 1. Controller parameters and control performance assessed for plant P_1.

	Pretuning	Fine Tuning	Method of Moments + AMIGO
controller parameters			
K_p	6.31	11.1	4.4
T_i [s]	0.49	0.37	0.49
control performance: IAE index			
step	7.33	7.44	7.09
trajectory	2.22	1.84	1.54
disturbances	3.38	3.82	3.14

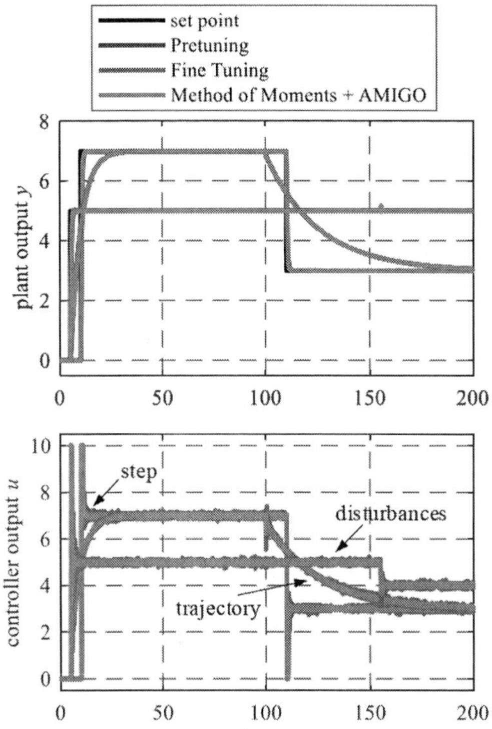

Fig. 4. Plant output and controller output for plant P_1 (curves are very close to each other).

Table 2. Controller parameters and control performance assessed for plant P_2.

	Pretuning	Fine Tuning	Method of Moments + AMIGO
controller parameters			
K_p	1.92	1.6	0.7
T_i [s]	5.72	5.48	2.27
control performance: IAE index			
step	60.3	54.5	49.5
trajectory	44.8	49.4	35.5
disturbances	30.2	27.4	26.5

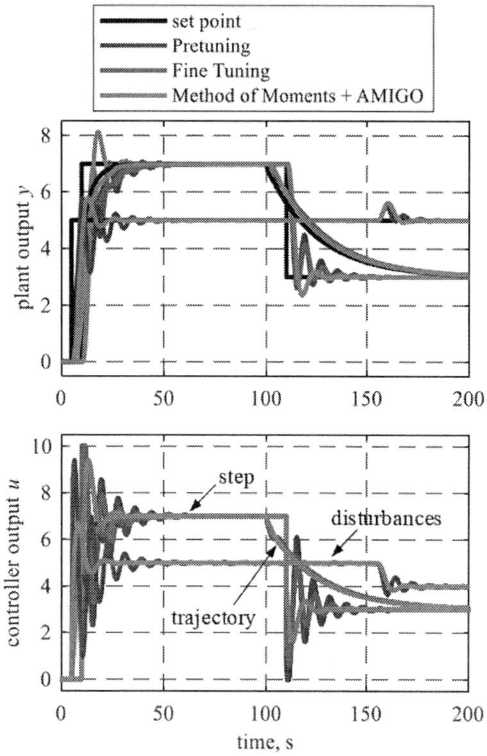

Fig. 5. Plant output and controller output for plant P_2.

DOI: 10.1051/itmconf/20171505009

Table 3. Controller parameters and control performance assessed for plant P_3.

	Pretuning	Fine Tuning	Method of Moments + AMIGO
controller parameters			
K_p T_i [s]	0.26 4.09	0.57 1.89	0.25 0.51
control performance: IAE index			
step trajectory disturbances	181 179 98.7	41.2 40.8 22.6	22.9 22.5 13

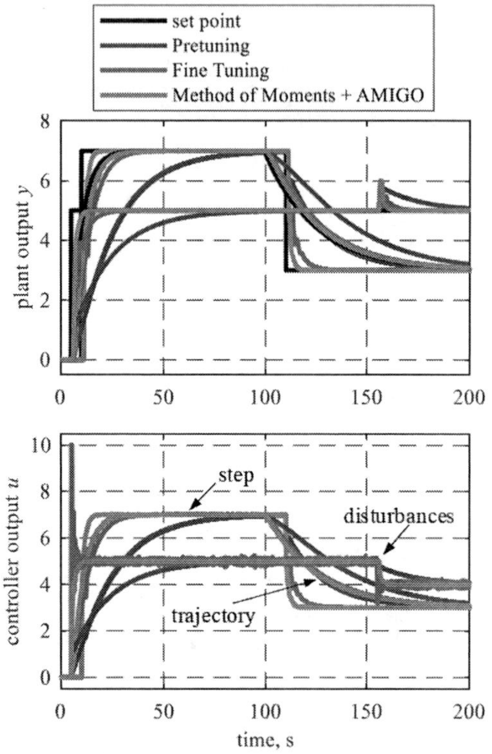

Fig. 6. Plant output and controller output for plant P_3.

All tested methods allow to tune the PID controller so that in each testing scenario tuned systems reached the steady state after set point change or load disturbance. There was no case of unstable system and the overshoots were small.

Results for lag-dominated plant (P_1) show the value of IAE index for Method of Moments + AMIGO was smaller by 3.3%, 30.6%, and 7.1% for step change in the set point, trajectory tracking, and load disturbances, respectively compare to Pretuning procedure. In the case of balanced plant (P_2) proposed procedure allows to achieve control performance index smaller by 9.2%, 28.2%, and 3.25% for step change in the set point, trajectory tracking, and load disturbances respectively compare to Fine Tuning procedure. However, the greatest advantage of described procedure can be seen for delay-dominated plant (P_3). The value of IAE index was smaller by 44.4% (step change in the set point), 44.7% (trajectory tracking), and 42.7% (load disturbances) in comparison to Fine Tuning procedure.

5 Conclusions

Our goal was to implement combination of Method of Moments and AMIGO as a new autotuning procedure to PLC controller. We also compared this method with two built-in PID autotuning procedures which were available in Siemens S7-1200 PLC controller. The results show that control loops tuned by proposed method has lower values of control quality index.

In future works we will focus on minimising the errors of integrals calculation caused by the rectangle rule. The trapezoidal rule should give better approximation. The source code in ladder diagram language should be optimised to reduce the controller memory occupation.

References

1. L. Desborough, R. Miller, AIChE Symp. Ser., **98**, 326, 169–189 (2002)
2. P. Van Overschee, B. De Moor, IFAC Proc. Vol., **33**, 4, 595–600 (2000)
3. A. O'Dwyer, *Handbook of PI and PID controller tuning rules*, London: Imperial College Press (2009)
4. Siemens, *SIMATIC S7-1200, S7-1500 PID control function manual* (2017)
5. J. Müller, *Controlling with SIMATIC: practice book for SIMATIC S7 and SIMATIC PCS7 control systems*, Berlin: Publicis Corporate Publishing (2005)
6. K. J. Åström, T. Hägglund, *PID Controllers: Theory, Design, and Tuning*, ISA (1995)
7. K. J. Åström, T. Hägglund, J. Process Control, **14**, 6, 635–650 (2004)
8. J. Berner, K. Soltesz, T. Hägglund, K. J. Åström, IFAC-PapersOnLine, **50**, 1, 5332–5337 (2017)
9. J. Berner, K. J. Åström, T. Hägglund, IFAC Proc. Vol., **47**, 3, 11288–11293 (2014)

ITM Web of Conferences 15, 05010 (2017) DOI: 10.1051/itmconf/20171505010
CMES'17

Indoor mobile robot attitude estimation with MEMS gyroscope

Radosław Cechowicz[1,*]

[1]Lublin University of Technology, Mechanical Engineering Faculty, Department of Automation, Nadbystrzycka 36, 20-618 Lublin, Poland

Abstract. A method of attitude estimation with a low-cost, strap-on MEMS sensor was proposed in the article. The method relies on dynamic gyroscope bias change estimation and updates during detected stops of the robot. The algorithm has been tested in laboratory with iRobot Roomba robot and should be improved to be useful in an industrial environment. The measurement of attitude of a mobile platform is necessary for correct navigation, especially in autonomous vehicles, which use dead reckoning between position updates from external sources. Since the system is intended to operate indoors, in industrial halls and shops, it cannot avail of GPS and Earth magnetic field sensors because of anomalies, which are common inside the steel constructions. Therefore, the accuracy of the gyroscope-based attitude estimation is significant. The proposed method aims to address the common problem of gyroscope bias drift, by dynamic update of sensor bias and simultaneous use of all gyroscope axes, to improve the quality of the measurements. A popular 3-axial gyroscope and 3-axial accelerometer sensors were used during the test runs. Obtained results suggest that it is possible to improve short-term accuracy of inertial dead reckoning, to get a system that could be of practical use in industrial AGV systems or intelligent vehicles.

1 Introduction

One of the impedimenta to the development of mobile platforms, designed to work in industrial halls and public buildings, are the difficulties in identifying the position of the platform (in relation to other elements of the environment/building). In industrial AGV systems, as well as those operating in warehouses, this problem is usually solved by installing fixed route markings or way-points. This can be done by using contrasting paint, magnetic tape, inductive circuits, or surface modifications (guiding canals, rails, kerbs, etc.). Precise position of the mobile unit can also be determined by reading or counting route markings — for example bar codes, QR codes, RFID tokens, etc. In some systems, the position of the moving object, in relation to fixed beacons, can be established by triangulation [1,2].

According to [3], industrial navigation systems share a common problem of limited flexibility. The mobile units in such systems are forced to follow fixed routes (or go over fixed way-points). In consequence, the mobile units can only follow a limited number of routes. Additionally, in some solutions, the space used by the mobile units (communication paths) cannot be shared or allocated to other tasks (such as temporary storage). Transport routes can only be changed by moving or installing new infrastructure (route markings) and reserving new space for the AGV system [4].

If the autonomy of mobile platforms was increased by detaching them (partially or completely) from permanently marked routes, it would allow [3]:

- to increase the flexibility of the system, allowing an adjustment of transport routes to the dynamic change of the environment (e.g. the emergence of obstacles, temporary route closure, etc.).
- to reduce the number of fixed way-points, decreasing the cost of the infrastructure.

Detaching the mobile units from fixed paths would be possible, if their local control systems are able to navigate autonomously over a (limited) time. In the areas where fixed markings are not available, the system could determine its position from dead reckoning (reading onboard sensors), and react to events nearby.

Orientation of mobile platform can be established from odometry, inertial sensors or a combination of these two. The possible solutions have been documented, for instance in [5–7].

2 The experiment setup

The purpose of the experiment was to determine the accuracy of the relative orientation angle measurement of a mobile platform with a low-cost, strap-on inertial gyroscope sensor.

The measurements were taken in the conditions similar to those of industrial warehouse mobile platforms – the sensor was mounted onto the robot (iRobot Roomba was used for this purpose), which was moving on an even surface, similar to those in clean warehouses (PVC cover).

* Corresponding author: r.cechowicz@pollub.pl

 DOI: 10.1051/itmconf/20171505010

The 6-axis IMU unit consisting of MEMS gyroscope (L3GD20H) and accelerometer (LSM303D) was used during the experiment. The sensor was mounted onto the modified iRobot Roomba robot (fig. 1). The sensor was attached to the Odroid C2 single board computer (control unit) via the I2C interface. The computer was also used to control the robot and to read sensor data via the Roomba serial interface. All data collected during the experiment was stored on-board and then sent to central system for processing.

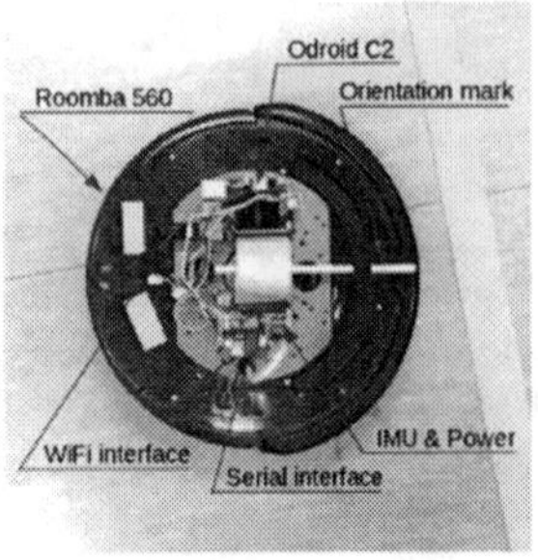

Fig. 1. Modified iRobot Roomba used in the experiment.

During the test runs, the robot was controlled by the single-board computer in the open-loop mode (without position feedback). The pre-programmed and timed sequence of movements was being sent to the Roomba, and the inertial sensor and odometry data were read.

Since the main purpose of the experiment was to test the gyroscope, a custom sequence of movements (fig. 2) was chosen, instead of standard square path (proposed by [8]), to simulate various working conditions of the warehouse platform, such as various turn radii and forward and backward movements. The real orientation angle of the robot was measured optically, in defined stop points along the path, from the picture taken by a camera above the robot. The optical measurements were taken by relating the position of the robot to the markings on the floor (vertical and parallel lines, markers). The accuracy of optical measurement was about ±1 deg. The area covered by the robot path fitted within the rectangle of 3 by 4 m. Time of a single round (including control stops) was 250 s.

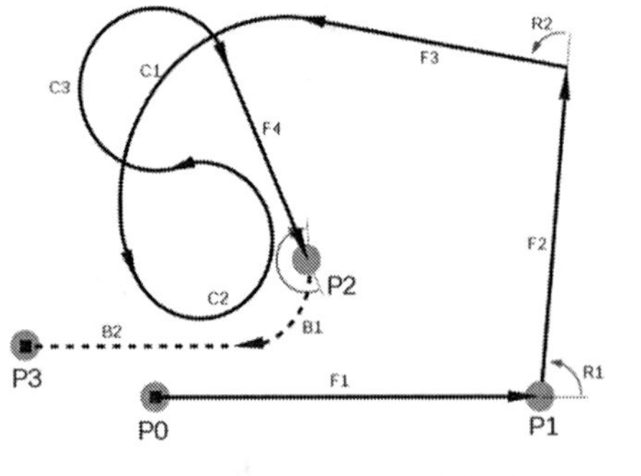

Fig. 2. Robot path during the tests (approximate): P0 - starting point, P1 - first control point, P2 - second control point, P3 - end point, Rx - in-place rotation, Fx, Cx - straight and curved path forward movement (continuous line), Bx - backward movement (dashed line).

3 Measurements

The IMU sensor sampling rate was 100 Hz. During the test runs, the IMU was fixed to the Roomba in a position where none of the axes were perpendicular to the floor. The final rotation in the floor plane was then calculated from vector (quaternion) transformation. A sample of recorded gyroscope signal is shown in fig. 3. In the figure, there are filtered (lowpass Butterworth $f_{cut}=0.05f_{sample}$) signals shown for each axis, and a raw signal from axis Z to illustrate the gyro noise. The noise level in the remaining two axes was similar. All gyroscopes were calibrated during the first 10s of the test run.

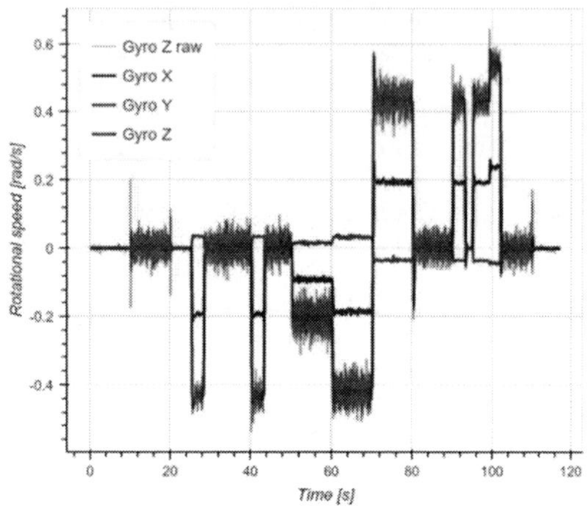

Fig. 3. Gyro signal recorded during test run.

Normalised vector A_v perpendicular to the floor was derived from accelerometer:

$$A_v = [A_x, A_y, A_z]/||[A_x, A_y, A_z]|| \tag{1}$$

where: A_x, A_y, A_z – stand for values of accelerometer readings while the robot was idle (not moving).

Rotation vector at the moment t can be obtained from gyroscope readings (scaled in rad/s), using quaternion mathematics, commonly applied to rotation, orientation and tracking problems in navigational systems [9–12]. For any sufficiently small dt, a rotation vector can be derived directly from 3-axis gyroscope reading:

$$r_v = [g_x, g_y, g_z] \tag{2}$$

where: r_v – rotation vector, g_x, g_y, g_z – gyroscope readings (axes X, Y and Z) scaled to rad/s. Rotation vector can be converted to quaternion Q (or angle-vector representation) with the following:

$$Q = q(||r_v||dt, r_v/||r_v||) \tag{3}$$

The (normalised) unit quaternions can be used for tracking the orientation of an object in space, especially, when the rotation axis changes frequently, such as in the case of wearable sensors [11]. For the purpose of the experiment, an assumption was made that the floor plane is horizontal (perpendicular to the Earth gravity vector) and does not change during the run time. The vertical

DOI: 10.1051/itmconf/20171505010

direction, obtained from (1) was assumed to be constant. The rotation about the axis A_v is then a dot product of:

$$R_v = dot(r_v, A_v) \tag{4}$$

The relative orientation angle is then a sum of R_v over time:

$$O = \int_0^t R_v dt \tag{5}$$

For a short time periods, orientation angle calculated with (5) gives good estimation of robot attitude. The obtained results (fig. 4) are similar to those documented in [5,6]. The attitude error from gyroscope signal is smaller than the odometry error.

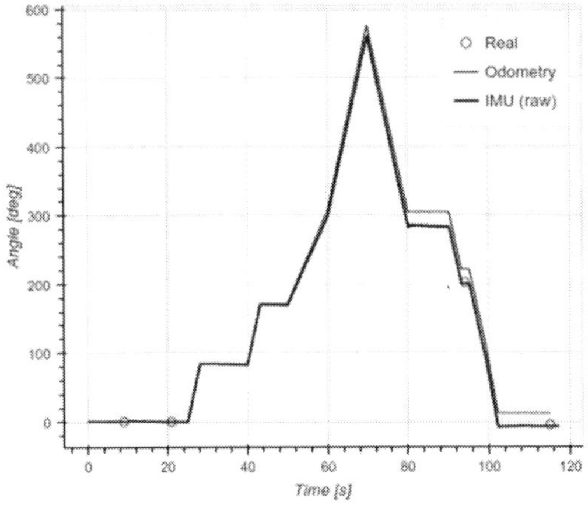

Fig. 4. Relative orientation angle (IMU) compared to orientation from Odometry and optical measurements in control points (Real)

The result presented in fig. 4 is valid only for short periods of time, for measurements taken immediately after gyroscope calibration, and for the sensor working in stable thermal conditions. For any longer integration periods, the attitude error would grow because of gyro bias fluctuations.

4 Sensor movement detection

It was shown in [3] that even in a static MEMS gyroscope, the bias changes (drifts) constantly over time. It is possible to compensate the integration error from bias drift using a filtering [13] or modelling the process [14] but this could be difficult to perform in an industrial system, consisting of a number of gyros or for a low-cost or mass produced system.

Periodic, runtime calibration could be used to follow the drift of the gyroscopic sensors and to keep the attitude error low. This can be done on a hardware level, as proposed by [15], or by updating sensor bias (as in [3,16]). Bias value can be updated while the sensor remains stationary. Therefore, in such system a movement detection procedure is required. In [3], a statistical filter was proposed for this purpose. For a mobile robot, a simple classification algorithm could be used, since the state of moving can be determined by measuring vibration noise.

In fig. 4, movement can be easily noticed by the increased noise in the sensor output. Following this observation, a simple measure of sample range could be probably used for detection. However, it provided only rough estimation of movement. It turned out that better and more exact rotational movement classification can be done with measure proportional to signal energy. In fig. 5, there are two histograms shown: one for the still robot (Still), the other for the robot moving forward (FWD, only leftmost part of the histogram shown) and their cumulative distribution functions calculated from data. It can be noted that slightly less than 20% of movement samples are in the overlap with still area. It is assumed that robot is not turning at these points. The remaining samples are classified as rotating and included into attitude estimation.

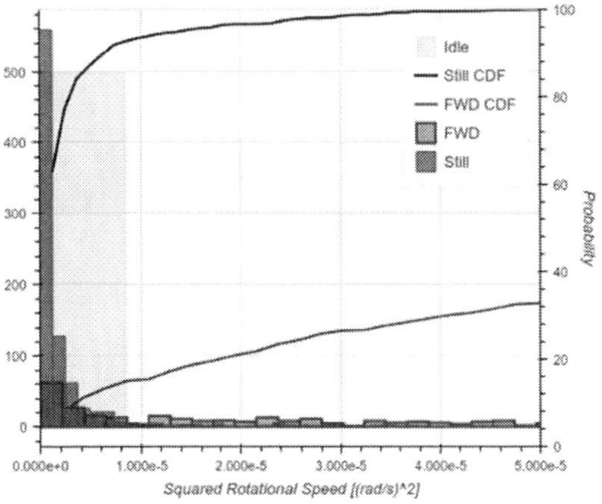

Fig. 5. Rotational speed histograms of the idle and moving robot (section F1) and cumulative distribution functions (CDF) from data. Idle area marked in light blue.

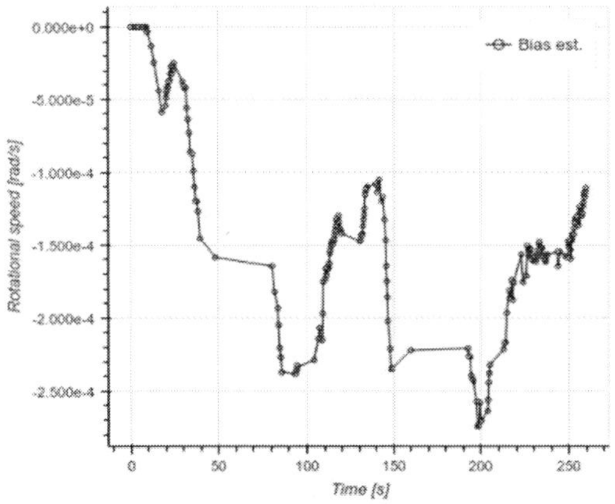

Fig. 6. Estimated sensor bias change during test run.

The robot was slightly turning during all the straight-line segments, due to unequal effective diameter of the wheels. The average rotational speed of the robot during the straight-line segments, was about 0.0005 rad/s.
The classification method was then used to update the sensor bias value during the test runs. In the points classified as still (no rotational movement) that bias of

the sensor was updated, using a complementary filter with parameters determined from experiment. The resulting bias change after first calibration is shown in fig. 6. The plot was made from a test run composed of two rounds along defined path (fig. 2), one immediately after the other. It can be noted that the general shape of bias changes is similar for both rounds, which can suggest a correlation of bias and movement history.

The bias drift estimation allowed to significantly decrease the attitude error. The comparison of the obtained results is shown in fig. 7.

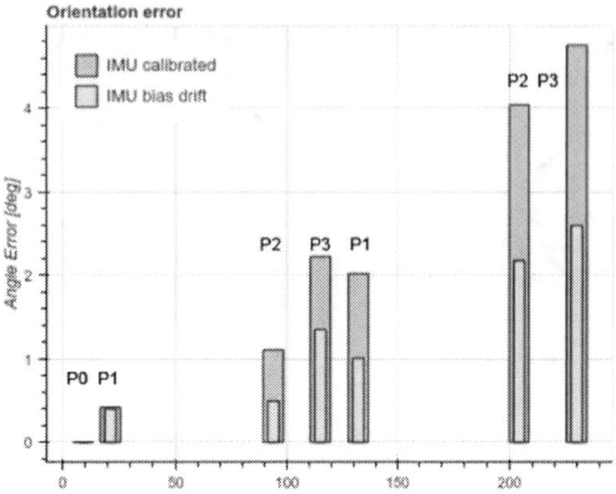

Fig. 7. Comparison of attitude estimation results for calibrated IMU and IMU with bias drift estimation, against real orientation of the robot in defined control points.

5 Conclusion

The proposed method of sensor bias change estimation allows to decrease the attitude error by about 40% in comparison to calibrated-only approach. Since the gyroscopic signal must be integrated to obtain orientation angle, bias change brings major contribution to attitude error. Bias changes registered during the 12 hours-long still test with the sensor used in the experiment are shown in fig. 8.

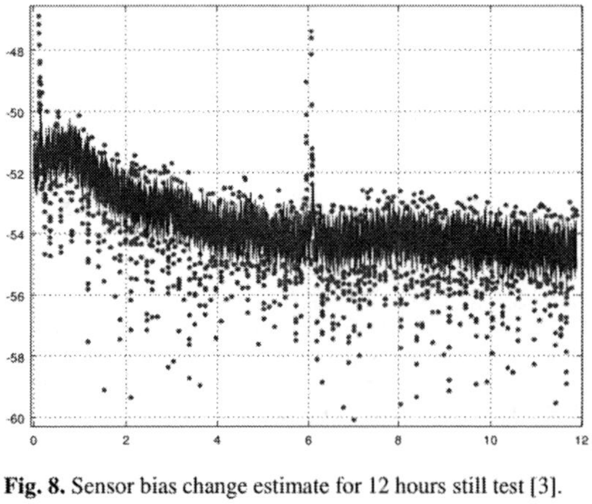

Fig. 8. Sensor bias change estimate for 12 hours still test [3].

The bias does not change linearly – it is influenced by the temperature (of the sensor itself and of the environment), fluctuations in power supply, reading procedure and other, less known factors.

The use of quaternions or rotation vectors (in simplified case) simplifies the calculations and makes possible to include measurements from more than one axis, which results in better statistical properties of the signal, or allows to easily incorporate inertial measurement in tracking applications like robot arm steering [17].

The research will be continued, the movement detection algorithm should be improved to increase its selectivity. The error distribution in the control points suggest that the slight drifting of the robot during the forward movement is not fully measured.

References

1. F. Ijaz, H. K. Yang, A. W. Ahmad, C. Lee, 15th International Conference on Advanced Communications Technology (ICACT), 1146–1150 (IEEE, 2013)
2. R. Mautz, Geod. Kartogr. **35**, 18 (2009)
3. R. Cechowicz, Acta Mechanica et Automatica. **11**, 2, 104–110 (2017)
4. D. Herrero-Pérez, J. J. Alcaraz-Jiménez, H. Martínez-Barberá, Int. J. Adv. Rob. Syst. **10**, 292 (2013)
5. M. Guráň, T. Fico, A. Chovancová, F. Duchoň, P. Hubinský, D. Dúbravský, 23rd International Conference on Robotics in Alpe-Adria-Danube Region (RAAD), 1–7 (2014)
6. N. Houshangi, F. Azizi, World Automation Congress, 1–8 (2006)
7. B. Gersdorf, U. Freese, ICINCO, **1**, 388–395 (2013)
8. J. Borenstein, L. Feng, IEEE Trans. Rob. Autom. **12**, 869 (1996)
9. A. Janota, V. Šimák, D. Nemec, J. Hrbček, Sensors. **15**, 7016–7039 (2015)
10. R. G. Valenti, I. Dryanovski, J. Xiao, Sensors. **15**, 19302–19330 (2015)
11. V. Renaudin, C. Combettes, and F. Peyret, IEEE/ION Position, Location and Navigation Symposium, 645–656 (PLANS 2014)
12. A. Kim and M. F. Golnaraghi, in Position Location and Navigation Symposium, 268–272 (IEEE, 2004)
13. X. Yuan, S. Yu, S. Zhang, G. Wang, S. Liu, Sensors. **15**, 10872–10890 (2015)
14. L. Xue, C. Jiang, L. Wang, J. Liu, W. Yuan, Micromachines. **6**, 266–280 (2015)
15. I. P. Prikhodko, C. Merritt, J. A. Gregory, J. A. Geen, J. Chang, J. Bergeron, W. Clark, M. W. Judy, Transducers – 2015, 18th International Conference on Solid-State Sensors, Actuators and Microsystems (TRANSDUCERS), 35–38 (IEEE, 2015)
16. S. K. Hong, S. Park, Sensors. **8**, 7287–7299 (2008)
17. M. M. Plechawska-Wójcik, P. K. Wolszczak, R. J. Cechowicz, K. Łygas, 9th International Conference on Human System Interactions (HSI), 348–354 (IEEE, New York, 2016)

DOI: 10.1051/itmconf/20171505011

Measurement system analysis for one-sided tolerance

Szemik Kamil[1,*], *Bogacz* Pawel[1]

[1]AGH University of Science and Technology, Faculty of Mining and Geoengineering, Department of Economics and Management in Industry, Al. Mickiewicza 30, 30-059 Cracow, Poland

Abstract. Measurement system analysis is carried out in order to determine if a capability to perform measurements in terms of product and process control is sufficient, indicating that the type I and the type II appraisal errors probability are acceptable. Statistical analyses for measurement system evaluation presented in the literature and the industrial manuals are not applicable for all complex and unusual applications. Therefore, the purpose of this study was to develop a robust statistical analysis method for measurement system variability analysis, in terms of product control scenario applied to one-sided tolerance. In the hereby presented study, the authors presented the theoretical principles of statistical techniques for measurement variations evaluation. Subsequently, the formula of gauge repeatability and reproducibility in terms of lower specification limit was proposed. The research hypothesis was tested using the statistical analysis.

1 Significance and determinants of measurement systems quality

The quality of product and process has become one of the most significant elements of the successful management process in high volume manufacturing facilities. Therefore, measurement systems are used for control of manufacturing process variability and location, as well as a criterion for product acceptance.

Due to the measurements variability, the results, which are close to the specification limit, do not allow for a proper appraisal of product conformance in relation to the design tolerance limits. Consequently, in terms of statistical process control (SPC), the measurements variation has a negative impact on the evaluation of the process performance and stability; thus, the measurement system shall allow for a reliable evaluation of the manufacturing process statistical properties.

If the performance of measurement process is unknown, then a manufacturer may be exposed to bear significant costs due to the customer complaints, warranty returns, scrap increase and special causes in ongoing process monitoring applied throughout the SPC.

First of all, capability indexes are one of the main statistical techniques applied in SPC methodology. Next, as a product is defined by specification limits upon which allowable deviation is limited through tolerances [1], thus, the variability of the characteristic with respect to the specification limits is evaluated throughout Cp and Cpk indexes. The Cp index compares the product tolerance zone to the 6 standard deviations of the manufactured products distribution. The Cpk index takes also distribution location into account [2]. Finally, the effect of measurement system variability on the process capability index [3] is demonstrated in figure 1.

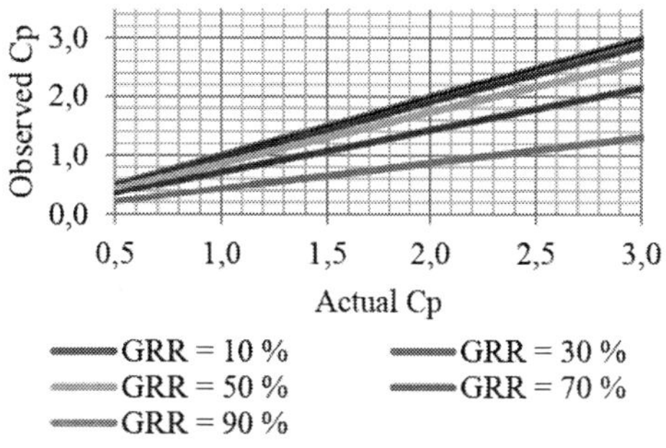

Fig. 1. Observed vs actual Cp value.

The importance of qualified measurement systems can be seen in figure 1. The repeatability and reproducibility of the measurement system has significant impact on the estimation of capability indices, thus, the actual capability index (Actual Cp) was introduced [3].

Respectively, based on collected measurements, the following crucial decisions are taken:

- To accept or reject (scrap or rework) manufactured product;
- To issue corrective action in a manufacturing system;
- To issue 100% control in running production;
- To qualify machine for production use.

The effect of excessive measurement system variability, in reference to the specification limit, is that the type I and type II classification errors are probable to occur. The probability of classification conforming part as non-conforming part is called type I error, producer's risk or false alarm.

The probability of type I error is shown in figure 2.

* Corresponding author: szemik@agh.edu.pl

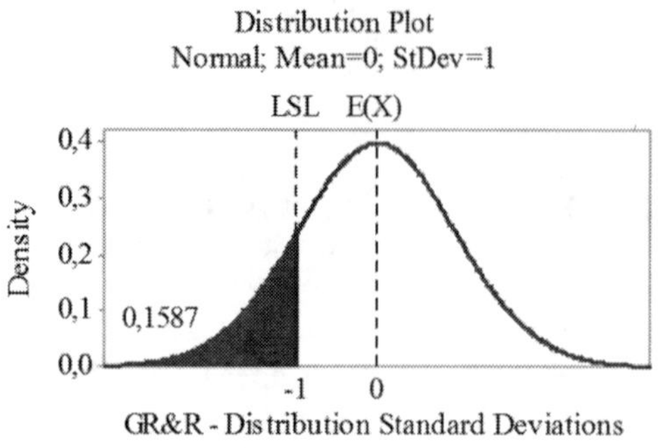

Fig. 2. The probability of type I error.

Probability chart demonstrated that in this particular scenario the likelihood of classification a conforming part as non-conforming is 0,1598. In this case, the variability of measurement process exposes the manufacturer to extra costs.

Respectively, the probability of classification non-conforming part as conforming is called miss rate, consumer's risk or type II error.

The probability of type II error is shown in figure 3.

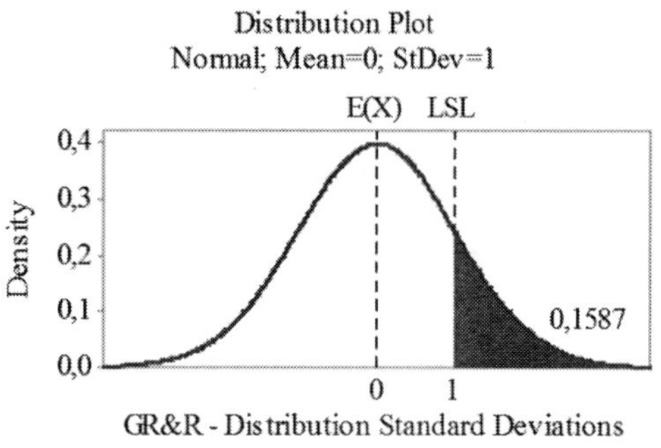

Fig. 3. The probability of type II error.

Probability chart demonstrated that in this particular scenario, the likelihood of classification a non-conforming part as conforming is 0,1598. In this case, the variability of measurement process exposes the customer to the risk of utilizing part out of the specification.

Thereby, the measurement systems analysis has become a widely used method for the gauges approval for production use.

The purpose of the two main industrial manuals *MSA 4th Edition* and *VDA 5 Capability of Measurement Process* respectively is to [3-4]:

- Present guidelines for assessing the quality of measurement system without intention to be a compendium of analyses for all measurement systems;
- To summarize the requirements and procedures in order to gain a standardized model for the estimation of expanded measurement uncertainty.

The unusual application of product control scenario applied to one-sided tolerance is not discussed in both manuals. Therefore, the aim of this study was to develop a method for measurement system variability analysis in the case of product tolerated by a single lower (minimum) specification limit.

2 Methodical assumptions for management of measurement systems

In order to facilitate reading of this paper, the summary of essential terms which are used in this study is presented in the following section.

A gauge is any equipment used to assign a measured value to the material thing [4]. A measurement system is "the collection of gauges, standards, operations, methods, fixtures, software, personnel and assumptions used in measurement process" [4]. In general, the measurement system is a complete process used to obtain the measurement.

As the variation in any process comes from the many sources, therefore, measuring the same part does not result in identical measurements [5].

The main variability sources influencing the measurement process are as follows [6]:

- Operator;
- Measured object;
- Measurement method;
- Environment;
- Gauge;
- Fixture.

Measurement uncertainty is a "parameter, associated with the result of a measurement that characterizes the dispersion of the values that could reasonably be attributed to the measurand" [7]. Discussed uncertainty is crucial if measurements are performed on a product with respect to a specification limit. If the measurement result, along with its uncertainty, does not exceed the specification limits, then the product can be evaluated as meeting the design criteria [8]. Each measurement is "biased, which means that measured value is never equal to true value" [9]. A true value is the value consistent with the definition of and observed measured characteristic [6]. The difference between estimating variation by measurement uncertainty and MSA methods is that MSA evaluates variation of the measurement system for product and process control [4].

Repeatability is "variation in measurements obtained with one measuring instrument when used several times by an appraiser while measuring identical characteristic on the same part" [10]. Consequently, reproducibility is "variation in the average of the measurements made by different appraisers using the same gauge when measuring a characteristic of a part" [10].

Gauge R&R (GRR) is "the combined estimate of measurement system repeatability and reproducibility" [10]. Studying the gauge variability determines the measurements variability and the part-to-part variation which is for instance the variability of measurements due to the significant part circularity [11].

A number of distinct categories is "the number of data categories that can be distinguished by the effective resolution of the measurement system and part variation from the observed process" [12].

Effective resolution is "the size of the data category when the total measurement system variation is considered" [12]. The Gaussian distribution properties are used to evaluate the above statistical parameters of the measurement systems.

3 Proposal of authorial changes in measurement system management methodology

According to the *MSA 4th Edition Manual*, the %GRR, when evaluating the quality of measurement process with respect to the product tolerance specification limits, is given by a formula (1) [4]:

$$\%GRR = 100 \cdot \frac{6 \cdot GRR}{T} \qquad (1)$$

Where:

- T = *tolerance zone = USL- LSL;*
- *USL = upper specification limit;*
- *LSL = lower specification limit.*

Therefore, to calculate %GRR when the product is tolerated by the single lower (minimum) specification limit, the authors proposed the formula (2):

$$\%GRR = 100 \cdot \frac{0,5 \cdot k \cdot GRR}{\overline{\overline{X}} - LSL} \; for\; \overline{\overline{X}} > LSL \qquad (2)$$

Where:

- k = *coverage factor = 6;*
- $\overline{\overline{X}}$ = *mean of all measurements.*

The coverage factor is to be further discussed in this paper. In the above formula, the measurement process variation is compared to the range between the lower specification limit and the average of all measured parts. The above formula is valid, if the mean of all measurements is greater than LSL.

The disadvantage of such an approach is that the capability of the measurement system is determined by the average of the manufacturing process, while the measurement system is evaluated with respect to tolerance specification limit.

Next, if the measurement result is in the area between LSL and half of the GRR distribution, then conformity with specification cannot be proved properly.

Moreover, a similar method has already been introduced in MiniTab 17 Statistical Software [13].

An alternative approach proposed by the authors to the following issue is that regardless the parts average value, the 3 standard deviations of measurement process (assuming the normality of measurement process and measured characteristic distribution) shall be smaller than the minimum average value of measured parts, which are representative to cover the entire manufacturing process.

The illustration of this scenario is demonstrated in figure 4. Probability chart shown in figure 4 demonstrated that in this case the likelihood of classification non-conforming part as conforming is 0,00135, thus, such probability of 3 standard deviations of measurement process is assumed by the authors as acceptable in accordance with the 6-sigma standard in quality engineering. Therefore, to evaluate GRR, when the product is tolerated by the single lower (minimum) specification limit, the authors proposed the formula (3):

$$P_{gk} = \frac{min\{\overline{X}_1,...,\overline{X}_n\} - LSL}{0,5 \cdot k \cdot GRR} \geq 1 \; for \; min\, \overline{X}_i > LSL \qquad (3)$$

Where:

- k = *coverage factor = 6;*
- $\min\{\overline{X}_1,...,\overline{X}_n\}$ – *minimum average value of the measured parts.*

The factor equal to 6 is sufficient to cover the probability of 0,9973 of a normal distribution. The graphical interpretation of the probability is shown in figure 4.

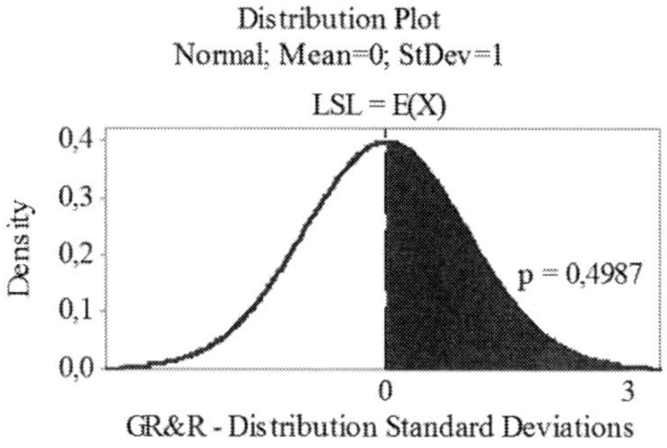

Fig. 4. The probability of type II error.

3.1 Research assumptions

The numeric study has been conducted by measuring 10 representative parts 3 times by 3 appraisers. According to the MSA 4th Edition Manual the detailed procedure of GRR study is as follows:

- Obtain a sample of n > 10 parts that represent the actual process variation;
- Calibrate the gauge;
- Measure the parts in random order three times for each part and appraiser combination;
- Analyze the results with respect to the acceptance criteria.

The acceptance criteria according to the manual are %GRR < 10% and a number of distinct categories (NDC) ≥ 5.

3.2 Used methods and research tools

The analysis of variance (ANOVA) method was used to analyze the results. In the study, the variation components of the measurement process can be distinguished in four categories (parts, appraisers, interaction between parts and appraisers, and measurement equipment).

The advantage of using analysis of variance method instead of average range method (ARM) is that estimation of variation components is more accurate, as well as more data related to the measurement process is provided (e.g. interaction between operators and parts).

On the other hand, the complexity of calculation as compared with ARM method is significant. Finally, the Gaussian probability distribution properties were used to evaluate formulas (2) and (3) respectively.

The statistical evaluation was performed using *MiniTab 17* software.

4 The results of applying the proposal on a selected example

In this section, authors presented the numeric results of gauge repeatability and reproducibility study for illustration of proposed formulas applicability and statistical validation. Collected data are shown in table 1.

Table 1. Measurement data for the numerical example.

Appraisal		A		
Trial		1	2	3
Part	1	15,572	15,568	15,568
	2	15,610	15,613	15,610
	3	15,609	15,610	15,608
	4	15,627	15,629	15,631
	5	15,609	15,612	15,608
	6	15,549	15,552	15,550
	7	15,555	15,557	15,555
	8	15,600	15,604	15,602
	9	15,572	15,572	15,573
	10	15,546	15,548	15,548
Appraisal		B		
Trial		1	2	3
Part	1	15,565	15,568	15,566
	2	15,612	15,611	15,613
	3	15,610	15,609	15,608
	4	15,630	15,628	15,626
	5	15,609	15,609	15,607
	6	15,551	15,550	15,550
	7	15,556	15,556	15,556
	8	15,602	15,603	15,602
	9	15,574	15,571	15,572
	10	15,548	15,547	15,548
Appraisal		C		
Trial		1	2	3
Part	1	15,568	15,568	15,566
	2	15,612	15,611	15,610
	3	15,609	15,609	15,609
	4	15,627	15,633	15,631
	5	15,608	15,607	15,611
	6	15,549	15,551	15,551
	7	15,556	15,555	15,556
	8	15,602	15,602	15,602
	9	15,571	15,573	15,573
	10	15,548	15,546	15,547

5 Discussion of obtained results

Gauge repeatability and reproducibility analysis was carried out using collected data. The result of the study using the analysis of variance method is shown in table 2.

Table 2. Measurement systems GRR study results.

Source	σ_{GRR}
Total Gage R&R	0,0013706
Repeatability	0,0013706
Reproducibility	0
Operator	0
Part-To-Part	0,0298342
Total Variation	0,0298656
NDC	30

Subsequently, the %GRR was calculated using the proposed formula (2):

$$\bar{\bar{X}} = 15{,}5854$$

$$\%GRR = 4{,}82\,\%$$

The graphical analysis of formula (2) is presented in figure 5.

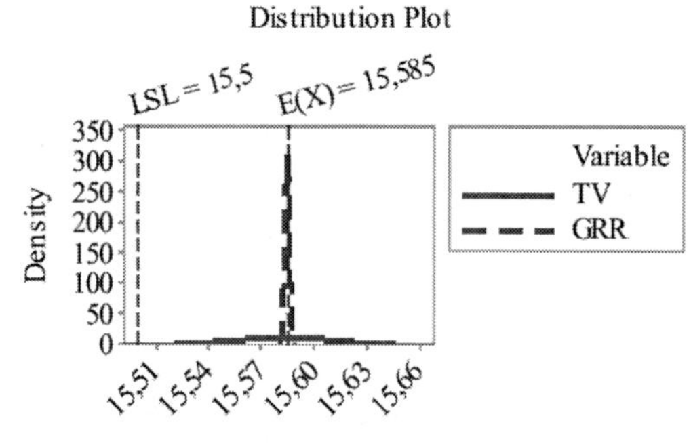

Fig. 5. % GRR with respect to LSL.

The analysis of the formula's (2) result shows that the variation of the measurement system, in comparison to the total variation (the combined estimate of the process and the gauge variation), equals 4,82%. This result is assumed as acceptable since according to the manual, the acceptance criterion is %GRR < 10%.

Subsequently, the number of distinct categories is greater than 5, which means that the measurement system can distinguish 30 number of groups from measurement data. This means that the measurement process is capable towards effective resolution, which is not necessarily required to be evaluated in analysed case study. However, if the effective resolution is not sufficient for the evaluation of the parts variability, it may be an input to improve the gauge and therefore lead to better process capability.

Moreover, with respect to the manufacturing system, if the process overall capability is greater or equal to 1,33, then the probability of type II error is ~ 0,001 assuming the %GRR equals 33%. The type II error probability in this case is shown in figure 6.

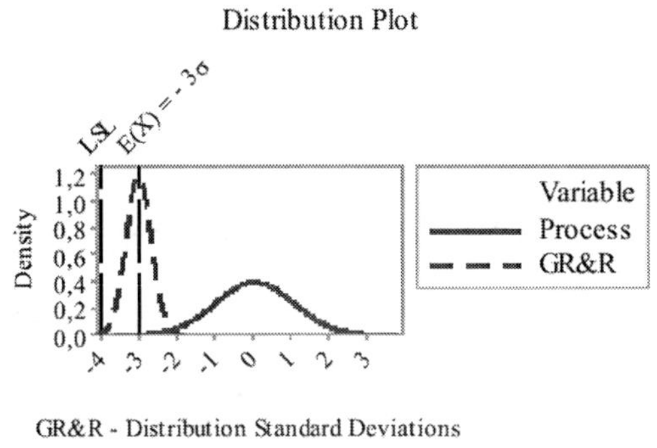

Fig. 6. % GRR impact on type II error with a capable process.

Subsequently, the P_{gk} was calculated using the proposed formula (3):

$$min\{\overline{X_1},...,\overline{X_n}\}=15,547$$

$$P_{gk}=11,51$$

The graphical analysis of P_{gk} index is presented in figure 7.

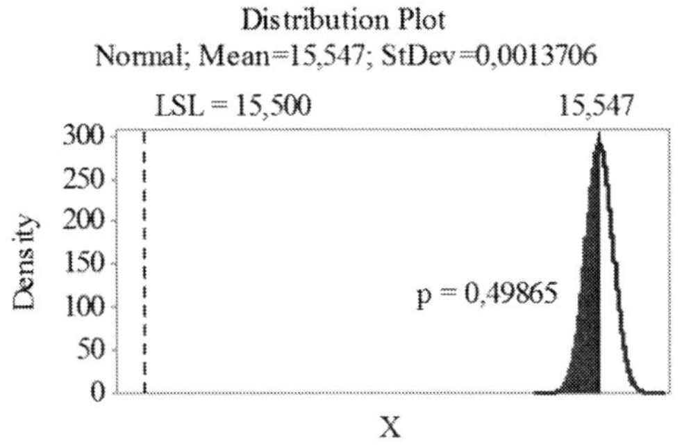

Fig. 7. % GRR with respect to LSL.

The analysis of the formula's (3) result shows that the percentile of 3 standard deviations of measurement process (which is equal to the probability of 49,865%) is over 11 times smaller than the zone between the minimum of observed parts average results. Therefore, the measurement process is considered as capable since $P_{gk} \geq 1$.

In a proper approach, a single index as the acceptance criterion for measurement system should not be taken. The final decision should consider different evaluations. For this reason, multiple studies should be used to determine the quality of measurements [14]. Therefore, if the overall process performance is less than 1, then the criterion for the product acceptance shall be the result in the conformance zone [15]. In this particular case, the conformance zone is the result $x \in C_z$, according to the formula (4):

$$C_z = \{LSL + 3\sigma_{GRR}; +\infty\}. \quad (4)$$

6 Conclusions

In the carried-out research authors have elaborated two robust, statistical methods for the gauge repeatability and reproducibility analysis, in terms of single lower specification limit scenario, as the additional technique of the analysis method presented in the *MSA 4th Edition Manual*.

The methods proposed by the authors are to be used for the evaluation of the measurement system variability and effective resolution regarding the product control. The elaborated formulas use statistical properties of studied parts.

Since in the studied scenario there is no upper specification limit, the elaborated formula (2) takes the mean of the studied parts as the upper specification limit for the tolerance zone determination. It may not be fully effective, as the quality of measurement system becomes dependent on the process location. Thus, the alternative approach was elaborated and presented in the formula (3). This method takes the minimum value of the parts average as the upper specification limit and compares the variation of the measurement process to the zone between the lower specification limit and calculated minimum value. The advantage of such method is that assuming the measured parts representative to the process 6 standard deviations, it prevents the manufacturer from the type II error.

Nevertheless, if the process capability is insufficient, the minimum mean value of measured parts may be smaller than lower specification limit, and then the method cannot be used. Therefore, the conformance zone based on the gauge repeatability and reproducibility was elaborated. Hence, the authors have referred to the type II probability error as the additional criterion in the acceptance of the complete measurement results.

Consequently, the proposed formulas for the GRR were used in the numeric study following the *MSA 4th Edition Manual* procedure, using the analysis of variance method.

Afterwards, the numeric study results were evaluated using the statistical properties of the Gaussian distribution. Additionally, the graphical interpretation of each formula was presented.

Moreover, the presented methods are applicable for all measurement systems which are normally distributed.

In addition, the influence of the measurement process variability on the process evaluation (capability index, control charts) and product qualification (type II, type II errors) was described.

Finally, the most important definitions used in this paper were discussed, including such terms as gauge, measurement system, measurements variability sources, measurements uncertainty, gauge repeatability and reproducibility, measurement system qualification criteria.

References

1. B. Colismo, N. Senon, *Geometric Tolerances* (2011)
2. AIAG, *Statistical Process Control (SPC) 2nd Edition* (2005)

3. T. M. Yeh, J. J. Sun, J Appl. Res. Technol. **11**, 780-786 (2013)
4. AIAG, *Measurement Systems Analysis (MSA) 4th Edition* (2010)
5. M. J. Flynn, IEEE Trans. Intrum. Meas. Mag. **11**, 10-15 (2008)
6. VDA, *VDA volume 5 Capability of Measurement Process* (2011)
7. International Organization for Standardization, *PN-EN ISO 14253-2: 2011: Specyfikacje geometrii wyrobów (GPS) -- Kontrola wyrobów i wyposażenia pomiarowego za pomocą pomiarów -- Część 2*, 9 (2011)
8. N. Ridler, B. Lee, J. Martens, K. Wong, IEE Microw. Mag. **8**, 44-53 (2007)
9. JCGM, *Evaluation of measurement data — Guide to the expression of uncertainty in measurement*, 5 (2008)
10. AIAG, *Measurement Systems Analysis (MSA) 4th Edition*, 13 (2010)
11. A M. Awad, T. P. Erdmann, Y. Shanshal, B. Barth, Qual. Eng. **21**, 300-305 (2009)
12. AIAG, *Measurement Systems Analysis (MSA) 4th Edition*, 214 (2010)
13. http://support.minitab.com/en-us/minitab/17/topic-library/quality-tools/measurement-system-analysis/gage-r-r-analyses/calculation-of-percent-tolerance-with-a-one-sided-tolerance/
14. C J. N. Pan, Qual. Quant. **40**, 499-518 (2006)
15. International Organization for Standardization, *PN-EN ISO 14253-1: 2014-02: Specyfikacje geometrii wyrobów (GPS) -- Kontrola wyrobów i wyposażenia pomiarowego za pomocą pomiarów -- Część 1: Reguły orzekania zgodności lub niezgodności ze specyfikacją* (2014)

DOI: 10.1051/itmconf/20171505012

Modelling and simulation of a hydraulic active heave compensation system

Arkadiusz Jakubowski[1,*], and *Arkadiusz* Kubacki[1]
[1]Institute of Mechanical Technology, Poznan University of Technology, Skłodowska-Curie Square 5, 60-965 Poznań, Poland

Abstract. The following work presents modelling, simulation and results of hydraulic motor and hydraulic cylinder position measurement for a hydraulic active heave compensation system (AHC). In order to perform experimental research the authors developed a test stand. A hydraulic cylinder is used to carry out a simulation of sea waves. The development of a test stand made it possible to compare the simulation with the experimental results. A hydraulic motor is used for compensation of sea waves. Optimization of design parameters for such systems can be done by analysing a model simulation. The main components of this system were modelled. This model was implemented in simulation software Mathlab Simulink and its dynamic performance was tested. The Authors presented the simulation and the experimental results for main components of a hydraulic active heave compensation system.

1 Introduction

An electrohydraulic servo-drive is used in many industrial applications. Hydraulic systems are well established in the marine industry. Hydraulic actuators provide the highest power to weight ratio of any actuator currently on the market. The electrohydraulic servo-drives can be controlled by two types of electro-valves: proportional and servo-valves. Basic information about electrohydraulic servo drives is found in [1] and [2].

The heave compensation unit is an extremely important piece of equipment. Drilling, load handling, observation of seabed during rough sea are difficult to execute. For this reason heave compensation systems are applied. There are three types of heave compensation: passive heave compensation (PHC), active heave compensation (AHC) and semi-active heave compensation (SAHC). The first two are more commonly used than the last one. In the last few years, both passive heave (PHC) and active heave (AHC) systems have been researched and developed in order to improve drilling and load handling on the ship during rough sea [3,4].

Examples of research on PHC systems include the work by Huster *et al.* who used a gas-backed accumulator as a spring for PHC [5]. Another work about PHC is by Kidera who found the solution to the problem of a cylinder stick where static friction was too large for the load. Hatleskog and Dunnigan researched passive heave systems and one of the conclusions they made was that the only way to reduce heave motion by over 80% is by using an active compensator [6]. In the active heave compensation systems, ship's heave motion is measured and sent to a controller which moves an actuator. In these systems hydraulic drivers such as a hydraulic motor and a hydraulic cylinder are used. Over the last few years the research focused on the development of a new control algorithm [7, 8]. Kyllingstad applied transfer function filters [9]. Another work about a new algorithm is by Kuchler et al. who used heave-prediction algorithms to predict heave motion [10].

Modelling of the AHC system can be divided into three main parts: control, hydraulic and mechanical system. In this work the Authors focused on a modelling hydraulic system. The hydraulic system consists of a hydraulic cylinder and a hydraulic motor which are controlled by two servo valves.

2 Hydraulic active heave compensation system – modelling and simulation

2.1. Modelling of the electrohydraulic servo drive with a hydraulic cylinder

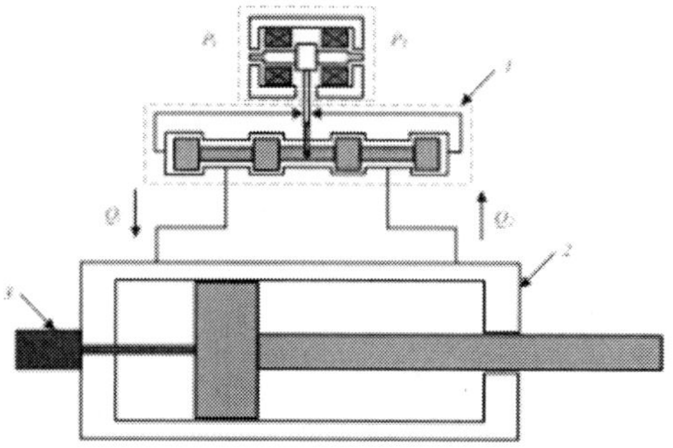

Fig. 1. Scheme of the electrohydraulic servo drive with a hydraulic cylinder.

* Corresponding author: arkadiusz.z.jakubowski@doctorate.put.poznan.pl

DOI: 10.1051/itmconf/20171505012

The first electrohydraulic servo drive consists of a servo valve *1* and a hydraulic actuator *2*. The servo valve is connected to the hydraulic cylinder as shown on Fig. 1. The actual position of a hydraulic cylinder piston (double acting cylinder) is measured via a displacement sensor *3*.

Scheme of the four edge electrohydraulic amplifier is shown on Fig.2.

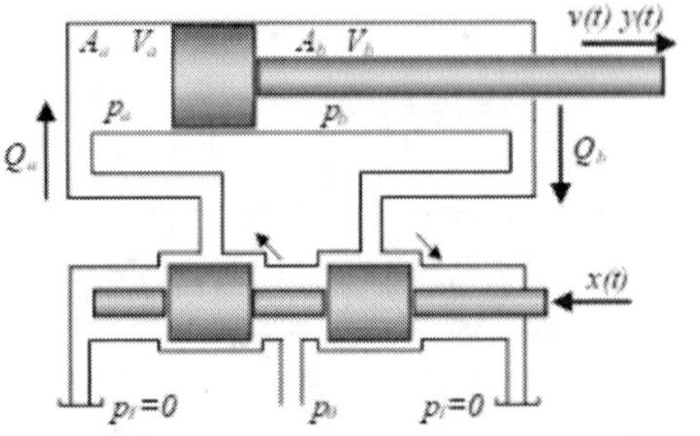

Fig. 2. Scheme of the four-edge electrohydraulic amplifier with a hydraulic actuator.

The equations describing the electrohydraulic servo drive model with a hydraulic actuator can be specified as follows [11, 12]:

$$Q_a(t) = Q_{sa}(t) + Q_{ha}(t) + Q_v(t) \tag{1}$$

$$Q_b(t) = Q_{sb}(t) + Q_{hb}(t) - Q_v(t) - Q_{vb}(t) \tag{2}$$

$$Q_a(t) = K_{Qp}x(t) - K_l[p_0 - p_a(t)] \tag{3}$$

$$Q_b(t) = K_{Qp}x(t) - K_l p_b(t) \tag{4}$$

$$Q_{ha}(t) = A\frac{dy(t)}{dt}, \quad Q_{hb}(t) = aA\frac{dy(t)}{dt} \tag{5}$$

$$Q_{sa}(t) = \frac{V_a}{E_o}\frac{dp_a(t)}{dt}, \quad Q_{sb}(t) = -\frac{V_b}{E_o}\frac{dp_b(t)}{dt} \tag{6}$$

$$Q_v(t) = K_v[p_a(t) - p_b(t)] \tag{7}$$

$$Q_{vb}(t) = K_{vb}p_b(t) \tag{8}$$

$$m\frac{d^2y(t)}{dt^2} + D\frac{dy(t)}{dt} = A[p_a(t) - ap_b(t)] \tag{9}$$

where: Q_a, Q_b – flow, Q_{ha}, Q_{hb} – absorption of the actuator chambers, Q_{sa}, Q_{sb} – flow of the covering losses due to compressibility, Q_{vb} – leakage flow on the piston rod, p_a, p_b – the pressure in the chambers of the actuator, A_a, A_b – active surfaces of the piston, V_a, V_b – the volume of liquid in the chambers of the actuator.

2.2 Modelling of the electrohydraulic servo drive with a hydraulic motor

The second electrohydraulic servo drive consists of a servo valve *1* and a hydraulic motor *2*. The servo valve is connected to the hydraulic motor as shown on Fig. 3. The actual position of the hydraulic motor is measured by an incremental encoder *3*.

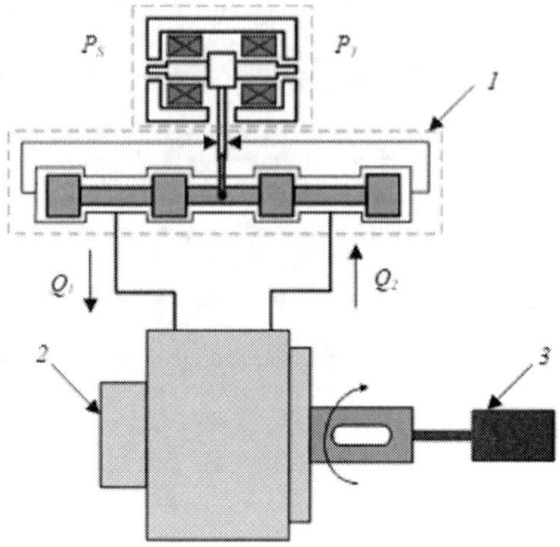

Fig. 3. Scheme of the electrohydraulic servo drive with a hydraulic motor.

The parameters of a hydraulic motor are needed for its modelling. The equations below are used to determine these parameters:

$$n_m = Q_t / V_m \tag{10}$$

$$\Delta P = \frac{2\pi}{V_m}T \tag{11}$$

where: n_m – motor speed, ΔP – applied pressure differences, V_m – geometric volume of motor, T – loading torque, Q_t – theoretical flow rate.

The theoretical motor flow rate is smaller than the real flow due to the internal leakage. The volumetric efficiency of the motor can be specified as follows:

$$\eta_v = \frac{Q_t}{Q} \tag{12}$$

or

$$n_m = \frac{Q\eta_v}{V_m} \tag{13}$$

The output mechanical power of the motor is smaller than the input hydraulic power. The power losses are evaluated by the total efficiency η_t:

$$Q\Delta P\eta_T = 2\pi n_m T \tag{14}$$

DOI: 10.1051/itmconf/20171505012

then:

$$\Delta P = \frac{2\pi}{V_m \eta_m \eta_h} T \tag{15}$$

where: Q – real motor flow, η_T– total motor efficiency η_m – motor mechanical efficiency, η_v – motor volumetric efficiency, η_m – motor hydraulic efficiency.

The mathematical model of the hydraulic motor describes its dynamics. The equations below describe the hydraulic motor [13, 14].

$$\dot{\omega}_k = J_m^{-1}[q_m f(q)(p_i - p_j) - (M_l / u_{mech} + b_\omega \omega_k + b_p |p_i - p_j| \cdot sign\,\omega_k + b \cdot sign\,\omega_k)] \tag{16}$$

$$\dot{\varphi}_k = \omega_k \tag{17}$$

$$Q_{i,j} = q_m f(q)\omega_k \pm k_{lea} p_{i,j} \tag{18}$$

where: ω_k – angular speed of the hydraulic motor shaft, J_m – moment of inertia, q_m – maximal geometric volume, $f(q)$ – parameter of regulation: $-1 \leq f(q) \leq 1$, M_l – loading moment, b_ω, b_p, b– coefficients of the hydraulic motor hydro mechanical losses, u_{mech} – transfer number of the working mechanism gear, k_{lea} – coefficient of the hydraulic motor volumetric losses.

3 Simulation

The simulation model was built based on the Simulink SimHydraulics tool. In Fig. 4 the simulation model of the electrohydraulic servo drive with a hydraulic cylinder is shown. The stroke of the double acting hydraulic cylinder was equal to 400 mm. The diameters of the piston and piston rod were respectively 60 mm and 100 mm. The value of the reduced mass was equal to 37, 3 kg (mass of the piston, piston rod and hydraulic motor, intermediate plate which is fixed at the end of piston rod).

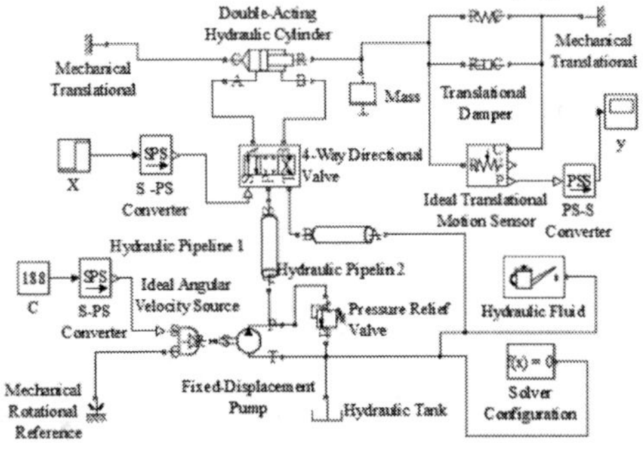

Fig. 4. Model of the hydraulic cylinder in Mathlab Simulink software.

The step response model of a hydraulic cylinder was checked by the Authors. Fig. 5 shows the simulation results with displacement of the piston rod. The simulation was performed for the supply pressure p_o equal to 5 MPa, 10 MPa and 15 MPa.

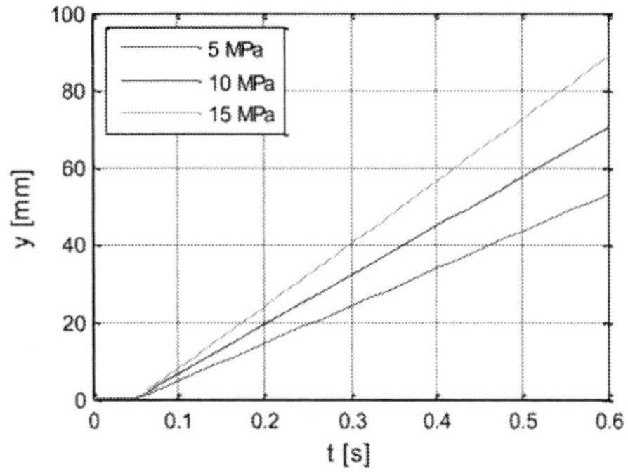

Fig. 5. Displacement of the piston road for a supply pressure of 5 MPa, 10 MPa and 15 MPa.

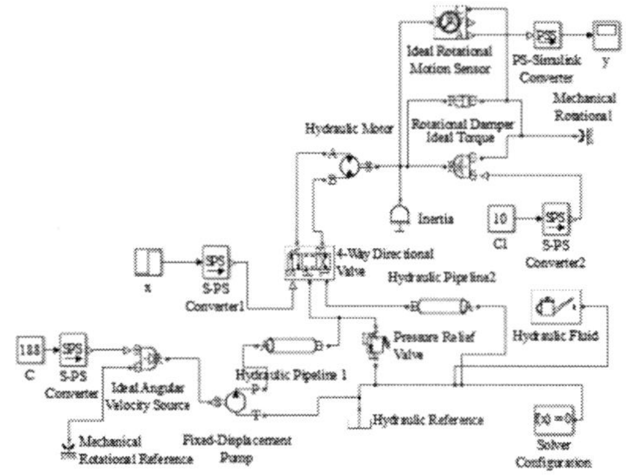

Fig. 6. Model of the hydraulic motor in Mathlab Simulink software.

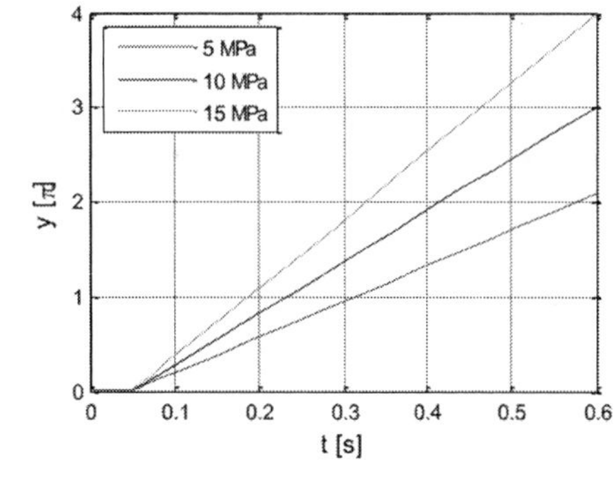

Fig. 7. Displacement of the shaft hydraulic motor for a supply pressure of 5 MPa, 10 MPa and 15 MPa.

In Fig. 6 the simulation model of the electrohydraulic servo drive with a hydraulic motor is shown. A steel load was fixed to the hydraulic motor via a steel cable. Mass of the steel load was equal to 5 kg.

The model of the electrohydraulic servo drive with a hydraulic cylinder and a hydraulic motor was tested by the step response signals. Fig. 7 shows the simulation results for the hydraulic motor shaft position. The model of the hydraulic motor was tested for different supply pressures such as: 5 MPa, 10 MPa and 15 MPa.

4 Experimental research

The electrohydraulic servo drives were tested by the Authors. In Fig. 8 the test stand is presented. The test stand consists of a hydraulic cylinder with a stroke range of 400 mm. The hydraulic cylinder is used for waves simulation. At the end of the piston rod the hydraulic motor with a maximum rotary speed range of 810 rpm is fixed. Maximum output power of the hydraulic motor was equal to 16 kW. The hydraulic cylinder and hydraulic motor were combined with servo valves. The hydraulic cylinder was equipped with the displacement sensor. The actual position of the hydraulic motor was measured via an incremental encoder.

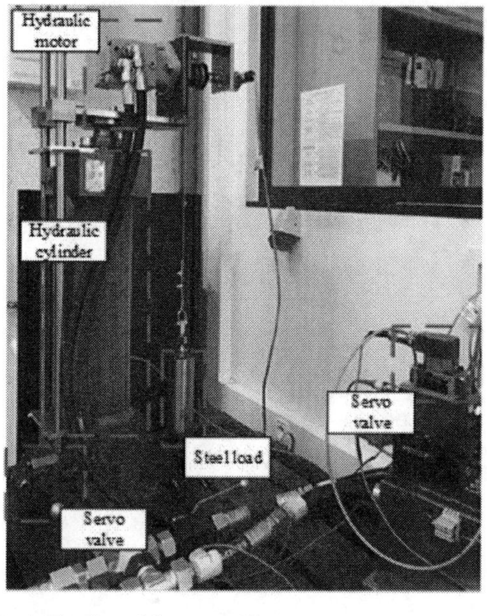

Fig. 8. The view of the test stand.

The hydraulic power supply used by the Authors is characterized by the following parameters: maximum flow rate= 100 dm3/min, maximum pressure p_0 = 40 MPa, motor power = 37 kW, filtration at 6 microns. The last step was to test a real object and compare the step response simulation model and a real hydraulic cylinder and a hydraulic motor. The supply pressure for the tested hydraulic cylinder and hydraulic motor was equal to 10 MPa. In Fig. 9 the displacement piston rod (hydraulic cylinder) for a real object and a model is presented. In Fig. 10 the displacement of the shaft hydraulic motor is illustrated by the Authors.

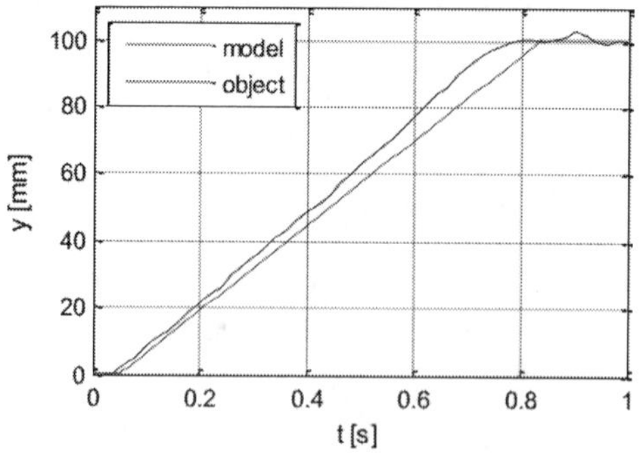

Fig. 9. Displacement of the piston rod for a model and a real object.

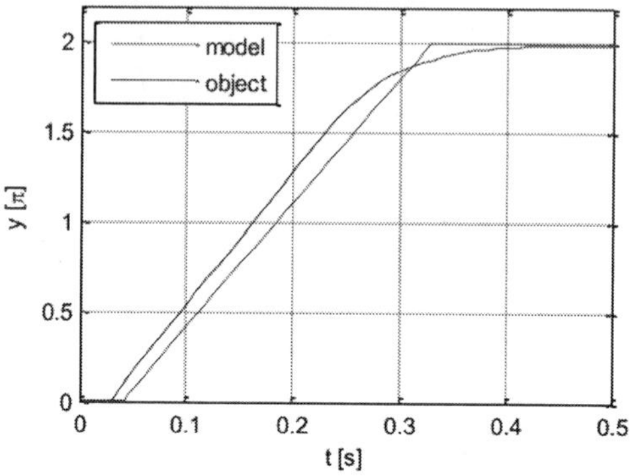

Fig. 10. Displacement of the shaft hydraulic motor for a model and a real object.

5 Conclusions

The article presents basic equations describing an electrohydraulic servo drive with a hydraulic cylinder and a hydraulic motor. The PLC was connected to the servo valves via dedicated control cards (valves amplifier). The simulation model was built by using the SimHydraulics tool. The hydraulic motor and hydraulic cylinder were controlled by the electrohydraulic servo valve. The test stand was built to compare the results of the simulation and the real objects. Further research aims to improve the simulation models of a hydraulic motor and a hydraulic cylinder.

The work described in this paper was funded from 02/22/DSMK/1408 (Sterowanie urządzeniami mechatronicznymi z zastosowaniem technik nieliniowych oraz bioelektrycznej aktywności mózgu).

References

1. J.S. Cundiff, *Fluid power circuits and controls* (CRC Press LLC, USA, 2002)
2. P. Chapple *Principles of hydraulic system design,* (Coxmoor , 2003)
3. J.K. Woodacre, R.J. Bauer, R.A. Irani, J. Phys. Conf. Ser. **104,** 140-154 (2015)

DOI: 10.1051/itmconf/20171505012

4. A.A. Walid, P. Gu, M.R. Hansen, G. Hovland, Y. Iskandarani, IJMECH. 291-296 (2011)
5. A. Huster, H. Bergstrom, J. Gosior, D. White, Wit. Trans. Ecol. Envir. 1-8 (2009)
6. J.T. Hatleskog, M.W. Dunningan, Wit. Trans. Ecol. Envir. 1-6 (2006)
7. L. Huang, Y. Zhang, L. Zhang, M. Liu, Petrol. Explor. Develop. **40(5),** 655-670 (2013)
8. J. Ni, S. Liu, M. Wang, X. Hu, Y. Dai, Appl. Mech. Mater. 5111-5116, (2009)
9. A. Kyllingstad, US. Patent 8,265,811 (2012)
10. S. Kuchler, T. Mahl, J. Neupert, K. Schneider, O. Sawodny, Trans. of Mechatronics. **16.2,** 297-309 (2011)
11. A. Milecki, D. Rybarczyk, Stroj. Vestn-J. Mech. E. **61(9),** 517-522 (2015)
12. D. Rybarczyk, D. Sędziak, P. Owczarek, A. Owczarkowski, *Progress in Automation, Robotics and Measuring Techniques*, 215-222 (2015)
13. E.Y. Malinovsky, L.B. Zaretsky, Y.G. Berengard, M.M. Gaitsgori, PJSC. (1980)
14. I.I. Bazhin, Y.G. Berengard, M.M. Gaitsgori, PJSC. (1988)

Recurrence analysis of regular and chaotic motions of a superelastic shape memory oscillator

Joanna Iwaniec[1], *Grzegorz* Litak[2,*], *Davide* Bernardini[3], and *Marcelo A.* Savi[4]

[1]AGH University of Science and Technology, Faculty of Mechanical Engineering and Robotics, Department of Robotics and Mechatronics, Mickiewicz Alley 30, 30-059 Krakow, Poland
[2]AGH University of Science and Technology, Faculty of Mechanical Engineering and Robotics, Department of Process Control, Mickiewicz Alley 30, 30-059 Krakow, Poland
[3]University of Rome Sapienza, Department of Structural and Geotechnical Engineering, via Antonio Gramsci 53, 00192, Rome, Italy
[4]Universidade Federal do Rio de Janeiro, COPPE – Mechanical Engineering, Centre for Nonlinear Mechanics, Rio de Janeiro - RJ - Brazil

Abstract. The recurrence analysis is a promising tool for diagnostics of periodic and chaotic solutions, as well as identifying bifurcations. This paper deals with the application of this analysis for the first time to identify regular and non-regular motions of a superelastic shape memory alloy oscillator. The numerical analyses show that the method is capable of distinguishing periodic and chaotic trajectories. Recurrence quantities are applied, showing that different approaches are possible to establish the distinction between periodic and chaotic signals. Basically, recurrence entropy, trapping time, and characteristic recurrence time are considered.

1 Introduction

Shape Memory Alloys (SMAs) exhibit a complex thermomechanical behaviour determined by the occurrence of phase transformation between solid phases, namely austenite and martensite, at the microscopic scale [1, 2]. This produces several functional properties, like superelasticity (or pseudoelasticity) and shape memory effects, which are largely utilized nowadays, for several applications in various fields. In particular, the superelastic behaviour is characterized by hysteretic loops without residual displacements (Figure 1).

The force-displacement curves (Figure 1a) are characterized by two plateaus associated with the occurrence of the Forward A→M transformation (FwT) and Reverse M→A transformation (RvT). The hysteresis loops can, in general, have different shape in tension and compression. Figure 1b also shows the temperature dependence of the four transformation forces that characterize the hysteresis loop, as well as the force-temperature path that corresponds to the pseudoelastic loop in (a). In particular, FwT is associated with temperature rises, while RvT to temperature decreases. The intersections of the various lines with the temperature axis in (b), show four transformation temperatures – M_S, M_f, A_S, A_f – that characterize a specific SMD. By and large, depending on the specific SMA, each of the main properties that characterize the pseudoelastic loops can assume different values in tension and compression.

When employed in dynamical systems, superelastic SMA elements may induce non-regular chaotic responses, as shown for example in [3-11].

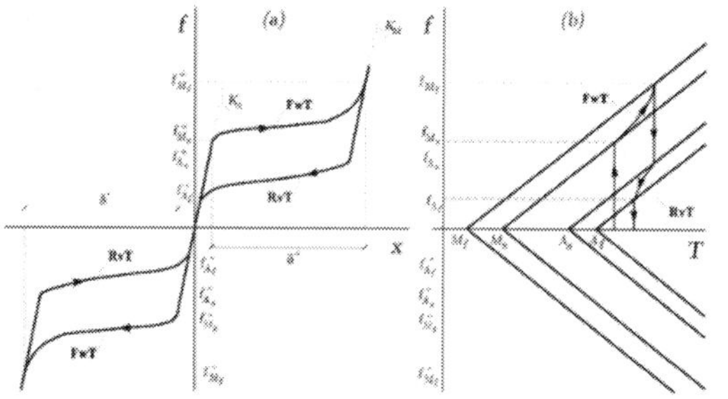

Fig. 1. General features of the superelastic behavior: a pseudoelastic loop (a), the intersections of the various lines with the temperature axis (b).

The eventual presence of such chaotic motions may be an important aspect considered in the design of application and, for this reason, it is very important to be able to detect them. Due to the hysteretic behaviour induced by the non-smoothness of underlying differential equations, the most typical tool for chaos diagnosis, namely Lyapunov exponents, is difficult to employed. Therefore, various, alternative diagnostic methods have been proposed, as for example, 0-1 test [12, 13] and the method of wandering trajectories [10, 11].

This work investigates the opportunity to apply recurrence analysis that has not been utilized yet for SMA oscillators. This approach includes Recurrence Plots (RP) and Recurrence Quantification Analysis (RQA) that have been of interest to characterize the non-linear dynamics of many dynamical systems [14], including solutions and bifurcation identification in

* Corresponding author: g.litak@pollub.pl

mechanical systems [15, 16] and smart material structures [17]. In this method, the periodicity of the underlying dynamics can be investigated by means of recurrences that are calculated for each point of the reconstructed trajectory. This method was originally developed by Eckmann et al. [18] and directly extended by Webber and Zbilut [18, 19], Casdagli [20], and finally, by Marwan et al. [14]. The application of recurrence analysis can be applied for both short deterministic and noisy experimental data (see [15, 21]).

2 Modelling of the shape memory alloy oscillator

An oscillator, composed by a mass m and a restoring force provided by a superelastic SMA device, is considered. The oscillator is subjected to a harmonic excitation represented by $F(t)$ (Figure 2).

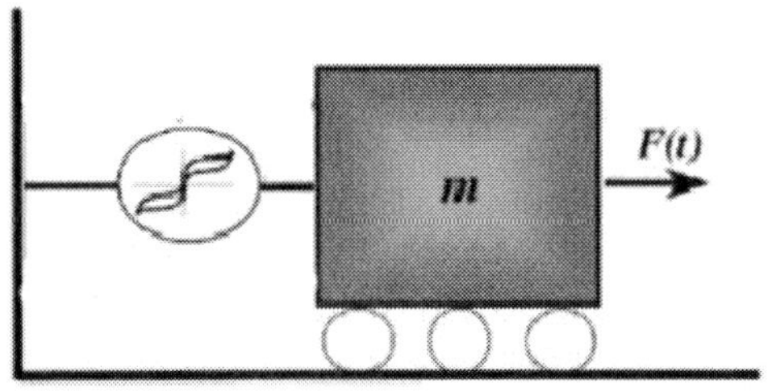

Fig. 2. Typical layout of a superelastic SMA oscillator.

The thermomechanical constitutive model, proposed in [22], has been adapted in [23] as a model for the restoring force of SMA devices, suitable for the use in the analysis of nonlinear dynamics of Shape Memory Oscillators (SMO) [10, 11].

This model is employed to form the SMO analysed in this paper, considering a dimensionless form discussed in the sequence. Specifically, quantities with the dimensions of a temperature are normalized with respect to a reference value θ_{ref}, greater than the transformation temperature A_f, typical for each specific SMA. On the other hand, quantities with the dimensions of time, length and force are normalized, respectively, to the elastic fundamental period $1/\omega$ of the oscillator and to the displacement x_{Ms} and force f_{Ms} at the beginning of the forward A $\rightarrow$ M transformation at the temperature θ_{ref}.
The SMO is subjected to a harmonic excitation $\Phi(t) = F\cos\Omega t$, and operates in a thermomechanical environment, where a convective rate of heat exchange $Q(t) = h(\theta_e\text{-}\theta(t))$ takes place. Here, F and Ω are the normalized excitation amplitude and frequency, θ_e the environment non-dimensional temperature and h the coefficient of convective heat transfer.

At each time t, the state of the SMO is described by displacement $x(t)$, velocity $v(t)$ and temperature $\theta(t)$ of the device, as well as the martensitic fraction $\xi(t) \in [0,1]$. Moreover, to model the complex hysteretic behaviour of SMA related to internal subloops, the state of the device also depends on the value $\xi_0(t)$, that represents the martensitic fraction ξ, attained at the beginning of the last phase transformation, occurred before time t. As discussed in [8] Bernardini and Rega (2005), time evolution of the SMO is described by the following system of 5 differential equations:

$$\begin{cases} \dot{x} = v \\ \dot{v} = -x + (\mathrm{sgn}(x)\lambda)\xi - \zeta v + \Phi \\ \dot{\xi} = Z[\mathrm{sgn}(x)v - JQ] \\ \dot{\theta} = ZL\left(\dfrac{\Lambda}{J\lambda} + \theta\right)[\mathrm{sgn}(x)v - JQ] + Q \end{cases} \quad (1)$$

The constitutive functions Λ and Z take different expressions, depending on the fulfillment of suitable transformation criteria that detect the occurrence of the phase transformations [8]. Subscripts F and R indicate the expressions relative to the upper and lower plateaus, respectively, that correspond to the forward (A $\rightarrow$ M) and reverse (M $\rightarrow$ A) transformations:

$$\begin{aligned} &\Lambda_F(\xi,\xi_0) = \lambda\frac{(J-1)(q_2-1)}{2} + \lambda(J-1)(1-q_1)\Psi_F(\xi,\xi_0) \\ &\Lambda_R(\xi,\xi_0) = -\lambda\frac{(J-1)(q_2-1)}{2} - \lambda(J-1)q_2(1-q_3)\Psi_R(\xi,\xi_0) \\ &Z_{F,R}(\vartheta,\xi,\xi_0) = \frac{1}{\lambda + JL\theta + \frac{L}{\lambda}\Lambda_{F,R} + \frac{1}{\lambda}\frac{\partial \Lambda_{F,R}}{\partial \xi}} \quad , \end{aligned} \quad (2)$$

where $\Psi_{F,R}$ are constitutive functions that determine the shape of the upper/lower pseudoelastic plateaus and the other quantities are model parameters. Whereas various choices at different levels of refinement are possible, the following one is adopted:

$$\begin{aligned} &\Psi_F(\xi,\xi_0) = \frac{1}{2}\left(1 + \frac{1}{b}\ln\frac{1-\xi+e^b(\xi-\xi_0)}{\xi-\xi_0+e^b(1-\xi)}\right) , \\ &\Psi_R(\xi,\xi_0) = \frac{1}{2}\left(-1 + \frac{1}{b}\ln\frac{\xi-\xi_0+e^b\xi}{-\xi+e^b(\xi-\xi_0)}\right) . \end{aligned} \quad (3)$$

Here b is a parameter that determines the smoothness of the transition between the elastic branch and the transformation plateaus.

The second law of thermodynamics expresses the non-negativity of the rate of energy dissipation Γ. In this framework $\Gamma=\Lambda\dot{\xi}$. Since $\Lambda=\Lambda_F$ when $\dot{\xi}>0$ and $\Lambda=\Lambda_R$ when $\dot{\xi}<0$, the second law requires: $\Lambda_F \geq 0$, $\Lambda_R \leq 0$. Such constraints impose limitations on the range of variations of model parameters. For example, while evaluating Λ_F at the beginning of the upper plateau, the following constraint among λ, J, q_2 is obtained:

$$\Lambda_F(0,0) = \lambda\frac{(J-1)(q_2-1)}{2} \geq 0 . \quad (4)$$

Considering (1) and (2), the system response depends on 7 model parameters, besides the above-mentioned b: q_1, q_2, q_3, λ, L, h, J.

The first four parameters determine the shape of the outer pseudoelastic loop. In particular: q_1 and q_3,

 DOI: 10.1051/itmconf/20171505013

respectively, influence the slopes of the isothermal upper and lower plateaus. Physically meaningful ranges of values may be identified with $q_1 \in [0.7, 1.0]$ and $q_3 \in [1.0, 1.5]$. On the other hand, q_2 determines the position of the isothermal lower plateau with respect to the upper one hence, determines the size of the hysteresis loop. Finally, λ directly influences the length of both plateaus. The remaining three parameters characterize the thermomechanical properties.

In particular, a parameter L determines the amount of heat which is produced or absorbed during mechanical loading, and the parameter h characterizes the rate at which the involved heat can flow out from the system to the environment by convection. Physically meaningful ranges of values may be identified with $L \in [0.0, 0.5]$ and $h \in [0.0 ,0.2]$. The limit case $L = 0$ corresponds to shape memory materials which transformations produce negligible amounts of heat, while the limit case $h = 0$ models an adiabatic environment in which all the heat produced remains in the system.

The *thermo-mechanical parameter* J determines the slope of the linear dependence of the transformation forces on the temperature. A physically meaningful range of values may be identified with $J \in [1.0, 4.0]$.

3 Recurrence analysis

In early beginning, the concept of recurrences was applied to study the nonlinear dynamics and bifurcations in terms of stroboscopic phase diagrams, namely Poincare maps [24]. The method of the Recurrence Plot (RP), introduced by Eckmann et al. [17], was generalization on all the states along the trajectory [14, 24]. RP is based on identification of the same states in the phase space of the dynamical system. The recurrence of the state $\mathbf{x}_i$ at time i, at a different time j, $\mathbf{x}_j \approx x_i$, is counted within a two-dimensional squared matrix $\mathbf{R}$ with ones and zeros, i.e.

$$R_{ij} = \Theta\left(\varepsilon - \left\|\mathbf{x}_i - \mathbf{x}_j\right\|\right), \quad i,j = 1,\ldots,N, \qquad (5)$$

where N is the number of considered state states $\mathbf{x}_i$, along the embedding space trajectory, defined in a discrete sampling time $t_i=i\Delta t$, ε is a threshold distance, $\|\ \|$ indicates a norm (usually the L_2 norm) and Θ (.) represents the Heaviside step function.

The graphical representation of such a matrix, where ones and zeros are represented by black and white dots respectively, is called recurrence plot (RP). In other words, information provided by the RP can be described as follows:

$$R_{ij} = \begin{cases} 1: & \mathbf{x}_i \approx \mathbf{x}_j \\ 0: & \mathbf{x}_i \neq \mathbf{x}_j \end{cases}, \quad i,j = 1,\ldots \qquad (6)$$

where $\mathbf{x}_i \approx \mathbf{x}_j$ indicate the points from the neighbourhood of radius ε (defined according to the applied norm).

In order to calculate the RP representing properties of the considered system, it is necessary to determine the embedding space and define values threshold ε. Embedding parameters include the time delay, τ, to introduce missing independent variables as vectors $\mathbf{x}_i = [x_i(1), x_i(2), x_i(3), \ldots, x_i(m)\ ]$, where $x_i(n) = x_{i-(n-1)\tau}$ with the appropriate embedding dimension, m, where $n \leq m$ [26].

To estimate the smallest sufficient embedding dimension m, the false nearest neighbours' algorithm is frequently used [27]. Appropriate time delay τ can be determined with the use of the auto-correlation or mutual information function [28, 29]. In the original definition of the method, the neighbourhood is specified with the use of the L_2 norm and its radius is chosen in such a way that it contains a fixed amount of states. For such a vicinity, the radius ε_i changes for each x_i ($i = 1, \ldots, N$). This neighbourhood is often denoted as *fixed number of nearest neighbours* (FAN). Other commonly used neighbourhood is that of a fixed radius, firstly used by Zbilut and Weber [30].

During research, carried out over the last few years [15, 31-33], it has been proven that the recurrence-based methods are particularly sensitive to changes in the dynamic behaviour of mechanical systems. In this paper, the methodology based on the recurrence plots (RP) method was applied to analyse regular and chaotic motions of the superelastic shape memory oscillator. In all computations, FAN definition of the neighbourhood was used [25]. The smallest sufficient embedding dimension ($m = 3$) was estimated with the application of the false nearest neighbours' algorithm [25]. Appropriate time delay ($\tau = 11$) was determined by searching for the first minimum of the mutual information function [29], while the phase space of the considered systems has been reconstructed by delay embedding [30, 34]. For the purposes of consistency, for both signals, the same value of threshold ($\varepsilon = 0.09$) was assumed. All the necessary computations were performed with the use of the CRP Toolbox for MATLAB [35].

Recurrence Quantification Analysis (RQA) [14, 30] is a method of nonlinear data analysis which quantifies the number and duration of recurrences of a dynamical system, represented by its state space trajectory. Definitions of the most popular RQA measures, such as: The Recurrence Rate (*RR*), Determinism (*DET*), Laminarity (*LAM*), Averaged diagonal line length (*L*), Trapping Time (*TT*), Longest diagonal line (L_{max}), Longest vertical line (V_{max}), Divergence (*DIV*), Entropy (*ENTR*) can be found in [30,36]. Extended discussion, including recurrence times of the first and second orders as T_1 and T_2, can be found in Marwan et al. 2007. On the other hand, the additional topological measures, as transitivity or causality (TRAN) and clustering coefficient (CC), were given in [37].

The simplest measure of the RQA analysis is the *recurrence rate* (*RR*):

$$RR(\varepsilon) = \frac{1}{N^2}\sum_{i,j=1}^{N} R_{i,j}(\varepsilon), \qquad (7)$$

defined as a ratio of all recurrence states (points) to all possible system states. Therefore, *RR* describes the probability that a state recurs in its ε - neighbourhood in phase space.

DOI: 10.1051/itmconf/20171505013

Definition of another measure – *DET* – is based on the observation that processes with uncorrelated or weakly correlated, stochastic or chaotic behaviour, cause none or very short diagonals, while deterministic processes result in longer diagonals. Therefore, the ratio of recurrence points forming diagonal lines to all recurrence points is treated as a measure for *determinism* (predictability) of the system:

$$DET = \frac{\sum_{l=l\min}^{N} lP(l)}{\sum_{l=1}^{N} lP(l)}, \tag{8}$$

where $l_{\min}$ often is fixed to the shortest line length $l_{\min}=2$ and $P(l)$ denotes a histogram of diagonal lines of the length l:

$$P(\varepsilon,l) = \sum_{i,j=1}^{N}\left(1-R_{i-1,j-1}(\varepsilon)\right)\left(1-R_{i+l,j+l}(\varepsilon)\right)\prod_{k=0}^{l-1} R_{i+k,j+k}(\varepsilon). \tag{9}$$

A diagonal line of length l denotes that a segment of the trajectory is rather close during l time steps to another segment of the trajectory at a different time; thus, these lines are related to the divergence of the trajectory segments. The average diagonal line length L:

$$L = \frac{\sum_{l=l\min}^{N} lP(l)}{\sum_{l=l\min}^{N} P(l)} \tag{10}$$

stands for the average time that two segments of the trajectory are close to each other. This parameter is the longest for periodic behaviour.

Another RQA measure refers to the length $L_{\max}$ of the longest diagonal line found in the RP (excluding diagonal line):

$$L_{\max} = \max\left(\{l_l\}_{l=2}^{Nl}\right), \tag{11}$$

where Nl is the total number of diagonal lines. Depending on the system character divergence of the dynamical system can be associated $1/L_{\max}$ or $1/L_{mean}$, where L_{mean} is an average length of diagonal lines, can be used to distinguish the chaotic and periodic systems. Periodic systems would have long lines and chaotic broad spectrum of line lengths resulting with increase of divergence.

The ratio between the recurrence points forming the vertical structures and all the recurrence points is called *laminarity* (*LAM*). A value of *LAM* decreases if the RP consists of more separated recurrence points than vertical structures:

$$LAM = \frac{\sum_{v=v\min}^{N} vP(v)}{\sum_{v=1}^{N} vP(v)}. \tag{12}$$

Trapping time, *TT*, expresses the average length of vertical structures:

$$TT = \frac{\sum_{v=v\min}^{N} vP(v)}{\sum_{v=v\min}^{N} P(v)}. \tag{13}$$

It can be interpreted as the time during which the system stays in a particular state.

Transitivity, *TRAN*, is a network topological parameter defined on the recurrence matrix, **R**, elements:

$$TRAN = \frac{\sum_{i,j,k=1}^{N} R_{ij}(\varepsilon)R_{jk}(\varepsilon)R_{ki}(\varepsilon)}{\sum_{i,j,k=1}^{N} R_{ij}(\varepsilon)R_{kj}(\varepsilon)}. \tag{14}$$

In general TRAN is a measure which refers to the binary relations between three given points *i*, *j*, and *k*, and it is sensitive to the dynamical system attractor geometry [16, 38]. On the other hand, the maximal length of the vertical structures in the RP, denoted as V_{max}:

$$V_{\max} = \max\left(\{v_l\}_{l=1}^{Nv}\right) \tag{15}$$

is a measure, which is the analogue to the standard RQA measure L_{max} (here, Nv denotes the absolute number of vertical lines).

In contrast to the RQA measures based on diagonal lines, the measures based on vertical lines can find chaos–chaos and chaos–order transitions.

Furthermore, the recurrence times of the first, T_1, and of the second, T_2, type define the average time distances between the recurrence structures. T_1 includes the average horizontal space between subsequent structures, in terms of the white points, while T_2 corresponds to the full periods including white and black points. Both parameters give an idea on system periodicity given in RP by the lengths between (basically diagonal) lines.

4 Results

As discussed in detail in [8], the SMO can undergo chaotic motions. For the purposes of the presented comparison, two specific trajectories are shown: one chaotic and the other – periodic. Both trajectories are obtained after numerical integration of the system for 500 periods by means of a standard 4th order Runge-Kutta algorithm with 4000 steps per periods. The output is then sampled to get time series of 200,000 points. Model parameters are summarized in Table 1 corresponding to devices with a hysteresis loop of medium-high size and low hardening. The latent heat of transformation is low so that thermal phenomena are small, and the resulting conditions are close to the isothermal ones.

Table 1. Model parameters.

q_1	q_2	q_3	λ	L	h	J	B
0.980	1.200	0.980	8.125	0.001	0.080	3.1742	0.030

The two trajectories differ for the features of the two-forcing excitation (Tab. 1). The simulation results using the above presented model are given in Figure 3. Time histories and corresponding spectra are plotted in Figures 3a, c, and 3b, d for periodic and chaotic signals, respectively.

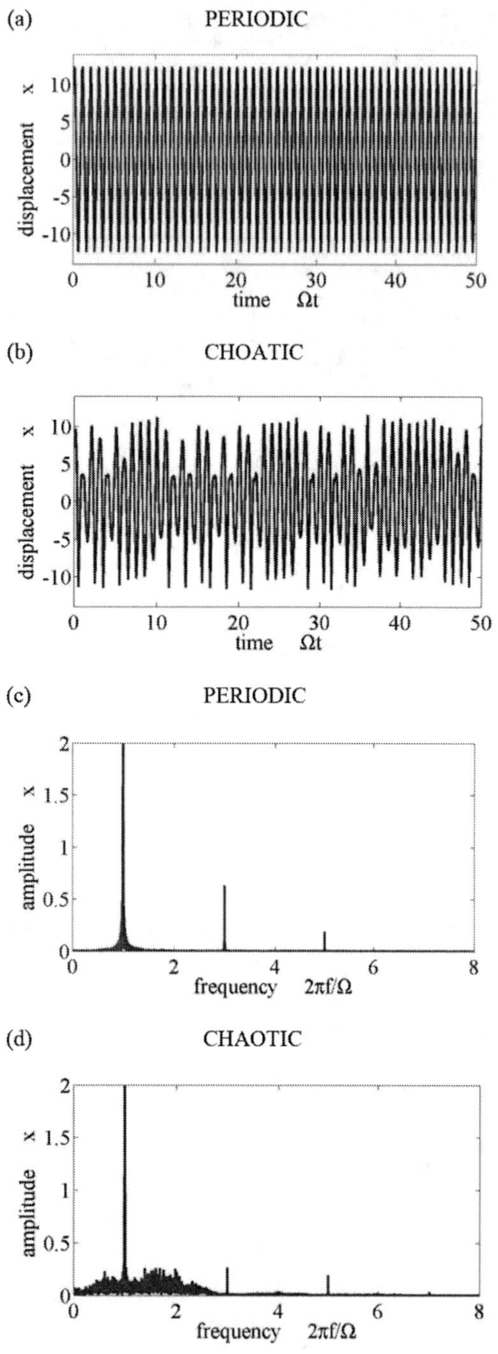

Fig. 3. Time histories and spectra of the considered periodic (a, c) and chaotic (b, d) signals. The sampling was 400 points per excitation period $2\pi/\Omega$.

Table 2. Forcing parameters related to different kinds of response.

	Ω	F
Chaotic	0.227	1.000
Periodic	0.400	1.000

Note that Fourier spectrum of the periodic response is given by a singular spectrum (Figure 3c), while the chaotic response is characterized by a continuous spectrum with additional singular peaks of higher amplitude. Occurrence of higher singular periods (with respect to excitation frequency $\Omega/2\pi$) in the spectrum is a consequence of system nonlinearities (see Eq. 1).

The spectrum criterion has a qualitative character. A standard way to categorize such solution would be using Lyapunov exponent or 0-1 test [12]. In the sequence, the recurrence analysis is performed, showing that it can be used as an alternative way to estimate the character of solutions [14, 15].

The recurrence analysis includes recurrence plots (RPs) and estimation of the recurrence quantifiers (Equations 7-15). The RPs are presented in Figure 4. Note, that periodic response is characterized by the collection equally spaced diagonal straight lines (Figure 3a). On the other hand, the chaotic response is based on the shorter curved lines with evident variable distances between them. This basic qualitative difference included in the RPs is mirrored in the recurrence parameters (Table 3). Namely, values of particular RQA measures, estimated for both cases, represent quantitative information sufficient to decide about the character of system response.

Table 3. RQA measures computed for the considered periodic and chaotic signal for RPs from Figure 4. Assumed parameters: $m = 3$, $\tau = 11$, $\varepsilon = 0.09$, FAN criterion. Calculations were done using the numerical package [35].

State	RR	DET	L_{max}	L_{mean}	$ENTR_L$	LAM
periodic	0.10	0.99	3977	38.02	1.42	1.00
chaotic	0.10	1.00	3977	35.18	4.34	1.00
State	TT	V_{max}	T_1	T_2	$TRAN$	
periodic	23.12	43	10.02	229.82	0.66	
chaotic	33.95	178	10.02	340.74	0.62	

Parameters RR, DET, L_{max}, LAM, T_1 are very similar. This occurs due to the chaotic solution being complemented by some periodic features. These features are related to the coexistence of the broad band of frequency and narrow peaks. This structure is mapped into RP (see Figure 4), where the line layout in Figure 4b follows, in some extend, the line structure of Figure 4b. However, it is possible to observe important deviations into the discontinuous and curved lines. This specific behaviour can show the intermittency on the system [16]. Different recurrence parameters support this possibility.

Firstly, we notice the difference in L_{mean}. As expected, chaotic response shows larger divergence (defined as $1/L_{mean}$). Secondly, entropy $ENTR_L$ is 3 times bigger for chaotic system than regular one. Thirdly, V_{max} is larger for chaotic system (4 times larger); this is due to the appearance of the curved lines characteristic for variable frequency responses. This result, together with the results of the trapping time, *TT*, in much larger for chaotic system support a scenario of system persistence around a specific state mediated by more periodic motion.

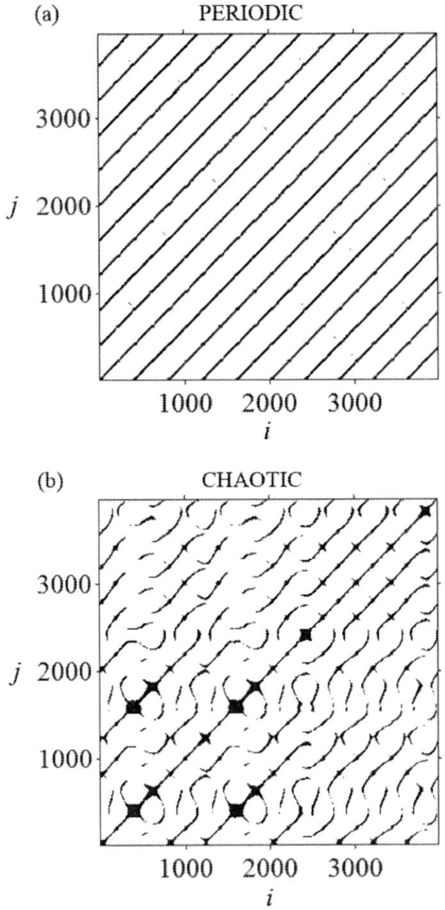

Fig. 4. Recurrence plots computed for periodic (a) and chaotic (b) signals. Assumed parameters: $m = 3$, $\tau = 11$, $\varepsilon = 0.09$, FAN criterion.

Finally, the additional arguments are provided by the T_2 and *TRAN* parameter. Obviously, it can be noticed that the ratio T_1/T_2 indicates on some smearing in response phase (local decelerations or accelerations in the system response). *TRAN* shows the information about the reproduction of binary related between black points on RP. It is obvious, that the complexity of the studied system responses is different.

5 Conclusions

We have studied nonlinear dynamics of a pseudoelastic shape memory alloy oscillator. The hysteretic behaviour of the SMA element is described by a phenomenological model with internal variables. Numerical simulations are performed showing periodic and chaotic responses. Diagnostic analysis of the kind of response is performed considering recurrence plots. Recurrence plots and recurrence quantification analyses are considered. Using various recurrence measures, a successful distinction of regular and chaotic solutions has been obtained using the following parameters: L_{mean}, $ENTR_L$, *TT*, T_2, V_{max}. Results are compared with Fourier spectra presenting coherent results. In general, it is possible to conclude that the recurrence analysis is capable of distinguishing periodic and chaotic responses from short intervals (ten cycles of excitation – see Fig. 4).

The authors would like to acknowledge the support of Brazilian Research Councils CNPq, CAPES and FAPERJ.

References

1. M. Schwartz, *Encyclopedia of Smart Materials*, vols. 1,2 (Wiley and Sons, New York, 2002)
2. D. C. Lagoudas, *Shape memory alloys: modeling and engineering applications* (Springer, Berlin, 2008)
3. M. A. Savi, P. M. C. L. Pacheco, Int. J. Bif. Chaos **12**, 645 (2002)
4. L. G. Machado, M. A. Savi, P. M. C. L. Pacheco, Shock Vibr., **11**, 67 (2004)
5. B. C. dos Santos, M. A. Savi, , Chaos Solit. Fract. **40**, 197 (2009)
6. D. Sado, M. Pietrzakowski, Int. J. Non-Lin. Mech. **45**, 859 (2010)
7. V. Piccirillo, J. M. Balthazar, B. R. Pontes jr, Nonlin. Dyn. **59**, 733-746. (2010)
8. D. Bernardini, G. Rega Math. Computer Model. Dyn. Syst. **11**, 291 (2005)
9. D. Bernardini, G. Rega, Int. J. Non-Linear Mech. **45**, 933 (2010)
10. D. Bernardini, G. Rega, Int. J. Bifur. Chaos **21**, 2769 (2011)
11. D. Bernardini, G. Rega, Int. J. Bifur. Chaos **21**, 2783 (2011)
12. G. Litak, D. Bernardini, A. Syta, G. Rega, A. Rysak, Eur. Phys. J.-Spec. Topics **222**, 1637 (2013)
13. D. Bernardini, G. Rega, G. Litak, A. Syta, Proc. Inst. Mech. Engineers Part K-J. Multi-Body Dyn. **227**, 17 (2013)
14. N. Marwan, M. C. Romano, M. Thiel, J. Kurths, Phys. Rep. **438**, 237 (2007)
15. G. Litak, M. Wiercigroch, B. W. Horton, X. Xu, ZAMM **90** 33 (2010)
16. G. Gorski, G. Litak, R. Mosdorf, A. Rysak, Eur. Phys. J. B **88**, 239(2015)
17. P. Harris, G. Litak, J. Iwaniec, C.R. Bowen, Appl. Mechan. Mater. **849**, 95 (2016)
18. J.-P. Eckmann, S.O. Kamphorst, D. Ruelle, Europhys. Lett. **5**: 973 (1987)
19. C. L. Webber jr., J.P. Zbilut, J. Appl. Physiol. **76**, 965 (1994)
20. M.C. Casdagli, Physica D **108**, 12 (1997)
21. J. Hunicz, M. Geca, A. Rysak, G. Litak, P. Kordos, J. Vibroengineer. **15**, 1093 (2013).

22. D. Bernardini, T. J. Pence, Int. J. Non-Lin. Mech. **37**, 1299 (2002)
23. D. Bernardini, T. J. Pence, Meccanica **40**, 339 (2005)
24. J.C. Sprott, *Chaos and Time-Series Analysis*, (Oxford University Press, Oxford 2003)
25. N. Marwan, *Encounters with neighbours—current developments of concepts based on recurrence plots and their applications*. (PhD Thesis, University of Potsdam, Germany 2003)
26. F. Takens *Detecting strange attractors in turbulence*, in: Lecture Notes in Mathematics, **898**, 366–38, (Springer, Heidelberg 1981)
27. A. M. Fraser, H. L., Phys. Rev. A **33**, 1134 (1986)
28. R. Hegger, H. Kantz, T. Schreiber, Chaos **9**, 413 (1999)
29. H. Kantz, T. Schreiber, *Nonlinear Time Series Analysis*, (Cambridge, University Press 1997)
30. J.P. Zbilut, C.L. Weber, Phys. Letter A **171**, 199 (1992)
31 J. Iwaniec, T. Uhl, W. J. Staszewski, A. Klepka, Nonlin. Dyn. **70**, 125 (2012)
32. J. Iwaniec, Int. J. Struct. Stabil. Dyn. 13, 1340008 (2013)
33. J. Iwaniec, P. Kurowski, J. Vibrat. Control **23**, 1133 (2017)
34. L. Cao, Physica D **110**, 43 (1997)
35. N. Marwan, *Cross Recurrence Plot Toolbox for Matlab, Reference Manual*, Version 5.15, Release 28.6. Available at: http://tocsy.pik-potsdam.de (2010)
36. E. J. Ngamga, A. Nandi, R. Ramaswamy, M.C. Romano, M. Thiel, J. Kurths, PRAMANA J. Phys. **70**, 1039 (2008)
37. R.V. Donner, J. Heitzig, J.F. Donges, Y. Zou, N. Marwan, J. Kurths, Eur. Phys. J. B **84**, 653 (2011)
38. G. Gorski, G. Litak, R. Mosdorf, A. Rysak, Eur. Phys. J. Plus **131**, 111 (2016)

 DOI: 10.1051/itmconf/20171506001

An experimental study of exploitation of materials used for prosthetic temporary restorations

Leszek Szalewski[1,*], *Elżbieta* Celej-Piszcz[1], *Małgorzata* Stodółkiewicz[2], *Radosław* Cechowicz[3], *Marcin* Bogucki[3], *Janusz* Borowicz[1], and *Jacek* Szkutnik[1]

[1]Medical University of Lublin, Department of Dental Prosthetics, Karmelicka 7, 20-081 Lublin, Poland
[2]Medical University of Lublin, Department of Functional Masticatory Disorders, Karmelicka 7, 20-081 Lublin, Poland
[3]Lublin University of Technology, Mechanical Engineering Faculty, Department of Automation, Nadbystrzycka 36, 20-618 Lublin, Poland

Abstract. Temporary restorations more often play their role for a definitely longer period of time, than the time necessary for implementation of final prosthetic work. Therefore, they are subjected to adverse effects of chemical, physical and thermal factors in a patient's oral cavity. Since loss of temporary prosthetic work can have negative consequences for an entire treatment process, it is important to identify a potential risk of damaging temporary restorations by factors derived from the diet of patients. The aim of this study was to evaluate the influence of components of the beverages, such as coffee and tea, consumed daily by the patients, on the mechanical properties of materials used in temporary restorations. The experimental design applied, as well as gained results, were justified by appropriate, chosen statistical methods. The conducted research demonstrated the change of mechanical properties of studied materials for their flexural strength, as well as microhardness. The tendency of changes in both result variables, compared with the samples kept in neutral environment – distilled water, indicates the decrease in mechanical strength and microhardness of materials conditioned in coffee and tea. It can be claimed that the components of beverages present in an everyday diet of patients have a considerably adverse effect on the quality of materials used for provisional crowns and bridges.

1 Background

New possibilities of prosthetic treatment sometimes require fixed temporary restorations to be used for longer than the time needed to prepare a final restoration.

During treatment, patients expect temporary restorations to be esthetic and comfortable thus, the requirements for temporary restorations are almost identical with those set up for final restorations [1]. Additionally, apart from being esthetic, temporary restorations must meet several conditions – protection of a living dental pulp against chemical, thermal and mechanical factors, assurance of stable occlusion, continuity of functions of stomatognathic system and support for soft tissues [2].

Currently, on the market, there are many materials for temporary prosthetic restorations. These are mainly materials based on acrylic mass or composite materials, and they can meet requirements for temporary restorations to a different extent.

A prosthesis in the oral cavity is constantly prone to mechanical, chemical and thermal factors deriving from the diet of patients. These factors can considerably influence mechanical and esthetic properties of prepared crowns and bridges [3]. A human should drink approximately 2-3 liters of liquids. The type of liquid depends on the age and individual patient's preferences. The adults, who constitute the majority of patients treated prosthetically, most often drink coffee and tea. These beverages can have a crucial impact on mechanical and esthetic properties of temporary and final restorations.

The materials for temporary restorations can change their mechanical and esthetic properties. Diaz-Arnold et al. found that the microhardness of 5 different temporary materials diminishes with time, regardless of chemical or thermal factors derived from the patient's diet [4]. In their studies, they kept samples in artificial saliva for 14 days and after that time they compared microhardness with primary samples measured after 24 hours since preparation. On the other hand, Balkenhol et al. while comparing the flexural strength, depending on the time that passed from polymerization, proved that for the first few hours materials showed lower flexural strength [3]. The effect can be explained by the fact that materials prepared on the basis of acrylic products need more time to undergo full polymerization. When comparing materials based on composites and acrylic resins, Balkenhol suggested the use of composite materials, which are characterized by higher values of flexural strength directly after their preparation [5]. The analysis of literature studies, related to the effect of dietary factors on the properties of materials used for temporary restorations, indicates several articles concerning the change of color of temporary restorations under the

* Corresponding author: leszek.szalewski@umlub.pl

influence of exogenous coloring agents [1,6,7]. Yet, there are few studies determining the influence of dietary factors on the mechanical properties of temporary materials. Akova et al. evaluated an influence of heptane, ethanol and citric acid on the mechanical properties of temporary material samples [8]. When analyzing the subject literature, publications describing the influence of beverages consumed by patients with temporal prosthesis, such as coffee, tea, fruit juices and alike, have not been found. Thus, it is highly justified to conduct such studies. An experimental design and statistical analysis of its outcomes could validate the expected premises about detrimental influences of the aforementioned factors. Moreover, results gained from this study might contribute to defining the recommendations for patients using temporary restorations, especially for a long period of time.

2 Objectives

The aim of this study was to evaluate the influence of components of the beverages, such as coffee and tea, consumed daily by patients on the mechanical properties of materials used in temporary restorations. In the research, an experimental design of univariate analysis was applied. The studied factors were compared with the influence of substances which are contained in coffee and tea; yet, each of the factors was studied individually.

The effects of interactions (in other words: cross-validation) of simultaneous interaction of two factors were not included in the study. The impact of the studied factors on the temporary restoration material was examined for three agreed systems. The first experimental system was treated as so-called control group and it corresponded to the samples of conditioned materials in a neutral environment. The rest of two experimental systems were related to the conditioning of samples in two water solutions respectively: coffee and tea. The studies tried to prove the effect of change in the mechanical properties of materials, namely flexural strength and microhardness under the influence of substances from the two solutions mentioned above.

Altogether, there were six experiments conducted during the study – two for each three tested materials used for temporary restorations.

3 Material and methods

Three materials for temporary restorations available on the market were used in the study: Structure 2 (VOCO), Unifast TRAD (GC) and Revotek LC (GC). 24 samples in the shape of a cuboid measuring 2 x 2x 25 mm and 30 disc-shaped samples measuring 5 x 2 mm were made from each of the materials. The samples were prepared in specially designed metal matrices. The metal form was layered thinly with vaseline, to avoid the material sticking to the matrix. The material was condensed with hand pluggers. Each sample, requiring polymerization with the use of light, was treated with radiation from LED Clear Blue lamp of 1200 mW/cm^2. Both sides of every sample were radiated for 20 seconds. The samples were polymerized via polyethylene foil, to eliminate the oxygen inhibition layer. After hardening, the samples were cautiously taken out of the matrix, so that the edges stayed intact. Possible uneven parts were polished with the Sof-lex (3M ESPE) disc. Then, the samples were thoroughly examined to eliminate samples with air bubbles or other inaccuracies that might have emerged during condensation and polymerization. For the first 24 hours, the samples were kept in distilled water in 370C. After that time, the samples were randomly divided into three subgroups and conditioned in fluids: coffee (Nescafe), tea (Lipton) and distilled water for 7 days. The coffee solution was prepared by pouring 2 spoons (10 g) of instant coffee Nescafe Classic with 200 ml of boiling water. On the other hand, the tea solution was made from 1 tea bag of Lipton Yellow Label Tea poured over with 200 ml of boiling water and left for 5 minutes. Next, the samples were rinsed with distilled water and dried. Then, they underwent the flexural and Vickers microhardness tests.

3.1 Flexural testing

Flexural testing was conducted with the use of materials testing machine (Zwick Roell) in the support spacing 20 mm and the speed of head 0,75 mm/min, with the initial load of 1 N. The width and thickness of every sample was measured with the use of a digital caliper to obtain possibly the most accurate results. The test finished when a sample was broken.

The tension determining the flexural strength was calculated on the basis of the following formula:

$$\sigma = 3F_m I /(2bh^2) \qquad (1)$$

where:

F_m – the maximum load applied at the moment of a sample braking,
I – the support spacing 10 mm,
b – the width of a sample,
h – the thickness of a sample.

3.2 Vickers microhardness study

In case of measuring microhardness of provisional restorative materials, the Vickers test is better than other methods of examining hardness, due to its low sensitivity to the material surface imperfections after polymerization. The microhardness testing was conducted with the Vickers method with the use of Micro Hardness Tester mhp 100 and the load of 20 g for 10 seconds.

The length of the diagonal imprint on the surface of a studied sample was measured by examining sample surface in the optical microscope Neophot 2. Each sample had 3 imprints done in different areas of its surface. While calculating, it was considered that the microhardness is proportional to the quotient of the load and the imprint surface area, which is assumed to have a shape of a regular pyramid with a square base and the same apex angle as its indenter.

3.3 Statistical analysis of the study results

The series of result variables measurements underwent the preliminary analysis with the exclusion of extreme values, which were observed in the case of flexural testing. The extreme values occurred due to the unrepeatable sample setting in the material testing machine and the impurity (defective structure) of a sample material. The measurement of result variables compared to the rest of experimental samples, subjected to the same environmental impact, demonstrated only random changes in an established mean value, which indicated maintaining repeatable study conditions.

The conducted research was analyzed statistically with the use of the following experimental data analysis [9]:

- Parametric variance analysis, when the results of trials observed in given experimental systems were characterized by homogenous variance, and when the distribution of the rest of the variance model was not different from the normal distribution, which was verified by an appropriate statistical testing;
- Non-parametric type of variance analysis – the rank-based Kruskal-Wallis test when the above-mentioned assumptions were not met;
- Post hoc comparison tests of the measures achieved in given experimental systems, which aim was to indicate reciprocal, statistically significant differentiation of the results in the range of the used restoration material. In the case of the parametric analysis, the HSD Turkey test was applied, otherwise, an appropriate type of non-parametric test was used.

The aim of the ANOVA analysis was to demonstrate the negative impact of dietary factors contained in the studied beverages. Due to the conciseness of the research paper, only the most significant results and comments of statistical analyses were enclosed in the publication – Box-Plot figures summarizing descriptive statistics of measurements achieved in given experimental systems (mean value with the standard error – illustrated with so-called *figure frame* and the highest and lowest value of measurement – so-called *whiskers*). In every statistical test, the probability $\alpha = 0.05$ was assumed as a threshold of the statistical significance.

4 Results

The measurements of the result variables: microhardness and flexural strength have been presented in Figs. 1 and 2.

The conducted research demonstrated the change of mechanical properties of the studied materials for their flexural strength, as well as microhardness. The tendency of changes in both result variables, compared with the samples kept in neutral environment – distilled water, indicates the decrease in mechanical strength and microhardness of materials conditioned in coffee and tea. The discussed changes are visible in Box-Plot figures (1 and 2), which was formally proven by the analysis of variation results and the results of post-hoc comparisons. In the case of flexural strength of materials, none of the experimental trials conducted for control group samples demonstrated variation in the level of mean values which was approximately 66 MPa.

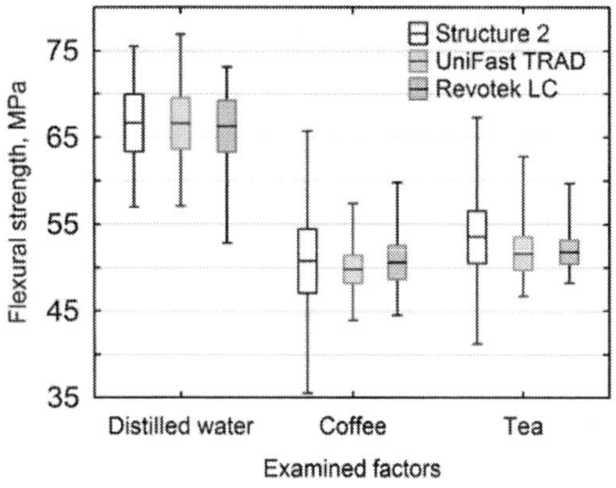

Fig. 1. The summary of flexural strength measurement results achieved in given experimental sets

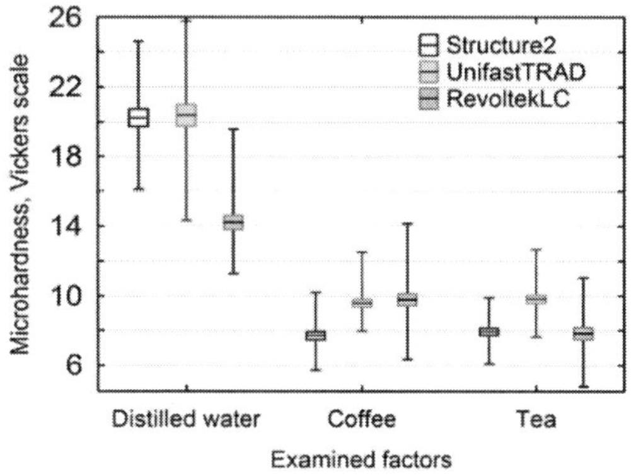

Fig. 2. The summary of microhardness measurement results achieved in given experimental sets

Similarly, statistically significant differences in the measurements of the flexural strength of the samples made from various materials conditioned in coffee or tea solution were not observed. This result proves a very similar influence of the studied solutions on the temporary restorations. The post-hoc tests showed that the influence of coffee solution components on the samples made from various materials was not considerably different from the impact of the tea solution ingredients, in a statistical and practical sense. The above-mentioned observations concern also the second group of studied variables – microhardness.

5 Discussion

It should be noted, that the 3-point bending test and Vickers microhardness test were used in the research. Flexural strength measured with 3-point bending test reflected processes occurring in the oral cavity, where the prosthetic bridge is supported by abutment teeth on both edges, and the destructive force occurs in the middle of restoration, in the span of a bridge.

In the studies, the samples were immersed in the tested fluids for 7 days, which seems to be a long period.

Despite the fact, that restorations in an oral cavity are constantly cleaned by saliva or hygienic procedures, the food can remain in unavailable places (such as overhangs of temporary prosthetic works or under the spans of the bridge) for a long time. The adverse factors derived from the diet of patients can also be absorbed by plaque and tartar, which additionally extends the contact of chemical factors with teeth and restorations.

All the materials tested in the control samples were characterized by the similar average flexural strength (66 MPa) and in the case of microhardness, the lowest value was achieved by Revotek material. This may be due to a lower filler content (15-35%), as compared to conventional composite materials with a filler content of 85% [8].

The microhardness of the samples after immersing in beverages decreased by approx. 50-60% for Structure 2 and Unifast TRAD materials, and approx. 35% for Revotek material. The observed effects can be explained by the lack of material surface polishing, which increases the value of microhardness of the material and resistance to chemical agents [7].

Similar results of the negative impact of food onto the materials used for prosthetic temporaries were achieved by Akova et al. [8], in the case of heptane and 75% ethanol use. Just in the case of 0,02 citric acid, the increase in Knoop microhardness values was observed. Analogous results were depicted in the studies concerned with the composite materials used for direct teeth restoration [10,11].

There are few papers describing the impact of diet on temporary restorations in the oral cavity; as it is commonly believed, because of a short period of use. However, due to diagnostic reasons, temporary restorations more and more often must remain in the oral cavity for a longer period (even for a few months). Hence, it seems important to examine what impact the chemical and/or thermal factors derived from the diet of patients can have on this kind of restorations.

6 Conclusions

Taking the results of the discussed studies into account, it can be claimed that the components of beverages present in the everyday diet of patients have a considerably adverse effect on the quality of materials used for provisional crowns and bridges. Thus, dietary recommendations for patients using temporary restorations for a longer period and the diligence of their preparation should be emphasized. Provisional restorations, as well as fixed restorations, cannot encourage the stagnation of food and should be available for self-cleaning and hygienic procedures. The durability of temporaries is also affected by the choice of material with suitable properties, which should be the subject of further experimental research.

References

1. D.K. Prasad, H. Alva, M. Shetty: J. Ind. Prost. Soc., **14**, 1 (2014)
2. S. Rosenstiel, M.F. Land, J. Fujimoto: *Contemporary Fixed Prosthodontics* (Elsevier Health Sciences, 1141, 2006)
3. M. Balkenhol, P. Ferger, M. Mautner, B. Wostmann, Dent. Mater. **23**, 12 (2007)
4. A.M. Diaz-Arnold, J.T. Dunne, A.H. Jones, J. Prosthet. Dent. **82**, 5 (1999)
5. M. Balkenhol, M.C. Mautner, P. Ferger, B. Wostmann, J Dent. **36**, 1 (2008)
6. M.D. Gross, J.B. Moser, J. Oral Rehabil. **4**, 4 (1977)
7. Y.K. Lee, J.M. Powers, J. Biomed. Mater. Res. Part B Appl. Biomater. **73**, 2 (2005)
8. T. Akova, A. Ozkomur, H. Uysal, Dent Mater, **22**, 12 (2006)
9. D. Montgomery, *Design and Analysis of Experiments*, (John Wiley and Sons, 2004)
10. H. Sazak-Öveçoğlu, M. Günday, M.L. Öveçoğlu, F. Özer, B. Tarçın,. K. Eng. Mater. **264-268** (2004)
11. A.U. Yap, J.S. Low, L.F. Ong, Oper. Dent. **25**, 3 (2000)

ITM Web of Conferences **15**, 06002 (2017)
CMES'17
DOI: 10.1051/itmconf/20171506002

Analysis of fracture roughness parameters of S355J2 steel and EN AW-2017A-T4 aluminium alloy

Wojciech Macek[1,*], and *Tomasz* Wołczański[1]

[1]Opole University of Technology, Faculty of Production Engineering and Logistics, Sosnkowskiego 31, 45-272 Opole, Poland

Abstract. This article presents an analysis of fracture surface for steel S355J2 and EN AW-2017A-T4 aluminium subjected to fatigue bending, torsion, and bending with torsion. Fracture surfaces were observed using the focus variation microscope, which allows for the acquisition of data sets with large depth of focus. The authors focus on presenting the features of profile roughness parameters on the example of differences between fracture propagation and rupture areas. The researchers are looking for a correlation or ratio between different profile roughness parameters, especially arithmetical mean deviations of the roughness profile Ra and maximum height of the roughness profile Rz, for this case.

1 Introduction

Fracture topography [1] is one of the basic investigations aimed at determining the cause of the damage. It allows determining what kind of the loading the material was subjected. Several typical macroscopic patterns of fatigue damage can be distinguished. Among them there are functions of type and magnitude of loading. The surface analysis reveals localization of initiation and crack path propagation.

Nowadays surface roughness parameters [2] are defined in a handful of international standards [3, 4]. There are more than a hundred parameters describing the surface topography that we divide into three groups, which are defined as amplitude parameters, spacing parameters, and hybrid parameters [5]. The two most popular surface roughness parameters are arithmetical mean deviation of the roughness profile Ra (1) and maximum height of the roughness profile Rz (2). Fig. 1 shows graphically how the Ra and Rz parameters are calculated.

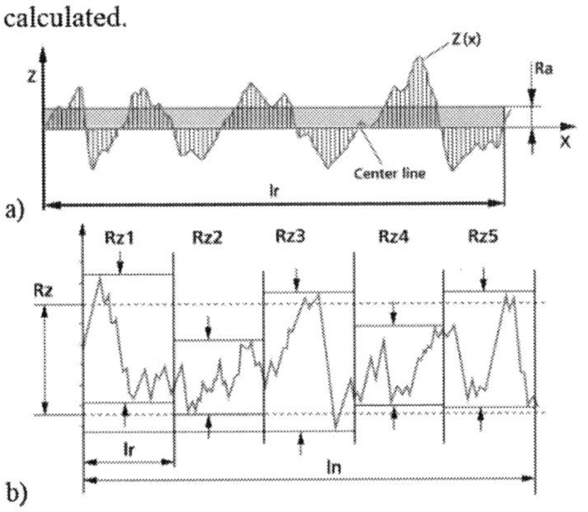

Fig. 1. The surface profiles and the method for calculating parameters a) Ra, b) Rz.

$$Ra = \frac{1}{lr}\int_0^{lr}|Z(x)|dx \quad (1)$$

$$Rz = \frac{Rz1 + Rz2 + Rz3 + Rz4 + Rz5}{5} \quad (2)$$

Ra averages all peaks and valleys of the roughness profile and then neutralizes the few outlying points so that the extreme points have no significant impact on the final results. Rz averages only the five highest peaks and the five deepest valleys, therefore extremes have a much greater influence on the final value.

The results of fracture topography studies, and in particular the roughness profile parameters Ra and Rz, were analysed in this paper.

2 Materials and methods

The following section gives a short description of two kinds of fatigue tests for two materials, both in bending, torsion, and bending with torsion loadings. The method of measuring fracture surface topography is also presented.

2.1 Fatigue tests

2.1.1 Steel S355J2 subjected to non-proportional bending with torsion

Table 1 shows the mechanical properties of the tested steel. Fatigue tests on hourglass specimens, as shown in Fig. 2, with a minimum diameter d = 8 mm, made of S355J2 steel, were performed on a MZGS-200PLZ test stand [6, 7]. In order to distinguish the type of loading

* Corresponding author: w.macek@po.opole.pl

 DOI: 10.1051/itmconf/20171506002

stress ratio $r = \tau_{max} / (\sigma_{max} + \tau_{max})$ was employed, amounting from 0 to 0.55.
Mean stress effect did not occur and the stress ratio in these studies were R = 0.

Table 1. The mechanical properties of the S355J2 steel.

σy (MPa)	σu (MPa)	E (GPa)	ν (-)
357	535	210	0.30

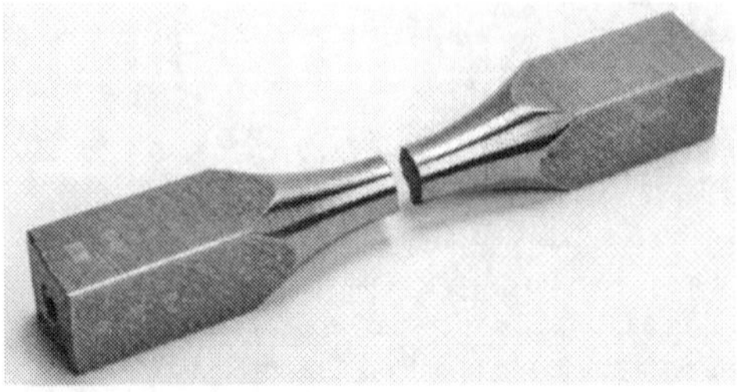

Fig. 2. The exemplary hourglass specimen after fatigue test.

2.1.2 EN AW-2017A-T4 aluminium alloy subjected to proportional bending with torsion

This material was characterized by the mechanical properties in table 2.

Table 2. The mechanical properties of the EN AW-2017A-T4 aluminium alloy.

σ_y (MPa)	σ_u (MPa)	E (GPa)	ν (-)
382	480	72	0.32

The specimens with rectangular cross-section dimensions 8x10 mm were tested. The specimens had an external, unilateral, sharp and blunt one-sided notches, which radius was ρ = 0.2 mm, 5 mm, 10 mm and 22.5 mm. The shape and dimensions of the specimen EN AW-2017A-T4 aluminium alloy for the smallest and largest notch radius, respectively a) ρ = 0.2 mm and b) ρ = 22.5 mm, are shown in Fig. 3.

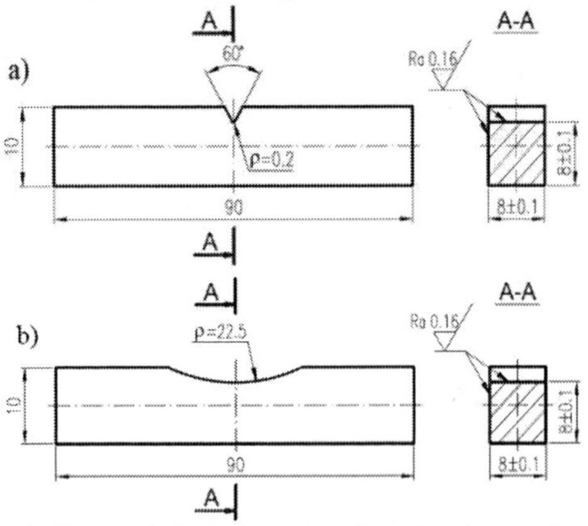

Fig. 3. Shape and dimensions of specimen for: a) ρ = 0.2 mm, b) ρ = 22.5 mm.

Fatigue tests were performed on a MZGS-100 test stand [8]. In order to distinguish the type of loading stress ratio $r = \tau_{max} / (\sigma_{max} + \tau_{max})$ was employed, amounting from 0 to 1.
The stress ratios for this research were R = -1, - 0.5, 0.

2.2 Fracture study

The study of fracture surfaces was carried out by employing the InfiniteFocus® IF G4 focus variation microscope, produced by Alicona Imaging, shown in Fig. 4. The lateral and vertical travel range (x-y-z axes) of the instrument was 100×100×100 mm^3. The measurement device was equipped with a motorized nosepiece using five dedicated microscopic objective lenses from 2.5× to 100× magnification.

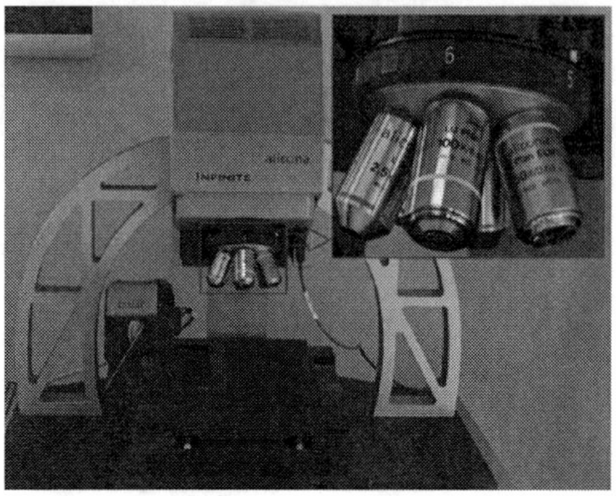

Fig. 4. Focus variation microscope InfiniteFocus® IF G4 produced by Alicona Imaging.

The InfiniteFocus® IF G4 focus variation [5, 9] microscope was supported by IF-MeasureSuit 5.1 software produced by Alicona Imaging and by advanced MountainsMap® Universal 7.4 software provided by Digital Surf.

3 Results and discussion

This section includes visualization of the fracture geometry along with an exemplary fracture profiles, and the presentation of roughness Ra and Rz results. The results were used for a brief analysis.

3.1 Fracture geometry

Fig. 5 demonstrates exemplary scanned fractures with marked propagation and rupture area, for 10x magnifications. All exemplified specimens shown in Fig. 4 were made from EN AW-2017A-T4 aluminium alloy and have the same notch radius ρ = 0.2 mm.

Fig. 6 shows exemplary 3D views of the surface topography of propagation and rupture area, for the 100x magnifications.

DOI: 10.1051/itmconf/20171506002

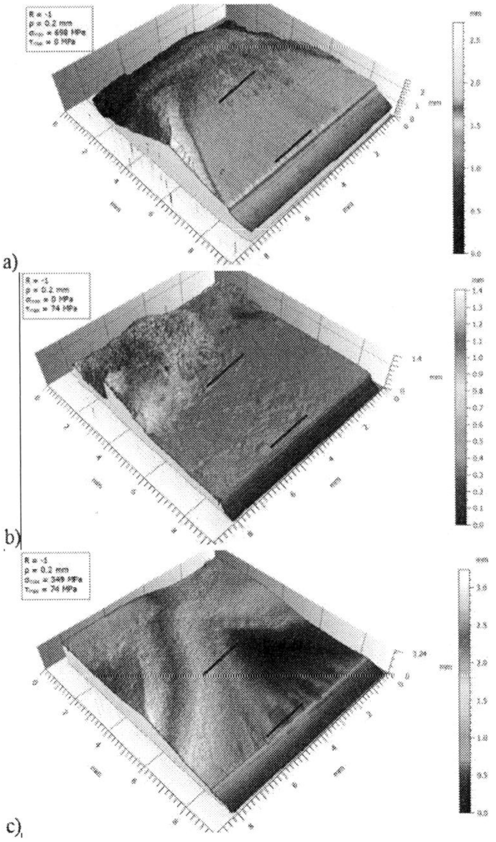

Fig. 5. EN AW-2017A-T4 specimens with marked crack propagation and rupture roughness profiles, for exemplary specimens loadings subjected to a) bending, b) torsion, c) bending with torsion.

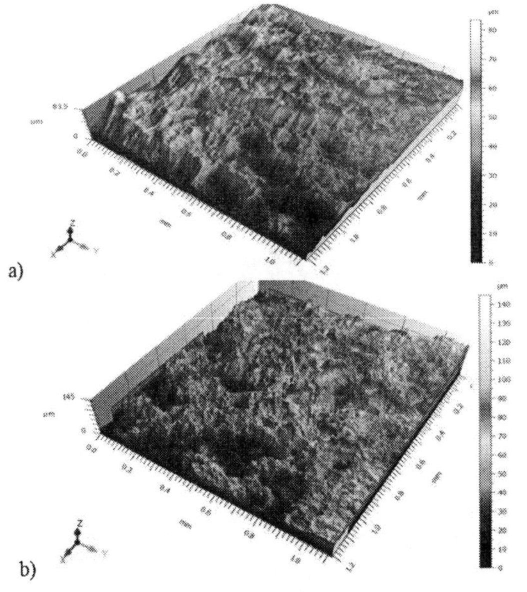

Fig. 6. Roughness surface steel S355J2 specimen subjected to combined bending-torsion loading, with magnitude 100x, for a) the propagation area, b) the rupture area.

Fig. 7 demonstrates exemplary fracture roughness profile a) propagation and b) rupture area, for EN AW-2017A-T4 specimen subjected to bending loading and r=0.

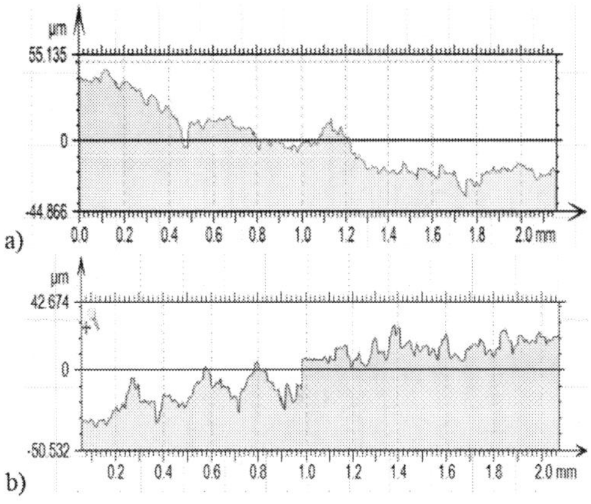

Fig. 7. The profile diagram of EN AW-2017A-T4 specimen subjected to bending loading, r=0, for a) the propagation area, b) the rupture area.

3.2 Roughness parameters analysis

Propagation and rupture profiles were measured for representative 2 mm profile length as marked in Fig. 5. For the sake of great impact of filter waviness λc, on roughness measurements [3, 4], constant value λc = 250 µm was used.

Rz and Ra roughness parameters were used for the analysis in the areas of propagation and rupture. The Rz/Ra factor relationship to r parameter was also checked for propagation and rupture area.

Fig. 8 shows the relationship between Rz and Ra of propagation area for both EN AW-2017A-T4 and S355J2 materials.

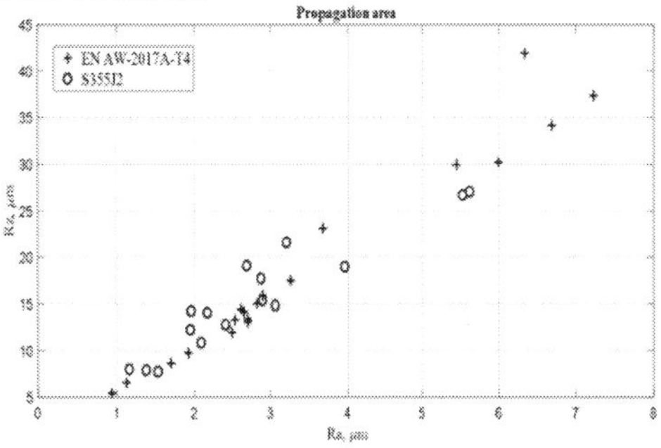

Fig. 8. The relationship between Rz and Ra of propagation area.

The mean value Ra for propagation amounted to Ra=3.46µm for EN AW-2017A-T4, and Ra=2.79µm for S355J2. On the other hand, respectively Rz=18.71µm and Rz= 15.53µm.

The dependence of Rz to Ra on the rupture area is shown in Fig. 9.

DOI: 10.1051/itmconf/20171506002

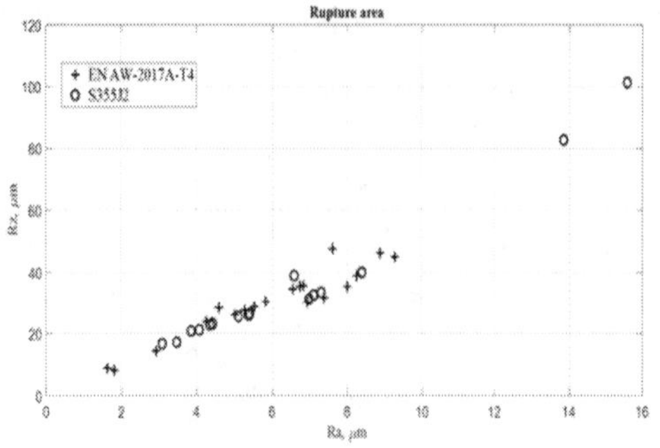

Fig. 9. The relationship between Rz and Ra of rupture area.

For rupture area, the mean value of roughness, is for EN AW-2017A-T4 Ra=5.96µm, Rz=30.40µm and by analogy for S355J2 Ra=6.56µm, Rz=34.93µm.

Comparing data from the graphs in Figures 8 and 9, we can infer that regardless of the method of loading and the type of notch, on average for EN AW-2017A-T4 the tested specimens Ra is 1.88 times bigger for rupture than for propagation area. Similarly Rz, for the same material, is 1.80 times bigger. While for S355J2 all tested specimens Ra and Rz are respectively 2.64 and 2.37 times bigger for rupture than for propagation area.

Fig. 10 and Fig. 11 show the relationship of Rz/Ra to r parameter respectively for propagation area and the rupture area, in a comprehensive overview, for all types of fatigue tests being analysed.

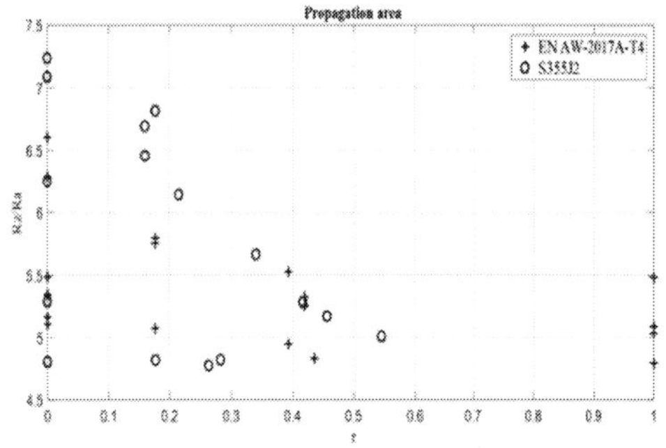

Fig. 10. The Rz/Ra factor relationship to r parameter, for initiation area.

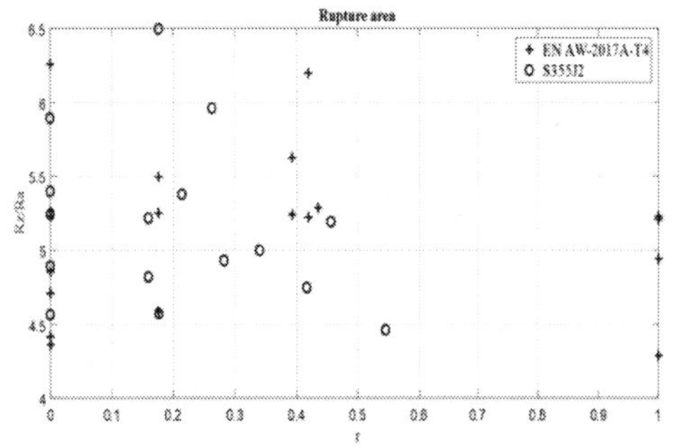

Fig. 11. The Rz/Ra factor relationship to r parameter, for rupture area.

Independent on stress ratio R, notch radiusρ, loading ratio r and form of fatigue, the following information was extracted:

- for propagation areas of S355J2 specimens, the mean value amounts to Rz/Ra=5.77,
- for rupture areas of S355J2 specimens, the mean value is Rz/Ra=5.17,
- for propagation areas of EN AW-2017A-T4, the mean value reaches Rz/Ra=5.38,
- for rupture areas of EN AW-2017A-T4, the mean value totals Rz/Ra=5.14.

The increase in the share of torsional loading components increases the roughness of the fracture surface, however, there was no similar dependence between Rz/Ra and r.

4 Conclusions

Different fatigue tests produce different surface characteristics. On the above experimental results, it can be stated that both roughness parameters Ra and Rz grow along with the increase in the proportion of torsional loadings. For steel tested the differences between the rupture and propagation areas were higher for Ra (1.88) rather than Rz (1.80), when for aluminium alloy, the results were 2.64 and 2.37 respectively. Rz/Ra ratio amounted from 4.29, for AW-2017A-T4 in rupture area to 7.23, for S355J2 in propagation area. Rz averages the highest peaks and the deepest valleys, therefore extremes have a much greater influence on the final value, than Ra parameter.

References

1. K. Slámečka, J. Pokluda, P. Ponížil, S. Major, P. Šandera, Eng. Fract. Mech., **75**(3-4), 760-767(2008)
2. E.S. Gadelmawla, M.M. Koura, T.M. Maksoud, I.M. Elewa, & H.H. Soliman, J. Mater. Process. Tech., **123**(1), 133-145 (2002)
3. ISO 25178-2:2012 - *Geometrical product specifications (GPS) — Surface texture: Areal — Part 2: Terms, definitions and surface texture parameters.*
4. ASME B46.1-2009, *Surface Texture, Surface Roughness, Waviness and Lay.*
5. Leach R. (Ed.): *Optical Measurement of Surface Topography,* Springer-Verlag Berlin Heidelberg (2011)
6. Z. Marciniak, D. Rozumek, E. Macha, Int. J. of Fatigue, **58**, 84-93 (2014)
7. W. Macek, E. Macha, Arch. Mech. Eng., **62**(1) 85-100 (2015)
8. S. Faszynka, D. Rozumek, Solid State Phenomena, **250**, 10-15, 2016,
9. N.I. Nikolaev, J.N. Petzing, J.M. Coupland, Appl. Optics., **55**(13), 3555-3565 (2016)

DOI: 10.1051/itmconf/20171506003

Application of computer image analysis software for determining incubation period of cavitation erosion – preliminary results

Mirosław Szala[1,*]

[1]Lublin University of Technology, Mechanical Engineering Faculty, Department of Materials Engineering, Nadbystrzycka 36, 20-618 Lublin, Poland

Abstract. The paper discusses the application of image analysis software ImagePro Plus and MetIlo for assessing cavitation erosion wear. The investigation of cavitation damage is performed on a vibratory test rig in compliance with the ASTM G32 standard using the stationary specimen method. Low-alloy steel grade 34CrNiMo6 is used as the test material. Cavitation wear is evaluated by gravimetric and roughness measurements, microscopic observations and computer image analysis in specified exposure times. The computer image analysis of selected surface areas during a predetermined exposure to cavitation time is performed using images captured with a metallographic microscope. Based on the results of scanning electron and stereoscope microscopic observations, the cavitation worn surfaces are qualitatively described. The relations between surface topography, gravimetric measurements and the microscopic results are discussed. The findings obtained by gravimetric and roughness measurements, image processing and microscopic observations are complementary. The results prove the suitability of image analysis for investigating incubation period of cavitation erosion. The quantification of cavitation erosion damage indicates that the incubation period of cavitation erosion of the tested steel lasts for 20 minutes. The results demonstrate that cavitation-worn surfaces at the initial stage of cavitation erosion can be examined using both ImagePro Plus and MetIlo.

1 Introduction

Cavitation consists in the formation and subsequent implosion of vapour bubbles due to local pressure fluctuations in a liquid. Bubbles start to grow even at an almost constant temperature in the vicinity of a low-pressure region, below the saturated vapour pressure of the liquid. When subjected to higher pressure, the bubbles implode.

In engineering systems, two types of cavitation usually occur: hydrodynamic and vibratory. These phenomena have effect on the performance, service and operation time of different types of fluid machinery such as pump systems [1], propellers [2], processing equipment [3], diesel engines [4] and even rocket engines [5]. Even though, cavitation can be beneficial in engineering as shown in the work [6], where ultrasonic processing of uranium desorption is studied, it also causes serious material damage. Components that undergo cavitation can be negatively affected by the cavitation erosion process, which may ultimately lead to the system's failure. Cavitation erosion consists in surface damage and progressive loss of material from a solid due to the action of bubbles in the liquid that systematically collapse at the surface. It is generally accepted that the impact pressure and shock waves generated by the implosion of cavitation bubbles lead to wear of materials [7–10].

Typical cavitation erosion curves plotted in laboratory conditions are shown in Fig. 1. However, as a survey of the literature [7–13] indicates, different nomenclature of periods and various duration times of erosion progress are used. Generally, the cavitation erosion process of metals can be divided into two characteristic stages of damage, distinguished by the amount of material loss. In one stage, surface layer properties are changed, and in the other stage material removal is started, Fig. 1.

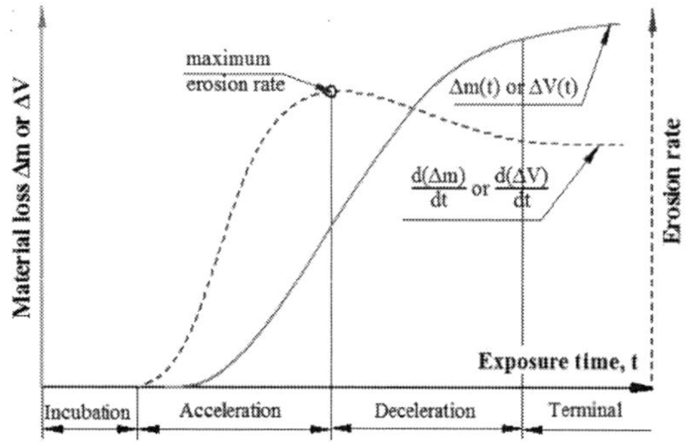

Fig. 1. Cavitation erosion curves with typical period names (Δm – mass loss, ΔV – volume loss).

Nevertheless, it is considered that the incubation period of cavitation erosion of metal alloys or overlay coatings can easily be identified [7, 10, 13–18]. At this stage of damage, material loss cannot be measured (Fig. 1) and wear can mainly be recognised because of surface deformation. Therefore, the application of topography

* Corresponding author: m.szala@pollub.pl

 DOI: 10.1051/itmconf/20171506003

measurement, microscopic surface investigation and worn surface image analysis are essential to describe qualitatively and quantitatively the incubation period duration and to determine the initial stage of the cavitation erosion process. On the other hand, the literature of the subject does not clearly state which from the above approaches is the most accurate for assessing cavitation wear. The ASTM standard [11] only mentions incubation period duration, but other methods for the evaluation of cavitation wear in incubation period are omitted. For this reason, it seems worthwhile to develop a more efficient method for wear resistance classification based on observations conducted at the initial stage of cavitation erosion.

The incubation period ends when the erosion rate starts to increase, hence the acceleration period begins, Fig. 1. The gravimetric measurement of material loss is a basic method of estimating cavitation erosion of materials. Unfortunately, this method relies on long-term testing in which indicators such as mean depth of erosion and maximum erosion rate of tested materials are determined and compared [11, 16]. In light of the above, it is vital to devise a method that would rely on the results obtained in the cavitation erosion incubation period (quick testing) and thus help to ensure long-term cavitation erosion resistance. Various method are being developed to shorten test time. In particular, computer image analysis of damaged surface is widely used, hence different approaches are reported [7, 18–20]. A survey of the literature of the problem indicates that despite the fact that Metllo or ImagePro Plus are widely used in materials science, there are no publications devoted to the application of this software for measuring cavitation erosion resistance. Therefore, this work focuses on this new application for the above software.

The aim of this study is to verify whether the Metllo and ImagePro Plus image analysis software can be used for evaluating cavitation erosion wear. This is done by comparing the image analysis results with the experimental findings of mass loss, topography measurements and microscopic observations.

2 Materials and methods

2.1 Specimen preparation

The test material for this study is low-alloy steel grade 34CrNiMo6 in untreated conditions with the following chemical composition: C – 0.3 to 0.38%, Si max. – 0.4%, Mn – 0.5 to 0.8%, P max. – 0.025%, S max. – 0.035%, Cr – 1.3 to 1.7%, Mo – 0.15 to 0.3% and Ni – 1.3 to 1.7%, according to the EN 10083 standard. The specimen for the cavitation erosion test was machined from a steel rod as a cylinder with a diameter of 25 mm and a height of 10 mm. The surface of the specimen was described by the following roughness parameters: Ra = 0.06 μm and Rt = 0.5 μm.

2.2 Cavitation erosion test

The cavitation test was performed using a vibratory apparatus by the stationary specimen method in compliance with the ASTM G-32 [11] test method recommendations. The test was conducted on a test stand schematically shown in Fig. 2 and described in detail in the works [17, 18]. In the present study the distance between the horn-tip and the specimen's working surface was set to 1 mm and distilled water with a temperature of 25±2°C was used as a working liquid. The surface of the specimen was examined in exposition times of 5, 10, 20, 40 and 60 minutes until 120 minutes. The specimen was weighed at each time interval with an accuracy of 0.1 mg. A cavitation curve of material mass loss due to cavitation erosion is plotted in Fig. 3.

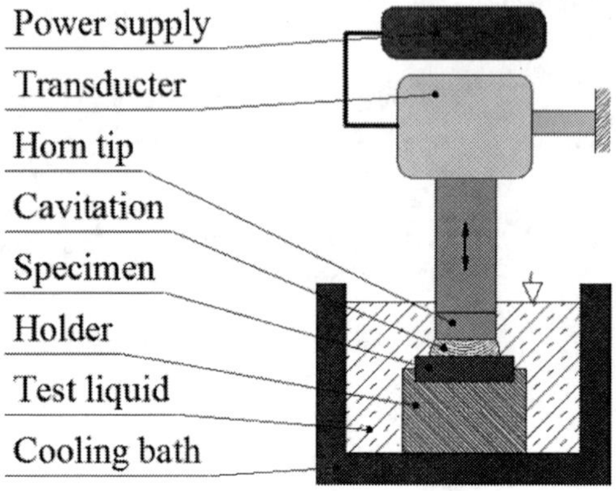

Fig. 2. Schematic design of cavitation erosion test.

Before and during the test, the surface of the specimens was systematically examined and captured in the selected specimen area of 2.21 mm^2 with a metallographic optical microscope Nikon MA 200. This was done by the bright field technique. Each time a large image composed of multiple image frames was captured because the size of the examined area exceeded the camera's field of view. The area of surface used for image analysis is shown in Fig. 4, and the development of the area due to cavitation is shown in Fig. 5.

Surface damage caused by cavitation was examined using a stereoscopic microscope Nikon SMZ 1500, Fig. 6. In addition, selected regions of the specimen surface in predetermined exposure times were examined with a scanning electron microscope Phenom ProX (Fig. 7, Fig. 8 and Fig. 9). The scanning electron microscopy (SEM) analysis was conducted using a backscattered electron detector (BSD) with topographical modes (Topo). Two roughness parameters: the arithmetic mean roughness Ra and the total height of the profile Rt, were measured with a Surtronic 3+ profilometer, and the results are plotted in Fig. 10.

2.3 Image analysis

The computer image analysis was conducted based on the images systematically captured with a metallographic microscope operated in the specified sample area. The surface development due to cavitation was analysed with the Metllo or ImagePro Plus image analysis software. Metllo is dedicated to metallographic stereology analysis

 DOI: 10.1051/itmconf/20171506003

[21] while ImagePro Plus provides state-of-the art imaging and analysis capability for acquiring, enhancing and analysing images [22]. Both programmes enable to detect and measure object attributes such as area, shape characteristic, perimeter, diameter, roundness and aspect ratio, ferret, and many more parameters.

The MetIlo analysis was performed according to the standard metallography procedure described in the works [21, 23, 24]. The tracing and counting of objects by ImagePro Plus was preceded by the assessment of bitmap image (hence the areas affected by cavitation are marked in white). Top and bottom binary thresholds were selected manually. The analysis was based on the images shown in Fig. 11 and Fig. 12. Parameters such as volume fraction, total count, area of each element, perimeter, maximal and minimal feret were measured by both image processing programmes, and the results are given in Fig. 13 and Table 1.

3 Results and discussion

The mass loss of 34CrNiMo6 steel due to cavitation was within a range of measuring error until 20 minutes of cavitation exposure (Fig. 3). The total mass loss after 2 hours of cavitation exposure is 15 mg, which means that the incubation period of cavitation erosion lasts until 20 minutes.

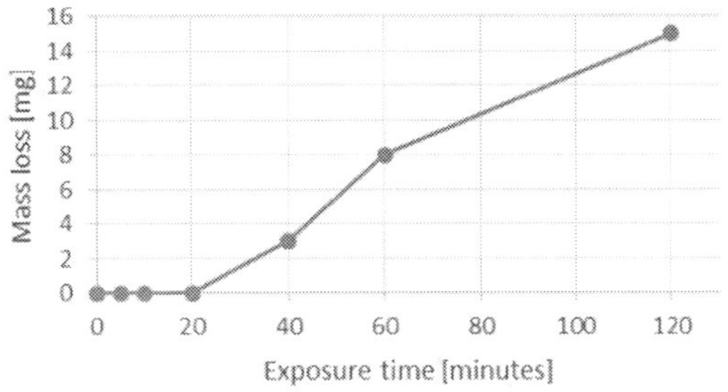

Fig. 3. Cavitation curve of specimen mass loss.

A development of the examined area (Fig. 4) exposed to cavitation is shown in Fig. 5.

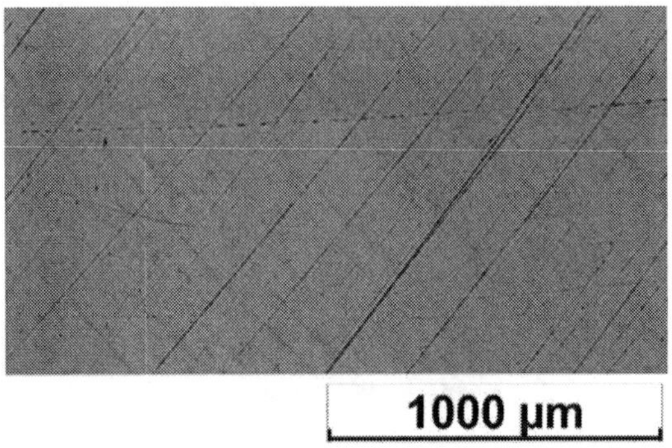

Fig. 4. Surface of 34CrNiMo6 sample before cavitation test – input image to analysis.

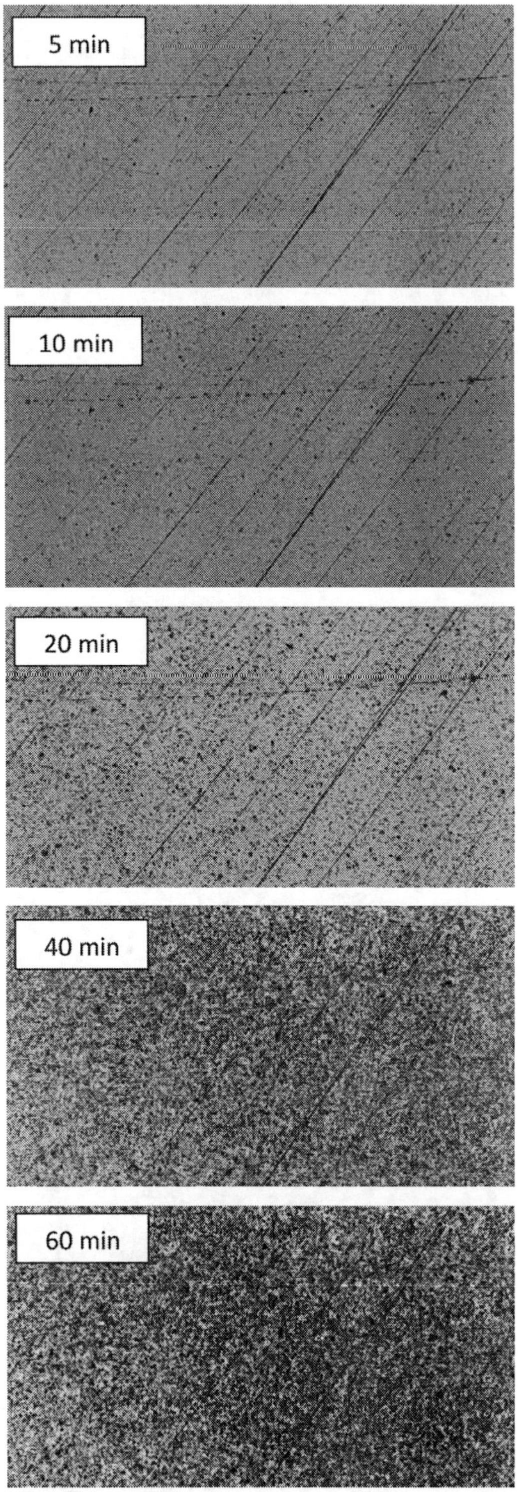

Fig. 5. Cavitation worn surfaces of 34CrNiMo6 steel in different exposure times, metallographic microscopy.

 DOI: 10.1051/itmconf/20171506003

Changes in the surface are clearly visible when using metallographic optical microscopy (bright light technique). At first, the surface of the specimen under cavitation load begins to deform. Finally, the areas affected by cavitation do not reflect and diffusing light as a result of surface deformation that leads to darker or dark areas in the captured image (Fig. 5).

In contrast, the stereoscope microscope examination visualizes the affected areas as brighter areas. Prior to the test only surface scratches are visible (Fig. 6a), but after 10 minutes of exposure to cavitation one can also notice worn areas (Fig. 6b). The SEM observation of the specimen prepared for the cavitation test confirmed the presence of surface non-uniformities (Fig. 7). Cavitation exerts impact on plastic deformation of the specimen surface (Fig. 8 and Fig. 9) and leads to an increase in the roughness parameters of the investigated steel (Fig. 10).

The qualitative examination of the worn surface indicates that the incubation period of cavitation erosion involves plastic deformation and the formation and growth of pits. The acceleration period begins after 20 minutes, so the growth of pits and removal of material are visible (Fig. 8 and Fig. 9). Material erosion starts in the vicinity of the scratches, in well-deformed areas and pits as well as in the vicinity to non-uniformities such as inclusions.

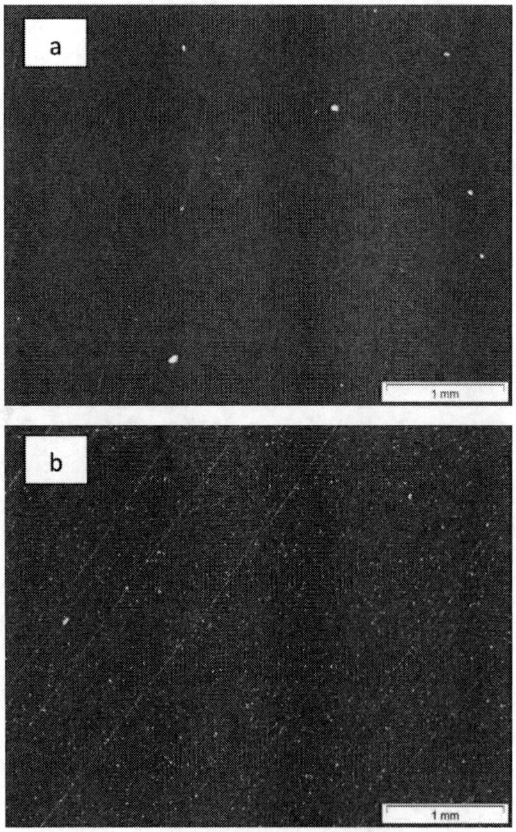

Fig. 6. Surface before cavitation (a) and after 10 minutes (b) of cavitation, stereoscopic microscope.

It can be observed that the roughness parameters Ra and Rt systematically increase with increasing the exposure time. In comparison with the untreated specimen, the surface roughness of the cavitation-exposed steel specimen almost doubled by the end of the incubation period. There is a correlation between mass loss (Fig. 3) and the Ra parameter, and similar results are reported in the literature [18, 25]. The quantitative erosion results are in agreement with the SEM results.

Surface roughness increases because of the peaks and valleys in the height direction caused by material erosion. Finally, after 120 minutes, Ra is 0.23 μm and Rt is 3.05 μm, which means that image analysis processing is impossible to perform in this particular moment using optical metallographic microscope images captured with the bright field technique. Consequently, the possibility of conducting dark-field examination must be verified

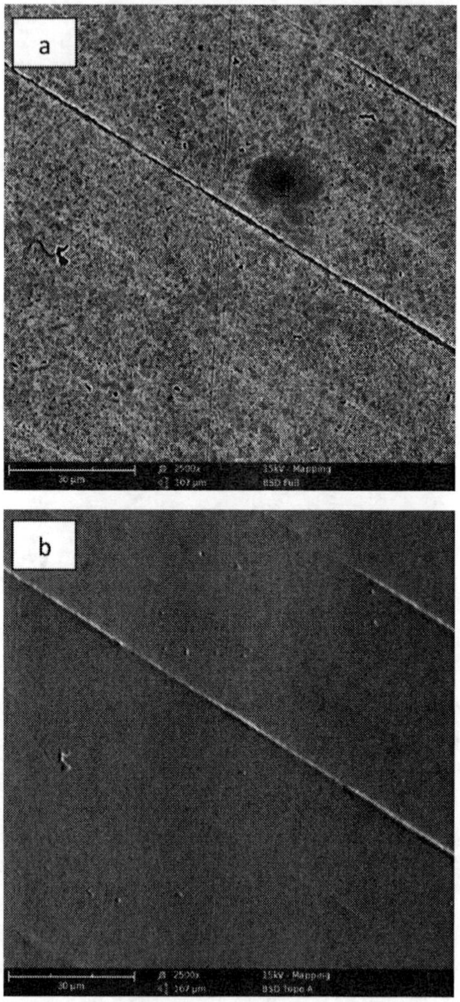

Fig. 7. Selected area of steel surface before cavitation tests, SEM-BSD (a)-and SEM-BSD-Topo (b) modes.

 DOI: 10.1051/itmconf/20171506003

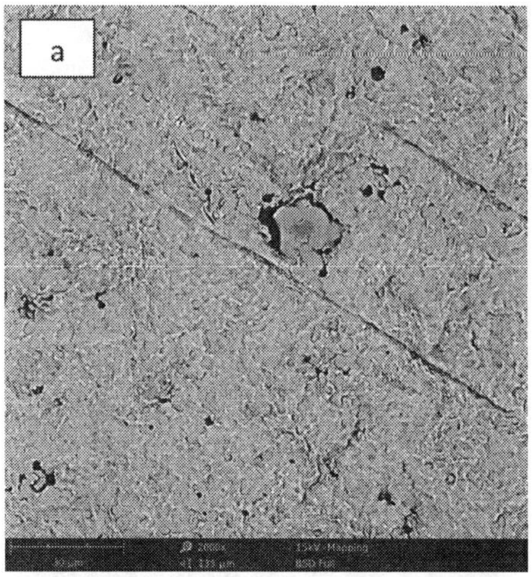

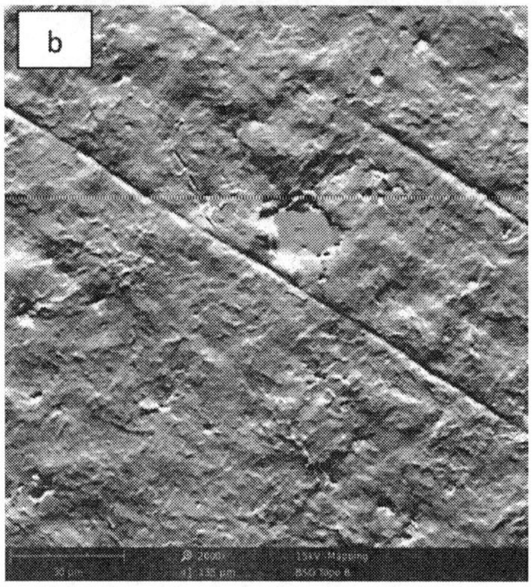

Fig. 8. Selected surface area after 40 minutes of cavitation, SEM-BSD (a), SEM-BSD-Topo (b) modes.

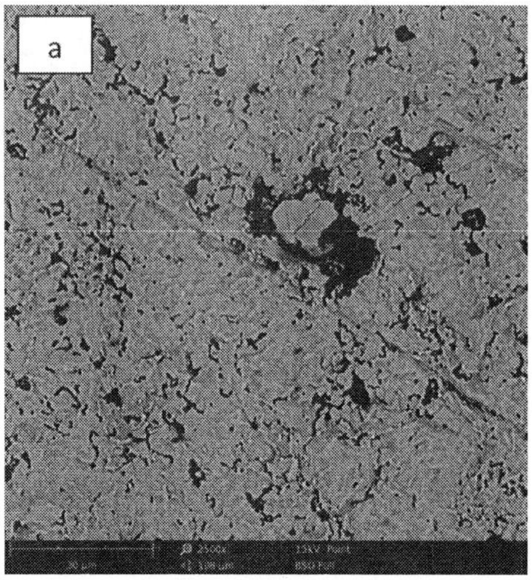

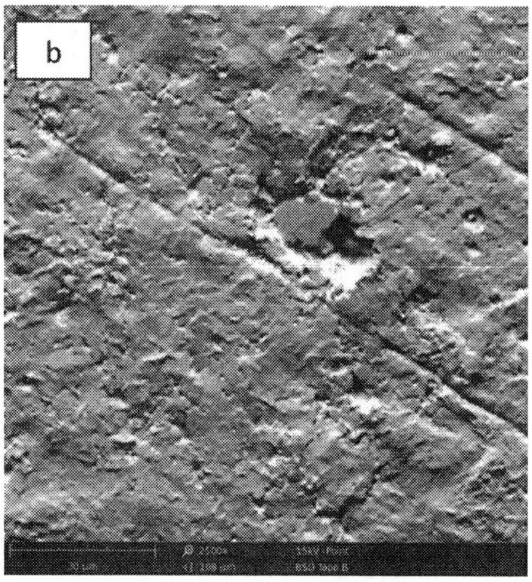

Fig. 9. Selected surface area after 120 minutes of cavitation, SEM-BSD (a), SEM-BSD-Topo (b) modes.

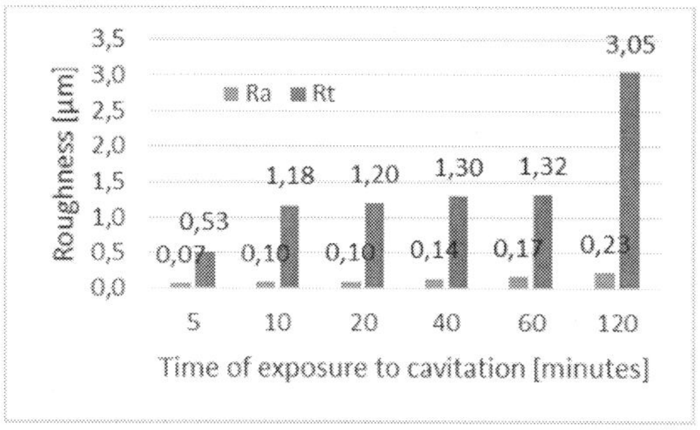

Fig. 10. Results of roughness measurement.

The images generated by computer image analysis are shown in Fig. 11 and Fig. 12, and the numerical results are given in Fig. 13 and Table 1. MetIlo and ImagePro Plus were used to assess wear in the incubation period of cavitation erosion. It can generally be observed that, irrespective of the programme used, every parameter tends to increase, decrease or remain constant depending on the exposure time (Table 1). Once the incubation period of cavitation erosion is over, the trend is reversed. Therefore, it must be statistically investigated whether ImagePro Plus yields more accurate results than MetIlo, and, to this end, more specimens must be statistically examined.

Despite the fact that the results obtained with the two image analysis programmes differ, it is possible to point to the exact moment when the incubation period ends by identifying the moment when parameter tendency starts to change. The end of the incubation period of cavitation erosion can be distinguished by estimating the moment when the investigated parameters change their tendency (from increasing to decreasing, or vice versa). Nevertheless, this problem requires further detailed studies.

The numerical results listed in Table 1 can be useful for describing erosion in terms of quantity and quality. Naturally, one must decide which parameter is the most representative to evaluate cavitation resistance of the material and hence could be used for predicting its

DOI: 10.1051/itmconf/20171506003

cavitation wear. The volume fraction of worn areas is not an adequate parameter to determine incubation period duration.

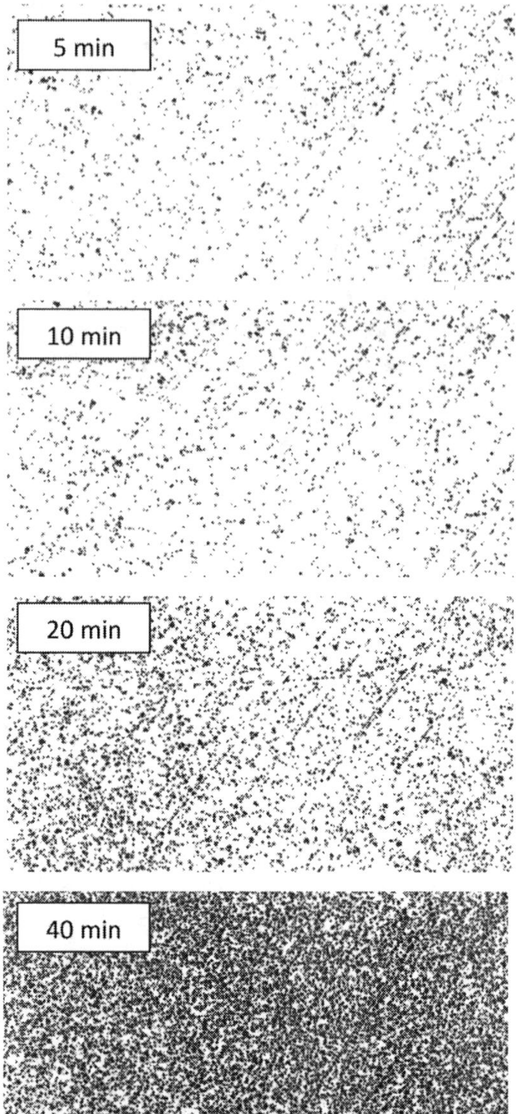

Fig. 11. Images used for image analysis with MetIlo.

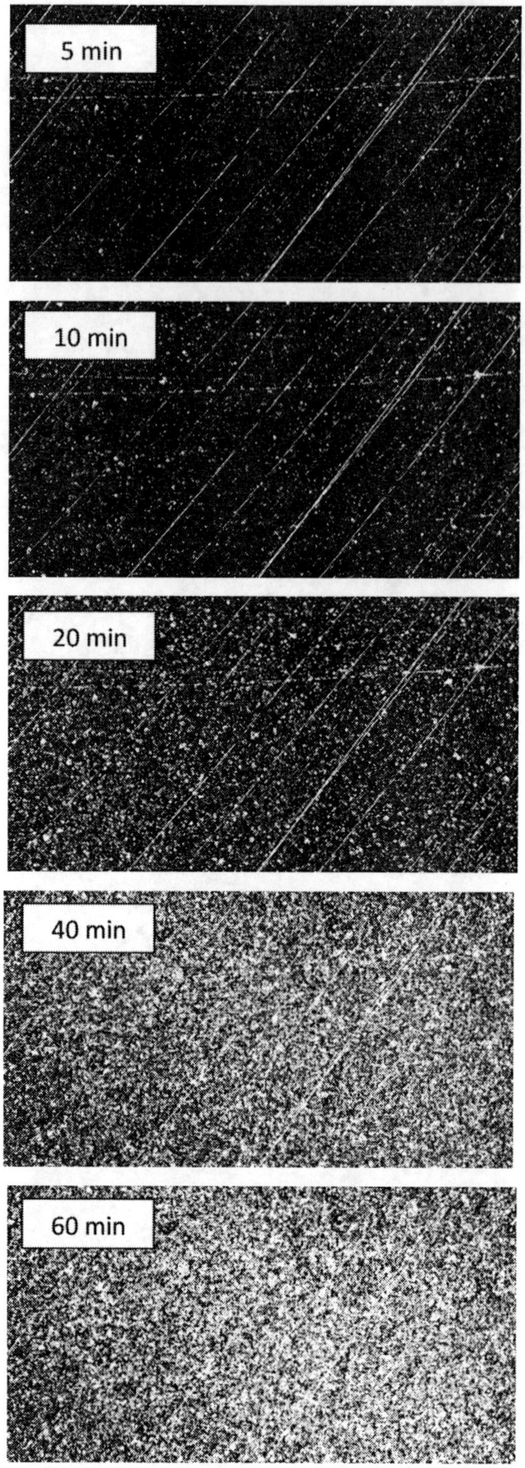

Fig. 12. Images used for image analysis with ImagePro Plus.

It seems that perimeter and ferret could be used to this end because their results indicate that the incubation period of cavitation erosion is 20 minutes, which complies with the results plotted in Fig. 3. In contrast, the volume fraction first shows a steady increase, irrespective of the image processing programme. However, after 40 minutes of exposure to cavitation, the Metllo-simulated volume fraction shows a decreasing tendency, while the ImagePro Plus results show that this parameter is still on the increase (Fig. 13). Examining the volume fraction results it is difficult to determine the exact moment when the acceleration period of cavitation erosion begins.

In addition to the above, statistical methods of results evaluation must be employed, too. It would be a good idea for future studies to compare these results with results obtained with other image analysis software. This would help verify accuracy of the applied procedure.

The scratches observed on the surface of the investigated area do not blur the computer analysis results. This means that the quality of specimen surface roughness preparation was sufficient to determine cavitation wear using computer image analysis.

Table 1. Parameters calculated with Metllo and ImagePro Plus (mean values) vs cavitation exposure time.

	Time [minutes]	5	10	20	40	60
Metllo	Volume fraction [%]	5.0	6.6	16.4	51.4	42.7
	Total count [-]	1555	2115	4360	8599	7233
	Elements per area [1/μm²]	704	958	1974	3893	3275
	Area of element [μm²]	70.5	68.4	82.8	132.0	130.0
	Perimeter [μm]	25.7	24.7	27.8	37.5	37.4
	Feret max [μm]	11.0	10.7	11.7	15.3	15.3
	Feret mean [μm]	9.7	9.5	10.5	13.2	13.3
ImagePro Plus	Volume fraction Vv [%]	5.7	7.3	19.6	50.4	60.2
	Total count [-]	5152	5267	10920	4562	2429
	Elements per area [1/μm²]	2342	2394	4964	2074	1104
	Area of element [μm²]	24.9	31.0	40.0	246.0	551.5
	Perimeter [μm]	17.3	20.5	24.9	45.5	42.8
	Feret max [μm]	7.1	8.1	8.8	9.0	9.0
	Feret mean [μm]	5.0	5.9	6.6	6.8	6.8

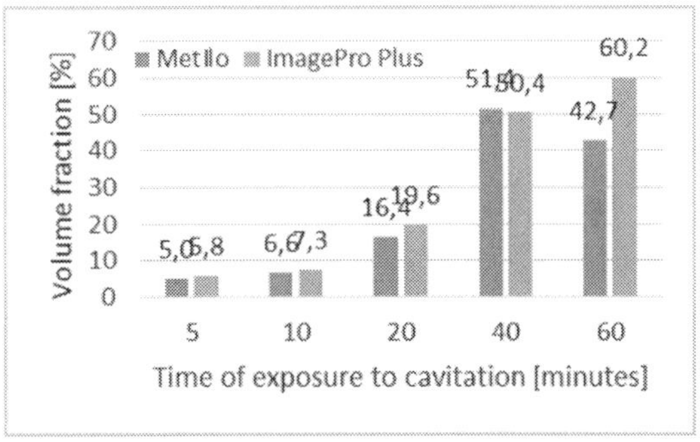

Fig. 13. Comparison of volume fractions of worn area calculated with ImagePro Plus and Metllo.

4 Conclusions

This study reported the preliminary results of study investigating, whether two image processing software programmes: Metllo and ImagePro Plus, can be used for assessing cavitation erosion wear. The numerical results produced by the above software was then compared with the image analysis results, mass loss calculations, topography measurements and microscopic observations. A new application for these image analysis programmes was proposed.

The results confirm that Metllo and ImagePro Plus can be effectively used for assessing wear during the incubation period of cavitation erosion.

The duration of incubation period for 34CrNiMo6 steel was determined using gravimetric, roughness and microscopic measurements and found to be equal to 20 minutes, which is in agreement with the results of computer image analysis. Both image analysis software programmes are suitable for determination of the incubation period duration of cavitation erosion.

The end of the incubation period of cavitation erosion can be determined by computer image analysis based on variations in the investigated parameters. This, however, requires further studies. Unlike parameters, such as perimeter or feret, the volume fraction of worn areas is not adequate to determine incubation period duration.

The results obtained with ImagePro Plus and Matllo allow for a complementary computer image analysis with gravimetric and roughness measurements and microscopic observations.

Based on the microscopic observations, it has been found that the wear during the incubation period for the investigated steel grade is characterised by plastic deformation and the formation and growth of pits, resulting from the changes in surface roughness. One can observe a systematic increase in the roughness parameters Ra and Rt with increasing the exposure time – from Ra=0.06 μm and Rt=0.5 μm for the specimen before cavitation, to Ra=0.23 μm and Rt=3.05 μm for the specimen after 120 minute-exposure to cavitation.

Future studies will focus on enhancing the computer image analysis procedure and evaluating suitability of its parameters, as well as devising a formula for long-term cavitation resistance of engineering materials based on data collected in the incubation period.

The author wishes to express his gratitude to Bartłomiej Dybowski (Silesian University of Technology, Poland) for carrying out the Metllo analysis.

References

1. M. Szala, D. Łukasik, *J. Technol. Exploit. Mech. Eng.* **2**, 40–44 (2016)
2. E. A. Weitendorf, On the History of Propeller Cavitation and Cavitation Tunnels (2001), (available at http://resolver.caltech.edu/CAV2001:sessionB9.001)
3. B. Liu, J. Zhao, J. Qian, *ENG FAIL ANAL.* **80**, 312–324 (2017)

4. M. Simionov, *Constanta Maritime University Annals*. **15** (2011) (available at https://trid.trb.org/view.aspx?id=1124082)
5. M. G. De Giorgi, D. Bello, A. Ficarella, *Measurement*. **47**, 971–981 (2014)
6. E. Panturu, A. Filcenco-Olteanu, A. D. Radu, M. Zlagnean, *E3S Web Conf.* **18**, 01034 (2017)
7. M. Dular, A. Osterman, *Wear*. **265**, 811–820 (2008)
8. J.-P. Franc, J.-M. Michel, *Fundamentals of Cavitation* (Kluwer Academic Publishers, New York, Boston, Dordrecht, London, Moscow, 2004), vol. 76 of *Fluid Mechanics and Its Applications*
9. C. E. Brennen, *Cavitation and Bubble Dynamics* (Oxford University Press, Oxford, 1995)
10. G. Gottardi, M. Tocci, L. Montesano, A. Pola, *Wear*. **394–395**, 1–10 (2018)
11. "ASTM G32-03 Standard Test Method for Cavitation Erosion Using Vibratory Apparatus" (ASTM International, 2003)
12. M. Szkodo, *Journal of Materials Processing Technology*. **164–165**, 1631–1636 (2005)
13. C. Haosheng, L. Jiang, C. Darong, W. Jiadao, *Wear*. **265**, 692–698 (2008)
14. B. Dybowski, M. Szala, T. J. Hejwowski, A. Kiełbus, *Sol. St. Phen.* **227**, 255–258 (2015)
15. S. Hattori, K. Maeda, *Wear*. **268**, 855–862 (2010)
16. J. Steller, *Wear*. **233–235**, 51–64 (1999)
17. M. Szala, T. Hejwowski, I. Lenart, *Adv. Sci. Technol. Res. J.* **8**, 36–42 (2014)
18. M. Szala, *Coatings for increasing cavitation wear resistance of machine parts and elements* (Lublin University of Technology, Poland, Poland, 2016), *Doctoral degree thesis*
19. M. Szkodo, *Problemy Eksploatacji*. **1**, 225–231 (2006)
20. A. Osterman, B. Bachert, B. Sirok, M. Dular, *Wear*. **266**, 945–951 (2009)
21. J. Szala, *Zastosowanie metod komputerowej analizy obrazu do ilościowej oceny struktury materiałów* (Silesian Univeristy of Technology, Poland, 2001), vol. 61 of *Zeszyty Naukowe. Hutnictwo*
22. Image Processing Software - Image-Pro Plus - Media Cybernetics, (available at http://www.mediacy.com/imageproplus)
23. H. Myalska, B. Dybowski, G. Moskal, *Adv. Sci. Technol. Res. J.* **11**, 220–231 (2017)
24. T. Rzychoń, B. Dybowski, A. Kiełbus, *Archives of Metallurgy and Materials*. **60**, 167–170 (2015)
25. S. A. Romo, J. F. Santa, J. E. Giraldo, A. Toro, *Tribology International*. **47**, 16–24 (2012)

 DOI: 10.1051/itmconf/20171506004

Computational image analysis of Suspension Plasma Sprayed YSZ coatings

Monika Michalak*[1], *Leszek* Łatka[1], *Patrycja* Szymczyk[1] and *Paweł* Sokołowski[1]

[1]Faculty of Mechanical Engineering, Wroclaw University of Science and Technology, Łukasiewicza 5, 50-371 Wrocław, Poland

Abstract. The paper presents the computational studies of microstructure- and topography- related features of suspension plasma sprayed (SPS) coatings of yttria-stabilized zirconia (YSZ). The study mainly covers the porosity assessment, provided by ImageJ software analysis. The influence of boundary conditions, defined by: (i) circularity and (ii) size limits, on the computed values of porosity is also investigated. Additionally, the digital topography evaluation is performed: confocal laser scanning microscope (CLSM) and scanning electron microscope (SEM) operating in Shape from Shading (SFS) mode measure surface roughness of deposited coatings. Computed values of porosity and roughness are referred to the variables of the spraying process, which influence the morphology of coatings and determines the possible fields of their applications.

1. Introduction

The paper aims at demonstrating the computational methods of image analysis, including porosity, morphology and topography assessment of plasma sprayed coatings.

Currently, one of the most extensively studied deposition methods is suspension plasma spraying (SPS). The method originates from conventional atmospheric plasma spraying (APS). SPS was developed in order to spray fine-grained powders that cannot be deposited by APS. On one hand, SPS is a very versatile technique (applied amongst others in energy, aerospace, mechanical and petrochemical industries) and provides a variety of coating microstructures [1–3]. However, on the other hand, the final coating's properties are highly dependent on the deposition parameters. APS spraying is already known to have 50 variables [4], but SPS method is even more complex (for example, the suspension properties highly influence process characteristics) [5–8]. Therefore, in this work, two types of SPS coatings with different microstructures (porous and dense) are investigated.

The evaluation of coating porosity, especially suspension plasma sprayed ones, is a challenging task. As already proven [1,8], different scaled porosity can be observed. In general, many characterization techniques may be used for porosity analysis. The basic ones are [1,9–16]:

(i) water adsorption (WA): inexpensive and single method; does not measure closed porosity;

(ii) mercury intrusion porosimetry (MIP): evaluates only open porosity but provides detailed information on pore size distribution;

(iii) helium pycnometry (HP): measures open porosity but does not provide the pore size distribution;

(iv) image analysis (IA): the object of interest of this paper; this method detects both open and close pores, allows to determine distribution of pore size; magnification and image contrast can influence the results;

(v) 3D reconstruction phase contrast by X-ray microscope (3D XRM): measures both open and close pores, with the size of pores up to 50 nm; a new and expensive technique;

(vi) NMR cryoporometry: measures pores down to nanometre size, approximates pore shape, a new technique.

The manner of porosity evaluation is classified differently by the literature [1,17–20]. Two main approaches can be distinguished: (i) shape-based [1,16–18], and (ii) size-related [1,16,17,20]. Notwithstanding a considered approach, the computations have to be carried out repetitively for a certain amount of images. Ganvir et al. [1] conducted such study for 20 images per sample. The ASTM Standard [12] defines the accuracy of results as dependent on the number of micrographs and 20 images is the minimum that should be taken to receive reliable results.

When investigating the porosity, pore shape should be considered. Deshpande et al. [17] distinguishes three types of pores: globular, interlamellar and cracks. The same is proposed by Ganvir et al. [1]. Authors of [18] evaluate the porosity at the following levels: (i) 0.25-1, (ii) 0.50-1 and (iii) 0.75-1 (where 1 stands for perfectly round shape).

Another important parameter is pore size. As suggested in [1], the porosity of SPS YSZ coatings can be divided into nano-, submicron- and micron- scale. Moreover, pores above 10 μm are considered to be

*Corresponding author: monika.michalak@pwr.edu.pl

globular pores, large inter-columnar spacing and/or larger vertical and horizontal cracks [1]. This assumption is consistent with the approach of Friis [20], which diversified pores into small and large groups (large were those of at least two splat thicknesses). Deshpande [17] evaluates the porosity in the subsequent limits: 3.0-10.0 µm for coarse globular pores, 0.1-3.0 µm for fine pores resulting from incomplete contact between lamellae during coating formation and 0.01-0.5 µm for fine cracks, derived from the relaxation of stresses occurred during splats cooling. The limitation of pore sizes depends on many factors.

An innovative approach of this paper is to study the relationships between the mentioned parameters by computational means. In this work, ImageJ public domain free software is the computational tool used for porosity evaluation. The reliability of this method is assumed to give 95% confidence level [21]. Nevertheless, the great attention has to be paid to this technique [1,2,5]. The total coatings porosity (including both nano- and micropores) is of significant meaning: it affects the wide range of coating properties, such as: elastic modulus, thermal conductivity, dielectric behaviour and many others [3,5,15,17,22].

The surface topography is also the important issue when characterizing coatings. The classic problem in computer imaging is the proper recovery of the shape of strongly irregular surfaces. The novel method, called *shape from shading* (SFS) is considered in this paper. In this technique, the 3D shape of a surface is computed from one or more 2D images [23,24]. The 3D views of the coating surfaces reconstructed by SFS are believed to give another insight on the surface profile and morphology [5].

2. Materials and method

2.1 Powders and suspension characterization

Yttria-stabilized zirconia (YSZ) powder (Tosoh TZ-8YS, Tokyo, Japan), with a chemical composition of ZrO_2 + 14 wt.% of Y_2O_3 was used for suspension preparation. The powder particle size (Zetasizer Nano ZS, Malvern Instruments Ltd., Worcestershire, UK) was: dv_{10} = 0.36 µm, dv_{50} = 0.53 µm and dv_{90} = 0.95 µm.

Each suspension was a mixture of: (i) 10 wt.% of solid powder, (ii) solvent (ethanol and water, in ratio 1:1) and (iii) dispersant (Beycostat C213, CECA, La Garenne-Colobes, France). The suspension formulation process was described in details in [8].

2.2 Substrate preparation

The 2 mm thick, 304L stainless steel coupons of 25 mm were grit-blasted prior to the spraying. The substrate roughness was equal to Ra = 7.8 µm (Fig. 1.). The details of substrate preparation were presented in work [8].

a

b

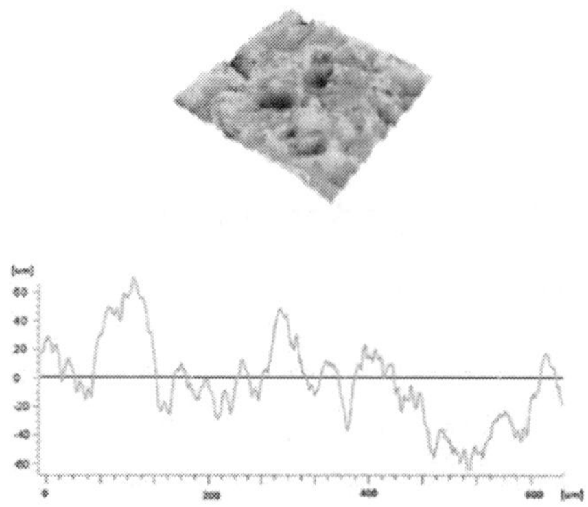

Fig. 1. The topography (a) and roughness profile (b) of substrate after grit blasting.

2.3 Spraying description

The coatings deposition was carried out with the use of two high-energy plasma torches: (i) Axial III (Mettech, North Vancouver, Canada) at University West (Trollhättan, Sweden) and (ii) WSP-H 500 at Institute of Plasma Physics CAS (Prague, Czech Republic). The torches were different in the terms of construction, plasma arc stabilization, feedstock injection, etc. (Table 1.). Consequently, all parameters were individually set for WSP-H 500 and Axial III. The aimed thickness of both kinds of YSZ coatings was 150 µm [8].

Table 1. The spray process details [8].

Torch	WSP-H 500 (Sample 1)	Axial III (Sample 2)
Electric power, kW	~150	~120
Working gases composition	Ar + H_2O decomposition products	Ar + H_2 + N_2
Total plasma gases flow rate, slpm	15 Ar + H_2O decomposition products	245
Stand-off distance, mm	100	75
Relative torch scan speed, mm/s	~900	~1000
Offset between plasma torch passes, mm	30	5
Nozzle injector internal diameter, mm	0.35	0.50
Suspension injection system	Pneumatic system; continuous stream injector	Peristaltic pomp; continuous stream injector
Suspension injection rate, ml/min	90–100	45
Injector type	Nozzle outside the torch, injection realized radially	Nozzle inside the torch, injection realized axially

DOI: 10.1051/itmconf/20171506004

2.4 Porosity analysis

In order to study the coatings microstructures, the specimens were firstly sectioned and embedded in a low viscosity epoxy resin. A standard procedure was applied to polish the samples manually. Gold sputtering was used prior to SEM investigations (Phenom G2 pro, Phenom-World B.V., Eindhoven, the Netherlands) to reduce the charging effect on ceramics.

Porosity analysis was carried out by the use of ImageJ software (ImageJ is free, open source software for images processing. It enables to calculate the area- or pixel- based statistics of user-defined settings [18]). The preceding stage before computational evaluation was the acquisition of SEM images in order to verify the scale of pores. Nanometre- (below 0.1 µm), submicrometre- (0.1-1.0 µm) and micrometre- sized (above 1.0 µm) pores are all observed in SPS coatings (Fig. 2.). The pores can be either round or irregular.

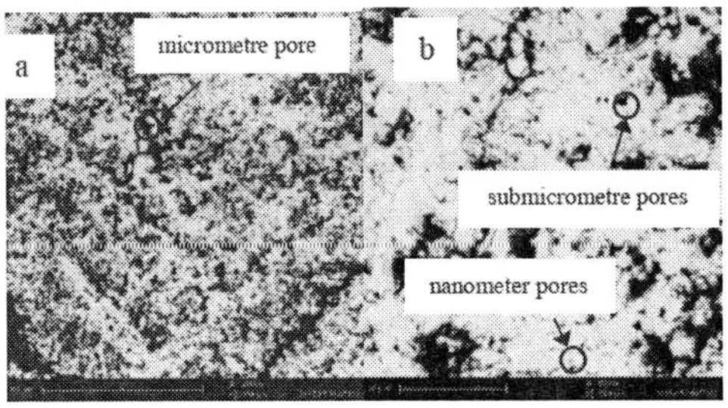

Fig. 2. SEM images of coating of sample 1 at various magnifications: × 2000 (a) and × 8000 (b), showing two different scales of porosity.

In this work the porosity was investigated with the use of micrographs taken at two magnifications (similarly as in [1]): (i) lower magnification for micrometre-sized pores, big cracks and column gaps, (ii) the higher magnification for submicrometre- and nanometre-sized pores. The images were taken at magnifications of × 2000 and × 8000. A total of 30 SEM micrographs were recorded across the 25 mm cross-section of coating in each case. All images were converted into binary files prior to the porosity calculations.

To sum up, the investigations cover the study of the following variables (Fig. 3. and Fig. 4.):

(i) Magnification: × 2000 and × 8000,

(ii) Circularity (which stands for the factor assessing the globularity of pores; its value is between 0 and 1; 1 stands for perfectly rounded pores): five circularity ranges were applied: 0.00-0.25; 0.26-0,50; 0.51-0.75; 0.76-1.00; 0.00-1.00,

(iii) Pore size: nanometre-sized (0.01-0.1µm), submicrometre-sized (0.1-1.0 µm), micrometre-sized (1.0-10.0µm), coarse (> 10.0µm) and pores with no size limits,

(iv) Structure type: porous (sample 1) and dense (sample 2) coatings.

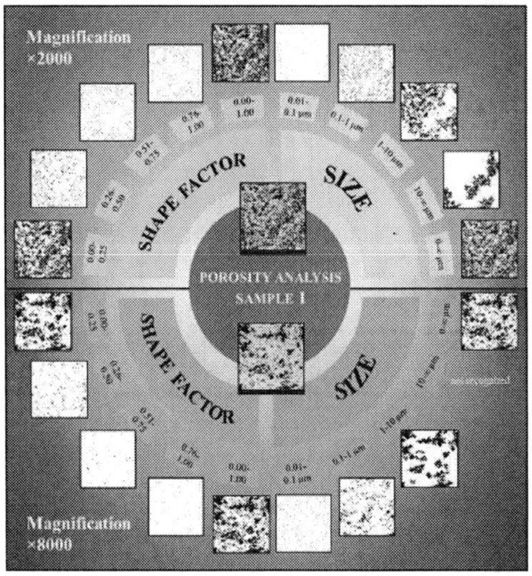

Fig. 3. The scope of investigations for the sample 1 (sprayed by WSP-H 500).

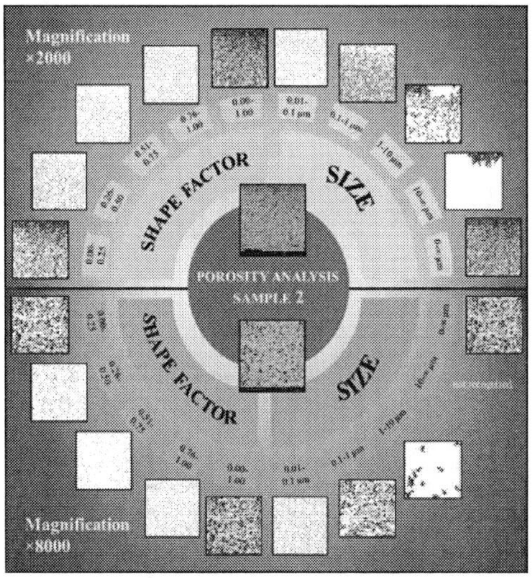

Fig. 4. The scope of investigations for the sample 2 (sprayed by Axial III).

For both shape- and size- related analyses, the porosity analysis covered the following steps. (Fig. 5.): (i) firstly, the image was opened (if bimodal grey value 8-bit was not set yet, it should be done here). (ii) Next, the image was calibrated by adjust to the scale bar. (iii) Subsequently, the image was cropped to the rectangle which comprises only the coating (e.g.: the substrate was not included; the intercolumnar areas between columns, visible at low magnification, were not excluded from the analysis). (iv) The threshold for the image processing was applied and (v) finally the porosity value was calculated.

DOI: 10.1051/itmconf/20171506004

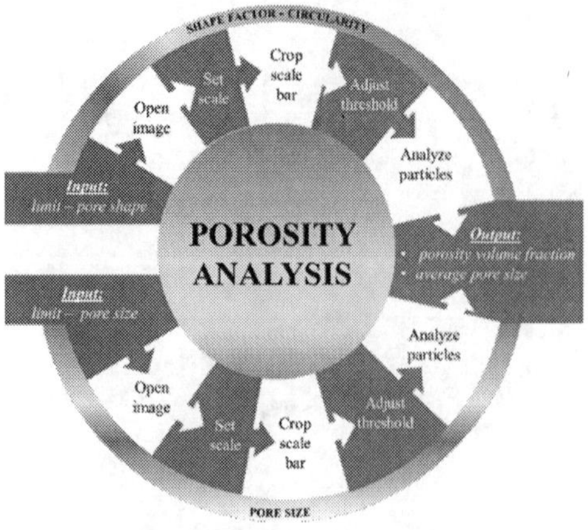

Fig. 5. The steps of image analyses.

The threshold is particularly important to correctly distinguishing the pores (Fig. 6.). There are many threshold methods (e.g.: *Intermodes*, *IsoData*, *MaxEntropy*, *Minimum*, *Percentile*, etc.). In order to set the suitable one, each method was tested in this study. The most appropriate threshold was assumed to be *Mean*.

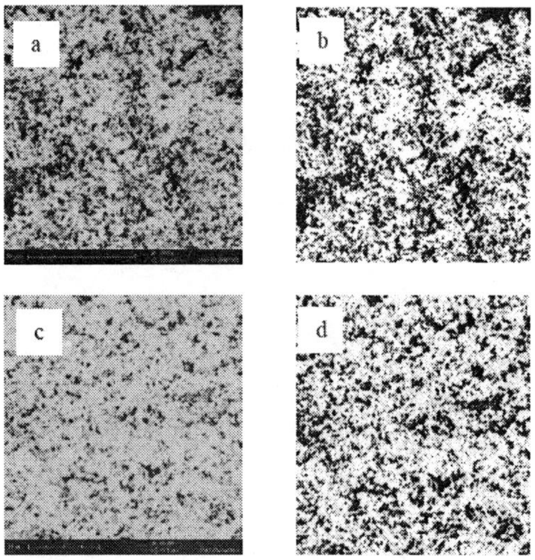

Fig. 6. SEM images before (a – sample 1, x 2000; c – sample 2, x 8000) and after threshold (b – sample 1, x 2000; d – sample 2, x 8000).

2.5 Characterization of coatings topography

In order to study the coatings topography, the roughness was measured with the use of: (i) non-contact confocal laser scanning microscopy (CLSM, Olympus LEXT OLS4000, Tokyo, Japan) and (ii) 3D Roughness Reconstruction application, dedicated for Phenom G2 pro (Phenom-World B.V., Eindhoven, The Netherlands) scanning electron microscope (SEM). SFS technique was applied to visualize coating free-surface in 3D mode. The views were created automatically by assembling four SEM micrographs, taken by symmetrically located BSD detectors (Phenom G2 pro, Phenom-World B.V., Eindhoven, the Netherlands). More details on SFS are provided in work [25].

3. Results

3.1 Morphology of coatings

The microstructure analysis of sample 1 reveals porous and columnar-like microstructure (Fig. 7a.). The coating thickness is in the range of 170 to 180 μm. The structure contains vertical voids but also a great number of closed pores. The pores aggregate reaches the total size of even a few micrometres. This coating has a typical cauliflower-like surface (Fig. 7b.).

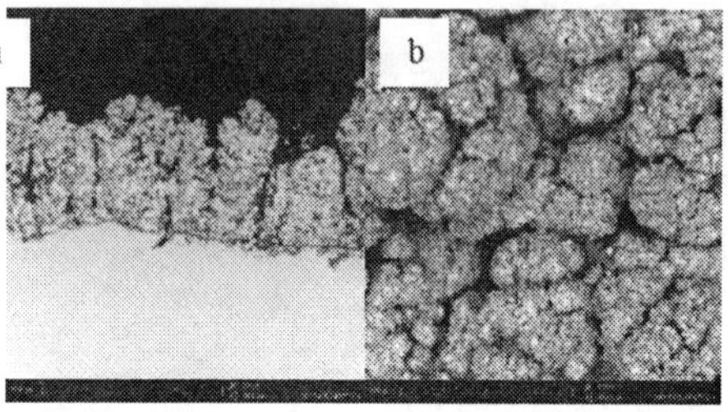

Fig. 7. The morphology of columnar-like coating of sample 1: a – columnar like structure of cross-section, b – cauliflower morphology of coating surface.

The coating labelled as sample 2 is 120-130 μm thick. The use of an Axial III plasma torch caused the formation of a very dense-looking coating and the columnar-like microstructure is not observed (Fig. 8a.). Nevertheless, although the coating seems to be dense, the pores are still present in the structure. Those pores are expected to be of a very small size (in the range of hundreds of nanometres) and uniformly distributed in the coating. They do not form bigger clusters as it was observed in the coating of sample 1. The top view of coating indicates the fine-grained nature of deposits (Fig. 8b.).

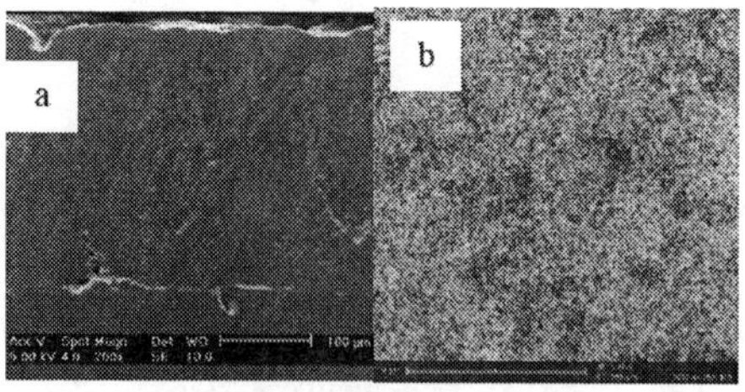

Fig. 8. The morphology of dense coating of sample 2: a – cross-section view, b – top view morphology.

More advanced topography features were provided by SFS technique (Fig. 9. and Fig. 10). The coating of sample 1 reveals columnar microstructure. The *Ra* roughness of this coating is two times higher in comparison with the dense and relatively smooth coating of sample 2 (Table 2.). The results were confirmed both by SEM (in SFS mode) and non-contact confocal laser scanning microscope (CLSM).

Fig. 9. *Shape from shading*: the partial 2D images and computed 3D shape of sample 1 (shown in Fig. 2); 650x.

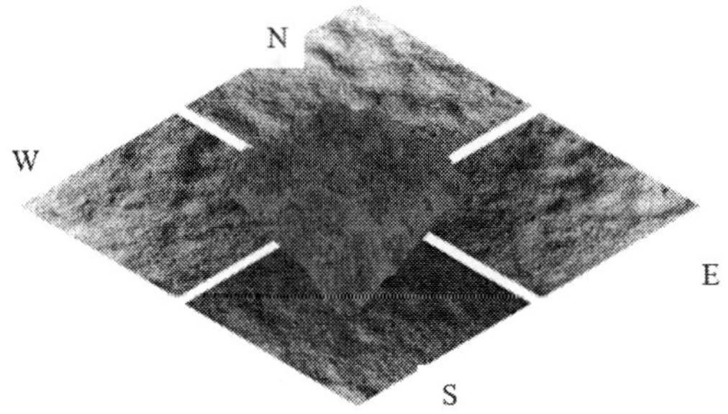

Fig. 10. *Shape from shading*: the partial 2D images and computed 3D shape of sample 2 (shown in Fig. 3); 650x.

Table 2. Roughness *Ra* of coatings measured by SEM (SFS) and CLSM

	Sample 1	Sample 2
SEM Phenom G2 pro	15.9 ± 1.5 µm	7.7 ± 0.7 µm
CLSM Olympus LEXT OLS4000	14.1 ± 1.6 µm	4.5 ± 0.5 µm

3.2. Porosity evaluation

3.2.1 Size influence

In the pore size-based studies, images were analysed for different size ranges, specified already in chapter 2.4. The results were obtained for: (i) total volume fractions of pores, (ii) percent relative accuracy, described in ASTM Standards [12] and (iii) the average size of pore.

3.3. Volume fraction

The computed pore volume fractions show different behaviour for coatings that were sprayed by Axial III and H-WSP 500 plasma torch (Fig. 11.). Pores in sample 1 are mainly micrometre in size. Coating of sample 2 contains both submicrometre - and micrometre-sized pores. The results confirm that the coating of sample 1 is porous, whilst the coating of sample 2 – more dense. It is recommended to distinguish the range of analysed pores. The results without the diversified size, although still valid (see next chapter), provides less detailed data.

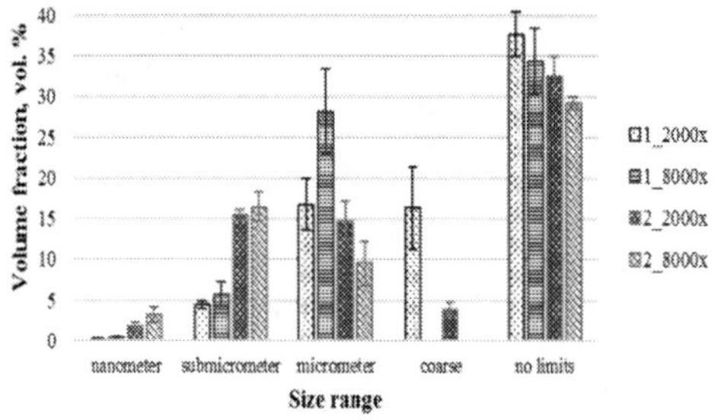

Fig. 11. The total volume fractions of pores, computed for different pore sizes.

3.4. Relative accuracy

The porosity determination is not complete without the calculation of its precision within normal confidence. According to ASTM E2109-01 (2014) [12], the common engineering practice assumes normal confidence as the expectation that the actual is within the stated uncertainty 95% of the time. One of the factors which assesses the reliability of the results, is the percent relative accuracy, *% RA* (equation 1), calculated for the 95% confidence interval (equation 2).

$$\% RA = \frac{95\% \, CI}{\bar{x}} \cdot 100 \tag{1}$$

$$95\% \, CI = \frac{t \cdot s}{\sqrt{n}} \tag{2}$$

% RA – the percent relative accuracy,
95% CI – 95% confidence interval,
$\bar{x}$ – the mean volume fraction of porosity,
t – constant, confidence interval multipliers,
s – standard deviation of total porosity volume fraction,
n – number of measurements.

ASTM E2109-01 (2014)[12] considers *RA* equal to 10% (or lower) as acceptable precision. The percent relative accuracy values, calculated for different ranges of pore sizes, magnifications and coating's types are summarized in Fig. 12. All results are below 10% *RA* value. The only inacceptable outcome is 1_2000x (sample 1, magnification × 2000) with the pore sizes above 10 µm. This suggests that the upper boundary pore size should be set as well.

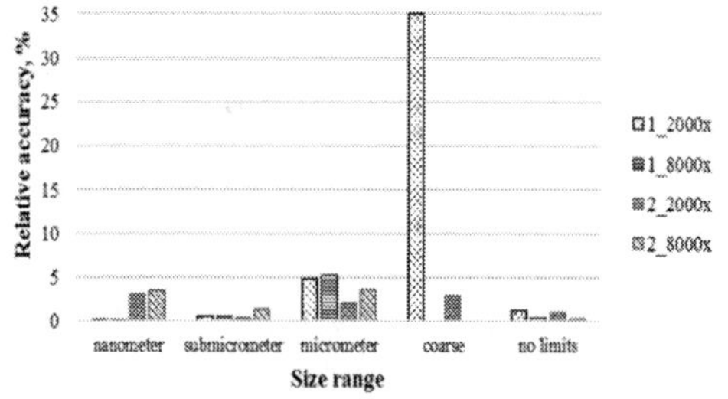

Fig. 12. The percent relative accuracy, depending on the pore size.

 DOI: 10.1051/itmconf/20171506004

3.5. Average pore size

When comparing the average pore size, depending on the limited size and applied magnification (Table 3.), it is observed that: (i) in each case, the computed results agree with restricted sizes. (ii) It is not possible to detect and measure the coarse pores (> 10 μm) when using higher magnification (× 8000), so for the assessment of coarse pores, the lower magnification is recommended. (iii) It is recommended to distinguish the range of analysed pores, otherwise the analysis provides more inaccurate results.

Table 3. The influence of assumed range of pore size on the computed average size of pore.

Image		Sample 1 × 2000	Sampl e 1 × 8000	Sample 2 × 2000	Sampl e 2 × 8000
Fine	nano, nm	94.0 ±5.0	45.0 ±8.0	97.0 ±4.0	29.0 ± 2.0
	submicr on, nm	639.0 ±51.0	599.0 ±45.0	578.0 ± 18.0	552.0 ± 18.0
	microm etre, μm	4.9 ±0.2	5.2 ±1.1	3.3 ± 0.2	2.9 ± 0.2
Coarse	μm	35.1 ±9.2	-	27.7 ± 3.1	-
Fine and coarse	μm	1.3 ±0.2	0.4 ±0.1	0.4 ± 0.1	0.1 ± 0.0

3.5.1. Shape (circularity) influence

The influence of the pore shape (circularity) on the volume of pores was also investigated (see chapter 2.4). The *circularity* command calculates the globularity of object(s), by the use of *area* and *perimeter* variables (see equation 3 and Fig. 13.) [26, 27]. The circularity value of 1.00 indicates a perfect circle. As the value approaches 0.00, it indicates an increasingly elongated polygon [26].

$$circularity = 4 \cdot \pi \cdot \frac{area}{perimeter^2} \qquad (3)$$

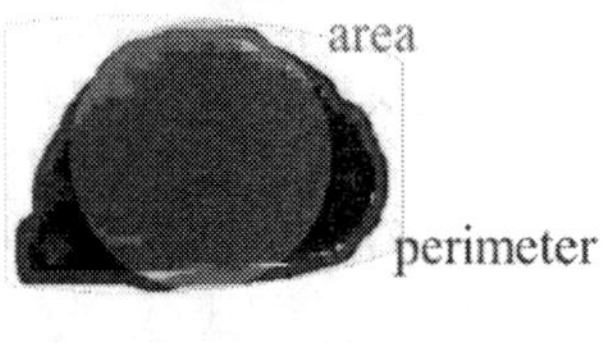

Fig. 13. Illustration of area and perimeter variables [27].

The results are calculated (for × 2000 and × 8000 magnifications) for: (i) total volume fractions of pores and (ii) percent relative accuracy, as specified in ASTM Standards [12].

3.6. Volume fraction

The dominating type of pores in both investigated samples are the ones of longitudinal shape (Fig. 14.), with the range of circularity between 0.00-0.25. However, the dense coating of sample 2 contains a slightly lower volume of non-circular pores. The other shape ranges do not differ significantly. The sum of calculated fractions with circularities: 0.00-0.25, 0.26-0.50, 0.51-0.75, 0.76-1.00 is equal to the amount of porosity identified for 0.00-1.00 range.

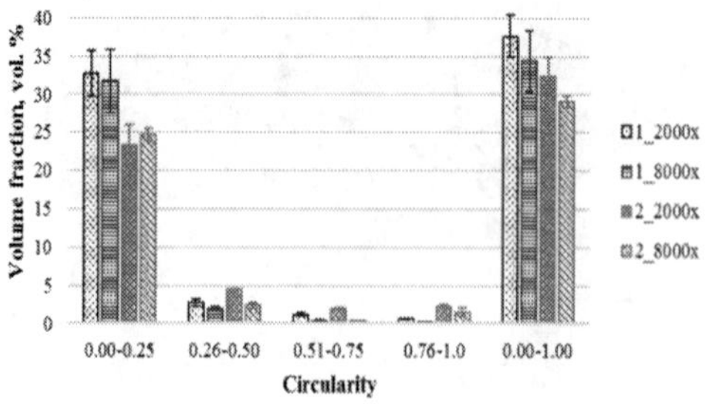

Fig. 14. The total volume fractions of pores, computed for different pore shapes.

3.7. Relative accuracy

Finally, the *% RA* was determined for shape-diversified volume fractions of pores. All calculated *95% RA* values are below 10 (Fig. 15.), so the precision of porosity evaluation is satisfactory.

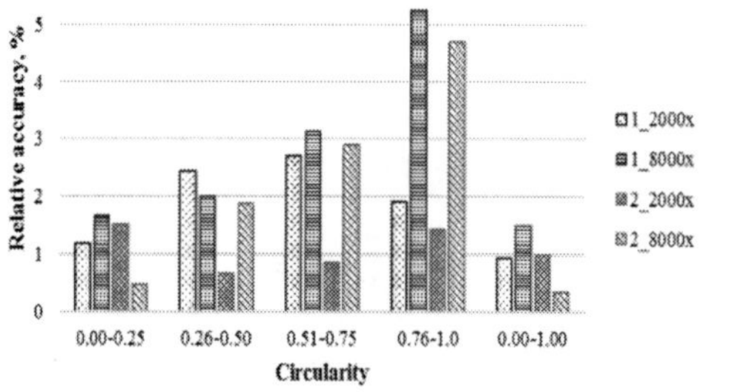

Fig. 15. The percent relative accuracy, depending on the pore shape

4. Discussion

The spraying of fine powder particles, like in SPS, results in coatings with a wide range of porosity. The distribution of pore size and the total number of pores depends also on the condition of spraying. For the investigated coatings, the total closed porosity (without visible inter-columnar gaps) is found to be in the range of 37 vol.% (without discontinuities) and 29 vol.%, for coatings of sample 1 and sample 2, respectively. Those results are consistent with the porosity investigated by Ganvir et al. [1], who compared axial suspension plasma sprayed YSZ coatings, having quite similar microstructures – the author observed both small-scaled porosity (at × 1000) and large-scaled porosity (at × 10 000), respectively in amounts of 24 and 42 vol.%. The investigations of pore size distribution show that the dominated pores in the coating of sample 1 are those of micrometre size (28 vol.% with × 8000 magnification). In the coating of sample 2, smaller (submicron) pores are dominating (16 vol.% with × 8000 magnification). The spraying parameters result in formulation of porous (sample 1) or dense (sample 2) types of coatings.

As introduced above, the presented study is focused on the coatings with different types of microstructure – dense and porous. The visual assessment, carried out by SEM, is consistent with the results of image analysis: the coating of sample 1 is much more porous than dense coating of sample 2.

The roughness of the coating originates from the method of spraying, spray process parameters and substrate topography. The strong correlation between coating roughness and morphology of coatings can be also observed (compare Fig. 9. and Fig. 10. and Tab.2.). The coating of sample 1 reveals higher roughness values which is caused by the columnar type of coating. On the contrary, the coating of sample 2 is much more homogeneous and therefore has lower roughness values. The visualisation of the surface topography may be regarded as a useful tool for the evaluation of coatings.

Finally, the image analysis of the porosity in the plasma sprayed coatings has to be carried out with attention to the specified problem. The brightness and contrast (necessarily at the same level for each image), magnification (lower – better for coarse pores, higher – for fine pores) and threshold (*Mean* mode) decided about the reliable results. In addition, the repetitive manner of image analysis is required for proper porosity computations. Properly conducted image analysis enables to characterize shape and size distribution of pores in user-friendly way, without expensive and advanced equipment (Fig. 16.).

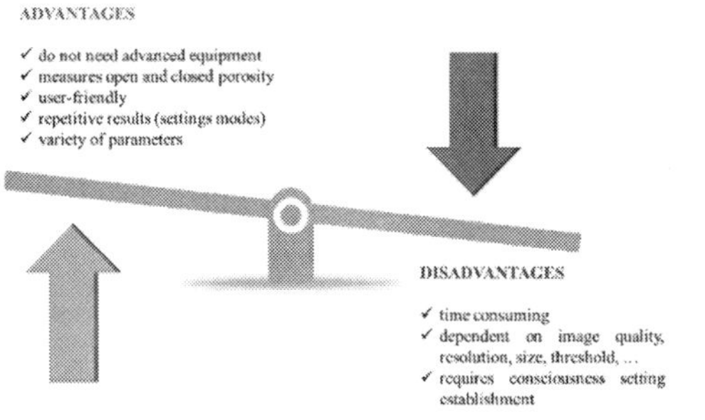

Fig. 16. The use of image processing for porosity measurements - advantages and disadvantages

5. Summary

As was presented in this work, the quantitative computational image analysis is an integral part of the microstructural characterization of thermally sprayed coatings. Both the literature and the article show that this tool may provide a reliable outcome – the obtained results are acceptable, especially in the terms of columnar coatings, which were tailored for thermal barrier coatings. Nevertheless, the capturing of images and their post-processing has to be carried out with the attention to detail, as many variables (among others: magnification, limits of pore shape and size, quality of the image and others) influences the final results.

The work was financed by the research project Sonata, UMO-2013/11/D/ST8/03400, awarded to the Faculty of Mechanical Engineering at the Wroclaw University of Science and Technology by the National Science Centre (Poland). The authors would like to acknowledge the help of research group of prof. Per Nylén from University West and research group of PhD Tomáš Chráska from Institute of Plasma Physics from Prague with the coating deposition process.

References

1. A. Ganvir, N. Curry, S. Björklund, N. Markocsan, P. Nylén, J. Therm. Spray Technol. **24** (2015)
2. A. Bacciochini, G. Montavon, J. Ilavsky, A. Denoirjean, P. Fauchais, J. Therm. Spray Technol. **19** (2010)
3. A. Ganvir, N. Curry, N. Markocsan, P. Nylén, S. Joshi, M. Vilemova, Z. Pala, J. Therm. Spray Technol. **25** (2016)
4. P. Fauchais, J.L. Coudert, B. Pateyron, La production des plasmas thermiques, Rev. Générale Therm. **35** (1996)
5. P. Sokołowski, P. Nylen, R. Musalek, L. Łatka, S. Kozerski, D. Dietrich, T. Lampke, L. Pawłowski, Surf. Coat. Technol. **318** (2017)
6. F. Gitzhofer, E. Bouyer, M.I. Boulos, Suspension plasma spray, US5609921 A, 1997. http://www.google.ch/patents/US5609921 (accessed December 20, 2016).
7. R. Rampon, O. Marchand, C. Filiatre, G. Bertrand, Surf. Coat. Technol. **202** (2008)
8. P. Sokołowski, S. Kozerski, L. Pawłowski, A. Ambroziak, Surf. Coat. Technol. **260** (2014)
9. ISO/TR 26946:2011 - Standard method for porosity measurement of thermally sprayed coatings. https://www.iso.org/standard/43929.html (accessed July 17, 2017).
10. B.D. Craig, L.F. Francis, L. Abrams, J. Am. Ceram. Soc. **79** (1996)
11. M. Lawrence, Y. Jiang, *Porosity, Pore Size Distribution, Micro-structure*, (Bio-Aggreg. Based Build. Mater., Springer, Dordrecht, 2017)
12. ASTM E2109-01(2014) Standard Test Methods for Determining Area Percentage Porosity in Thermal Sprayed Coatings, ASTM International, West Conshohocken, PA, 2014.
13. U. Klement, J. Ekberg, S.T. Kelly, 3D Analysis of Porosity in a Ceramic Coating Using X-ray Microscopy, J. Therm. Spray Technol. **26** (2017)
14. J. Ekberg, L. Nordstierna, U. Klement, Surf. Coat. Technol. **315** (2017)
15. U. Klement, J. Ekberg, A. Ganvir, Mater. Sci. Forum. **879** (2017)
16. A. Ganvir, N. Curry, S. Govindarajan, N. Markocsan, Int. J. Appl. Ceram. Technol. **13**, (2016)
17. S. Deshpande, A. Kulkarni, S. Sampath, H. Herman, Surf. Coat. Technol. **187** (2004)

18. W. Shen, J. Wan, T. K. Tokunaga, K. Yongman, i L. Xizhe, EJGE **20** (2015)
19. F. Andreola, C. Leonelli, M. Romagnoli, P. Miselli, Am. Ceram. Soc. Bull. **79** (2000)
20. M. Friis, C. Persson, J. Wigren, Surf. Coat. Technol. **141**, 115–127 (2001)
21. D.B. Fowler, W. Riggs, J.C. Russ, Adv Mater Proc., **11**, 41–52 (1990)
22. A. Ganvir, N. Curry, N. Markocsan, P. Nylén, F.-L. Toma, Surf. Coat. Technol. **268**, 70–76 (2015)
23. E. Prados, O. Faugeras, *Shape From Shading* (Handb. Math. Models Comput. Vis., Springer, Boston, 2006)
24. R. Zhang, P.-S. Tsai, J.E. Cryer, M. Shah, IEEE Trans. Pattern Anal. Mach. Intell. **21**, 690–706 (1999)
25. P. Sokołowski, L. Pawłowski, D. Dietrich, T. Lampke, D. Jech, J. Therm. Spray Technol. **25**, 94–104 (2016)
26. W. Rasband, Circularity, (2000). https://imagej.nih.gov/ij/plugins/circularity.html (accessed July 30, 2017)
27. Public Lab: Passive Particle Monitoring Analysis, https://publiclab.org/wiki/passive-pm-analysis (accessed July 30, 2017)

 DOI: 10.1051/itmconf/20171506005

Effect of porosity on physical properties of lightweight cement composite with foamed glass aggregate

Marzena Kurpińska[1,*] and *Tomasz* Ferenc[1]

[1]Gdansk University of Technology, Faculty of Civil and Environmental Engineering, Narutowicza 11/12, 80-233 Gdańsk, Poland

Abstract. This paper reports on a study of physical properties of lightweight cement composite. We investigate the possibility of replacing traditional aggregate with Granulated Ash Aggregate (GAA) and above all with Granulated Expanded Glass Aggregate (GEGA). For this purpose, 15 specimens of different percentage share of each aggregate in total aggregate volume were tested: 0%, 25%, 50%, 75% or 100% of foam glass aggregate (GEGA) partially replaced by ash aggregate (GAA) content in the cement composite. The water-cement ratio was constant and equal to w/c=0.5. Three grain sizes were analyzed: 2mm, 4mm (both GEGA) and 8mm (GAA). Numerical simulations of concrete specimen behavior under static loading were conducted with the implementation of elastic plastic model of each component. The study shows a significant impact of grain type and size on physical properties of lightweight concrete. Due to lower density of foamed glass aggregate, specimens shows various apparent density and porosity, which affect concrete properties. Compressive strength of concrete decreases with the increase in foam glass aggregate content; however specimens show different workability and in consequence porosity of lightweight concrete.

1 Introduction

In order to meet the increasing requirements for engineers to face more and more challenging structural designs new solutions need to be sought. On the one hand, structures need to be greater, more durable and reliable, on the other hand, standards for material itself are more demanding, *e.g.* for materials thermal conductivity. Hence, considering concrete as a part of material engineering [1] [2], the use of new components like lightweight foamed aggregate is investigated [3].
Lightweight concrete (LWC) with foamed aggregate has been a subject of study *e.g.* [4], [5]; however, our study presents a new approach to lightweight cement composite design considering Granulated Expanded Glass Aggregate (GEGA) as a concrete component instead of commonly used Granulated Ash Aggregate (GAA).

2 Composite components

In LWC design the following foamed aggregates were subjected to analysis: GEGA (Fig. 1a) of two grain sizes: 2 mm and 4 mm, and GAA (Fig. 1b) grain size 8 mm. The purpose of the study was to investigate the influence of aggregate share and its grain size on physical properties of LWC. In the tested specimens the aggregates always constituted about 70% of the whole lightweight composite volume. Due to high impact of aggregate parameters on the physical properties of the whole composite they were defined and listed in Table 1.

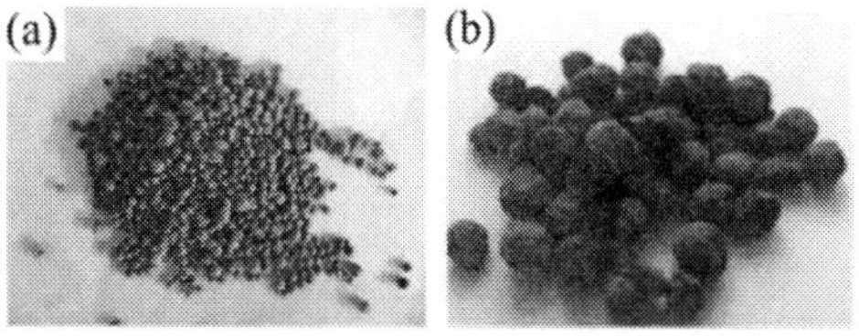

Fig. 1. Aggregates: (a) Granulated Expanded Glass Aggregate GEGA, (b) Granulated Ash Aggregate GAA.

Table 1. Physical properties of aggregates.

Property	GAA 8 mm	GEGA 4 mm	GEGA 2 mm
Water absorption WA_{24} [%]	16.5	17.8	15.2
Volume density ρ_a [Mg/m^3]	1.35	0.35	0.38
Density of dried grain ρ_{rd} [Mg/m^3]	1.25	0.31	0.34
Density of saturated grain ρ_{ssd} [Mg/m^3]	1.29	0.33	0.36.
Porosity p [%]	37	42	37
Crumble indicator X_r [%]	17.8	25.9	22.3
pH after 24h	11.1	11.9	11.9
Bulk density in loose state ρ_b [Mg/m^3]	0.68	0.18	0.20

* Corresponding author: marzena.kurpinska@pg.edu.pl

 DOI: 10.1051/itmconf/20171506005

The Portland cement CEM I 42.5R was applied as a binder according to EN 197-1. The morphological characteristics of components were investigated by means of Scanning Electron Microscope (SEM).
The main component of lightweight aggregates is silica SiO_2 (from 52% to 63%) and less significant components are Na_2O, CaO, MgO and Al_2O_3. Regarding cement, CaO is the main component, amounting to 63%, while the content of silica SiO_2 is about 22%. All content is listed in Table 2.

Table 2. The chemical compositions.

Component	Content [%]		
	GAA	GEGA	Cement
SiO_2	52.82	63.33	21.7
Al_2O_3	24.28	0.74	6.1
Fe_2O_3	7.5	–	3.1
CaO	4.5	14.19	63.4
MgO	3.19	2.98	1
SO_3	0.43	0.32	3.9
K_2O	0.2	0.57	0.64
Na_2O	–	13.35	0.16
Loss of roasting	7.1	4.53	–
Na_2Oeq	–	–	0.7
$(Na_2Oeq) = Na_2O + 0.658\ (K_2O) = 0.7$			

3 Porosity

Lightweight concrete (LWC) contain pores (empty spaces inside material), which may correspond to anywhere between a dozen and several dozen percent of total volume. Porosity is the result of the use of porous lightweight aggregate, where the amount of pores can be up to 67 %. Two types of pores can occur in LWC: open pores (o) and closed ones (c) (see Fig. 2). Open pores (o) are connected with each other and with material surface, thus they are permeable to liquids and gases. Closed pores (c) are isolated and not connected. LWC containing only closed pores are impermeable for liquids and gases, thus is commonly use as thermal and acoustic insulation.
Pores in material have a considerable impact on its mechanical properties, which increase with increasing porosity. Hence, density and porosity investigation is a crucial parameter in manufacturing process control and quality of the final product. Differences between LWC with various porosity concern also different behavior of material after immersion in fluid.

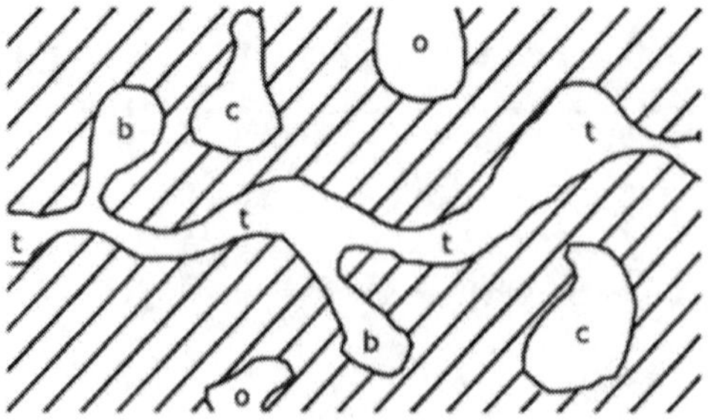

Fig. 2. Types of pores in lightweight concrete: o - open, c - close, t – transportation, b – blind.

The amount and the size of space between grains depend on the pores (voids content) of aggregate skeleton (composition) and on the amount of mortar surrounding the aggregate grains and filling the pores. Porosity of LWC increases when the void content of the aggregate skeleton is higher and when pores fulfillment is lower. The highest porosity is obtained when only one or a maximum of two fractions of large aggregate is applied and the decreased amount of cement grout is used instead of cement mortar. The role of the cement grout is to merge individual grains. When its amount increases, the voids of aggregates are being filled. Thus the void content is the property of lightweight concrete, which is independent of the porosity of the grains itself. The porosity of lightweight aggregate is bound with an aggregate skeleton structure and with grain shape.
The grain shape (Fig. 3a) affects physical properties of the aggregates listed in Table 1, while properties of aggregates have a great impact on the volumetric weight and on physical properties of hardened LWC.

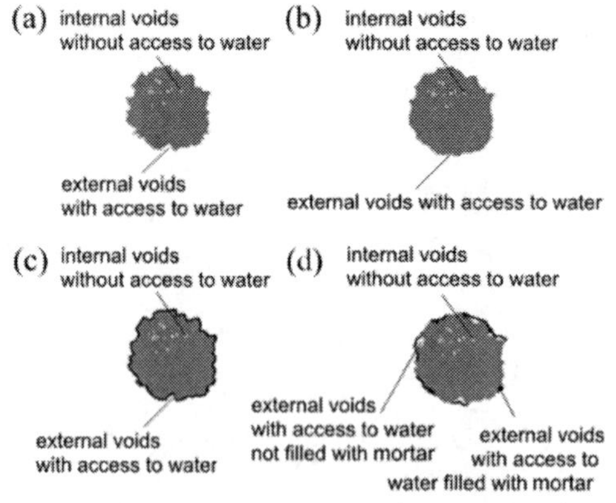

Fig. 3. Grain density: (a) apparent, (b) volume, (c) dried grain, (d) effective.

The aggregates volume density ρ_a was determined according to EN 1097-6 as a ratio of a dried aggregate sample mass to the volume that it takes in water with internal voids without access to water and external voids

 DOI: 10.1051/itmconf/20171506005

with access to water (Fig. 3b). According to EN 12697-5 it is called apparent grain density. The density of dried grain ρ_{rd} was calculated according to PN-EN 1097-6 as a ratio of dried aggregate sample mass to the volume it takes in water with internal void without access to water and with external void with access to water (Fig. 3c). According to PN-EN 12697-5 it is denoted as the density of saturated grain ρ_{pdd}. The effective grain density is a ratio of dried aggregate mass to the volume with internal voids without access to water, external voids with access to water and with the volume of empty spaces (Fig. 3d).

4 Experiments

4.1 Description

In order to examine physical properties of LWC with different aggregates fifteen specimens were investigated. Samples were prepared according to the authors' triangle rule (Fig. 4).

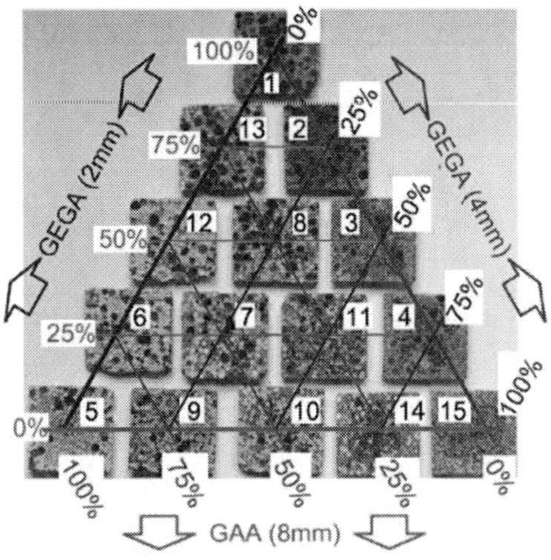

Fig. 4. Investigated specimens in the authors' triangle.

The percentage share of each aggregate in total aggregate volume is assumed in intervals 0%, 25%, 50%, 75% and 100% and listed with specimens' designation in Table 3.
The water-cement ratio is constant at w/c=0.5, thus 250 kg of water and 500 kg of Portland cement CEM I 42,5R were used. No admixtures or supplements were added.
A series of experiments was carried out in order to investigate physical properties of lightweight cement composite: strength, porosity and density.
Compressive strength was determined by means of Controls Advantest 9 machine with capacity of 300 kN (Fig. 5) on cubic specimens and dimensions 5x5x5 cm.
During concrete mix density examination, air content and consistency was investigated using a shock table inquiry. Afterwards, the concrete mix was placed into steel mold and stored in 20°C. After demoulding the specimens were kept in water. Experimental tests were carried out 28 days after the microscopic examination (Fig. 6).

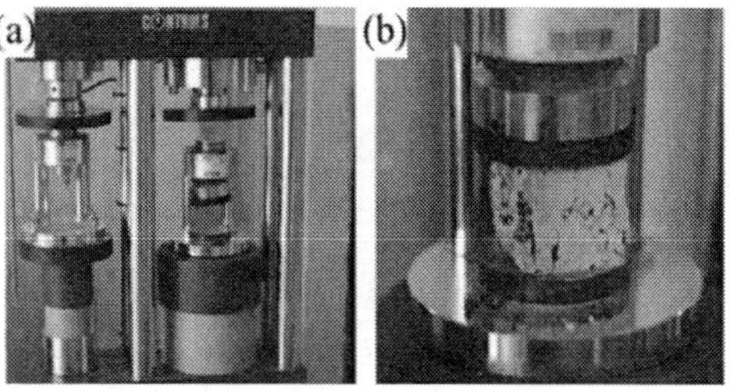

Fig. 5. Compressive machine Controls Advantest 9: (a) overall view, (b) investigated specimen.

Table 3. Participation of aggregates.

Specimens designation	GEGA 2 mm		GEGA 4 mm		GAA 8 mm	
	[%]	[kg/m³]	[%]	[kg/m³]	[%]	[kg/m³]
LWC 1	0	0	0	0	100	574
LWC 2	25	65	0	0	75	431
LWC 3	50	130	0	0	50	287
LWC 4	75	194	0	0	25	144
LWC 5	0	0	100	245	0	0
LWC 6	0	0	75	184	25	144
LWC 7	25	65	50	123	25	144
LWC 8	25	65	25	61	50	287
LWC 9	25	65	75	184	0	0
LWC 10	50	130	50	123	0	0
LWC 11	50	130	25	61	25	144
LWC 12	0	0	50	123	50	287
LWC 13	0	0	25	61	75	431
LWC 14	75	194	25	61	0	0
LWC 15	100	259	0	0	0	0

Porosity was determined on specimens with dimensions 2x2x2 cm. After removing dust, the samples were dried and weighed (m_1). Afterwards, the air was removed and the specimens were saturated with kerosene by keeping them in boiling fluid for two hours. Next, the specimens

DOI: 10.1051/itmconf/20171506005

were set to cool in kerosene to the temperature of 20°C was achieved. The prepared samples were hydrostatically weighed (m_3) in fluid. The next step was to determine the mass immediately after removing the specimens and wiping them with a cloth (m_2). Accuracy of all weighing amounted to 0.02 g. Apparent density (ρ_p) and opened porosity (p_o) were calculated from equations, respectively:

$$\rho_p = \frac{m_1 \rho_c}{m_2 - m_3}\left[\frac{g}{cm^3}\right] \quad (1)$$

$$p_o = \frac{m_2 - m_1}{m_2 - m_3} 100\% \quad (2)$$

where m_1 is mass of dried samples, m_2 is mass of samples saturated is kerosene, m_3 is mass of samples saturated in kerosene and weighted in it, ρ_c is density of kerosone used for saturation and for hydrostatic weighting (ρ_c=0.8241 g/cm^3).

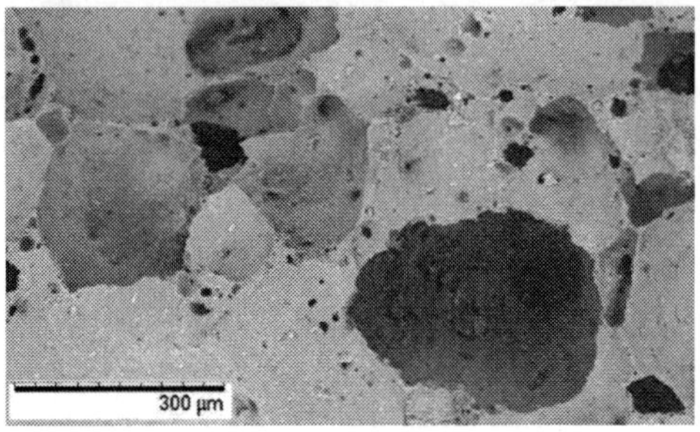

Fig. 6. Microscope picture of specimen.

4.2 Results

The experimental results were listed in Table 4, containing compressive strength of specimens, their apparent density and porosity. The list order was assumed according to specimen strength.

As shown, material properties are always related to its porosity. LWC containing lightweight aggregates made from foam glass (GEGA) is characterized by high porosity and relatively low apparent density. Specimens with porosity in the range of 15 % and 20 % (LWC 1, LWC 2, LWC 3) obtained the highest compressive strength 13-21 MPa. Concrete with porosity in the range of 45 %-65% (LWC 6 and LWC 9) show compressive strength of about 4 MPa.

However, specimen properties are not only affected by porosity alone, but the selection of lightweight aggregate does as well. The highest compressive strength was obtained for specimens consisting of lightweight aggregates GEGA 2 mm and GAA 8 mm in various proportions.

Specimens that were characterized by the highest porosity 67 % (LWC 5) containing only aggregate GEGA 4 mm show compressive strength of approximately 13 MPa and apparent density 1078 kg/m^3.

Table 4. Specimen properties.

Specimens designation	Apparent density [kg/m3]	Porosity p_o [%]	Compressive strength [MPa]
LWC 2	1378	17.7	21.35
LWC 1	1560	20.8	18.65
LWC 3	1177	16.0	13.43
LWC 5	1078	67.0	12.49
LWC 8	1117	24.0	10.10
LWC 13	1304	25.6	8.92
LWC 11	1041	20.1	6.99
LWC 15	1002	15.2	6.86
LWC 14	1060	19.3	6.37
LWC 10	903	26.4	5.44
LWC 12	1059	36.6	5.38
LWC 6	1028	45.0	4.59
LWC 7	1058	27.4	4.29
LWC 9	929	65.9	4.21
LWC 4	877	22.1	3.72

5 FEM modeling

Fifteen specimens were modeled using Finite Element Method by means of shell elements. The specimen geometry was assumed according to a section obtained through the middle of the cubes with dimensions 5x5x5 cm, *i.e.* square 5x5 cm was received.

The results are presented on specimen 13 (Fig. 7).

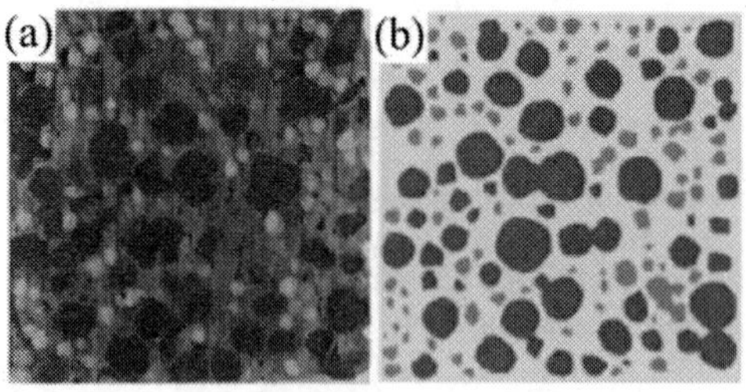

Fig. 7. Specimen 13: (a) geometry with aggregate decomposition, (b) FEM model.

FEM models were created using shell quadratic and triangle elements of size around 0.25 mm. For specimen 13 the total number of elements is 16959 while the total

DOI: 10.1051/itmconf/20171506005

number of nodes is 16081. With the assumption of isotropy, material parameters (stiffness modulus, Poisson ratio and strength) were taken from [1] and are listed in Table 5. For nonlinear analysis ([6], [7]) the elastic plastic models were taken into account, where the value of stress in each component cannot exceed its strength.

Table 5. Material parameters.

Material parameters	**Material**			
	GAA 8 mm	**GEGA 4 mm**	**GEGA 2 mm**	**Mortar**
Stiffness modulus [GPa]	18.9	4.7	3.8	23.9
Poisson ratio	0.1	0.1	0.1	0.167
Strength [MPa]	9.8	1.5	1.8	52.8

The specimens were acted upon by force representing the real one converting load value to 1 mm thickness. Hence, for specimen 13 instead of 22.31 kN (see strength in Table 3) the force equal to 0.4462 kN was applied. The HMH stress map is presented in Fig. 8.

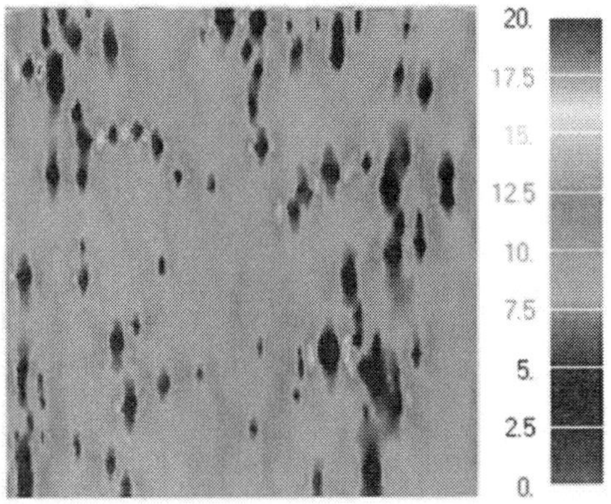

Fig. 8. HMH stress map for specimen 13 [MPa].

As it can be observed, the analyzed specimen failure occurs due to capacity exhaustion of the aggregate (GAA and GEGA 4 mm), which are the weakest element in the composite with strength, for GEGA 4 mm, equals 1.5 MPa (blue color in Fig. 8). It is a confirmation of what was observed during experiments (e.g. Fig. 5b).

6 Conclusions

The purpose of the paper was to study the possibility of applying lightweight concrete (LWC) in such structures as structural elements or extenders and insulation material. Hence, mechanical properties of 15 different material and aggregate grain size specimens were investigated.

Considering mechanical properties, the lightweight aggregate is the weakest component in concrete, nevertheless, analyzing other physical properties such as density (mass) or porosity it was revealed that the material offers multiple advantages, *e.g.* high thermal or acoustic insulation. Low density and in consequence low self-weight allow reducing the total weight of a structure by up to 35%. Furthermore, it can decrease cross-section are or the amount of necessary reinforcement; hence the cost of construction can be reduced. The manufacturing and the curing of lightweight concrete can be investigated analogously to the traditional one [8].

Calculations were carried out at the Academic Computer Center in Gdańsk. Tomasz Ferenc is supported with grant for the development of young scientists from the Faculty of Civil and Environmental Engineering, Gdańsk University of Technology.

References

1. L. Domagała, Cem. Wapno Beton **2** (2011)
2. B. Grzyl, A. Kristowski, Czas. Tech. Bud. **2-B**(6), (2014)
3. M. Kurpińska, T. Ferenc, Shell Structures: Theory and Applications Volume 4, pp.549-552 (CRC Press/Balkema Taylor & Francis Group, London 2018)
4. N. Ciak, M. Ciak, J. Harasymiuk, Effects of the lightweight concrete's composition's modification on its properties (IX Conference Days of Concrete Wisła, Poland, 2016)
5. K.-Ch Thienel, Materialtechnologische Eigenschaften der Leichtbetone aus Blähton (Conference: TFB Seminar "Leichtbetone im konstruktiven Ingenieurbau" Berlin, Germany 1997)
6. J. Pamin, R. de Borst, Arch. App. Mech. **68**(9) (1998)
7. J. Pamin, R. de Borst: Arch Mech **51** (3-4) (1999)
8. J. Chróścielewski, A. Mariak, A. Sabik, B. Meronk, K. Wilde, Adv. Sci. Technol. Res. J. **10** (32) (2016)

DOI: 10.1051/itmconf/20171506006

Evaluation of scratch resistance of nitride coatings on Ti grade 2

Kamil Pasierbiewicz[1,*], *Anna* Rzepecka[1], and *Mariusz* Walczak[1]

[1]Lublin University of Technology, Faculty of Mechanical Engineering, Department of Material Science, Nadbystrzycka 36, 20-618 Lublin, Poland

Abstract. Hard coatings limiting direct contact of material susceptible to wear are applied in order to improve resistance to abrasive wear. The key importance in using anti-wear coatings has their adhesion to substrate and differences in mechanical properties of the coating and substrate, particularly their Young's module. Studies have analyzed merits of applying nitride coatings TiAlN, AlTiN and hybrid coating TiAlN/TiSiN on titanium substrate ASTM grade 2. For that purpose, nanohardness and scratch tests for each coating were conducted. Apparatus of Anton Paar Company was used in these studies – ultra-high resolution nanoindenter and micro combi tester. Due to the fact, that surface roughness is a significant parameter affecting properties of the surface layer, the first appointed parameters were Sa roughness with a use of optical profilometer Bruker Contour GT. Obtained results suggest the best adhesive grip of hybrid coating TiAlN/TiSiN compared with other tested. The weakest adhesion to titanium grade 2 has TiAlN coating, on which in scratch test has been observed significantly faster total damage, in comparison with other tested. Research results will provide the input data for numerical analysis, enabling fast receiving effects of applying tested coatings without the need to carry out time-consuming research.

1 Introduction

Titanium and titanium alloys, especially Ti6Al4V, found application not only in the aerospace or petrochemical industries, but also in biomedical industry. They have several beneficial features, which must be fulfilled by the materials used in such conditions. These features are: high load resistance factors, low degrees of corrosion and oxidation, low toxicity and beneficial biocompatibility [1]. In addition, a relatively small Young's module causes that titanium and titanium alloys are very good materials for bone implants. However, their low abrasion resistance and metallosis effect caused by alloys, poses a challenge. The most commonly used alloy in medicine – Ti6Al4V – can cause side effects due to formation of cytotoxic vanadium oxides [2]. For this reason, researchers are looking for solutions by applying appropriate coatings, such as nitrides. This is very important, because medical implants are used to improve the quality of patients' lives by reducing the pain and restoring a function of infected structures [3]. In recent years, TiN, TiO_2 and N-TiO_2 were successfully used as protective coatings against wear and corrosion. It was observed, that this increases the life span of surgical implants and prostheses. Coatings, among others with PVD method, enable to obtain a gradient diffusion layer of sufficient thickness and adhesion, which ensures long-lasting abrasion-resistant surface [4].

In this research, it was decided to study the usefulness of AlTiN TiAlN coatings and TiAlN/TiSiN applied on titanium ASTM grade 2 to improve its abrasion resistance. For that purpose, applied coatings should have high hardness, as well as a good connection with substrate. Only two conditions allow for their use in mentioned above medical applications. Comparison of properties of three tested coatings enables to assess, which one is the best to protect titanium substrate against wear. Comparison with other similar tested coatings allows for general evaluation of merits of their application. Appointed hardness values of coatings and Young's module enable their use as the input in computer modelling of coatings and simulating their behaviors, as well as cooperation with the substrate.

2 Studied materials

TiAlN, AlTiN and hybrid coating TiSiN/TiAlN were the subject of research. The substrate material was technically pure titanium – ASTM grade 2. The sample in a form of discs ø 25 mm and thickness of 6 mm were prepared in the industrial technological line. At first, studied surface were polished mechanically to achieve surface roughness below Ra 0.5 µm. Pre-cleaning and degreasing in biological parts washer machine was followed by blasting in water corundum suspension of 300 µm gradation at a pressure of 0.4 MPa through 10 seconds. The last preparation stage of samples was a bath in washing line in the water subjected to reverse osmosis process with ultrasounds and gas bubbles flowing through the liquid. Coatings were placed physically from the gas phase PVD in the magnetron sputtering process. Targets placed on the cathode in appropriate proportions, as described in Table 1, were the source of coating material. After placing the samples

* Corresponding author: pasierbiewicz.k@gmail.com

DOI: 10.1051/itmconf/20171506006

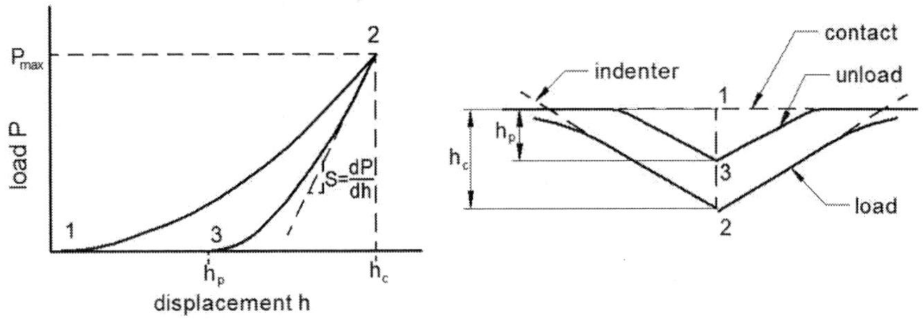

Fig. 1. Scheme of loading relation – offloading in nanohardness studies with Olivier Phar method; 1 – contact with studied surface; 2 – maximum deflection h_c with P_{max} loading and maintenance of the force for 5 seconds; 3 – residual deflection h_p after removal of the load.

in working chamber, the sputtering process was preceded by preliminary heating to 400° C, ion etching with direct current of 100 V with intensity 20 A, and then with modulated current with frequency of 240 kHz. Sputtering process was carried out in vacuum below 10^{-7} Pa in argon and krypton atmosphere.

Table 1. Proportion of coating materials and predicted thickness

Pos.	Coating	Coating thickness	Proportion of targets with coating material
1	AlTiN	3.0 - 4.0 µm	3 x TiAl70 (70 points Al) 1 x Ti
2	TiAlN	2.0 - 3.0 µm	3 x TiAl60 (60 points Al) 1 x Ti
3	TiAlN/TiSiN	4.0 - 5.0 µm	2 x TiSi23 (23 points Si) 2 x TiAl60 (60 points Al)

3 Surface characteristics

Geometrical measurements of surface roughness parameters were conducted in the first research stage with a use of optical profilometer Bruker Contour GT, with parameters using white light interference. Samples without coating were measured up before washing in industrial washing line and right after that coating was applied. Measured values of roughness parameters Sa from the surface with dimension 1 mm x 1 mm were described in Table 2. Measured surfaces were characterized by a uniform geometrical structure with local defects in a form of scratches.

Table 2. Parameters of surface roughness Sa.

Sample	Sa [µm]
Ti grade 2 without coating	0.209
AlTiN	0.217
TiAlN	0.234
TiAlN/TiSiN	0.223

4 Nanohardness studies

Nanohardness study of coatings was conducted in the next research stage using Olivier Phar method [5]. Berkovich diamond indenter of angle cut cube shape was pressed on ultra-high resolution nanoindenter of Anton Paar Company into tested surface with small P_{max} load (Figure 1) with value 2 mN and 5 mN. Sample surface was loaded gradually from the moment of contact to maximum adopted load in time 2 mN/min. Before the surface was loaded with the same velocity, P_{max} force maintained at a constant level for 5 seconds. At that time, indenter reached the maximum deflection h_c. After total removal of load, the indenter remained at residual deflection h_p, dependent on the quantity of plastic strain. 30 attempts were conducted for each coating with maximum load of 5 mN and 2 mN. Only the smallest load enabled to carry out study without excessive deformation of coatings. Hardness (H) was set from the relation:

$$H = \frac{P_{max}}{A(h_c)} \quad (1)$$

where A is a contact area of indenter with tested surface resulting from measured h_c values and applied indenter geometry. The effective Young's E_{eff} module was appointed considering stiffness S, which is tangent to the initial load curve. Both the deformation of studied material and indenter deformation according to the relation are affecting E_{eff} [5].

$$\frac{1}{E_{eff}} = \frac{1-\nu^2}{E} + \frac{1-\nu_i^2}{E_i} \quad (2)$$

The study was conducted with assumption that Young's module of indenter was E_i = 1140 GPa and its Poisson number ν_i = 0.07. Poisson's rate of coating was assumed as ν_i = 0.031. Measurement results of hardness and appointed effective Young's module were presented in Table 3. The use of tested nitride-coatings clearly increased the hardness of surface layer by approximately 7 times compared with substrate material. The largest hardness showed AlTiN coating, and the smallest layered TiAlN/TiSiN coating, however given the overlapping confidence intervals of results it is possible to state that coatings have similar properties in this respect. Similarly appointed values of the effective Young's module did not show major differences and analyzing all received load curves – it will not be possible to state, which of all coatings is more elastic-plastic. Clearly, we can see that study of technically pure titanium shows properties characteristic of plastic material (Figure 2a), and after applying the coating of studied nitrides, curves take a form of more elastic materials (Figure 2b). The surface of crosshatch area on Figures 2a and 2b indicates plastic strain, and it is clearly greater for pure titanium.

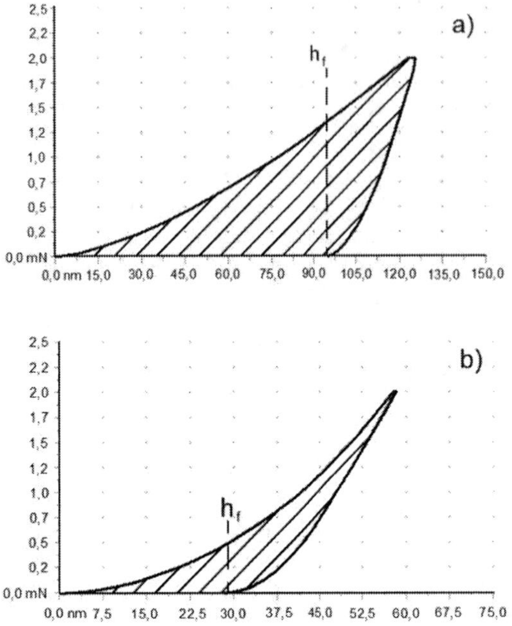

Fig. 2. Load curves – displacement in nanohardness studies; a – titanium grade 2; b – AlTiN.

Table 3. Results of measurements carried out with nanohardness.

Sample	Nanohardness [GPa]	Young's modulus E_{eff} [GPa]	Friction coefficient μ
Ti grade 2 without coating	4.99 ± 0.42	186 ± 13	–
TiAlN / TiSiN	33.91 ± 10.55	515 ± 175	0.16
TiAlN	36.08 ± 7.84	622 ± 185	0.10
AlTiN	36.49 ± 8.06	610 ± 162	0.29

5 Scratching tests

Scratching tests of coatings were carried out using Rockwell's indenter with a radius of 100 μm with a growing load of 0.5 N up to 30 N on 6 mm. The velocity of normal forces to the surface was 10 N/min. There were analyzed damages of coatings on scratches signs with a use of the optical microscope built-in micro combi tester presented in Figure 3a, b, c. In each test, three distinctive limits were indicated i.e. Lc_1, Lc_2 and Lc_3, which corresponded to the normal indenter forces respectively described in Table 4. Lc_1 was accepted as the normal force by which first damage of coating was visible usually with fracture along direction of the test. These fractures were peculiarly visible on hybrid coating TiAlN/TiSiN and provided cohesiveness of coating material [6]. AlTiN and TiAlN films presented more plastic initial deformations without distinct fractures (Fig. 3f, j). As Lc_2 limit adopted forces by which first wear of adhesion connection of the coating with substrate with crosswise fractures (Figure 3 e, i, l) and chips in the lateral parts of scratch sign. With Lc_3 limit, places of total coating rupture were indicated (Figure 3d, g, k). On TiAlN coatings, spreads of coating chips were often visible much earlier than clear limit of total coating destruction (Figure 3h). In carried-out scratching tests, the best adhesive properties to the substrate had hybrid coating TiAlN/TiSiN. Figure 3n presents visible transition between TiAlN and TiSiN coating. It is not possible to explicitly highlight the place of first coating wiping off. Noticeable fluent color change of scratch sign provides very good connection of both coatings.

Table 4. Results obtained in scratching tests.

Sample	Lc_1 [N]	Lc_2 [N]	Lc_3 [N]
TiAlN / TiSiN	0.68 ± 0.31	9.90 ± 3.29	23.90 ± 4.23
TiAlN	0.91 ± 0.54	7.23 ± 1.19	15.38 ± 3.43
AlTiN	0.91 ± 0.61	8.87 ± 2.40	18.93 ± 4.76

6 Discussion

Scratching tests were conducted to evaluate the adhesion. This method was often used by other researchers – among others [6, 7]. The following limits, appropriately indicating indenter load in marked places, were adopted in the evaluation [7]: Lc_1: cohesive failure to indicate edge or parallel cracking, Lc_2: adhesive failure to indicate delamination of film from its substrate, and Lc_3: total coating failure by the total exposure of the substrate surface. In the current work, Lc_2 was identified to mark the critical load that was required to break the bonding between the film and its substrate. As it is known from [7] and own studies, the results of scratching tests depend largely on such parameters as load rate, maximum test force and its duration. Other factors, determining the different test results, are associated with the same sample. It can be a different thickness of coating, local defects, different surface roughness and surface defects, such as scratches. In studies of this work, despite preserving identical test conditions, discrepancies were obtained in results of total destroying of coating presented in Table 4. What is interesting, on each sample conducted, 5 identical scratching tests and their results were often different. This indicates the stochastic nature of the phenomenon of adhesion breaking the coatings, and it is impossible to accurately determine this limit. Successfully, it is possible to relate above statements to the research on nanohardness where indenter enters the material on a very small depth of 50 – 150 nm while the average surface irregularities in analyzed samples were more than 200 nm.

DOI: 10.1051/itmconf/20171506006

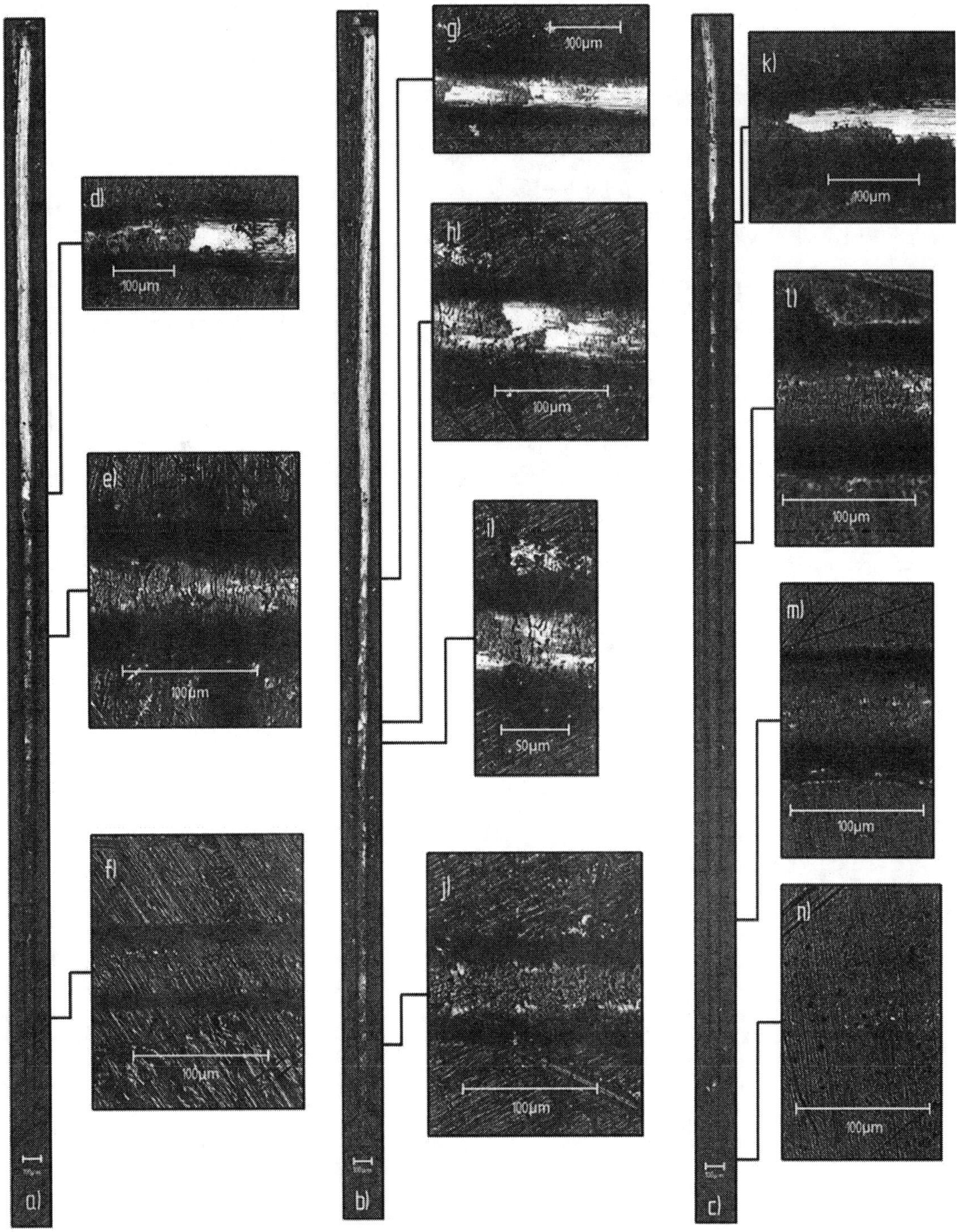

Fig. 3. Scratching test with Rockwell's indenter; a – scratching sign of AlTiN coating; b – scratching sign of TiAlN coating; c – scratching sign of TiAlN/TiSiN coating; d, g, h, k – surface of total coating rupture; e, i, l – surface of transverse fracturing caused by adhesive wear of coating; f, j, n – initial decohesion surface of coating material; m – transition from TiAlN to TiSiN coating.

DOI: 10.1051/itmconf/20171506006

7 Conclusions

In this study, AlTiN, TiAlN and TiSiN coatings were deposited on titanium grade 2. The surface roughness, nanohardness, as well as adhesion of the coatings were evaluated. According to the results provided, the following conclusions can be drawn:

- Modification of the surface layer of titanium materials using PVD magnetron sputtering, with studied nitride materials, allows for application in biomedical implants. Studies have shown that their use significantly improves exploitation properties, such as hardness and abrasion resistance compared with pure technical titanium grade 2.
- Application of analyzed coatings on titanium grade 2 increased the hardness of substrate surface layer by about 7 times, and the effective Young's module by 3 times.
- Based on conducted studies, similar hardness of analyzed coatings on titanic substrate was observed. Relatively great dispersion of test results on nanohardness does not enable explicitly to assess which coating is the hardest. Large disparity in results can be caused by too big surface roughness on tested samples. It is necessary to consider further research on more precisely polished samples.
- Hybrid coating TiAlN/TiSiN presented much better adhesion properties to the substrate as compared to the TiAlN and AlTiN.
- Friction coefficients in scratching tests are presented in the following order: AlTiN > TiAlN/TiSiN > TiAlN.

References

1. Q. Yao, J. Sun and others, Vacuum. **142**, 45-51 (2017)
2. M. Tarnowski, K. Kulikowski and others, Diam. Relat. Mater. **75**, 123-130 (2017)
3. P. Yi, L. Peng and others, Mater. Sci. Eng. C. **59**, 669-676 (2016)
4. H. Wang, R. Zhang, Ceram. Int. **41**, 11844-11851 (2015)
5. W.C. Oliver, G.M. Pharr, J. Mater. Res. **19**, 3–20 (2004)
6. M. Łępicka, M. Grądzka-Dahlke, D. Pieniak, K. Pasierbiewicz, A. Niewczas, Wear. **382–383**, 62–70 (2017)
7. A.R. Bushroa, H.H. Masjuki and others, Surf. Coat. Tech. **206**, 1837-1844 (2011)
8. Wo P.C., Munroe P.R. and others, Mater. Sci. Eng. A. **527**, 4447-4457 (2010)
9. Deng Y., Tan Ch. And others, Ceram. Int. **43**, 8721-8729 (2017)
10. Chen J., Li H. and others, Surf. Coat. Tech. **308,** 289-297 (2016)
11. B. Surowska, J. Bieniaś, M. Walczak, K. Sangwal, A. Stoch, App. Surf. Sci., **238**, 288-294 (2004)
12. J. Bieniaś, B. Surowska, A. Stoch, H. Matraszek, M. Walczak, Dent. Mater. **25**, 1128-1135 (2009)

ITM Web of Conferences **15**, 06007 (2017)
CMES'17
DOI: 10.1051/itmconf/20171506007

Experimental and numerical assessment of the characteristics describing superelasticity in shape memory alloys – influence of boundary conditions

Jakub Bryła[1] and *Adam* Martowicz[1*]

[1]AGH University of Science and Technology, Faculty of Mechanical Engineering and Robotics, Department of Robotics and Mechatronics, Mickiewicza 30, 30-059 Krakow, Poland

Abstract. Relatively recent discovery of shape memory alloys (SMA) justifies ongoing research on their properties and an attempt to explain the physical phenomenon responsible for the characteristic behaviour of SMA. Moreover, there have been reported many successful commercial SMA applications to medical cases, mostly based on superelasticity. Even though a wide application range is confirmed, its further contribution growth is currently not seen - mostly due to deficiency of reliable modelling techniques. Recently, lively discussion in the SMA academic community is observed, which deals with modelling issues and numerical implementation.
Considering the current trends, the authors of the work make an attempt at qualitative analysis of the material properties for superelasticity. The material characteristics – found using static stretching tests – are sensitive to the variation of local stresses induced in the area where a SMA sample is mounted in a fatigue testing machine. As shown, the phenomena present at the clamping area seem to initiate and govern the process of the solid phase transformation within the entire SMA body. The overall objective of the presented research is to assess the influence of the above stated boundary conditions on the properties of selected types of SMA, using both experimental and numerical results.

1 Introduction

Shape memory alloys (SMA) belong to smart materials. Their unique mechanical properties reveal through the two-way martensitic phase transition activated by force and thermal excitations [1-3]. This nanoscale phenomenon introduces the specific macroscale behaviour of SMA, namely one- and two-way shape memory effects, and superelasticity. In general, the mentioned effects cause the nonlinear mechanical response of SMA. During each of them the transformation between solid phases occurs, where martensitic and austenitic structures have the main role. The differences in material properties of both phases cause unique behaviour of SMA. Nature of structure transformations is briefly described below for better understanding of the physical background of the SMA behaviour. Next, the superelasticity is considered due to its adoption in many practical applications. Further details regarding the one- and two-way shape memory effects may be found in the authors' former works [4-7].

As previously mentioned, the martensite and austenite phases have different material properties. The most important, from practical reasons, are their values of the Young's modulus. Martensite is much less stiff than austenite. Moreover, the existence conditions of both phases are clearly identified. Therefore, the martensite phase appears in SMA at low temperature, contrary to the conditions when austenite may be found. Note that, the temperature below/above which only one solid phase in material is observed (martensite/austenite) varies according to the direction of temperature changes. Hence, a hysteresis in the material response is observed. The relation between phases percentage contribution rates and ambient temperature is schematically shown in Fig. 1.

Therefore, the four temperatures describe the conditions for the phase transformations between martensite and austenite in SMA: M_s, M_f, A_s and A_f, respectively marked in Fig. 1. As a matter of fact, these temperatures are functions of stress. It means that the above explanation refers to the case when mechanical constraints are not applied to an element made of SMA. The values of the mentioned characteristic temperatures rise along with greater stress generation in the material. This phenomenon has a significant impact on the SMA behaviour at specified temperature range, for which only austenite exists in a non-loaded element: $M_d > T > A_f$. Consequently, the values of the characteristic temperatures rise after applied mechanical constraints to a SMA element. Generation of a fairly high stress field causes the situation when $T < M_s$, which finally leads to initiation of phase transformation process. The phenomenon is called as superelasticity.

*Corresponding author: adam.martowicz@agh.edu.pl

DOI: 10.1051/itmconf/20171506007

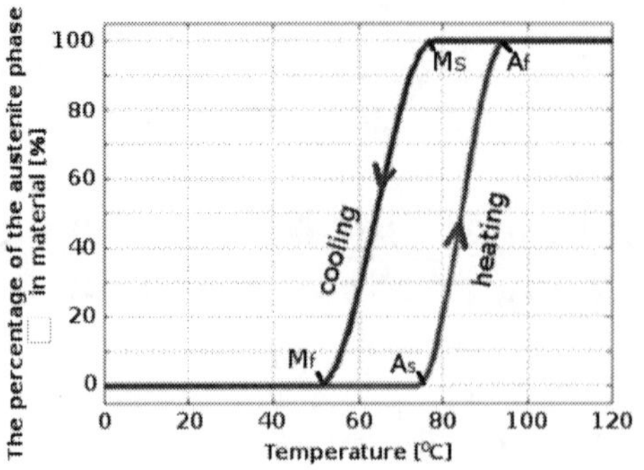

Fig. 1. Relation between phases contribution rates and temperature changes.

The superelasticity is a reversible effect. It means that martensite transforms back to austenite when a SMA component is unloaded. The full load cycle is shown in Fig. 2. One can observe that the stress-strain path depends on the constraint direction [8]. As a result the hysteresis occurs. This fact is widely applied when designing damping systems [9]. Application possibilities for a bumper made of SMA were also verified by the authors of the present work by performing uncertainty analysis [4-5]. Another application area is medicine, showing how crucial the existing features for popularization the SMA devices are.

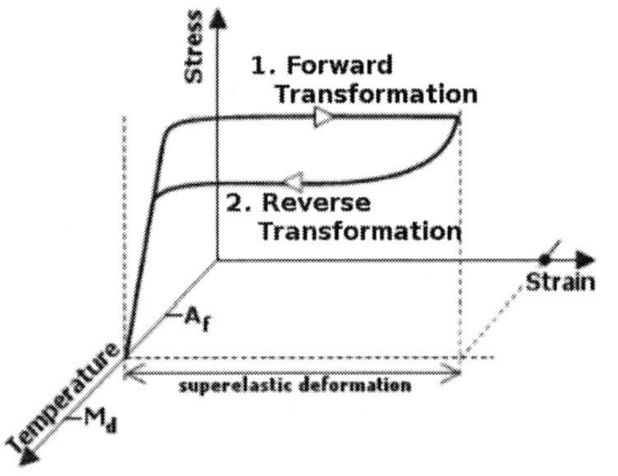

Fig. 2. Stress-strain characteristics exhibiting the superelasticity effect [8].

To understand the origin of high significance of the SMA properties, one should focus on the changes of characteristics shown in Fig. 2. The most crucial is the region where the stress-strain characteristics appoints constant stress field under wide strain field changes during phase transformations (a plateau region). Therefore, a SMA device can generate constant force while the displacement varies in the assumed range. Thanks to this phenomenon, SMA was successfully implemented in commercial applications such as medical staples and dental braces [10].

The unique behaviour of SMA is very attractive to be employed in unconventional cases where rigorous demands related to the number of movable parts and occupied space are required. Therefore the SMA devices are successfully developed in both aerospace and spacecraft applications [11]. However, SMA as smart materials are still at the research stage of their development. There is a group of existing phenomena with incomprehensible physical explanations. Presently, there is a great deal of effort to implement an efficient SMA model which would be able to accurately consider the numerous physical observations. The number of existing models shows that there is a lot of work to reach the aim. More information about different approaches to numerical modelling of SMA can be found in [12].

The present paper consists of 5 sections. After introductory Section 1, Section 2 provides description of the experiments carried out to obtain material properties for the tested SMA wire. Section 3 addresses a numerical model developed by the authors used to study influence of boundary conditions, followed by the results of numerical analyses presented in Section 4. Final Section 5 summarizes the paper and draws the concluding remarks.

2 Experiment: material properties determination

Before the details regarding carried out experiments are shown, the authors wish to introduce the background for SMA model validation issues, highlighting their specificity. Practical aspects of SMA modelling in the context of experimental activities are addressed in the following.

Most of commercial Finite Element Method (FEM) software packages have implemented SMA models. These models are often able to simulate only the phenomenon of superelasticity – as it is accessible in the ANSYS [13]. On the other hand, the MSC.Software MARC library contains two models of SMA, where the one-way memory effect is also implemented. However, one of them is the Aurrichio's model from 2001 [14], while the second one is the Saeedvafa's and Asaro's model proposed in 1995 [15]. It should be, however, noted that, a lot of new models have been proposed over more than 15 years, some of them employ more complex framework. Nevertheless, none of them was implemented in the mentioned FEM software packages. This can be seen as a determinant that reflects the difficulty of implementing such complex material models. Moreover, one should note that among all widely used approaches related to the SMA modelling [16], the FEM programs have implemented only the phenomenological models.

2.1 Experimental uncertainty for material parameters identification

One of the disadvantages of the phenomenological models is necessity to define numerous input parameters. In general, these parameters are divided into 3 groups:

- elastic parameters of both martensite and austenite such as: Young's modulus, Poisson's number, thermal expansion coefficient etc.,
- functions governing phase transformation caused by mechanical constraints,

- functions governing phase transformation caused by thermal constraints.

The most popular SMA material is Nitinol, which is a nickel and titanium based alloy. Another example is Flexinol, which has the same chemical composition as Nitinol, but different manufacturing process causes that smaller wire diameters are possible to be achieved. In both cases the literature review provides a wide range of the elastic parameters values, e.g. the Young's modulus for austenite is found to be in the range of 60-83GPa for Nitinol [17]. The reason for this fact may be related to various alloy additions – used in order to change some material properties such as values of the characteristic temperatures. The side effect of the alloy composition variation can be a significant change of the elastic properties. From the other side, the above mentioned observations may lead to necessary adaptation and use of different experimental procedures. Lack of datasheets on the SMA properties as well as lack of appropriate norms and standards are the factors which aggravate the problem of the material parameters uncertainty. In case of tensile tests, one should note the existence of the ASTM norm [18], which was created mainly due to medical restriction reasons. This 6-page document contains only the description of methods used to determine the material parameters for the superelasticity at room temperature. Moreover, most of researches use standard grips during the tensile tests. These components hold specimens in position by generation the compression force. The aim of this paper is to prove that this kind of constraints could have negative impact on reliability of experimental results.

2.2 Experiment: superelasticity in Flexinol wire

Considering the above statements, a uniaxial tensile test was performed for a Flexinol wire with diameter of 0.51mm and length of 50mm for which the temperature A_f equals 90°C. The experimental test stand contains a fatigue testing machine Instron8872 with maximal load 10kN and a heating chamber Instron 3119-505 with temperature control in the range -70°C to 300°C. However, the most important, from the research point of view, is the information about used grips. The scheme and schematically presented grip operation are shown in Fig. 3 and 4, respectively.

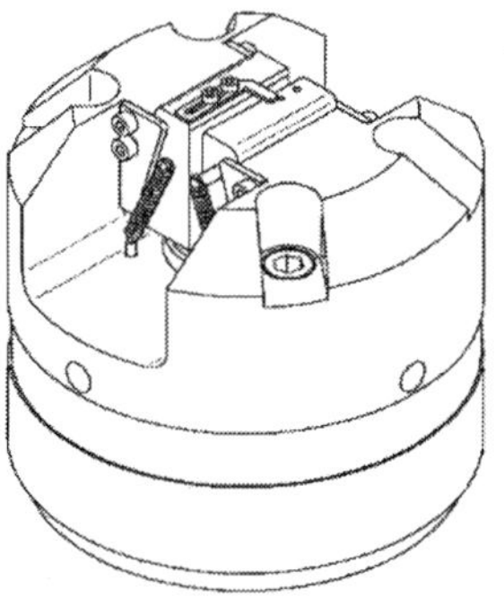

Fig. 3. Scheme of the grips [19].

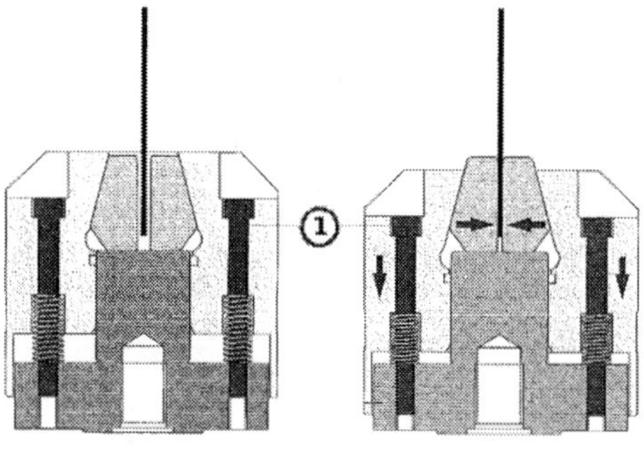

Fig. 4. The principle of the grip operation [19].

As shown in Fig. 4, tightening the screws 1 leads to closing the jaws. This method allows using a single grip for wide range of the specimen thicknesses. Moreover, the jaws are exchangeable and the various variants of their contact surfaces may be provided, which gives the opportunity to fit the experimental condition to demanded requirements. Taking into account the small diameter of the used Flexinol wire the authors used the most convenient grip configuration, which uses the jaws with serrated contact surfaces, as shown in Fig. 5.

Fig. 5. Serrated jaws used to hold Flexinol wire.

The intention of the authors was to perform the test following the most common condition applied by the research community. The similar system's configuration was used in many top works as [8], even for the high strain rates tests [20].

For the prepared measurement system, the test procedure was next defined. The used configuration gives an opportunity to control both the displacement of the grips as well as the force exerted on a tested specimen. The value change paths with respect to time obtained for the two mentioned parameters for three tests performed at 100°C are shown in Fig. 6. The displacement path is recorded as relative value where 0 is assigned to the initial position of the top grip. According both characteristics presented in Fig. 6, one can realize that the experiment contains six intervals:

1. pause for 5 seconds;
2. preloading sample with force of 10N in order to cancel the slacks – load rate: 10N/30sec;
3. pause for 5 seconds;
4. elongation of the sample up to 2.5mm – load rate: 0.05mm/s;
5. pause for 5 seconds;
6. return of the top grip to the position from the step 3 – the same load rate applies as in the step 4.

DOI: 10.1051/itmconf/20171506007

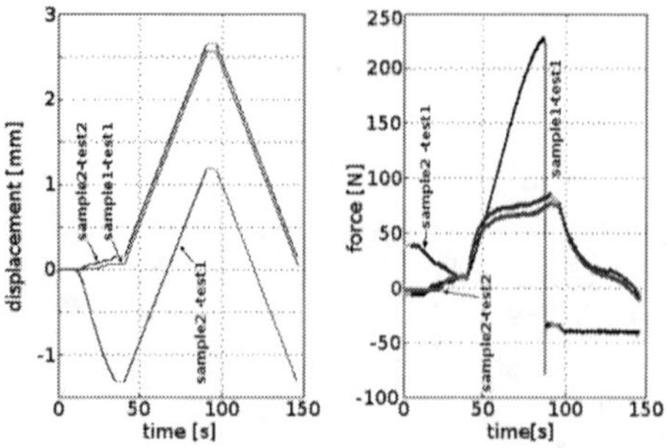

Fig. 6. The displacement of the grips (left) and the force exerted on a tested specimen (right).

As one can observe, the essential part of the test is the control process utilized by the displacement path – the mentioned steps 4-6. However, the initial position of the sample is set by the 10N force first. Therefore, the displacement path with respect to time is presented again according to the proper absolute initial position (reached at the end of step 2). The characteristics are shown in Fig. 7. Now, it is clear that for each experiment the top grip reaches the assumed displacement values from the steps 4 and 6.

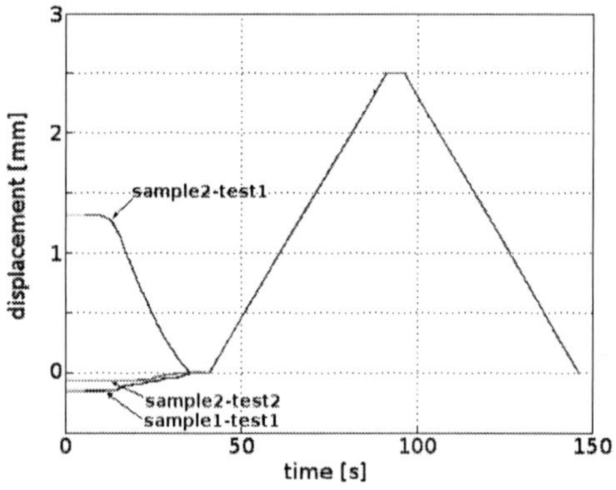

Fig. 7. The displacement path with respect to time according to the absolute initial position.

In the force-time characteristics in Fig. 6, it can be observed that the first sample was damaged during the test. This is one of the reasons why the essential parts of the tests were achieved by displacement controlling – due to the safety reasons. There should be also noted that the first sample was used earlier during development of the test procedure. It means that the Flexinol wire was subjected to cyclic loading (as in case of fatigue), during which the local stress field propagates within the whole body of the specimen. Moreover, the cut off part of the sample does not lose the shape memory features like the one-way shape memory effect. According to the above presented statement, the most likely reason of the specimen damage was the influence of the sharp edge of the jaw, which is located between serrated and a flat surface of the contact area – see Fig. 5.

The results for the second sample are analysed according to the force-displacement characteristics shown in Fig. 8. The hysteresis does not close because of two reasons: too high initial force as well as generation of compression stress at the end of the test sequence. Moreover, the hysteresis paths are slightly different for the two following tests. These two, recently mentioned phenomena interfere and lead to the two-way shape memory effects manifestation [8]. Summarizing, reduction of the initial force and usage of the force control instead of the displacement control during the step 6 should be sufficient to improve the observed results.

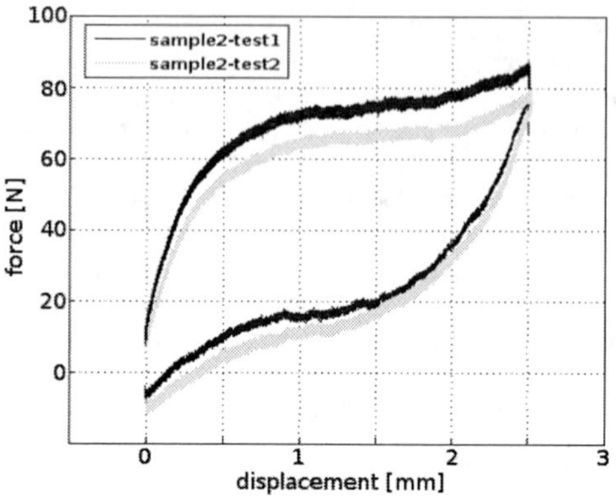

Fig. 8. The force-displacement characteristics of the experiments performed on the second specimen.

In the authors' opinion, the most important fact confirmed in the work, is that the presented results correctly project the superelasticity effect. Fig. 9 presents the stress-strain characteristics for the two considered tests. The values of the Young's modulus were calculated from the plots and they are 30.5GPa and 43GPa respectively for martensite (E^M) and austenite (E^A). The values fall within the range given in the literature, thereby confirming the correctness of the results.

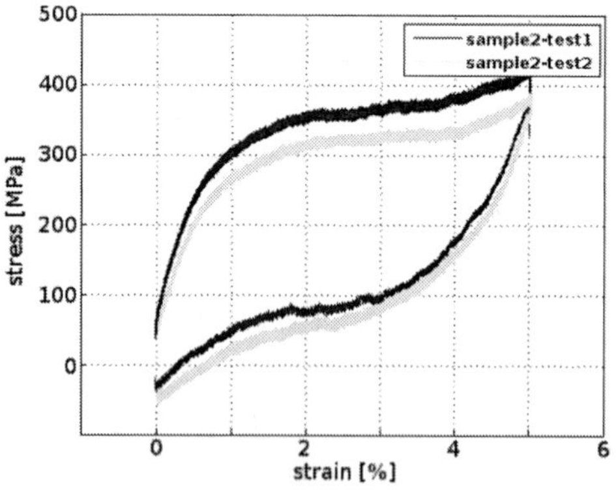

Fig. 9. The stress-strain characteristics of the experiments performed on the second specimen.

The second sample was subjected to 12 test cycles and all of the material responses convergence. On the other hand, the first sample was damaged after third test cycle. Both specimens were clamped in the same manner and with the same range of force and displacement values. The compression force generated by the grips is the only parameter out of the control, which can affect the test conditions considerably. As stated above, the hold strength is controlled by tightening the screw shown in Fig. 4. The disadvantage of this method is lack of a feedback on the reached value of the compression force. Thereby, the impact of this parameter on the material response is analysed in the simulation stage, presented in next sections.

3 Model of the tested system: SMA sample and grips of fatigue testing machine

The goal of the work is to reflect the experimental test conditions in the simulation software – MSC.Software MARC. Various parts of the measurement system were taken into account. There should be noted that presented SMA analysis considers strong nonlinearity because of specificity of the material behaviour as well as contact conditions. As a matter of fact, the number of FEs should be controlled to reach both result's convergence and acceptable computational time. Therefore, only half of the system is modelled based on symmetry of the model. Moreover, only the jaws and specimens are modelled as deformable bodies. The rest of the machine parts are assumed as rigid bodies due to their large masses and volumes compared to the parameters of the tested Flexinol wire. The final model (shown in Fig. 10) was created using the interface MSC.Software MENTANT. The way how the jaws were built is also presented in the plot to show the effort of minimisation the number of FEs. Finally, the model contains over 24500 FEs, where 17000 FEs represent the Flexinol wire.

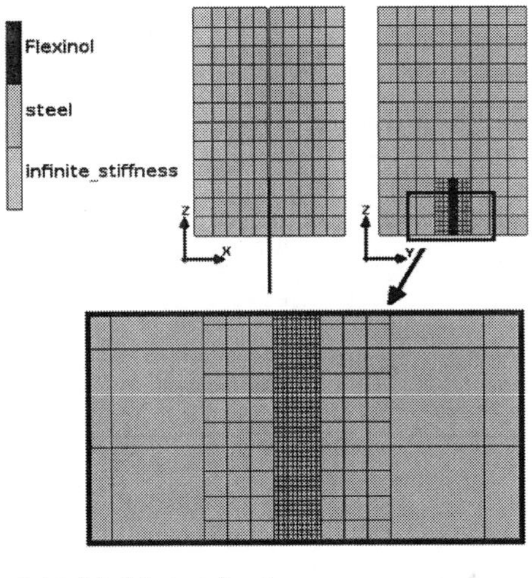

Fig. 10. Model of the tested system.

Simulation of the experimental work requires definition of two load cases: first one to analyse the jaws compression and second one to reflect the sample elongation. Therefore, proper boundary conditions must be defined. The functions used to describe the two forces, responsible for wire compression by the jaws and its further elongation, are crucial when obtaining required outcomes - which would confirm a hysteretic character of the material response. There should be mentioned that quasistatic analyses were performed. The constraint paths referring to the simulation steps for both above mentioned boundary conditions are shown in Fig. 11. Compression of the jaws was controlled with the displacement constraint, following the principle of grip operation – as seen in Fig. 4. The maximum value of this parameter, i.e. the displacement while compression of the jaws, is changed at each simulation, while the force path remains unchangeable. One can observe that the first load case lasts for 10 steps, while the second covers 400 steps. The reason of this situation is imposed by the convergence requirement of the analysis.

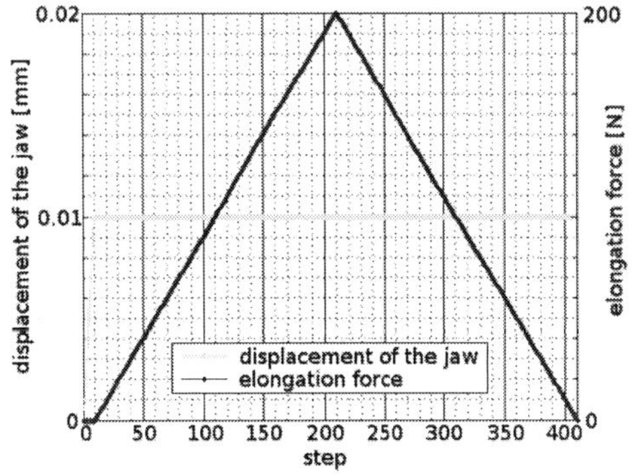

Fig. 11. The simulation constraints: displacement of the jaw and elongation force.

The simulation part of the paper is a theoretical one – a validation process was not performed yet. Therefore, the material parameters and input model data took the recommended values - after [17]. Only the Young's modulus values were changed according to the validated values, as described in Section 2.2, to improve the results. By default, these parameters are equal in the MARC software. Table 1 presents the most important input data used in the model.

Table 1. Input parameters

Parameter	Value
E^A Young's modulus for austenite	43.297GPa
E^M Young's modulus for martensite	30.455GPa
ν Poisson's number	0.3
σ_s^{AS}	500MPa
σ_f^{AS}	500
σ_s^{SA}	300
σ_f^{SA}	300
ε Transformation Strain	0.07

 DOI: 10.1051/itmconf/20171506007

The four stress parameters present in Table 1 mean respectively:

- σ_s^{AS} – tensile stress for which transformation from austenite to martensite begins;
- σ_f^{AS} – tensile stress for which transformation from austenite to martensite ends.
- σ_s^{SA} – tensile stress for which transformation from martensite to austenite begins;
- σ_f^{SA} – tensile stress for which transformation from martensite to austenite finishes.

4 Numerical results

In this section the superelasticity effect is numerically simulated in tensile tests. The studies are divided into two parts: first one presents the results for a theoretical elongation, while second one is devoted to the experiments in fatigue testing machine.

4.1 Theoretical results

The theoretical elongation case was studied by including only the SMA wire in the simulation. Therefore, the first load case described in Section 3 was omitted and the specimen was only strengthened by forces applied to its ends. The simulation was performed for a 10mm-long SMA wire. Finally, the force and displacement data was obtained from the result file for each step. The identified characteristics are shown in Fig. 12. The obtained results are considered as reference for further analysis. Thereinafter, the mentioned relation will be recalculated to presents stress-strain characteristics due to their more general form, which better reflects the material behaviour.

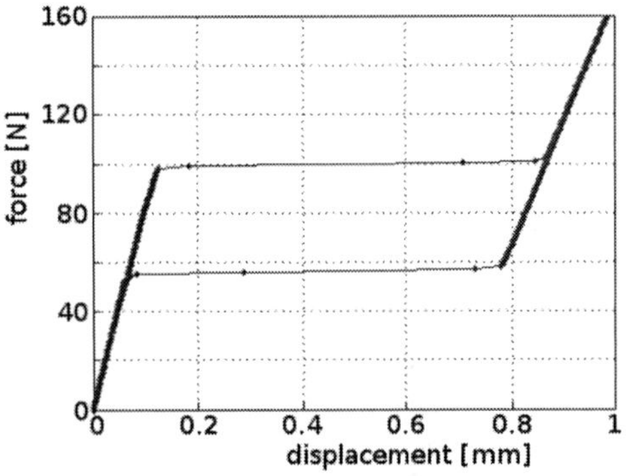

Fig. 12. The force-displacement characteristics of the theoretical elongation – simulation results.

4.2 Simulation of elongation experiments with use of fatigue testing machine

Five simulations were performed for different maximum values of the compression force applied during the first load step. As mentioned in Section 3 the boundary condition's constraint is defined by the displacement path. The maximum values for the performed analyses are in the range [0.01, 0.05mm] with the increment step 0.01mm. The force path defined for the second load step remains the same for all cases and is shown in Fig. 11.

The obtained results are compared with the referential ones, for which a theoretical case was solved. The simulations were performed for a 200mm-long SMA wire, which is clamped between jaws along the longitudinal distance of 10mm. Therefore, in all simulations the 10mm-long part of the specimen is elongated which gives opportunity to perform a quantitative analysis. Fig. 13 presents the stress-strain characteristics from all simulations.

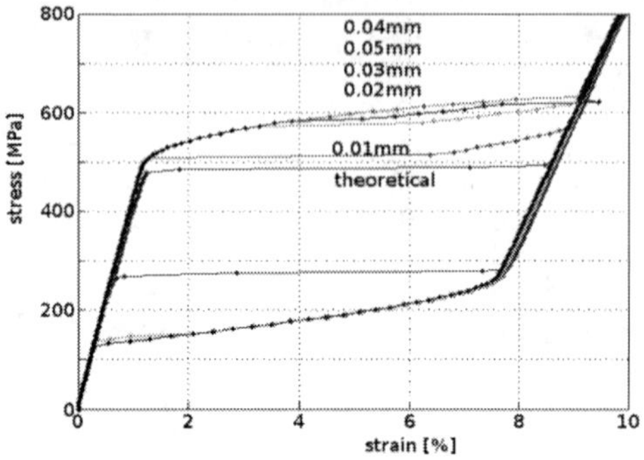

Fig. 13. The stress-strain characteristics for all simulations – node located at the specimen central longitudinal axis.

Starting from the initial point of the curves seen in Fig. 13, i.e. at the coordinates (0,0), we can observe that the Young's modulus for elastic strain is a little higher than the ones found for the theoretical case and the value of σ_s^{AS}. The last four simulations show a nonlinear behaviour for the transformation path and much higher stress value where it finishes. The exception is the first case where the maximum displacement of the jaws is 0.1mm. In this simulation the transformation path is similar to the theoretical one for major part of the path. Therefore, the behaviour of the specimen is changed during the transformation process. The stress-strain characteristics freezes at a specific level of the stress value equal 519MPa to be almost parallel to the strain axis, what takes place in the 115-th step of the simulation. From this purpose, Fig. 14 presents the contribution percentage of martensite in the material for steps from 114 to 122. Fig. 14 contains information about values of the parameters for the SMA wire close to the contact area. The results are presented for two planes: XZ and YZ – one of the grips was hidden to plot FEs of SMA wire at the second plane.

As one can observe, at the 115-th step the phase transformation is started in the whole body of specimen. This phenomenon appears on the stress-strain path by changing the slope of the curve. However, the phase transformation actually starts earlier. Fig. 14 presents the distribution of the martensite in material which is heterogeneous. The phase transformation begins at the contact area of the SMA wire and grips. The shown force-displacement as well as stress-strain characteristics present the results for node located on the longitudinal axis of the wire, i.e. for the coordinates (0, 0, 10mm). Next, the analysis extends to the points placed at the wire circumference, i.e. (0, 0.255, 10mm) as well as

(0.255, 0, 10mm). The stress-strain characteristics for both nodes are respectively presented in Fig. 15 and 16, for all simulations.

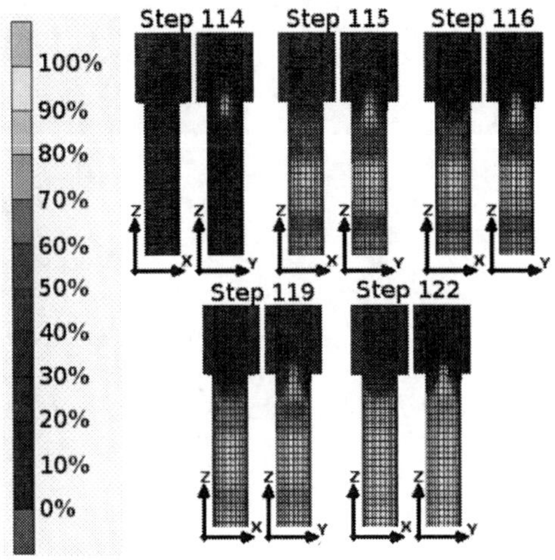

Fig. 14. The simulation results for first case (0.01mm) – MENTAT visualisation of results presented in Fig. 13.

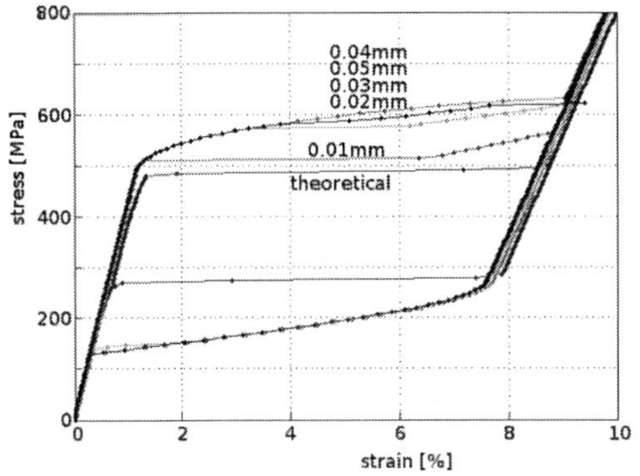

Fig. 15. The stress-strain characteristics for all simulations – node located at the wire circumference and X axis.

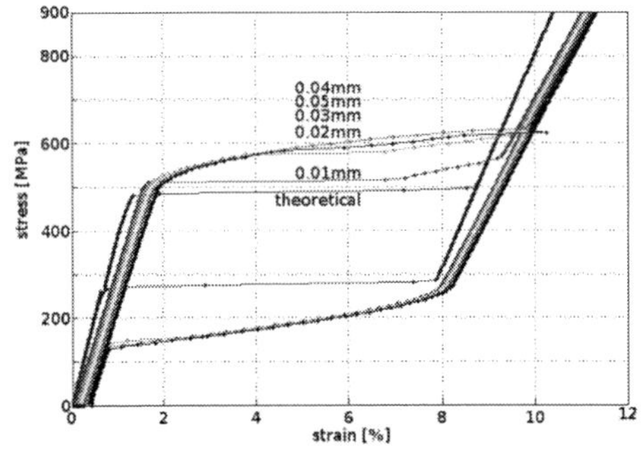

Fig. 16. The stress-strain characteristics for all simulations – node located at the wire circumference and Y axis.

By comparison of Fig. 13 and 15-16 one can see that the first two mentioned characteristics are almost the same, while the last one is significantly different from the previous ones. In the last shown relationship (Fig. 16) the initial strains are observed due to compression force generated by the jaws. Moreover, different stiffness value is observed for elastic strain at the area of the highest stress value (as shown in Fig. 14).

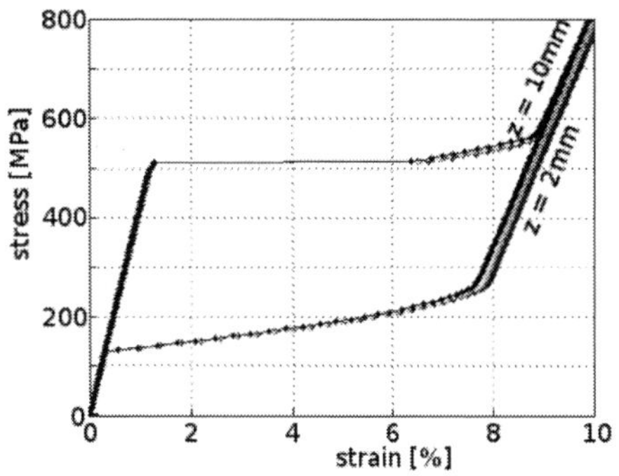

Fig. 17. The stress-strain characteristics for first simulation (0.01mm displacement of jaws) – nodes located at the specimen Z coordinates: {2, 4, 6, 8, 10mm}.

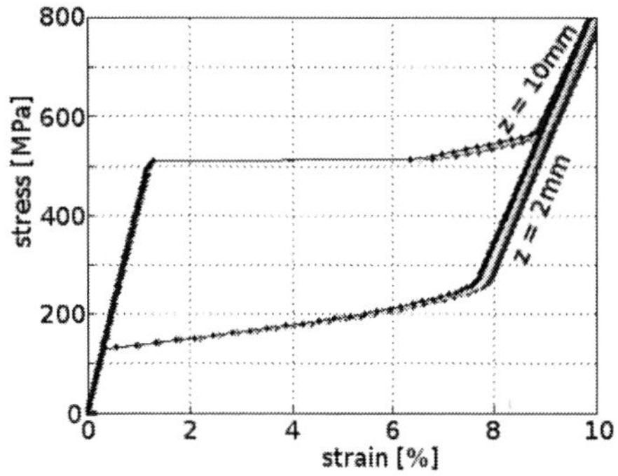

Fig. 18. The stress-strain characteristics for first simulation (0.01mm displacement of jaws) – nodes located at the wire circumference and X axis with Z coordinates: {2, 4, 6, 8, 10mm}.

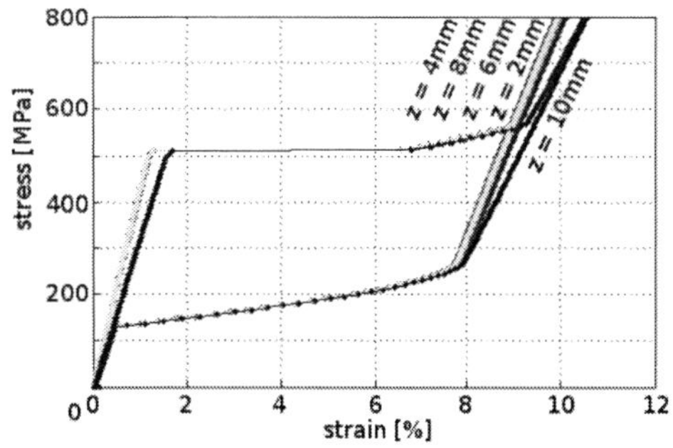

Fig. 19. The stress-strain characteristics for first simulation (0.01mm) – nodes located at the wire circumference and Y axis with Z coordinates: {2, 4, 6, 8, 10mm}.

Fig. 14 confirms that the martensite distribution is heterogeneous not only for the points placed on the wire circumference, but also along Z axis. Therefore the results for other nodes placed in the whole body of SMA wire are also presented. Exemplary characteristics are shown only for the simulation with 0.01mm displacement of the jaws. Hence:

- Fig. 17 presents the results for nodes with coordinates (0, 0) on the XY plane;
- Fig. 18 presents the results for nodes with coordinates (0, 0.225mm) on the XY plane;
- Fig. 19 presents the results for nodes with coordinates (0.225mm, 0) on the XY plane;

Each figure contains the characteristics for nodes with the coordinate along Z axis from the set: {2, 4, 6, 8, 10mm}.

5 Summary and concluding remarks

The purpose of the study is to verify the influence of boundary conditions on the superelasticity effects occurring in SMA. Hence, the tensile test procedure was modelled in FEM software. The simulation includes contact conditions between the jaws of the grips and wire made of SMA as well as implementation of boundary conditions reflecting real constraints which appear during experimental tests. The theoretical elongation was simulated to obtain the referential data for the other models. Finally, five models were defined with different values of compression force generated by the grips. Therefore, six simulation cases were studied.

As expected, the compression force affects the local phase transformation in the SMA specimen significantly. Initially, the authors expected that martensite fraction occurrence at the beginning of the simulation would lead to decrease of the wire stiffness value. However, the Young's modulus for elastic region of the stress-strain curve is slightly higher than the one for austenite for all performed simulations. On other hand, in the martensite occurrence region the elastic modulus is lower than for the mentioned phase. The authors consider the influence of geometry of the deformed wire on the stiffness factor as one of possible explanation for the above mentioned observation. However, the analysis should be continued to verify the proposed origin of the phenomenon. Moreover, other aspects in superelasticity behaviour which require explanation are indicated:

- significant increase of dissipation energy opportunity which is proportional to the size of the area enclosed by hysteresis (seen in Fig. 15 and 16);
- changes of the values of σ_f^{AS} as well as σ_f^{SA}, while two others (σ_s^{AS} and σ_s^{SA}) remain almost constant;
- nonlinear shape of transformation curve (a plateau region for the stresses, e.g. seen in Fig. 15 and 16).

There should be noted that the results obtained for the case with prestress (due to jaws compression) are more similar to the experimental response, presented in Section 2, than the theoretical one. That should be considered as motivation to continue the study. However, future validation is essential to gain quantitative dependencies which could confirm the above stated ideas.

This work has been supported by the AGH University of Science and Technology, WIMiR, research grant no. 15 .11.130.633 and 11.11.130.560.

References

1. M. Jianzuo, H. Haolei, H. Jin, Adv. in Mat. Sc. And Eng., **1-6** (2013)
2. F. Auricchio, E. Bonetti, G. Scalet, F. Ubertini, Int. J. Plast., **59** (2014)
3. D. Lagoudas, P. Entchev, P. Popov, E. Patoor, L. Brinson, X. Gao, Mech. Mater., **38** (2006)
4. J. Bryła, A. Martowicz, *Projektowanie mechatroniczne: zagadnienia wybrane*, (AGH, 2016)
5. A. Martowicz, J. Bryła, T. Uhl, *Uncertainty quantification for the properties of a structure made of SMA utilising numerical model* (ISMA2016 & USD2016)
6. J. Bryła, P. Palenica, A. Martowicz, *Projektowanie mechatroniczne: zagadnienia wybrane* (AGH, 2016)
7. J. Bryła, A. Martowicz, *Mechanics&Control*, 34:**3**, (2015) (in press)
8. D. Lagoudas, *Shape Memory Alloys: Modeling and Engineering Applications* (Springer, 2008)
9. F. Aurcchio, D. Fugazza, R. DesRoches, J. Eng. Mat. and Tech., **128** (2006)
10. L. Lecce, A. Concilio, Shape Memory Alloy Engineering, *Butterworth* Heinemann, Chapter 11, (2014)
11. O. Godard, M. Lagoudas, D. Lagoudas, Smart Struct. and Mat., **5056** (2003)
12. C. Cisse, W. Zaki, T. Zineb, *Smart Mater. and Struct.*, **10**, (2016)
13. S. Imaoka, ANSYS Tech. Supp. Gr., **4** (2014);
14. F. Auricchio, Int. J. Pl., **17** (2001);
15. M. Saeedvafa, R. Asaro, Los Alamos Report (1995)
16. A. Khandelwal, V. Buravalla, Int. J. Struct., **1**,111-148 (2009)
17. Marc Volume A: Theory and User Information. Release Marc2015 (2015);
18. ASTM International, Standard Test Method for Tension Testing of Nickiel-Titanium Superelastic Materials (2006)
19. Fatigue Testing Machine documentation (2017)
20. A. Vitiello, G. Giorleo, R. Morace, Smart Mater. Struct., **14** (2015).

Structure and tribological properties of MoS_2 low friction thin films

Agnieszka Paradecka[1,*], *Krzysztof* Lukaszkowicz[1], *Jozef* Sondor[2], and *Mieczysław* Pancielejko[3]

[1]Silesian University of Technology, Institute of Engineering Materials and Biomaterials, 44-100 Gliwice, Poland
[2]LISS, a.s., Dopravni 2603, 756 61 Roznov p.R., Czech Republic
[3]Koszalin University of Technology, Department of Technical Physics and Nanotechnology, 75-453 Koszalin, Poland

Abstract. The main aim of the studies was the deposition of the AlCrN film, covered by molybdenum disulphide (MoS2) – based lubricant, on the austenitic steel substrate. The AlCrN and MoS2 layers were deposited by PVD lateral rotating ARC-cathodes (LARC) and magnetron sputtering technology on the X6CrNiMoTi17-12-2 respectively. Structural characterizations of the MoS_2 thin films have been carried out using SEM (scanning electron microscopy) and AFM (atomic force microscopy) to determine the surface topography as well as HRTEM (high-resolution transmission electron microscopy) and Raman spectroscopy for structural investigations. The tribological wear relationships using ball-on-disc test were specified for surface layers, determining the friction co-efficient and mass loss of the investigated surfaces. Tests of the coatings' adhesion to the substrate material were made using the scratch test. HRTEM investigation shows an amorphous character of the MoS_2 layer. In sliding dry friction conditions, the friction co-efficient for the investigated elements is set in the range between 0.4-0.5. The investigated coating reveals high wear resistance. The coating demonstrated a dense cross-sectional morphology as well as good adhesion to the substrate. The good properties of the PVD AlCrN+MoS_2 coatings make them suitable in various engineering and industrial applications.

1. Introduction

The modern way of countering the harmful effects of friction for the elements of machines working under fixed tribological load is the use of solid lubricants in the form of thin films or coatings on sliding surfaces. These films provide protection in trying conditions, chemical stability, low coefficient of friction and abrasion resistance [1, 2].

One of the most interesting low-friction coatings is molybdenum disulphide (MoS_2), mainly used as a friction reducer in dry, vacuum and inert gases [3, 4]. The fundamental problem in the use of MoS_2 as a solid lubricant is a deterioration of its tribological properties in a humid environment. The reaction of MoS_2 and H_2O formed MoO_3 and H_2S (gaseous). MoO_3 shows a higher shear strength, which contributes to increased coefficient friction and the reduced lifetime of coated components. However, the sensitivity to the humidity of pure MoS_2 coatings can be reduced by, e.g. adding Ti, Cr, C, Al, Au and TiN, or increasing their density [5, 6]. Currently, MoS_2 coatings with the addition of Ti can be used not only in a vacuum or zero relative humidity but also at up to 50% humidity without losing their tribological properties, which allowed the extensive use of terrestrial applications [6].

In the category solid lubricant coatings MoS_2 is classified as soft. The anisotropic crystal structure is influenced by the good tribological properties [7]. These coatings can be applied to the steel substrate by bonding, burnishing, and more modern methods such as physical vapour deposition, which allow better adhesion to the substrate and amorphous structure of the coatings [3, 4, and 7]. The MoS_2 coating lubrication system is derived from a honeycomb structure that consists of a hexagonal array where the molybdenum atoms are surrounded by six sulphur atoms placed at corners of a triangular prism. Strong covalent bonds occur between atoms in layers (atom S and atom Mo_2), and weaker van der Waals forces between adjacent layers (S-Mo-S) allow for a low friction coefficient. The triple layer (S-Mo-S) can exist in three variants: the hexagonal (2H-MoS_2), the rhombohedral (3R-MoS_2) and the trigonal (1T-MoS_2) of which only hexagonal exhibit a low coefficient of friction [8]. In addition, the lubrication effect is extended by transferring the MoS_2 molecules to ablated areas [9]. Particularly important here is film-substrate adhesion, where the stronger it is, the longer the shelf life of the film [6]. MoS_2 coatings are widely used in space applications and industry (e.g. bearings, blades and vanes of turbine engines, HCCI engines, aircraft engines). The development of nanomaterials has allowed them to use their physical properties in nanocomposites,

* Corresponding author: agnieszka.paradecka@polsl.pl

DOI: 10.1051/itmconf/20171506008

for next-generation electronic devices and nano-electro-mechanical systems and optoelectronics [5, 8, 10-13].

The aim of this paper was to examine the microstructure, and tribological properties of MoS_2 films with additional hard AlCrN interlayer deposited on the austenitic steel X6CrNiMoTi17-12-2.

2. Experimental details

The examinations were made on the X6CrNiMoTi17-12-2 austenitic steel (substrates are cylindrical with 30 mm base diameter and 5 mm thickness) disk covered by AlCrN+MoS_2 coating.

The AlCrN film was deposited in the unit π1511 based on PLATIT's rotating and planar ARC-technology. The coatings deposition was carried out in an Ar and N_2 atmosphere. Cathodes containing pure metal Cr and Al were used for coatings deposition. After pumping the chamber to a base pressure below 5×10^{-5} Pa the substrate was heated to an appropriate temperature. Then, the substrates were cleaned by argon ion at a pressure of 2 Pa and applied bias voltage of 600 V/400 V for 20 min. To improve the adhesion of the coatings, after heating and both cleaning steps the LARC cathode (Cr) was burnt and formed of CrN adhesion layer. The MoS_2 topcoat was deposited by the magnetron sputtering method in the unit PLATIT Pl200. The deposition conditions are summarized in Table 1.

Table 1. Coating deposition parameters.

Parameter	AlCrN	MoS_2
Working pressure [Pa]	3	-
Substrate bias voltage [V]	60	-
Arc of target/magnetron [A]	250/235	1.2
Process temperature [°C]	450	80
Time [s]	4200	1640

The cross section of the deposited coatings was mapped with the use of the SUPRA 35 scanning electron microscope (SEM). The detection of secondary electrons was used for generation of fracture images with 4 kV bias voltage.

Diffraction and thin film microstructure was tested with the use of the TITAN 80-300 ultrahigh resolution scanning/transmission electron microscope. The thin cross-section lamellas for Transmission Electron Microscopy (TEM) observations were prepared by Focused Ion Beam (FIB) technique using Quanta 200i instrument with gallium ions.

Raman measurements were performed at room temperature with an in Via-Raman microscope (Renishaw) with a 633 nm Ar^+ laser, with a resolution 1 cm^{-1}.

The topography of the samples was examined using the atomic force microscope (AFM) Park System X E-100.

Adhesion of the coating to the substrate material was verified by the scratch test on the CSEM REVETEST device, by moving the diamond indenter along the examined specimen's surface with gradually increasing the load. The tests were made using the following parameters: load range: 0–100 N, load increase rate (dL/dt): 100 N/min, indenter's sliding speed (dx/dt): 10 mm/min.

The friction coefficient and wear rate of the coating were determined in the ball-on-disc test. The tests were carried out on the THT (CSM Instruments) device with the following parameters: sliding speed: 0.2 m/s; normal load: 10 N; counterpart: Al_2O_3 of 6 mm diameter; sliding distance: 1000 m; temperature: 23°C (±1°C); relative humidity: 26% (±3%).

3. Results and discussion

The cross-section of the investigated coating is presented in Fig. 1. The coating presents a compact structure; there is no trace of any coating delamination. The morphology of the MoS_2 layer is characterized by a porous microstructure. The clear boundary line is visible between AlCrN and MoS_2 layers. The adhesion layer between AlCrN and steel substrate is not as well defined but still visible on the SEM image. The measured thickness values of the particular layers MoS_2, AlCrN and adhesion layers (CrN) are approximately 1.2 μm, 2.0 μm and about 130 nm, respectively.

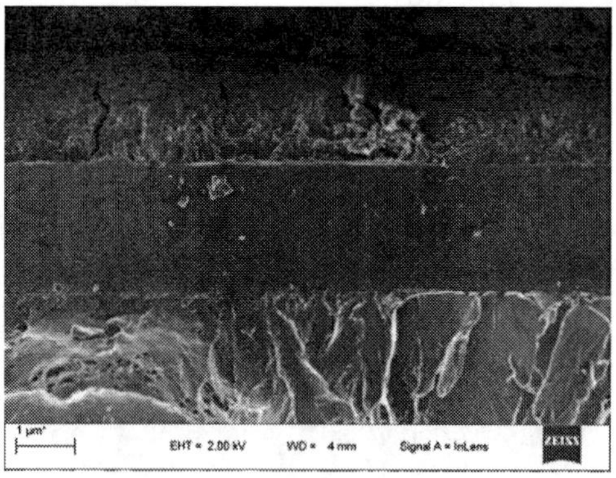

Fig. 1. The cross-section image (SEM) of AlCrN+MoS_2 coating deposited on the steel substrate.

Subsequently, for the coating structure characterization, TEM and HRTEM microscopes were used. The images, presented in Fig. 2, were obtained from selected regions. The bright-field images (Fig. 2a) and the HRTEM micrograph (Fig. 2b) show an amorphous character of the MoS_2 films. As expected, the electron-diffraction patterns obtained show a considerable broadening of diffraction rings (the inset image in Fig. 2a).

DOI: 10.1051/itmconf/20171506008

In the case of the AlCrN layer, it can be concluded, based on HRTEM images, that the investigated layer has a nanocrystalline structure (Fig. 3). It was found, by examining thin lamellas from the cross section of AlCrN layer produced by the arc method, that the layer features a compact structure with high homogeneity and grain size is less than 10 nm.

The observations of the analysed coatings surface topography using the AFM revealed that the AlCrN/MoS_2 is characterised by a high inhomogeneity relating to multiple droplet-shaped or nearly ball-shaped particles existing on the surface (Fig. 4).

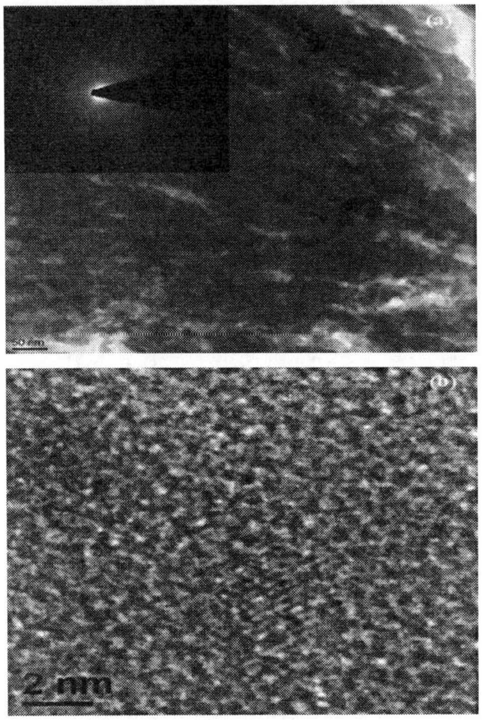

Fig. 2. MoS_2 top layer: a) TEM bright-field image and electron diffraction pattern, b) HRTEM micrograph.

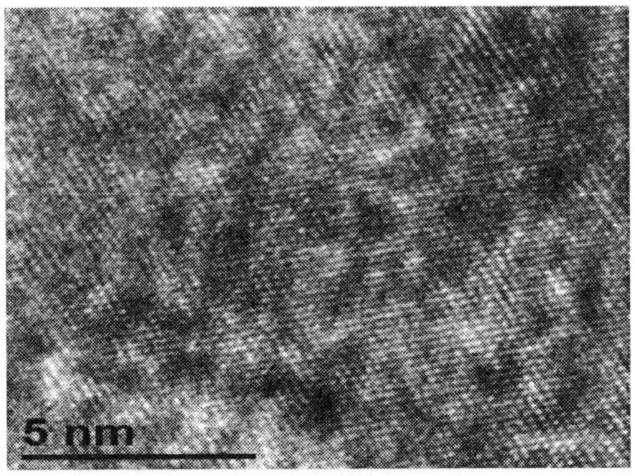

Fig. 3. HRTEM micrograph of the AlCrN layer.

An additional surface development analysis performed determining a ratio of the actual area measured for a given specimen according to its orthogonal projection on the XY plane revealed a large development for hard nitride layers. Surface development increases by about 10 % according to the coating type as a consequence of depositing a low-friction layer onto the surface of nitride layers. The measurements taken also allowed us to determine the layers surface roughness (by R_a).

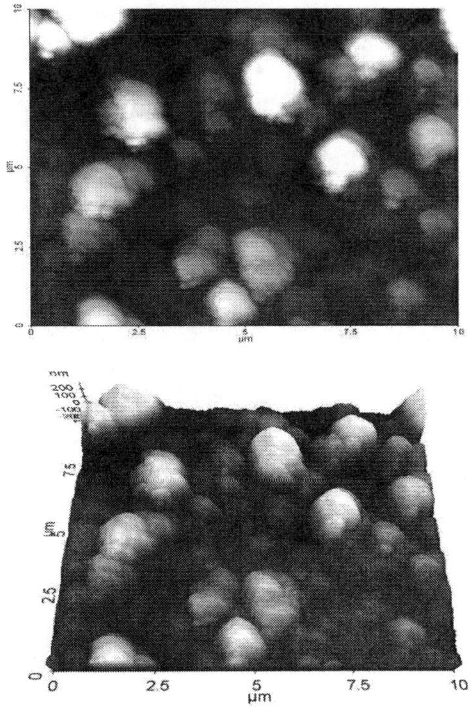

Fig. 4. Two- and three-dimensional AFM images of AlCrN+MoS_2 thin films

Raman spectroscopy shows first-order bands a result of vibrational modes within the S–Mo–S layer (227 cm^{-1}, 313 cm^{-1}, 354 cm^{-1}, 453 cm^{-1}) (Fig. 5). Additional recorded peaks 489 cm^{-1} and 566 cm^{-1} belong to the second-order lines derived from resonance type of Raman Effect. The noticed peaks 202 cm^{-1} and 726 cm^{-1} come from MoO_2 bonds. Visible broadening further on in the spectrum is due to the amorphous structure of layer [14, 15].

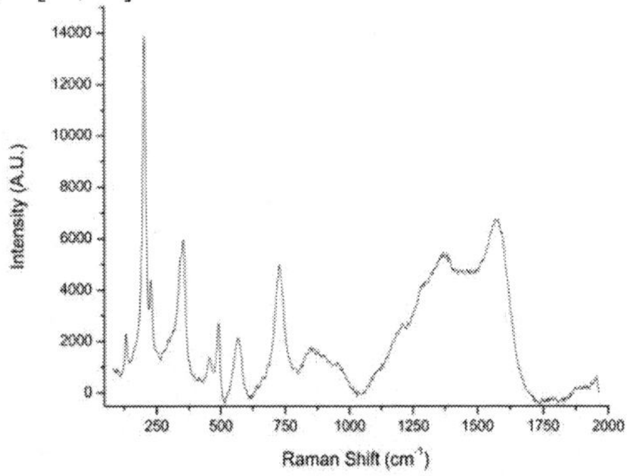

Fig. 5. Raman spectra of AlCrN+MoS_2 coating.

Critical load values L_{C1} and L_{C2} were determined using the scratch-test method with a linearly increasing load, characterizing the adhesion of the investigated coatings to the steel substrate (Fig. 6). The load at which the first coating defects appear is known as the first critical load, L_{C1}. The first critical load, L_{C1}, corresponds to the point, at which the first damage is observed (Fig. 7a); the first appearance of micro-cracking, surface flaking outside or inside the track without any exposure of the substrate material – the first cohesion – indicates a failure event. The second critical load, L_{C2}, is the point at which a complete delamination of the coating starts; the first appearance of cracking, chipping, spallation and delamination outside or inside the track with an exposure of the substrate material – the first adhesion – indicates a failure event (Fig. 7b). The investigated coating shows relatively good values of the critical load. The first failure occurs at a value of ~14 N. The second critical load, L_{C2}, occurs at 48 N. This may be caused by a significantly lower hardness of the X6CrNiMoTi17-12-2 steel substrate (in compare of tool steel).

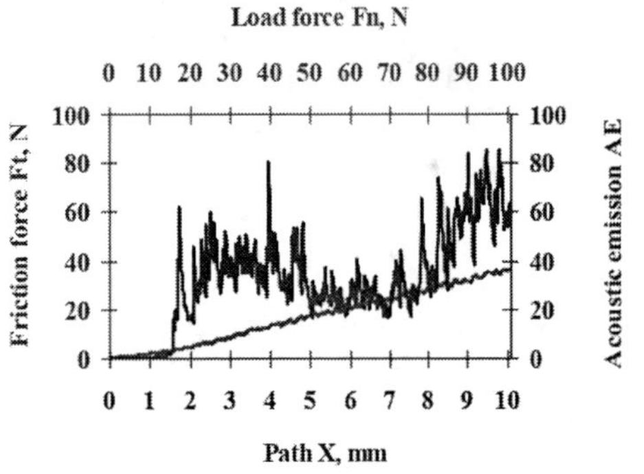

Fig. 6. Acoustic emission (AE) and friction force F_t as a function of the load Fn for AlCrN+MoS_2 coating.

To determine the tribological properties of the AlCrN+MoS_2 coating, an abrasion test under dry slide friction conditions was carried out by the ball-on-disk method. Fig. 8 presents the graph of friction coefficient μ changes obtained during wear tests in relation to counterpart with Al_2O_3. The friction curve has an initial transitional state of an un-stabilized course, during which the friction coefficient is reduced along with the growth of sliding distance up to obtaining the stabilized state, which normally occurs after a distance of about 100 m.

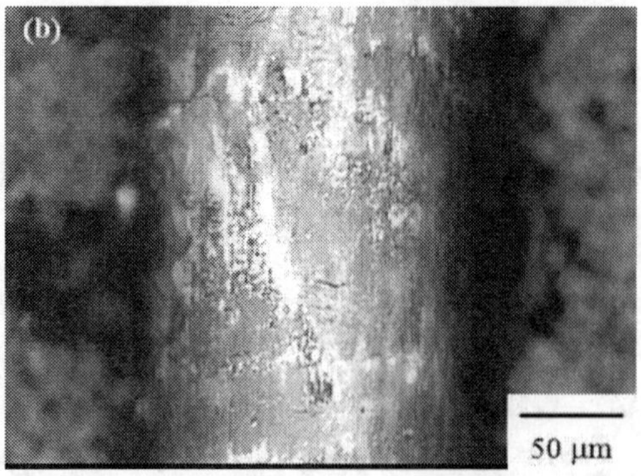

Fig. 7. Scratch-failure images of the AlCrN+MoS_2 coating at: a) L_{C1}, b) L_{C2}.

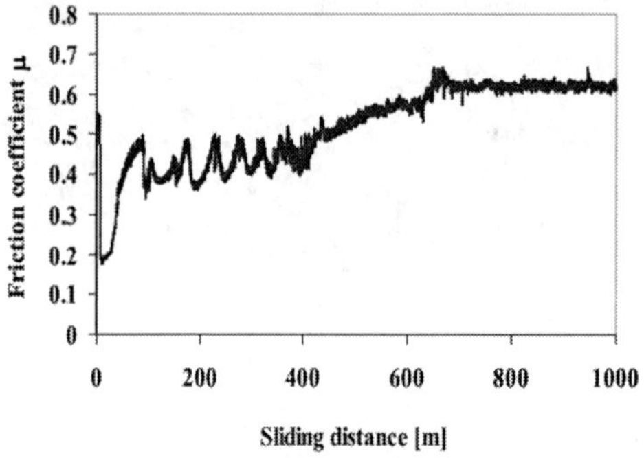

Fig. 8. Dependence of the friction coefficient on the sliding distance during the wear test for the AlCrN+MoS_2 coating.

Under technically dry friction conditions, after the wearing-in period, the friction coefficient recorded for the associations tested is stabilized in the range 0.4–0.5. There was no case of the coatings wearing through completely because the maximum wear-in depths were below their thicknesses.

4. Conclusion

The AlCrN+MoS_2 coating was deposited successfully on the X6CrNiMoTi17-12-2 steel substrate. The dense microstructure of the coating without any visible delamination was observed in the scanning electron microscope. Tests using the TEM confirmed an amorphous character of a low-friction MoS_2 layer. Under the technically dry friction conditions, the friction coefficient for the associations tested is within the range 0.4-0.5 for the investigated coating. The scratch tests on coating adhesion reveal the cohesive and adhesive properties of the coatings deposited on the substrate material.

References

1. T. Hu, Y. Zhang, L. Hu, Wear **278-279** (2012)
2. K. Lukaszkowicz, J. Sondor, K. Balin, J. Kubacki, Appl. Surf. Sci. **312** (2014)

3. T. Kubart, T. Polcar, L. Kopecky, R. Novak, D. Novakova, Surf. Coat. Tech. **193** (2005)
4. L. Rapoport, A. Moshkovich, V. Perfilyev, I. Lapsker, G. Halperin, Y. Itovich, I. Etsion, Surf. Coat. Tech. **202** (2008)
5. B. Vierneusel, T. Schneider, S. Tremmel, S. Wartzack, T. Gradt, Surf. Coat. Tech. **235** (2013)
6. N.M. Renevier, J. Hamphire, V.C. Fox, J. Witts, D.G. Teer, Surf. Coat. Tech. **142-144** (2001)
7. S. Sivarajan, R.P. Padmanabhan, Mater. Today-Proc. **3** (2016)
8. A. Abinaya, B.G. Jeyaprakash, Mat. Sci. Semicon, Proc. **31** (2015)
9. B. Behrens, H.J. Maier, S. Hubner, C.Bonk, A. Almohallami, C. Lummer, P. Scheint, H. Scheland, M. Micke-Camuz, Procedia Eng. **183** (2017)
10. C. Wang, H. Li, Y. Zhang, Q. Sun, Y. Jia, Tribol. Int. **77** (2014)
11. G. Colas, A. Saulot, E. Regis, Y. Berthier, Wear **330-331** (2015)
12. R.B. Pujari, A.C. Lokhande, A.R. Shelke, J.H. Kim, C.D. Lokhande, J. Colloid Interf. Sci. **496** (2017)
13. J. Theerthagiri, R.A. Senthil, B. Senthilkumar, A. Polu, J. Madhavan, M. Ashokkumar, J. Solid State Chem. **252** (2017)
14. B. C. Windom,W. G. Sawyer, D W. Hahn, Tribol. Lett. **42** (2011)
15. M. Dieterle, G. Mestl, Chem. Phys. **4** (2002)

DOI: 10.1051/itmconf/20171506009

The tensile strength test of thermoplastic materials based on poly(butylene terephtalate)

Anna Rzepecka[1,*], and *Monika* Ostapiuk[1]

[1]Lublin University of Technology, Department of Materials Engineering, Nadbystrzycka St. 36, 20-618 Lublin, Poland

Abstract. Thermoplastic composites go toward making an increasingly greater percentage of all manufacturing polymer composites. They have a lot of beneficial properties and their manufacturing using injecting and extrusion methods is a very easy and cheap process. Their properties significantly overtake the properties of traditional materials and it is the reason for their use. Scientists are continuously carrying out research to find new applications of composites materials in new industries, not only in the automotive or aircraft industry. When thermoplastic composites are manufactured a very important factor is the appropriate accommodation of tensile strength to their predestination. Scientists need to know the behaviour of these materials during the impact of different forces, and the factors of working in normal conditions too. The main aim of this article was macroscopic and microscopic analysis of the structure of thermoplastic composites after static tensile strength test. Materials which were analysed were thermoplastic materials which have poly(butylene terephthalate) – PBT matrix reinforced with different content glass fibres – from 10% for 30%. In addition, research showed the necessary force to receive fracture and set their distinguishing characteristic down.

1 Introduction

Thermoplastic composites determine a small percentage in all produced polymer composites, because they have only relatively recently been in the market. Scientists observe a steady increase in the interest of these materials, whereas duroplastic composites are no longer used. When we compare both types of materials we predict that the first type of materials will displace duroplastic composites because they have better properties. Manufacturing thermoplastic composites using injecting and extrusion methods are easier and cheaper methods than manufacturing duroplastic composites. Other benefits are easier recycling, bigger impact strength, improved fracture toughness, easier moulding material and low density, so low mass too – all these properties mean they are used in different industries, for example the automotive or aircraft industry [1].

Because they are used in very responsible parts in machines, and in different elements, a very important aspect is tensile strength. Thermoplastic composites can be destroyed in two ways. Fibres, which are used to reinforce materials, can be broken so different fibres can be overloaded and this situation leads to broken composites. The second way is the stretching of certain fibres that have small adhesion to matrix. The reason for this can be defects in boundary fibre/matrix or mistakes during manufacturing of materials [2].

So during planning and manufacturing methods of thermoplastic composites it is very important is consort with a lot of factors and aspects to create materials with the best properties. Not only the kind of force, but also the structure of boundary layers too and the defects that exist there can influence stability and quality of the connection fibre/matrix. Composites during exploration are exposed to different types of outside strains which can be begin to form micro-cracks which develop from existing defects [3]. This behaviour can lead to previous broken composite elements so to generate a good connection is one of the biggest problems during examination composite materials.

Banik K. [4] the effect of mould temperature variation on the short and long-term behaviour of injection melded PBT were presented. The results indicated that the influence of mould temperature variation has little manifestation from the tensile and impact properties of the moulded PBT parts, but while considering their creep and dynamic mechanical behaviour, they show a significant effect. This deformation behaviour could be explained by the influence of processing conditions on the semi-crystalline morphology, particularly crystallinity as well as molecular packing density of the moulded PBT parts [4]. Regarding the lack of knowledge in terms of the precise understanding on the failure basis on the interface between fibre and thermoplastic, there have not been many publications on the degradation mechanisms especially after tensile strength of PBT with short glass fibres.

So in this paper, a PBT/short fibre fillers composite has been introduced trying to show and discuss the

* Corresponding author: annarzepecka92@gmail.com

phenomena and structure characterization after tensile strength test.

2 Materials and methods

The chosen polymer was glass fibre reinforced PBT (poly(butylene terephthalate)) - Celanex® (Celanese, USA) containing 10%, 20% and 30% glass fibres. Corresponding material symbols i.e. PBT10% for PBT20% and PBT30% in order to simplify the notation. Injection moulding processing conditions as suggested in Processing Guide for Celanex® (product of Celanese Corporation) were presented in Table 1. It was taken for tests 5 samples from one kind of materials. However, the graph was only showing for 3 samples.

Table 1. Parameters of injection process for materials CELANEX 10, 20, 30%.

Parameters	Value
Melting point [°C]	260-270
Temperature of cast [°C]	75-85
Speed of inject	fast
Returnable pressure [bar]	10-30
Inject pressure [bar]	600-1000
Holding pressure [bar]	400-800

Macro-structural analysis was carried out on stereoscopic microscope (Nikon SMZ100, Japan). Micro-structural analysis was performed with using an optical microscope (Nikon MA200, Japan) and scanning electron microscope (NovaNano SEM450, FEI). Metallographic micro-sections have been prepared in accordance with the methodology of the author and with methodology created by Buehler Company using materials supplied by Buehler and Struers. Testing samples were not subjected to etching.
Tensile strength test were used with the specifications of PN EN ISO 527–2 and were carried out on ZWICK100 testing machine. These tests were undertaken at a crosshead displacement of 5 mm/min.

3. Results

3.1 Tensile strength test

In Figure 1 (a,b,c) are charts for three select samples and their dependence of force from displacement for PBT samples 10%, 20% and 30%. The sample is the lowest for PBT 10% whereas the biggest for PBT is 30% and run value about 6000 N. Experiments were conducted to complete destruction samples what is the end of curve on the Figures.

Table 2 shows selected parameters of tensile strength samples PBT 10, 20, 30%. It shows what the lowest force to broken sample is for material with 10% glass fibres and the biggest for 30% glass fibres. The difference in force needed to break the samples between these two types of material was about 2000 N. PBT 20% shows intermediate value. Similarly as in case of value of yield strength and Young's modulus.

a)

b)

c)

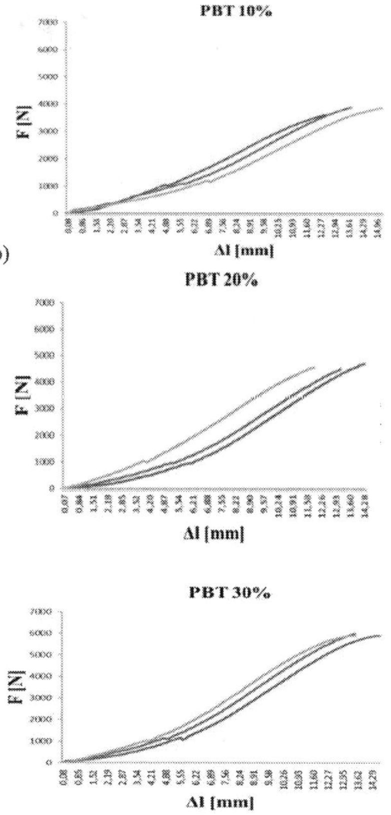

Fig. 1. Dependence of force F from displacement Δl for a) 10% glass fibres b) 20% glass fibres c) 30% glass fibres.

Table 2. Select parameters of tensile strength for sample PBT.

Kind of material	F [N}	R_m [MPa]	Young's Modulus [GPA]
PBT10%	3820±250	92 ±3,21	4,87±0,75
PBT 20%	4560±176	120±5	6,95±1,04
PBT 30%	5900±40	147±6	9,76±1,45

3.2 Macroscopic observation

In Figure 2 are examples of microscopy images samples after being broken. It shows that for sample PBT10% breakthrough is the least fragile character whereas for

DOI: 10.1051/itmconf/20171506009

samples 20% and 30% fragile breakthroughs are more visible. All samples were broken in middle part of paddles.

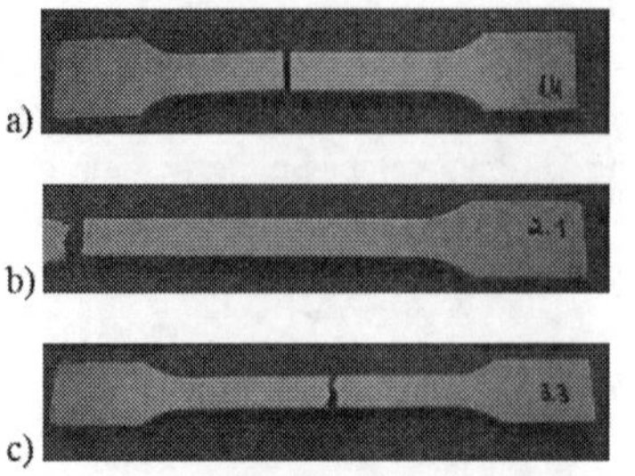

Fig. 2. Breakthroughs of samples after tensile test a) 10% glass fibres b) 20% glass fibres c) 30% glass fibres.

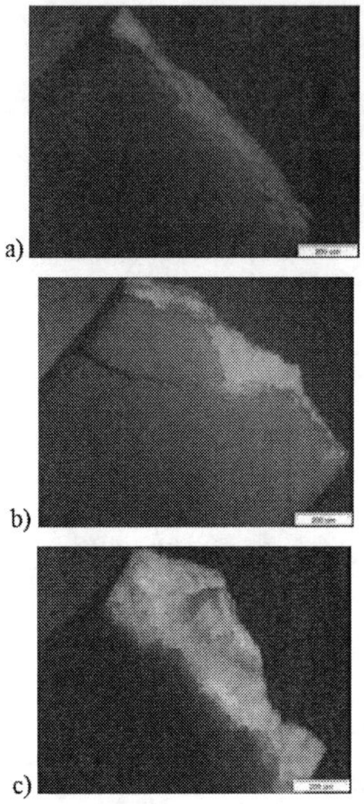

Fig. 3. Breakthroughs of samples a) 10% glass fibres b) 20% glass fibres c) 30% glass fibres; stereoscopic microscope.

Using scanning microscopy, we get the character of breakthroughs. We can see that the fragile character of samples shows interim breakthroughs. Furthermore, we can see individual fibres project from breakthroughs. For samples PBT 10% we cannot see destruct matrix but in the case of PBT 20% and 30% it is present, which is shown in Figure 3 (a,b,c).

3.3 Microstructural observation

Figures 4, 5 & 6 shows panoramic photos with photos developed from the assembly sequence. In Figure 4 we can singularise inner part and core that is typical for injection methods. In inner part is different composition of fibres than in core. Composition of fibre is chaotic enough. The dominate type of composition is composition at an angle bigger than 0° on inboard profile in the core but inner part direction aim at to 0°. On microstructure photos, we can see different content of fibres for every type of sample.

Fig. 4. Panoramic view of microstructure for PBT 10%; optical microscope.

Panoramic photo from Figure 5 shows the inner part and core too. In the case of sample PBT 20% a larger amount of glass fibres is visible. The composition of fibre is still chaotic but there exists a bigger aggregation. In microstructure, we cannot see individual cracks in matrix like in case of PBT 10%. We can see visual degradation of matrix in the inner part of the sample and it means lower tensile strength fibres in this composition.

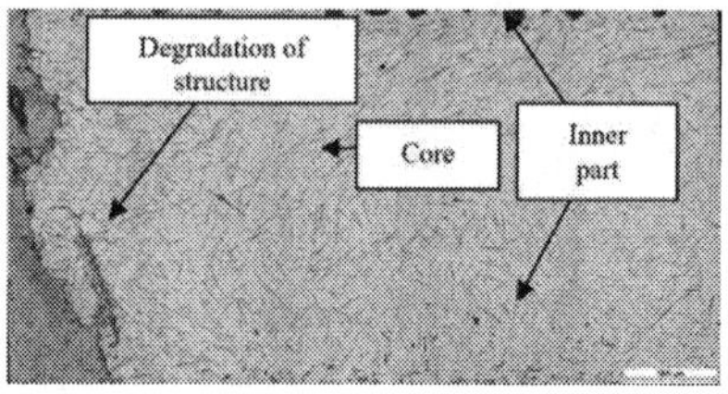

Fig. 5. Panoramic of microstructure for PBT 20%; optical microscope.

In Figure 6 we can see the biggest aggregation of fibres and this means the biggest contest of glass fibres in sample (PBT 30%). The biggest degradation of sample exists in the core. In this case, the inner part has the biggest tensile strength. The reason for this situation is the different direction of composition fibres. Fibres that exist in the core mark significant elongating. On microstructure 7,8 we can distinguish small cracks on board of fibres, cracks of resin and cracks between fibres.

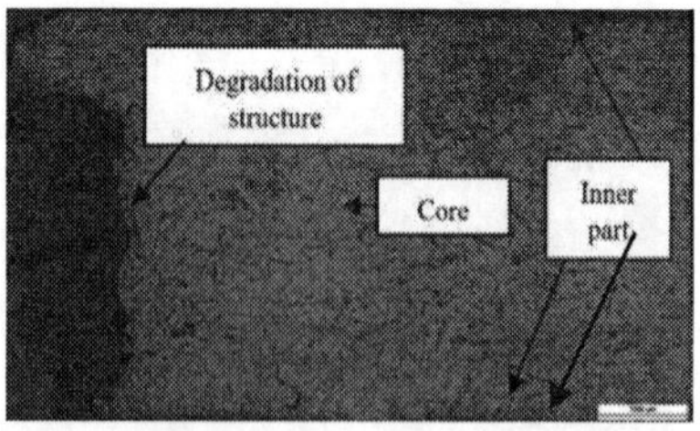

Fig. 6. Panoramic of microstructure for PBT 30; optical microscope.

Figure 9 (a, b) shows microstructures in which the characteristic are a large amount of cracks not only on board of the breakthrough but also deep into the core. These cracks exist in opposite directions than operate tensile force.

DOI: 10.1051/itmconf/20171506009

After detailed microstructural analysis we can see not only cracks of the matrix but also cracks of fibres, cracks on board the matrix/fibre, debonding with separation between the matrix and fibre [5,6].

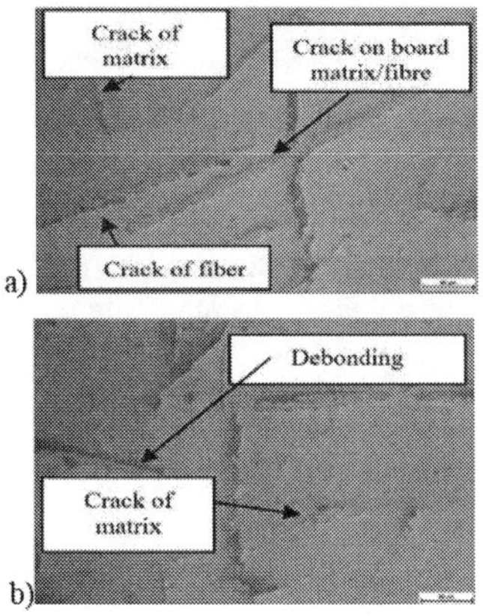

Fig. 7. Microstructure of PBT 10% a) different type of cracks b) debonding; optical microscope.

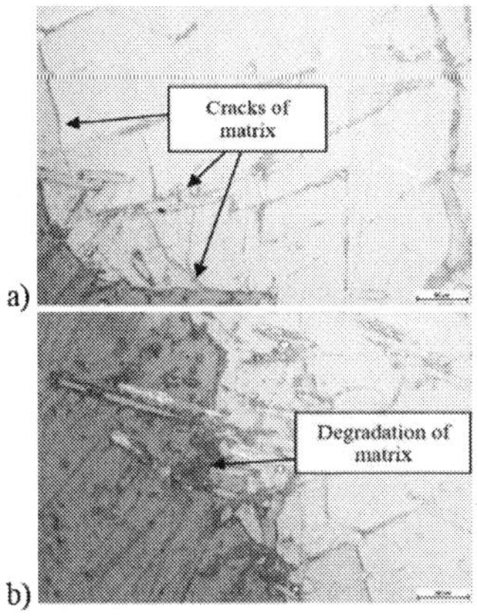

Fig. 8. Microstructure of PBT 20% a) different type of cracks b) debonding; optical microscope.

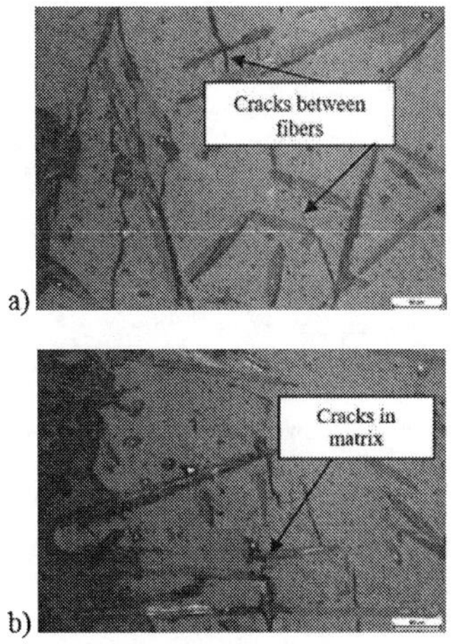

Fig. 9. Microstructure of PBT 30% a) different type of cracks b) debonding; optical microscope.

Figure 10 (a,b,c) shows images of breakthrough using scanning microscope.

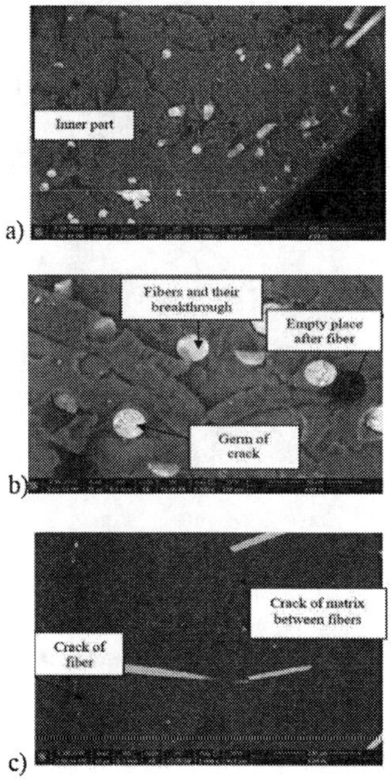

Fig. 10. PBT 10% a),b) breakthroughs c) microstructure; SEM.

In the case of Figure 10, composition indicates inner part of sample. After analysis, we can gather that fibres were broken or were hoick out of place from where they were. Breakthrough shows that here exists the appearance of shortcrust crack. We can see cracks line from fibre too. The reason of this is because of the fact that fibre have a larger strength than matix. One of the basic qualities of polymer composites is the characteristic surface of matrix (textured microflow). We can see cracks of fibres and cracks between fibres [7,8].

Figure 11 shows core of samples. Concentration of fibres testifies to the larger amount of glass fibres in structure. On breakthroughs, the amount placed after draw out of fibres is greater too. In this case, matrix is kept in a different way than sample PBT 10%. There are fewer lines shown in the texture of matrix. The appearance does not exist in PBT 10% but what exists here is fibre imprints with trails after fibres in matrix. Fibers that stick out from the matrix aren't the last of the matrix and is possible when we are using different types of fibres. On microstructure we can see cracks of matrix and cracks of fibres [7,9,10].

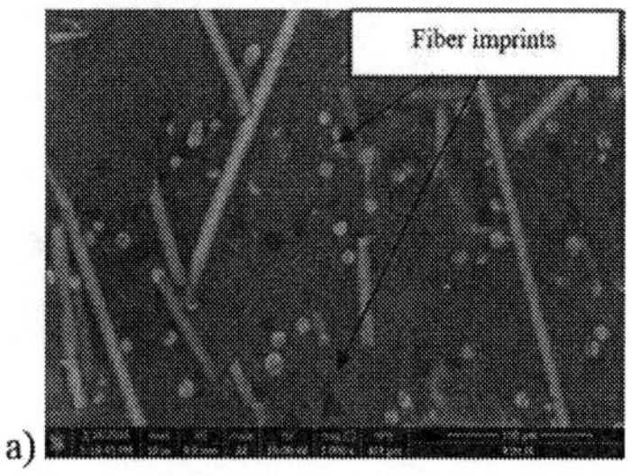

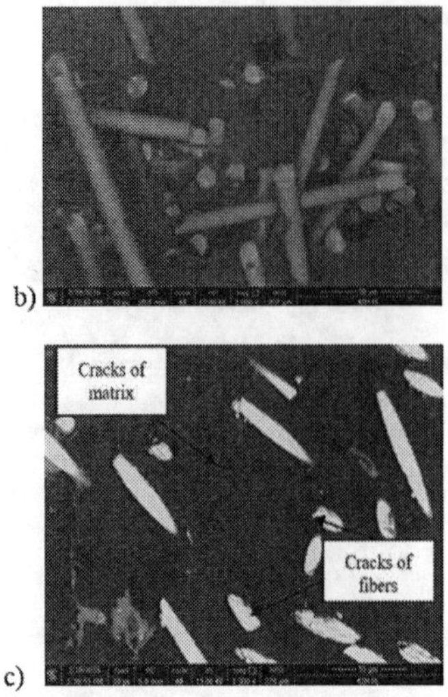

Fig. 11. PBT 20% a), b) breakthroughs c) microstructure; SEM.

Degradation of sample PBT 30% is different than samples with 10% and 20% glass fibres. The reason for this fact is that in PBT 30% we have a greater amount of fibres in the structure. The new appearance is the apparition of debris with the last of matrix. In Figures 11 and 12, we can see cracks of matrix and cracks of fibres too. Whereas on microstructure cracks on matrix and destroyed fibres [7,11].

Hull [12] argued that a further factor which influences the fracture is the fibre/ matrix interfacial strength. This parameter dictates whether matrix/fibre debonding or fibre fracture controls the failure processes and material toughness.

Hahn [13] said that the reinforcement architecture significantly affects the manner in which the side composite breaks. This variation of orientation within the material can lead to localized weakening, highly anisotropic regions, which are often associated with failure orientation in components [14].

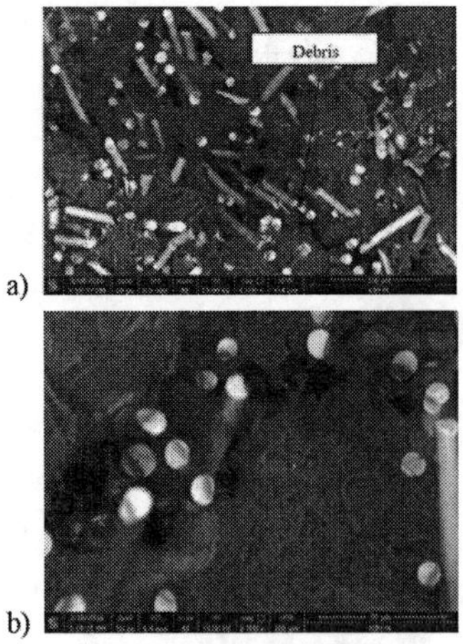

Fig. 12. PBT 30% a),b) breakthroughs c) microstructure; SEM.

Wang [15] concluded that a key parameter that is often the contributing factor to the failure of short fibre composites is the local fibre volume fraction. In regions where the fibre volume fraction is low, the material is locally weaker and brittle. Consequently, matrix features can be used to interpret the fracture directions at these sites [14].

In short fibre composites cracks often initiate from voids or micro-cracks at the fibre and propagate in the matrix until they are arrested by neighbouring fibres. Failure can then propagate through a number of mechanisms such as fibre/ matrix debonding, in-plane delamination, fibre fracture and fibre pullout [14].

4 Conclusion

Polymer composites are used, with good grace, in many types of industry. It is the reason to do research in order to get to know the attitude of this material and the influence on different forces and factors and to imitate an environment of work. There still exists work to make composite materials better and better, for example to make new combinations of fibre/matrix and examine their properties. One of these examples is the examination of tensile strength. In this article we show, agreeably with the main aim, results after statistic tensile strength and structural analysis of polymer composites – poly(butylene terephthalate) with different content of glass fibres – 10, 20 and 30%.

Degradation of samples were the biggest in the core part and it means that composition of fibres in the inner part have the largest tensile strength. In materials, there are many defects: cracks of fibres, cracks on board matrix/fibre, cracks between fibres and debonding. On breakthrough, we can see the growth of cracks and lines from this place so it means that breakthrough has a brittle character. This way of breaking is very dangerous due to flow being very fast after the advanced designated value of strains. The structure of the matrix shows its characteristics in places, in which we can see fibre imprints and debris.

I would like to say thank you to Mr. Adam Bartoszek for his cooperation and sharing the samples to perform this research (Account Development Manager Poland, Celanese Engineering Materials).
The present study was supported by grants S2/MN/2017 from the Faculty of Mechanical Engineering.

DOI: 10.1051/itmconf/20171506009

References

1. J. M. Hodgkinson, Woodhead Publishing Limited (2000)
2. L. St-Pierre, N.J. Martorell, S.T. Pinho, Composite Structures **168** (2017)
3. D. D. L. Chung, Materials Science and Engineering : R : Reports **113** (2017)
4. K. Banik, G. Mennig, Mech-time Dependent Mater **9** (2006)
5. J.Liu, G. Zhang, Composites Science and Technology **98** (2014)
6. U. Chaduvula B.V.S. Viswanadham, J. Kodikara, Applied Clay Science **142** (2017)
7. E. S. Greenhalgh, Woodhead Publishing Limited (2009)
8. S. Mortazavian, A. Fatemi, Composites: Part B **72** (2015)
9. B. Klimkeit, S. Castagnet, Y. Nadot, A. El Habib, G. Benoit, S. Bergamo, C. Dumas, S. Achard,, Materials Science and Engineering A **528** (2011)
10. B. Madsen, Mustafa Aslan, Hans Lilholt, Composites Science and Technology **123** (2016)
11. Z. Jiang, L. A. Gyurova and others, Composites Science and Technology **68** (2008)
12. D. Hull, Cambridge University Press (1999)
13. H. Hahn, K. Jerina, F. Burrett, ASTM Special Technical Publication **183** (1989)
14. M. Ostapiuk, Materials Engineering **5** (2016)
15. L. Wang L., W. C. Tjiu, S.F. Teh, C.B. He, T.X. Liu, Polymer Composites **26** (2005)

ITM Web of Conferences **15**, 07001 (2017)
CMES'17

DOI: 10.1051/itmconf/20171507001

Analysis of dynamic stiffness of arch bridges by means of the first natural frequency

Slawomir Karas[1,*], *Radoslaw Wrobel*[2]

[1]Road and Bridge Department, Lublin University of Technology, Nadbystrzycka 40, 20-814 Lublin, Poland
[2]Budimex S.A., Stawki 40, 01-040 Warszawa, Poland

Abstract. Arch bridges have always boasted a strong aesthetic value. However, their design is challenging due to dynamic excitations. It regards, above all, an area up to ¼ of a bridge length. In bridge dynamics numerical analyses, the basic importance is attached to the natural vibration frequency corresponding to the first bending mode. The multiple first mode is connected to the horizontal shift of a platform or arches, additionally coupled with the deformation corresponding to torsion. In the bridge standards, the first bending mode was adopted as default followed subsequently by higher rank modes of torsional or transversal shapes which do not always occur. This paper describes fourteen numerical models of various arches used in bridge design. The arch bridge, of vertical arches, constitutes a reference. Other models are their varieties through a change of the arch height, a method of connecting platforms and arches at a head and their resulting interaction. Furthermore, variability of arch tilts from a vertical plane was discussed. Conclusions are not unanimously final from the technical point of view; nevertheless, certain beneficial solutions, corresponding to the models of structures of higher natural frequencies, can be indicated.

1 Introduction

Arch bridges come in the variety of forms [1]. Their types can be classified according to:
- a form of an arch and their number,
- a static scheme of an arch, including several joints and how they are attached to a platform,
- a proportion of stiffness between an arch and a - platform, e.g. flexible arch – stiff platform,
- tied arch and arch,
- a type of hangers used,
- a location of an arch in relation to a platform,
- other criteria.

Even in terms of statics, arch bridge design, regardless whether for the purposes of railways or roads, may face numerous problems in stresses distribution. In practice, a numerical analysis is indispensable.

Dynamics is much more complex. Here, the modal analysis as implementation of FEM is applied. At present, the Fourier's harmonic analysis is rarely in use. In case of dynamics, apart from examining limit states, i.e. ULS and SLS in relation to strain/stress amplitudes, deformations and accelerations, also damping vibrations and possible excitations are important. In the face of such complexity, it is necessary to look for a relatively simple measure for purposes of preliminary assessment of a model of bridge to be designed. Most often, the measure is the shape and value of the first natural vibration frequency [2-4]. Such an approach is to be found among the European standards, where it is a simplified criterion for simple (non-complex) structures, simple beams [5]. Even the best design cannot replace a proofing test load performed on a constructed bridge, both in static and dynamic terms. Nevertheless, it is assumed that results of the numerical model analysis should be consistent with in situ measurements.

The paper discusses fourteen numerical models of bridges where all of them are modifications of the basic reference model, namely, a real railway bridge structure (Fig. 1). An assumed scope of the research of arch bridges is not exhaustive, nevertheless, it sheds light on the question of dynamic stiffness.

2 Assumptions and the range of the research

The paper discusses fourteen numerical models of bridges where all of them were modifications of the basic reference model, namely, a real railway bridge structure (Fig. 1).

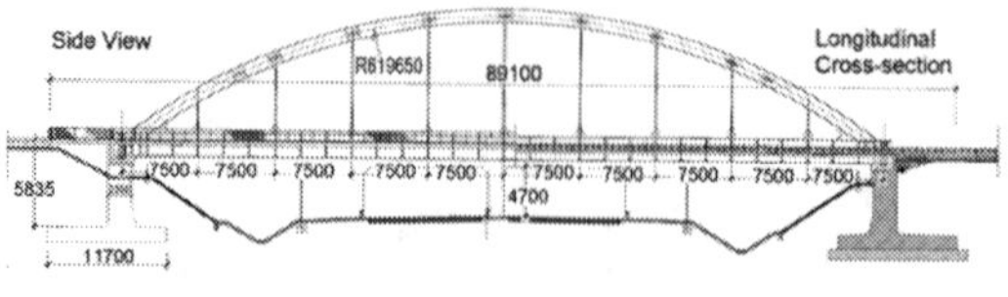

Fig. 1. Reference bridge, load class LM71, α=1.21; [3].

The assumed scope of the research of arch bridges was not exhaustive, nevertheless, it shed light on the

* Corresponding author: s.karas@pollub.pl

DOI: 10.1051/itmconf/20171507001

question of dynamic stiffness. The reference bridge characteristics were: steel S355; cross-section of welded arch box h=1,200 mm, width b=600 mm, and thickness t=30/22 mm; hanger rods -- d=100 mm. Bearing in mind further adaptations, an original steel orthotropic platform was replaced by RC plate of the height h= 45 cm, made with concrete of C30/37 strength class.
Massive abutments, enough to carry the thrust, were also assumed. Therefore, it was not a tied arch scheme. Characteristics of bridge geometry were: span L= 75 m; segment of circular arc with radius R= 61.95 m (angle α=71°); arch elevation f =11 m; arch axle spacing 5.5 m; Langer system of hangers fixed at their ends. Steel transverse stiffeners between arches had a box form of h=300 mm and b=600 mm.

The analyses were carried out regarding the following groups of models.

2.1 Group 1. Testing an influence of hinge joints

Model 1. It differed from the original arch by: a parabolic arch shape, elevation amounted to f_{M1}=12 m and hinged connection between arches, plate and abutment. The hangers were clamped at the ends.

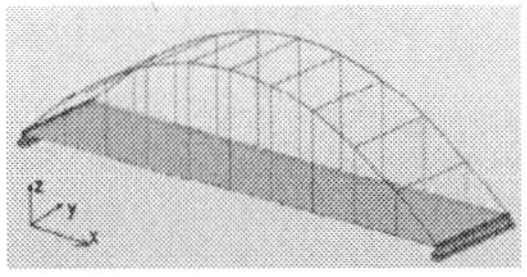

Fig. 2. Model 1.

Model 2. Hangers had hinges at their ends.

Model 3. Hangers were clamped. The arch was circular and its elevation was fixed f_{M3}= f_{M2}= 12 m.

Model 4. As in the Model 1; however, hangers had an articulation at the ends, f_{M4}= 12 m.

2.2 Group 2. The influence of arch elevation

Additional horizontal stiffeners, HEB 200, were used. In this case, the stiffening system had a form of a truss. The arch had a parabolic scheme and hangers had an articulation at the ends.

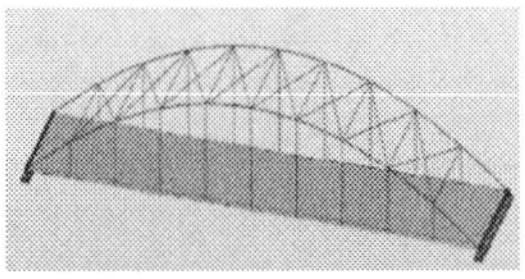

Fig. 3. Model 5.

Model 5. f_{M5}=10 m.
Model 6. f_{M6}=12 m.
Model 7. f_{M7}=15 m.
Model 8. f_{M8}=18.75 m.

2.3 Group 3. Arches tilted from the vertical

Model 9. The 1 m tilt at the keys of arches (4.5 deg)
Model 10. The tilt by 9 deg.

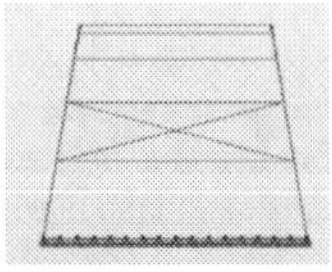

Fig. 4. Model 10.

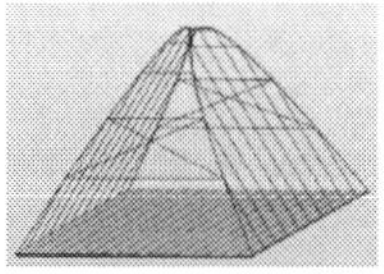

Fig. 5. Model 11.

Model 11. The horizontal distance between the keys equalled 1 m i.e. the tilt was of 10.2 deg.

2.4 Group 4. Analysis of Nielsen hanger systems

Arches were tilted inwardly by 9 deg. A stiffening frame was reduced to a bar system where each bar was perpendicular to the arches. The arch was parabolic with the elevation f= 12.5 m.

Model 12. Spacing at the concrete plate level equalled 15 m.

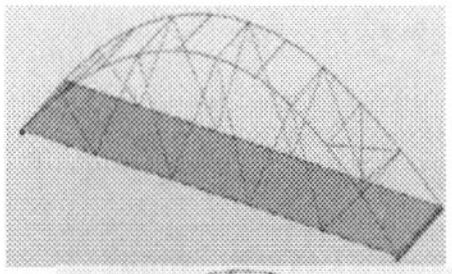

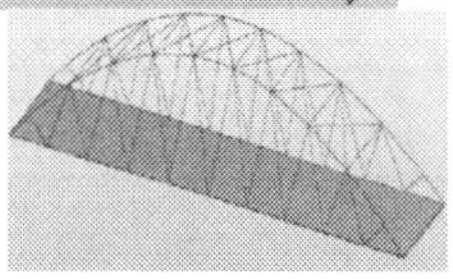

Fig. 6. Model 12. **Fig. 7.** Model 13.

Model 13. Hangers' spacing was broadened to 7.5 m. Again, the stiffening system had the truss form.

2.5 Group 5. Model 14.

A model of the cathedral type of an arch bridge, with an upper platform for traffic, was constructed. Concrete was used. The arch cross-section amounted to 60x120 cm. Elevation -12.5 m. Arches tilted by 10 deg.

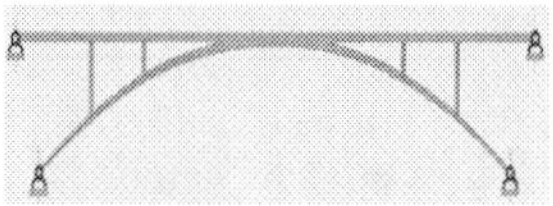

Fig. 8. Model 14.

3 Method of the analysis

The purpose of the paper was a relative comparison of the models.

Calculations were carried out in Robot Structural Analysis Professional 2017 using the spatial bar element

 DOI: 10.1051/itmconf/20171507001

(e1, p3) and the shell element (e3, p3) for the plate. In all models, regarding both cases where FEM elements were used, the plate and the beam, the max. distance between mesh vertices was max. 50 cm, in relation to the real structure dimensions.

Other models were also analysed, especially the ones allowing for a horizontal movement over one of the supports. However, in such cases the influence of freeing movement in the longitudinal direction on the arch was insignificant.

The results were presented in the tables, although the mode visualisations did not include all the models – some of them were only described.

The low values of the natural vibration frequencies resulted from adopted assumptions. In particular, that was due to the omission of: the weight of removable loads, the weight and stiffness of node plates, additional ties in bottom chords etc.

In the paper the following abbreviations were used:

- B-y - bending in y plane,
- B-x - bending in z plane or the translation of an arch in its keys in y direction,
- T-x - torsion about x axis,
- M.1 – Model 1, for instance.

4 Numerical results

4.1 Results obtained for Group 1

Table 1. Model 1.

No.	Mode	Value [Hz]	Depiction
n_1	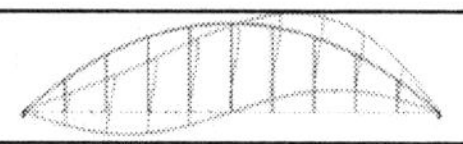	0.74	B-y
n_2	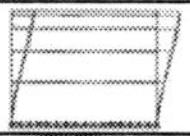	0.79	B-x
n_3		1.49	T-x
n_4	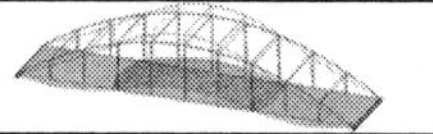	1.56	B-y
n_5	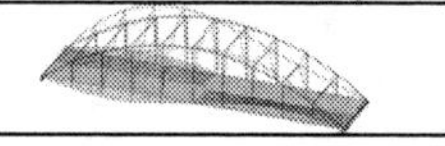	1.88	T-x

The nodes n_1 and n_4 were of bending type. They corresponded to the first and second modes of a simple beam. Both, arches and the deck have 1 full sinusoid (function domain: <0; 2π>) and 2 sinusoidal shapes, respectively. However, the analysed structure was significantly more complex than the simple beam. In the case of the beam, the first mode had a shape of ½ of sinusoid, which indicates the occurrence of greater stiffness in case of arch bridge. The uniform translation of the upper arch parts was a warning signalising the lack of transverse stiffness in this part of the structure. Also, it should have been noticed that the ratio of bending frequencies amounted to

$$n_4/n_1=2.11 \rightarrow \sim 211\% \qquad (1)$$

In the qualitative sense, the obtained results overlapped with those in the Table 1. The presence of two bending modes directly successive, n_2 and n_3, indicated better dynamics of the structure (cf. tables 1.2-4).

Table 2. Model 2.

No.	Mode	Value [Hz]	Depiction
n_1	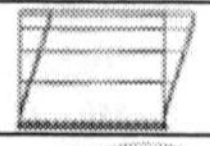	0.76	B-x
n_2	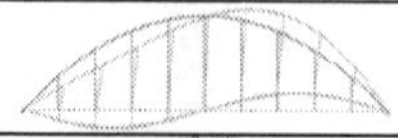	0.8	B-y
n_3	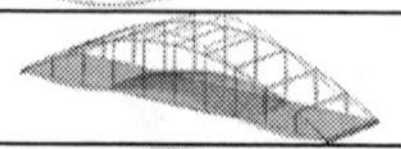	1.6	B-y
n_4	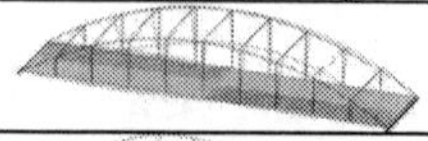	1.66	T-x
n_5		1.94	T-x

Table 3. Model 3.

No.	Mode	Value [Hz]	Depiction
n_1	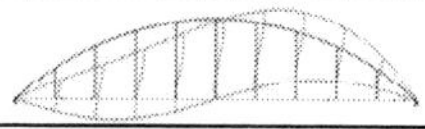	0.74	B-y
n_2	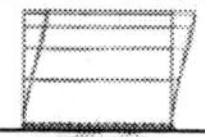	0.78	B-x
n_3		1.46	B-y
n_4		1.47	T-x
n_5	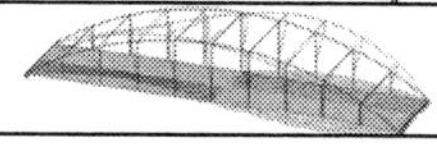	1.89	T-x

Table 4. Model 4.

No.	Mode	Value [Hz]	Depiction
n_1		0.73	B-x B-y
n_2	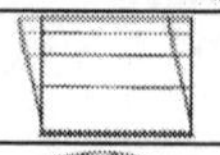	0.73	B-x
n_3		1.43	T-x
n_4		1.46	B-x B-y
n_5		1.86	T-x

Commenting on the results obtained from the analysis of the models in the Group 1, a conclusion that at the assumed span of 75 m the shape of the parabolic arch, and the arch in the form of a circle arc are similar can be drawn. Also, the differences between the stiff and articulated fastening of the hangers were insignificant. According to the Table 4, the modes corresponding to n_1 and n_3 corresponded to the bending effect on two planes, x and y.

More specifically, the highest values of bending vibrations were noted for the Model 2 – the parabolic arch with the articulated fastening of hangers.

4.2 Group 2

Here, the inflow of the arch elevation was investigated.

Table 5. Model 5; H=10 m; H/L=0.13 m.

No.	Mode	Value [Hz]	Depiction
n_1	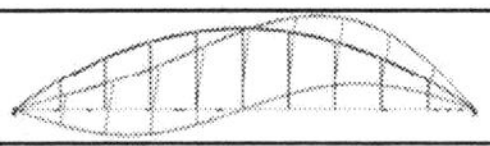	0.8	B-y
n_2	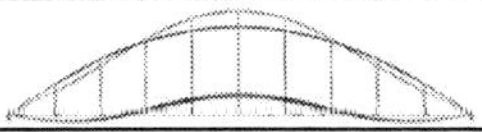	1.58	B-y
n_3	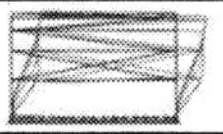	1.64	B-x
n_4	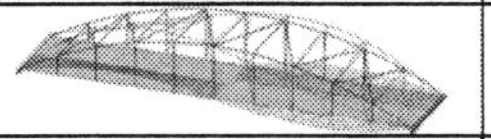	1.87	T-x
n_5	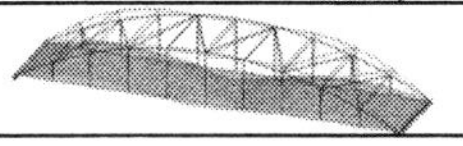	2.14	B-y

Table 6. Model 6; H=12 m; H/L=0.16 m.

No.	Mode	Value [Hz]	Depiction
n_1	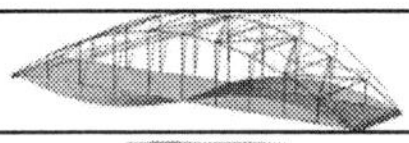	0.79	B-x B-y
n_2	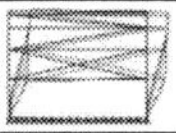	1.57	B-x
n_3	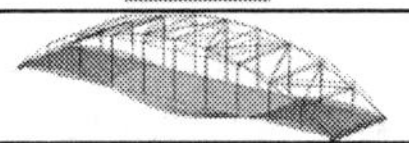	1.6	B-y
n_4	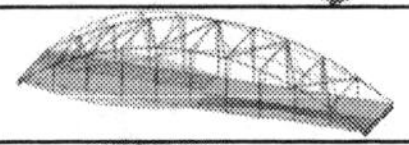	1.85	T-x
n_5	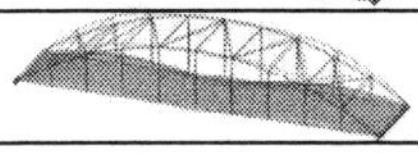	2.24	B-y

Table 7. Model 7; H=15 m; H/L=0.2 m.

No.	Mode	Value [Hz]	Depiction
n_1	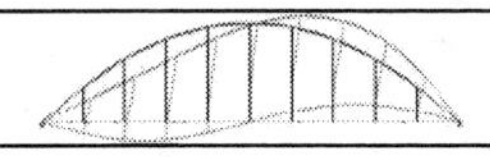	0.78	B-y
n_2	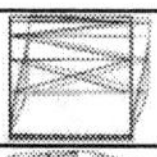	1.46	B-y
n_3		1.6	B-y
n_4	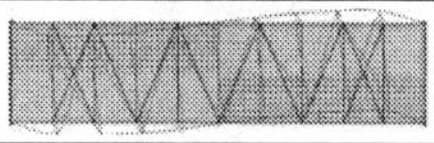	1.79	T-x
n_5		2.34	B-y

Table 8. Model 8; H=18.75 m; H/L=0.25 m.

No.	Mode	Value [Hz]	Depiction
n_1	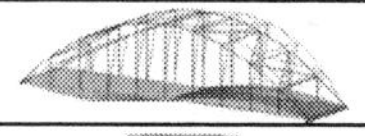	0.75	B-y
n_2	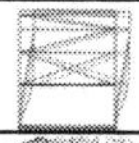	1.32	B-y
n_3	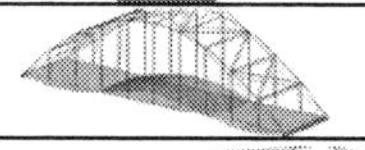	1.58	B-y
n_4	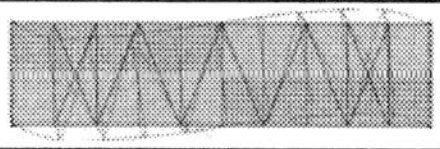	1.69	T-x
n_5		2.39	B-y

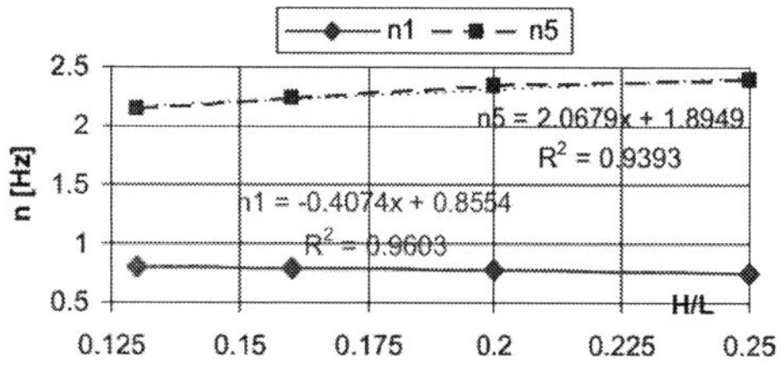

Fig. 9. Distribution of n_1 and n_5 values for models of the Group 4 .

Drawing conclusions at this stage regarded, above all, the variability of values n_1 and n_5 to which bending modes corresponded. n_1 decreased with the increase of the elevation of the arch, while n_5 increased, which was depicted in the Fig. 9. Even though lines in both cases were curved, their variability was insignificant, (2-3), therefore linear regression was applied.

The highest value of n_1 occurs for the Model 5, where

$$n_1(M.8) / n_1(M.5) = 0.9375 \rightarrow \sim 94\% \quad (2)$$

and appropriately n_5 rose somewhat

$$n_5(M.8) / n_5(M.5) = 1.1168 \rightarrow \sim 112\% \quad (3)$$

4.3 Group 3

The influence of the tilt of the arch planes towards the inside of the bridge was analysed. The examined parabolic arch was elevated to 12.5 m.

Table 9. Model 9; α=4.5°.

No.	Mode	Value [Hz]	Depiction
n1		0.80	B-y
n2		1.50	B-x T-x
n3	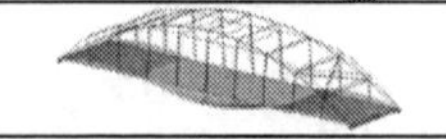	1.61	B-y
n4	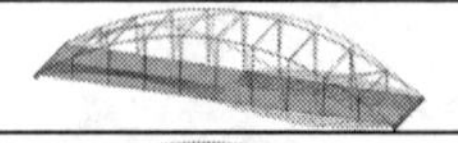	1.90	T-x
n5	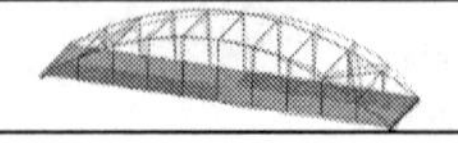	2.20	B-x T-x

Table 10. Model 10; α=9°.

No.	Mode	Value [Hz]	Depiction
n1	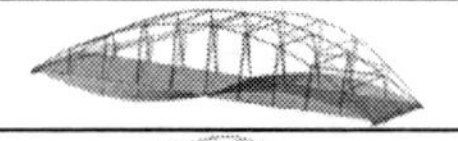	0.80	B-y
n2		1.61	B-x T-x
n3	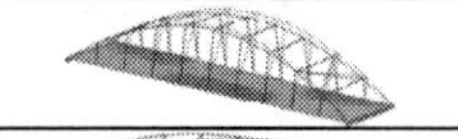	1.67	B-x T-x
n4	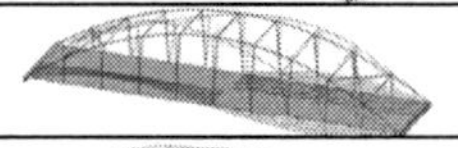	1.99	T-x
n5	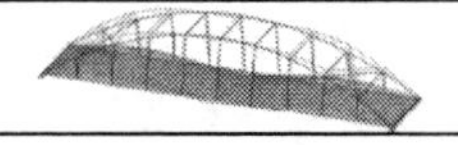	2.26	B-x B-y

Table 11. Model 11; α=10.2°.

No.	Mode	Value [Hz]	Depiction
n1	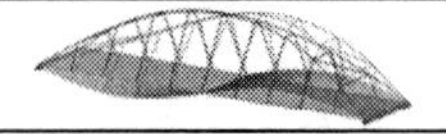	0.78	B-y
n2	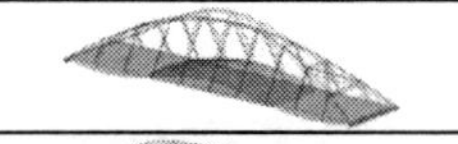	1.57	B-y
n3		2.20	B-y
n4	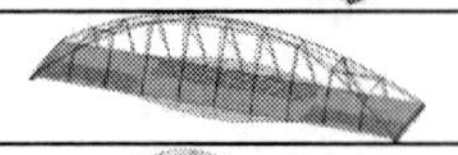	2.70	T-x
n5	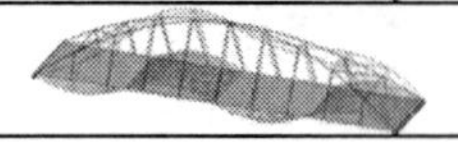	2.93	B-y

First modes were of the bending form and were characterised by an almost constant value of natural frequencies. In the cases of Models 9 and 10, apart from bending forms, there were modes of torsion shapes. However, in the last case of the Model 11, there was only one torsion mode, the forth one. The other ones were bending modes. This indicated that the bridge would work mainly on bending, which was a very beneficial result.

4.4 Group 4

Till this moment the bridges with vertical suspension systems were considered, the Langer system. Now, results obtained from the models with Nielsen hangers should be examined. The bridge arches were parabolic (non-circular) and elevated to 12.5 m. Each of the arches was tilted inwardly by 9 deg. This arrangement has been further analysed.

Table 12. Model 12; spacing of 15 m.

No.	Mode	Value [Hz]	Depiction
n1	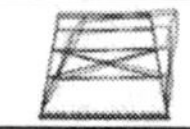	1.68	B-x
n2	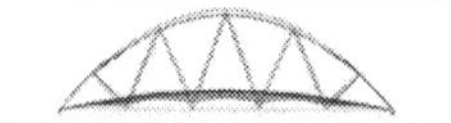	1.87	B-y
n3		2.16	B-x T-x
n4		2.4	T-x
n5	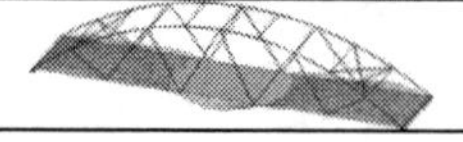	2.63	B-y

Table 13. Model 13; spacing of 7.5 m.

No.	Mode	Value [Hz]	Depiction
n1	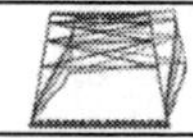	1.34	B-x
n2	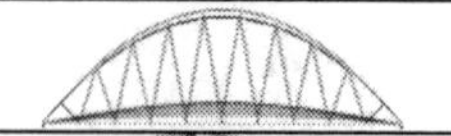	2.14	B-y
n3	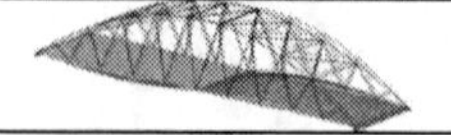	2.35	B-y
n4	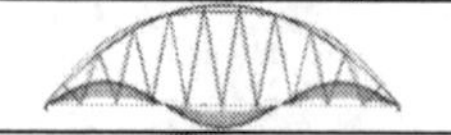	3.14	B-y
n5	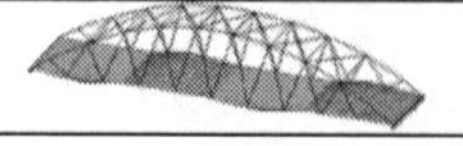	4.05	B-y

What was worrying in this case, was that the first mode corresponded to arches bending in plane x, although, its value was almost twice higher than in the previous cases, admittedly. The most favourable result of the analysis of the modes and the corresponding values

 DOI: 10.1051/itmconf/20171507001

of natural vibration frequencies was obtained from the Model 13. Apart from the first mode, the remaining ones were bending in plane y, while n_5 had value of 4.05 Hz. Therefore, it could be inferred that it was the most appropriate scheme of the arch bridge. The obtained result can also be extrapolated to the network arch suspension scheme which has not been analysed due to its relatively greater complexity.

4.5 Group 5

Here, the RC cathedral arch with open-spandrels was the subject of the analysis.

Table 14. Model 14.

No.	Mode-y	Mode-x	Value [Hz]	Depiction
n_1			1.79	B-y
n_2			2.65	B-y B-x
n_3	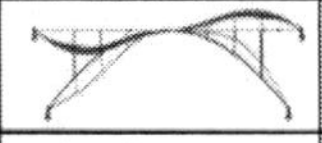		2.67	B-y B-x
n_4	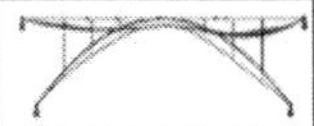		2.83	B-y B-x
n_5		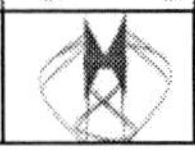	2.93	T-x

This bridge varies from the afore-analysed structures, although it has been popularly used. At the design stage, the most important task was to transfer horizontal forces – thrust. Model 14 used arches tilted from the vertical plane by 10 deg. Apart from the column supports of the platform, there was an additional support located in the bridge key. The better the bridge support is, the better its dynamic features are – this assumption has been illustrated by the subsequent modes. In the variant of wall supports of the width equal to the width of the platform, which was not represented in this analysis, the values of the natural vibration frequency increase additionally. Similarly, when the width of the arch equals the width of the platform, i.e. replacing a beam arch with a cylindrical shell strip [6].

As far as dynamics is concerned, the arch bridge consistent with the Model 14 is similar to the bridge with diagonal hangers, Model 13.

5 Conclusions

The carried out comparative analysis has both qualitative as well as quantitative assessments.

14 models of 75 metre span arch bridges were constructed where a typical railway arch bridge served as a reference. The models were modified to include: parabolic and circular arch shapes, method of fixing hangers to the arch and the platform, variable elevation of the arch, increased stiffening of the arches in the span element. Furthermore, Langer and Nielsen hanger systems were discussed taking into consideration arch tilts from the horizontal plane, and finally, the model of the arch bridge with the upper traffic platform.

For the purposes of the conducted comparative analysis, the first modes and corresponding natural vibration frequencies were used.

No influence of the method of fixing hangers was established.

Moreover, the influence of the type of the arch curve was negligibly small.

An increase in the values of the natural vibration frequency in the cases where arches were tilted from their initial horizontal planes was noted, as well as the noticeable influence where the concentration of the Nielsen system hangers was greater.

The stiffest models in terms of dynamics were the Model 13 and the Model 14, where the conclusions from the first model examinations were applied.

References

1. M. Dębiński, M. Jukowski, S. Karaś, M. Kowal, A. Leniak-Tomczyk (Ed.: Karaś S.): *Aesthetics of road structures*, (Lublin University of Technology, 2016).
2. S. Karaś, A. Krasnowski, Roads and Bridges - Drogi i Mosty, **4** (2013).
3. F. Magalhães, Á. Cunha, E. Caetano, Eng. Struct. **30** (2008).
4. Wei-Xin Ren, Tong Zhao, I.E. Harik, J. Struct. Eng. **4** (2004).
5. EN 1991-2 Eurocode 1 : *Actions on structures* – Part 2: *Traffic Loads on Bridges.*
6. S. Karaś, P. Jamińska, *Historyczne lubelskie mosty żelbetowe*, W: *Aktualne zagadnienia budownictwa komunikacyjnego* : wydanie jubileuszowe z okazji 75-lecia Profesora Wacława Szcześniaka, (Oficyna Wydawnicza Politechniki Warszawskiej, 2015).

Dental implant stress analysis with selected prosthetic crown overhangs sizes

Agnieszka Łagoda[1*], and *Adam* Niesłony[2]

[1]Opole University of Technology, Faculty of Mechanical Engineering, Department of Machine Technology and Production Automation, Mikołajczyka 5, 45-271 Opole, Poland
[2]Opole University of Technology, Faculty of Mechanical Engineering, Department of Mechanics and Machine Design, Mikołajczyka 5, 45-271 Opole, Poland

Abstract. Dental implantology is one of burgeoning methods of teeth loss reconstruction. Because each patient is different, there are still many problems to be solved. One of them is occurrence of prosthetic crown overhangs. The overhangs increase the stress value and risk of the glue shattering, causing a crown fall out and requiring another implantologist's appointment. The aim of the study is to assess the stress value generated in the dental implant using Finite Elements Method and simplified models.

1 Introduction

Nowadays, a continuous increase in demand for dental implantology services is observable. Dentistry which is dealing with the methods of replenishing teeth loss, is one of the fastest flourishing area of medicine. The first world-spread information about tooth implants have been realised merely 30 years ago, when the scientist named Per-Irvin Brånemark published a scientific paper about this subject. As a first person describing this issue, he focused on osseointegration problems and trouble connected with metal implant screw integration with animal bone [1]. The paper concerned not only esthetical values of implantology, which still are the main reason for any patient to opt for a dental implant as a method of dental treatment. It has also begun research in osseointegration and became the initiation of works on biology aspects of implant installation. Additionally, there is a group of mechanics, carrying out their research on the mechanical aspects of teeth loss in oral cavity treatment process. This is their contribution to constant development of dental implantology and introduction fundamentals of new discipline named biomechanics.

There are still many problems to solve for dentist and biomechanics around the world. The methodology of dental implant installation process is sufficiently advanced to guarantee the reliability and safety of implant use [2], but there is a lot of unquestioned issues, especially with individual patients' cases.

The success of implantation therapy depends on many factors. The most important are [3-5]:
- the progress of the installation process affecting implantation stabilisation,
- the progress of the osseointegration process (that is the integration between the implant and the patient's bone),
- the quality of patient's bone,
- appropriately selected dimensions of the implant screw.

* Corresponding author: a.lagoda@po.opole.pl

The aim of this study is to present the stress level generated into a dental implant construction. The Finite Element Method was used with a simplified model of dental implant with a selected range of prosthetic crown overhangs sizes.

The use of FEM in silico research is becoming increasingly popular [6]. This method works well in the very first step of teeth replenishing – treatment planning. Finite Elements Method supports the initial evaluation of the treatment success level, before the operation on the patient [7]. At the computerisation stage it is possible to eliminate the possible problems related to choosing the wrong implant geometry or unsuitable prosthetic crown overhangs size and shape. It also enables optimisation of the implantation system [8].

Commonly used implant system consists of following elements: abutment, connecting screw, implant screw (fig. 1). Prosthetic crown is connected to abutment by adhesive.

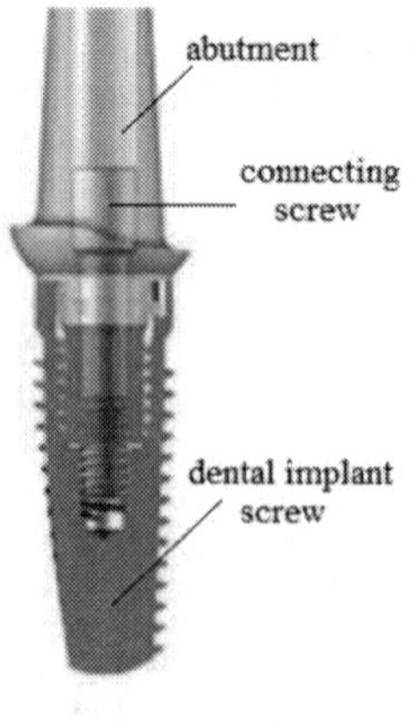

Fig. 1. Dental implantation system with its parts [9].

 DOI: 10.1051/itmconf/20171507002

The level of success in treatment depends on plenty of dependences and factors, of course. First, there is an implant stability problem and every trouble connected with the patient bone and the implant installation. The Finite Element Method could be helpful assuming that every process is proceeding correctly. It can be only used to help in assessment of the decision, either using selected implant geometry or changing it at the very primary stage of treatment.

2 Methods and materials

2.1. CAD Modelling

Dental implant is a screw usually made from a titanium alloy. It is screwed directly into the patient mandible or maxilla. After the recovery process (approximately 3-6 months), the implant becomes a support for the prosthetic crown.

There are two kinds of dental implant: one-piece and two-pieces. They are different in terms of their construction and installation processes. Two-pieces implant consists of implant screw and separated abutment, which is connected to implant by connecting screw and becomes a pillar for the crown [10]. The one-piece implant consists of one element – implant and abutment belong together and are inseparable.

One model was used to the Finite Elements Method analysis. Two-pieces implant model was common in screw length and diameter. Model was designed such as real implants manufacturing by CAMLOG® Company. Autodesk Inventor Professional 2017 has been used to CAD modelling.

Model length and diameter are: l=16mm and ø=6mm (fig. 2). Selected geometry is an image of screw geometry offered by CAMLOG® Company in their implantation systems [10].

The overhang is a part of crown which is beyond the implant screw. There are two kinds of overhangs:
- one-side – the overhang is only on one side of implant screw,
- two-side – the overhang is on two sides of implant screw.

Premolars and molars teeth as five, six, seven or eight are quite large and much larger even than the ø6mm which is the biggest diameter among CAMLOG® implants [10]. It is a natural matter for those teeth. There is a problem when the overhang is too wide – especially on one side. Then the impact of force in jaw is strong enough to shatter the adhesive. This is the reason why the crown can fall out, which leads to a necessity to visit dentist again and change the prosthetic crown shape and adhesive it on abutment one more time. Repeated visit does not ensure a successful treatment. New tooth crown very often falls again and there is a need to see a doctor for several times. Overhangs are amongst the most problematic issues in implantology, as well as osseointegration trouble, implant stabilisation or implant installation problems, or lack of bone to qualify the patient to operation.

Research includes one-side overhang in sizes of 0mm and 3mm. Fig. 3 presents the parts of implantation system for crown without overhang. Fig. 4 shows the parts of implant system for crown with 3mm overhang. Furthermore, the parts are divided for elements and prepared to the Finite Elements Method analysis.

Fig. 2. Dental implant model without tooth crown.

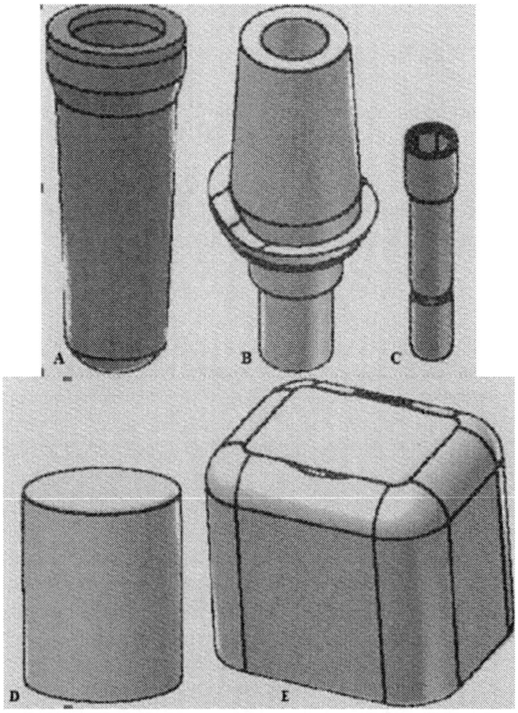

Fig. 3. Dental implant system parts, where A – implant screw, B – abutment, C – connection screw, D – adhesive layer, E – prosthetic crown without overhang [11].

DOI: 10.1051/itmconf/20171507002

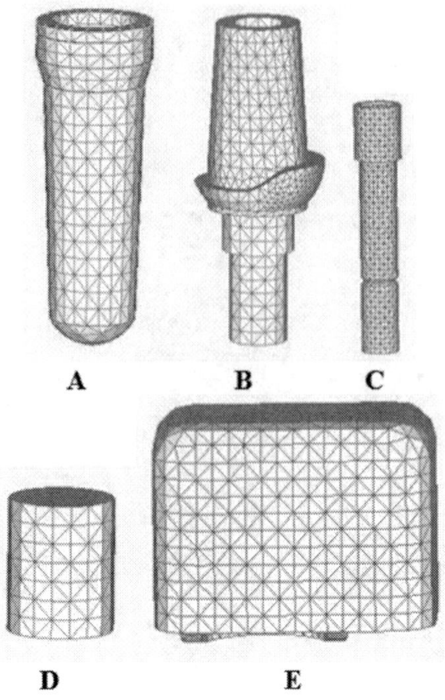

Fig. 3. Dental implant system parts with elements, where A – implant screw, B – abutment, C – connection screw, D – adhesive layer, E – prosthetic crown with 3mm one-side overhang [12].

2.2. Material properties

It was assumed that the implant was made of titanium and more specifically, the titanium alloy named Ti6Al4V. This material is very popular and used by CAMLOG® because of good strength properties and no negative impact on the human body [10]. This alloy is characterised by greater resistance to fatigue compared to pure titanium. It is also indifferent to the difficult conditions prevailing in human oral cavity [13]. Those are very eligible properties for biomaterial.

However, at present titanium alloy Ti4Al4V is less often used due to the content of vanadium in its composition. There are several reports on hazardous effects of this material on a human body [14]. As replacement for this type of alloy, the Ti6Al4V ELI alloy is becoming increasingly popular. The content of harmful elements is reduced in this alloy [15]. In terms of material properties, both alloys do not differ from each other. Their modulus of elasticity and Poisson's ratios needed for research remain at the same level [16]. Any change of material does not affect obtained results. It has an impact on patient's condition.

Cobalt-chromium alloy is the most commonly used metal material in metal construction, for instance abutment and connection screw. Young's module is on high level which is 220 GPa [7]. The material is mainly valued for its anti-corrosive properties and no negative impact on the human body. The chromium content increases the hardness of steel and has a significant role in the alloys used for prosthetic tools. This application of chromium in larger amounts (12 ÷ 30%) provides corrosion resistance. Cobalt increases hardness of steel [17].

The RelyX Unicem dental adhesive is the most commonly used adhesive by implantologists. It is a composite cement characterised by the absence of etching of the glued surface. It only needs rinsing and drying. This is a self-etching adhesive, so it is less durable than bonding materials requiring special surface preparation [18]. However, the lack of eugenol in glue content and zinc oxide base make the product extremely versatile. It can be used not only for adhesiving a tooth crown on an abutment, but also as a precise replacement for cavities. The adhesive provides a strong bond to tooth and dental materials [19].

The most frequently used material for dental crowns is zirconium oxide. Research on its use in dentistry dates back to the 90s of the twentieth century [20]. Zirconium oxide has many positive aspects that are used in dentistry. The most important of these are: corrosion resistance, mechanical strength, high hardness and the highest bending strength of dental and prosthetic materials [21]. Due to zirconium oxide high costs, standard dental ceramic becomes more and more popular. Its properties are eligible enough to perform an operation. It is cheaper and this material is adopted to analysis.

Each item of implant system was assigned to the appropriate material. The properties are shown in the table 1. Ti6Al4V ELI is material for dental implant screw, cobalt-chromium alloy for abutment and connection screw, adhesive for adhesive connection between abutment and tooth crown and dental ceramic is for tooth crown.

Table 1. Material properties [7, 13, 22].

Material	Young's Modulus [GPa]	Poisson's ratio
Ti6Al4V ELI	E=110	υ=0.32
Cobalt-chromium alloy	E=220	υ=0.3
Adhesive	E=3.9	υ=0.35
Dental ceramic	E=61.2	υ=0.19

2.3 Boundary conditions

The analysis was carried out for few forces [23]:
- 20N – force for chewing food,
- 100N – popular force repeated in literature,
- 157,7N – force acting on premolar tooth no. 35 [5],
- 196N – force acting on molar tooth no. 36 [5],
- 200N – maximum bite force in healthy human jaw,
- 500N – maximum bite force in unhealthy human jaw (bruxism).

There were six analyses of crown without overhang model and consecutive six analyses for crown with overhang model. Force was set perpendicular to the corner of overhang at implant models with overhang crown. For tooth crown without overhang, force was set perpendicular to the crown surface (green arrows in the fig. 4).

DOI: 10.1051/itmconf/20171507002

Glued contacts were used as a connection between elements of FEM model.

The linear analysis was carried out in FEMAP program with NX Nastran modulus implemented inside. The nonlinearity geometry was not taken into consideration, because there was no need for that. Discretisation of each object for finite elements was conduct in FEMAP program also. There were 37296 tetra-shaped elements for model without the overhang and 38298 for model with overhang.

Both models were constrained along the dental implant screw surface. Models were fixed to facilitate research and keep behaviour of dental implant after correct osseointegration process with the bone. Black triangles correspond with fixed constrain of each model in the fig. 4.

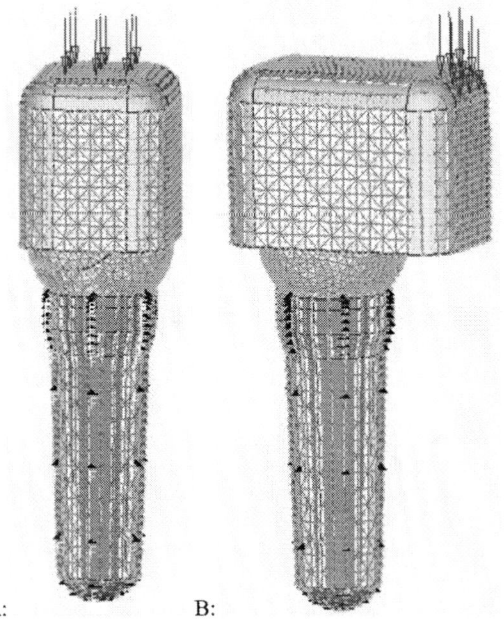

A: B:

Fig. 4. Models boundary conditions, where green arrows show the load place and direction and the black triangles represent the fixed constrains. A – model without the overhang, B – model with one-side overhang.

3 FEM stress analysis

3.1. Model without overhang

In the fig. 5, the stress analysis of tooth crown section without overhang is presented.

The maximum reduced stress, according the Huber-Mises hypothesis, is in an upper part of model – between the abutment and adhesive layer. There are also results for side part of connection between implant screw and abutment. It comes from the specific shape of abutment characteristic for CAMLOG® implant system.

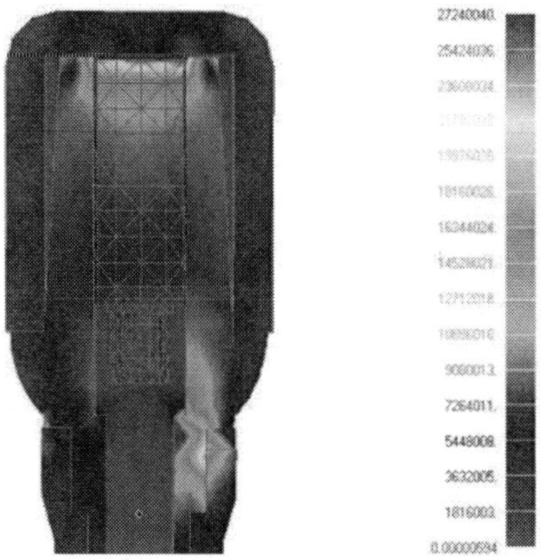

Fig. 5. Result for section of dental implant without overhang for 200N.

3.2. Model with one-side overhang

In the fig. 6, the stress analysis of tooth crown with overhang is depicted.

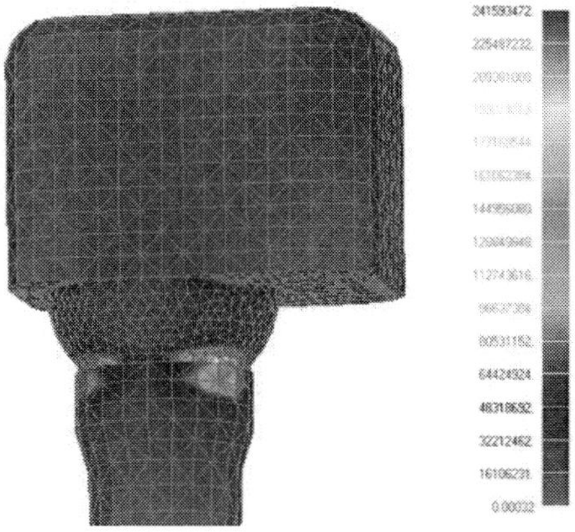

Fig. 6. Result for dental implant with 3mm one-side overhang for 200N.

The maximum stress, according the Huber-Mises hypothesis, is in a part of model between the abutment and implant screw.

4 Results and conclusion

The results of linear stress analysis are compared in the table 2. The most precarious results are for 500N force. This force is assign for ill people with occlusion parafunctions, such as bruxism. Generally, this group of patients is not qualified for dental implant treatment; nevertheless, after medical consultation and positive opinion, they could be considered for replacement of the teeth loss using implants.

 DOI: 10.1051/itmconf/20171507002

Table 2. Results for stress analysis.

Force	**Maximum reduced stress according to Huber-Mises hypothesis**	
	without overhang	with overhang
20N	2.7 MPa	24 MPa
100N	13.5 MPa	120 MPa
157,5N	21.3 MPa	190 MPa
196N	26.5 MPa	236.2 MPa
200N	27 MPa	241 MPa
500N	67.5 MPa	602.5 MPa

There is maximum reduced stress according to Huber-Mises hypothesis bar graph with comparison of stresses between no overhang and overhang model shown in the fig. 7. Clearly, the value of stress generated under 500N force impact is almost ten times greater with overhang application than without.

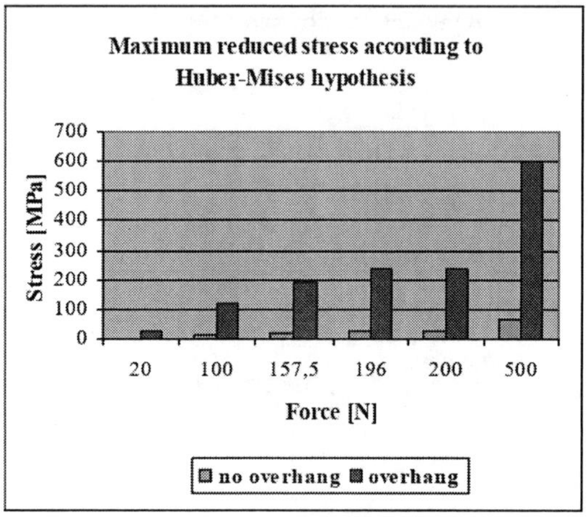

Fig. 7. A bar graph with comparison of model without overhang stress and model with one-side overhang stress.

The overhangs are very typical problems to struggle with in dentistry. FEM can be helpful to assess the level of treatment success, assuming that every installation and osteointegration problems proceed fortunately. The model was constrained along with the implant screw surface. It reflected the implant set inside the bone. In these conditions, the research revealed that 3mm overhang for person with occlusion parafunction could be dangerous and there was a need to assess the treatment success more precisely. More than 600 MPa stress level can be the reason for tooth crown falling out or even the abutment damage.

References

1. P-I. Brånemark, *Osseointegration Book: from Calvarium to Calcaneus*, 1. edit, Berlin; Chicago: Quintessence Pub. Co. (2006).
2. G. Wang, X. Gao, E. Lo, J. of Dent. **43** (2015).
3. W. Chladek, *Biomechanika Inżynierska Narządu Żucia. Zagadnienia Wybrane*, Gliwice (2008)
4. R. J. Lazzara, JIRD **1** (2012).
5. G. Milewski, *Wytrzymałościowe Aspekty Integracji Biomechanicznej Tkanka Twarda – Implant Stomatologiczny*, Kraków (2002).
6. J. Geng, K. Tan, G. Liu, J. Pros. Dent. **85** (2001).
7. O. Kayabasi, E. Yuzbasioglu, F. Erzincanli, Adv. Eng. Soft . **37** (2006).
8. A. Dobosz, H. Panek, P. Napadłek, Dent. Med. Probl. **42** (2005).
9. Camlog Compedium, *1. Surgery*, Thieme Verlagsgruppe (2007).
10. Camlog, *Conelog ® Implant System*, Product Catalog (2012).
11. A. Materac, A. Niesłony, APB **11** (2016).
12. A. Materac, A. Niesłony, APB **9** (2016).
13. B. Jedynak, E. Mierzwinska-Nastalska, Dent. Forum **1** (2013).
14. J. Marciniak, *Biomateriały*, Wyd. Pol. Sl., Gliwice (2002).
15. Y. Okazaki, S. Rao, Y. Ito, T. Tateishi, Biomaterials **19** (1998).
16. ISO 5832-3, *Implants for surgery – Wrought titanium-6 aluminium-4 vanadium alloy. ASTM F-136 – Specification for Titanium –6Al-4V ELI*, Alloy for Surgical Implant Applications.
17. Z. Bańkowski, *Mały poradnik mechanika*. WNT, Warszawa (1976).
18. M. Prylicki, P. Deręgowska-Nosowicz, H. Shaw, E. Kaczmarek, Dent Med Prob **43** (2006).
19. 3M Espe, *RelyX product catalog*.
20. H.J. Conrad, W.J. Seong, I.J. Pesun, J. Prosthet. Dent. **98** (2007).
21. K. Lasek, P. Okoński, E. Mierzwińska-Nastalska, Prot Stom **6** (2009).
22. M. Marzec-Gawron, S. Michalska, B. Dejak. Prot Stom **3** (2012).
23. M. Idzior-Haufa, W. Hędzelek, Sz. Rządowski, P. Gajdus, M. Rychlik, Prot. Stom. **6** (2013).

 DOI: 10.1051/itmconf/20171507003

Dynamic diagnostics of moving ferromagnetic material with the linear induction motor

Krzysztof Szewczyk[1,*], and Tomasz Walasek[2]

[1] Częstochowa University of Technology, Faculty of Electrical Engeneering, Str. Armii Krajowej 17,42-200 Częstochowa, Poland
[2] Częstochowa University of Technology, Faculty of Mechanical Engineering and Computer Science, Str. Armii Krajowej 17,42-200 Częstochowa, Poland

Abstract. The paper presents the application of a three-phase induction motor as a sensor measuring the force of the electromagnetic field connection between the engine and produced sheet steel. The force interaction between the engine and the manufactured sheet metal treated as a treadmill for a linear motor may be an indicator of damage to the material. Detection of places where the sheet does not meet the quality requirements may be very useful in the production process. FEM calculations were performed in the ANSYS MAXWELL environment. The results suggest the possibility of using this type of construction to test the quality of produced materials. The computational results and their analysis are presented in this article.

1 Introduction

The process of producing materials such as wire and sheet metal requires material control. As the product is reeled in during the final phase, a possibility to inspect the produced material disappears [1]. The top layer covers the bottom layer. Repeated control requires rewinding the material [2]. The testing method often requires cutting off a sample such as an Epstein frame.

The authors attempted to design and simulate an inspection device for the final phase of sheet metal production.

2 The measuring principle

The idea of measuring principle is illustrated in Figure 1. The part of the drawing on the left shows the metal sheet production area. The produced sheet metal is moved to the right in the direction of the winder.

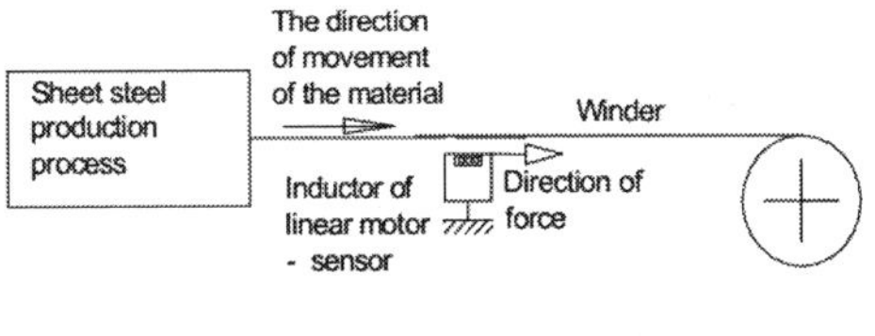

Fig. 1. Idea of diagnostic system.

A sheet metal sensor was placed before the winder. It can be connected to a device that identifies the position of the sheet. The identification device may be helpful in finding a broken piece of sheet metal later. The identified sheet is wound onto the winding shaft.

A liner electrical motor is used as a sensor. Before winding, the metal sheet passes the inductor of the electrical line motor. Depending on the direction of field movement and the direction of product movement, the sensor will show the force value. The force acting on the exciter depends on the magnetic and mechanical properties of sheet steel. With the constant sheet metal moving, any change in the magnetic properties of the sheet will change the parameters acting between the sheet and the sensor.

The main purpose of this article is to identify phenomena occurring in the sensor and around it. Special attention will be paid to the impact of an air gap as a material defect. All the calculations and their analyses are presented in the paper.

It is the authors' belief that the conducted analysis and the conclusions drawn from the project will encourage further research and contribute to the development of diagnostics of this kind.

3 Space: 2D or 3D

The simulation of the electromechanical system was performed in the Ansys Maxwell environment. It is an ideal environment for simulating different electromechanical converters and electromagnetic field distribution. Electrotechnical metal sheet constitutes the main part of the simulated system, being simulated as both the tested product and a sensor. Sensor construction is based on the principle of linear induction electric motor. All these elements can be simulated in the Ansys

* Corresponding author: szewczyk500@gmail.com

DOI: 10.1051/itmconf/20171507003

environment with the FEM method and the Stress Tensors Maxwell calculations. This procedure allows you to calculate the distribution of the magnetic field in the simulated system, and further to calculate the forces and torques generated by the system under test.

It is possible to analyse this kind of problems in 2D and 3d in Ansys Environment. The 3D model is capable of analysing the problem without the limitation of the current flow range. In Figure 2, basic differences between 2D and 3D models have been presented. In the 2D model, on the left, the discussed currents can flow perpendicularly to the surface. Treating the computer screen as the analysed surface the currents can flow in and out in the perpendicular direction to the screen. In the 3D model we can analyse additional currents flow parallel to the screen surface. This enables analysis of boundary effects and their impact on the studied phenomena. 3D modelling of the discussed problem requires enormous computing power, large memory areas and time consuming calculations. Reducing the space of analysis to a 2D model diminishes the prospect of simulation and limits the flow of currents only in a direction perpendicular to the plane of the analysed space (Fig. 2).

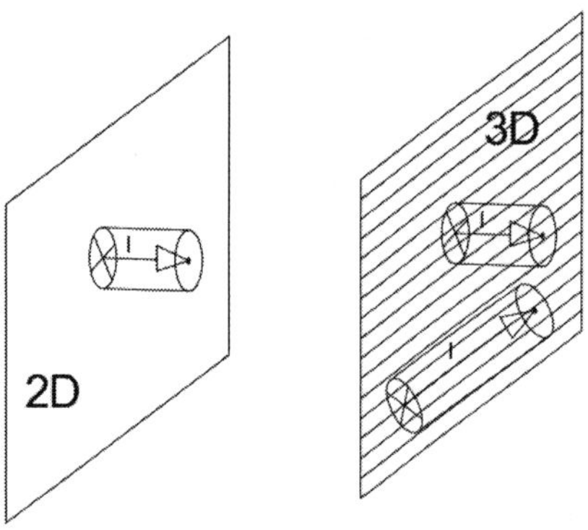

Fig. 2. Direction of currents flow in 2D and 3D simulations.

Reducing the analysed space to the two-dimensional model accelerates computations and simplifies the analysis of phenomena. However, it requires a precise design of the experiment so as not to lose the opportunity to draw the right conclusions as a consequence of limited area of analysis.

The problem discussed in the paper does not require modelling of the influence of boundary phenomena on the system. It is intended to present the idea of using a typical construction of an inductive linear motor for dynamic ferromagnetic product identification. As the boundary phenomena occurring in the construction of the linear motor drive are not the subject of this article, the 2D space was used for simulation. Presented linear phenomena in linear motors were based on common knowledge.

4 Assumption for the Project

The tested system consists of two basic components: the inductive linear motor and the test plate. The linear motor, shown in Fig. 3, consists of magnetic circuit and windings. The rotor of the linear motor is a sliding steel sheet. It is a treadmill for the linear engine. In a three-phase motor for the field to move the engine should be composed of 15 coils. The movement of the field will generate eddy currents in the treadmill, which in turn will generate forces between the treadmill and the engine. The position of the motor working as a defect sensor for the controlled sheet is shown in Fig. 3.

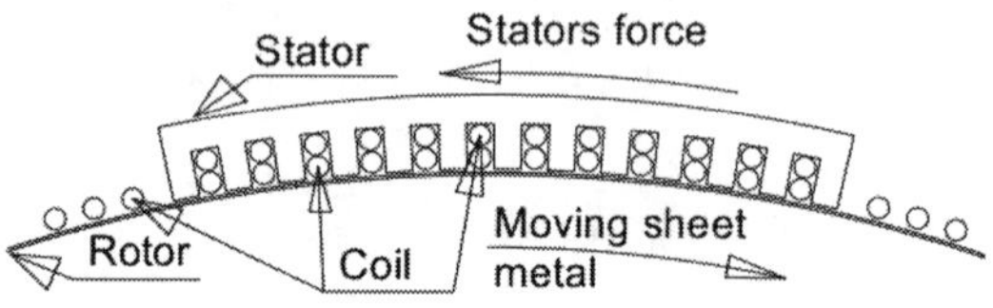

Fig. 3. Three-phase inductive linear engine. Sensor.

Winding the linear motor requires additional explanation. In the case analysed by the authors, the principle is analogous to that of an induction arc motor. In this type of engines magnetic circuit is built on the basis of an open arc rather than a closed circle shape. Due to edge effects the specific winding of the induction motor is required. Windings at the beginning (on the left) and at the end (on the right) of the magnetic circuit have coils not covering the magnetic circuit. The shape of the construction shown in Fig. 3, based on the radius, is explained in the next sections of the paper. It is related to the limited space of analysis resulting from memory constraints of the processor. The presence of additional windings outside the magnetic circuit is required by the occurrence of longitudinal boundary phenomena resulting from the limited length of the inductor of the induction motor. Longitudinal boundary phenomena consist of the formation of an alternating field (in addition to the moving component) and the input and output at the ends of the inductor, also described as marginal phenomena.

The phenomenon of induction changes B described in the function of time can be presented in the form of two components: the alternating field and the moving field as in equation (1):

$$B = B_t(\cos(\omega t - \frac{\pi}{T}z) + B_p \sin(\omega t - \frac{\pi}{2}) \quad (1)$$

where: B_t, B_p are the induction of moving and alternating field, T is the period of induction distribution along the axis of length Z of the driver in the direction of the moving sheet. Explanation of the field compensation principle B_p is shown in Fig. 4.

 DOI: 10.1051/itmconf/20171507003

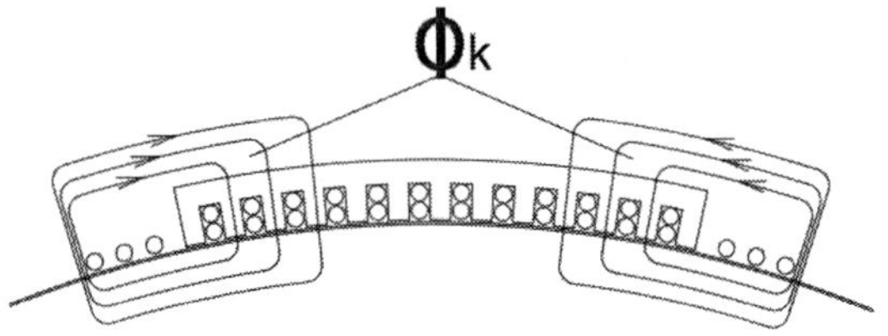

Fig. 4. Streams compensating the boundary phenomena of a linear motor.

Stream Φ_k, consisting of two components located on the edges of a linear motor is compensated in the magnetic circuit. Compensation of the component B_p is important because of the additional power losses it generates and the braking forces occurring in the absence of compensation. In the design of real arc-line systems, two-layer windings with the sides of the coils outside the magnetic circuit are used to compensate the boundary effect as shown in Fig. 4.

Under such assumptions, the flow of currents in individual coils was routed according to the pattern of a linear motor winding with 12 coils of the magnetic circuit (Fig. 5).

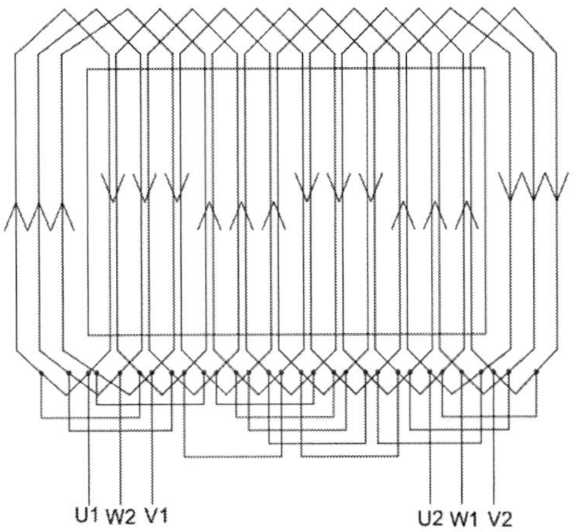

Fig. 5. Winding of the linear engine with direction of the currents flow.

Such a design of the linear motor generates a running field that interacts with the treadmill at the levelled pulsating field. The parasitic effect of the boundary effect disappears at the extension of the linear motor inductor. For relatively short structures, it should be compensated by coils located outside the magnetic circuit as proposed in the article.

5 Simulation of infinite metal sheet length

Simulation processes do not tolerate infinity. Computer memory, the limited area of the simulation of the phenomenon, does not allow the use of the notion of infinity, hence the produced sheet is "only" very long. Simulation of length dimension makes it difficult to model such an object. We propose a solution to this problem: a special design of the simulated system. Figure 6 shows a proposal for the analysis of a practically infinitely long sheet. Taking into consideration the ratio of sheet width to length, *e.g.* 1 m to 10 km, which can be easily wound on a single roll of the output product, the term infinity can be considered in the simulation of the real system. The authors propose to simulate the interaction of forces and torques of the electromechanical system as shown in Fig. 6.

The sheet is a circle with the diameter of 1 m and the width of 1 m. The sensor, located at the top of the circle, is a linear motor. Because of the shape of the sheet metal, it is a section of the induction motor (arc motor). The axis of rotation is located in the centre of the sheet metal. Analysis of the forces generated by that engine is reduced to the analysis of torque versus the rotation axis of the sheet metal. The fragment of the analysed sheet contains delamination.

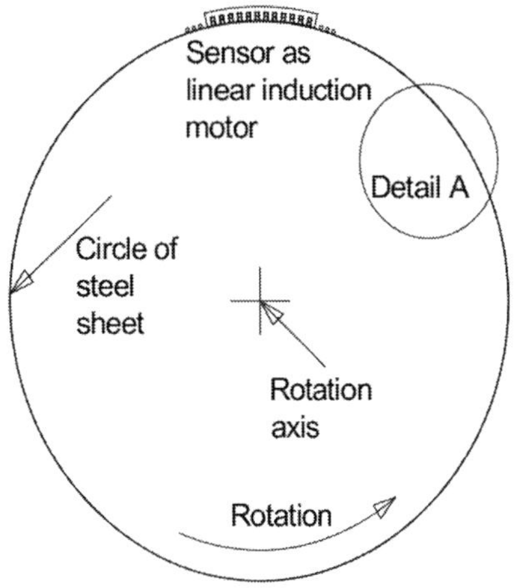

Fig. 6. Schematic process adopted for the calculation model.

The detail A shown in Fig. 7 shows the defect modelled in the article. The air gap formed during the roll of sheet metal disqualifies the product. It can be formed at the stage of the air bubble hidden in the material rolled in the production process.

Due to the place of occurrence (inside the sheet) finding the defect is impossible without special test methods and equipment (*e.g.* Epstein apparatus). Identifying defects requires complex tests and is not possible at the production stage. If no defect is detected, the sheet is wound on the reel drum. Finding a piece of sheet with a flaw after the end of production is practically impossible. There is a possibility to correlate the sensor system with the measurement of sheet length. This kind of electronic identification allows, at a later stage, during the product distribution, to separate the defective material.

 DOI: 10.1051/itmconf/20171507003

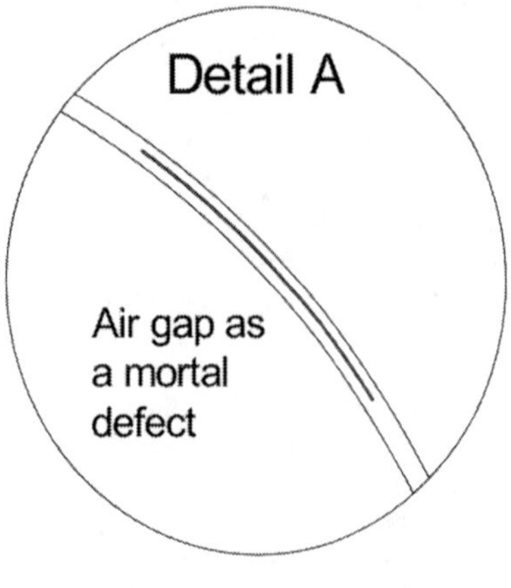

Fig. 7. Airgap. Mortal defect in produced material.

In simulations for the calculation of flux density, forces and torques have been performed by the program Ansys Maxwell. An object shown in Fig. 6 has been taken for the calculations. A circle made of metal sheet is turning round on its own rotary axis. At the top there is the sensor based on induction linear motor. All details of the sensor are explained in Fig. 3. In real conditions there is no ring of steel sheet, but simply a produced steel sheet. The simplification applied in the model is the easiest way to describe this kind of object in the program. In Fig. 7 in the middle of the steel sheet there is an air gap which symbolizes the defect of material. The air gap inside of material is moving with the steel sheet along the sensor [3,4].

AC sensor works as a linear induction motor whose raceway moves the metal sheet. In addition, the metal sheet as a ferromagnetic material induces eddy currents effect. It is the source of forces acting on the sensor. By measuring the force, torque, and eddy current effect acting on the sensor we can identify disturbances in the magnetic field resulting from product defects.

6 Computing methodology

All simulations were performed with the Ansys Maxwell program. This piece of software uses the Maxwell Stress Tensors (MST) methodology [5,6,7]. Zero boundary conditions have been assigned around the machine. The area around the motor has been divided into about 300.000 triangular elements. The elements are of different dimensions, depending on the importance of the considered area. The sensitive space around the air gap, the sharp shapes around the poles, was analysed with finer mesh. The software performing the simulation is capable of modelling both dynamic and static elements. The motor was divided into the stator with 15 winding poles as the static part of the motor, and the rotor, raceway made of metal sheet as the dynamic element. The shape of the air gap between the stator and the rotor is very important [8]. The rotor axis is placed perpendicularly to the surface of screen. Considering the formulas for computing force:

$$F = \int \left[\frac{1}{\mu_0} B(B \cdot n) - \frac{1}{2\mu_0} B^2 \cdot n \right] dC \tag{2}$$

and torque

$$T = r \cdot F \tag{3}$$

Values in formulas (2) and (3) are as follows:
B - momentary value of induction in the air gap B [T]
N - unit vector, perpendicular to the surface of rotor,
μ0 - magnetic permeability of vacuum μ0= 4 π·107 [H/m].

Value of torque is calculated from the formula (3) taking the radius of rotor as r [m].

It is important to assume a very big difference between magnetic permeability (minimum 1/1000) in those two analysed phases (air and iron). In the calculations the normal part of flux vectors in the air gap, and big difference of magnetic permeability between the air and electromagnetic steel were assumed. The presence of the air gap in the analysed region fulfils these conditions.

7 Computational results

The sensor system along with the sheet metal was modelled in the Ansys Maxwell program. Linear motor is powered according to the electrical diagram shown in Fig. 8.

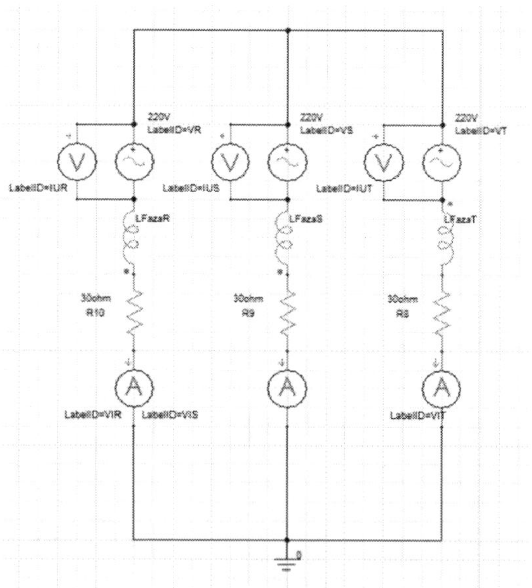

Fig. 8. Electric scheme for supplying linear induction motor.

The diagram shows three voltage sources. Sinusoidal voltage phase of each of the three sources is shifted by 120° comparing to the voltage of the adjacent phases. A voltmeter was connected at each phase for voltage control. Coils of the linear motor are serially connected

DOI: 10.1051/itmconf/20171507003

to the power source. It should be noted that, according to the scheme of the windings of the linear motor shown in Fig. 5, the coil of the phase T is connected in the reverse direction to the phases R and S. The change of phase is necessary to correlate the direction of the moving sheet and the moving field generated by the induction motor. Serial-connected resistances limit the coil current. The ammeter connected at the end of each phase circuit controls the current connected to the linear motor. In the simulation the source of AC power at 50Hz and 220V was used. Simulation results are shown in the figures below.

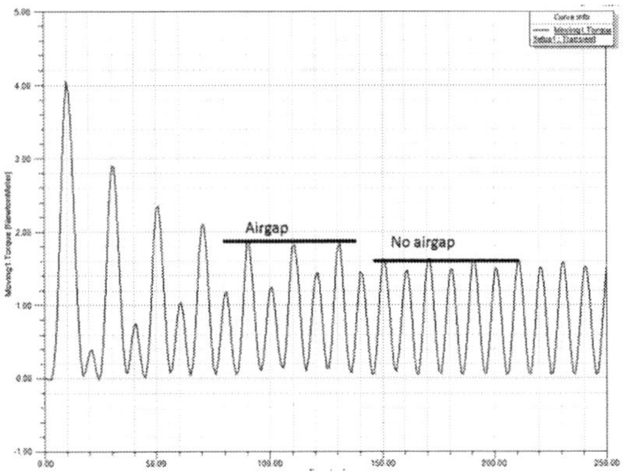

Fig. 9. Dynamic torque waveform in dependence on time.

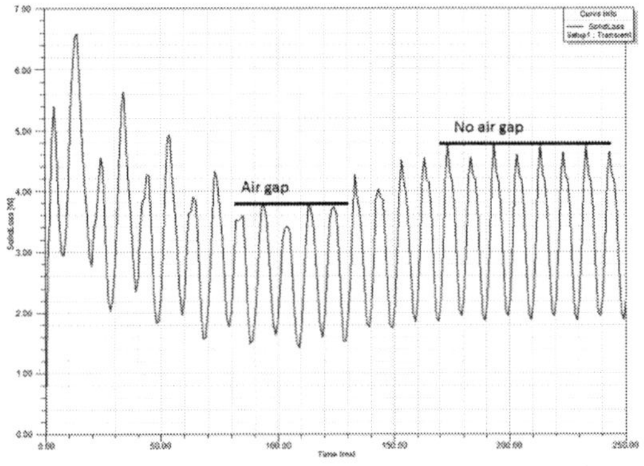

Fig. 10. Dynamic solid loss in dependence on time.

Computer simulation was made assuming the sheet thickness equal to 2 mm and the air gap of 0.2 mm. The fixed state of the system was considered, which in the case shown in Fig. 9 appeared after about 80 ms from the time t = 0. The segment marked "air gap" is faulty, the further part of the "no air gap" is a material without defects. A comparison of the numerical results of torque over time is presented in Table 1.

Table 1. Comparison of the moving torque values with and without air gap.

Air-gap 0.2 mm	**Moving Torque**
With air gap [Nm]	1.82
Without air gap [Nm]	1.6
Difference [Nm]	0.22

Additional simulation of losses occurring in the produced sheet is also highly interesting. Due to the fact that losses from the linear motor, affecting the sheet are active, they can be distinguished by simple measurements. The losses (Eddy Current) show significant differences in size. In Fig. 10 the pass of material with and without air gap near the sensor was shown.

The simulation results presented in the time chart show a significant difference. The gap appearing in the linear motor decreases the loss by more than 20%. Comparison of the results of the simulation for changes in the solid losses occurring in the test sheet is presented in Table 2.

Table 2. Value comparison for solid loss with and without air gap.

Air-gap 0.2 mm	**Solid loss**
With air gap [W]	3.8
Without air gap [W]	4.8
Difference [Nm]	0.22

8 Conclusion

The results of the simulation show that the construction of a sensor based on the design of an inductive linear motor for testing material quality in continuous production processes is possible. By measuring the change of torque (or thrust force) of a linear motor and changing the quality of the treadmill it is possible to generate an output signal at low air gap thicknesses. For the simulation presented in this paper, the change of torque acting on the motor at an air gap of 0.2 mm represents 12% of the generated torque in the steady state. It generates a signal informing the operator about the change in the quality of the material produced.

Increasing the torque acting on the sensor is explained in the second graph. It shows that in the area of the delaminated material the active loss in treadmill material is decreased. This is natural because of the reduction of eddy current losses in layered materials. Both increased torque (strength) and reduced eddy current losses can be easily measured. The noise signal given to the electronic memory system can easily be used for the distribution of the produced material. After distribution, you can verify the quality of the highlighted material to determine its suitability.

Qualitative study of materials is a broad field that uses a variety of methods. Penetration testing is a non-destructive method which employs Barkhausen effect, radiometry, x-ray, echolocation, enables material testing resulting in higher quality of products and better performance of production machinery. This article shows the possible application of the theory of electrical machines for the construction of sensors and their use in production processes, and for the diagnostics of ferro and diamagnetic materials. The work aims to inspire designers to use existing designs in material diagnostics.

The authors intend to conduct further research by examining the effects of linear engine parameters, ferromagnetic thickness, and air gap on accuracy of measurements, thanks to which better sensor parameters of this type can be found.

References

1. R. Zimroz, R. Błażej, P. Stefaniak , A. Wyłomańska, J. Obuchowski, M. Hardygóra, Fundamental Problems of Conveyor Transport **21**(2), 99–109 (2014)
2. T. Machula, *Opracowanie metody ciągłej analizy stanu technicznego taśm przenośnikowych z linkami stalowymi*, Doctoral dissertation (AGH Kraków, 2011)
3. J. Jonak, J. Gajewski, Maintenance and Reliability **4**, 74-78 (2006)
4. T. Piech, *Badania magnetyczne, Wykorzystanie efektu Barkhausena* (Biuro Gamma, Warszawa, 2003)
5. Maxwell 15.0 *Application User's Guide*
6. Ansys 14.0 *Application User's Guide*
7. WorkBench 14 *Application User's Guide*
8. M. Lis, *Modelowanie matematyczne procesów nieustalonych w elektrycznych układach napędowych o złożonej transmisji ruchu* (Wydawnictwo Politechniki Częstochowskiej, Częstochowa, 2013)

FEM simulation research of natural frequency vibration of crankshaft from internal combustion engine

Paweł Magryta[1,*], *Konrad* Pietrykowski[1] and *Krzysztof* Skiba[1]

[1]Faculty of Mechanical Engineering, Department of Thermodynamics, Fluid Mechanics and Aviation Propulsion Systems, Lublin University of Technology, ul. Nadbystrzycka 36, 20-618 Lublin, Poland

Abstract. The article presents vibration simulation studies of a crankshaft used in internal combustion engine. The simulation was performed using FEM method. The study was conducted in Abaqus software, and a shaft model was implemented in Catia v5. The influence of a mesh element size on the results of obtained calculations was analysed. A shaft which will be used in the ultralight aircraft engine was the subject of research. Results show the first 10 frequency modes and 12 grid examples for various element sizes, from 9 to 1.3 mm. Moreover, the effect of mesh size on the obtained results is presented. It has been proven that the maximum error for two extreme results (the densest and sparsest grid) is approximately 1.4%; therefore, it is justified to carry out calculations of own vibration on a fatal grid. Results presented herein will be used in the future work on the crankshaft geometry modifications.

1 Introduction

Internal combustion engines use a crank-piston system to convert reciprocating motion to rotary motion. The crankshaft is a key element of this system. Its torque is not constant over time but varies in complexity, depending on the position of the crankshaft for each cylinder [1]. In many cases the vibration phenomena can be decisive for the proper functioning of the shaft. Vibration can cause an engine as well as an aircraft's work disturbance. The resonance amplification of the vibration amplitude threatens to damage of the shaft and its associated components. Regardless of the dynamic system in which the engine works, the greatest danger to its crankshaft are torsional vibrations. Among of many forces acting in the crank-piston system, crankshaft rotational force is caused by tangential force around the crank. Force which is one of two component forces acting along the connecting rod axis is the periodic force [2]. Its period for a two-stroke engine is 2π, and for four-stroke engines 4π [3]. Variation of force induces acceleration in the crankshaft rotation of the engine, causing the formation of torsional vibrations, which additionally change with the change in shaft speed. The resonance may occur when the frequency of the forced oscillation equals or is multiple of the vibration frequency of the shaft. The vibration frequency of the shaft depends on the size and distribution of the masses, how the shaft is supported and its geometrical properties [4].

Very often design support processes are used in the form of computer simulation. Modern design methods allow for a considerable reduction in design time [5]. In the field of mechanics and machine construction, the use of computing power allows for faster verification of proposed technical solutions [6]. This approach avoids undesirable design errors as well as shortens design time [7, 8]. With the development of computer systems, increased computing power is available. This allows for more complex simulation processes. In the field of mechanics, simulations are mainly used for testing the strength of the tested parts of machines or devices. Most of simulations are based on the so-called FEM ¬ Finite Elements Method. The main aspect influencing the accuracy of calculations using this method is a size of the primary computing element. The choice of its shape and size obviously affects the time of calculation but also the accuracy of results [8, 9].

This article describes the impact of computational grid selection on results obtained during the simulation of the crankshaft of a diesel engine. Presented methodology will be useful for making new calculations for new crankshafts. The paper presents differences of results, that can be used for study of next engine parts [10, 11, 12]. This engine is a research engine with opposed pistons. It is a propulsion unit that will be used in ultralight aviation. Abaqus FEM software was used to calculate the crankshaft frequency. This program allows to specify a user-defined number of forms of the vibration frequencies of any model. Crankshaft geometry was imported into the environment as a .stp file (Figure 1).

* Corresponding author: p.magryta@pollub.pl

DOI: 10.1051/itmconf/20171507004

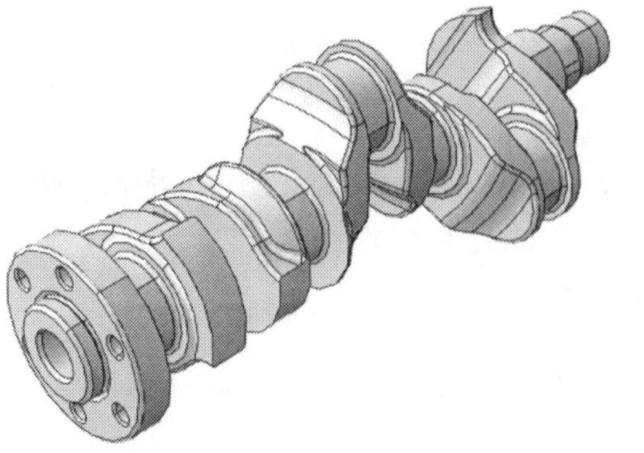

Fig. 1. Solid model generated in Catia v5 software.

2 Simulation

The first step in preparing the simulation was to determine a type of simulation. In order to calculate a form of the vibrations, a frequency simulation was selected and *Lanczos* solver was used. Because Abaqus software allows to determine the number of calculated vibration patterns or to determine the lower and upper limits of vibration frequencies, a frequency range between 10 Hz and 6000 Hz was set after initial calculations. The analysis of the crankshaft's own vibration was performed to determine its own frequency, so the model was not fixed in any place. Due to the complex geometry of the crankshaft model, *Free Tet* mesh grid was used for its discretisation. Material properties were defined by choosing the material as cast iron. The material properties adopted for the calculation are shown in the Table 1.

Table 1. Material properties (cast iron) of the analysed shaft.

Name	Value
Young Modulus	178 GPa
Poisson Number	0.33
Density	7197 kg/m^3

During simulation tests, mesh grid optimisation was performed by varying the size of a single element from an initial value of 9 mm to final size of 1.3 mm. In the Figure 2 a ready-made model with mesh grids for cases 9 and 1.3 mm is presented. Numbers of elements for models are also included. Based on grids constructed in such a way, crankshaft frequencies were measured. The legends show the overall magnitude of the elements scaled in mm.

The Figure 3 shows a visualisation of all 10 modes of engine crankshaft vibration for the two extreme versions of the grid: 9 and 1.3 mm. These are the first 10 free frequency modes between the assumed range. This range is assumed for the specific part – crankshaft from the opposed piston engine. The frequency value for every mode is calculated automatically by Abaqus software. The colours indicate the displacements. Because at given frequencies of the oscillation, the system falls into resonance, the displacements take the values corresponding to the destruction of the element. For a better visualisation and evaluation of the behaviour of the model, a uniform multiplication coefficient for all visualisations of 0.033 and the same legend of 0-1.35 mm were adopted. The Table 2 below, shows results of the received vibration values for the two extreme cases of grid: 9 and 1.3 mm.

Table 2. The crankshaft's own vibration frequencies for the 9 mm and 1.3 mm grid.

Mode number	Frequency [Hz] for 9 mm grid	Frequency [Hz] for 1.3 mm grid
Mode 1	764.88	754.83
Mode 2	838.54	828.19
Mode 3	1445.7	1428.3
Mode 4	1715.1	1705.6
Mode 5	1950.1	1928.4
Mode 6	2563.7	2560.5
Mode 7	3300.2	3262.5
Mode 8	3872.2	3847.7
Mode 9	4731.1	4687.8
Mode 10	5179.7	5125.5

9 mm 27885 elements

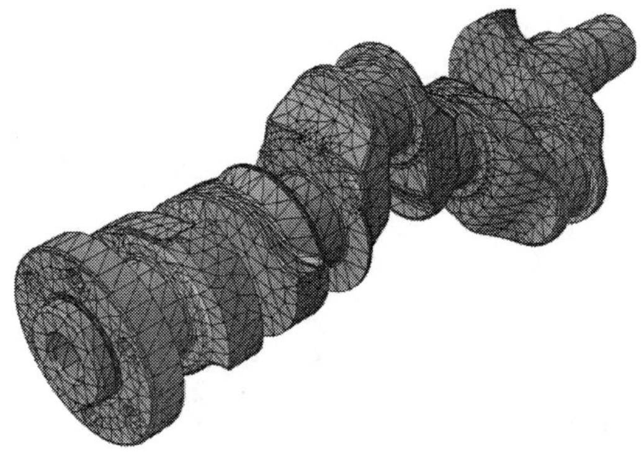

1.3 mm 1888341 elements

Fig. 2. Mesh grid of crankshaft.

DOI: 10.1051/itmconf/20171507004

<table>
<tr><td>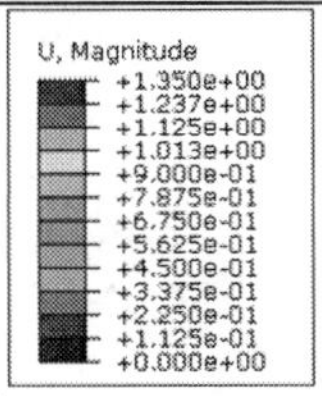
9 mm</td><td>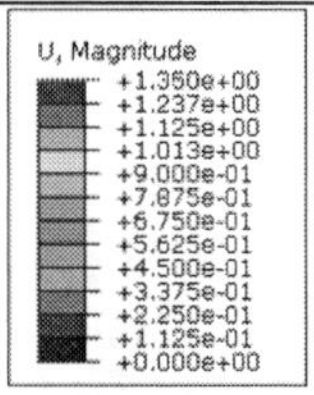
1.3 mm</td></tr>
<tr><td>Mode 1
</td><td>Mode 1
</td></tr>
<tr><td>Mode 2
</td><td>Mode 2
</td></tr>
<tr><td>Mode 3
</td><td>Mode 3
</td></tr>
<tr><td>Mode 4
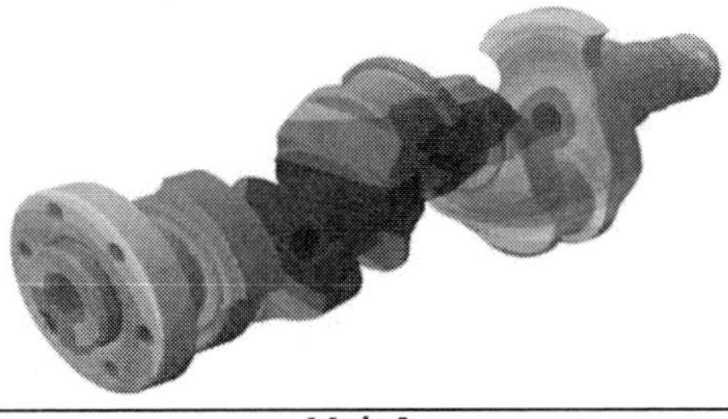</td><td>Mode 4
</td></tr>
<tr><td>Mode 5
</td><td>Mode 5
</td></tr>
</table>

DOI: 10.1051/itmconf/20171507004

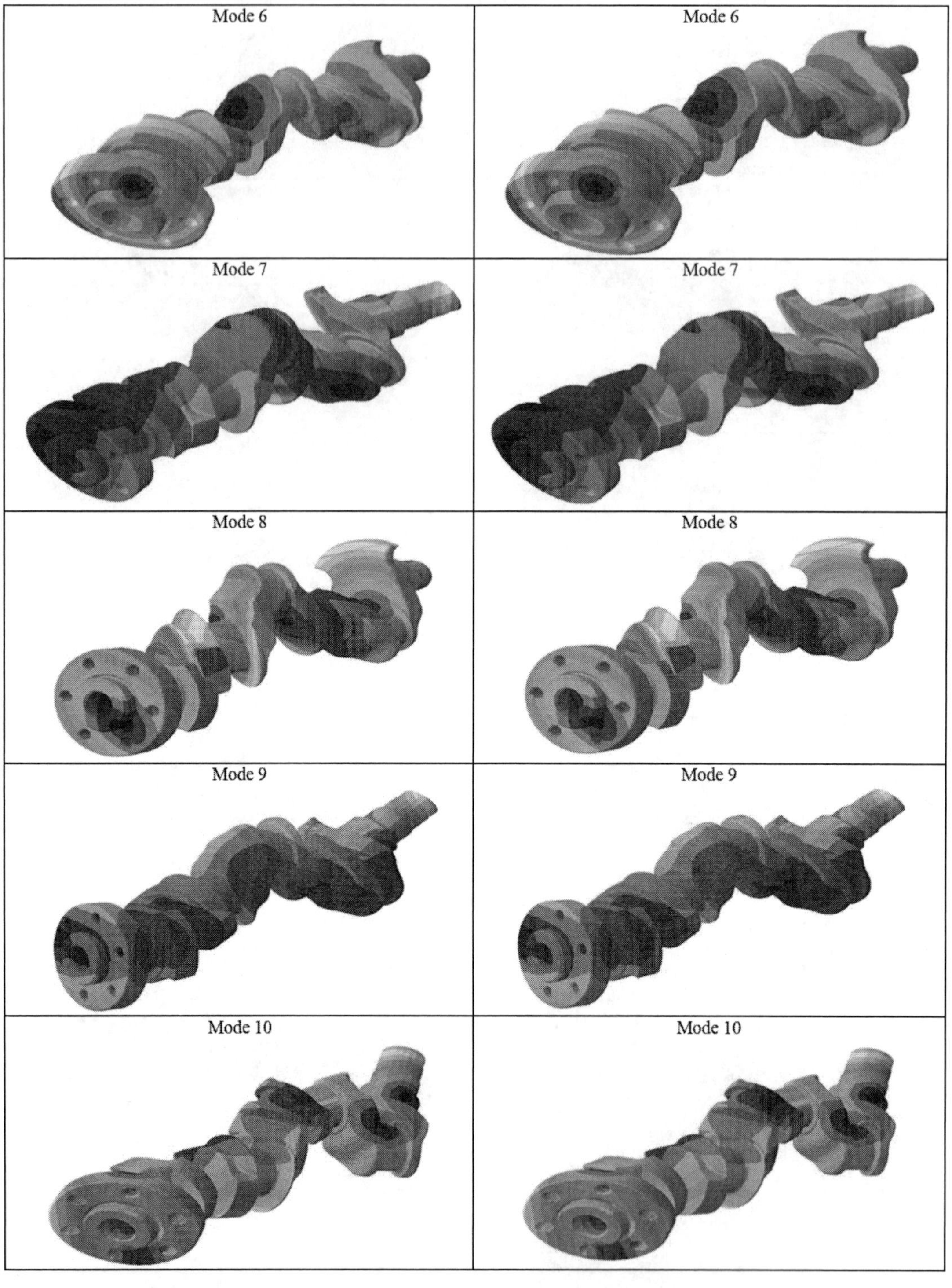

Fig. 3. Calculation results for 9 mm and 1.3 mm grids.

 DOI: 10.1051/itmconf/20171507004

3 Results

In addition, a range of changes in the frequencies value of received native vibration for each grid in the form of figures 4-8 were presented. Modes 1, 3, 5, 7 and 9 are referred to the left axis, modes 2, 4, 6, 8 and 10 are referred to right axis.

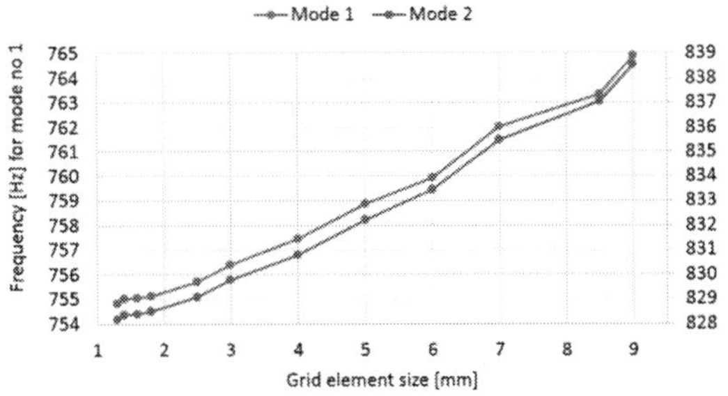

Fig. 4. Frequency values depending on grid size for the mode 1 and 2.

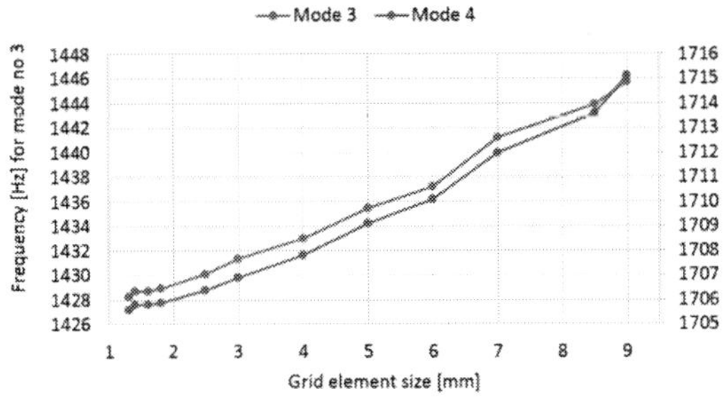

Fig. 5. Frequency values depending on grid size for the mode 3 and 4.

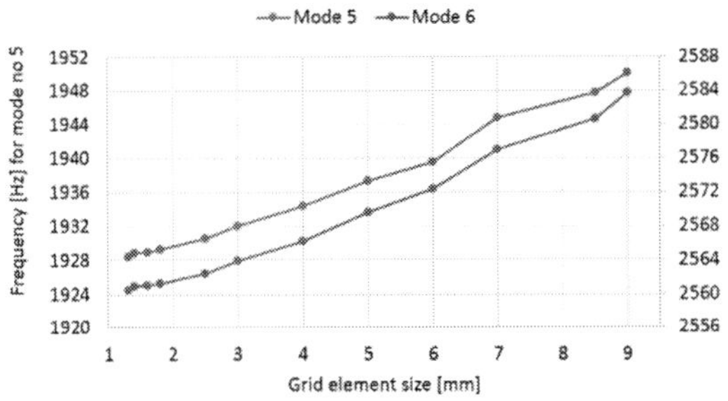

Fig. 6. Frequency values depending on grid size for the mode 5 and 6.

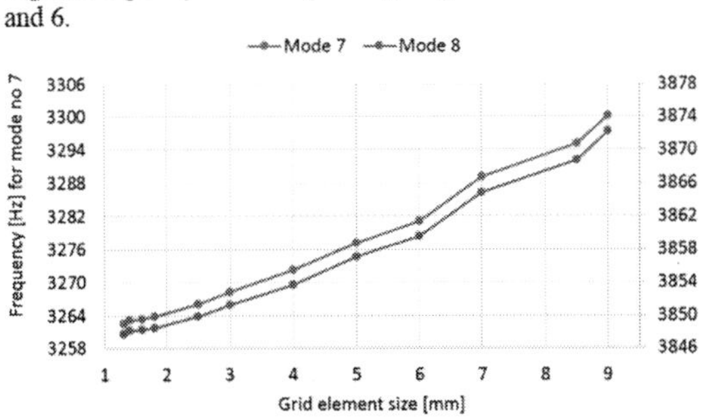

Fig. 7. Frequency values depending on grid size for the mode 7 and 8.

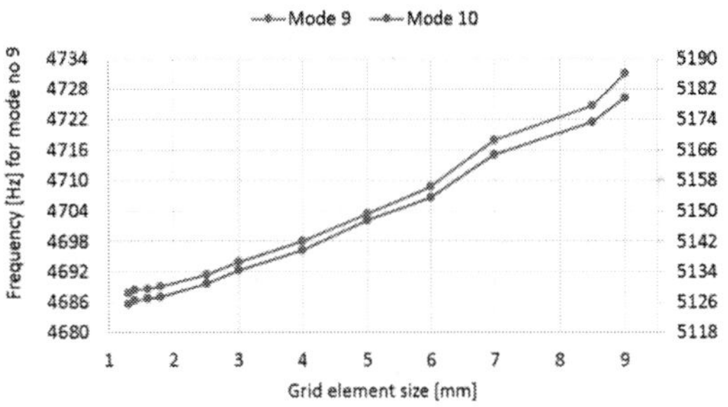

Fig. 8. Frequency values depending on grid size for the mode 9 and 10.

To check the mesh size quality for all cases, the dimensionless factor of $V_{i,j}$ was adopted, which is described by the equation:

$$V_{i,j} = m_{i,j} / m_{mix\,j} \tag{1}$$

Where:
V – coefficient,
$m_{i,j}$ – frequency value for i-grid and j-mode,
$m_{mix\,j}$ – minimum frequency value for j-mode,
i – number of grid,
j – number of mode.

This factor was calculated for each of shaft vibration frequencies received, and is shown in the Figure 9.

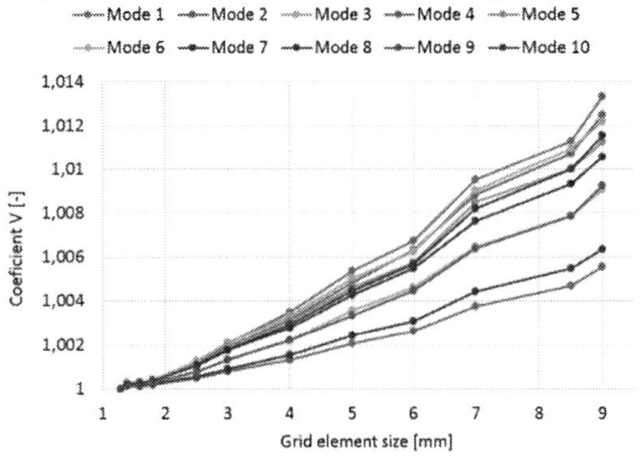

Fig. 9. Characteristics of factor $V_{i,j}$ for all frequency modes.

The most likely to occur in reality are the first four forms of self-oscillation. They correspond visually to the figures below. As it is shown in the Figure 3, they have the same shape as presented below.

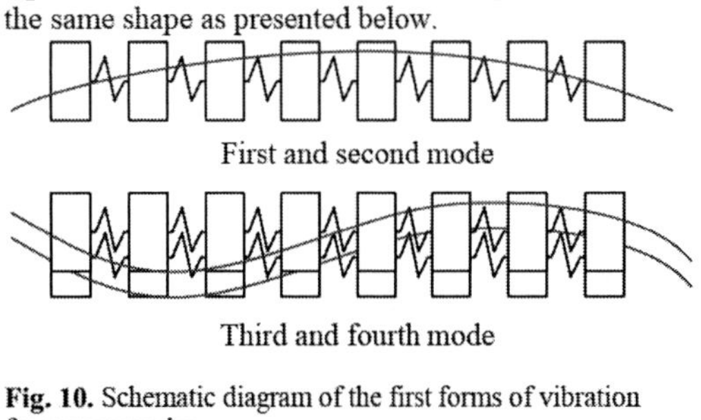

Fig. 10. Schematic diagram of the first forms of vibration frequency modes.

4 Summary and conclusions

As can be seen based on presented results of the simulation calculations in the Table 2, the values of vibrations in the individual modes differ among themselves. These differences reach as much as 50 Hz for higher vibration values, but dimensionally they reach a maximum of approximately 1.3% (as shown in the Figure 9). According to the authors, these derogations are mainly due to the lack of reflection of the roundedness of the analysed solid in the case of worse quality of mesh grid. The grid of 9 mm (Figure 2) causes distortion of the crankshaft geometry, which affects other mass distributions in some parts of the crankshaft, consequently having an impact on results of the simulation. However, it should be noted that the visual representations of the form of the oscillation shown in the Figure 3, for both extreme grids look almost identical.

The grid with a single element size of 1.3 mm generates 1,888,341 elements. Such a large number of elements results in a considerable prolongation of the simulation time. Further reduction in size of a single element is pointless: if the element is reduced to a size close to 0 mm, results obtained (considering the downward trend of the curves in the Figures 4-8) would not significantly differ from results obtained for the 1.3 mm mesh.

In case of superficial calculation of the value of the oscillation, less accurate grid can be used. However, it should be noted that the selection of grid elements and its compaction must be in line with the geometric model. Nevertheless, to obtain very precise calculations, the grid should be compacted as accurately as possible. It should be considered in the future in case of calculating new engine parts.

This allows the authors to say that if vibrations are calculated for visualisation purposes only, a mesh with a bigger size of elements can be applied.

Acknowledgement: This work has been realised in the cooperation with The Construction Office of WSK "PZL-KALISZ" S.A." and is a part of Grant Agreement No. POIR.01.02.00-00-0002/15 financed by the Polish National Centre for Research and Development.

References

1. W. Mitianiec, K. Buczek, Technical Journal, Cracow University of Technology Publish House, **z. 8-M/2008** (2008).
2. A. S. Mendes, Proceedings of the Institution of Mechanical Engineers Part K Journal of Multi-body Dynamics **222**, 2 (2008).
3. W. Homik, Pol Marit Res, **no 3**, 43-47, (2011).
4. S.S. Mohtasebi, H. Afshari, H. Mobli, J of Appl Scien, **6**, 3 (2006).
5. T. Łodygowski, Kąkol W., *Finite element method in selected problems of mechanics engineering structures* (in polish), Poznań University of Technology Publish House, (1994).
6. R. Neugebauer, C. Scheffler, M. Wabner, CIRP Journal of Manufacturing Science and Technology **4** (2011).
7. A. Zielińska, P. Wyciślok, High-speed Tracked Vehicles (in polish), **18**, 2 (2003).
8. P. Magryta, M. Biały, K. Siadkowska, Logistyka **3** (2015).
9. T. Markowski, G. Budzik, J. Pacana, Engineering Modeling (in polish) **39** (2010).
10. Ramesha C. M., Abhijith K. G., Abhinav S., Abhishek R., Chetan S. N., IJETAE **5**, 6 (2015).
11. Deshbhratar R. J., Suple Y. R., IJMER, **2**, 5 (2012).
12. Horváth P., Égert J., Dynamic, Acta Technica Jaurinensis **8**, 4 (2015).

DOI: 10.1051/itmconf/20171507005

Influence of gas inlet angle on the mixing process in a Venturi mixer

Mathias Romańczyk[1,a]

[1]Czestochowa University of Technology, Institute of Thermal Machinery, al. Armii Krajowej 21, 42-201 Czestochowa, Poland

Abstract. In this paper numerical analysis were performed to investigate the influence of gas inlet angle on mixing process in a Venturi mixer. Performance of an industrial gas engine depends significantly on the quality of mixing air and fuel; therefore, on the homogeneity of the mixture. In addition, there must be a suitable, adapted to the current load of fuel, air ratio. Responsible for this fact, among others, is the mixer located before entering the combustion chamber of the engine. Incorrect mixture proportion can lead to unstable operation of the engine, as well as higher emissions going beyond current environmental standards in the European Union. To validate the simulation the Air-Fuel Ratio (AFR) was mathematically calculated for the air-fuel mixture of lean combustion gas engine. In this study, an open source three-dimensional computational fluid dynamics (CFD) modelling software OpenFOAM has been used, to investigate and analyse the influence of different gas inlet angles on mixer characteristics and their performances. Attention was focused on the air-fuel ratio changes, pressure loss, as well as improvement of the mixing quality in the Venturi mixer.

1 Introduction

The performance of industrial gas engines strongly depends on the quality of mixing air and fuel, and therefore homogeneity of the mixture. Rapid variations between the concentration of fuel and air called "Air-Fuel Ratio (AFR) changes" have a very negative impact not only on environmental emissions, but also on the fuel consumption in industrial gas engines. With increasing the requirements in line with European Union environmental standards for high efficiency combustion process in the gas engine, the need for a precise, homogeneously mixed air-fuel mixture, become more and more essential for better engine performance [5]. Therefore, an ideal Venturi mixer prepares a mixture with an appropriate air-fuel concentration, depending on the demand of the gas engine, both from no load during its start-up, to full load conditions during its constant operation [3]. Improper air-gas mixture can lead to unstable operation of the entire gas engine and excessive emissions, which are going beyond the applicable environmental standards. It is worth noting that for industrial gas engines, the anomalies of combustion processes are very different from those of conventional combustion processes. These processes can negatively affect the operation of the entire gas engine, which is associated with the occurrence of many kinds of undesirable phenomena [6]. The mixture proportion of air and fuel has a great impact on how the complete combustion process develops. When the air-fuel mixture is optimally prepared, there occurs an optimal combustion process. This situation changes when the air-fuel mixture is too lean (i.e. it has more air than necessary), then this can lead to a non-ignition mixture that results with misfire. As a result, there are unburned gas-fuel fractions in the exhaust duct. An additional undesirable phenomenon can also be a slow combustion process. This involves high CO emissions and consequently high instability of the entire combustion process. On the other hand, when the air-fuel mixture is too rich (i.e. it has more fuel than necessary) there may occur sudden, violent undesirable processes such as pre-ignition and glow-ignition. Most often, they occur during engine operation at maximum performance. Glow-ignition causes a strong instability of the entire combustion process, higher exhausts emissions and is, additionally, very dangerous for the gas engine itself. Various reasons for pre-ignition and glow-ignition have been analysed and investigated over the past few years, but the most common cause of self-ignition has been stated as the inadequate proportion of the air-fuel mixture [4]. Another symptom of a too rich air-fuel mixture is the knocking phenomenon. This is the most undesirable phenomenon, because it affects not only very badly on exhaust emissions, but primarily on the life of the entire gas engine. Consequently, it can lead to complete destruction of the whole industrial gas engine. To provide an efficiency combustion process in an industrial gas engine, the Venturi mixer should be designed to allow the best possible mixing of the two components. Additionally, it should be compact, with minimum of pressure loss, and moreover, should have a good suction pressure in the throat due to the Venturi principle. Different analyses were

[a] Corresponding author: mathias@imc.pcz.czest.pl

DOI: 10.1051/itmconf/20171507005

performed to improve the efficiency of the whole mixing process in the Venturi mixer. The effectiveness of the mixing process in the Venturi mixer depends on many factors, such as the Venturi throat diameter and its position, gas inlet area, as well as the gas inlet position. As was shown by Danardono et al. [2], by decreasing the throat diameter in the Venturi mixer, the mixture is richer in the end, and the Air-Fuel Ratio decreases significantly. Additionally, they showed that by increasing the size of the gas inlet area, the air-gas mixture is richer, there would be a better mixing process and a lower pressure loss. The influence of gas inlet position was investigated by Romańczyk and Elsner [8]. They showed that the optimal gas inlet position is located at an angle of inclination directly into the flow stream of air, where the mixing process takes place without much resistance from the air itself, what leads to a better mixing process and in addition to a lower pressure loss. They also paid attention to the gas inlet located indirectly into the flow stream of air (i.e. in the opposite direction to the flow), which results in a poorer mixing process and automatically – a bigger pressure loss through the Venturi mixer. To show how the whole mixing process depended on the gas inlet location, it was decided, by using the computational fluid dynamics simulation software OpenFOAM, to investigate the influence of different gas inlet angles in 5° intervals, which were located directly into the flow stream of air.

2 Mixer design and numerical modelling software OpenFOAM

In modern industrial gas engines, the manufactures more often try to design mixers based on the Venturi principle. The Venturi principle describes the drop in fluid pressure that occurs when a fluid flows through a throat in a pipe. In fluid dynamics, a fluid's velocity increases as it passes through a throat, in accordance with the principle of mass continuity, while its static pressure decreases, in accordance with the principle of conservation of mechanical energy. By decreasing in some points, the cross-sectional area of the air supply (see Figure 1), in accordance with Bernoulli's law and the continuity condition, creates a vacuum resulting in gas being sucked into the Venturi mixer. In this way, gas is being mixed with air.

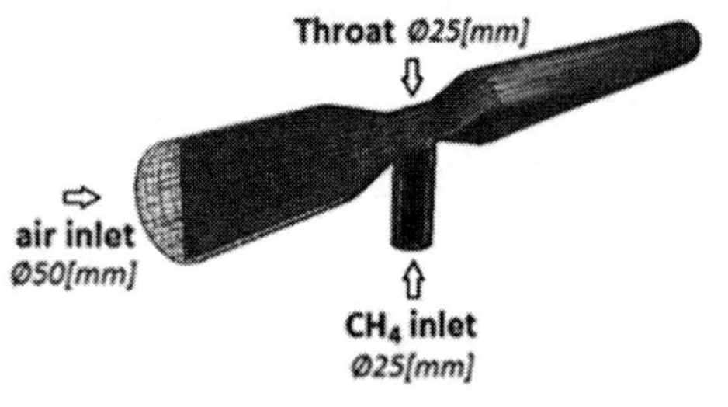

Fig. 1. Venturi mixer design with dimensions [mm].

In previous paper the optimal localisation of gas inlet position was investigated. It was shown that the optimal gas inlet position was located at an angle of inclination, directly into the flow stream of air, where the mixing process was taking place without much resistance from the air itself. Therefore, resulting in better mixing process and lower pressure loss. To demonstrate how the whole mixing process depends on the gas inlet location, it was decided to investigate the influence of different gas inlet angles located directly into the flow stream of air, by using the computational fluid dynamics simulation software OpenFOAM. As a result, seven gas inlets were analysed at different angles from 0°-30° with 5° intervals.

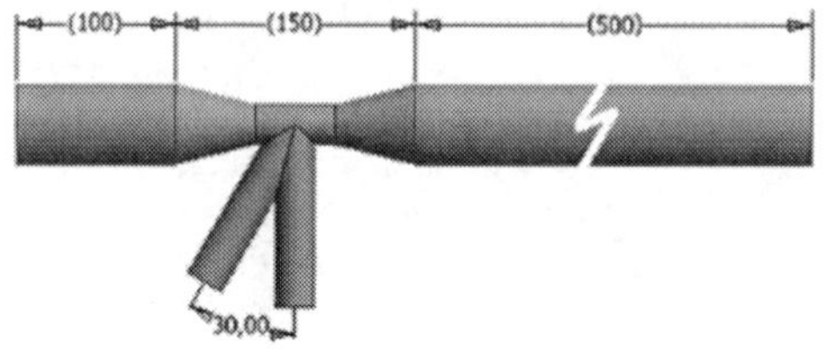

Fig. 2. Analysed cases (0°-30°) with the dimensions [mm].

The Venturi mixer models were designed with software Autodesk Inventor. The dimensions of the Venturi mixer were adopted in all analysed cases at the same level (see Figure 1 and Figure 2). Differences were in the gas inlet angles, only. The air inlet was equipped with a diameter of → 50 [mm], and length of 100 [mm]. The throat point, as well as the gas inlet (CH_4), were provided with a diameter of → 25 [mm]. The throat point was centred on a length of the next 150 [mm] beyond the air inlet, while the gas-air mixing pipe after the Venturi throat was set on 500 [mm] in length to achieve an appropriative convergence of the whole simulation. For this reason, the total length of the analysed Venturi mixers was set on 750 [mm]. Numerical calculations were performed for a lean combustion mixture, in which the most common industrial gas engines operate, with an air excess coefficient of λ = 1.6. In numerical analysis, OpenFOAM software was used. OpenFOAM (Open Field Operation and Manipulation) is open source CFD software with a package for solving a wide range of engineering issues, from complex CFD calculations, including chemical reactions and turbulence flows. The flow through the Venturi mixer is a turbulent flow, so it was necessary to use a turbulence model in numerical calculations. For this purpose, a two-equation turbulence model k-ε was used to analyse the Turbulence Kinetic Energy (TKE), along with the flow through the Venturi mixer. That turbulence model was selected, because it is one of the most widely used, good validated model in terms of consistency and reliability. It also consumes less computer time [9]. To analyse the efficiency of the gas-air mixing process, the standard OpenFOAM solver *reactingFoam* was used. This model is based on the VOF (Volume of Fluid) methodology, which allows to analyse different mixing reactions. In this paper, the mixing of methane → CH_4 with air was analysed. During numerical calculations, the composition of air was adopted as standard in 21% of oxygen → O_2 and 79% of nitrogen → N_2. The next chapter in this paper discusses results of numerical calculations of the analysed cases with different gas inlet angles. Particular attention has been paid to the Air-Fuel Ratio changes (AFR), by analysing the concentrations of methane mass fraction → CH_4 and air through the Venturi mixer. In addition, the distribution of the Turbulence Kinetic Energy (TKE) through the Venturi mixer, and the velocity changes with the pressure loss (which is one of the most qualitative parameters to describe the quality and efficiency of a gas mixer) were analysed.

3 Results and discussion

The results presented in this paper show a comparison between seven different gas inlet angles, located directly into the flow stream of air from 0°-30° with 5° of intervals. Analysing the whole mixing process in a Venturi gas mixer, it is very important to know, how the concentrations of methane mass fraction → CH_4 and air changes through the whole mixer section. Therefore, the mixing characteristics in the cross-sectional view of the Venturi mixer were firstly analysed in the Figure 3 after passing the Venturi throat, and in the Figure 4 at the outlet of the Venturi mixer.

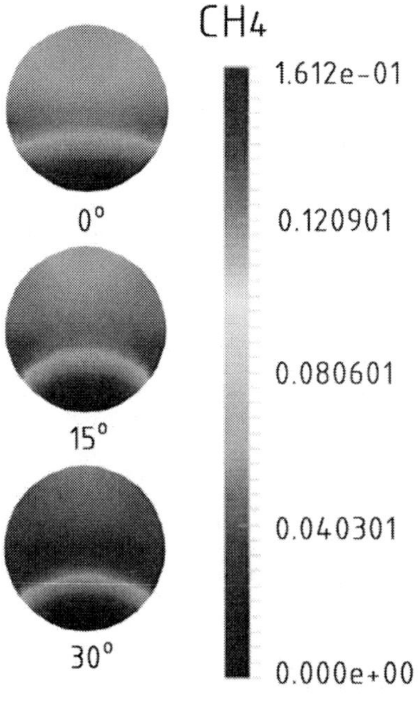

Fig. 3. Distribution of methane mass fraction in the cross-sectional view between analysed cases – 0°,15, 30° after passing the Venturi throat at 0.25 [m] from the inlet.

Analysing contours of distributions of methane mass fraction in the cross-sectional view, after passing the Venturi throat in the Figure 3, following assumption can be made: the greater the inclination of the gas inlet, the more methane → CH_4 is sucked into the Venturi mixer.

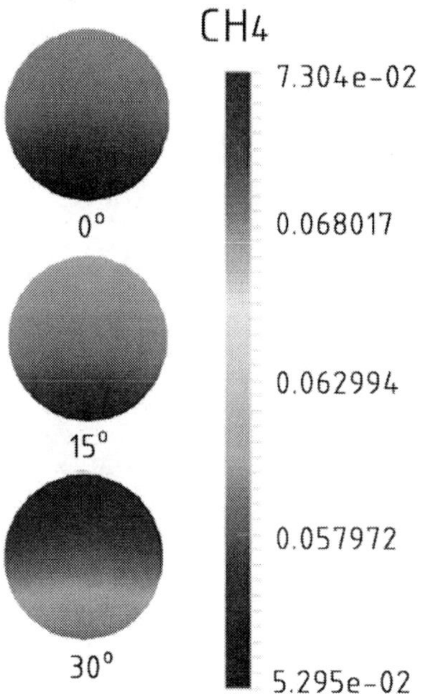

Fig. 4. Distribution of methane mass fraction in the cross-sectional view between analysed cases – 0°,15, 30° at the outlet of the Venturi mixer at 0.74 [m] from the inlet.

As a result, there would be a richer air-gas mixture at the outlet of the Venturi gas mixer, as presented in the Figure 4, that for 0° gas inlet angle the methane content is much lower than the methane content for a 30° angle. Analysing Figure 4, it could be seen that the greater the angle of inclination of the gas inlet, the higher is the concentration of methane mass fraction at the outlet of the Venturi mixer. These contours of methane mass fraction in the cross-sectional view, presented above, are very important, but unfortunately, they cannot fully describe the phenomenon occurring in the whole Venturi mixer section. Therefore, detailed analysis between the different gas inlet angles with average numerical results from the whole Venturi mixer were performed. Each numerical modelling analysis should be validated in the form of experimental data or mathematical calculations, using book definitions. For this reason, in the paper mathematical calculations, according to the book definition, were made to verify the correctness of this numerical calculations. The Air-Fuel Ratio (AFR) was calculated for a lean air-gas combustion mixture for which the most common industrial gas engines operate, with an air excess coefficient ratio of → λ=1.6, where the stoichiometric (i.e. λ=1.0) Air-Fuel Ratio AFR_{stoich} for methane CH_4 equals → 9.52 $\frac{m^3 air}{m^3 CH_4}$ [7]. The mathematical calculation was presented below:

$$AFR = \frac{m_{air}}{m_{fuel}} \tag{1}$$

$$AFR_{stoich} = 9.52\ \frac{m^3 air}{m^3 CH_4} \tag{2}$$

where the air excess coefficient ratio → λ:

DOI: 10.1051/itmconf/20171507005

$$\lambda = \frac{AFR}{AFR_{stoich}} \tag{3}$$

for $\rightarrow \lambda = 1.6$:

$$1.6 = \frac{AFR}{9.52} \Rightarrow AFR = 15.232 \frac{m^3 air}{m^3 CH_4}$$

and as a result:

$$for\ AFR = 15.232 \frac{m^3 air}{m^3 CH_4} \Rightarrow \frac{0.9384\, m_{air}\, [-]}{0.0616\, m_{CH_4} [-]}$$

The mathematical calculations gave the following results. For an air excess coefficient ratio of $\rightarrow \lambda = 1.6$, in which the most common industrial gas engines operate, the concentration of methane equals $CH_4 \Rightarrow 0.0616\, m_{CH_4}$ while the concentration of $air \Rightarrow 0.9384\, m_{air}$. In the next step, detailed analysis between different gas inlet angles were made, with average numerical data, to show if numerical results match the mathematical calculations.

Table 1. Comparison between mathematical and numerical calculations with specified Approximation error at the outlet of the Venturi mixer.

	Gas inlet angle	Concentration of air	Concentration of CH_4	Air excess coefficient $\rightarrow \lambda$	Approx. error
Units	[°]	[-]	[-]	[-]	[%]
Math. calculations	0°	0,93840	0,06160	1,60000	0,00
Numerical calculations	0°	0,93835	0,06165	1,59888	0,07
	5°	0,93790	0,06210	1,58635	0,86
	10°	0,93747	0,06253	1,57481	1,60
	15°	0,93708	0,06292	1,56446	2,27
	20°	0,93679	0,06321	1,55675	2,78
	25°	0,93485	0,06515	1,50727	6,15
	30°	0,93177	0,06823	1,43449	11,54

Accordingly to data in the Table 1, describing the concentrations of air and methane mass fraction $\rightarrow CH_4$ through the Venturi mixer, the results calculated with theoretical mathematical formulas show full compatibility with the numerical calculations performed using OpenFOAM software. The approximation error for 0° of gas inlet angle is only 0.07 [%], so it can be concluded that the numerical calculations were performed correctly for these analysed cases. The greater the gas inlet angle, the higher is the approximation error, because more methane $\rightarrow CH_4$ is sucked into the Venturi mixer which causes a richer mixture proportion at the outlet of the Venturi mixer as evidenced by a smaller air excess coefficient ratio $\rightarrow \lambda$, from $\lambda = 1.6$ up to $\lambda \approx 1.44$. In the Figure 5 were presented the concentrations of methane mass fraction changing through the Venturi mixers for different gas inlet angles.

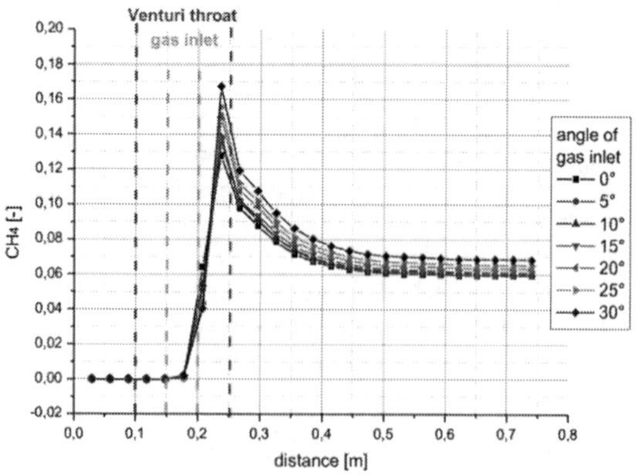

Fig. 5. Concentration of methane mass fraction CH_4 through the Venturi mixer for different gas inlet angles.

Analysing the concentration of methane mass fraction CH_4 through the Venturi mixer in the Figure 5, it could be seen that for mixers equipped with a greater gas inlet angle than > 20° the suction of methane appears a little bit later and initially is slower, but in the end more methane CH_4 is sucked into the mixer, which results in a better mixing process of the two components – air and gas. After passing the gas inlet location (marked with orange dashed lines), there is a rapid increase of the concentration in the Venturi throat, while after passing the throat at 0.25 [m] a drastic drop in the concentration can be observed. However, at 0.6 [m] there occurs a stabilisation of the whole distribution of methane concentration. If more methane CH_4 is sucked in, better and intensified mixing process of two components, air and gas, can occur. Automatically, there is a richer mixture proportion at the outlet of the Venturi mixer. Here, we could observe the Air-Fuel Ratio (AFR) changes, which could be explained with the angle of the gas inlet flow, which causes more suction of gas into the flow stream of the Venturi mixer, at identical inlet conditions in the simulations for all analysed cases.

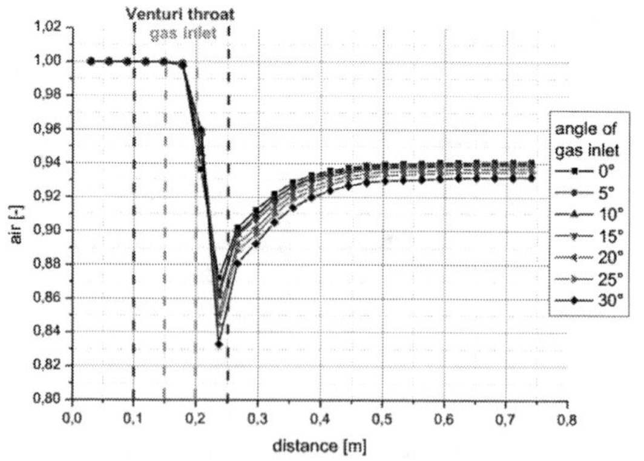

Fig. 6. Concentration of air through the Venturi mixer for different gas inlet angles.

These drops in concentration changes are also seen in the Figure 6, while analysing the concentration of air through the Venturi mixer, because in the total flow volume there is always a 100 [%] air-gas mixture totalling $\rightarrow 1$. Thus, the distribution of air concentration is the mirror image of the distribution of methane concentration. As was shown previously, the result of numerical calculations matches very well with the theoretical mathematical calculations. By analysing the concentration changes of methane CH_4 (Figure 5) and air (Figure 6) along the flow, it can be observed that the greater the gas inlet angle, more methane $\rightarrow CH_4$ is sucked into the Venturi mixer which causes a richer mixture proportion at the outlet of the Venturi mixer. This results with a smaller air excess coefficient ratio → λ, changing from λ = 1.6 up to λ ≈ 1.44. Having the simulated concentration changes (m_{air}, m_{fuel}), in the next step, additional mathematical calculations for the air excess coefficient → λ were made, to show how the Air-Fuel Ratio changes for the different analysed cases in the flow through the whole Venturi mixer. For example, calculation is presented below for $\rightarrow \lambda = 1{,}6$:

Knowing that:

$$AFR = \frac{m_{air}}{m_{fuel}} \Rightarrow \frac{0{,}9384\, m_{air}}{0.0616\, m_{fuel}}$$

$$AFR_{stoich} = 9.52\, \frac{m^3 air}{m^3 CH_4},$$

It can be calculated the air excess coefficient $\rightarrow \lambda$:

$$\lambda = \frac{AFR}{AFR_{stoich}} \Rightarrow \frac{\frac{m_{air}}{m_{fuel}}}{9.52} \Rightarrow \frac{\frac{0{,}9384\, m_{air}}{0.0616\, m_{fuel}}}{9.52} = 1.6$$

The calculated distributions of the air excess coefficient $\rightarrow \lambda$ for these analysed cases is presented in the Figure 7.

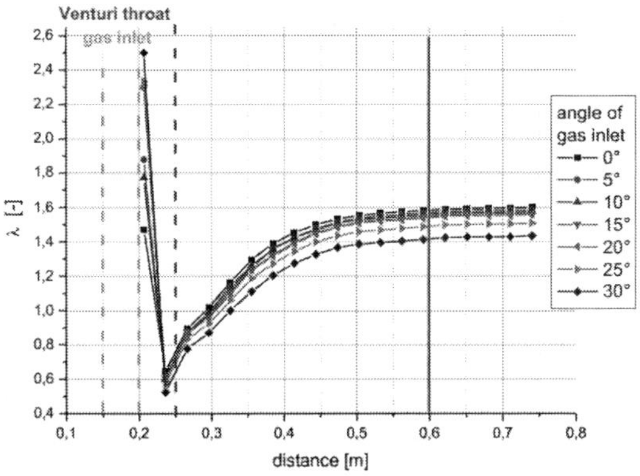

Fig. 7. Distribution of air excess coefficient → λ, through the Venturi mixer for different gas inlet angles.

Analysing the distributions of the air excess coefficient ratio it could be seen that after passing the Venturi throat of the mixer, the coefficient rapidly increases and stabilises at about 0.6 [m] for all analysed cases. The greater the gas inlet angle, the faster is the drop after passing the gas inlet, while the grow up in the Venturi throat is faster. More methane $\rightarrow CH_4$ is sucked into the Venturi mixer which causes a richer mixture proportion at the outlet. This results with a smaller air excess coefficient ratio → λ, changing from λ = 1.6 for a standard gas inlet at 0° up to λ ≈ 1.44 for a gas inlet located with an angle of 30°. As it was mentioned before, the flow through the Venturi mixer is a turbulent flow, so it was necessary to analyse the Turbulent Kinetic Energy (TKE) along the flow through the Venturi mixer. Figure 8 shows the distribution of Turbulent Kinetic Energy (TKE). In fluid dynamics, the Turbulent Kinetic Energy is defined as the mean kinetic energy per unit mass, associated with eddies in the turbulent flow. Physically, TKE is characterized by measured root-mean square (RMS), therefore velocity fluctuations. There is a drastically grow up of the turbulent fluctuations if the flow rate increases. The biggest whirls are affecting and drawing energy from the main flow. In this case, the forces of inertia dominate, while the viscosity forces are negligible. In Reynolds-averaged Navier Stokes equations (RANS), the turbulence kinetic energy (TKE) can be calculated based on the closure method, i.e. a turbulence model. As it was mentioned before, in the numerical calculations the turbulence model k-ε was applied. Turbulence kinetic energy (TKE) can be generated by fluid shear, friction or buoyancy, or through an external force at low-frequency eddy scales [1].

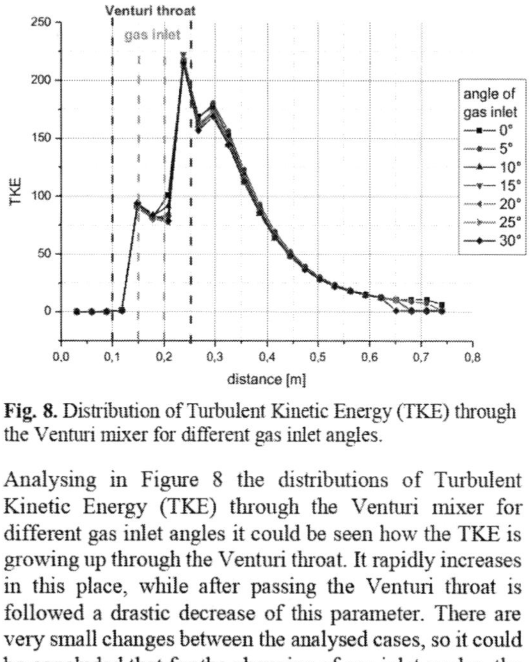

Fig. 8. Distribution of Turbulent Kinetic Energy (TKE) through the Venturi mixer for different gas inlet angles.

Analysing in Figure 8 the distributions of Turbulent Kinetic Energy (TKE) through the Venturi mixer for different gas inlet angles it could be seen how the TKE is growing up through the Venturi throat. It rapidly increases in this place, while after passing the Venturi throat is followed a drastic decrease of this parameter. There are very small changes between the analysed cases, so it could be concluded that for the changing of gas inlet angles, the Turbulent Kinetic Energy (TKE) is almost constant. In the next step were analysed in detail velocity changes in the flow through the Venturi mixer. As was mentioned before, the flow through a Venturi mixer is a turbulent flow, so there could be expected drastic increases of the velocity after passing the Venturi throat of this mixer. For a reminder, in fluid dynamics, a fluid's velocity increases as it passes through a throat in accordance with the principle of mass continuity, while its static pressure decreases in accordance with the principle of conservation of mechanical energy. The initialised velocity at the air inlet was set on 12 [m/s] in the numerical calculations, which caused the gas being sucked into the Venturi mixer by an

DOI: 10.1051/itmconf/20171507005

under-pressure generated in the throat point of the Venturi mixer. Therefore, firstly the contours of velocity magnitude between the analysed cases for three different gas inlet angles → 0°, 15° and 30°, were analysed in the Figure 9.

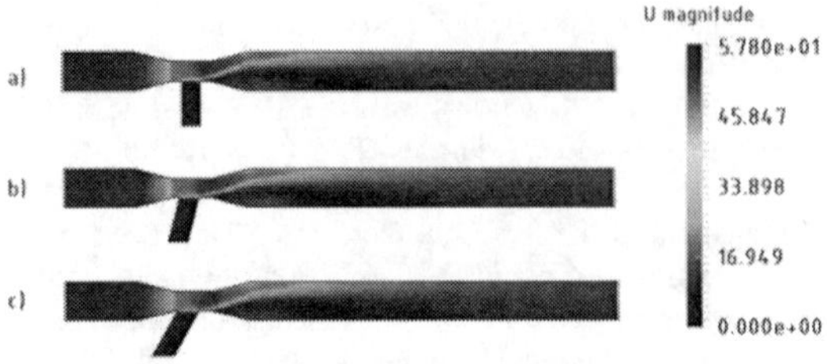

Fig. 9. Contours of velocity magnitude between analysed cases – a) 0°, b) 15° and c) 30° [m/s].

The expected velocity was increasing nearly to 58 [m/s] during the flow through the Venturi throat. During the flow through the Venturi mixer, there are strong turbulences, which also have a positive effect on the overall mixing process. Analysing the contours of velocity magnitude (Figure 9), it could be seen that the differences between them are so minor, that it cannot be described correctly by only analysing these contours. Therefore, better view of the flow rate will give the analysis of the average distribution of velocity magnitude [m/s] through the Venturi mixer for different gas inlet angles presented in the Figure 10.

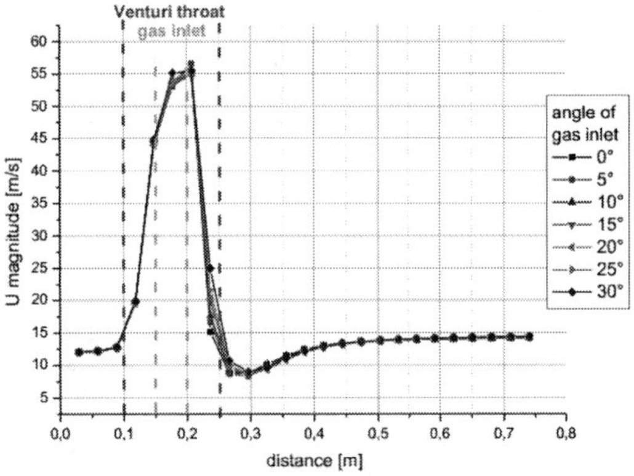

Fig. 10. Distribution of velocity magnitude [m/s] through the Venturi mixer for different gas inlet angles.

Analysing the average distribution of velocity magnitude [m/s] through the Venturi mixer (presented in the Figure 10) for different gas inlet angles, it could be seen that the biggest changes happened in the Venturi throat section of the gas mixer. Initially, in the Venturi throat inlet the velocity was drastically growing up nearly to 58 [m/s], while after passing the gas inlet section it started to fall to a smaller velocity, as set in the inlet of the Venturi mixer. The greater the gas inlet angle, the lower is the velocity drop after passing the gas inlet section. But it is worth noting, that the differences in drops of velocity between the analysed cases with different gas inlet angles are very small. Finally, the most significant parameter of the Venturi mixer – the pressure loss was analysed. This is one of the most qualitative parameters to describe the quality and efficiency of a gas mixer. Firstly, the contours of pressure loss were analysed in the Figure 11, between three analysed cases for a gas inlet angle of 0°, 15° and 30°.

Fig. 11. Contours of pressure loss between analysed cases – a) 0°, b) 15° and c) 30° [Pa].

Analysing the contours of pressure loss between these three cases, it could be noticed which of the following Venturi gas mixer has a lower pressure loss, by analysing the colour scale bar of the pressure contour located at the right-hand side in the Figure 11. Here, the differences are much greater than in the previous analysis of the velocity magnitude contours. Supposing, the brightest red colour at the Venturi gas inlet angle of 30° should provide the lowest pressure loss. In addition, after passing the Venturi gas inlet the largest vacuum zone (marked in blue) is in the mixer with the greatest gas inlet angle for 30°. This results in bigger under-pressure which causes more gas being sucked into the Venturi mixer. Therefore, in the end there would be a richer air-fuel mixture proportion which results with a smaller air excess coefficient ratio → λ. This contour analysis shows a good comparison between the three analysed cases. On the other hand, here too, for a more detailed analysis, it was necessary to analyse the average numerical data of the pressure loss through the Venturi mixer for different gas inlet angles. This analysis was shown in the Figure 12.

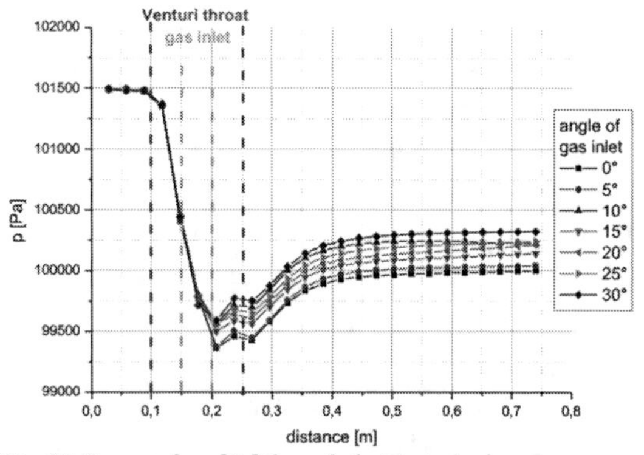

Fig. 12. Pressure loss [Pa] through the Venturi mixer for different gas inlet angles.

Analysing the pressure loss [Pa] through the Venturi gas mixer for different gas inlet angles (Figure 12), it could be noticed that the greater the gas inlet angle, the smaller the pressure loss through the whole Venturi mixer was. It could be also seen that the whole pressure loss occurs in the Venturi throat of the gas mixer, and after that it was growing slowly again until the distance of 0.6 [m] from the

Venturi mixer inlet where it stabilises and keeps at a constant level for each analysed case.

4 Conclusions

The detailed analysis presented in this paper showed that the greater the inclination of the gas inlet, the more methane → CH_4 is sucked into the Venturi gas mixer. As a result, there would be a richer air-gas mixture at the outlet of the Venturi mixer. This also significantly affects the increase of the efficiency of the whole mixing process. Moreover, smaller pressure loss occurs through the whole Venturi mixer. In this paper, detailed analysis between seven different gas inlet angles was made, with average numerical result from the whole Venturi mixer performed with the numerical OpenFOAM software. As it is commonly known, each numerical modelling analysis should be validated in the form with experimental data or mathematical calculations. In this paper mathematical calculations were made accordingly to the book definition, to verify the correctness of this numerical calculations. As was shown previously, the result of numerical calculations matches very well with the theoretical mathematical calculations.

This research is a part of a project that has received funding from the European Union's Horizon 2020 research and innovation programme under grant agreement No.691232–Knocky–H2020–MSCA–RISE–2015/H2020-MSCARISE–2015 as well as a part of the personal Czestochowa University Grant BS/MN 1-103-302/2017/P.

References

1. R. Absi, J. Appl. Mech., **75**, 1 (2008)
2. *D.* Danardono, K. S. Kim, S. Y. Lee, J. H. Lee, J. MECH. SCI. TECHNOL. **25** (9), 2285 (2011)
3. R. Devarajan, Comparative Evaluation of a Two Stroke Compressed Natural Gas Mixer Design using Simulation and Experimental Techniques. *International Conference on Association of Science and Technology for Development, Power and Energy Systems*, Langkawi: 359 (2008)
4. M. Günther, Y. Uygun, F. Kremer, S. Pischinger, *MTZ Worldw*, **74**, 46 (2013)
5. M. Gorjibandpy, M. K. Sangsereki, World Academy of Science, Engineering and Technology. International Journal of Mechanical, Aerospace, Industrial, Mechatronic and Manufacturing Engineering, **4** (2010)
6. K. M. Han, *Lichtleiterbasierte Methoden zur optischen Analyse von räumlichen Verbrennungsprozessen und Verbrennungsanomalien in Ottomotoren*. Technische Universität Karlsruhe, Dissertation (2010)
7. K. Rażniewić, *Tablice cieplne z wykresami* (Warszawa WNT 1966)
8. M. Romańczyk, W. Elsner, *23rd International Conference Engineering Mechanics 2017, Book of full texts: ISBN 978-80-2014-5497-2*, Svratka, Czech Republic, 15 (2017)
9. S. Sundararaj, V. Selladurai, International Journal of Applied Engineering Research, **7**, 891 (2008)

 DOI: 10.1051/itmconf/20171507006

Influence of patient position and implant material on the stress distribution in an artificial intervertebral disc of the lumbar vertebrae

Robert Karpiński[1,*], *Łukasz* Jaworski[2], *Mirosław* Szala[3], and *Monika* Mańko[4]

[1]Lublin University of Technology, Faculty of Mechanical Engineering, Department of Machine Design and Mechatronics, Nadbystrzycka 36, 20-618, Lublin, Poland
[2]Poznan University of Technology, Faculty of Mechanical Engineering and Management, Piotrowo 3, 60-965, Poznań, Poland
[3]Lublin University of Technology, Faculty of Mechanical Engineering, Department of Materials Engineering, Nadbystrzycka 36, 20-618 Lublin, Poland
[4]Lublin University of Technology, Faculty of Electrical Engineering and Computer Science, Nadbystrzycka 38A, 20-618, Lublin, Poland

Abstract. The aim of this paper was to determine the effect of using cobalt and titanium-based alloys as implant materials for the lumbar vertebrae with an artificial intervertebral disc on the stress distribution. The lumbar vertebrae were chosen for the study because they carry considerably higher loads, especially while standing or sitting. Finite element method (FEM) simulations were conducted for three standard loads reflecting three patient's positions: recumbent, standing and sitting. The FEM analysis was performed using the SolidWorks Simulation module. Artificial units containing a pair of vertebrae with a prosthesis between them were designed by the Solid Edge software, based on micro-computed tomography CT scans of the patient's spine. The implant model was designed with its shape based on the geometry of surrounding vertebrae, consisting of an upper pad, a bottom pad and an insert (intervertebral disc). Two implant material configurations were studied. One involved the use of titanium alloy for the upper and bottom pads, while in the other, these pads were made of cobalt alloy. In both cases, a polyethylene insert was used. The FEM results demonstrate that both material configurations meet the requirements for prosthesis design. In both material configurations, the maximum stresses in each prosthesis element are almost twice higher in a sitting posture than in a recumbent position.

1 Introduction

Spinal diseases pose a serious problem both from a point of view of diagnosis and treatment as well as biomechanics. The mechanisms of development of an abnormal posture of the human body are not fully understood yet, which makes rehabilitation and treatment difficult.

The range of segmental motion of the spine is a standard clinical assessment of motor organs. It is reported in literature [1–4] that computer-aided methods (CAD) are necessary in biomechanics, so the applications of the finite element method (FEM) are constantly developed. Modern computer technology enables the development of increasingly advanced numerical models not only of the human spine [1], but also dentate prostheses [2], knee or hip joints [3, 5], dentinal microtubules [4], which – combined with engineering and medical knowledge – provides an excellent basis for conducting biomedical research and analyses that ensure optimal choice of treatment or rehabilitation methods for the spine. Nowadays it is possible to simulate models of entire biomechanical systems and individual elements tailored to the needs of a particular patient. This is in agreement with the current trend for implant personalisation in biomedical engineering. Micro-computed tomography (CT) is useful in solving biomedical problems [5–8]. Based on images obtained from the CT scans of the human spine, it is possible to perform said personalised treatment in the area of intervertebral disc replacement. In such cases, one of the first stages of treatment is to design an artificial intervertebral disc. However, various biomaterials are available that could be considered for implant applications [1, 3, 9–13], and FEM simulations are helpful in proper material selection.

The aim of this study is to determine the effect of using cobalt and titanium-based alloys as implant materials for lumbar vertebrae with a polyethylene intervertebral disc on stress distribution. FEM simulations are conducted for three standard loads reflecting the following positions of the patient: recumbent, standing and sitting.

2 Structure of the spine

The spine provides support for the upper body. Depending on the location, the vertebrae are divided into: 7 cervical vertebrae, 12 thoracic vertebrae, 5 lumbar vertebrae, 5 sacral vertebrae and 4 or 5 coccygeal vertebrae [14]. The cervical, thoracic and lumbar vertebrae are the movable presacral part of the spine, and are called true vertebrae.

* Corresponding author: r.karpinski@pollub.pl

DOI: 10.1051/itmconf/20171507006

The intervertebral disc is a connector between the true vertebrae. There are 23 intervertebral discs in the spine which connect the surfaces of two adjacent vertebral bodies [15, 16]. The adjacent vertebrae are mutually stabilized by a strong ligament apparatus. Two connected vertebrae with an intervertebral disc are shown in Fig. 1.

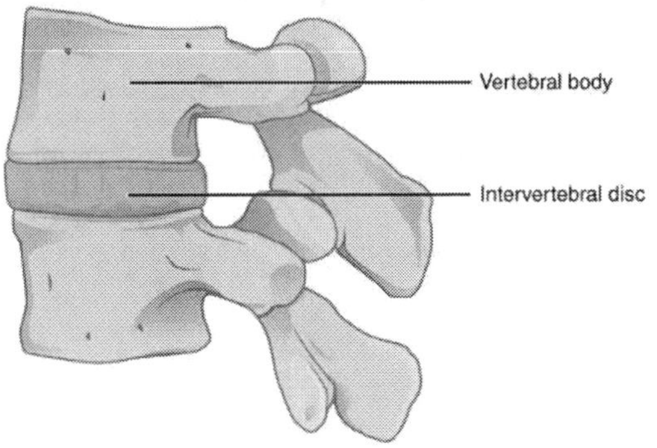

Fig. 1. Connection of two vertebrae, lateral view [17].

The lumbar vertebrae is characterised by its size and mass [18]. The lumbar spine as seen from above is shown in Fig. 2.

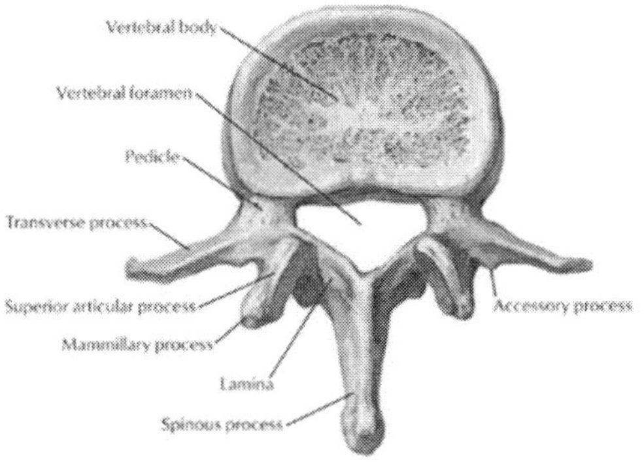

Fig. 2. Lumbar spine as seen from above [14].

Typical discs are made up of boundary plates, fibrous rings and atherosclerotic nuclei. The fibrous ring consists of a dozen layers of variable material characteristics and encapsulates an incompressible gelatinous atherosclerotic nucleus. The stiffness of the fibrous ring layers decreases with the distance to the canter of disc, the outer layers have the highest rigidity and are the least susceptible to deformation [19].

The discs on each level have a similar shape but vary in dimensions. The height and width of the lumbar discs are almost twice as high as those of the discs in the neck. As is the case with the articular cartilage, the performance of the intervertebral disc depends on the degree of saturation by water and the degree of degeneration. With age, they flatten and lose some elasticity. The main tasks of the intervertebral discs are to ensure the stability of the spine, damp vibration of the vertebrae and act as a link enabling the adjacent vertebrae to move [20, 21].

3 Materials and methods

For our study, we chose lumbar vertebrae due to the fact that they carry considerably larger loads, especially while standing or sitting. The patient's vertebrae were measured using CT (micro-computed tomography), and selected results are given in Fig. 3. Artificial units containing a pair of vertebrae with a prosthesis between them were modelled with Solid Edge ST8 (CAD software) based on CT scans of the patient's spine. The basic shape of the bottom pad and polyethylene insert were designed. The insert's dimensions had to exactly match the size of the bottom pad, as these two elements are closely intertwined [5]. When assembled, these elements form an endoprosthesis. The original implant model had a shape based on the geometry of the surrounding vertebrae and consisted of an upper pad, a bottom pad and an insert (intervertebral disc), shown in Fig. 4.

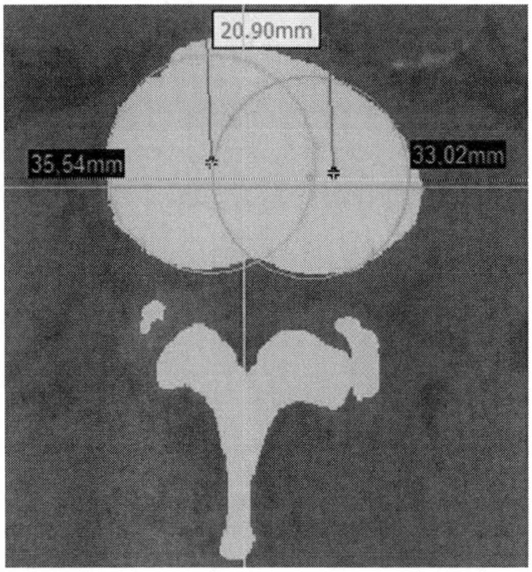

Fig. 3. Measurements performed with CT scans of the patient.

For the purpose of the study, three construction materials were used: ultra-high-molecular-weight polyethylene for the insert, titanium alloy Ti-6Al-4V and cobalt alloy Co-Cr-Mo for the bottom and upper pads of the prosthesis. The mechanical properties of these materials are given in Table 1.

Table 1. Mechanical properties of tested materials.

Material	Element	Young's modulus [MPa]	Poisson's ratio [-]	Density [g/cm^3]
Titanium alloy Ti-6Al-4V	Upper, bottom pad	$1.05 \cdot 10^5$	0.31	4.43
Ultra-high-molecular-weight polyethylene	Insert	1070	0.41	0.95
Cobalt alloy Co-Cr-Mo	Upper, bottom pad	$2.2 \cdot 10^5$	0.22	8.4

DOI: 10.1051/itmconf/20171507006

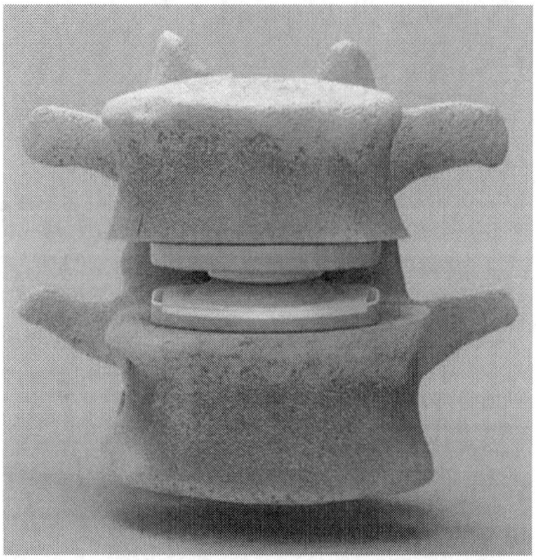

Fig. 4. Original model of artificial intervertebral disc with dimensions corresponding to CT result presented in Fig. 3.

The FEM modelling was performed using the SolidWorks Simulation module. The model was fixed rigidly on the underside of the bottom plate, imitating the connection between the fourth lumbar vertebra. By analogy, the loads were applied to the surface of the upper pad where the implant is attached to the third vertebra. Two configurations of the prosthesis materials were investigated: (I) upper and bottom pads made of titanium alloy, and (II) upper and bottom pads made of cobalt alloy, both configurations with a polyethylene insert.

The FEM analysis of the units containing a pair of vertebrae with a prosthesis between them was conducted under various load conditions, depending on the body position. The value of load was intended to represent the force acting on the intervertebral disc of a 100-kg body weight (BW) patient. We studied different human body positions: recumbent, standing and sitting, with the BW percentage being 75%, 100% and 140%, respectively [22].

Constraints and loads are shown in Fig. 5. The simulation module generated a triangular mesh with 16533 nodes and 10082 elements. The meshed model is shown in Fig. 6.

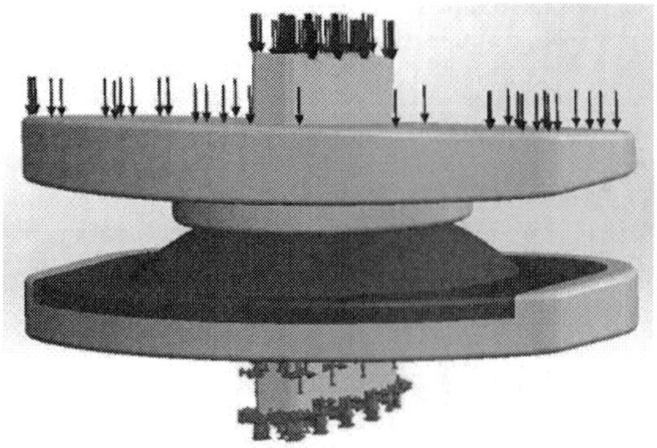

Fig. 5. Constraints and loads acting on the model.

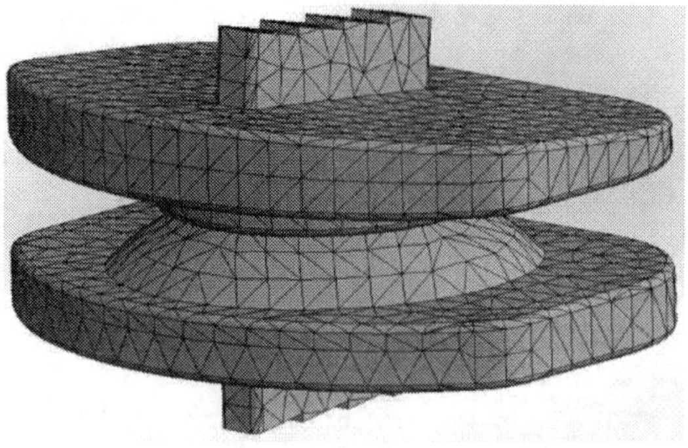

Fig. 6. Triangle mesh generated for the simulation.

4 Results

The FEM analyses resulted in obtaining the maps of stress distribution in the artificial vertebral disc, as shown in Figs 7-9. The maximal values of von Mises stress for each element depending on the load and structural material of both pads are listed in Table 2.

It can be observed that in each material configuration the maximum stresses acting on each element of the prosthesis are almost twice higher in the sitting position than when the patient is recumbent.

In general, the results demonstrate that despite a considerable difference between the mechanical properties of titanium and cobalt alloys, the stresses acting on the prosthesis under given loads have similar values in both cases. The greatest difference in stresses of approximately 2% can be observed when the prosthetic insert is pressed by the upper pad. Hence, in this case, it seems that titanium alloy characterised by a suitable weight-to-strength ratio and good biological performance [9] is more suitable for the prosthesis than cobalt alloy (Table 1).

However, not only mechanical properties must be taken into account in biomaterials selection, factors such as microstructure, corrosion resistance and wear properties are crucial, too [12, 13]. This paper offers preliminary results of a study intended to optimise the proposed prosthesis design in compliance with the requirements of modern bioengineering. To this end, ex vivo testing is often employed [6]. However, ex vivo tests must be preceded by FEM simulations.

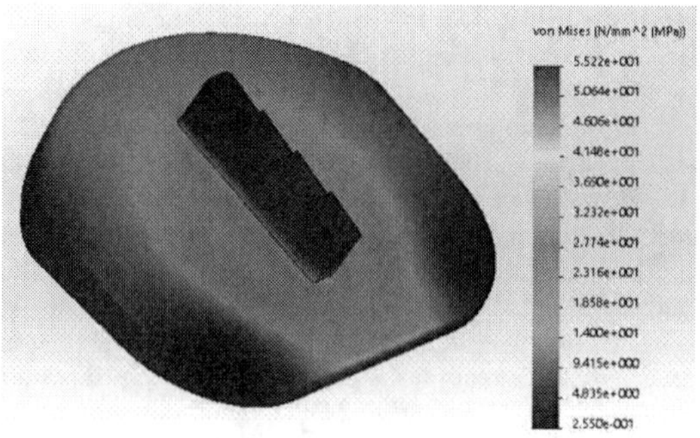

Fig. 7. Stress distribution on the surface of the upper pad.

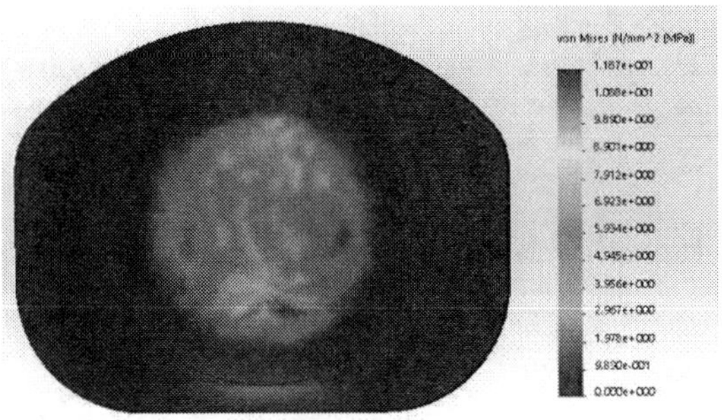

Fig. 8. Stress distribution on the spherical surface of the insert.

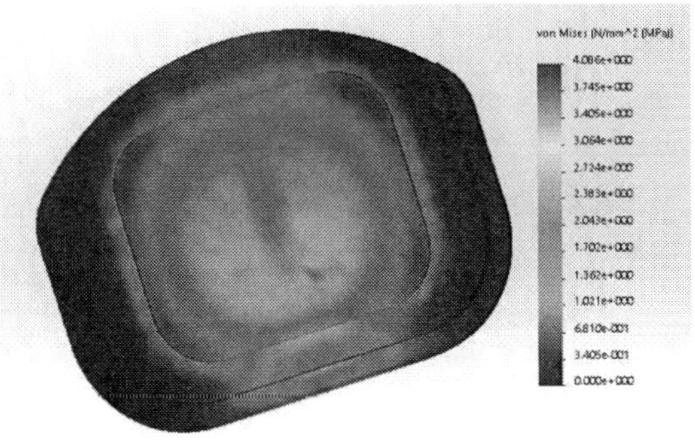

Fig. 9. Stress distribution on the inside of the bottom pad.

Table 2. Maximum stresses acting on prosthesis elements for two material configurations.

Element	Material	75%BW	100%BW	140%BW
Material Configuration I				
Upper pad	Ti-6Al-4V	41.41 MPa	55.22 MPa	77.30 MPa
Bottom pad	Ti-6Al-4V	30.64 MPa	40.86 MPa	57.20 MPa
Insert	polyethylene	8.91 MPa	11.87 MPa	16.62 MPa
Material Configuration II				
Upper pad	Co-Cr-Mo	41.43 MPa	55.24 MPa	77.33 MPa
Bottom pad	Co-Cr-Mo	31.22 MPa	41.63 MPa	58.28 MPa
Insert	polyethylene	9.09 MPa	12.13 MPa	16.98 MPa

In the work [8], the authors conducted a structural analysis of the knee joint endoprosthesis, finding that the weakest element of the model was a polyethylene liner. Therefore, in the present study, a polyethylene disc is used to prevent its damage under the action of stresses.

With the use of titanium alloy, the maximum stresses slightly decrease compared to those observed for the cobalt alloy configuration (Table 2). This is due to mechanical properties of the investigated metal alloys. It is obvious that both Young's module and Poisson's ratio have impact on the FEM results. Moreover, the determined stress values are lower than the nominal strength of the tested materials, therefore the polyethylene insert disc is prevented from damage.

Based on the FEM results, it is possible to claim that the prosthesis design proposed in this paper can withstand loads generated during daily activities, as the maximum stresses acting on the prosthesis elements are significantly under the yield point of each tested material, and the displacements are marginal. However, in order to optimise the proposed design, it is crucial to perform further simulations using different materials like porous titanium and titanium–nickel (shape memory) alloys or implants with modified surface properties [1, 3, 9–11, 13].

5 Conclusions

The objective of this paper was to determine the effect of using cobalt and titanium-based alloys as implant materials for lumbar vertebrae with a polyethylene intervertebral disc on stress distribution. The FEM (finite element method) analysis was conducted for three standard loads reflecting three patient's postures: recumbent, standing and sitting.

The model of an artificial intervertebral disc was designed with the dimensions corresponding to the CT (micro-computed tomography) scans of the patient's spine.

The FEM results demonstrate that the prosthesis proposed in this paper can withstand loads generated during daily activities, as the maximum stresses acting on the elements are significantly under the yield point of each tested material, and the displacements are marginal.

It has been found that in both investigated material configurations, the maximum stresses acting on each element of the prosthesis are almost twice higher in the sitting posture than in the recumbent position.

The FEM simulations of stress distribution have confirmed that both material configurations yield lower values than the nominal strength of the tested materials, which prevents the polyethylene insert disc from damage.

To optimise the prosthesis design, it is crucial that further simulations be performed for different materials and loads.

References

1. Y.-H. Lee, C.-J. Chung, C.-W. Wang, Y.-T. Peng, C.-H. Chang, C.-H. Chen, Y.-N. Chen, C.-T. Li, *Comput. Biol. Med.* **71**, 35–45 (2016)
2. Z. Gümrükçü, Y. T. Korkmaz, F. M. Korkmaz, *Comput. Biol. Med.* **86**, 47–54 (2017)
3. L. Zach, L. Kunčická, P. Růžička, R. Kocich, *Comput. Biol. Med.* **54**, 53–60 (2014)
4. M. M. Shirazi, O. Abouali, H. Emdad, M. Nabavizadeh, H. Mirhadi, G. Ahmadi, *Comput. Biol. Med.* **89**, 1–17 (2017)
5. R. Karpiński, Ł. Jaworski, J. Zubrzycki, *Adv. Sci. Technol. Res. J.* **10**, 240–246 (2016)
6. J. P. Caffrey, E. Cory, V. W. Wong, K. Masuda, A. C. Chen, J. P. Hunt, T. M. Ganey, R. L. Sah, *J. Biomech.* **49**, 4090–4097 (2016)
7. S. M. Moon, J. H. Yoder, A. C. Wright, L. J. Smith, E. J. Vresilovic, D. M. Elliott, *Eur. Spine J. Off. Publ. Eur. Spine Soc. Eur. Spinal Deform.*

Soc. Eur. Sect. Cerv. Spine Res. Soc. **22**, 1820–1828 (2013)
8. J. Zubrzycki, R. Karpiński, B. Górniak, *Appl. Comput. Sci.* **12** (2016) (available at http://yadda.icm.edu.pl/yadda/element/bwmeta1.element.baztech-69507fc1-0eca-4b84-b42a-0c1edb67506c)
9. R. A. Gittens, R. Olivares-Navarrete, Z. Schwartz, B. D. Boyan, *Acta Biomater.* **10**, 3363–3371 (2014)
10. A. Dudek, L. Adamczyk, *Opt. Appl.* **43** (2013) (available at http://yadda.icm.edu.pl/baztech/element/bwmeta1.element.baztech-6bff853a-6c74-46c0-bd04-81008281c5f6)
11. S. Sobieszczyk, P. G. Wydawnictwo, *Rozwój bioaktywnych implantów porowatych na osnowie stopów tytanu* (Wydawnictwo Politechniki Gdańskiej, 2013)
12. M. Walczak, *Wpływ wybranych zabiegów technologicznych na trwałość użytkową układów metal-ceramika stosowanych w protetyce stomatologicznej* (Politechnika Lubelska, Lublin, Poland, 2014; http://bc.pollub.pl/dlibra/docmetadata?id=8653)
13. S. Taksali, J. N. Grauer, A. R. Vaccaro, *Spine J. Off. J. North Am. Spine Soc.* **4**, 231S–238S (2004)
14. F. H. Netter, *Atlas of Human Anatomy* (Elsevier Health Sciences, 2010)
15. K. Lipscomb, N. Sarigul-Klijn, E. O. Klineberg, in press, doi:10.1115/IMECE2016-66384
16. R. Będziński, *Biomechanika inżynierska: zagadnienia wybrane* (Oficyna Wydawnicza Politechniki Wrocławskiej, 1997)
17. Classification of Joints, (available at http://oerpub.github.io/epubjs-demo-book/content/m46383.xhtml#fig-ch09_01_02)
18. M. A. Adams, P. Dolan, *J. Biomech.* **38**, 1972–1983 (2005)
19. J. P. G. Urban, S. Roberts, *Arthritis Res. Ther.* **5**, 120–130 (2003)
20. K. Luoma, H. Riihimäki, R. Luukkonen, R. Raininko, E. Viikari-Juntura, A. Lamminen, *Spine.* **25**, 487–492 (2000)
21. F. Paulsen, J. Waschke, *Sobotta Atlas of Human Anatomy, Vol. 3, 15th ed., English/Latin: Head, Neck and Neuroanatomy* (Elsevier,Urban&FischerVerlag, 2013)
22. Ł. Jaworski, R. Karpiński, *J. Technol. Exploit. Mech. Eng.* **3**, 8–12 (2017)

DOI: 10.1051/itmconf/20171507007

Numerical analyses of adhesive-bonded joints in steel I-beams reinforced with CFRP strips

Maciej Kowal[1,*], and *Mateusz* Hypki[1]
[1]Lublin University of Technology, Faculty of Civil Engineering and Architecture, Poland

Abstract. Numerical analyses represent a useful structure-dimensioning tool and are indispensable for the analyses of structures featuring varied materials or complex geometric layouts. Modelling reinforced structures by gluing their elements is complicated as frequently at least two elements, made of different materials, are joined. This influences their behaviour under a load. Numerical analyses can provide invaluable support in this respect. To compare the laboratory test results obtained by the authors referring to reinforcing bent steel I-beams with adhesive-bonded composite-CFRP strips, and theoretical analyses (which are not the subject of this study), a preliminary FEM analysis was carried out. FEM analysis models were created with the same geometries and material parameters of the strengthening composite strip, the adhesive, and the reinforced I-beam, as in the laboratory tests. The distribution of stresses along the bond end, and their maximum values at the strip end, were specified in detail. ABAQUS®6.12-3 software was used to verify the results. The compliance of the FEM analysis with the laboratory measurements was confirmed in terms of the identified values and shearing-stress diagrams.

1 Introduction

Designing structural reinforcements for bridges is a difficult yet crucial aspect of engineering. Using conventional methods of strengthening steel structures (cutting out and inserting a new element, increasing structure sections, external compression, or using rope-expansion structures) can turn out to be cost-ineffective or technically difficult to perform. The structure's deadweight can be considerably increased in the process, and static patterns can be disrupted. The reinforcing structure can be difficult to apply, as well as susceptible to corrosion and wear; therefore, the result may not be completely satisfying.

Using CFRP (Carbon Fibre Reinforced Polymer) composites is a promising alternative to the currently used methods of reinforcing steel structures. CRRP composites are resistant to corrosion, and have low density and very high tensile strength. Additionally, the method of adhesive-bonding of composite elements is easy to apply.

Aerospace engineering is a pioneer in composite application. Mainly, concrete structures are reinforced with composites. However, the number of applications of composite strips to increase the bearing capacity of steel bridges is also on the rise. In Poland the bridge over the Vistula in Chełmno was reinforced using this method [1]. Elsewhere in Europe, mainly in the United Kingdom, the number of reinforcements of bridges with composites is more common, and focuses mainly on metal structures [2-6].

The major purpose of the article was to compare laboratory test results, obtained referring to bent steel I-beams reinforced, with adhesive-bonded composite-CFRP strips with the FEM analysis results.

2 Reinforcing steel structures with FRP composites in the literature

An important stage in reinforcing steel structures with composite materials is preparing the contact surfaces of the reinforced element and the adhesive-bonded element for joining, and determining stresses in the adhesive layer. The strength of the joint has a considerable impact on the bearing capacity of the entire reinforcement system. The significance of this stage for structure reinforcement is described, i.a., in [7-11]. Exceeding border stresses in the adhesive, causing the disjoining of the reinforcement at the steel-adhesive bond, is the main, but not the only feasible form of damage to the structure.

Scientists, dealing with the reinforcement of steel structures with composite materials, seek solutions enabling a possibly permanent reinforcement with high efficiency. For this purpose, various solutions are considered for increasing the strength of the joint [12-14].

Reinforcing steel elements with FRP composites aims to restore or increase the structure's bearing capacity. In bridge structures, the main structural elements are usually the main girder bars. The steel bar tests to be found in the literature on the subject usually

* Corresponding author: m.kowal@pollub.pl

present three- or four-point bending bars with an I-beam [15-19].

3 The laboratory tests presented by the authors

The authors carried out tests at the Civil Engineering Laboratory of the Lublin University of Technology. The main objective was to confirm the possibility of increasing the strength of steel structure reinforced with adhesive-bonded composite-CFRP strips.

The tests were divided into three stages. In the first stage, various geometries of the end of the strip and the adhesive bond at the joint were compared. 78 samples were subjected to axial tension. Each sample consisted of two flat steel bars joined with carbon plates with specified dimensions, adhesive-bonded on both sides. The second stage consisted of verifying the results of the first. For this purpose, 18 reinforced samples were prepared with three types of strip ends, and subjected to three-point bending. The third stage tested the influence of composite-strip adhesion on increasing the bearing capacity of reinforced I-beams. Adhesives with three degrees of thickness (0,6mm; 1,2mm; 1,8mm) and three anchoring lengths were considered (65mm, 165mm, 265mm), composite strips with two varying elastic moduli (168GPa and 210GPa). In the third stage of the tests, undamaged I-beams were used – 3 for comparison, and 16 reinforced steel I-beams INP 140 with a theoretical span of 1.90m – to perform 4-point bending. Two types of joints were checked -- the square end (square) and the one providing the best possible strength of the bond end selected from the ones used in the first stage, i.e. reverse tapered with an adhesive fillet (reverse tapered). Fig. 3.1 presents an outline of the sample with the load. More about the laboratory tests in [20].

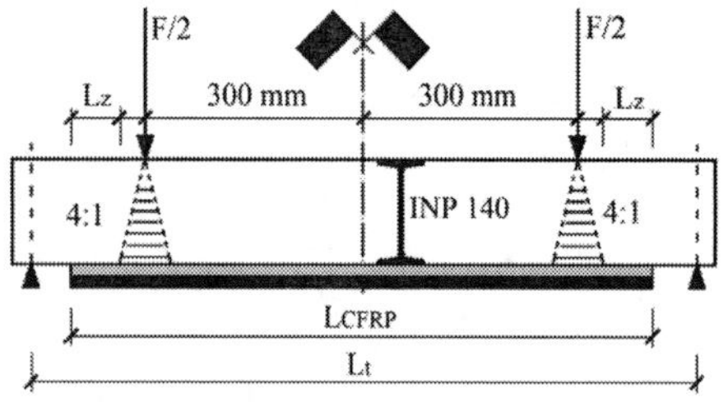

Fig. 3.1. The sample in the second stage of tests

All the reinforced samples were covered with electric-resistance tensometers. The adhesion patterns were changed depending on the type of the samples. Samples 200.120.Z.A (square, 1,2m long strip, 210GPa elastic module) and 200.120.F.A (reverse tapered, 1,2m long strip, 210GPa elastic module) were covered with nineteen tensometers arranged symmetrically around the centre of the span, as in the Fig. 3.2. In the case of these samples, Tenmex foil electric resistance tensometers type TFs10/120 were used, with a resistance of 120±0.2%Ω.

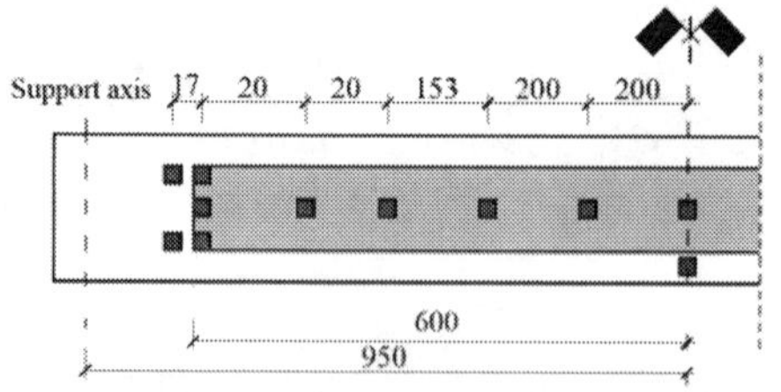

Fig. 3.2. Diagram of the distribution of tensometers for samples 200.120.F.A and 200.120.Z.A.

The 4-point bending tests were stopped after tearing the composite strip or reaching the force of 95 kN on the press stroke, which ensured the plastic softening of the steel I-beam reinforced with the CFRP strip in the area where the forces concentrated on the upper flange were applied. The forces damaging the non-reinforced steel I-beams oscillated around 76 kN.

Two main forms of sample damage were observed. The first was the disjoining of the composite strip at the steel-adhesive bond. This form was observed in 12 samples. The second observed form of damage was the plastic softening of the I-beam's steel, not accompanied by strip disjoining.

4 Numerical analysis

Numerical analyses are useful structure dimensioning tools, indispensable for analyses of structures with complicated geometric layouts or using varied materials. Depending on the purpose of the reinforced structure, the proper selection of the reinforced material dimensions, the elastic moduli, and the strength of the reinforcing material and the adhesive, are of key importance to the durability and strength of the reinforcement. A FEM analysis can bring invaluable support in this respect.

Among a broad range of numerical analyses described in the literature on the subject those dealing with the FEM models of steel beams reinforced with composite strips are few in comparison to other types of supports of steel elements with composites, i.a., [21-25]. The studies involve the verification of theoretical analyses of stresses at the bond end for various types of load on the beams.

To compare the obtained results of the authors' laboratory tests with the theoretical analysis, a preliminary MES analysis had been conducted. The tests were carried out to verify the accuracy of the developed theoretical model for determining shearing and peel stress, which is not the subject of this study. The secondary goal was to verify the shearing stress model against the laboratory test results. Almost all the experimental measurements of stresses described in the

 DOI: 10.1051/itmconf/20171507007

literature refer to shearing stresses and are based on the measurement of strains on the outer surface of the adhesive. The calculation of shearing stress at the bond end is based on the difference in strains between two neighbouring tensometric sensors, the values depending on the distance between the sensors and the thickness and the elastic modulus of the composite tape. This leads to including in the measurements at least two obvious approximate stress values. First, due to neglecting the impact of strip bending on the stress value, which is important in the strip end area. Second, the shearing stress between the adjacent tensometric sensors is averaged, and hence high shearing stress values in the strip end area might not be recorded. Furthermore, the average shearing stresses obtained because of such a measurement are closer to the stress values at the strip-adhesive bond than to those of the adhesive-steel bond, which is more prone to disjoining. Similar conclusions can be found, i.a., in [22]. The authors are not familiar with any existing peel stress measurement methods.

4.1. Numerical model

FEM analysis models were created with the same geometries and reinforcement parameters as the samples used in the laboratory tests. ABAQUS®6.12-3 software was used to verify the results. A three-dimensional model was created for a two-dimensional space, using shell-type elements, and taking into consideration the entire length of the bar. The attachment points at the end of the bar represented the support conditions in the vertical direction. A measurement grid for the beam model is presented in the Fig. 4.1. The steel bar, the adhesive, and the composite strip were modelled with 4-node doubly curved shell elements without the reduction of integration points (S4). To obtain accurate results for stresses in the strip end area, the grid was condensed in this area.

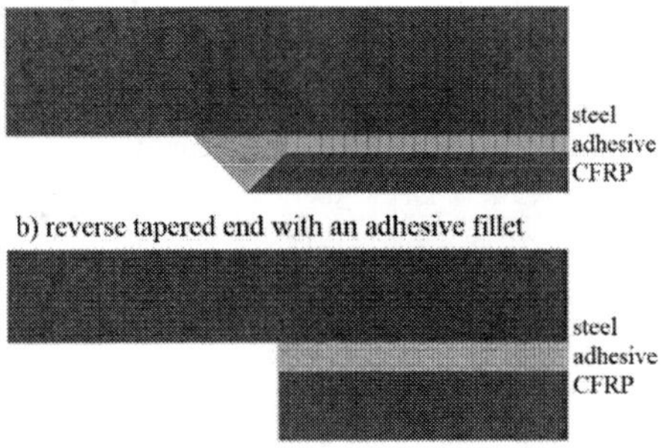

Fig. 4.1. The model of the grid - a section of the reinforcement in the bar with the adhesive strip end area:
a) a reverse-phased end with adhesive outflow,
b) a typical end.

Sample type a):

-- I-beam (60 elements in the flanges and the web heightwise, lengthwise 10 elements per millimetre in the joint end area, normally 0.2 elements per millimetre);
-- adhesive (10 elements heightwise, lengthwise 10 elements per millimetre in the joint end area, normally 0.2 elements per millimetre);
-- strip (10 elements heightwise, lengthwise 10 elements per millimetre in the joint end area, normally 0.2 elements per millimetre);
Sample type b):
-- I-beam (60 elements in the flanges and the web heightwise, lengthwise 5.9 elements per millimetre in the joint end area, normally 0.2 elements per millimetre);
-- I-beam (7-22 elements heightwise, normally 7 elements, lengthwise of 5.9 elements per millimetre in the joint end area, normally 0.2 elements per millimetre);
-- strip (15 elements heightwise, lengthwise 5.9 elements per millimetre in the joint end area, normally 0.2 elements per millimetre);

The geometrical model used in the numerical analysis was identical with the one applied in the mathematical analysis and the laboratory tests. The numerical analysis involved bar INP140 with a span of 1.90 m, a reinforced strip 1.4 x 50mm with a length of 1.20 m with a typical strip end (type "a" in the Fig. 4.1), and a reverse-phased end (45-degree shearing) with an outflow and thickness of the bond end of 0.6 mm (type "b" in the Fig. 4.1).

As this was the primary verification of the theoretical model, the properties of the reinforced bar within the elastic range was considered, together with the isotropic behaviour of all the materials. The following material properties were assumed: steel – elastic modulus 205GPa, Poisson's coefficient 0.3; adhesive – elastic modulus 7.1GPa, Poisson's coefficient 0.35; shear modulus 2.63GPa; composite – elastic modulus 210GPa, Poisson's coefficient 0.3.

To verify the consistency of the FEM module with the laboratory tests, the strain values on the lower surface of the composite strip were compared.

4.2 Numerical analysis results

4.2.1 The compliance of the FEM model results with the laboratory tests

It was assumed that the compliance of the numerical module with the laboratory tests would be confirmed if the values of strains measured along with the lower edge of the composite strip had been compliant. To verify the results obtained in the FEM analyses, the values of the calculated strains were confirmed with those identified in the laboratory tests, in points consistent with the measurement points; see Figs. 4.2÷4.3.

The comparison was carried out for bars loaded with two concentrated forces with a value of 20 kN. The blue points mark the values of the non-dilatational strains found in the laboratory tests for all the samples (OBT).

The solid red line marks the outline of the strains identified based on the numerical analysis (FEM). The dotted green line presents the outline of the stresses from the numerical analysis reduced (FEM.R) to the strains obtained in the laboratory measurements.

The lowest square difference error for samples 200.120.ZA was obtained by reducing the numerical analysis value to 102.1%, for samples type 200.120.FA by reducing the value to 99.8%, and for samples 165.120.FA – to 95.4% of the value. As a result, it can be concluded that the difference between the values measured and determined using the numerical analysis was 4.6% (165.120.FA), 2.1% (200.120.ZA) and 0.2% (200.120.FA). The value of the difference between the numerical analysis and measurements in this stage makes it possible to conclude that the proper numerical model was assumed for the calculations.

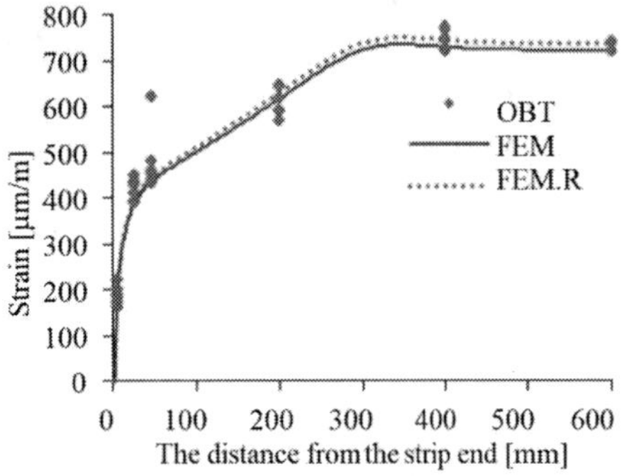

Fig. 4.2. A comparison of the strain diagrams at the lower edge of the composite strip obtained from the FEM analysis and the laboratory tests on sample 200.120.ZA.

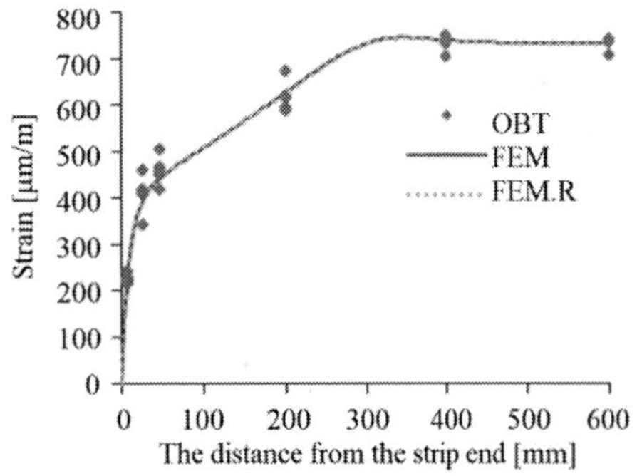

Fig. 4.3. A comparison of the strain diagrams at the lower edge of the composite strip obtained from the FEM analysis and the laboratory tests on sample 200.120.FA.

4.2.2 Parametric tests

To verify the results obtained in the laboratory tests, a FEM analysis was carried out on the impact of the following elements on reinforcement: the FRP-laminate thickness, the CFRP strip elastic modulus, thickness of the bond end layer, and the length of the bond. The properties of the material and the load conditions were the same as those applied in the experimental tests.

Based on the numerical analysis results, it was found that using a strip with an elastic modulus of 168GPa, instead of a tape with a modulus of 210GPa, lowers both shearing and peel stresses. The shearing stress value dropped by 12.8%, and the peel stress value by 5.9%. Similar conclusions were drawn from analyses not described in the present study. However, the values of higher destructive forces were obtained for the samples reinforced with a strip with a modulus of 210GPa. Based on the numerical analysis presented in the form of stress diagrams in Figs. 4.4 & 4.5, it can be concluded that using a reverse tapered strip with adhesive fillet instead of a strip with a square end has a significant impact – it lowers the values of the shearing and peel stresses. The shearing stress value dropped by 29.4%, and the peel stress value by 71.8%. This leads to an increase in the bearing capacity of the reinforcement system, which complies with the results obtained in the laboratory and analytical tests.

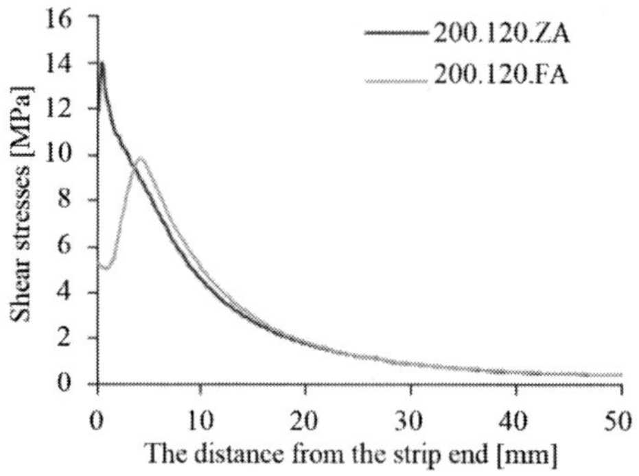

Fig. 4.4. Diagrams of the shearing-stress tests at the bond end of the bar, reinforced with a strip with a length of 1200mm with a modulus of 210GPa, thickness of 1.4mm with a square end, and a reverse tapered strip with adhesive fillet, with bond end thickness of 0.6mm.

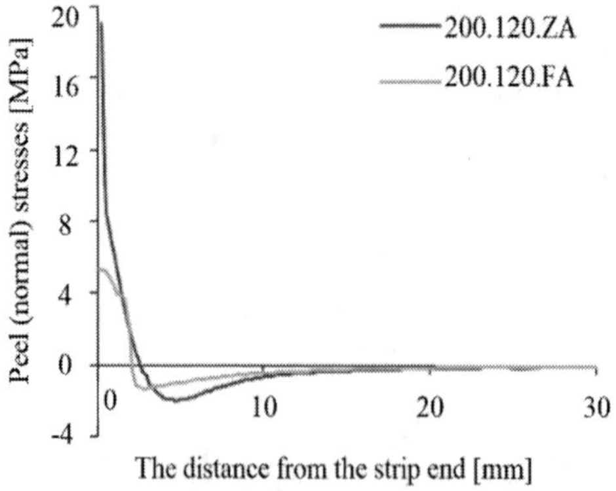

Fig. 4.5. Diagrams of peel stress tests at the bond end of the bar, reinforced with a strip with a length of 1200mm with a modulus of 210GPa, thickness of 1.4 mm with a square end, and a reverse tapered strip with adhesive fillet, with bond end thickness of 0.6mm.

DOI: 10.1051/itmconf/20171507007

4.2.3 A comparison of stress diagrams from the tensometric measurements and the numerical analysis

Based on the values of shearing stresses obtained from strain measurements in the laboratory tests and from the numerical analyses, the values and the distribution of stresses at the bond end were compared. Figs. 4.6÷4.8 present the distribution of shearing stresses at the bond end lengthwise after loading the bars with two concentrated forces with a value of 20kN.

The descriptions of the Figures have the following meanings:

OBT – the values of shearing stresses obtained in laboratory tests.

FEM.CA – the value of stresses from the FEM analysis at the strip-adhesive bond.

FEM.SA – the value of stresses from the FEM analysis at the steel-adhesive bond.

THEO - the value of stresses from the analytical analysis at the strip-adhesive bond.

The presence of two types of shearing and peel stresses in the FEM analyses is caused by their difference depending on the contact line.

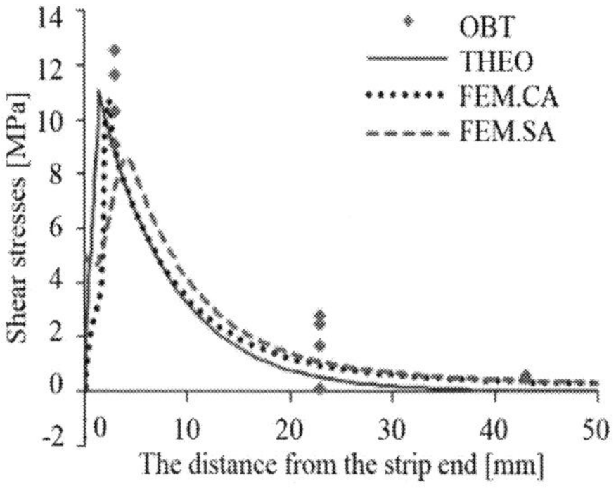

Fig. 4.6. A comparison of the values and distribution of shearing stresses at the bond end of reinforced bar type 165.120.FA.

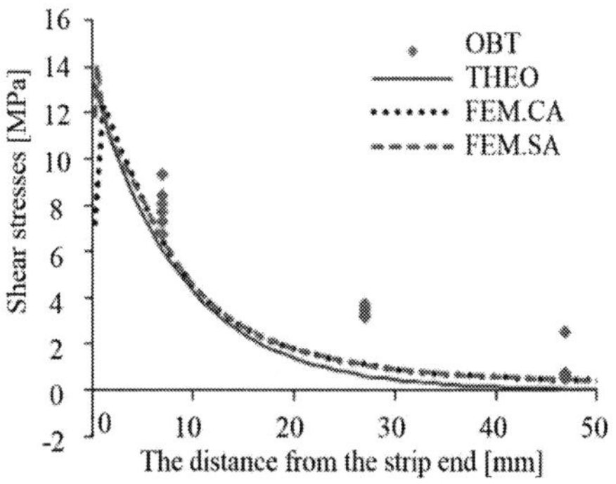

Fig. 4.7. A comparison of the values and distribution of shearing stresses at the bond end of reinforced bar type 200.120.ZA.

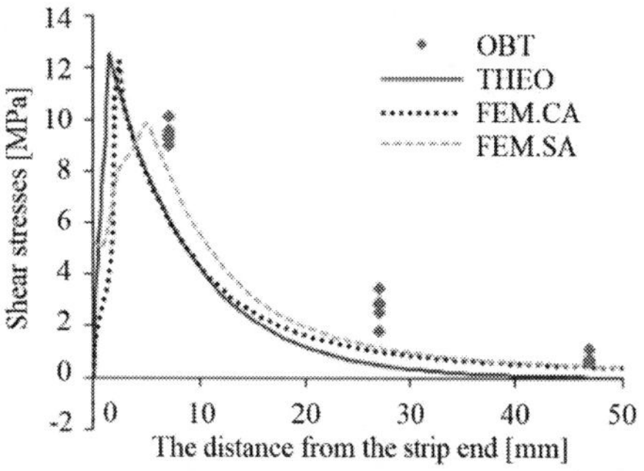

Fig. 4.8. A comparison of the values and distribution of shearing stresses at the bond end of reinforced bar type 200.120.FA.

Analysing Figs. 4.6÷4.8, it can be noted that the diagrams of shearing stresses determined on the basis of the FEM analysis and the maximum values are consistent with the laboratory test results.

5 Conclusions

On the basis of the laboratory tests and the numerical analysis, the following conclusions can be drawn.

-- There is an acceptable compliance with the results for the strains measured in the laboratory tests and those determined based on the numerical models. Based on the three types of sample it was found that the differences between the laboratory measurements and the numerical models were below 5%.

-- Shearing stress diagrams were calculated using the numerical method and their maximum values are consistent with the laboratory test results.

-- The change in stress values at bond end thickness is of some significance. The peel stress in the upper edge of the bond end has a positive value, and at the lower edge the value is negative. In fact, the negative stress in the lower edge does not contribute to bond damage. Peeling is a result of positive peel stresses concentrated in the upper edge of the bond end, as described in [25]. Changes in stress values at bond end thickness is more important in the case of peeling stresses.

As the described numerical analyses were of a preliminary nature and were simplified (the elastic range, the isotropic properties of materials) and their main purpose was to verify the analytical model, which is not described in this study, and to compare them with the laboratory test results, future studies should focus on:

-- Developing a model for elastic and plastic properties taking into consideration the anisotropy of adhesive materials (including at least the adhesive and the CFRP strip).

-- Developing models for other types of sample which was not included in the numerical analysis and described in this study (samples with strips with a modulus of 210GPa, a length of 80cm and adhesive thickness 0.6mm, and strips with a length of 100cm, and bond end

DOI: 10.1051/itmconf/20171507007

thickness of 0.6mm, 1.2mm and 1.8mm, all with a reverse tapered strip end with adhesive fillet).
-- A more detailed analysis of the impact of peel stresses on the bearing capacity of the reinforcement system.

References

1. G. Łagoda, M. Łagoda. 33rd IABSE Symposium, Bangkok, Thailand, Sust. Infrastr. Env. Friendly, Safe and Res. Eff. Vol. **96**,September 9-11, 2009.
2. S. Luke. Oxford and Qafco Prill tower, Qatar, NGCC first annual conference and AGM – Composites in construction, through life performance, BRE, Watford, UK, 30-31 October 2001.
3. S. Luke. International Conference on FRP Composites in Civil Engineering, J. G. Teng, ed., Hong Kong, China, v. **2**.
4. Moy S. S. J., Bloodworth A. G. ICE Proceedings – Struct. and Build. 160/SB2 (2007).
5. N. Farmer, I. Smith. Proc. 9th Int'l Conference on Structural Faults and Repairs (2001), 4th – 6th July 2001, London.
6. Church D.G., Silva T.M.D. Proceedings Advanced Polymer Comp. for Struct. Appl. in Constr., Thomas Telford, London 2002.
7. D. Schnerch, K. Stanford, E. Sumner, S. Rizkalla. Proceedings of the International Symposium on Bond Behaviour of FRP in Structures (BBFS 2005) Chen and Teng (eds)© 2005 International Institute for FRP in Construction 435.
8. D. Schnerch, M. Dawood, S. Rizkalla, E. Sumner, K. Stanford. Adv. in Struct. Eng. **9**, 6 (2006).
9. A.F. Harris, A. Beevers. Int. J. of Adh.& Adh. **19** (1999).
10. D.E. Packham. Int. J. of Adh. & Adh. **23** (2003) 437–448
11. M. Łagoda, M. Kowal. Bud. i Arch. **12,** 2 (2013).
12. M.Y.Tsai, J.Morton. Comp. Struct. **32** (1995).
13. M.D. Fitton, J.G. Broughton. Int. J. of Adh. & Adh **25** (2005).
14. M. Kowal, M. Łagoda. Struct. and Env. **6**, 4 (2014).
15. P.Colombi, C.Poggi. Comp.: Part B **37** (2006).
16. J.Deng, M.M.K.Lee. Comp.Struct.**78** (2007).
17. D. Linghoff, M. Al-Emrani, R. Kliger. Comp.: Part B (2009), doi:10.1016/j.compositesb.2009.05.008.
18. N. Ochi, M. Matsumura, N. Hisabe. Proc. Engi. **14** (2011).
19. Y. J. Kim, G. Brunell. Comp. Struct. **93** (2011).
20. M. Kowal, M. Łagoda. R.a.B. – D.i.M. **16**, 2 (2017), 85-99, DOI: 10.7409/rabdim.017.006.
21. J.Deng, M.M.K. Lee, S. S.J. Moy. Comp. Struct. **65** (2004).
22. J. Yang, J.F. Chen, J.G. Teng. Constr. and Build. Mat. **23** (2009).
23. M. Bocciarelli, P. Colombi. Eng. Struct. **49** (2013).
24. S.P. Chiew, Y. Yu, C.K. Lee. Comp.: Part B **42** (2011).
25. Y. Yu, S.P. Chiew, C.K. Lee. Comp.: Part B **42** (2011).

Numerical analysis of the influence of particular autogyro parts on the aerodynamic forces

Zbigniew Czyż[1,*], *Paweł* Karpiński[1], *Tomasz* Łusiak[2], and *Tomasz* Szczepanik[3]

[1]Lublin University of Technology, Faculty of Mechanical Engineering, Department of Thermodynamics, Fluid Mechanics and Aviation Propulsion Systems, Nadbystrzycka 36 Str., 20-618 Lublin, Poland
[2]Polish Air Force Academy, Department of Airframe and Engine, 35 Dywizjonu 303 Str., 08-521 Dęblin, Poland
[3]Institute of Aviation, Aerodynamics Department, Krakowska 110/114, 02-256 Warsaw, Poland

Abstract. This paper describes the research gyrocopter and presents the methodology of numerical calculations for this type of research objects. There are also the results of an ANSYS Fluent three-dimensional simulation of an airflow around a gyrocopter, excluding an airflow around its main rotor and propeller blades. The calculations enabled basic aerodynamic characteristics. The analysis of the numerical calculations focuses on the impact of gyrocopter individual sections like a fuselage, tail, tail boom, etc., on aerodynamic forces and moments, as well as coefficients of aerodynamic forces and moments. The research scope covers an angle of attack α ranging between -20° and 25° at a sideslip angle equal to 0.

1 Introduction

Gyrocopters are widely used for sports and leisure activities and they have recently been gaining popularity as they can be used to travel fast short distances, i.e. up to hundreds of kilometres and to perform patrol and rescue missions. Their advantage is undoubtedly their short take-off and low speed horizontal landing. This type of aircraft is also cheaper to purchase, maintain and operate than small standard-configured helicopters.

There are few publications on research into the gyroplane aerodynamic construction so the publications on helicopters should be referred to if investigating the state of the art. Most of the works on gyroplanes investigates and discusses modelling of a main rotor. Based on the characteristics of the NACA profile, drag acting on the main rotor during the prerotation of a gyrocopter main rotor is defined in [1]. The value of this force was used to calculate the power required to propel the rotor to achieve the assumed speed. Another example of a work that studies aerodynamics of the rotor is [2], where the numerical model to calculate aerodynamic forces and moments acting on the gyrocopter rotor is defined.

The common methods to aerodynamically study the aircraft are computational fluid dynamics (CFD) and wind tunnel experiments. Both were applied in [3] to study drag that occurs on the model of a helicopter fuselage. Particle Image Velocimetry (PIV) was applied in wind tunnel experiments. The validation of both methods was followed by specifying drag coefficients of the fuselage and elements connected to it to estimate interference drag. Furthermore, it has been found that the rear fuselage section is largely responsible for the generated drag. It was shown that the optimisation of the shape leads to a reduced drag coefficient. A numerical analysis of a helicopter fuselage and its wind-tunnel validation is described in [4]. The authors optimised there the mesh of a model of a fuselage and applied a hybrid model of turbulence-detached eddy simulation (DES).

A non-standard approach in an experimental investigation of an airflow around a gyrocopter was followed by the authors in [5]. To investigate how the drag force that generates on a model of a gyrocopter is impacted by speed, a research object was mounted to a car roof. The experimental results were compared with the CFD results and it was found that the numerical analysis differs from real measurements because of the ideal-conditioned simulation, i.e. steady-state, ideal gas as well as the vehicle-disturbed airflow. To fast study and optimise aerodynamics of a compound gyroplane, methods other than CFD can be applied. They are discussed in [6] where a fast method of aerodynamic calculation (FMAC), combining the advantages of the theory of the blade element moment and the estimation method, is applied.

The authors [7] created their own statistical-engineering-model-based method. Such a model enabled them to define, with the required accuracy, aerodynamic coefficients of a helicopter fuselage for varied angles of attack.

* Corresponding author: z.czyz@pollub.pl

DOI: 10.1051/itmconf/20171507008

Similar studies are discussed in [8] where aerodynamic forces occurs on a standard-configured helicopter fuselage, depending on an angle of attack, a sideslip angle and a side flow.

CFD is applied in [9] to determine drag of a helicopter fuselage according to the Reynolds number. The numerical calculations enabled them to specify the volume of interference drag of a wind tunnel prop with drag of the fuselage.

Aerodynamic forces and moments acting on the aircraft have an impact on its performance and stability. The work [10] investigates gyrocopter flight longitudinal stability. The analysis of the mathematical model and experiments enabled the conclusion that rotor speed is an important parameter for stability. Moreover, a vertical position of centre of gravity in relation to a rotor drag axis may affect longitudinal stability. A preliminary analysis of aerodynamics shall be carried out for the largest aircraft surfaces such as a fuselage or a tail. Components of forces and moments that have extremal values are also important. The fuselage of both helicopter and gyrocopter is the largest part of its entire construction, so its aerodynamic characteristics impacts most on overall aircraft aerodynamics [11].

While investigating the characteristics of coefficients of aerodynamic forces and moments in rotary wing aircraft, the main rotor surface area is usually considered, however, the correlation of reference values of similar aircraft occurs then. To facilitate the comparison of individual types of aircraft, the *f* parameter, i.e. equivalent flat-plate area is applied, given as [11]:

$$f = \frac{D}{0{,}5 \cdot \rho V^2} \qquad (1)$$

where D – drag, ρ – air density, V – airflow speed

This paper investigates what portion of total drag, lift and pitching moment is generated on individual elements of the gyrocopter construction, depending on an angle of attack at a constant sideslip angle equal to 0. Accordingly, a three-dimensional computational model of the aircraft was formed to create an Ansys Fluent tetrahedral mesh, to define boundary conditions, specify a model of turbulence, and finally perform simulations for varied angles of attack. Later, aerodynamic characteristics were created, and conclusions were formulated referring to the obtained results. The impact of the operation of the main rotor and propeller blades on the airflow around the gyrocopter construction was ignored.

2 Research object and boundary conditions

The numerical research is based on the Ansys Fluent. Our calculations use the k-SST turbulence model. Aerodynamic forces and moments are specified in relation to the coordinate system associated with the gyrocopter as in Figure 1. It is a right-handed system. The moment along the longitudinal axis *x* defines roll, along the lateral axis *y* defines pitch, along the perpendicular axis *z* defines yaw. It was assumed that a positive moment causes a clockwise rotation around the z axis, seen from the beginning of the system in the direction of the positive return of the axis. Angle of attack α and sideslip angle β are calculated clockwise as positive values.

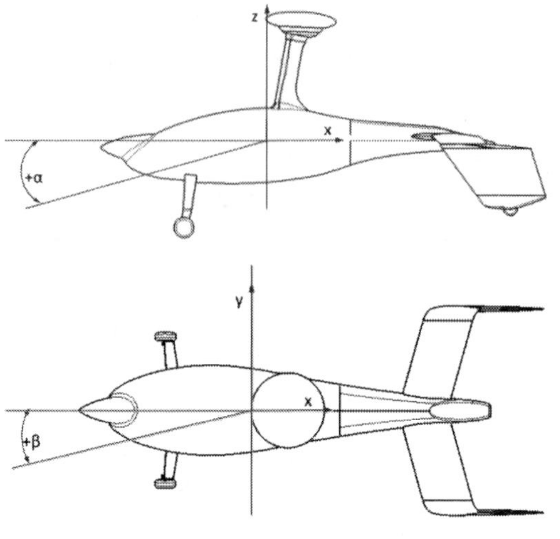

Figure 1. Coordinate system and directions of angles of attack α and a sideslip angle β applied in the numerical calculations.

The measuring area of the research object is a cuboid (Figure 2) which walls are 20 m away from the research object on the *x* axis, 15 m on the *y* axis and 15 m on the z axis (at both sides). The gyrocopter is 6,348 mm x 2,802 mm *x* 3,148 mm (length x width x height).

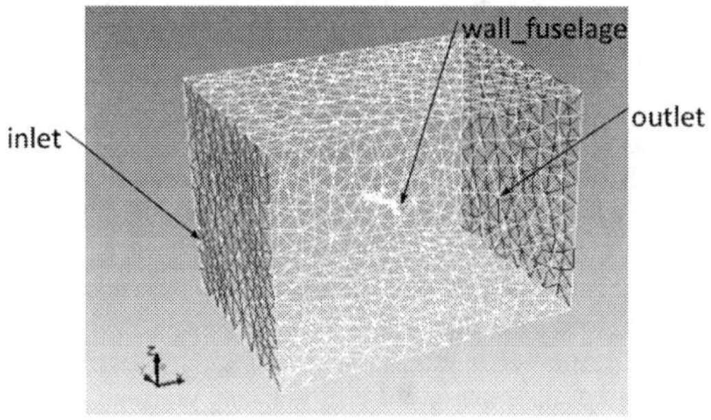

Figure 2. Boundary conditions.

On such a defined calculation domain, there were applied the conditions of *velocity inlet* and *pressure outlet*. The remaining walls, including the surfaces of the gyrocopter fuselage are defined as the wall, however, they are distinguished and named basic elements of the gyrocopter construction, i.e. a head, left leg, right leg, fuselage, mast, nose, left vertical stabiliser, right vertical stabiliser, left horizontal stabiliser, right horizontal stabiliser.

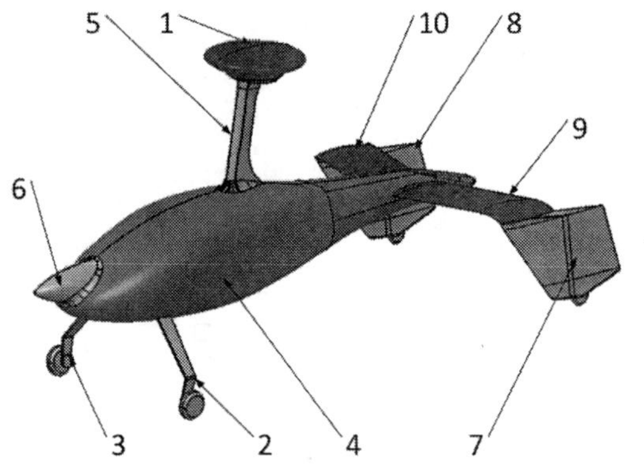

Figure 3. Sections of the gyrocopter construction in the numerical research: 1—head, 2—left leg, 3 – right leg, 4 – fuselage, 5 – mast, 6 – nose, 7 – vertical stabiliser (left), 8 – vertical stabiliser (right), 9 – horizontal stabiliser (left), 10 – horizontal stabiliser (right).

The research conditions were pressure-based and steady. Air is a fluid material. A compressible flow was assumed, and a defined material was *ideal-gas*. In this case, for velocity v = 100 km/h and temperature T = 288 K, the boundary speed is 367.4 km/h (speed of sound for these conditions is 340 m/s, i.e. 1,224.6 km/h). *Turbulent intensity* is assumed as 1% and *turbulent viscosity ratio* as 5. The basic settings of the numerical research are given in Table 1.

Table 1. General settings of the numerical calculations.

General	Type	Pressure-based	
	Time	Steady	
Models	Viscous	k-ω (2-eqn) SST	
Materials	Fluid materials	Air	
	Density	Ideal-gas	
	Viscosity	Constant	$1{,}7894\cdot10^{-5}$ kg/ms
Boundary conditions	Inlet	Velocity	28 m/s
		Turbulent intensity	1 %
		Turbulent viscosity ratio	5

3 Research plan and methodology

The research was performed for the research gyrocopter for angles of attack ranging from -20° to + 25° every 5° at a constant sideslip angle β = 0°. Consequently, there were 10 calculation cases for which were read the values of aerodynamic forces and moments acting on the individual sections of the gyrocopter construction. Then, they were summed up to achieve total aerodynamic forces and moments acting on the whole gyrocopter construction. These, in turn, enabled basic aerodynamic characteristics, i.e. plotted against the correlation of coefficients of aerodynamic forces and moments as a function of an angle of attack α. These coefficients were calculated from the following formulas:

$$C_i = \frac{P_i}{0{,}5\cdot\rho V^2\cdot\pi R^2} \qquad i = x, y, z \qquad (2)$$

$$C_{mi} = \frac{M_i}{0{,}5\cdot\rho V^2\cdot\pi R^3} \qquad i = x, y, z \qquad (3)$$

The values of reference parameters applied to determine the above coefficients are in Table 2.

Table 2. Reference values of the parameters.

Parameter	Symbol	Value	Unit
Radius of the main rotor blade	R	5	m
Flight velocity	V	28	m/s
Air density	ρ	1.226	kg/m^3
Ambient temperature	T	288	K

The geometry of the gyrocopter, to be researched, required to be adapted to CFD calculations. The repair of the geometric model, i.e. connecting individual components, complementing certain shortcomings, discontinuity and holes, was done in the CATIA v5. The next step was to import the model into the Design Modeler of the Ansys Workbench. The model was simplified further on, mainly to reduce the number of its fragmented surfaces, to repair them, as well as to remove any acute angles, fragmented edges, etc. The repairing tools were: Repair, Hard Edges, Edges, Sems, Holes, Silvers, Spikes, Faces. The fragmented surfaces were connected with the Merge tool. Having completed any necessary simplifications of the geometry, a fluid domain (i.e. a research area) was created with the Enclosure tool by closing the research object in the cuboid. It was decided to create a universal geometric model, see Figure 4, for different position angles of the gyrocopter.

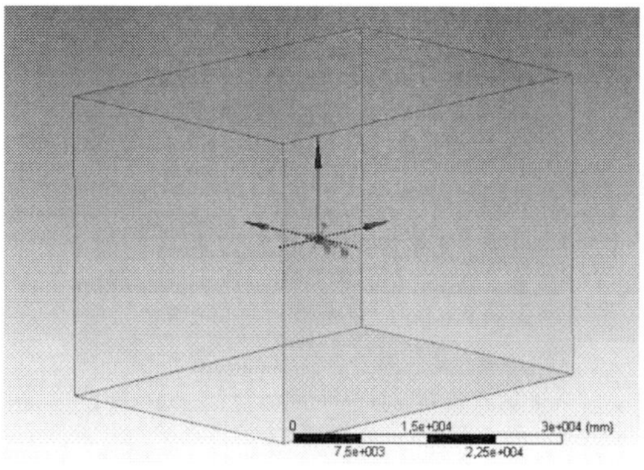

Figure 4. Positioning the model of the gyrocopter in the research area – isometric view.

It consisted in isolating the research object with cylinders, which enabled us to change the position of the gyrocopter in relation to any axis. Defining the angles of attract and sideslip angles was followed by adding the previously isolated cylinders with the embedded shape of the gyrocopter to the rest of the geometry, so that a uniform and coherent shape was created. This method can guarantee that operations, even such as giving a virtual topology, can be recorded at the level of the Ansys/Meshing. While changing, for instance, the angle of

 DOI: 10.1051/itmconf/20171507008

attack, this program does not lose the location and is still capable of recognising corresponding each other characteristic quantities of the geometry.

The geometries of this type are most often discretised by means of tetrahedral cells. Cell size and quantity and quality can significantly affect calculations, but on the other hand, defining them can facilitate mesh creation. Figure 5 shows the created mesh of the research model of the gyrocopter.

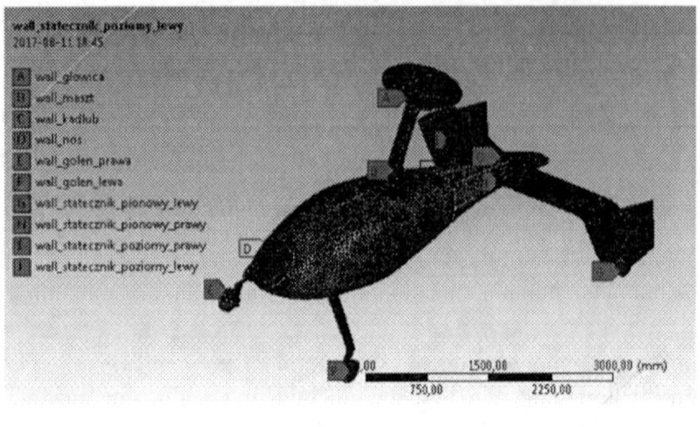

Figure 5. Mesh of the model of the research gyrocopter – isometric view.

This is the mesh created by the *Tetrahedrons* method with an advanced function of the *Curvature* dimension. The *Inflation* with the *Smooth Transition* with 7 layers and a *growth rate* of 1.15 was applied on the fuselage surface. The mesh was about 6.7 million cells in total.

4 Results and discussion

The simulation performed for 10 simulation cases (for varied angles α) enabled us to achieve the values of the aerodynamic forces and moments acting on the individual sections of the gyrocopter construction. Having summed up these individual values, the values of total forces and moments acting on the whole construction of the gyrocopter were achieved. Table 3 shows the values of the aerodynamic forces and moments, whereas Table 4 the values of the coefficients of aerodynamic forces and moments in relation to each axis for the research range of angles of attack α and a sideslip angle β = 0.

Table 3. Aerodynamic forces and moments in relation to each axis for the research range of angles of attack α and a sideslip angle β = 0°.

α (°)	P_x (N)	P_y (N)	P_z (N)	M_x (Nm)	M_y (Nm)	M_z (Nm)
-20	544,72	0,55	-995,55	3,10	934,68	-0,33
-15	412,34	3,63	-943,98	-3,49	1049,45	-4,87
-10	273,70	3,60	-783,02	-3,19	913,80	-0,59
-5	208,04	1,59	-383,78	-0,60	418,48	-0,01
0	184,74	-3,28	47,67	1,71	-100,71	-1,24
5	198,14	2,55	450,85	-5,39	-611,08	-0,91
10	244,13	2,50	776,25	-3,33	-1121,29	-0,45
15	365,93	0,84	956,04	-4,18	-1279,95	-11,60
20	509,58	3,53	1101,11	-4,13	-1421,13	-3,96
25	660,29	1,53	1208,53	-3,94	-1493,59	-1,44

Table 4. Values of the coefficients of the aerodynamic forces and moments in relation to each axis for the research range of angles of attack α and a slip angle β = 0°.

α (°)	C_x (-)	C_y (-)	C_z (-)	C_{mx} (-)	C_{my} (-)	C_{mz} (-)
-20	0,01444	0,00001	-0,02639	0,00002	0,00496	0,00000
-15	0,01093	0,00010	-0,02502	-0,00002	0,00556	-0,00003
-10	0,00725	0,00010	-0,02076	-0,00002	0,00484	0,00000
-5	0,00551	0,00004	-0,01017	0,00000	0,00222	0,00000
0	0,00490	-0,00009	0,00126	0,00001	-0,00053	-0,00001
5	0,00525	0,00007	0,01195	-0,00003	-0,00324	0,00000
10	0,00647	0,00007	0,02058	-0,00002	-0,00594	0,00000
15	0,00970	0,00002	0,02534	-0,00002	-0,00679	-0,00006
20	0,01351	0,00009	0,02919	-0,00002	-0,00753	-0,00002
25	0,01750	0,00004	0,03203	-0,00002	-0,00792	-0,00001

Due to numerous results, the values of the forces and moments acting on the individual sections of the gyrocopter for all the cases (for each of the angle of attack α) are not given. Figures 6 and 7 show the forces acting on the individual sections of the gyrocopter for angles of attack α =-20°, α = 0° α = 25° and a slip angle β = 0°. Similarly, Figure 8 shows pitching moment *My* acting on the individual sections of the gyrocopter for angles of attack α = -20°, α = 0°, α = 25° and a slip angle *β* = 0°.

The values of the lateral force *Py*, as well as rolling moment *Mx* and yawing moment *Mz* are not given. Their values are negligible due to the symmetry of the gyrocopter construction and the symmetry of the airflow (sideslip angle β = 0°) – the forces and moments generated on each half balance each other out. The values of drag *Px* and lift *Pz* are meaningful in terms of aircraft performance, and the value of pitching moment *My* has an impact on aircraft longitudinal stability.

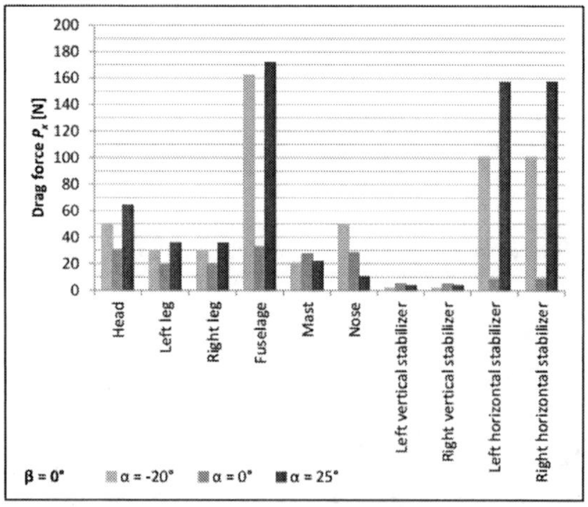

Figure 6. Drag Px acting on the individual sections of the gyrocopter for angles of attack α = -20°, α = 0°, α = 25° and a sideslip angle β = 0°.

The research shows that at the angle of attack equal to 0, the sections like the head, fuselage, mast and nose (at least 15% share) have the largest share in the generation of drag *Px*; the legs also have a significant share, i.e. more than 10%. For large and small angles of attack, the share of drag force, generated on the fuselage and horizontal stabilisers, increases a lot. The force generated on the legs and head also increases.

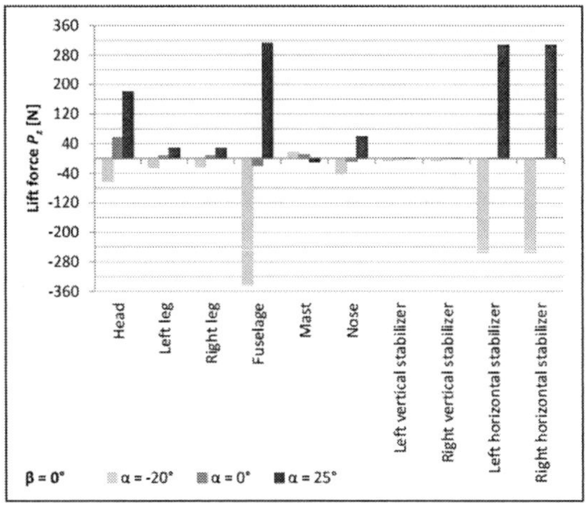

Figure 7. Drag *Pz* acting on the individual sections of the gyrocopter for angles of attack α = -20°, α = 0°, α = 25° and a sideslip angle β = 0°.

At the angle of attack equal to 0, the largest positive lift *Pz* is generated on the gyrocopter head. The mast and legs also show a small share. In the case of a large negative angle of attack, a large negative lift force is generated on the fuselage and horizontal stabilisers. The forces generated on the head, legs and nose are negative. For a large angle of attack, the forces generated on the fuselage, horizontal stabilisers, nose and legs are positive. The forces generated on the head, nose and horizontal stabilisers significantly increase.

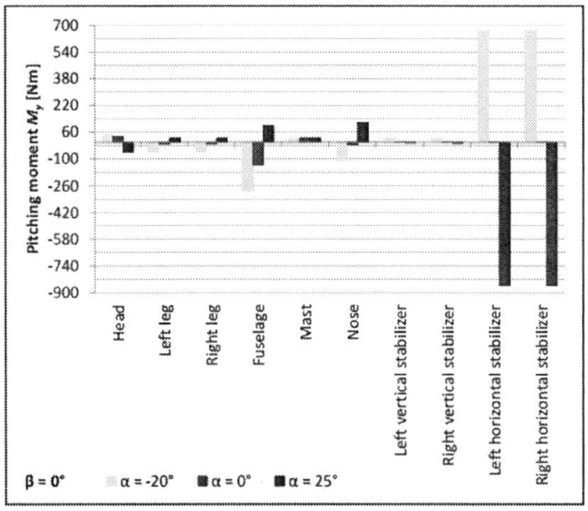

Figure 8. Pitching moment *My* acting on the individual sections of the gyrocopter for angles of attack α = -20°, α = 0°, α = 25° and a sideslip angle β = 0°.

For the angle of attack equal to 0, the pitching moment *My* is negative on the fuselage, nose and legs. Its largest absolute value is generated on the fuselage. Highly positive pitching moments are generated on the head and mast. For a large negative angle of attack, the pitching moments generated on the fuselage, head and nose increase their negative values. Positive pitching moments on horizontal stabilisers rapidly increase. For a large positive angle of attack, a large positive pitching moment is generated on horizontal stabilisers. The moments on the fuselage, nose and legs change their values into positive, whereas the ones on the legs, horizontal and vertical stabilisers change into negative.

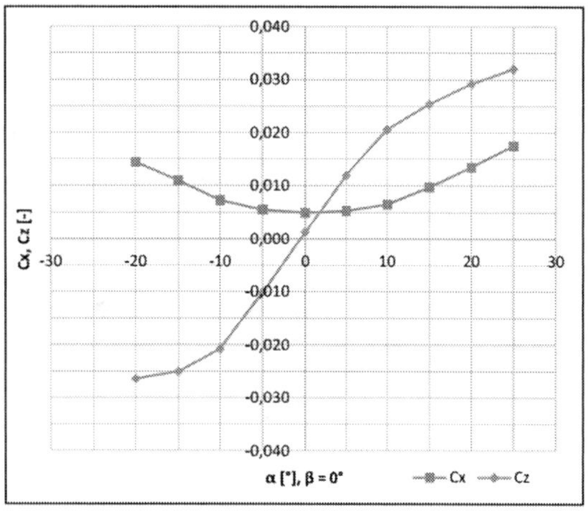

Figure 9. Drag coefficient *Cx* and lift coefficient *Cz* as a function of an angle of attack α for a sideslip angle β = 0°.

Drag coefficient *Cx* is the lowest for the angle of attack equal to 0. With increasing and decreasing the angle of attack, this coefficient increases to achieve its maximum value at an angle α = 25°. Lift coefficient *Cz* reaches 0 at the angle of attack of approx. 0.5°. This coefficient changes approximately linearly within the angles of attack ranging from -10° to 10°.

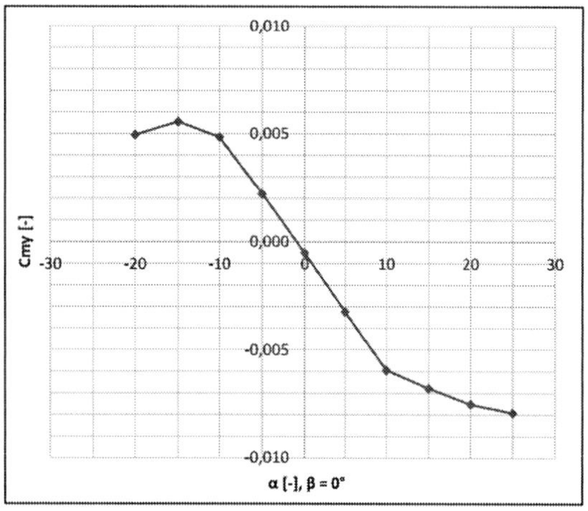

Figure 10. Coefficient of aerodynamic pitching moment *Cmy* as a function of an angle of attack α for a sideslip angle β = 0°.

Coefficient of aerodynamic pitching moment *Cmy* changes linearly in the angles of attack from -10° to 10°. Its highest value is at α =-15°, whereas the lowest one is at α = 25°. One of the criteria of aircraft longitudinal stability is that a derivative from a coefficient of aerodynamic pitching moment after an angle, i.e. $dC_{my}/d\alpha < 0$ should be negative (a descending function). In the range of angles between -10° and 10°, this derivative is constant as -0.0005 so the gyroplane can maintain longitudinal stability.

5 Conclusions

The paper shows the results of the numerical simulation of the airflow around the model of the gyrocopter for different angles of attack and a constant sideslip angle equal to 0. The research focused on how individual sections of the construction of the gyroplane, i.e. its head, left and right legs, fuselage, mast, nose, left and right vertical stabilisers, left and right horizontal stabilisers are impacted by the generated aerodynamic forces and moments. There were calculated coefficients of aerodynamic forces and moments to create basic aerodynamic characteristics.

This work has been financed under Grant Agreement No. POIR.01.01.01-00-0056/16

References

1. Z. Czyż, T. Łusiak, D. Czyż, D. Kasperek, Adv. Sci. Technol. Res. J., **10**(31), 169 (2016)
2. B. McCormick, AIAA **2002-5950**, 1 (2002)
3. A. Batrakov, L. Garipova, A. Kusyumov, S. Mikhailov, J. AIRCRAFT, **52**(5), 1634 (2015)
4. F. Vogel, Ch. Breitsamter, N. Adams, NNFM, **112**, 603 (2010)
5. Z. Czyż, P. Magryta, M. Szlachetka, Adv. Sci. Technol. Res. J., **9**(26), 89 (2015)
6. T. Ma, S. Hao, P. Xue, G. Li, W. Gan, ICMEIS 2015, Adv. Eng. Res., 91 (2015)
7. H. Sheikhi, A. Saghaie, Chin. J. of Aeronaut., **30**(1), 175 (2017)
8. A. Filippone, Aeronaut. J., **111**(1117), 175 (2007)
9. V. Gleize, AIAA, **2001**(0999), 1 (2001)
10. S. Houston, J. Guid. Control. Dynam., **21**(3), 391 (1998)
11. J.G. Leishman, *Principles of helicopter aerodynamics* (Cambridge University Press 2006)

Numerical simulations of crashworthiness performance of multi-cell structures considering damage evolution criteria

Quirino Estrada[1*], *Dariusz* Szwedowicz[2], *Alejandro* Rodriguez-Mendez[3], *Jesús* Silva-Aceves[1], *Javier S.* Castro[1], *Lara C.* Wiebe[1], *Elifalet* Gonzalez[1], *Julio* Vergara-Vazquez[2] and *Andrea E.* Chavez[1]

[1]Institute of Technology and Engineering, Autonomous University of Juarez City (UACJ), Ciudad Juárez, Chihuahua, México.
[2]Mechanical Engineering Department, National Centre for Research and Technological Development, Cuernavaca, Morelos, México.
[3]Department of Mechanical Engineering, University of California, Berkeley, CA 94720, USA.

Abstract. In this paper finite element software Abaqus was used to analyse the effect of cross-sectional shape on the crashworthiness performance of multi-cell profiles. An emphasis was placed on the modelling of the damage initiation criteria and its evolution during the crash event. The structures evaluated included square and circular multi-cell cross-sections fabricated with aluminium alloy EN AW-7108 T6. During the crash simulations, the structures were subjected to axial impact loads using a 500-kg rigid body striker with an initial velocity of 10 m/s. Accordingly to our results, profiles with circular cross-section base presented better crashworthiness performance than square profiles. An increase in crush force efficiency to 36.9% and specific energy to 35.4% was observed when a circular cross-section has been reinforced in the transversal and longitudinal directions. Finally, it was corroborated that the addition of the damage initiation criteria allowed for more reliable crash simulations of the structures.

1 Introduction

The use of thin-walled structures, as energy absorbers increases, improves the safety of passenger during car collisions [1]. Based on crashworthiness requirements, the thin-walled structures should minimise the peak load at the beginning of the crash event, maximise the energy absorption by plastic deformation and minimise weight [2]. Structures made of aluminium alloys are widely implemented in the automotive industry to satisfy the above requirements. Due to the lightness of aluminium alloys relative to steel, the total weight of a vehicle can be reduced by 28% [3]. This allows to reduce fuel consumption and consequently the emission of harmful contaminant agents into the atmosphere [4]. Several numerical and experimental studies have been conducted with the purpose of maximising the crashworthiness performance of thin-walled structures [5,6]. Hence, the study of multi-cell structures is developing. A multi-cell profile is obtained when a single cross-section is reinforced by plates at different angle of connectivity. Nia and Parpasour [7] conducted a comparative study of energy absorption of single and multi-cell structures. Several cross-sections based on triangular, square, hexagonal and octagonal shape were analysed. They observed that multi-cell profiles showed better energy absorption in comparison to single-cell profiles. Likewise, larger energy absorption per unit mass was obtained by hexagonal and octagonal multi-cell configurations. This observation was confirmed by Pirmohammad and Marzdashti [8] who analysed the energy absorption capacities of single and multi-cell profiles under axial and oblique quasi-static load. From their numerical results, it was determined that multi-cell structural members with an inner tube had larger energy absorption (E_a), notably when the inner and outer profiles were at a scale of 0.5. They found that multi-cell profiles with circular cross-sections allowed the best E_a performance. Tang et al [9] determined that the crashworthiness characteristics of circular multi-cell configurations depends mainly on the thickness and the number of cells along the radial and transversal directions. They concluded that the double layer cylindrical multi-cell structure was the most suitable to increase crashworthiness performance. Continuing with circular multi-cell profiles, Ahmed et al. [10] investigated the energy absorption capacities of several tube arrangements with curved stiffeners under dynamic axial load. Their numerical results evidenced the effectiveness of curved stiffeners to improve the energy absorption of columns. However, relative to circular single-cell profiles, bi-tubular arrangement with curved stiffeners showed better stability to the lowest peak load. On the other hand, to get reliable numerical results some researches have added theoretical damage models into their simulations [11-13]. Therefore, Hooputra et al. [14] presented a compressive macroscopic strain and stress model. The failure mechanism, considered in this approach, involves ductile, shear and necking instabilities. The effectiveness of this approach was shown by comparing numerical results with experimental data for a double chambered profile under

* Corresponding author: quirino.estrada@uacj.mx

DOI: 10.1051/itmconf/20171507009

axial compression and three point bending tests. Based on this approach Estrada et al. [13] evaluated the effect of circular discontinuities on the crashworthiness performance of square profiles under an impact load. In all cases, the addition of discontinuities improved the energy absorption capacity of the square profiles relative to a structure without holes. However, the best Ea performance equal to 17.61 kJ has been obtained when the holes were drilled at the top of the walls of the structure. As mentioned, significant efforts have been put to get more accurate numerical results, however the use of damage criteria and its evolution were less reported on numerical simulation of multi-cell structures. The current paper analyses the effect of cross-sectional shape on the crashworthiness response of multi-cell profiles. For this purpose, we considered multi-cell structures made with aluminium alloy EN AW-7108 T6 under a dynamic compression test. We tested multi-cell cross-sections with square and circular shapes. In order to get a reliable outcome, great emphasis was placed on the modelling of damage and its evolution.

1.1 Crashworthiness indicators

The parameters used to estimate the energy absorption capabilities of thin-walled structure should not dependent on geometry, material and mass. The energy absorption due to elastic and plastic deformation can be calculated from Equation 1. This refers to work done by the external force (F) along to the displacement (δ).

$$E_a = \frac{1}{2}\sum_{i}^{n-1}\left[\left(F(\delta)_{i+1} + F(\delta)_i\right)\cdot\left(\delta_{i+1} - \delta_i\right)\right] \quad (1)$$

The rate of the energy absorbed (E_a) to the displacement is described as the mean crush force (P_m) and it is calculated from:

$$P_m = E_a / \delta \quad (2)$$

The crush force efficiency (CFE) allows better comprehension of crashworthiness performance of structures. This parameter is given by Eq. 3. A value close to 1 indicates the optimal performance of the columns.

$$CFE = P_m / P_{max} \quad (3)$$

Also, when the structure has different masses (m), we define the specific energy absorption (SEA) by:

$$SFA = S_a / m \quad (4)$$

1.2 Damage criteria and its evolution

Finite element simulations are widely used to analyse the mechanical behaviour of structures. However, when these structures are subjected to dynamic loads, the mechanical behaviour becomes increasingly complex. Phenomena such as failure and fracture by ductile and shear mechanisms are presented. In order to correctly capture the behaviour of the structures, the article considers the ductile and shear damage initiation criteria proposed by Hooputra et al. [14] available in Abaqus/explicit. Experimental data for all models was obtained from [14,15] for aluminium alloy EN AW-7108 T6 and is shown in the Table 1. The ductile criterion, used in the simulations, assumes the equivalent plastic strain at the onset of damage to be a function of stress triaxiality and strain rate. Analogously, the shear criterion was specified by considering the equivalent plastic strain at the onset of shear damage to be a function of shear stress ratio and strain rate. Simplified analytical expressions presented by Hooputra et al. [14,15] are presented in Eqs. (5) and (6) for ductile and shear criteria, respectively.

$$\bar{\varepsilon}_D^{pl}\left(\eta, \dot{\bar{\varepsilon}}^{pl}\right) = \frac{\varepsilon_T^+ \sinh\left[k_0\left(\eta^- - \eta\right)\right] + \varepsilon_T^- \sinh\left[k_0\left(\eta - \eta^+\right)\right]}{\sinh\left[k_0\left(\eta^- - \eta^+\right)\right]} \quad (5)$$

$$\bar{\varepsilon}_S^{pl}\left(\theta_S, \dot{\bar{\varepsilon}}^{pl}\right) = \frac{\varepsilon_S^+ \sinh\left[f\left(\theta_S - \theta_S^-\right)\right] + \varepsilon_S^- \sinh\left[f\left(\theta_S^+ - \theta_S\right)\right]}{\sinh\left[f\left(\theta_S^+ - \theta_S^-\right)\right]} \quad (6)$$

where ε_T^+ and ε_T^- are equivalent plastic strains for equibiaxial tensile and equibiaxial compressive deformation, respectively. For isotropic materials, the stress triaxility in equibiaxial tensile deformation (η^+) is equal to 2/3, and for equibiaxial compression deformation, η^- is -2/3. In Abaqus, η is defined as the ratio of the equivalent mean stress to the Misses equivalent stress. The values of $k_{0,}\varepsilon_T^+$ and ε_T^- were taken from [15]. For Eq. (6), ε_S^+ is the equivalent plastic strain at the shear damage initiation for equibiaxial tensile and ε_S^- corresponds to equibiaxial compressive deformation. θ_S^+ and θ_S^- are the values of θ_S at $\eta = \eta+$ and $\eta = \eta-$, respectively. In this case, $\theta_S = \dfrac{1 - k_s\eta}{\phi}$ where $\phi = \tau_{max} / \sigma_{eq}$. The parameters $k_{s0}\varepsilon_S^+\varepsilon_S^-$ and f are determined experimentally.

DOI: 10.1051/itmconf/20171507009

Table 1. Failure parameters for aluminium alloy EN AW-7108 T6 at a strain rate of 250 s-1 [14, 15].

Parameter	Ductile failure parameters	Shear failure parameters
ε_T^2	0.44	-
ε_T^-	1.494	-
k_0	8.6304	-
ε_S^+	-	0.35
ε_S^-	-	1.2
f	-	2.05
k_s	-	0.3

2 Validation of numerical model

To evaluate the effectiveness of damage criteria on the progressive damage of a structure, a numerical model was developed using Abaqus/explicit. For this purpose, some guidelines given in [15] were considered. The discrete model consisted of a double-chambered profile under axial compression impact load. The profile was made of aluminium EN AW-7108 T6, with length L~396.5 mm, height H=68 mm, width W=95 mm and thickness t = 2.5 mm. The impact was done by a rigid body with a mass of 500 kg. The velocity applied was 10 m/s. We used a Young modulus of 70 GPa, Poisson ratio of 0.33 and density of 2700 kg/m^3. Likewise, ductility, shear, and Müschenborn-Sonne forming limit diagram (MSFLD) damage initiation criteria with evolution were considered. An element sized 5 mm was used to discretise the double-chambered structure. This value was validated using experimental data from [14,15]. The model and boundaries conditions are shown in the Figure1.

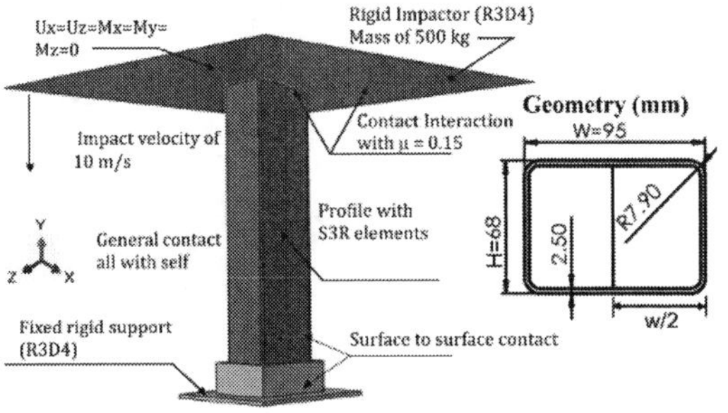

Fig. 1. Discrete model and boundary conditions.

Numerical results were compared with experimental data obtained by Hooputra et al. [14,15] and are presented in the Figure 2. The effect of damage modelling was studied, as well as its evolution, using a second numerical model without damage criteria. From the force vs displacement curve a P_{max} ~180 kN was computed; this value represented a difference of 5% relative to experimental data. The model with damage and its evolution properly represented the mechanical behaviour of the structure, especially after P_{max} was reached. At that point, a drop-in crushing force was registered, and the fracture of the profile could be visible. The opposite case was obtained by the model without damage criteria, where no cracks and progressive damage were observed (see in the Figure 3).

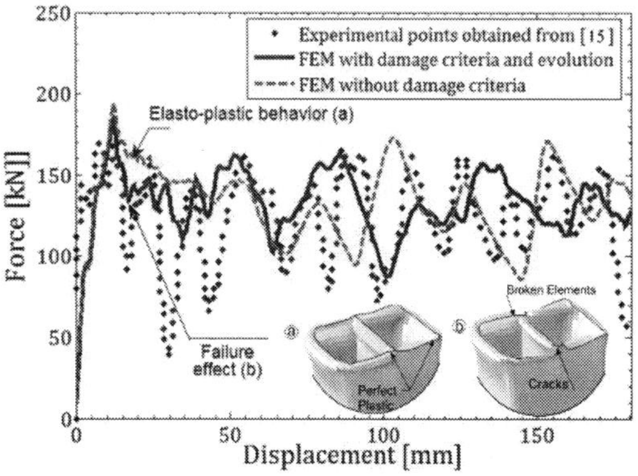

Fig. 2. Force vs displacement curves for the double-chambered profile.

Figure 3 presents the final deformation state of the structure, where it can be observed how the damage criteria and its evolution effectively reproduce the post-failure behaviour of the structure.

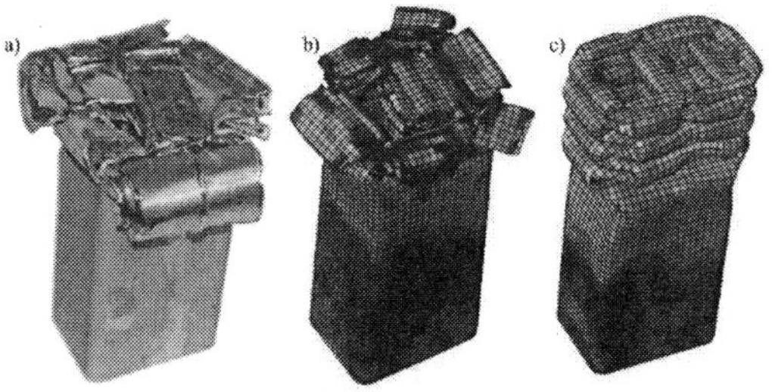

Fig. 3. Final deformation state, where: a) experimental [14,15], b) numerical with damage criteria and c) numerical without damage criteria.

Comparing the numerical and experimental results, the feasibility of the discrete model was validated. Thus, we can proceed to the crashworthiness analysis of multi-cell profiles.

3 Finite Element simulations

The article analyses the effect of cross-section on the crashworthiness response of multi-cell profiles. For this purpose, six structures with square and circular cross-sections were selected. The structures were made of aluminium alloy EN AW-7108 T6 with a length of 400 mm. The conditions of the numerical test were the same as those described in Section 2. The evaluated geometry of the specimens is displayed in the Figure 4 and the Table 2.

DOI: 10.1051/itmconf/20171507009

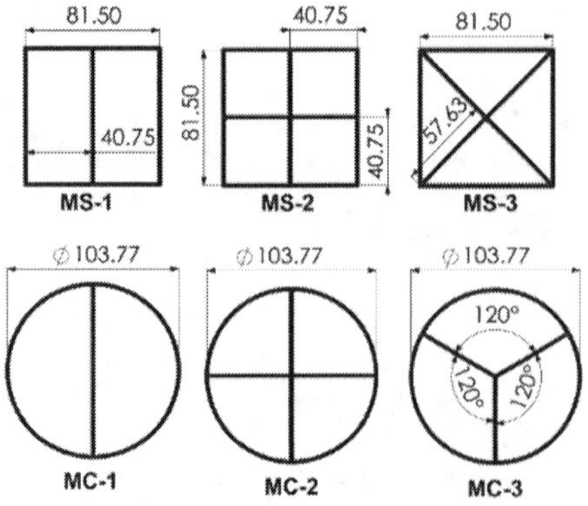

Fig. 4. Cross-sections analysed (mm).

Table 2. Details of multi-cell structures.

Code	Geometric base	Thickness (mm)	Length (mm)	Mass (gr)
MS-1	Square	2.5	400	1100
MS-2	Square	2.08	400	1100
MS-3	Square	1.83	400	1100
MC-1	Circular	2.37	400	1100
MC-2	Circular	1.91	400	1100
MC-3	Circular	2.12	400	1100

4 Results and Discussion

4.1 Square multi-cell profiles

Force vs displacement curves were obtained for all evaluated structures. To get a real comparison between profiles, the curves were delimitated to 160 mm. From the Figure 5, an approximate value of P_{max} of 350 kN can be interpreted. Three different behaviours were observed. The MS-3 profile showed the best stability to deformation. The curve presented straight zones which indicated a controlled deceleration. The opposite case was observed from the MS-2 structure.

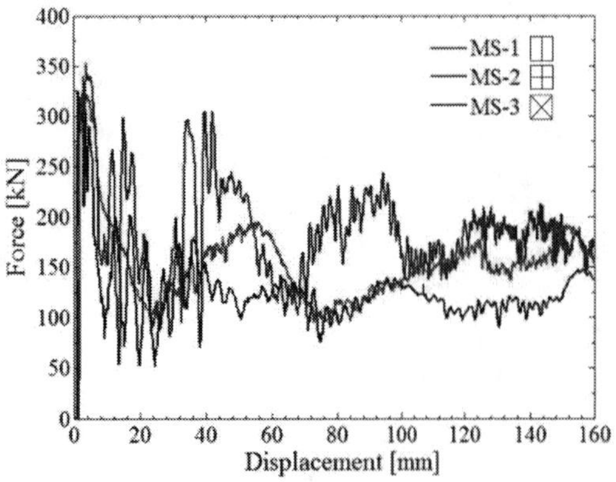

Fig. 5. Force vs displacement curves for square multi-cell columns.

The modelling of damage and its evolution becomes important when the final deformation of the structure is required with accuracy. Figure 6 shows the final deformation state of the profiles at 160 mm of displacement. In all cases, the failure and fracture of some elements located on the corners and ribs were observed. The deformation pattern of the columns was characterised by a symmetrical deformation mode described in [1].

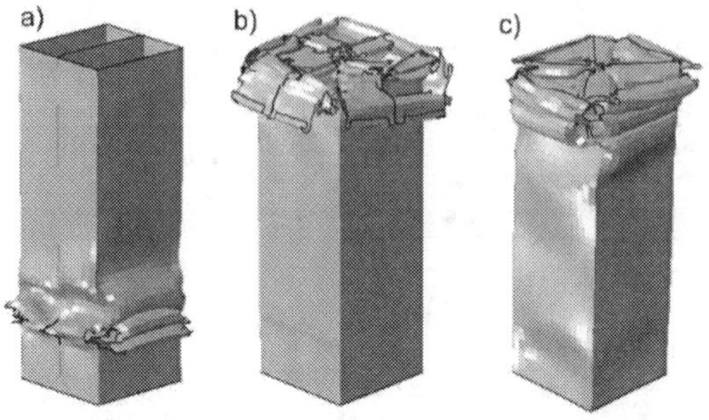

Fig. 6. Final deformation state for square multi-cell columns.

The mode of collapse and energy absorption performance depended on the failure of the structure by shear and ductile mechanism. Figure 7 shows the ductile damage initiation indicator. Failure by tension and compression effects were registered on some elements of the outer cylindrical surface on both the profile and the ribs. A value of 1 indicated that damage which has been initiated.

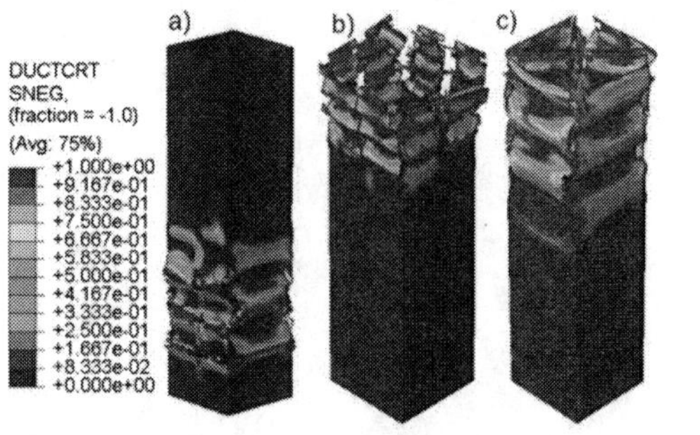

Fig. 7. Ductile damage initiation indicator.

Failure by shear band mechanism was more visible than failure by ductile effects. During the plastic wrinkle formation, failure by shear mechanism was observed on some elements used to form conical and toroidal surfaces. Additionally, shear effects were observed on the T joint formed by the ribs with the profile as seen in the Figure 8.

DOI: 10.1051/itmconf/20171507009

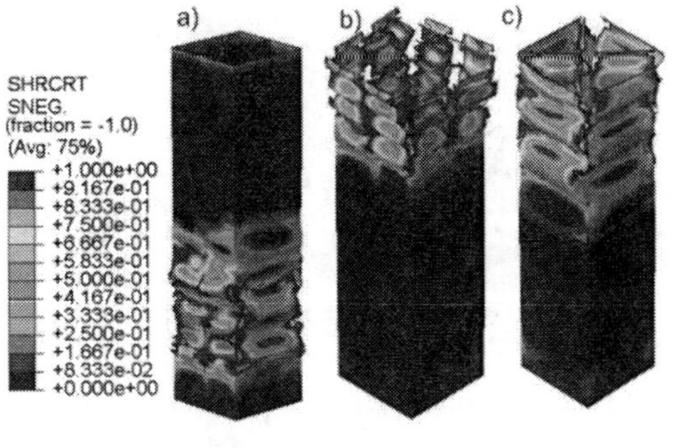

Fig. 8. Shear damage initiation indicator.

A summary of the results calculated for 160 mm of displacement is presented in the Table 3. An increase on energy absorption (E_a) in the rage of 21.66 to 39.31% was obtained for MS-1 and MS-2, relative to the worst energy absorption performance (MS-3). Since all structures had the same mass, the best energy absorption per unit of mass (SEA) and CFE value was for the MS-2 profile, with a value of 25.54 J/gr and 0.497, respectively. This condition indicated that the cross-section of this profile contributed to a better collapse mode of the structure. Thus, a large quantity of energy deformation was obtained (28.10 kJ).

Table 3. Summary of numerical results I.

Code	P_{max} (kN)	P_m (kN)	E_a (kJ)	SEA (J/gr)	CFE
MS-1	339.70	153.37	24.54	22.30	0.451
MS-2	353.01	175.62	28.10	25.54	0.497
MS-3	325.60	126.06	20.17	18.33	0.387

Moreover, it was confirmed that the energy absorption capabilities of profiles were mainly determined by the arrangement of the ribs in the profile, more than their number.

4.2 Circular multi-cell profiles

The crashworthiness performance of circular multi-cell profiles was obtained by the crash force vs displacement curves and they are presented in the Figure 9. The comparison of curves was realised at 160 mm of displacement. The curves were characterised by a controlled deformation along the crushing process. Thus, no abrupt changes in acceleration were registered. The curves presented similar mechanical behaviour, however the formation of wrinkles was out of phase, caused by the arrangement of the ribs in the cross section.

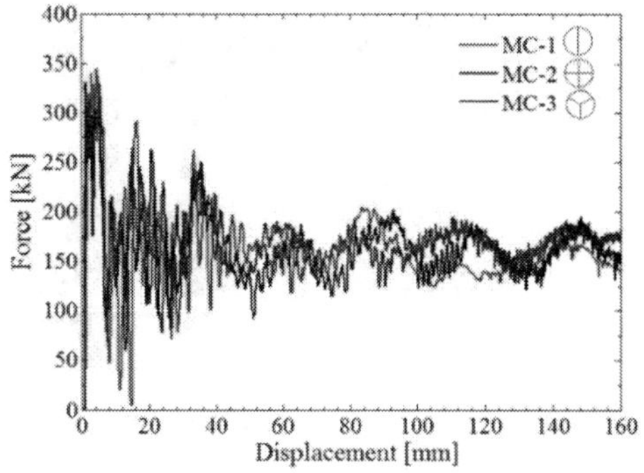

Fig. 9. Force vs displacement curves for circular multi-cell columns.

Figure 10 shows the final deformation state at a displacement of 160 mm. For all cases, the plastic deformation occurred at the top of the columns and was propagated in the downward direction. Even though the damage initiation criteria and its evolution modified the deformation mode of the structure, the columns were deformed in concertina and diamond mode. The fracture was located on the joints formed by the ribs and the outer profile. The largest damage effects were obtained by the MC-2 profile.

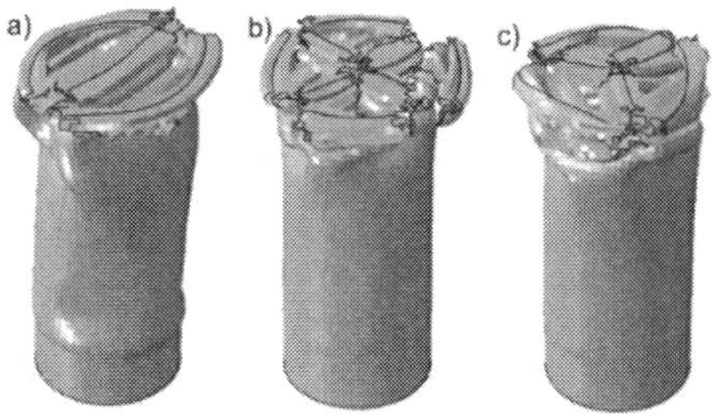

Fig. 10. Final deformation state for circular multi-cell columns.

The collapse mode was mainly determined by the cross-section arrangement and by failure of the elements. Thus, it was confirmed that the failure of circular multi-cell columns was due to ductile fracture (void nucleation, coalescence and growth) and shear band fracture. Figure 11 and 12 show ductile and shear damage initiation criteria, respectively, where a value of 1 indicates that the criterion has been satisfied. For all cases, the ductile criterion was met by some elements used to form the hinge line during the formation of the triangular folds. Meanwhile, the shear criterion was satisfied on elements near the T joint.

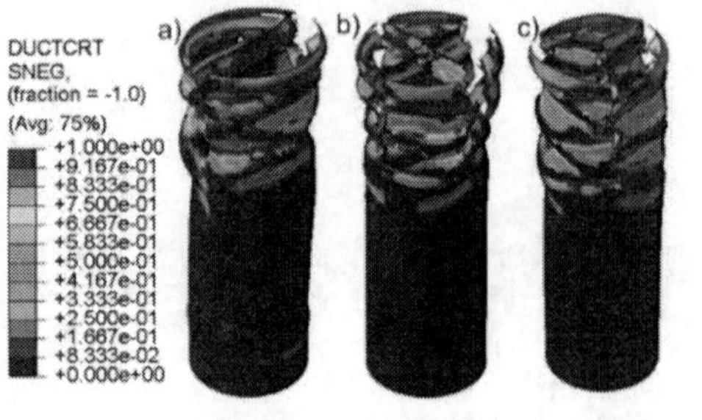

Fig. 11. Ductile damage initiation indicator.

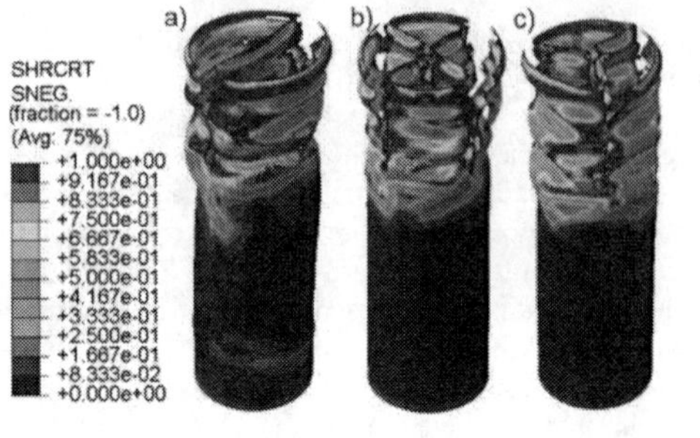

Fig. 12. Shear damage initiation indicator.

The addition of damage models and its evolution into the discrete models contributed to determine the correct mechanical behaviour of the structures. The numerical results are presented in the Table 4. The best energy absorption (E_a) was obtained by MC-2 profile (27.56 kJ) due to its larger deformation compared to all other profiles. Likewise, a greater external work was required to deform the walls of the column. The increase of the number of ribs contributed to improve the E_a performance in a rage of 1.65-6.88% relative to the worst E_a performance. The most favourable cross-section is the MC-2 profile, as demonstrated for higher values of SEA (24.83J/gr) and CFE (0.53).

Table 4. Summary of numerical results II.

Code	P_{max} (kN)	P_m (kN)	E_a (kJ)	SEA (J/gr)	CFE
MS-1	344.80	159.75	25.56	23.23	0.46
MS-2	321.01	170.75	27.32	24.83	0.53
MS-3	331.10	162.37	25.98	23.61	0.49

4.3 Comparing results

To understand the direct effect of the cross-section on the crashworthiness performance, a comparative analysis of SEA and CFE is shown in Figures 13 and 14, respectively. We observed the increase of SEA when the cross-sections have been reinforced in the longitudinal and traversal directions (MS-2 and MC-2). However, the best SEA performance of 25.54 J/gr was obtained by the MS-2 profile. That value represented the increase of 39.3% relative to the MS-3 profile. The worst CFE behaviour (18.33 J/gr) was obtained by the MS-3 structure.

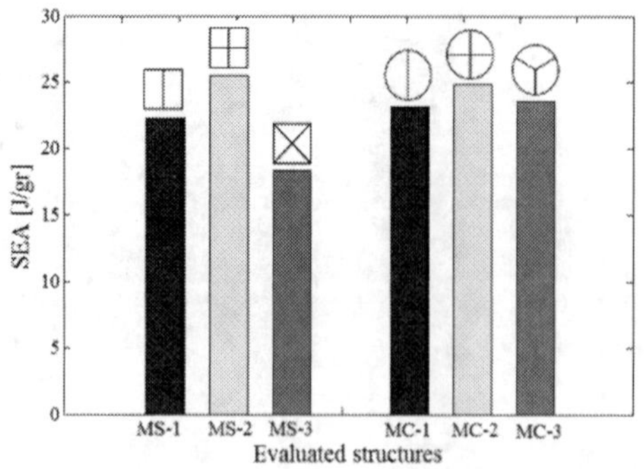

Fig. 13. SEA values of the structures evaluated.

A comparison of CFE value is presented in the Figure 14. A value close to 1 indicates the most desirable CFE value. The improvement of 36.95% was obtained for the MC-2 structure relative to the lowest CFE. For this case a maximum value of CFE around 0.58 was obtained. This value suggested a reasonable low peak load with a large quantity of energy absorption when the circular multi-cell cross-section has been reinforced in the transversal and longitudinal direction. In general, circular multi-cell profiles present better CFE performance than square columns in a range of 21.05-36.95%.

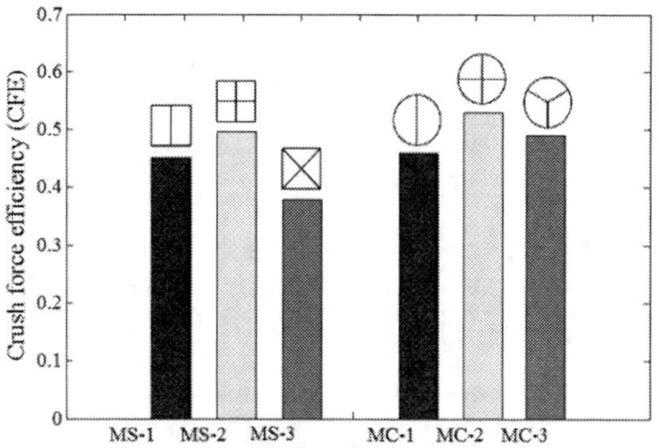

Fig. 14. CFE values of the structures evaluated.

From the Figure 13 and 14 the best crashworthiness performance was obtained for the circular multi-cell profile MC-2. Thus, this kind of cross-section arrangement should be considered in the design of energy absorption systems as an effective and low-cost solution for the automotive engineers.

5 Conclusions

A numerical study was conducted to evaluate the effect of the cross-sectional shape on the crashworthiness performance of multi-cell profiles with certain emphasis put on the modelling of progressive damage. For all cases, the use of ribs allowed better stability of the structure, minimising the buckling effect and maximising the energy absorption capacity. Regarding square multi-cell profiles, an improvement of CFE

(28.57 %) and SEA (39.31%) was obtained when the cross-section has been reinforced by 4 ribs in the transversal and longitudinal directions. That condition was also valid for circular multi-cell profiles, where an increase of E_a of 6.88% relative to MC-1 was obtained by the MC-2 profile. Of all the multi-cell cross sections, the circular shapes showed better crashworthiness performance when compared with square cross sections. The best performance was obtained by the MC-2 profile with a SEA value of 24.83 J/gr and CFE value of 0.53. It was determined that the crashworthiness performance of the structures depended mainly on the geometric base of the cross-section followed by the arrangement of the ribs and finally by the number of ribs. Likewise, it was confirmed the accuracy of numerical results to incorporate damage initiation criteria and its evolution. Finally, the MC-2 profile should be considered in the design of energy absorption devices to reduce the injuries to passenger during car collisions.

This work was supported by the Programa para el Desarrollo Profesional Docente para el tipo superior (PRODEP), through project No. PROMEP/103.5/16/10473, UACJ-PTC-349.

References

1. Q. Estrada, D. Szwedowicz, T. Majewski, E. Martinez, A. Rodriguez-Mendez, EIN-Maintenance and Reliability **18** (2016).
2. Q. Estrada, D. Szwedowicz, T. Majewski, M Baltazar, C Cortes, T Majewski, CA Estrada, Int. J Adv. Manuf. Technol. **84** (2016).
3. B. Simhachalam, K. Srinivas, R. Lakshmana, Int. J. Crashworthiness **19** (2014).
4. S. Wenlong, C. Xiaokai, W. Lu, Energy Proc. **88** (2016).
5. D. Szwedowicz, Q. Estrada, C. Cortes, J. Bedolla, G. Alvarez, F. Castro, Lat. Am. J. Solids Struct. **11** (2014).
6. M. Ferdynus, EIN-Maintenance and Reliability **15** (2013).
7. A.A Nia, M. Parsapour, Thin-Walled Struct. **74** (2014).
8. S. Pirmohammad, S. E. Marzdashti, Thin-Walled Struct. **108** (2016).
9. Z. Tang, S. Liu, Z. Zhang, Thin-Walled Struct. **62** (2013).
10. N. Ahmed, P. Xue, M. Kamran, N. Zafar, A. Mustafa, M.S. Zahran, Lat. Am. J. Solids Struct. **14** (2017).
11. J.S. Qiao, J. H. Chen, H. Y. Che, Thin-Walled Struct. **44** (2006).
12. J. Marzbanrad, A. Keshavarzi, F. H. Aboutalebi, Int. J. Crashworthiness **19** (2014).
13. Q. Estrada, D. Szwedowicz, J. Silva-Aceves, T Majewski, J. Vergara, A. Rodriguez-Mendez, Int. J. Mech. Sci. **131-132** (2017).
14. H. Hooputra, H. Gese, H. Dell, H. Werner Int. J. Crashworthiness **9** (2004).
15. Abaqus 6.10. Analysis User's Manual. (Dessault Systémes, 2010).

ITM Web of Conferences **15**, 07010 (2017)
CMES'17
DOI: 10.1051/itmconf/20171507010

Open-end tube dynamic flow model with an oscillatory extortion

Tytus Tulwin[1,*], *Monika* Klimek[2], and *Piotr* Piech[2]

[1] Lublin University of Technology, Mechanical Engineering Faculty, ul. Nadbystrzycka 36., 20-618 Lublin, Poland
[2] Medical University of Lublin, Department of Human Anatomy, al. Racławickie 1, 20-059 Lublin, Poland

Abstract. This paper presents a derivation of dynamic 2d mathematical model for open end tube with oscillatory extortion in the region of the closed end. The aim the research is to investigate possible uses of the increased pressure in the enclosed tube chamber, especially for energy efficient lift generation. The mathematical model allows to test and predict how flow modifications impact the resultant lifting force. A derivation of the proposed mathematical model is shown. The mathematical model is then compared to the computational fluid dynamics discrete model. The results prove the accuracy of the mathematical physical model.

1 Introduction

In fluid dynamics it is difficult to model complex scenarios, especially when turbulence and viscosity are involved. Omitting those terms can lead to a simplified mathematical model that can represent real world scenarios with good accuracy. This paper presents a derivation of dynamic 2d mathematical model for open end tube with oscillatory extortion in the region of the closed end. The other end of the tube is open. An open-air tube is often used in musical instruments as it can intensify the sound pressure level at resonant frequencies. This phenomenon known as acoustic resonance. This paper presents the results of the resonant pressure build-up experiment. The aim of the research is to investigate possible uses of the increased pressure in the enclosed tube chamber. Especially for energy efficient lift generation. The mathematical model allows to test and predict how flow modifications impact the resultant lifting force. There are studies investigating dynamic lift force acting on tube arrays [1][2][3][4][5].

2 Materials and methods

The experiment is based on the square profile 0.5 m long tube, with extra 0.1m pressure chamber with two oscillatory membranes. The tube width is 0.1 m. The investigation consists of three approaches: theoretical, CFD simulation and experimental. The frequency is set to 400 Hz with oscillatory membrane displacement close to 1 cm. The Computational Fluid Dynamics model is based on the same geometry, with hemispherical open field conditions. The theoretical model is simplified to the one-dimensional problem, basing on Euler governing energy equation. The experimental setup is visible in figure 1. The dimensions and parameters for theoretical model are presented in figure 2.

Fig. 1. The open resonance tube experimental setup.

The boundary condition of pressure P2 is set to 0 Pa relative to the atmospheric pressure, because it is a standing wave node. P0 is the maximum pressure extorted by the oscillatory membranes.

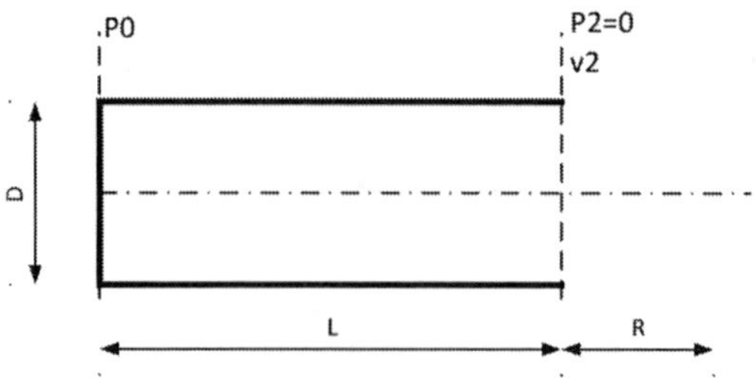

Fig. 2. The opn-tube dimensions and parameters.

A well-known Euler equation is used and simplified to the one-dimensional form:

[*] Corresponding author: t.tulwin@pollub.pl

$$\frac{\partial \bar{v}}{\partial t} + (\bar{v}\ grad\ \bar{v}) = g - grad\, p \qquad (1)$$

which can be written in differential form:

$$\frac{\partial v}{\partial t} + v\frac{\partial v}{\partial s} = \frac{1}{\rho}\frac{\partial p}{\partial s} \qquad (2)$$

Dividing both sides of the equation 1 by ∂s leads to equation 3. The first term corresponds to the transient inertial effect.

$$\frac{\partial v}{\partial t}ds + v\, dv = \frac{1}{\rho}dp \qquad (3)$$

At this point every term in the equation 3 can be integrated with known integration limits as shown in equation 4:

$$\int_0^{L+R}\frac{\partial v}{\partial t}ds + \int_{v_0}^{v_2} v\, dv = \int_{p_0}^{p_2}\frac{1}{\rho}dp \qquad (4)$$

The dimensional limits correspond to the tube location along its axis as shown in figure 3. The distance R corresponds to the area of effect in open field region. For simplicity it was assumed that the velocity profile decays in linear way with distance. For higher accuracy this profile could be represented by inverse distance equation.

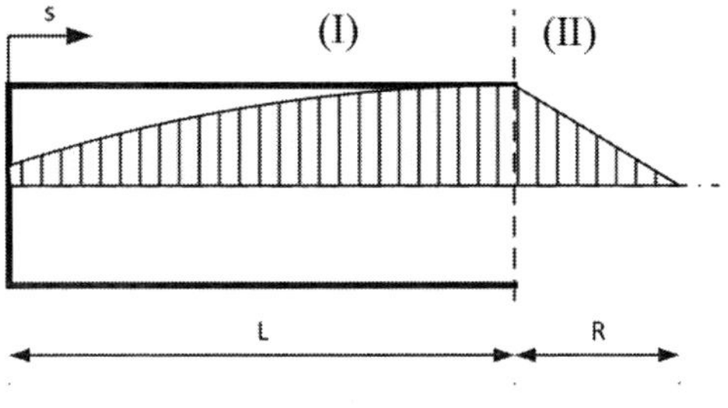

Fig. 3.. A predicted velocity profile along the tube's axis required for acceleration integration in expanded bernouli equation.

This leads to the modified unsteady Bernoulli equation 5 [6][7][8].

$$\int_0^{L+R}\frac{\partial v}{\partial t}ds + \frac{1}{2}(v_2^2 - V_0^2) = \left(\frac{k}{k-1}\right)\frac{p_2 - p_0}{\rho_0} \qquad (5)$$

The transient term needs to be integrated along the axial distance with limits covering whole region of its effect. The air outside the pipe has still significant influence on forces present. It is crucial to predict the shape of the acceleration profile in respect to the axial distance. Inside the pipe the shape is sinusoidal as the generated extortion signal is also sinusoidal. Outside the pipe air enters hemispherical open field region. The profile of acceleration still has a sinusoidal nature but decays quickly inversely proportional to distance (Fig. 4,5). It can be assumed that outside the tube air pressure and velocity are inversely proportional to the axis distance with sinusoidal term. For a frequency of 400Hz the sinusoidal term can be omitted for simplification.

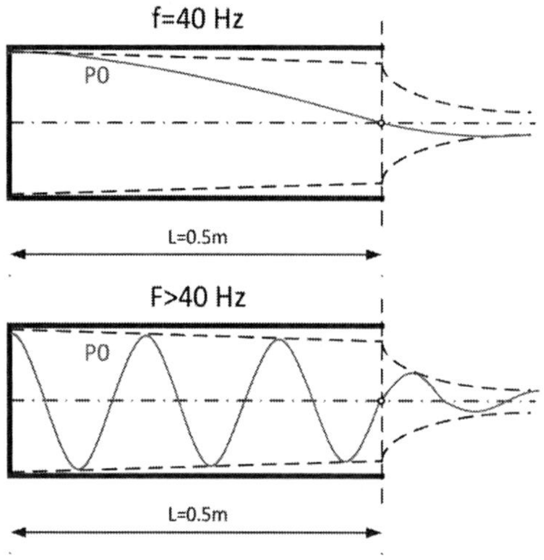

Fig. 4. Pressure distribution along the tube's axis for two different frequency levels.

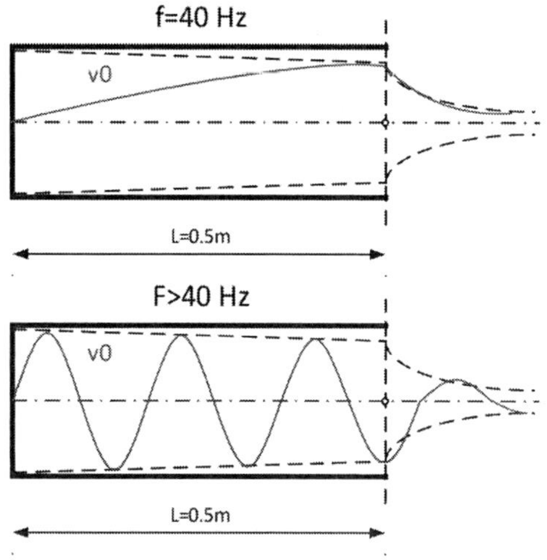

Fig. 5. Velocity distribution along the tube's axis for two different frequency levels.

In order to integrate the transient term of equation 5 it is split to two regions: I amd II. Region I represent the control volume inside the tube and region II represent the control volume outside the tube.

$$\int_0^{L+R}\frac{\partial v}{\partial t}ds = \int_0^{L}\frac{\partial v_{(I)}}{\partial t}ds + \int_0^{L+R}\frac{\partial v_{(II)}}{\partial t}ds \qquad (6)$$

Inside region I the velocity follows sinusoidal path, that can be written as:

$$v_{2(I)}(s) = -\cos\left(\frac{s-L}{c}\,2\,\pi\, f\right)v_2 \qquad (7)$$

Inside region II the velocity profile is considered as linear decaying to 0 at a distance value of R:

$$v_{2(II)}(s) = \frac{(R-(s-L))}{R} v_2 \tag{8}$$

Combining equation 7 and 8 gives:

$$\int_0^{L+R} \frac{\partial v}{\partial t} ds = \int_0^{L+R} \frac{v_2 - v_{2(t-1)}}{\Delta t} ds = \int_0^{L} \frac{v_{2(I)} - v_{2(I)(t-1)}}{\Delta t} ds + \int_0^{L+R} \frac{v_{2(II)} - v_{2(II)(t-1)}}{\Delta t} ds \tag{9}$$

Equation 9 is then integrated:

$$\int_0^{L+R} \frac{\partial v}{\partial t} ds = \frac{1}{\Delta t}\left(v_2 - v_{2(t-1)}\right)\left[\frac{c}{2\pi f}\left(-1 + \sin\left(\frac{L}{c} 2\pi f\right)\right) + \frac{1}{2} R + \frac{L^2}{2R}\right] \tag{10}$$

Equation 10 can now be inserted to equation 5, giving the final governing equation of the mathematical model:

$$\frac{1}{\Delta t}\left(v_2 - v_{2(t-1)}\right)\left[\frac{c}{2\pi f}\left(-1 + \sin\left(\frac{L}{c} 2\pi f\right)\right) + \frac{1}{2} R + \frac{L^2}{2R}\right] + V_2^2 \frac{1}{2}\left(1 + \lambda \frac{L}{D} + \xi\right) + \left(\frac{k}{k-1}\right)\frac{p_2 - p_0}{\rho} - \frac{1}{\Delta t}\left[\frac{c}{2\pi f}\left(1 + \sin\left(\frac{L}{c}\frac{c}{2\pi f}\right)\right) + \frac{1}{2} R + \frac{L^2}{2R}\right] v_{2(t-1)} = 0 \tag{11}$$

The viscosity loss is added in form of Darcy–Weisbach loss coefficients. Useful one-dimension governing equations can be found in paper [9] In this paper the author presented a different pressure wave equation [10].

3 Results and discussion

The results show that 2D model follows closely the CFD simulated pressure and velocity distribution along the tube axis (Fig 6).

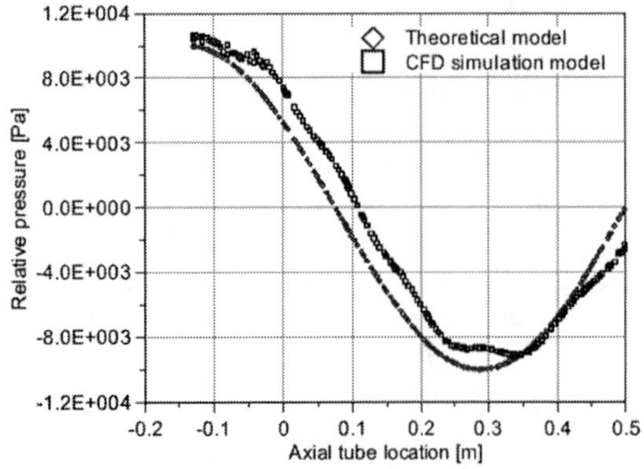

Fig. 6. Relative pressure distribution along the tube axis. The open end is on the right side.

Highest pressure is observed near the oscillation generator membranes on the left side of the tube. On the second antinode of the standing wave the amplitude drops slightly because of the viscous wall losses. A last pressure node of the confined tube is located at the end of the tube. A good representation of standing waves was found in paper [11]. CFD and theoretical models seem to overlap accurately with slight shift along x axis. This shift may be caused by differences in model geometry. The mathematical model covers exact 0.5m ideal tube, whereas the CFD model covers elongated tube with pressure membranes chamber. The velocity distribution along the tube axis follow a sinusoidal pattern, similar to the pressure distribution. Because the pressure and velocity quantities are coupled by the conservation of energy equations the energy transforms form kinetic form to potential pressure form and back.

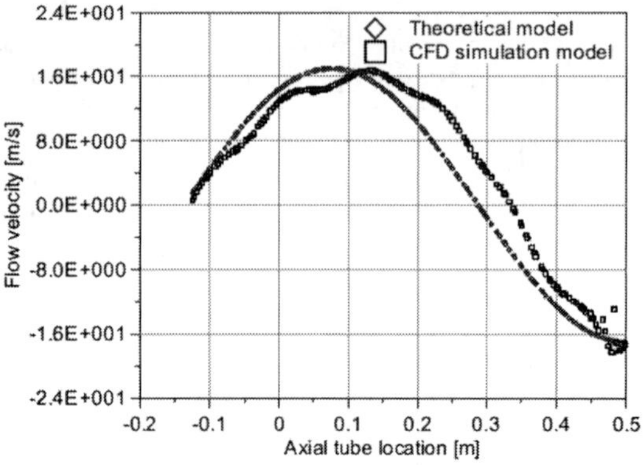

Fig. 7. Flow velocity distribution along the tube axis. The open end is on the right side.

This causes the velocity distribution to be shifted in phase when compared to the pressure distribution.

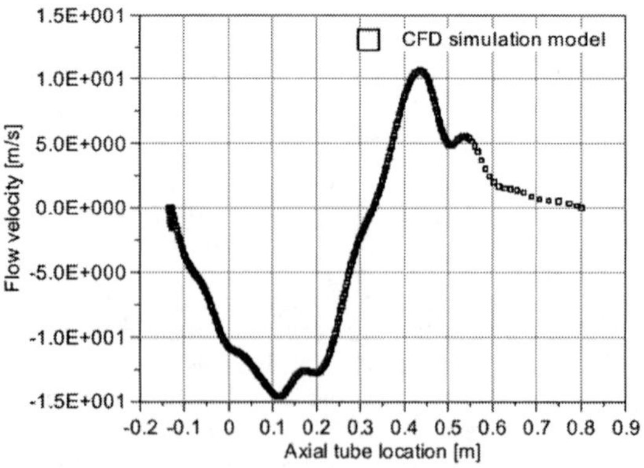

Fig. 8 Flow velocity distribution along the tube axis extended to the open field region. The open end is on the right side at 0.5m.

The pressure node is velocities antinode and opposite. The presented mathematical model follows this rule with good accuracy. When the wave leaves the tube the pressure and velocity amplitude drop is inversely proportional to the distance (Fig. 8). The outside conditions are hemispherical open field. Some noise is visible along the

DOI: 10.1051/itmconf/20171507010

path with higher frequency than the extortion frequency. It is caused by interference and geometry variations [12]. Fig. 9 shows how the velocity changes outside the tube in CFD simulation model. The fit was performed with the equation Y = A/X+B where A = 0.294 and B = -0.711. The residual sum of squares = 0.248936 and coefficient of determination R-squared = 0.963533.

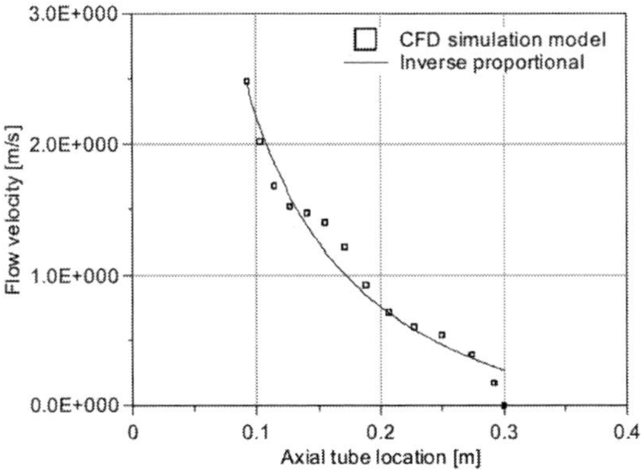

Fig. 9. A flow velocity ploted against distance from the tube exit. Inverse square fit is shown.

Figure 10 presents the CFD simulation pressure field during 400Hz membrane extortion operation. On the top of the tube two membranes generate pressure fluctuations. The pressure and velocity are than recorded along the tube's axis.

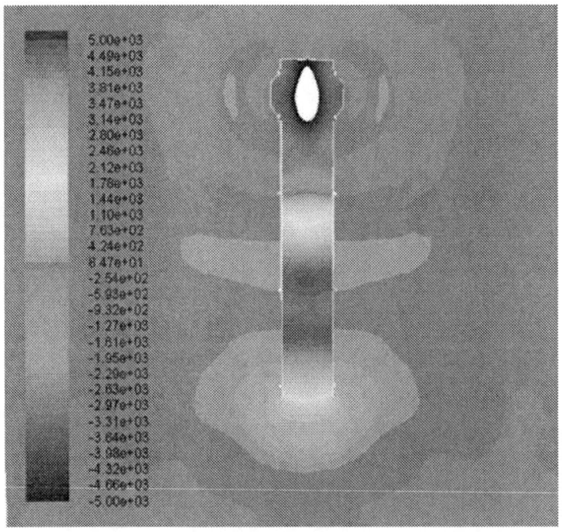

Fig. 10. Relative pressure field for an open end tube CFD simulation.

The research results indicate that the theoretical mathematical model accurately represents the pressure-velocity fluctuations inside the open-end tube. This is possible thanks to transient term in the Bernoulli equation together with the change of density term. The assumptions regarding the acceleration profiles are proven to be correct for analysed case. For increased accuracy the open-field part of the integral 8 can be modified to follow inverse distance equation. Further investigations are required to accurately estimate the R distance integral limit for different conditions. Although the pressure accumulated through the resonance phenomenon is relatively high for the energy input, it could not be used to generate work for every cycle without losing intensity. Any disruption of the pressure like adding one-way flow valve causes the pressure to drop significantly or act in opposite direction during next cycles. That is because the pressure builds up through many cycles and cannot be used and restored by one cycle.

4 Conclusions

The research provides several insights and phenomena descriptions:

- The theoretical mathematical model accurately represents the pressure-velocity fluctuations inside the open-end tube.
- The assumptions regarding the acceleration profiles are proven to be correct for analysed case.
- The linear representation of the open-field pressure decay seems to represent the case accurately, but for increased accuracy the integral 8 can be modified to follow inverse distance equation.
- CFD and theoretical models seem to overlap accurately with slight shift along x axis.
- The pressure accumulated through the resonance phenomenon is relatively high for the energy input, but it could not be used to generate work for every cycle without losing intensity

References

1. P. A. Feenstra, D. S. Weaver, and F. L. Eisinger, J. Fluids Struct., **21**, 89–101, (2005).
2. R. Hanson and S. Ziada, *J. Fluids Struct.*, **27**, 367–382, (2011).
3. A. Mohany and S. Ziada, *J. Fluids Struct.*, **25**, 461–478, (2009).
4. A. Oengören and S. Ziada, *J. Fluids Struct*, **6**, no. 3, 1992.
5. A. Mohany and S. Ziada, *J. Fluids Struct.*, **27**, 1149–1164, (2011).
6. S. Chen, D. Zhao, H. K. H. Li, T. Y. Ng, and X. Jin, *Appl. Therm. Eng.*, **112**, 1070–1082, (2017).
7. D. Zhou, Z. Luo, M. Fang, M. Lu, J. Jiang, H. Chen, and M. He, *Appl. Energy*, **185**, 2245–2250, (2017).
8. J. J. Heys, N. Holyoak, A. M. Calleja, M. Belohlavek, and H. P. Chaliki, *Open Biomed. Eng. J.*, **4**, 123–128, (2010).
9. R. C. Stevenson, Z. G. Yang, and V. Jairazbhoy, *Appl. Math. Model.*, **32**, 123–140, (2008).
10. MIT course, *Unsteady Bernoulli Equation* https://ocw.mit.edu, (2003), 11.2017.
11. S. Fontana, D. Moeller, I. Liu, and K. Robins, *Acoustic Levitation : A Theoretical Exploration*, (2016).
12. D. M. Howard and J. A. . Angus, *Acoustics and Psychoacoustics Fourth Edition*. (2009).

ITM Web of Conferences 15, 07011 (2017) DOI: 10.1051/itmconf/20171507011
CMES'17

Optimisation of liquid flow in cavitation tunnel using CFD method

Robert Jasionowski[1,*], and *Waldemar* Kostrzewa[1]

[1]Maritime University of Szczecin, Institute of Basic Technical Sciences, 2-4 Willowa Str., 71-650 Szczecin, Poland

Abstract. Computational Fluid Dynamics CFD is a type of software using a numerical modelling (finite element methods and finite volume methods) based on Navier-Stokes equations. This approach allows determining distributions of pressure, velocity, temperature and other features of a flowing media. In the present work. CFD software was used to a development of three-dimensional model of a cavitation tunnel, used for an examination of cavitation resistance of structural materials, as well as to simulate a liquid flow through the tunnel. The simulation was carried out for various assumed flow rates, to determine an optimal value leading to a more extensive area of low pressure assisted by a cavitation phenomenon.

1 Introduction

A progress in computer-based methods in the 20th century has led to an increase of popularity of the CFD as engineering software applied in various industry branches (i.e. aviation, automotive, medicine, petrochemical, energy etc.). The CFD has been applied to simulate lamellar and turbulent flow of liquids, a gas transport in combustion engines, a gas flow upon designing of aircrafts and sea crafts. Furthermore, the CFD allows verifying a nature of interaction between a flowing medium and an involved solid body [1-6].

A development of complete flow analysis giving many parameters (e.g. a stream course, local pressure drops, changes in flow rates, temperature variations) must be preceded by:

- building three-dimensional model of the analysed component (including a proper materials selection);
- assessing geometry of flowing medium (including its physical and chemical properties);
- setting boundary conditions of the flow by considering all necessary parameters;
- defining nodes and a mesh density in the most crucial parts of the working space;
- setting of parameters of analysis and a number of iterations;
- discussing obtained results.

Furthermore, the CFD allows performing the same analysis for a few different scenarios, i.e. for various initial parameters or model geometries.

In the present paper, the results of analysis of liquid flowing through a cavitation tunnel, under various flow rates, were discussed. The main purpose of presented simulation was to optimise flow parameters, in terms of formation of the largest low-pressure area leading to a cavitation phenomenon. An establishment of optimal flow conditions will allow conducting cavitation resistance test in shorter time, as well as give information about the best distribution of measurement and controlling devices.

2 Numerical model

2.1 Physical model

In the present work, a cavitation tunnel (Fig. 1), being a part of cavitation resistance measurement device, has been subjected to the analysis.

Fig. 1. Cavitation tunnel.

A PML2 80/200 type pump, supplied by 15kW engine, and LG iG5A inverter push water through the tunnel.

In previous experiments a rotor inside the PML2 80/200 type pump had been rotating with 2900 rpm, giving a flow rate of 9,95 m/s and delivery of a pump 200 m^3/h (measured in the tunnel inlet). The application of LG

* Corresponding author: r.jasionowski@am.szczecin.pl

iG5A inverter allowed setting a freely chosen rotational speed of the rotor; thus, controlling the flow and delivery rate. The cavitation induction system in the tunnel consisted of barricade and counter-barricade systems, and was similar to those located in University of Hannover, Hiroshima University and Institute of Fluid-Flow Machinery of the Polish Academy of Sciences.

2.2 Flow field model and boundary conditions

The geometric model of cavitation tunnel was designed in AutoDesk Inventor Professional 2018 software, by using a dimensional accuracy of 0,1 mm (Fig. 2).

Fig. 2. The geometric model of cavitation tunnel.

The geometric model of cavitation tunnel was made of six own-designed components (body, cover, upper and lower barricades, two flanges and 53 components from pre-existed libraries (Fig. 3)).

Fig. 3. The geometric model of cavitation tunnel - components.

The tunnel was made of 0H18N9 stainless steel, so the same material was also used upon the simulation of flow.

Volume mesh model (Fig. 4) and numerical analysis of liquid flow in cavitation tunnel were also designed in AutoDesk CFD 2017 software.

In order to increase the precision of measurements, tenfold concentration of mesh nodes between upper and lower barricades and under test sample (size adjustment parameter was 0,1) was performed. Additionally, components having no direct contact with the flowing liquid, as well as the roundness and chamfers of edges below 0,1 mm, were excluded from the analysis. Consequently, the achieved volume mesh model contained 960 000 nodes.

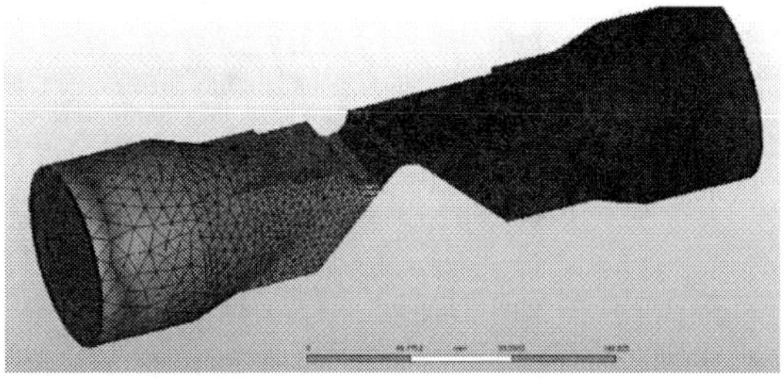

Fig. 4. Volume mesh model.

The following chemical and physical parameters of flowing water were used upon the analysis: density of 0,9982 g/cm3, viscosity of 0,001003 Pa·s, temperature of 25 °C, thermal conductivity of 0,0006 W/mm·K, bulk modulus of 2,18565 GPa.

The numerical analysis was carried out by using a liquid temperature of 25 °C and the following five different values of the velocity in the tunnel inlet: 7,74 m/s, 8,85 m/s, 9,95 m/s, 11,06 m/s and 12,16 m/s obtained for delivery rate: 160 m^3/h, 180 m^3/h, 200 m^3/h, 220 m^3/h and 240 m^3/h (maximum pump delivery rate).

3 Analysis of simulation and test results

The conducted analyses for five different velocity values allowed receiving a distribution of flow rate and pressure inside the cavitation tunnel.

To precisely determine positions of low pressure areas, additional 7 measurement points were added to the tunnel axis. The points separated by 5 mm distance were placed parallelly to the surface of sample at the distance of 0,5 mm (Fig. 5).

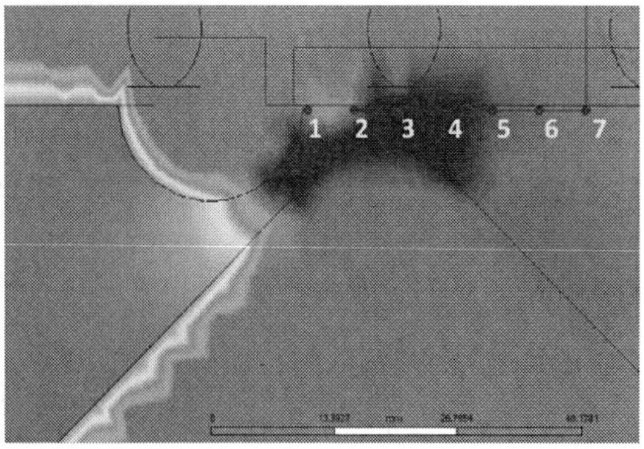

Fig. 5. Measurement points.

The results of the velocity are summarised in the Fig. 6, and pressure results in the Fig. 7.

DOI: 10.1051/itmconf/20171507011

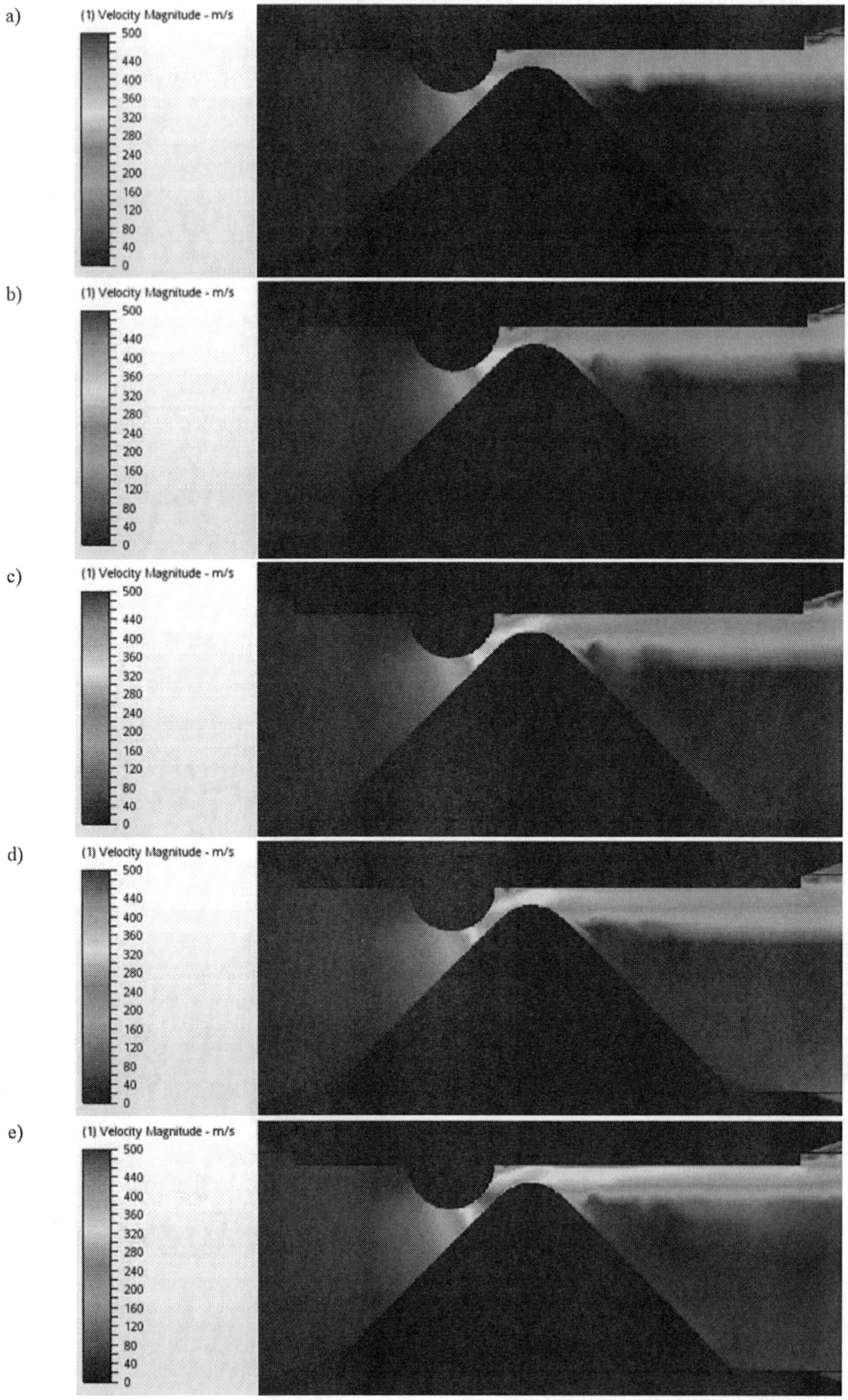

Fig. 6. The results of the velocity in the tunnel for the flow rate in the tunnel inlet: a) 7,74 m/s, b) 8,85 m/s, c) 9,95 m/s, d) 11,06 m/s, e) 12,16 m/s.

DOI: 10.1051/itmconf/20171507011

a)

b)

c)

d)

e)

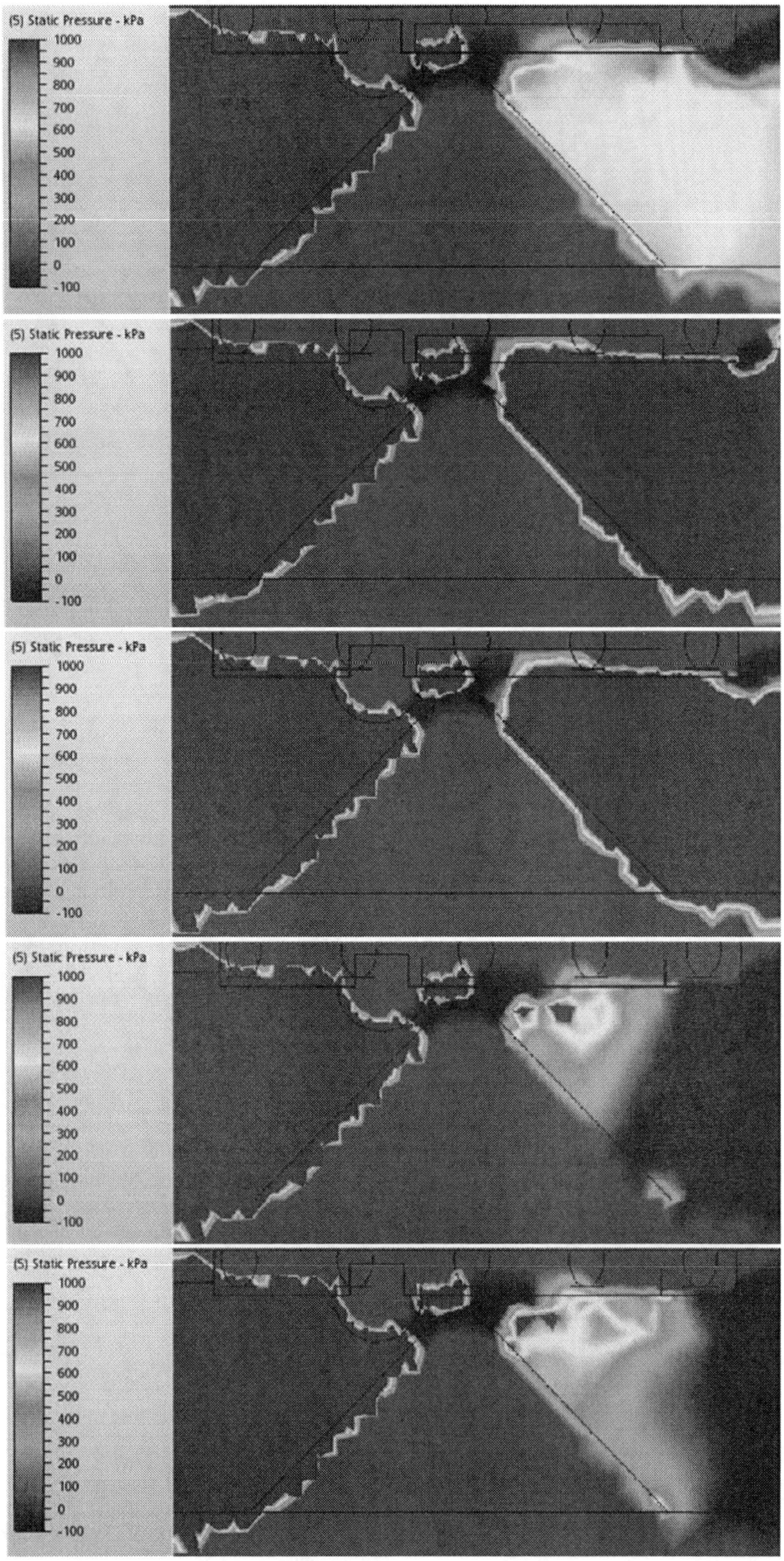

Fig. 7. The results of the pressure in the tunnel for the flow rate in the tunnel inlet: a) 7,74 m/s, b) 8,85 m/s, c) 9,95 m/s, d) 11,06 m/s, e) 12,16 m/s.

The velocity and pressure were analysed in these measurement points. The velocity and pressure were analysed in these measurement points. The results of velocity in the tunnel for all points are summarised as a diagram presented in the Fig. 8.

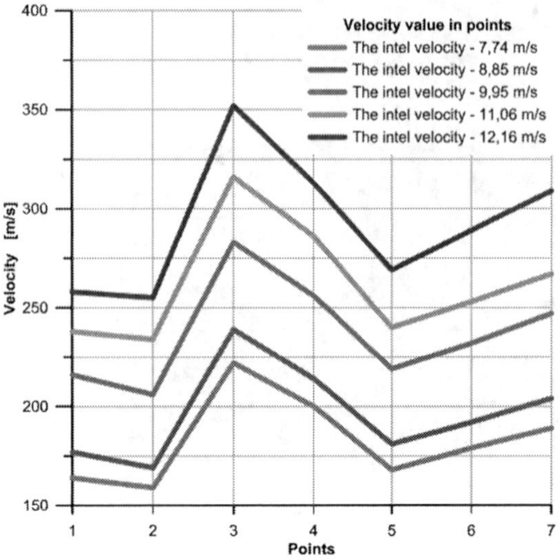

Fig. 8. The results of velocity in the tunnel for measurement points.

The results of pressure in the tunnel are summarised as a diagram presented in the Fig. 9 for all points, and in the Fig. 10 for points from 3 to 7.

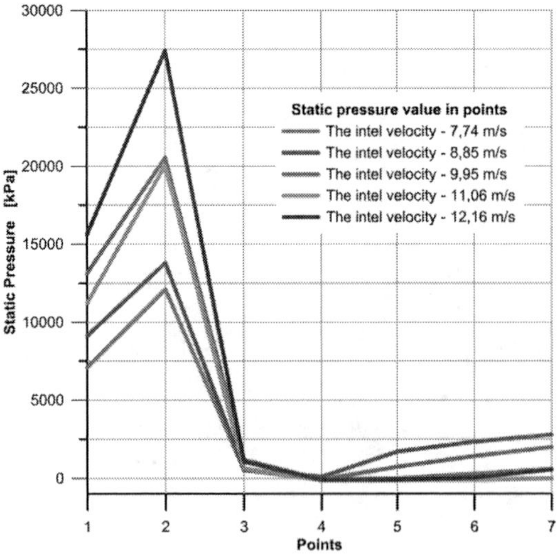

Fig. 9. The results of pressure in the tunnel for all points.

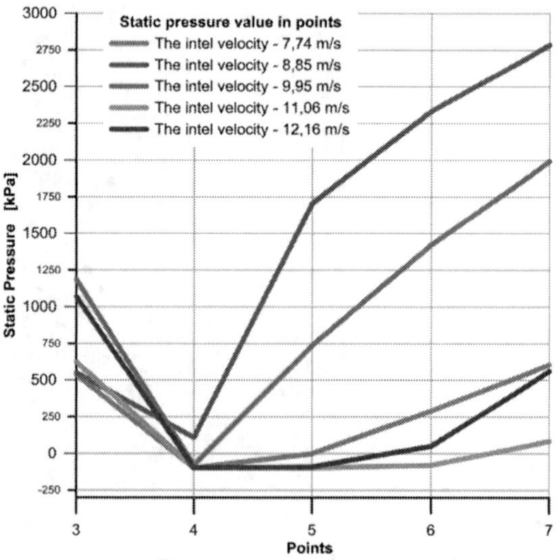

Fig. 10. The results of pressure in the tunnel for points 3 to 7.

4 Conclusions

The analysis of liquid flow with five different flow rates (within the range of 7,74 o 12,16 m/s) through the cavitation tunnel, was carried out in the present work. The conducted numerical calculations allowed determining the flow rate and pressure distribution near the surface of investigated materials. Following conclusions may be drawn from the obtained results:

- the highest values of flow rate (up to 350 m/s) were observed in the third measurement point located in the tunnel, i.e. in its smallest cross-section. The maximum value of flow rate was similar to that achieved in University of Hannover laboratories [7];
- the lowest pressure value between the fourth and fifth point of the measurement, and for all cases, was similar to -98986 Pa;
- the area of lowered pressure increased, raising the initial flow rate to its maximum for 11,06 m/s;
- by analysing all obtained results, it might be stated that the optimal velocity in the tunnel inlet was 11,06 m/s, for which the area of lowered pressure was between the fourth and sixth measurement point.

To sum up, the CFD method has confirmed its usefulness in conducting of simulations in terms of a proper design of machines, devices and laboratory stands. An implementation of this kind of software allows a better understanding of phenomena related to flowing liquids behaviour, as well as to reduce costs associated with experimental verifications. Nevertheless, further experiments regarding various flow geometry will be a valuable supplementary of the presently considered case of cavitation tunnel simulations.

Scientific work funded by the Polish Ministry of Science and Higher Education for statutory activities No. 1/S/IPNT/16.

DOI: 10.1051/itmconf/20171507011

References

1. J. Blazek, *Computational Fluid Dynamics: Principles and Applications*, (Elsevier, 2001)
2. J. Chung, *Computational Fluid Dynamics*, (Cambridge University Press, 2002)
3. T. Cebeci, J.R. Shao, F. Kafyeke, E. Laurendeau, *Computational Fluid Dynamics for Engineers*, (Springer, 2005)
4. A. Sayma, *Computational Fluid Dynamics*, (Ventus, 2009)
5. H. Lomax, T.H. Pulliam, D.W. Zingg, *Fundamentals of Computational Fluid Dynamics* (Springer, 2001)
6. J. D. Anderson Jr. *Computational Fluid Dynamics. The Basics with Applications* (McGraw-Hill, 1995)
7. J. Steller, B.G. Gireń, *International Cavitation Erosion Test. Final Report*, (Institute of Fluid-Flow Machinery Polish Academy of Sciences, 2015)

Selection of appropriate concrete model in numerical calculation

Krzysztof Śledziewski[1,*]

[1]Lublin University of Technology, Faculty of Civil Engineering and Architecture, Department of Road and Bridges, Nadbystrzycka 40, 20-618 Lublin, Poland

Abstract. Modelling of composite and reinforced concrete structures requires very precise determination of material parameters and constitutive relations between strain and stress. Erroneous selection of the dependencies and incorrect modelling, in particular, of the performance of concrete in tension may generate results in finite element method programs, that are far from the results obtained in an experiment. Using the example of a concrete damage plasticity model, based on fracture mechanics, this paper describes the physical interpretation and the method of the selection of parameters necessary for the appropriate modelling of concrete in a complex stress state. The correctness of the assumed description of concrete was verified on the basis of results of laboratory tests. A comparative analysis of the experimental and numerical results showed that the application of the concrete damage plasticity model allowed correct determination of the concrete element damage mechanisms for each level of strain.

1 Introduction

The appropriate modelling of composite structures [1] and reinforced concrete structures [2] is a major challenge, in particular due to the material properties of concrete. Unlike steel, concrete is a material that – in compression – behaves in a non-linear way from the very beginning. In addition, it undergoes degradation in a much faster manner [3] when in tension. Hence, the correct description of the concrete failure is the primary issue when considering the behaviour of concrete in a structure with advanced strains. As a result of the development of existing micro-cracks in concrete, the material softens. On the grounds of mechanics of continuous mediums, this phenomenon is defined as damage during which the tangent stiffness matrix ceases to be defined positively. In the case of concrete tension, due to the development of the damage, the tangent stiffness matrix is negatively defined, which leads to the location of the strains. From the mathematical point of view, a change occurs in a type of partial differential equations, which govern the process, from elliptic to hyperbolic. The problem is ill-conditioned, and consequently, the obtained solution depends on the density of the applied discretisation [4].

The presented problem can be regularised via different methods, among which one can distinguish two main groups. The first group concerns the regularisation at the level of a mathematical formulation, while the other regularises the problem at the level of a numerical formulation. The first group includes, above all, viscoplastic models in which the stress state depends on the rate of the strain [5]. They are used mainly in dynamics. This group also includes the Cosserat and micro-polar models which are used for soils and granular media [6]. In the case of modelling a failure in concrete, non-local models are also used [7], as well as the so-called higher-order, or gradient, models [8]. The non-local models introduce an averaging function which converts local variables into non-local ones, according to specific weights.

In turn, the possibility of regularisation at the numeric formulation level consists in the explicit introduction of the width of the plastic strain location zone into a finite element [4]. In addition, an automatic refinement of finite element mesh is used, based on local errors caused by large gradients of internal variables, or a displacement field discontinuity within the finite element is introduced [9]. This leads to the regularisation of the problem. All the above regularisation methods bring some characteristic internal length scale, which provides information on the width of the strain location zone in the material model.

2 Concrete damage plasticity model

2.1 Mathematical description

The numerical analysis of the initial boundary value problem with the location of failure requires a comprehensive, constitutive modelling of the material. In this model, the scalar damage parameter is used for modelling of the failure of concrete separately in compression and tension. The scalar description of the material damage [10] was initially used to describe the phenomenon of creep. The primary issue in the numerical analysis of concrete structures is to define the correct mechanism of structure failure and its load carrying capacity. This requires the identification of

* Corresponding author: k.sledziewski@pollub.pl

DOI: 10.1051/itmconf/20171507012

constitutive parameters of the model based on specific laboratory experiments [11].

The constitutive equation [10] of the material with scalar damage is as follows:

$$\sigma = (1-d)D_0^{el} : (\varepsilon - \varepsilon^{pl}) = D^{el} : (\varepsilon - \varepsilon^{pl}) \quad (1)$$

where: σ – the Cauchy stress tensor, d – scalar damage parameter (representing stiffness degradation), ε – strain tensor, D_0^{el} – initial tensor of elastic constitutive stiffness, D^{el} – degraded tensor of elastic constitutive stiffness.

It is necessary to specify an effective stress tensor based on the equation (2):

$$\bar{\sigma} = D_0^{el} : (\varepsilon - \varepsilon^{pl}) \quad (2)$$

where: ε^{pl} – plastic strain tensor, and the evolution of the damage parameter d as follows:

$$d = d(\bar{\sigma}, \bar{\varepsilon}^{pl}) \quad (3)$$

This, variable d is assumed to be a function of the effective stress $\bar{\sigma}$ and the equivalent plastic strain $\bar{\varepsilon}^{pl}$. In the concrete damage plasticity model, the stiffness of the material is initially isotropic, but during the process of stiffness degradation, it is determined by two variables: d_c in the compression and d_t in the tension. Ultimately, the Cauchy stress tensor σ is proportional to effective stress tensor $\bar{\sigma}$, and the coefficient of proportionality is $(1 - d)$:

$$\sigma = (1-d)\bar{\sigma} \quad (4)$$

The damage of the material in compression and tension is determined independently by two variables, $\bar{\varepsilon}_c^{pl}$ and $\bar{\varepsilon}_t^{pl}$, respectively, which specify the equivalent strains (in compression and tension):

$$\bar{\varepsilon}^{pl} = \begin{bmatrix} \bar{\varepsilon}_c^{pl} \\ \bar{\varepsilon}_t^{pl} \end{bmatrix} \quad (5)$$
$$\dot{\bar{\varepsilon}}_t^{pl} = h(\bar{\sigma}, \bar{\varepsilon}^{pl})\dot{\bar{\varepsilon}}^{pl}$$

Cracking (in tension) and crushing (in compression) in concrete are governed by increasing hardening (softening) variables. These quantities control the evolution of the loading surface F and the material stiffness degradation.

The loading surface determines the strain state as a function of stress and effective plastic strain. In the case of non-viscous damage plasticity model, the state of stress and strain must satisfy the condition:

$$F\left(\sigma, \bar{\varepsilon}^{pl}\right) \leq 0 \quad (6)$$

Plastic flow is determined by the plastic potential function $G(\sigma)$ and a non-associated flow rule in the form:

$$\dot{\varepsilon}^{pl} = \dot{\lambda}\frac{\partial G(\sigma)}{\partial \sigma} \quad (7)$$

2.2 Strength hypothesis and its parameters

The failure of concrete can be divided into two types. The first type refers to cracking in tension, which is associated with the occurrence of a dominant crack and the decline in load-carrying capacity in the direction perpendicular to it [12], while the other type is related to crushing in compression, during which many small tight cracks develop and evolve as a result of the loss of most of the concrete strength related to spalling.

The strength of concrete in a complex stress state is dependent on the complexity of the stress state present in the material. Thus, it cannot be determined only based on simple tests, such as uniaxial compression or tension. The strength of concrete can be determined only by considering the interaction of various stress state components.

Therefore, if one desires to describe the strength of concrete with a triaxial state equation, one must present its plane in a three-dimensional stress space, where on the surface there are stress states corresponding to the failure of the material, while inside there are safe performance states. In addition, inside the space there is the so-called plastic potential surface, after exceedance of which two situations occur [13]:

- an increase in strains without changing the stress (ideal plasticity),
- material softening – destruction of the material.

One of the most widely used hypotheses regarding the strength of concrete is the Drucker-Prager hypothesis [14], under which the failure depends on the shear strain energy and the failure surface in the stress space takes the shape of a cone (Fig. 1 [15]). An advantage of using such a failure criterion is the surface smoothness and thus no complications with numerical applications, but a disadvantage is the incomplete compliance with the real behaviour of concrete.

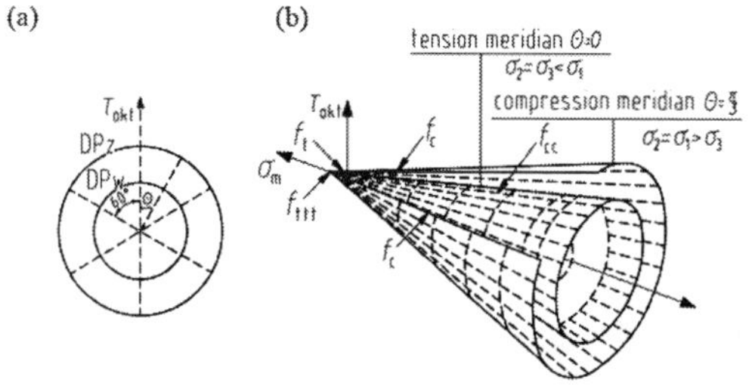

Fig. 1. The Drucker-Prager failure surface in deviatoric cross-section (a) and view (b) [15].

The CDP model used in Abaqus software is a modification of Drucker-Prager strength hypothesis. In the recent years, subsequent modifications have been made in it [16, 17] according to which the failure surface in the deviatoric cross-section does not need to be a circle and is regulated by parameter K_c. The physical interpretation of parameter K_c is the ratio of the distances from the hydrostatic axis to the compression and tension meridians in the deviatoric cross-section. This ratio is always larger than 0.5. If its value is 1, the deviatoric cross-section of the failure surface becomes a circle, as is the case in the classic Drucker-Prager hypothesis (Fig. 2 [15]).

In the paper [13], in accordance with the results of the experimental studies, it is observed that this value is 0.6 and slowly increases with the decrease in tension in the case of normal average stress of equal to zero. In accordance with [15], the recommended value K_c is 2/3. This shape is comparable to the strength criterion formulated by William and Warnke [18], which combined three mutually tangent ellipses. It is a theoretical-experimental criterion built based on the results of tests in a triaxial stress state.

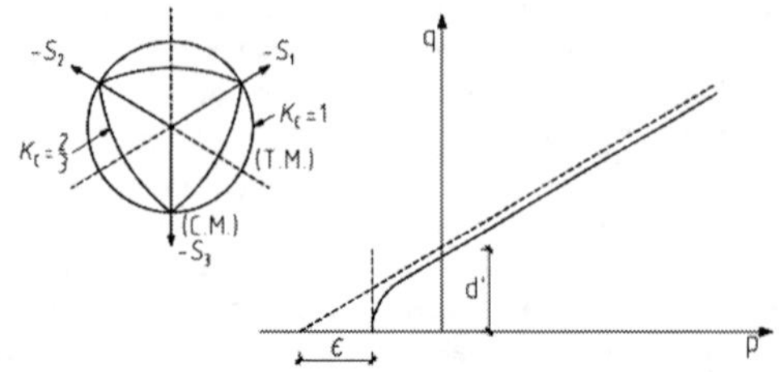

Fig. 2. Deviatoric cross section of failure surface in concrete damage plasticity model and hyperbolic surface of plastic potential in meridional plane [15].

A similar change applies to the shape of the meridians of the surface in the stress space, which are curved lines. In the CDP concrete model, the plastic potential surface in the meridian plane takes the form of a hyperbole (Fig. 2 [15]). The adjustment of the shape is done via the eccentricity plastic potential parameter. This is the length of the segment measured along the hydrostatic axis between the hyperbole vertex and the intersection of the asymptotes of this hyperbole (the centre of the hyperbole). The *eccentricity* parameter can be calculated as the ratio of the tensile strength to the compressive strength [19]. For the CDP model, the recommended value is $\varepsilon = 0.1$. In a limit case where it is equal to 0, the surface becomes a straight line in the meridian plane – the classic Drucker-Prager hypothesis.

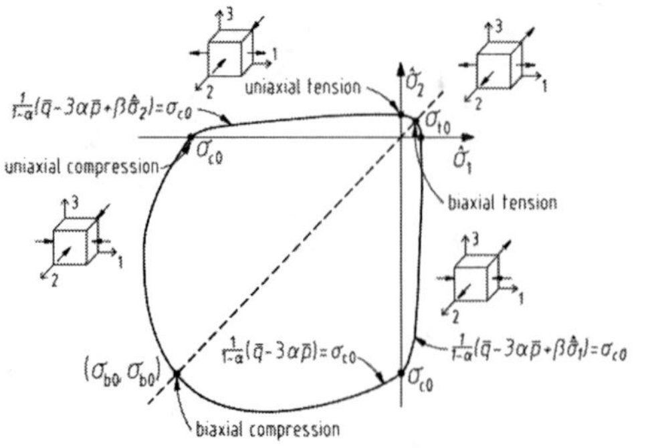

Fig. 3. Strength of concrete under biaxial stress in CDP [15].

Another parameter that describes the behaviour of concrete is the point at which it fails in biaxial compression (Fig. 3 [15]). The value σ_{b0}/σ_{c0} (f_{b0}/f_{c0}) determines the ratio of the strength of concrete in a biaxial state to the strength in the uniaxial state.

The most reliable research in this area is the study carried out by Kupfer in 1969. After its approximation with an equation of an ellipse, the uniform biaxial compressive strength f_{cc} = 1.16248 [13] or by default 1.16 according to [15].

The last parameter describing the behaviour of concrete in a complex stress state is the *dilation angle*, that is the angle of the inclination of the asymptote of the failure surface in relation to the hydrostatic axis, measured in the meridian plane (the angle of internal friction in the concrete). Numerical analyses usually assume the value of $\psi = 36°$ or $\psi = 40°$.

2.3 The stress-strain curve in uniaxial compression and tension

The most accurate method to describe the relationship between stress and strain is to perform uniaxial compression testing for the applied concrete and then transform the variables from the obtained graph. In the CDP model, one must determine the so-called inelastic strain $\tilde{\varepsilon}_c^{in}$ by deducting the elastic portion that corresponds to the undestroyed material from the total strain (recorded as part of a uniaxial compression test):

$$\tilde{\varepsilon}_c^{in} = \varepsilon_c - \varepsilon_{0c}^{el} \tag{8}$$

$$\varepsilon_{0c}^{el} = \frac{\sigma_c}{E_0} \tag{9}$$

During the transformation of strains, one must determine the moment from which the material is defined as non-linearly elastic (Fig. 4 [15]).

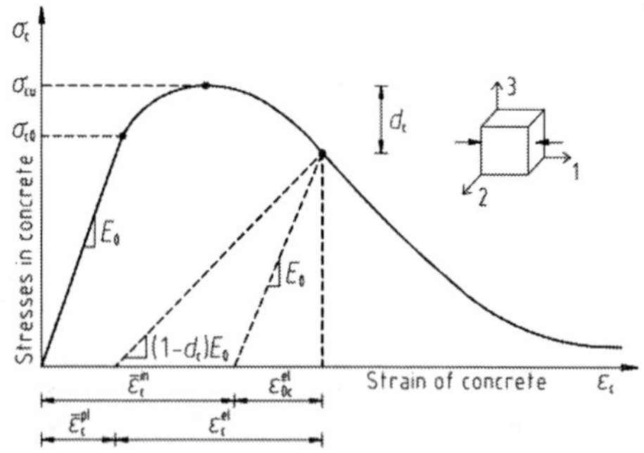

Fig. 4. The stress-strain curves of concrete in uniaxial compression [15].

Laboratory tests – uniaxial tests – show that such behaviour of material occurs almost from the beginning of the process of compression, but it is negligible in the initial stage for most numerical analyses. In the paper [13], the author writes that the end of linear elasticity should grow with the increase in concrete strength and should be more accepted from there, rather than determined on the basis of tests. In addition, the author described it as a percentage of stress in relation to the strength of concrete, according to the relationship:

 DOI: 10.1051/itmconf/20171507012

$$e_{\text{lim}} = 1 - \exp\left(\frac{-f_c}{80}\right) \quad (10)$$

A simpler method for determining this ceiling is to arbitrarily adopt it as $0.4 \cdot f_{cm}$. In accordance with [20], the modulus of elasticity of concrete is defined as secant within the range of 0–$0.4 \cdot f_{cm}$. It is good at this stage to adopt such a ceiling of the beginning of the inelastic phase so that the initial value of the Young's modulus and the secant value are consistent with each other. Most numerical analyses usually examine the phase of reaching the strength of the material, rather than the initial phase of its performance. Adoption of such a level gives greater certainty regarding the convergence of the solution.

When defining the stress-strain variables, one must specify the degradation variable d_c, which is 0 for an undestroyed material and 1 for a total loss of the ability to carry stress:

$$d_c = d_c(\bar{\varepsilon}_c^{\text{pl}}, f_i) \text{ if } 0 \le d_c \le 1 \quad (11)$$

These values can also be read from the uniaxial compression tests by calculating appropriately the ratio of stress values of the declining branch of the graph to the compressive strength of concrete. Hence, the CDP model specifies strains as:

$$\bar{\varepsilon}_c^{\text{pl}} = \bar{\varepsilon}_c^{\text{in}} - \frac{d_c}{(1-d_c)} \cdot \frac{\sigma_c}{E_0} \quad (12)$$

In Eq. (5), E_0 denotes the initial modulus of elasticity of the undestroyed material. Based on the plastic strain values, it is possible to determine the stress σ_c for uniaxial compression and their effective values $\bar{\sigma}_c$ describing the size of the flow and failure surfaces:

$$\sigma_c = (1-d_c) \cdot E_0 \cdot (\varepsilon_c - \bar{\varepsilon}_c^{\text{pl}}) \quad (13)$$

$$\bar{\sigma}_c = \frac{\sigma_c}{(1-d_c)} = E_0 \cdot (\varepsilon_c - \bar{\varepsilon}_c^{\text{pl}}) \quad (14)$$

The course of the material behaviour can also be determined without uniaxial compression testing, knowing only the average compressive strength of concrete f_{cm}. Then, the second value necessary to start the analysis of the variability of the stress-strain curve is the modulus of longitudinal elasticity of concrete E_{cm}. Its value can be estimated based on dependencies available in the code of practice [20]:

$$E_{\text{cm}} = 22 \cdot (0.1 \cdot f_{\text{cm}})^{0.3} \quad (15)$$

Other values indicating the location of the characteristic points on the graph are the strains while reaching the strength (ε_{c1}) and the limit strains at failure (ε_{cu1}).

In turn, the tensile strength of concrete in the uniaxial tension state is rarely determined in a direct tension test due to implementation difficulties and large result variability. Therefore, one usually uses indirect methods, such as splitting samples or a beam bending method. In the absence of laboratory tests, it is possible to derive tensile strength from the compressive strength in accordance with [20]:

$$f_{\text{ctm}} = 0.30 \cdot f_{\text{ck}}^{(2/3)} \quad (16)$$

The CDP model uses the notion of cracking strain $\bar{\varepsilon}_t^{\text{ck}}$. This allows to consider the phenomenon called tension stiffening effect [21]. Concrete in tension is not treated as an elastic-brittle body, but one takes into account the effects of phenomena such as the aggregate interlock in a crack and the adhesion of concrete to steel in the segment between the cracks. This assumption has a reason during the analysis with a fuzzy cracking image. In this way, the decrease in the stress in concrete in the tension zone does not occur suddenly but gradually. The post-cracking strains are defined as the difference between the total strains and elastic strains of the undestroyed material:

$$\bar{\varepsilon}_t^{\text{ck}} = \varepsilon_t - \varepsilon_{0t}^{\text{el}} \quad (17)$$

$$\varepsilon_{0t}^{\text{el}} = \frac{\sigma_t}{E_0} \quad (18)$$

In turn, the plastic strains $\bar{\varepsilon}_t^{\text{pl}}$ are calculated in the same way as for compression after defining the degradation parameter d_t:

$$d_t = d_t(\bar{\varepsilon}_t^{\text{pl}}, f_i) \text{ if } 0 \le d_t \le 1 \quad (19)$$

In order to build the stress-strain curve of concrete in uniaxial tension, one must determine the form of the softening function (Fig. 5 [15]).

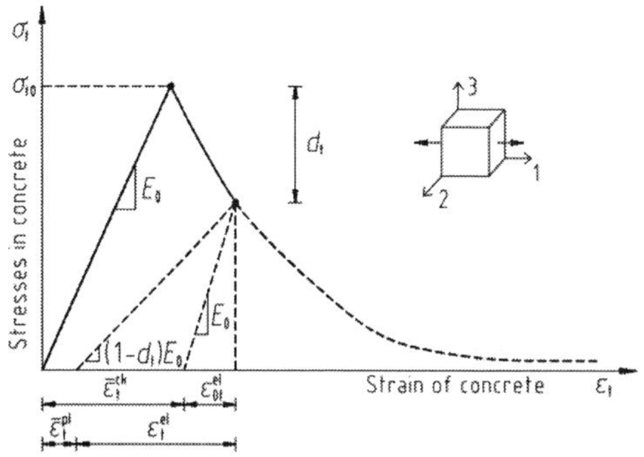

Fig. 5. The stress-strain curves of concrete in uniaxial tension [15].

In the absence of data, the stress can be reduced linearly down to zero from the moment of reaching the tensile strength in the case of total strains that are 10 times higher than at the time of reaching the value f_{ctm} [15]. A detailed description of this function, however, requires model calibration with experimental results in a specific analysis case [22]:

DOI: 10.1051/itmconf/20171507012

$$\left.\begin{aligned} \sigma_t &= E_c \cdot \varepsilon_t && \text{if } \varepsilon_t < \varepsilon_{cr} \\ \sigma_t &= f_{ctm} \cdot \left(\frac{\varepsilon_{cr}}{\varepsilon_t}\right)^{0,4} && \text{if } \varepsilon_t > \varepsilon_{cr} \end{aligned}\right\} \tag{20}$$

In Eq. (13), ε_{cr} denotes the strain at concrete cracking. As the effect of stiffening can significantly affect the analysis results and additionally this relationship requires calibration during a simulation, it is recommended to use a modification of the formula by Wang T. and Tsu for the softening function:

$$\sigma_t = f_{ctm} \cdot \left(\frac{\varepsilon_{cr}}{\varepsilon_t}\right)^n \quad \text{when } \varepsilon_t > \varepsilon_{cr} \tag{21}$$

where n denotes a parameter that determines the softening rate.

3 Numerical analysis

3.1 Test specimens

The correctness of the adopted concrete model was verified on the basis of the results of experimental tests of continuous composite beams subjected to bending [23].

Test specimens were made from steel girders (European I-beam, IPN 360 type) connected with a concrete slab along its entire length using two rows of pin fasteners. The fasteners with the diameter of ϕ16 mm and height of 75 mm were welded to the top flange of the beam with a 20 cm spacing. Such a connection was intended to ensure the rigidity of the joint in the entire load range. The primary reinforcement of the slab comprised twelve ϕ10 mm smooth reinforcement bars, laid in two rows upside and downside, along the entire length, in groups of 6 bars with the spacing of 8 cm (the reinforcement degree was 2%). The transverse reinforcement consisted of stirrups made of ϕ4.5 mm bars, arranged with a 20 cm spacing (Fig. 6).

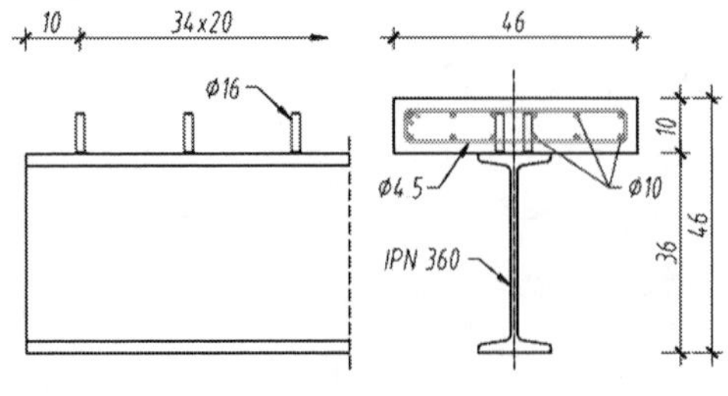

Fig. 6. Tested beam parameters.

The places of application of external loads were chosen in such a way that the test piece included both the zone of negative moments causing tension of the upper fibres of the concrete slab and the zones of positive moments causing their compression (Fig. 7). The beams were loaded gradually by applying incremental concentrated forces in four full load-unload cycles. The first cycle was always carried out in the range from 0 kN until the appearance of the first crack. In the subsequent cycles, the force was increased by approx. 200 kN until the maximum assumed load of 700 kN was reached. The maximum value was reached in each cycle by the gradual increase of the force, introducing intermediate load values. The load increase rate was on average 10 kN/min.

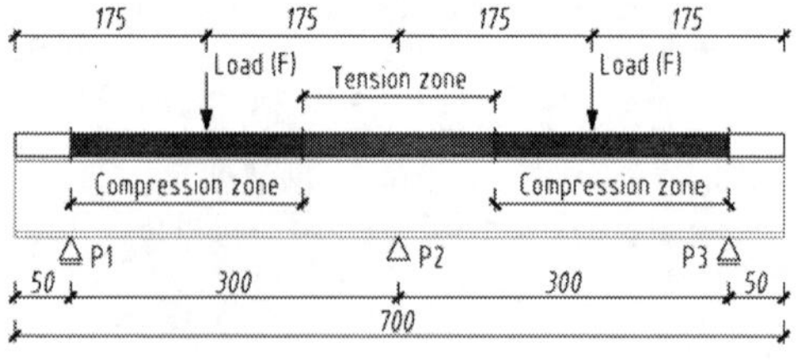

Fig. 7. Test piece load scheme together with the slab damage measurement area.

All computational tasks were solved in Abaqus/Standard software using the incremental iterative Newton-Raphson method [15]. The problems related to achieving the convergence of the solution, caused by the non-linearity of the material model, were solved by viscosity stabilisation μ [24]. The size of the load increment was also reduced (0.01 – 1E-12) and the maximum number of the load steps (max. 12000) was increased. The selection of the parameter μ was made iteratively after analysing the size of its impact on the results of the task. Finally, μ=0.0001 was adopted, allowing one to solve the task in more than 1.200 load increments formed in approx. 4.000 iterations. The analysis of the behaviour of the test configuration shows that such a value of the viscosity parameter makes it possible to reach a compromise between the computational magnitude of the task and the accuracy of the obtained results.

3.2 Geometrical models

Due to the nature of the structure being modelled, which consists of two different materials with clearly different geometries, the numerical model included different types of finite elements that best described the individual components of the beam.

Hence, the concrete slab was modelled with an 8-node brick element with reduced integration (C3D8R), and the I-beam with 4-node shell elements with reduced integration (S4R) and the primary reinforcement bars and stirrups with 2-node linear beam elements (B31).

Joining of the concrete slab with the upper flange of the steel beam varied depending on the distribution of bending moments along the length of the beam (Fig. 7). In particular, the joint was modelled in the tension zone as a discrete connection representing the occurrence of rigid connectors in the elements subjected to the laboratory tests. On the other hand, in the positive moment zones where the slab was compressed, the full joining of the upper surface of the steel beam with the lower surface of the reinforced concrete slab by means of a "continuous" (tie) connection was considered.

DOI: 10.1051/itmconf/20171507012

3.3 Identification of constitutive parameters of the concrete model

While identifying the constitutive parameters, it was assumed that concrete behaved in a linear-elastic manner up to $0.4{\cdot}f_c$ in compression and up to f_t in tension. For compression, the elastic limit is the point described by the stress value of 28.06 MPa and the strain value of 0.0006450. Concrete hardens in compression from the stress value of 28.06 MPa to the value of 70.15 MPa. In this range, any strain relief occurs along a straight line, whose slope is E_{cm}. Upon reaching stress that is equal to 70.15 MPa, concrete in compression softens, which is accompanied by the degradation of stiffness as well. After reaching the critical stress for concrete, the stiffness is reduced.

The stress-strain dependence of the non-linear behaviour of concrete under uniaxial compression (Fig. 8) was assumed in accordance with the recommendations [20]:

$$\sigma_c = f_{cm} \cdot \frac{k\eta - \eta^2}{1+(k-2)\eta} \quad \text{when } 0 < \varepsilon_c < \varepsilon_{cu1} \tag{22}$$

$$k = 1.05 \cdot E_{cm}\, \varepsilon_{c1} / f_{cm} \tag{23}$$

$$\eta = \varepsilon_c / \varepsilon_{c1} \tag{24}$$

The strain values corresponding to the greatest stress and the limit strain were determined from:

$$\varepsilon_{c1} = 0.7 \cdot f_{cm}^{0.31} \text{ and } \varepsilon_{c1} \le 2.8 \tag{25}$$

$$\varepsilon_{cu1} = 2.8 + 27 \cdot \left[0.01 \cdot (98 - f_{cm})\right]^4 \tag{26}$$

Then, the variables were transformed into inelastic strains $\tilde{\varepsilon}_c^{in}$, according to Eqs. (8)-(9) and degradation variables d_c were calculated as the ratio of stress values in the descending branch of the graph to the compressive strength of concrete. In this way, based on Eq. (12), plastic strains $\tilde{\varepsilon}_c^{pl}$ are determined during the numerical analyses.

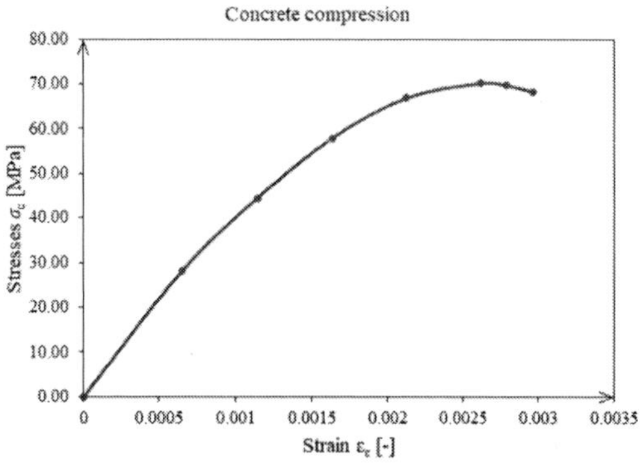

Fig. 8. Determined stress-strain curves of compressed concrete.

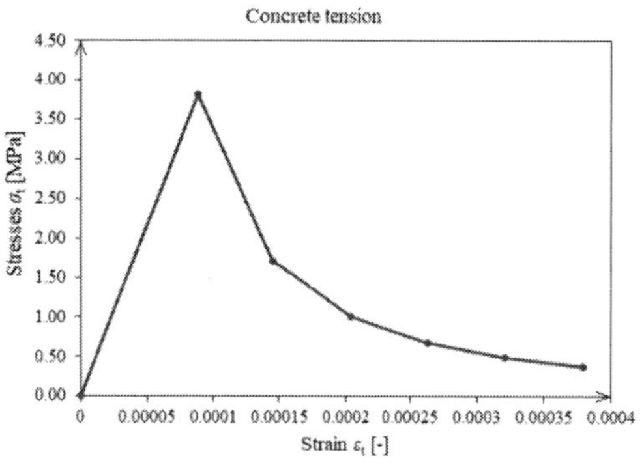

Fig. 9. Determined stress-strain curves of tensioned concrete.

Similar deliberations were made for tension. In accordance with the earlier assumptions, it was assumed that concrete behaved elastically up to the stress value of 3.82 MPa. According to Eqs. (17)-(18), cracking strains $\tilde{\varepsilon}_t^{ck}$ were determined, on the basis of which plastic strains $\tilde{\varepsilon}_t^{pl}$ are calculated taking into account the degradation parameter d_t (the maximum concrete degradation equal to 0.99 was adopted). Based on the model calibration, the form of the softening function was determined with the assumption that stress can be reduced linearly from the moment of reaching the tensile strength (Fig. 9).

Table 1 shows the material parameters that determine the behaviour of concrete in a complex stress state.

Table 1. Adopted CDP model parameters.

The parameters of CDP model			
Dilatation angle ψ	36°	f_{b0}/f_{c0}	1.16
Eccentricity ϵ	0.10	K_c	0.667
Concrete compression hardening		**Concrete compression damage**	
Stress [MPa]	Crushing strain [-]	DamageC [-]	Crushing strain [-]
28.06	0.000000	0.0	0.000000
44.38	0.000113	0.0	0.000113
57.87	0.000292	0.0	0.000292
66.84	0.000576	0.0	0.000576
70.15	0.000990	0.0	0.000990
69.69	0.001164	0.007	0.001164
68.24	0.001339	0.027	0.001339
Concrete tension stiffening		**Concrete tension damage**	
Stress [MPa]	Cracking strain [-]	DamageT [-]	Cracking strain [-]
3.82	0.0	0.0	0.0
0.38	0.00032	0.99	0.00032

4 Results and discussion

Identification of the slab cracking was done on the basis of the analysis of the maps of damage defined by the changes in the size of the DAMAGE parameter, i.e. the degradation of stiffness, which illustrates the material damage.

It should be remembered that the Concrete Damage Plasticity model of the material does not enable the formation of cracks in a discrete way taking into account the material spalling (its losses). It results only in the gradual exclusion of finite elements from the interaction. In this way, their "bonding" occurs, as well as their further involvement in the transfer of strain to the adjacent elements. This is one of the imperfections in the model, which has no significant effect on the behaviour of the entire test configuration, though.

The maps of damage defined by parameter d_t in the tension zone and parameter d_c in the compression zone, presented in the Fig. 10 and Fig. 11, can be identified with the spots where cracks appeared in the concrete slabs of the tested composite beams. It can be observed that the image of damage obtained in the numerical analysis corresponds qualitatively to the picture of the distribution of cracks obtained during the experimental tests.

The analysis of the maps of damage defined by parameter d_t also allows for tracing the process of the formation and development of cracks with an increasing load. In the case of the analysed beams, the first damage to the slab concrete appeared at the support axis, with the load causing tensile stress in concrete that was equal to the tensile strength. In the initial phase, as the load continued to be increased, more cracks began to appear simultaneously on both sides of the support, spreading towards the centre of the span. The subsequent increase in the load caused the density of the damage zones (cracks) to increase. A similar phenomenon was observed during the empirical research [23].

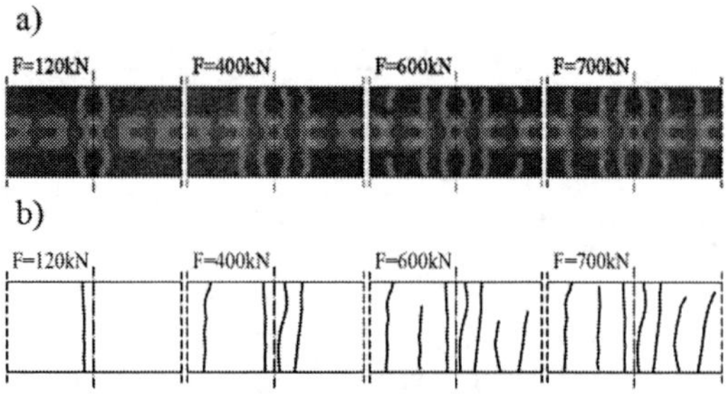

Fig. 10. The comparison of crack patterns for bending beam in support section: a) beam with CDP numerical model, b) the fracture path, which is observed in experiment.

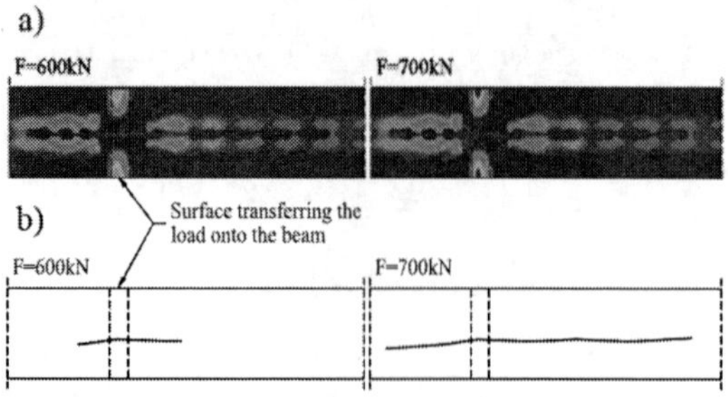

Fig. 11. The comparison of crack patterns for bending beam in span section: a) beam with CDP numerical model, b) the fracture path, which is observed in experiment.

In addition, the verification of the numerical model was carried out based on the comparison of the obtained displacements of the tested beams and the numerical model (Fig. 12 and Fig. 13). The deflections were measured at two points where the external load was applied. In addition, readings of vertical displacements were taken in the longitudinal beam axis, above each support.

The analysis of the relationship between the load and the displacement confirms the good agreement of the solutions. The mean percentage differences between the computer calculation results and the measured values oscillate between ~5% and ~8%.

The graphs also show that the beam stiffness in the experiment and the stiffness of the numerical model are similar to each other. This confirms the correctness of the selection of degradation variables d (parameter d_t, in particular) in the concrete model.

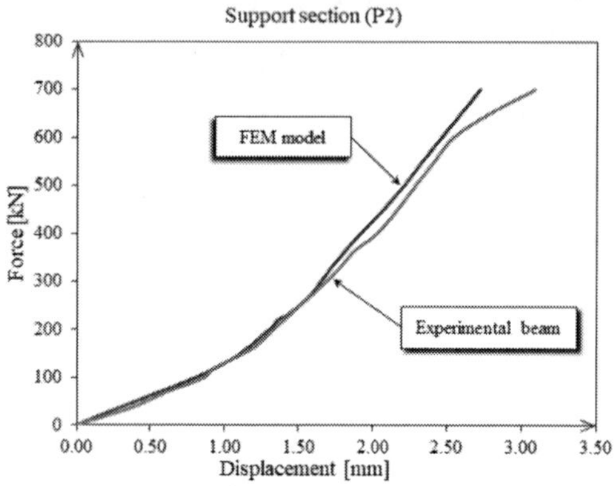

Fig. 12. Comparison of the vertical beam displacements measured and calculated in support section.

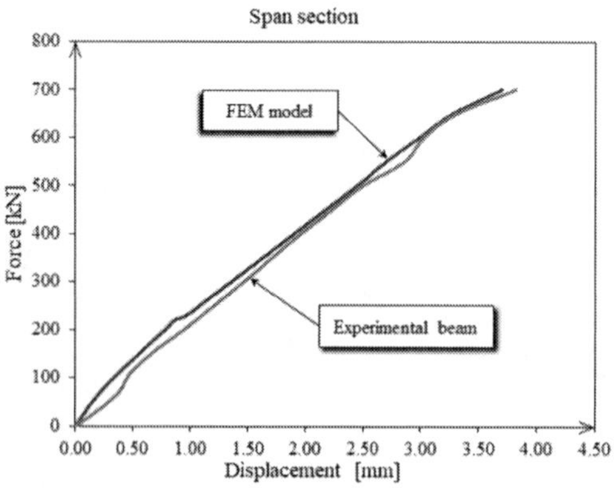

Fig. 13. Comparison of the vertical beam displacements measured and calculated in span section.

5 Summary and conclusions

The comparative analyses of the experimental for a numerical study results show that the application of the concrete damage plasticity model allows correct determination of the concrete element damage mechanisms. Considering the full non-linearity of concrete along with its progressive degradation as the stress values increase, however, can cause some problems related to achieving the convergence of the solution [24]. These problems are mainly related to the tensile response. FEM techniques based on reducing the load increment size or increasing the maximum number of steps while solving the problem using the Newton-Raphson approach may prove insufficient. Hence, the Concrete Damage Plasticity model includes the viscosity parameter μ, which is used for constitutive equation regularisation (allows to slightly exceed the plastic potential surface in some sufficiently small steps). The very idea of viscoplastic regulation consists in such a parameter selection so that the ratio of the time step of the task to the value of μ would approach infinity. Such a regulation method forces the parameter μ to be selected multiple times to check its impact on the results of the problem and thus, determine its minimum value.

The results of the studies make it possible to formulate the following conclusions:

- The material damage plasticity model is one of many possibilities for modelling concrete [25], which can be used for computer simulation of the behaviour in advanced load states of the entire structure and its individual elements. However, it is intended primarily for the analysis of reinforced concrete structures and composite structures of various types, such as steel-concrete and even concrete-concrete.
- Before the start of the analysis of a structure, one should determine the parameters of individual materials and constitutive relations between the strain and the stress.
- Due to the nature of the damage of concrete, some quantities should be adopted rather than determined as part of laboratory tests. Hence, the correctness of these assumptions should be checked by comparing other parameters, e.g. the deflection of the modelled structural element. Therefore, the results of the numerical analysis often require that the calibration of the model parameters be performed several times.

The author would like to thank the Gotowski Budownictwo Komunikacyjne i Przemysłowe Sp. z o.o. company for providing the testing materials free of charge. Laboratory studies were co-funded under the statutory measures of the Ministry of Science and Higher Education No. S-50/B/2012.

References

1. H. Hernández, J. Bonilla, G. Rodríguez, Revista Ingeniería de Construcción **29**(1), 5 (2014) http://dx.doi.org/10.4067/S0718-50732014000100001
2. H.M. Gomes, A.M. Awruch, Advances in Engineering Software **32**(4), 257 (2001) http://dx.doi.org/10.1016/S0965-9978(00)00093-4
3. K. Śledziewski, Maintenance and Reliability **4**(18), 578 (2016) http://dx.doi.org/10.17531/ein.2016.4.13
4. S. Pietruszczak, Z. Mróz, International Journal for Numerical Methods in Engineering, **17**, 327 (1981) http://dx.doi.org/10.1002/nme.1620170303
5. W. Wang, L. Sluys, R. de Borst, International Journal for Numerical Methods in Engineering **40**, 3839 (1997) http://dx.doi.org/10.1002/(SICI)1097-0207(19971030)40:20<3839::AID-NME245>3.0.CO;2-6
6. J. Tejchman, G. Gudehus, International Journal for Numerical and Analytical Methods in Geomechanics **25**, 1 (2001) http://dx.doi.org/10.1002/1096-9853(200101)25:1<1::AID-NAG115>3.0.CO;2-8
7. J. Bobinski, J. Tejchman, Archives of Civil Engineering **52**(1), 7 (2006) https://www.scopus.com/record/display.uri?eid=2-s2.0-33845642530&origin=resultslist&zone=contextBox
8. J. Pamin, *Gradient-Enhanced Continuum Models: Formulation, Discretization and Applications* (Cracow, 2004)
9. T. Belytschko, T. Black, International Journal for Numerical Methods in Engineering **45**, 601 (1999) http://dx.doi.org/10.1002/(SICI)1097-0207(19990620)45:5<601::AID-NME598>3.0.CO;2-S
10. L. Kachanow, Izv. Ak. Nauk CCCP, Otd. Techn. Nauk, **8**, 26 (1958)
11. T. Jankowiak, T. Łodygowski, Foundations of Civil and Environmental Engineering, **6**, 53 (2005) http://www.fcee.put.poznan.pl/sites/issues/2005.06/full/fcee_2005-06_053-069.pdf
12. A. Chen, W. Chen, Journal of the Engineering Mechanics Division **101**(4), 465 (1975) https://www.scopus.com/record/display.uri?eid=2-s2.0-0016543486&origin=resultslist&zone=contextBox
13. S. Majewski, *The mechanics of the structural concrete in elastic-plastic perspective* (Gliwice, 2003)
14. D. Drucker, W. Prager, Quarterly of Applied Mathematics **10**(2), 157 (1952) http://www.jstor.org/stable/43633942
15. Abaqus User's Manual (2012)
16. J. Lubliner, J. Oliver, S. Oller, E. Oñate, International Journal of Solids and Structures **25**(3), 229 (1989) doi:10.1016/0020-7683(89)90050-4
17. J. Lee, G.L. Fenves, Journal of Engineering Mechanics **124**(8), 892 (1998) http://dx.doi.org/10.1061/(ASCE)0733-9399(1998)124:8(892)
18. K.J. Willam, E.P. Warnke, Concrete Structures Subjected To Triaxial Stresses **19**, 1 (1974) http://doi.org/10.5169/seals-17526
19. Y. Sümer, M. Aktaş, Challenge Journal of Structural Mechanics **1**(3), 149 (2015) http://dx.doi.org/10.20528/cjsmec.2015.07.023
20. Eurocode 2, Part 1-1 (2004)

DOI: 10.1051/itmconf/20171507012

21. S.M. Allam, M.S. Shoukry, G.E. Rashad, A.S. Hassan, Alexandria Engineering Journal **52**(2), 163 (2013) http://dx.doi.org/10.1016/j.aej.2012.12.005
22. T. Wang, T.T.C. Tsu, Computers and Structures **79**, 2781 (2001) http://dx.doi.org/10.1016/S0045-7949(01)00157-2
23. K. Śledziewski, ET Sciences **5**, 106 (2017)
24. P. Kmiecik, M. Kamiński, Archives Of Civil And Mechanical Engineering **XI**(3), 623 (2011) http://dx.doi.org/10.1016/S1644-9665(12)60105-8
25. L. Jason, A. Huerta, G. Pijaudier-Cabot, S. Ghavamian, Computer Methods in Applied Mechanics and Engineering **195**(52), 7077 (2006) http://dx.doi.org/10.1016/j.cma.2005.04.017

Stability of rectangular plates with notch using FEM

Katarzyna Falkowicz[1,*]

[1]Faculty of Mechanical Engineering, Department of Machine Design and Mechatronics, Lublin University of Technology, Nadbystrzycka 36, 20-618 Lublin, Poland

Abstract. Buckling behaviour analysis of thin-walled composite plate under an axial compressive force is presented. The plate with central notch is made of a carbon-epoxy composite – a laminate consisting of eight symmetrically oriented plies. This paper addresses an influence of notch on buckling behaviour of laminated composite plates. In this analysis, FEM was applied to perform parametric studies on various plates based on stacking sequences, shape of notch, and its size. Buckling behaviour of laminated composite plates under an axial compression load is studied using ABAQUSS software.

1 Introduction

A composite material, given the high strength and rigidity as well as low weight, has been increasingly used not only in the aerospace and automotive sectors, but also in designs, where low structure weight is crucial. An in-depth understanding of their structural behaviour is essential, especially concerning buckling load, distributions of stresses and strain, deflection behaviour or influence on composite layers sequence.

Problems with plate are often a part of much larger overall stiffened or built-up structure, an aircraft wing or a ship. The mentioned elements happen to be weakened by any type of holes. The need of notching is typically actuated by weight optimisation and practical concern. There is a question about a carrying capacity of this type of structures in concrete work conditions, as well as the influence of the size and type of notch on the load capacity. In general, the analysis of composite laminated plates is more complex due to their heterogeneous and anisotropic nature. Studies on buckling analysis of plates and thin-walled structures under compressive loads are relatively pioneering as not many have been conducted, so far [1,2,3,4].

The buckling problem of a simply supported rectangular plate subjected to an axial load was studied in 1961 by Timoshenko [5]. Azizian and Roberts [6], Rockey et al. [7], Shanmugam and Narayanan [7], and Sabir and Chow [9] used the FEM to study the buckling of perforated square plates subjected to uniaxial and/or biaxial compression. Shanmugam et al. [10] used the FEM to develop a design formula to determine the ultimate load carrying capacity of axially compressed square plates with centrally located perforations. Ouinas and Achour [11] analysed the buckling of square plates with elliptical notch, investigated the impact of the size and location of the notch, as well as the influence of unsymmetrical layer of layout on critical force. Dinesh, Kumar and Singh [12] studied the influence of boundary conditions and post-buckling behaviour of quasi-isotropic laminate [45/-45/0/90]2s with different cut shapes. Based on the literature review, the conclusion that the buckling analysis of plates with notch is significant can be drawn.

In this paper, the influence of different shapes of the notch on the buckling load of thin composite laminate plates are taken into consideration. This study also contains the impact of size of the notch and of lay-orientation on buckling behaviour of the plate. The scope of research will include issues of linear stability composite plates structures subjected to uniform compression. The research will be lead on a few plate models using numerical calculations basing on the finite element method.

2 Material and Geometric Properties

In this study, a thin rectangular plate of length A=160mm and width B=80mm (Fig.1) with notch is used. To reduce the effect of stress concentration on shorter edges of the plate, above the central notch, technological cut-outs were used. The ratio between the length a=100 and the width b=30mm of the plate notch is R=5mm, and the thickness of each composite layer equals t=0,131mm.

The laminate consists of carbon fibres as reinforcement material and epoxy as matrix material, based on the previous study [13,14,15].

* Corresponding author: k.falkowicz@pollub.pl

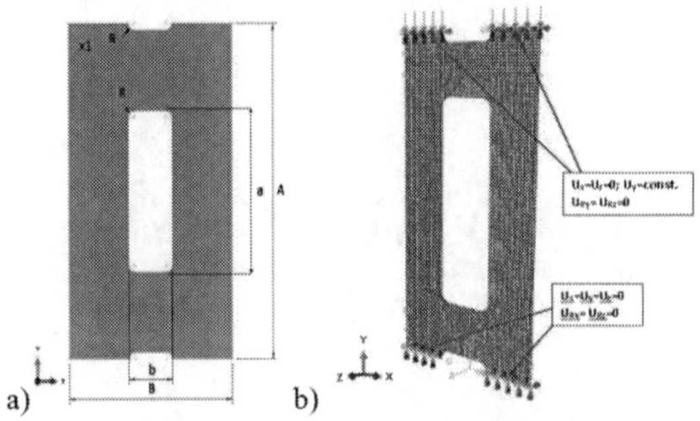

Fig. 1. Dimensions of plate containing notch (a) with FE-model of the plate (b)

The mechanical properties of used material were determined according to ISO standards and shown in the Tab.1.

Table 1. Mechanical properties of carbon-epoxy laminate.

Young's Modulus [GPa]		**Poisson's Ratio [-]**	**Shear Modulus [GPa]**
E_1	E_2	ν_{12}	G_{12}
0^0	90^0	0^0	$\pm 45^0$
131.7	6.36	0.32	4.18

A discrete model of the plate was made of eight-node shell elements with reduced integration. Known as S8R, these elements have six degrees of freedom at each node. A reduced integration allows for removal of false forms of deformation of finite elements by using higher order polynomials in the description of the function of a given element [16]. A Layup-Ply technique was used to model the laminate structure, by which using the composite layer configuration of plate thickness was mapped. Properties of the composite material were described by defining a model of orthotropic material in a flat state of stress. Boundary conditions of the numerical model were defined by locking kinematic degrees of freedom of nodes on upper and lower edges of the plate. These conditions of the numerical model, which plotted the articulated support of the plate, were defined by blocking kinematic degrees of freedom of nodes located on upper (i.e. Ux = Uz = 0 and URy = URz = 0) and lower edges of the plate (Ux = Uy = Uz = 0 and URy = URz = 0, leaving only the possibility of free rotation relative to the edge of the plate). The compression load was applied uniformly along the upper edge. In the analysed case, based on the criterion of minimum potential energy, the eigenvalue issue was solved. As a result of the calculations, the lowest form of plate stability loss and corresponding critical load values were determined. Finite element software Abaqus was preferred as numerical tool for present study.

3 Results

Detailed calculations have been made for three variants. Variant 1 dealt with the influence of composite layers layout on the critical force value. Variant 2 included the influence of the notch shape on the buckling load. Variant 3 showed how the buckling load value can be affected by changing the notch geometry. Calculations were aimed at determining the influence of the presented construction solutions on the form of buckling and the lowest critical force value, relative to the form of buckling. In all variants, the dimensions of the plate, the boundary conditions and the compression load were the same.

3.1 Influence of Lay-Orientation

To observe the buckling behaviour of composite laminate plate, subjected to axial compressive loading, the calculations were performed for composite plates made of laminate layers in a symmetrical arrangement, according to the median plane of the plate in the following variants: [90/-45/45/0]s, [45/-45/90/0]s, [0/-45/45/90]s, [45/-45/45/-45]s, [0/90/0/90]s.
The critical load values presented in the Tab. 2 for all tested laminate layers corresponded to the lowest flexural buckling form of the plate. Results showed that changing the layout of the laminate layers, in the case of the discussed configurations, resulted in changing critical buckling load about 82%.

Table 2. Buckling load results with different ply orientation

Ply orientation	**Buckling load [N]**
[90/-45/45/0]s	36.74
[45/-45/90/0]s	43.01
[0/-45/45/90]s	156.82
[45/-45/45/-45]s	31.77
[0/90/0/90]s	171.91

The highest critical stiffness had the plate of configuration [0/90/0/90]s, whereas lowest buckling stiffness was characteristic for a plate with [45 / 45-45-45]s configuration. Consequently, the layers in the 0^0 direction, parallel to the direction of the compressive load, caused the increase of stiffness of the structure, increasing the critical force of the structure.

3.2 Influence of Notch Shape

In this section the influence of different shapes of the notch are considered. It is assumed that a notch is located at the centre line of the rectangular with ply orientation [0/90/0/90]s, and subjected to uniaxial loading. The overall plate dimensions and the boundary conditions are the same as previous sections. The different shapes and their dimensions are presented in the Fig.2.

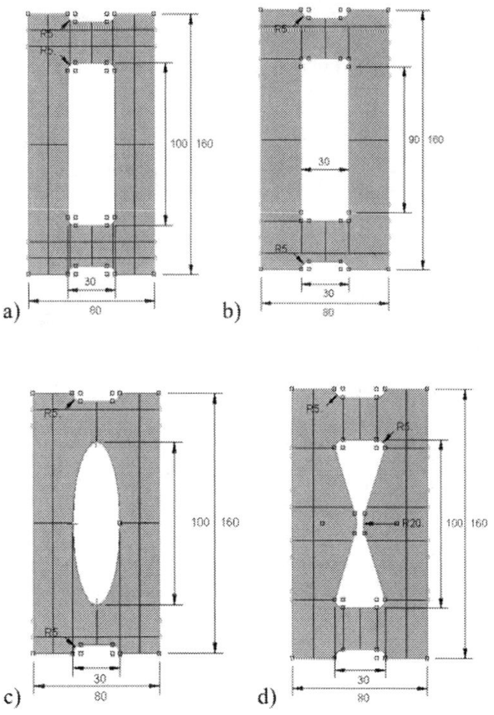

Fig. 2. Geometrical dimensions of the plate with notch: a) rectangular with sharp corners b) rectangular with rounded corners c) ellipse d) hourglass

Table 3 presents the impact of notch shapes on buckling behaviour of plates.

Table 3. Impact of notch shape on the critical value

Shape od notch	Buckling load [N]
Rectangle with sharp corners	171.51
Rectangle with rounded corners	171.91
ellipse	183.62
hourglass	204.85

Buckling load shows the highest value for hourglass notch and the lowest value for rectangular shape notch, which is a difference of 16%. The buckling load for rectangular shape notch with sharp corners and rectangular shape notch with rounded corners having nearly closer value.

3.3 Influence of Notch Size

This section deals with the buckling behaviour of perforated [0/-45/45/90]s laminated composite plate with plate's aspect ratio i.e. A/B =2, where 'A' is length and 'B' is width of plate. The notch is positioned in the centre of plate. Plates are provided with vertical rectangular shape notch, having different aspect ratio of rectangle notch a/b = 10; 5; 2.5, where a is length and b is width of rectangular notch. Buckling analysis is carried out by increasing the area of notch. Results of buckling load are depicted in the Tab.4.

Table 4. Buckling load for [0/-45/45/90] s plate with different aspect ratios of notch

a/b (a=100)	Buckling load [N]
2.5	125.53
3.3	156.82
5	224.15
a/b (a=80)	**Buckling load [N]**
2.5	160.67
3.3	188.40
5	188.40

This shows that aspect ratio of notch has some impact on buckling load. For plate with notch aspect ratio 5 provide greater value of buckling load than that of same plate but with notch aspect ratio 2.5. Results show that the buckling load increases with decreasing the notch size.

4 Conclusions

Based on present study, which dealt with the buckling behaviour of rectangular laminated composite plate under compression axial load, the following conclusion are made:

- Buckling behaviour of laminated composite plate varies for different ply orientation (stacking sequence). Changing the layout of the laminate layers in the case of considered configurations results in a change of the critical load value by 82%.
- Layers in direction 0^0, parallel to the direction of the compressive load acting, cause increasing stiffness of the structure, increasing the critical force value.
- Laminated composite plate with simply supported boundary condition, with ply orientation [0/90/0/90] s subjected to axial compression shows the highest value of buckling load for hourglass notch and the lowest value for rectangular shape notch.
- The performed analysis has shown the significant influence of the geometric parameters of the plate central notch on the buckling load value. The significant impact on the changing of the critical load value has a notch aspect ratio.

Obtained results confirm the possibility of shaping the rigidity of the thin-walled structure by changing the laminate layer configuration, geometrical parameters and shape of the central notch [8,9,10].

 DOI: 10.1051/itmconf/20171507013

References

1. M. Darvizeh et.al., Compos. Struct. **63**, 69–74 (2004).
2. M. Komur Aydin et.al., Adv. Eng. Softw. **41**, 161–164 (2010).
3. S. Wadhonkar, K. Kadam, IJIRSET **6**, 423-427 (2017).
4. D. Banat, Z. Kolakowski, R.J. Mania, Thin Wall. Struct. **107**, 335-344 (2016).
5. S.P. Timoshenko, J.M.Gere, *Theory of Elastic Stability*. 2nd ed. (New York (NY): McGraw-Hill Book Company 1961).
6. Z.G. Azizian, T.M., Proc. of the Int. Conf. on Instability and Plastic Collapse of Steel Struct., London: Granada Publishing, 392–8 (1983).
7. K.C. Rockey, R.G. Anderson, Y.K. Cheung, In: Proc. of the Int.l Conf. on Thin Wall Struct., Swansea (London), Crosby Lockwood, 148–69 (1967).
8. N.E. Shanmugam, R. Narayanan, In: Proc. of the Int. Conf. on FEM, Shanghai, August, (1982).
9. A.B. Sabir, F.Y. Chow, In: Morris LJ, editor. Proc. of the Int.Conf. on Instability and Plastic Collapse of Steel Struct., London: Granada Publishing, 311–21 (1983).
10. N.E. Shanmugam, V. Thevendran, Y.H. Tan, THIN WALL. STRUCT. **34**, 1–20 (1999).
11. D. Ouinas, B. Achour, Comp. Part B **55**, 575-579 (2013).
12. K. Dinesh, S.B. Singh, Compos. Struct. **92**, 769-779 (2010).
13. K. Falkowicz, H. Debski H., Adv. Sci. Technol. Res. J. **11**, 186-193 (2017).
14. K. Falkowicz., P. Mazurek P. Rozylo, P.Wysmulski, W. Smagowski, Adv. Sci. Technol. Res. J. **10**, 177-184 (2016).
15. P. Rozylo, A. Teter, H. Debski, P. Wysmulski, K. Falkowicz, Appl Compos Mater, **24**, 1251-1264 (2017).
16. O.C. Zienkiewicz, R.L. Taylor., *Finite Element Method* (5th Edition), (Vol.**2** - Solid Mechanics, Elsevier, 2000).

DOI: 10.1051/itmconf/20171507014

The analysis of distributions of effective strain and flow stress in longitudinal sections of cold backward extruded copper cans for different punch-face shapes

Tomasz Miłek[1,*]

[1]Kielce University of Technology, Faculty of Mechatronics and Mechanical Engineering, Department of Applied Computer Science and Armament Engineering, al. Tysiąclecia Państwa Polskiego 7, 25-314 Kielce, Poland, UE

Abstract. The paper presents computer modelling results of researches on cold backward extrusion of copper cans. The calculations were carried out using the commercial code QFORM-2D, based on the Finite Element Method (FEM). The simulation of cold backward extrusion process was performed for different punch-face shapes (flat; flat and conical with conical angle 90^0 and 150^0; as well as concave). On the basis of obtained results, the analysis of distributions of effective strain and flow stress in longitudinal sections of cold backward extruded copper cans was conducted.

1 Introduction

The process of backward extrusion of cans still plays an important role in the manufacturing industries [1]. Copper cans are traditionally produced in multi-stage deep drawing processes which, however, have some drawbacks, namely the design cost, material waste and inconsistent wall thickness [2,3]. Backward extrusion could be considered an alternative manufacturing process to produce cans [4]. When compared with other manufacturing processes, backward extrusion offers many advantages, including lower material consumption, higher dimensional accuracy and surface quality, adequate mechanical and microstructural properties and reduction or complete elimination of machining. Products after extrusion have better mechanical properties than those of the original material due to favourable grain flow [5-7]. Although backward extrusion has significant capabilities in production, it also shows some limitations. The unsteady deformation zone, which causes different strain distribution through the extruded part, is one of the problems [5].

The backward extrusion process, in which metal flows in the opposite direction to that of the punch movement, is relatively more energetically efficient. In the process, friction is considerably reduced, because the friction along the chamber walls does not need to be overcome. The material coming through the orifice formed by the punch land and the die wall undergoes no strain after this point. The remaining volume between the punch and the die bottom (or the ejector) is only partly located in the deformation zone. Depending upon the geometrical and friction conditions, a dome-shaped rigid plastic zone is formed in this region. The factors that affect the zone mentioned above include the can bottom thickness and the reduction in the area [2].

The deformation ratios of material in the literature [2,3,8] was defined as relative: strain of can bottom thickness ε_h, reduction in the area ε_A and equivalent strain ε. The strain ε_h, is calculated according to the following formula [2,3]:

$$\varepsilon_h = \frac{\Delta h}{h_o} = \frac{h_0 - h_1}{h_0} \tag{1}$$

where:
Δh – the punch displacement in mm,
h_0 – billet height in mm,
h_1 – thickness of cup bottom in mm,
The reduction in the area ε_A can be estimated from the relationship given below [2,3]:

$$\varepsilon_A = \frac{A_0 - A_z}{A_0} = \frac{d_0^2 - d_z^2}{d_0^2} \tag{2}$$

where:
A_0 – cross sectional area of the billet in mm^2,
A_1 - cross sectional area of the die stamping in mm^2,
d_0 – diameter of billet in mm,
d_s – diameter of punch in mm,

The unit operation of work of plastic deformation in industrial practice is determined by the homogeneous equivalent strain ε. It is given by a formula [8]:

$$\varepsilon = \ln \frac{A_0}{A_1} \tag{3}$$

1.1 State of the art

Cold backward extrusion of circular-shaped parts made from copper and aluminium and the design of the tooling have been reported in some studies [1,4,5,7,9-20]. Those covered both experimental and computer modelling investigations.

* Corresponding author: tmatm@tu.kielce.pl

DOI: 10.1051/itmconf/20171507014

Lee and Kwan [9] proposed a kinematically modified admissible velocity field for the backward extrusion of internally circular-shaped tubes from arbitrarily shaped billets. From the proposed velocity field, the upper-bound extrusion load and average extruded height for regular polygonal-shaped billets were determined with respect to the chosen parameters.

Zasadziński [10] evaluated the usefulness of different methods for continuous extrusion of aluminium and its alloys, based on the analysis of active friction forces possible to achieve in these methods. The Linex, Extrolling, Conform processes, as well as two methods proposed by the author were analysed.

Y.H. Kim and Park [11] studied backward extrusion process with low die rotation. The objective of investigations was to address problems related to conventional backward extrusion process, namely the necessity of using a large forming machine, the difficulty in selecting the die material due to high surface pressure, costs related to the reduction in noise and vibration generated from the forming machine, and others. They analysed experimental results for torsional and conventional backward extrusions. Results were compared using two methods, i.e. upper bound technique for computing the velocity field, and FEM simulation with DEFORM-3D.

Thomas [12] conducted a series of investigations into cold backward extrusion of cans for copper before and after heat treatment. He analysed the experimental change of the force for copper cans at the reduction in area ε_A=0,360-0,639.

Żmudzki A. et al. [13] presented numerical (FEM) and experimental results of the analysis of tests, which were performed to determine the friction coefficient in two types of metal forming processes (ring compression and combined forward-backward extrusion). Experiments were performed for copper deformed at room temperature. Żmudzki A. et al. demonstrated that friction coefficient calculated from the ring compression tests was slightly lower that determined from the direct-indirect extrusion test.

In their study, Shatermashhadi V. et al. [1], proposed a method of backward extrusion using small diameter billet. Their die setup consisted of three major components, namely the fix-punch, the moveable punch and the matrix. They demonstrated that the load was reduced to about less than a quarter when compared with the conventional backward extrusion process. Shatermashhadi V. et al. also conducted numerical investigations, in which the DEFORM 3D software was used for FE simulations. In their another work [7], they discussed the applications of the modified process to commercially pure aluminium.

Kim et al. [4] evaluated the effects of lubricants in backward extrusion of a rectangular aluminium case with large aspect ratio. The analysis of backward extrusion of a rectangular aluminium box for electrical battery casing was performed using MSC SuperForge, commercial finite volume software. The numerical analysis showed clearly that a friction factor of 0.2 yields an optimal forming shape, which was independently confirmed by experimental results.

Farhoumand and Ebrahimi [5] examined the effects of geometrical parameters including die corner radius and gap height, and of the process conditions, such as friction, on the radial-backward extrusion process.

In Wang et al. [14], recent developments in the friction testing techniques for aluminium extrusion processes were discussed and detailed comparisons of these techniques were made.

The most papers are concerned with the analysis on the forming of copper in micro-backward can extrusion process [15-20]. Ch. Chang et al. [15] discussed the effects of temperature and grain size on the deformation, dimension variation, and change of microstructure of copper from combined backward and forward extrusion at the micro scale. They conducted a series of investigations into extrusion of copper micro cups with 0.1 mm thickness. Three forming temperatures: 25, 200 and 400 °C, were considered in the micro extrusion experiments. Their studies show that the grain refinement improves material flow and thus lead to a better die filling with less variation in the rim height and wall thickness of the cup portion of the extruded part at relatively lower forming temperature.

Wang et al. [16] demonstrated that forming process of dispersion strengthened copper welding electrode consists of a forward extrusion and a backward extrusion. They analysed the characteristics of metal flow and the effect of different friction factors. Wang et al. [16] carried out the simulation of the upsetting-extruding process using Deform-2D finite element analysis software.

Geisdorfer et al. [17] investigated a possibility of using an ultrafine grained copper for micro-extrusion. The microforming process of backward extrusion was carried out at room temperature using half cylindrical billets. The extrusion force, grain flow, shape representation and surface quality of the extruded micro-components were compared.

Chan et al. [18] studied a size effect in micro-extrusion process of pure copper. The size effect on material deformation behaviours was characterised by grain size, part feature size, forming material size and interfacial condition. Chan et al. [18] performed research on micro-forward, backward, combined forward rod-backward can and double cup extrusions.

Bazaz et al. [19] evaluated the microstructure evolution of a pure copper processed by accumulative back extrusion (ABE) method at room temperature. The analysis of microstructure and hardness showed outstanding homogeneity improvement throughout the workpieces.

L. Yi et al. [20] proposed theoretical model to predict the backward extrusion force of copper-chromium alloy based on Projection Pursuit Regression (PPR).

1.2 The aim of investigations

The paper presents computer modelling (FEM) results of investigations on cold backward extrusion of copper cans. The different punch-face shapes were used for cold extrusion (flat; flat and conical with conical angle 90^0

DOI: 10.1051/itmconf/20171507014

and 150^0 and concave). Based on obtained results at relative strains ε_h=0,69 and ε_A= 0.67 and ε= 1.1, the analysis of distributions of effective strain and flow stress in longitudinal sections of cold backward extruded copper cans was conducted. Pure copper was selected as the testing material in these investigations due to its excellent formability and wide applications in industry.

2 Models and assumptions in numerical modelling with QFORM-2D

Computer modelling of metal forming processes is nowadays widely used for optimisation and shape prediction of newly designed parts during their manufacturing. At present there are many FEM systems to numerically investigate metal forming processes. Programs related to issues of forging and extrusion include: QForm MSC Superform, Deform, Forge, MSC Marc. QForm-2D/3D (Quantor-Form Ltd., Russia) uses the flow formulation for material behaviour. The estimation of the QForm simulation results was carried out in several works [21-24]. The analysis of these papers shows the good compatibility between the results of numerical simulation and experimental data.

Calculations of metal flow and study distributions of effective strain, flow stress and changes in loading were carried out with commercial code QFORM2D based on Finite Element Method (FEM).

The extruded material was incompressible rigid-plastic continuum and elastic deformations were neglected. The system of governing equations included the following [22,25]:

- equilibrium equations

$$\sigma_{ij,j} = 0 \tag{4}$$

- compatibility conditions

$$\dot{\varepsilon}_{ij} = \frac{1}{2}(v_{i,j} + v_{j,i}) \tag{5}$$

- constitutive equations

$$\sigma'_{ij} = \frac{2\bar{\sigma}}{3\dot{\bar{\varepsilon}}}\dot{\varepsilon}_{ij} \tag{6}$$

- incompressibility equation

$$v_{i,i} = 0 \tag{7}$$

- expression for flow stress

$$\bar{\sigma} = \bar{\sigma}(\bar{\varepsilon}, \dot{\bar{\varepsilon}}, T) \tag{8}$$

where σ_{ij} and $\dot{\varepsilon}_{ij}$ – components of stress and strain-rate tensors, v_i – velocity components, σ'_{ij} – deviatoric stress tensor, $\bar{\sigma}, \bar{\varepsilon}, \dot{\bar{\varepsilon}}$ – effective stress, strain and strain-rate, respectively, T – temperature.

In Eqs 4-8, summation convention was used. The prime denoted a derivative with respect to the axis following it. The indexes i and j for two-dimensional problems varied from 1 to 2, and repeated subscript represented summation.

The friction model proposed by Levanov et. al [22,25] was used for the contact region of workpiece surface. Eq. (9) could be considered as a combination of the constant friction model and the Coulomb friction model. The formula combined the advantages of both models [22,25]:

$$F_t = m\frac{\bar{\sigma}}{3}(1 - \exp(-1{,}25\frac{\sigma_n}{\bar{\sigma}})) \tag{9}$$

where m was the friction factor, σ_n was the normal contact pressure.

The flow stress characteristics were given as a function of the strain rate:

$$\bar{\sigma} = C\bar{\varepsilon}^n \tag{10}$$

Based on literature data [26] Eq. (10) was adopted for numerical modelling of cold backward extrusion of copper cans:

$$\bar{\sigma} = 420\bar{\varepsilon}^{0.3} \tag{11}$$

The investigations into backward extrusion involved the use of circular sectioned copper segments of rods with diameter d_0 = 24.5 mm and height h_0 = 16 mm (h_0/d_0=0.65). A tool for backward extrusion was equipped with replaceable punches with different punch shapes and constant diameter d_s= 20mm (Fig. 1). Initial workpiece and tools temperatures were 20^0C and 10^0C, respectively.

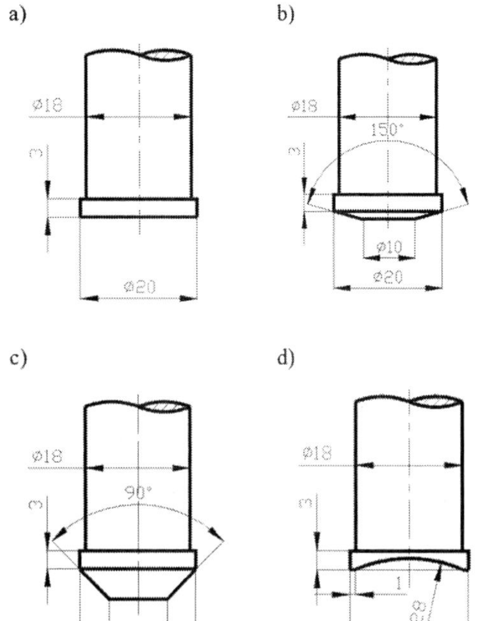

Fig. 1. The different punch-face shapes used in experiment and modelling of cold backward extrusion of copper cans: a) flat; b) flat and conical with a conical angle 150^0; c) flat and conical with a conical angle 90^0 d) concave.

DOI: 10.1051/itmconf/20171507014

3 Results and analysis

The numerical investigations produced copper cans in cold backward extrusion with strain of can bottom thickness ε_h=0.69 and reduction of the area ε_A= 0.67 and homogeneous equivalent strain ε=1.1 for different punch-face shapes (flat; flat and conical with conical angle 90^0 and 150^0 and concave). Results of the simulation process showed that the model of boundary conditions, presented in previous chapter, proved adequate. To analyse metal flow in the computer program, the flow lines were imposed. They formed a grid that made it possible to view the displacement and distortion of the metal selected volumes in deformations. In the simulation, ten inner flow lines along the OX and OY axes were assumed. Numerically calculated last stages of backward extrusion for different punch-face shapes are presented in the Fig. 2.

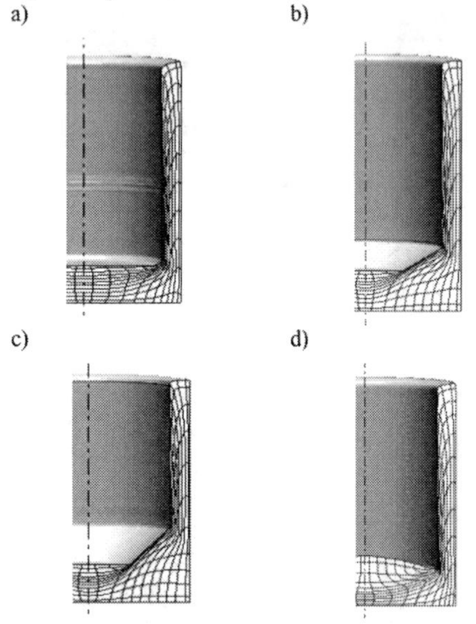

Fig. 2. Numerically calculated last stages a.- d. (which corresponded to equivalent strain ε=1.1 and ε_A= 0,67) of backward extrusion of copper can for different punch shapes: a) flat, b) flat and conical flat with conical angle 150^0, c) flat and conical with conical angle 90^0, d) concave.

The analysis of the flow lines confirmed the literature conclusions [2,3,8,23,27] concerning the occurrence of characteristic areas in the cross-section of extruded stampings at flat punch-face shape. The flow lines below the punch exhibit a curvature about the axis and the streamlines are curved toward the outside [3]. There are four zones of deformation in the cross-section of cans [2,3,8,23,27]. The lowest hardening area is located outside of the bottom part and the upper part of can. The intermediate area includes the inner part of the bottom and the zone adjacent to the outer wall of the product. The strongest hardening area occurs at the inner surface of the can wall. The described deformation zones have also been observed in investigations of cold backward extrusion of copper cans for different punch-face shapes (Fig. 2a-d). The most deformed flow lines were observed in longitudinal sections of cold backward extruded copper cans at flat punch-face and flat and conical punch-face shapes with conical angle 90^0 (Fig. 2a and Fig. 2b).

The analysis of the process was conducted based on results of numerically computed effective strain and flow stress distributions at intersections of backward extruded die stampings, too (Fig. 3 and Fig. 4).

Based on the analysis of numerically calculated effective strain distribution (shown in the Fig. 3) at the intersection of backward extruded cans made from copper, it can be stated that in the different stages of the modelling, the maximum effective strain values were found in the die stamping inner wall forming and area of the inner radii of the bottom transition to the wall. The maximum value of the effective strain was found in the last stage of simulation of cold backward extrusion for copper can extruded with concave punch shape. It occurred in inner walls and reached the value of 5.652 (Fig. 3d). In all stages of simulation of cold backward extrusion of copper cans for concave punch shape considered values of the effective strain were higher than values obtained for extruded die stampings by using other punch shapes (flat; flat and conical with conical angle 90^0 and 150^0 and concave, respectively). In cold backward extrusion of copper cans with strain of can bottom thickness ε_h = 0.69 and reduction of the area ε_A = 0.67 and homogeneous equivalent strain ε=1.1, maximum value of the effective strain obtained in longitudinal section extruded die stamping for flat and conical punch-face shape with conical angle 150^0 (Fig. 3b) did not differ much from maximum value of the effective strain for can formed at conical punch-face shape (Fig. 3c) with conical angle 90^0 (3.919 and 3.870, respectively). The maximum value of the effective strain obtained for copper can with flat punch-face shape (3.306) was smaller by approx. 41 % than the value obtained for die stamping with concave punch-face shape (Fig. 3d). The lowest values of effective strain occurred slightly outside the projection of the punch edges on the bottom surface for all types of punches (Fig. 3a-d).

Numerically obtained flow stress distributions in longitudinal sections of cold backward extruded copper cans for different punch shape are presented in the Fig. 4a-d. No significant differences were found while analysing the maximum values of flow stress for cold backward extrusion of copper cans for different punch-face shape. The highest value of the flow stress was noted in last stages of simulation for all punch shapes. It amounted to 420MPa (Fig. 4a-d). The main differences were observed in changes of distribution of flow stress in intersection of cold backward extruded cans. The area of maximum flow stresses occurred in inner walls of copper cans in all cases of punch shapes (Fig. 4a-d) and in the bottom of die stampings for backward extrusion at flat punch-face (Fig. 4a). The zone of relative lower flow stresses was observed outside of the bottom part (corners of the die) for extrusion at flat and conical punch-face shapes with conical angle 90^0 and 150^0 (Fig. 4b and Fig. 4c). Besides, it was located the inner part of the bottom for die stamping with concave punch-face shape (Fig. 4d).

DOI: 10.1051/itmconf/20171507014

a)

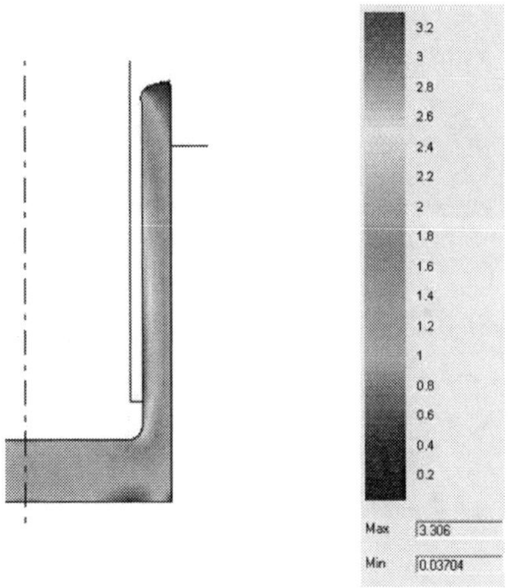

b)

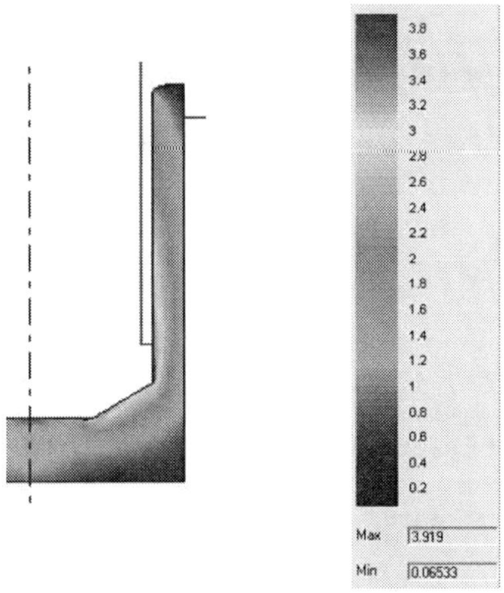

c)

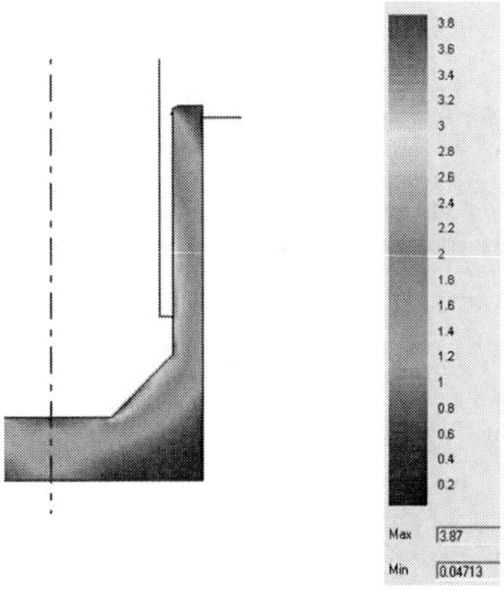

d)

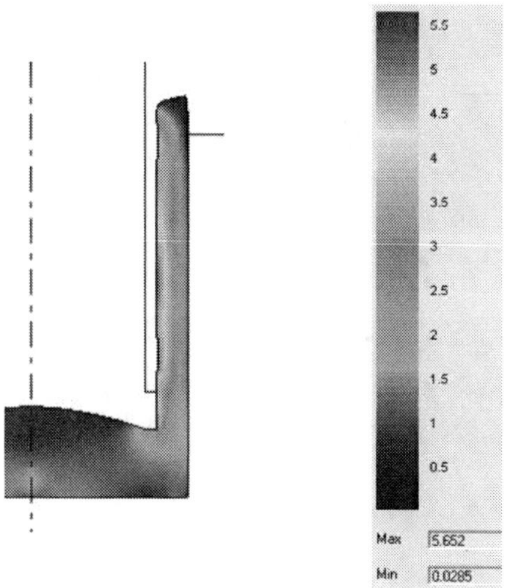

Fig. 3. Numerically computed effective strain distribution at the intersection of backward extruded can for the last stage of computer modelling ($\varepsilon_h = 0.69$; $\varepsilon_A = 0.67$; $\varepsilon = 1.1$) and for different punch shapes: a) flat, b) flat and conical flat with conical angle 150^0, c) flat and conical with conical angle 90^0, d) concave

a)

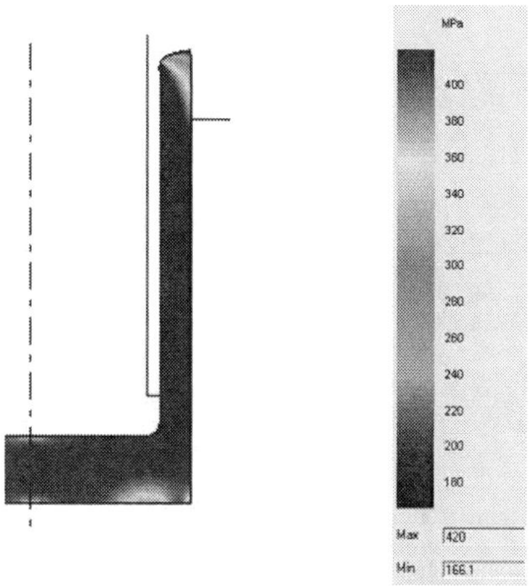

b)

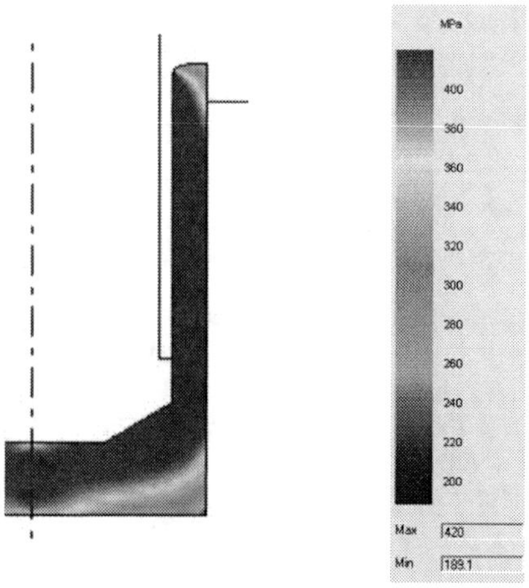

DOI: 10.1051/itmconf/20171507014

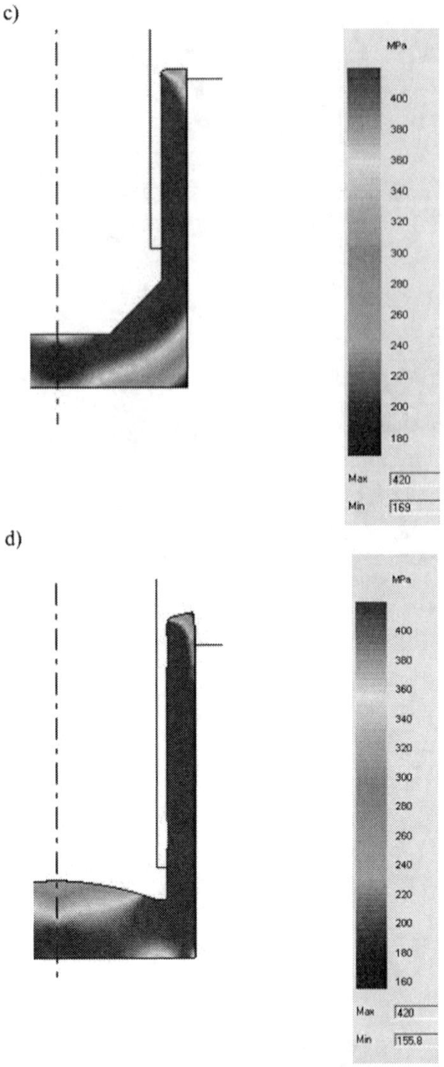

Fig. 4. Numerically computed flow stress distribution at the intersection of backward extruded can for the last stage of computer modelling ($\varepsilon_h = 0.69$; $\varepsilon_A = 0.67$; $\varepsilon = 1.1$) and for different punch shapes: a) flat, b) flat and conical flat with conical angle 150^0, c) flat and conical with conical angle 90^0, d) concave

Results of calculations of cold backward extrusion of copper cans process were validated against experimental data in terms of changes in forces for different punch shape. For the assumed maximum punch displacement Δh=11mm (which corresponded to the ε_h = 0.69; ε_A = 0.67; ε = 1.1) successful experimental tests were conducted (Fig. 5). The 2D FE models successfully described the backward extrusion of copper cans. Comparing changes in forces in cold backward extrusion for different punch shape, a conclusion that the load P_w increases with an increase in displacement Δh and strain of can bottom thickness $\varepsilon_{h\ can}$ [23].

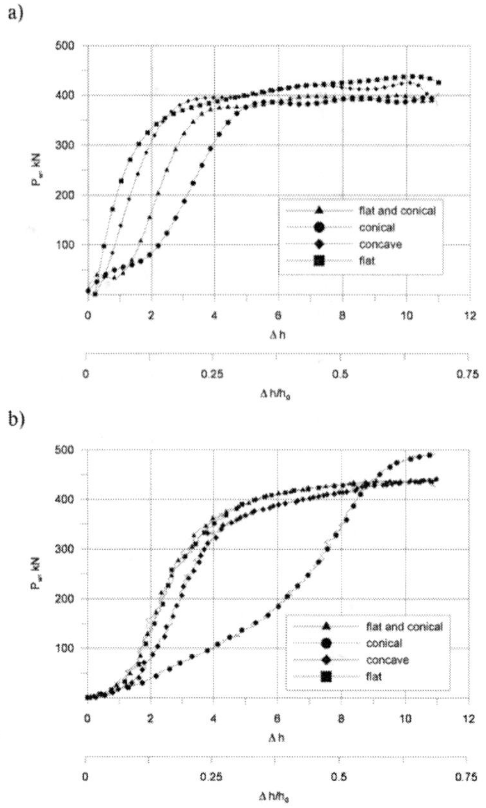

Fig. 5 The influence of punch shape on the cold backward extrusion force for copper cans at ε_A= 0,67 and ε= 1,1: a) numerical results, b) experimental investigations [23].

4 Summary

Although commercial package of QForm2D has been especially designed for numerical simulation of forging processes, the 2D FE model successfully described the backward extrusion of copper cans. It was possible to conduct cold backward extrusion of copper cans with strain of can bottom thickness ε_h=0.69 and reduction of area ε_A= 0.67 and homogeneous equivalent strain ε=1.1 for different punch-face shapes (flat; flat and conical with conical angle 90^0 and 150^0 and concave). It is confirmed by results of successfully performed computer modelling in these paper and experimental tests presented in previous author's studies [23]. The maximum value of the effective strain obtained for copper can with concave punch-face shape was higher by approx. 70 % than the value obtained for die stamping with flat punch-face shape. No significant differences were found while analysing the maximum values of flow stress for cold backward extrusion of copper cans with for different punch-face shape. The highest value of the flow stress was noted in last stages of simulation for all punch shapes.

DOI: 10.1051/itmconf/20171507014

References

1. V. Shatermashhadi, B. Manafi, K. Abrinia, G. Faraji, M. Sanei, Mater. Des. **62**, 361 (2014).
2. K. Lange, *Handbook of metal forming* (McGraw-Hill, Inc. USA,1985).
3. W. A. Golowin, A. N. Mitkin, A. G. Rieznikow, *Cold extrusion of metals* (WNT Warszawa, 1975).
4. S.H. Kim, S.W. Chung, S. Padmanaban, J. Mater. Process. Technol. **180**, 185 (2006).
5. A. Farhoumand, R. Ebrahimi, Mater. Des. **30**, 2152 (2009).
6. M. Plancak, M. Brameley, A. Osman F. J. Mater. Process. Technol. **34**, 465 (1992).
7. S.H. Hosseini, K. Abrinia, G. Faraji, Mater. Des. **65**, 521 (2015).
8. K. Kuczyński, E. Erbel, *Metal forming – lab* (Wyd. Polit. Warszaw., Poland, 1984).
9. R.S. Lee, C.T. Kwan, J. Mater. Process. Technol. **59**, 351 (1996).
10. J. Zasadziński, Rudy i Met. Nieżelazne **47**, 26 (2002).
11. Y.H. Kim, J.H. Park, J. Mater. Process. Technol. **143-144**, 735 (2003).
12. P. Thomas, Maint. Reliab. **2**, 63 (2003).
13. A. Żmudzki, R. Kuziak, M. Papaj, M. Pietrzyk, Obr. Plast. Met. **3**, 69 (2004).
14. L. Wang, J. Zhou, J. Duszczyk, L. Katgerman, Tribol. Int. **56**, 89 (2012).
15. Ch. Chang, J. Lin, Ch.-P. Siao, Steel Res. Int., spec. edit.: 14th Int. Conf. Met. Form. 467 (2012).
16. M. Wang, Y. Zhang, D. Huang, X. Liu, Trans. Nonferrous Met. Soc. China **17**, 449 (2007).
17. S. Geisdorfer, A. Rosochowski, L. Olejnik, U. Engel, M. Richert, Int. J. Mater. Form. **1**, 455 (2008).
18. W. Chan, M. W. Fu, B. Yang, Mater. Des. **32**, 3772 (2011).
19. B. Bazaz, A. Zarei-Hanzaki, S. Fatemi-Varzaneh, Material Science and Engineering A **559**, 595 (2013).
20. L. Yi, Y. Dong, M. Lin, L. Zhang, X. Wang, 2013 Int. Conf. on QR2MSE, 1181 (2013).
21. T. Miłek, B. Kowalik, B. Kuliński, Arch. Metall. Mater. **4**, 3043 (2015).
22. N. Biba, S. Stebunov, A. Vlasov, Steel Res. Int. **79**, 611 (2008).
23. T. Miłek, Mechanik **5/6**, 536 (2016).
24. T.Miłek, Int. Conf. Metal 2014, 285 (2014).
25. QFORM 2D/3D *Metal forming simulation program ver. 4.2. 2D simulation* (User's Guide, QuantorForm Ltd. 2007).
26. A. Mazurkiewicz, L. Kocur, *Metal forming – laboratory* (Wyd. Polit. Radom. 2001).
27. W. Muzykiewicz, A. Rękas, Rudy i Met. Nieżelazne **49**/10-11, 542 (2004).

 DOI: 10.1051/itmconf/20171507015

The design and structural analysis of the endoprosthesis of the shoulder joint

Robert Karpiński[1], *Łukasz* Jaworski[2], Jarosław Zubrzycki[1,*]

[1]Lublin University of Technology, Faculty of Mechanical Engineering, Department of Machine Design and Mechatronics, Nadbystrzycka 36, 20-618, Lublin, Poland

[2] Poznan University of Technology, Faculty of Mechanical Engineering and Management, 60-965 Poznań, Poland

Abstract. The shoulder joint is a crucial element of the upper limb and is necessary to maintain full mobility during daily activities. Similar as in the hip, it is an example of a ball and socket joint, enabling articulation between the head of the humerus and the glenoid cavity of the scapula. By studying kinematics of upper limbs, it is possible to distinct several possible movements in the shoulder joint. Due to many painful diseases and medical conditions, it may be crucial to perform a pain reducing procedure, like shoulder joint replacement. In order to preserve the function of the joint, the endoprosthesis should be designed exclusively for the patient. To assess a scale of damage in the joint or its specific structure, the shoulder was scanned using the computed tomography procedure. Results of the scan were processed with the use of Materialise Mimics software, which converts standard 2D images into 3D CAD models. Necessary analysis and measurements were taken leading to the beginning of the designing process. The prosthesis was created using Solid Edge software, developed especially for the purpose of rapid prototyping. After determining the physical properties of structural materials, the Finite Elements Analysis of the model was conducted using SolidWorks Simulation software under various load conditions.

1. Introduction

The shoulder joint is a crucial element of the upper limb and is necessary to maintain full mobility during daily activities. Unfortunately, due to pathological changes in the joint, it is often partially or completely impossible to fully exploit its capabilities. The ageing of the population, as a cause of growing demand for medical services dedicated to older people, poses a challenge to engineers trying to design solutions fully corresponding with the demand from both patients and physicians, especially in orthopaedics, and endoprosthesis in particular.

2. The anatomy of a shoulder joint

A shoulder girdle is connected with the upper limb (arm) via the shoulder joint (or glenohumeral joint - articulatio humeri) shown on Figure 1. The glenohumeral joint is a multiaxial, synovial ball and socket joint, one of the largest joints of this type in the human body. It guarantees movements such as flexion and extension, abduction and adduction, lateral and medial rotation, horizontal abduction and adduction, and a combination of mentioned movements called circumdiction. The articular surfaces of the joint consist of the head of the humerus and the glenoid cavity of the scapula. The head of the humerus is a section of a ball, and its surface is almost four times larger than the surface of an acetabulum (socket) of the joint. The surface of the socket is slightly recessed, oval-shaped and perpendicular to the surface of a scapula. Articular surfaces are covered with hyaline cartilage, which on the peripheries of the socket transforms into a few (up to 6) millimetres thick fibrocartilaginous rim called a glenoid labrum. The articular cartilage transfers and spreads loads between bones, while preserving appropriate stress distribution on the surface of the joint, reduces friction, absorbs sudden overstrains and protects bones from surface wear. Fully developed cartilage does not contain blood vessels or nerves. A very loose connection between the socket and the ball, along with a strong and large articular capsule, guarantees great mobility in the joint [1, 2, 3, 4, 5].

Beside muscles and tendons strengthening the joint capsule, there are also articular ligaments stabilizing the shoulder joint, such as coracohumeral ligament, glenohumeral ligaments and coraco-acromial ligament [1, 2, 3].

Coracohumeral ligament is the strongest ligament in the shoulder joint, and is located on the upper surface of the joint, between a coracoid process and the lesser and greater tuberosities of the humerus. Glenohumeral ligaments are located on the anterior side of the glenohumeral joint; the coraco-acromial ligament does not directly participate in strengthening the joint, but plays a crucial part in its biomechanics and protection [1, 2].

* Corresponding author: robert.karpinski@pollub.edu.pl

DOI: 10.1051/itmconf/20171507015

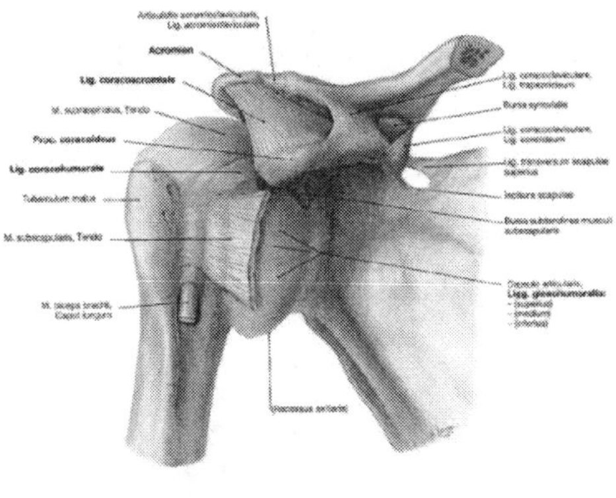

Fig. 1. Right shoulder joint (front view) [2].

3. Replacement of the shoulder joint

Alloplasty is a procedure that consists of implementing elements acting as a replacement for damaged parts of a joint into the human body, which can reinstate lost functions to these parts of the body. An endoprosthesis is an implant made of biocompatible materials, which acts as a replacement for damaged joints. Alloplasty of the shoulder joints, thanks to the implementation of the endoprosthesis in place of the defective joint, allows radically reducing pathological changes and pain, as well as recreating functions of the shoulder [4, 5, 6].

Endoprosthetics of the shoulder joint, or the shoulder joint replacement, is mainly performed on patients with osteoarthritis. Osteoarthritis is a group of overlapping medical conditions, which in spite of having different etiologies; cause similar biological, morphological and medical effects. A disease process affects not only the articular cartilage, but also the joint itself, including the subchondral bone, ligaments, the joint capsule, a synovial membrane and muscles around the joint. The osteoarthritis is one of the most frequently occurring diseases of the motion system, especially with elder people. It is often caused by biological and mechanical occurrences disordering a synthesis and triggering the defeneration of the articular cartilage, in chondrocytes and an extracellular matrix as well as in the subchondral layer of the bone [5, 6, 7, 8].

There are several conditions causing limitations in the mobility of the shoulder joint, including [7, 9]:

• changes in an acromioclavicular and a sternoclavicular joint, effecting in the limitation of movements of the scapula and a collarbone,
• inflammation of synovial bursae,
• weakening of muscles of the shoulder girdle,
• rotator cuff damage,
• increase of muscle tension, due to the pain in particular,
• constriction of the articular capsule,
• articular cartilage damage, effecting in a deficiency of a slide movement in the joint.

4. The design of the endoprosthesis of the shoulder joint

4.1 Software tools used in the study

Materialise Mimics allows the user to process and edit 2D images from CT or MRI scans in order to generate highly accurate 3D models. Moreover, it is possible to make precise measurements (dimensions, density) of bones, tissues or organs. The patient-specific models can be implemented in a variety of engineering applications directly in Materialise Mimics or Materialise 3-matic, or exported to 3D models and anatomical landmark points to third party software, like statistical, CAD, or FEA packages [10].

Solid Edge ST is a complex, parametric, hybrid 2D/3D CAD system using synchronous technology. Implementation of two highly-efficient graphic modellers - Parasolid and D-Cubed allowed to combine direct modelling with precise control of geometry, which gives engineers the opportunity to conduct the process of designing with speed and simplicity on a level that has never been seen before. It provides tools for solid modelling, assembly modelling and 2D orthographic view functionality for mechanical designers. Through third party applications it has links to many other Product Lifecycle Management technologies [10].

4.2 The design process

The alloplasty of each ball and socket joint, such as the shoulder joint, requires the best possible fitting of the implant within anatomical structures of the patient's body. Due to the increasing precision of medical imaging methods, processing of given data and many features of the CAD software, it is possible to design the prosthetic implant individually fitted for the patient.

Probably the most important step in a process of designing the endoprosthesis is obtaining measurements of structures that are going to be replaced. Based on 3D models of a patient's shoulder girdle, it was possible to find approximate measurements of the glenoid cavity of the scapula (length, width – Fig. 2) and the head of the humerus (diameter, angle between the axis of the head and axis of the shaft etc. – Fig. 3). Additionally, a parameter was determined on how much of the glenoid cavity could be excised to properly mount the implant.

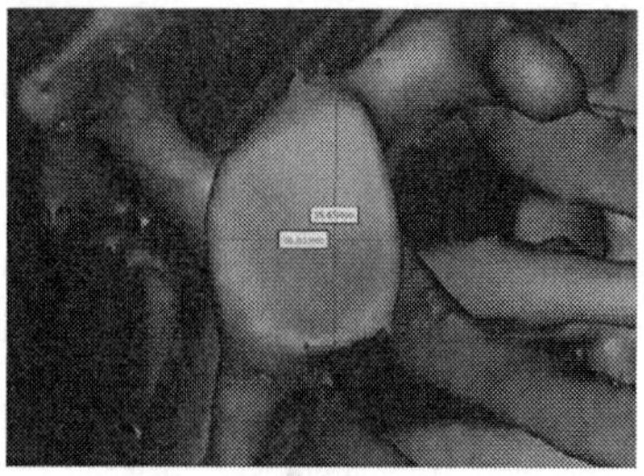

Fig. 2. Measurements of the glenoid cavity of the scapula.

 DOI: 10.1051/itmconf/20171507015

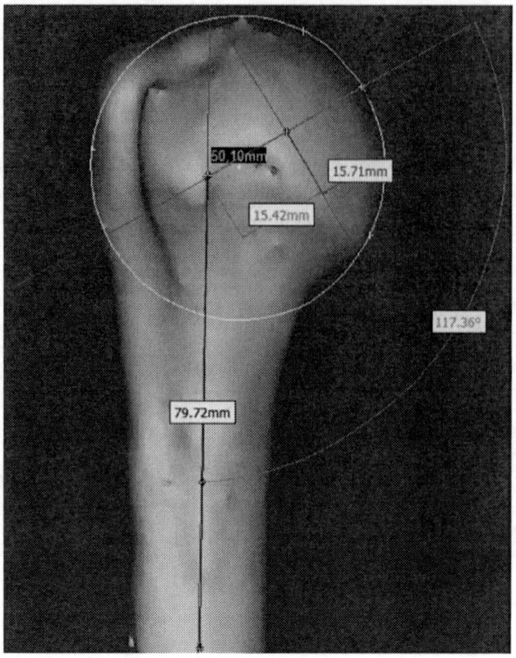

Fig. 3. Measurements of the humerus.

These values became a base for designing actual elements: a shell – replacing the glenoid cavity, a cup – acting as the articular cartilage, and the head on the shaft – substituting the top part of the humerus.

The shell and the cup (Fig. 4) are shaped like a section of a sphere. The shell is supposed to be mounted to the dissected articular cavity using a ø 4.5 mm orthopaedic screw and is stabilized by two pins. The cup is designed to tightly fit the shell. Its inner diameter (ID) is 45 mm while the outer diameter (OD) is 60 mm, which leads to an overall thickness of the set equal to 7.5 mm. The length and the width are respectively 35 mm and 25 mm, equivalent to measured values.

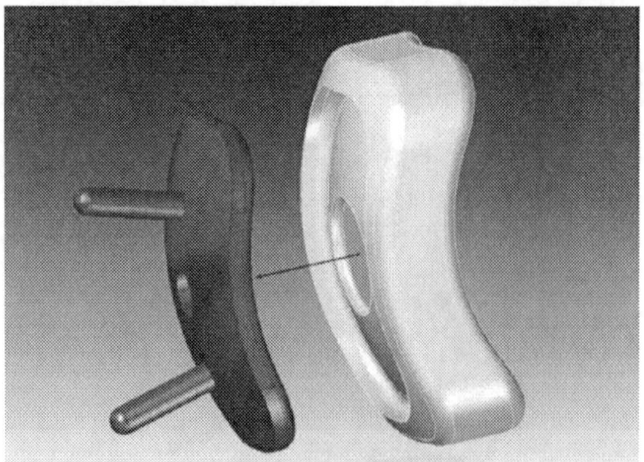

Fig. 4. The shell (left) and the cup (right) of the endoprosthesis.

The humerus' part of the endoprosthesis (Fig. 5) consists of the spherical cap shaped head and the shaft. The diameter of the head matches inner diameter of the cup, making it possible to perform expected movements. The shaft (or stem) is the element directly inserted into the humerus, hence the specific shape tampering to the downside with additional protrusions on the surface to ensure stability of the stem and maximize the process of bone healing, considering that the planned operation is going to be performed without the use of bone cement. The head is tightly mounted on the neck of the shaft with a mounting hole and is prevented from rotating with a protrusion on the neck matching an indentation in the mounting hole.

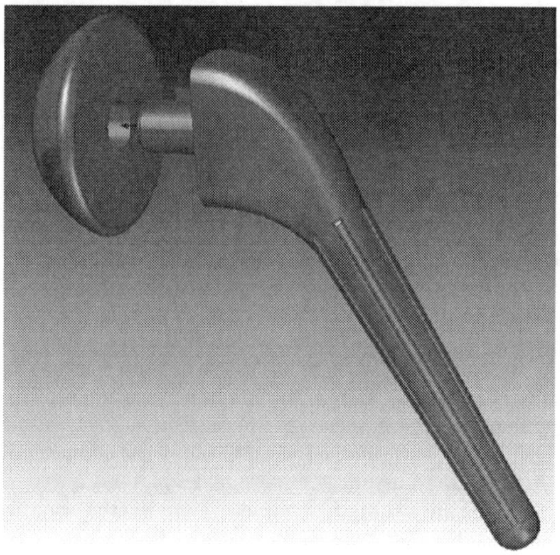

Fig. 5. The head and the shaft of the endoprosthesis.

5. Study on stress distribution in the endoprosthesis

The designed model was used to perform a series of preliminary studies including stress distribution in the endoprosthesis with the use of Finite Element Analysis method.

5.1 FEA method

Finite Element Analysis is one of the basic methods of conducting computer aided engineering calculations. It is one of the techniques of discretization of geometric systems, i.e. dividing a continuum into a finite amount of subareas.

The main principle of FEA is to divide geometric models into elements uniting in nodes, which effects in creating a discrete geometric model, split in simply shaped subareas, called the finite elements. During performing of calculations with the use of FEA, other physical quantities are being discretized, like loads, tensions, restraints or other examples represented in the system with the use of continuous function. While performing the process of discretization the software aims at maximally approximation of discreet and continuous form using approximation methods. After converting the data analysis follows, consisting in uniting individual elements as a whole using equilibrium conditions and displacement compatibilities, which results in receiving a set of algebraic, simultaneous equations, posing as a mathematical description of the analysed problem. Afterwards, the mentioned equations are being solved using values of equilibrium conditions, and their outcome used to compute sought quantities, i.e. tensions.

5.2 Methodology of studies

Studies were performed in the environment of Solidworks Simulation, which provides core simulation tools to test designs and to make the decisions to improve their quality. Full integration creates a short learning curve and eliminates the redundant tasks required with traditional analysis tools. Component materials, connections, and relationships defined during design development are fully understood for simulation. Its main advantage is its cost, which is much lower than any other FEA software. In order to achieve credible results, materials (of which the element would be made) were assigned to each element. Properties of materials are presented in Table 1.

Table 1. Table of materials

Material	Element	Young's modulus [MPa]	Poisson's ratio [-]
Titanium alloy Protasul Ti	Prosthetic shell	110000	0.3
Titanium alloy Protasul 100	Shaft, head	100000	0.3
Polyethylene	Socket inlay (cup)	1070	0.41

The force was applied to the shaft of the endoprosthesis with values matching the value of the body weight (measured in the unit of Newton [N]) while performing different activities. In performed studies, the value of the human body weight was set on 784.8 N (or 80 kg), which was then assumed as 100% of body weight (%BW). Three possible scenarios were considered with the upper limb in abduction to 30°, 65° and 90°, where reaction forces influencing the endoprosthesis would have approximate values of, respectively, 30, 50 and 60%BW [11].

5.3 Results

The analyses of the simulations resulted in obtaining maps of stress distribution in elements of the endoprosthesis, shown below.

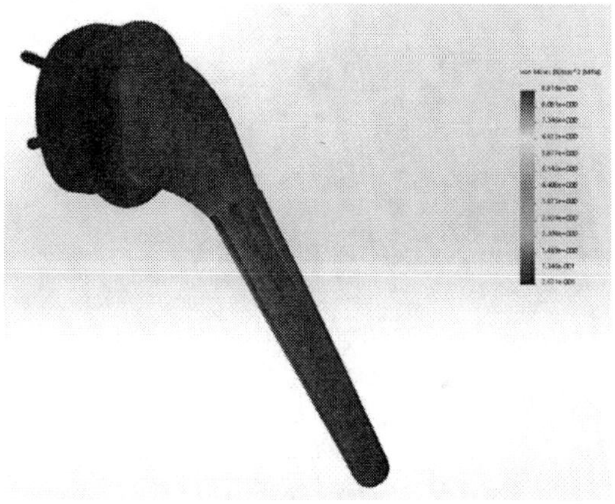

Fig. 6. Stress distribution in the endoprosthesis in 30° abduction.

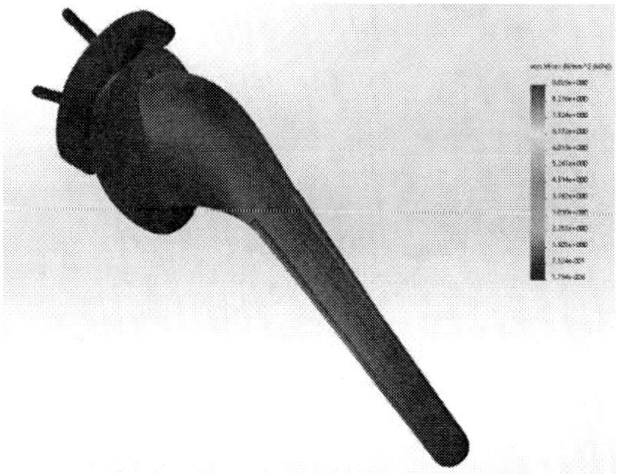

Fig. 7. Stress distribution in the endoprosthesis in 65° abduction.

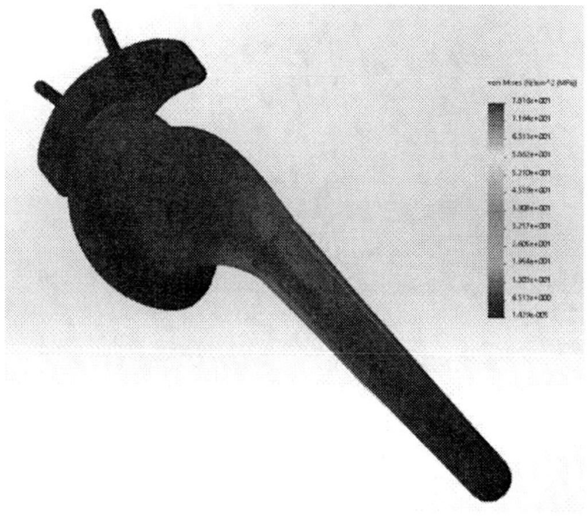

Fig. 8. Stress distribution in the endoprosthesis in 95° abduction.

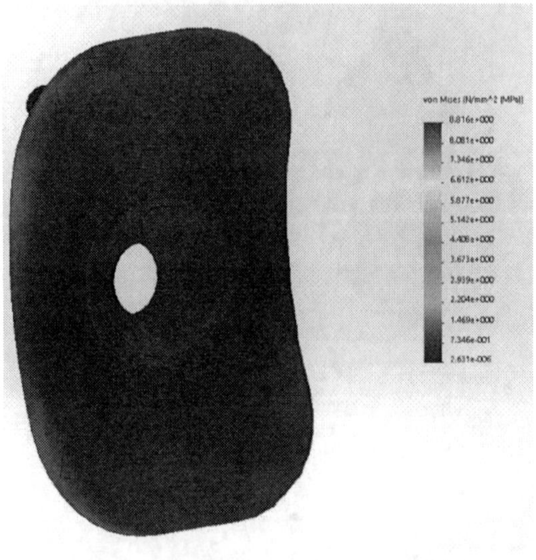

Fig. 9. Stress distribution in the prosthetic shell in 30° abduction.

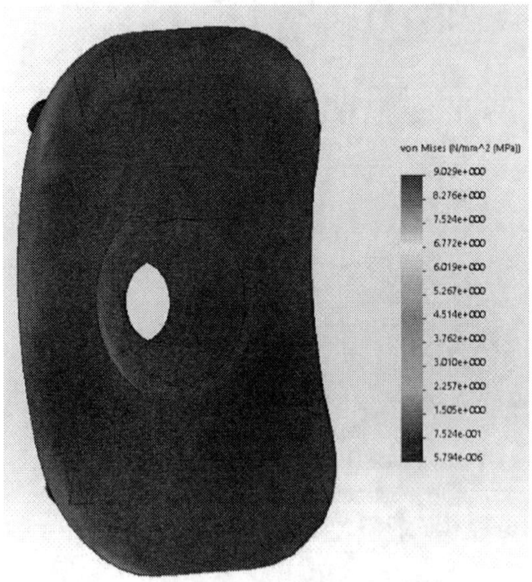

Fig. 10. Stress distribution in the prosthetic shell in 65° abduction.

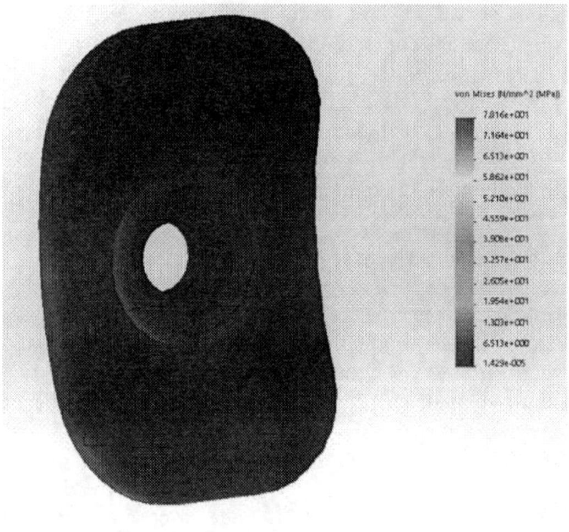

Fig. 11. Stress distribution in the prosthetic shell in 95° abduction.

Performed analyses show that forces acting upon the endoprosthesis lead to different stress distributions depending on the value of the force and orientation of the prosthesis. As expected, the maximum value of von Mises stress, in each case, is located on the shaft of the endoprosthesis, mostly near protrusions.

More interesting is the stress distribution in the prosthetic shell, since it is more likely to be damaged because of its structure. In each scenario simulation indicated, that the applied force would have greater impact on the edges of the shell. With the change in the angle of abduction of the endoprosthesis, the stress distribution changes – in 30° abduction the force acts almost evenly on the top and bottom edges of the shell, in 65° it influences both edges along the upper half of the shell, while in 95° most of the stress is located near the lower edge with small changes in the centre of the shell, which may indicate the possibility of the bending of the shell of the endoprosthesis.

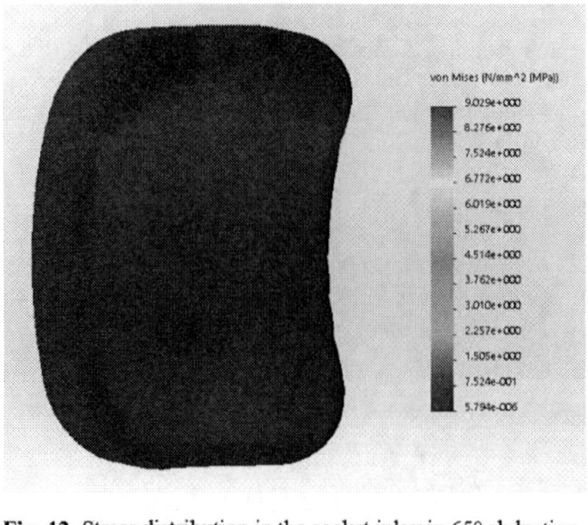

Fig. 12. Stress distribution in the socket inlay in 65° abduction.

Fig. 13. Stress distribution in the head of the endoprosthesis in 65° abduction.

In the case of the 65° abduction it is distinctive that most of the force affecting the socket inlay (or the cup) accumulates in the area between the centre and the top edge, as seen in Fig. 12. The corresponding area on the head of the prosthesis, acting upon the socket inlay, is shown in Fig. 13.

6. Conclusions

Modern technology gives engineers and physicians tools to implement solutions designed solely for the purpose of saving or at least easing the life of the specific patient. Due to the fact that dimensions of the endoprosthesis are based on a patient's physiognomy, it is possible to precisely plan and perform the implantation procedure. Thus, introducing more individual ways of treating the patient could lead not only to a decrease of the time required for convalescence, but in effect also reducing overall costs, mostly associated with the hospitalization of the patient.

Implementation of the FEA simulations is a crucial part of the designing process, since it is possible to predict and counteract possible damages, which may occur as a result of the forces influencing the part. Based on the performed simulations for designed endoprosthesis it can be assumed, that under given loads the elements of the endoprosthesis would maintain their structural integrity since the elastic limit (or yield strength) wasn't exceeded. For more accurate data it would be beneficial to perform a series of studies on the properly manufactured endoprosthesis.

References

1. Z. Ignasiak, *Anatomia układu ruchu* (Elsevier Urban & Partner, Wrocław, 2007)
2. A. Kontaxis, G. R Johnson, Clin. Biomech. 24, 254-260 (2009)
3. W. Woźniak, K.S. Jędrzejewski, *Sobotta. Atlas anatomii człowieka* (Elsevier Urban & Partner, Wrocław, 2012)
4. C.S. Neer, K.C. Watson, F.J. Stanton, . J. Bone Joint Surg. Am. 64 (3), 319–337 (1982)
5. R. Karpiński, Ł. Jaworski, J. Zubrzycki, Adv. Sci. Technol. Res. J. **10**, 240–246 (2016)
6. R. Będziński, *Biomechanika Inżynierska* (Oficyna Wydawnicza Politechniki Wrocławskiej, Wrocław, 1997)
7. T.S. Gaździk, *Biomechanika,* 641-664 (IPPT PAN, Warszawa, 2011)
8. P.A Kliniuk, A. Kuryliszyn-Moskal, Reumatologia **50**, 162-165 (2012)
9. A. Prusinowska, P. Turski, E. Przepiórski, P. Małdyk, K. Księżopolska-Orłowska, Reumatologia **51**, 108-112
10. R. Karpiński, Ł. Jaworski, J. Szabelski, Appl. Comput. Sci. **12**, 87–95 (2016)
11. P . Wodarski, R. Michnik, J. Jurkojć, A. Bieniek, M. Gzik, Modelowanie Inżynierskie, **29,** 74-79 (2016

The dual-fuel CFD combustion model with direct and indirect CNG injection

Tytus Tulwin[1,*], and *Rafał* Sochaczewski[2]

[1] Lublin University of Technology, Mechanical Engineering Faculty, ul. Nadbystrzycka 36, 20-618 Lublin, Poland
[2] Pope John Paul II State School of Higher Education, Department of Mechanical Engineering, ul. Sidorska 95/97, 21-500 Biala Podlaska, Poland

Abstract. A proper design of the compression ignition engine fuel system requires various fuel injection methods analysis. Direct and indirect CNG injection process simulation research was conducted. The results allow a mixture homogeneity analysis and its impact on combustion process and heat transfer. For an indirect injection the injector is placed in intake channel and the injection is performed during an intake stroke. For a direct injection the specially designed injector is installed in the glow plug position. The analysed case is set for 20% energetic ratio of diesel fuel and 80% of methane fuel. In order to perform the calculations a special combustion model has been implemented that allows duel-fuel combustion in a compression ignition engine.

1 Introduction

CNG gas is well suited as an additive fuel to a compression ignition engine. Methane autoignition occurs at the temperature of 537 °C, decreasing the chance of premature ignition during the compression stroke. Additionally, methane has good anti-knock combustion properties with equivalent octane number of 120. CNG is relatively ecological when compared to the gasoline or diesel, because the reaction produces mainly water and carbon dioxide. There aren't many additives to this fuel so there are fewer toxic products. The main problems of CNG fuel are storage and transport. The CNG fuel requires high pressures above 20 MPa, what requires appropriate installation. Table 1 presents the comparison of different combustion engine fuels [1].

Table 1. CNG, gasoline and diesel properties comparison.

	CNG	Gasoline	Diesel
Lower heating value MJ/kg	48.5	42.5	41.7
Higher heating value MJ/kg	53.9	46.7	44.8
Autoignition temperature °C	630-640	320-360	550-600
Octane number	115-135	90-98	25
Density kg/m3	0.73	700-750	800
Air-fuel ratio kg/kg	17.2	14.8	14.3

An internal combustion engine with high pressure direct fuel injection can be described with high thermodynamic efficiency thanks to stratified lean mixture combustion. Air excess that doesn't participate in combustion help to isolate the hot combustion gases from engine walls, therefore increasing efficiency. Additionally, the CO, CO2 and HC emissions are lower. In the following research the CNG injection is analysed for two cases: direct and indirect injection method. The direct injection method is characterised by the specially designed injector located in the combustion chamber, in the place of the glow plug. The indirect injection method is characterised by the injector located in the intake manifold wall outside the combustion chamber, as a multi-point injection. Both methods increase the homogeneity of the fuel-air mixture prior to the combustion – this effect is analysed in this paper [2].

2 Materials and methods

Accuracy of the simulation results of the simulation model is proved on the basis of the actual engine pressure at the same operating point - at partial load of 170 Nm and rotational speed of n = 3200 rpm. It is often difficult to take measurements inside the combustion engines, because of limited confined space. Often, required data must be indirectly approximated. A lot of combustion information can be obtained from the pressure readings. For the verification of the combustion model the heat generation rate was analysed. Its value as a function of time is derived from equation 1.

$$\frac{dQ_n}{dt} = \frac{\gamma}{\gamma-1} \cdot p \cdot \frac{dV}{dt} + \frac{1}{\gamma-1} \cdot V \cdot \frac{dp}{dt} \quad (1)$$

* Corresponding author: t.tulwin@pollub.pl

DOI: 10.1051/itmconf/20171507016

where: dQn - amount of heat change, γ – adiabatic index for chamber mixture, p – instantaneous mean cylinder pressure and V - instantaneous combustion chamber volume. The value of dQn / dt is expressed in J / s. The adiabatic index of gases in the combustion chamber is unknown in this equation. It is mainly temperature dependent, so no large error has been made assuming its constant value in the typical range of its occurrence (between 1.3 and 1.4). By combining the heat generation rate (equation 1), the heat generated during the combustion process was obtained without taking into account the heat loss via cooling:

$$Q = \int_{t_1}^{t_2} \frac{dQ}{dt} dt = m_f \cdot W_t \qquad (2)$$

where: Q –overall heat dissipated, mf – fuel mass, Wt – lower heating value and t_1, t_2 – combustion process integral time limits. Figure 1 shows the heat generation rate in respect to the crankshaft position, for the actual and computed case. Values for both cases were calculated using equation 1 based on the actual cylinder pressure and simulated cylinder pressure. The verification case considers a full dose of diesel fuel injection without CNG injection.

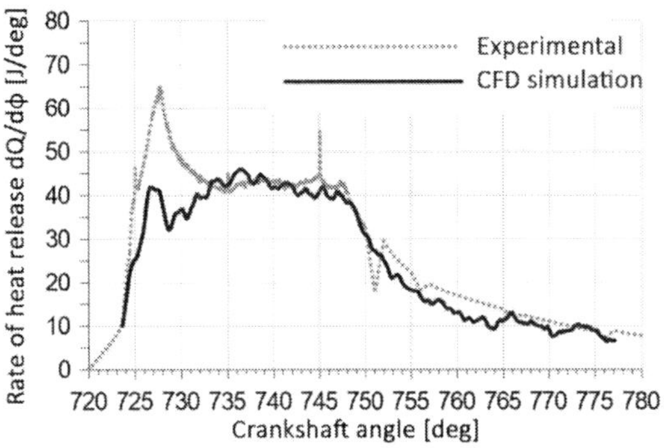

Fig. 1 Rate of heat release for experimental and CFD simulation results.

As shown in figure 1, both curves are closely aligned in the middle and at the end of the injection phase (> 735 °CA). This is the period of the diffusion combustion. The evaporation rate reaches plateau, causing a constant rate of heat release up to 748 °CA, where the injector closes. Fuel residues burn up for a longer time in relation to the injection time. An exemplary fuel injection flame visualisation is shown in figure 2. During the initial injection phase (<735 °CA) the fuel atomizes already in the injector nozzle, increasingly as penetration through the dense air in the combustion chamber. At first, the fuel mixture too rich to ignite. When the air is mixed sufficiently with the fuel there is a rapid combustion burst characterized by a peak in the rate of heat release graph. Figure 1 shows a small difference between real and computational cases, due to insufficient refinement of the computational mesh at the injection site, which is justified by the limitation of the calculation time of the workstation. However, for comparative purposes, the injection phase has been sufficiently reflected [3].

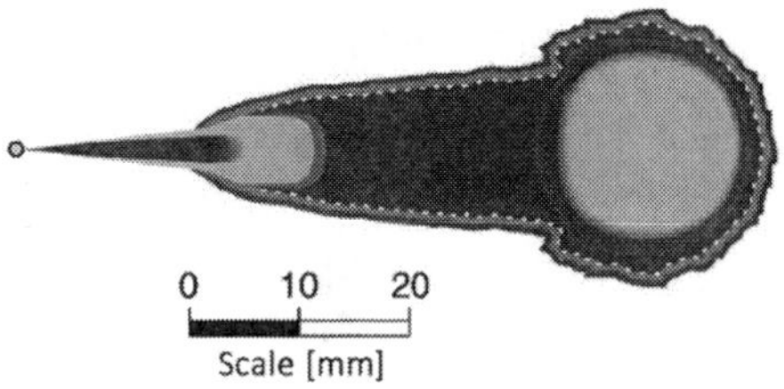

Fig. 2 Developed fuel injection for a typical diesel engine.

Information about the injector diameter and injection pressure allows to calculate the injection angle. High in-cylinder pressures empirical coefficients were assumed based on the results of the study. The injection stream angle was determined basing on equation 3:

$$\propto = 2 \cdot tan^{-1}(0{,}00751 \cdot W_e^{0{,}32} \cdot M^{-0{,}07} \cdot \delta^{0{,}5}) \qquad (3)$$

where δ is the ratio of mixture density to the fuel density. The Mach number is represented by the equation:

$$M = \frac{\mu^2}{\rho \cdot d \cdot \sigma} \qquad (4)$$

where: d is the injector nozzle diameter, σ is the surface tension, μ is the dynamic viscosity of the fuel. We is a Weber number, represented by the equation:

$$W_e = \frac{\rho \cdot w^2 \cdot d}{\sigma} \qquad (5)$$

The fuel stream velocity can be calculated by using Bernoulli equation for incompressible fluids:

$$w = \sqrt{2 \cdot \frac{\Delta p}{\rho}} \qquad (6)$$

where Δp is an injector pressure difference and ρ is a fuel density. Based on the maximum stream flow velocity an adequate injector mass-flow was formulated [4][5][6].

3 Results and discussion

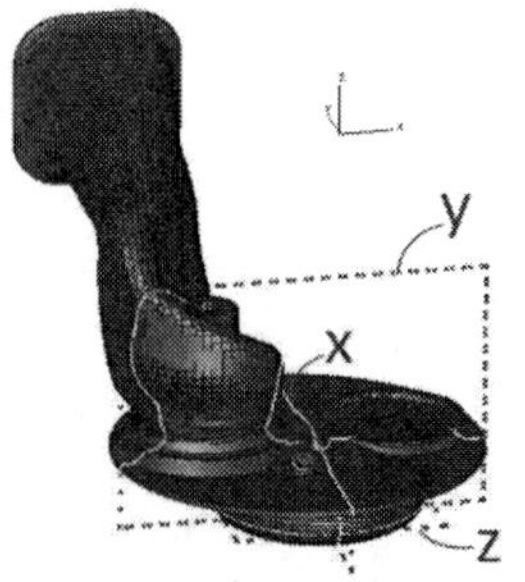

Fig. 3 Computational grid for CFD simulation model.

Figure 3 shows the ADCR diesel engine computational mesh. Reference planes for later visualisations are indicated. Two methods of supplying the CNG gas have been tested: indirect gas injection into the intake manifold and direct injection of gas into the combustion chamber. Figure 1 shows the locations of the injectors. Figure 4 shows the effect of the gasoline engine feed method on the decomposition of the mixture just before ignition (720 ° CA). Indirect gas injection causes a homogeneous mixture formation. A long mixing path and time resulted in a uniform CH4 distribution in the combustion chamber. Direct injection of methane results in uneven distribution of methane. At the bottom of the combustion chamber, the CH4 concentration is only 1%, when the uppermost part of the chamber is filled with a mixture containing up to 4% methane. The main advantage of direct injection is the fact that the entire fuel dose goes directly into the combustion chamber and takes part in combustion process during the same cycle. Indirect injection does not guarantee that the entire injected dose will reach the combustion chamber during the same cycle. This is important in states of transient engine operation.

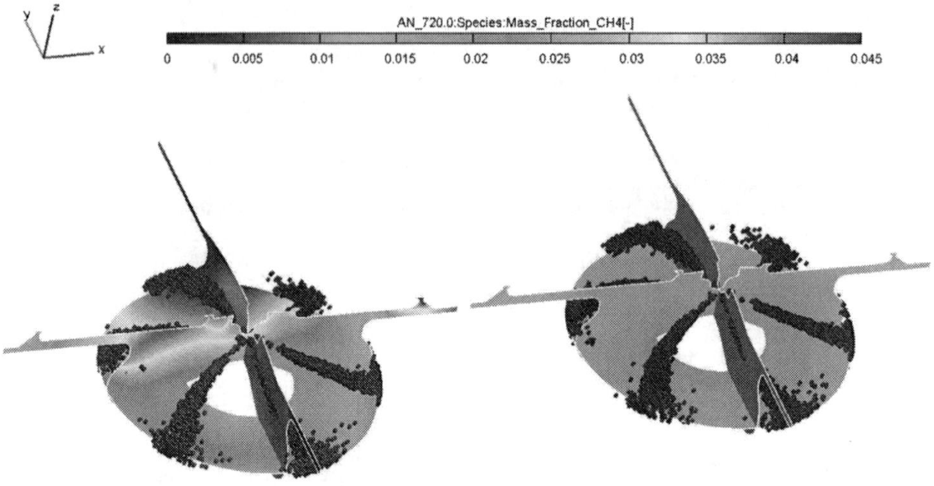

Fig. 4 Methane mass fraction and diesel droplets in case of direct and indirect gas injection.

The influence of the injection method on the cylinder pressure is shown in Figure 5. The direct injection occurs after the intake valve closes, increasing the mass of the mixture in the cylinder. This translates into increased maximal compression pressure (up to 0.4 MPa).

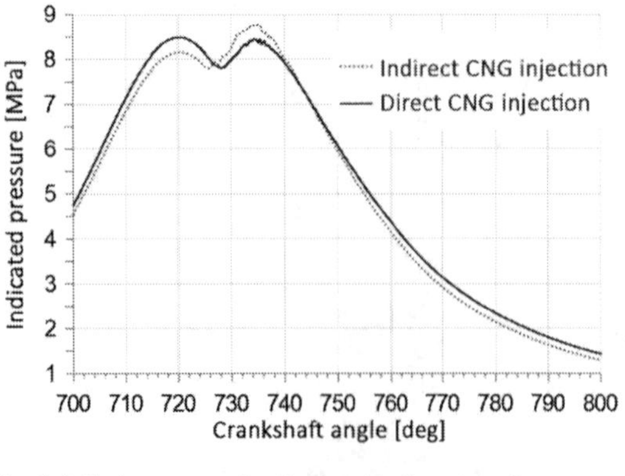

Fig. 5 Cylinder pressure for direct and indirect injection case.

Despite increased pressure, the ignition occurs later. In the case of indirect injection methane and air mixture is homogeneous. Ignition of the initial injection phase is accelerated. Pressure has significant influence on the beginning of self-ignition through chemical, physical phenomena and the local composition of the combustion mixture. It is difficult to predict the unambiguous influence when the above-mentioned phenomena are nonlinear. Local enrichment of the combustion mixture results in a reduction of the self-ignition delay. Increasing pressure (at pressures > 7 MPa) increases the delay caused by chemical effects by decreasing the diffusion coefficient. However, the impact on the delay is influenced by the injection phenomena. These include: reduced jet penetration, reduced droplet evaporation and lower fuel flow rates through the injector. Figure 6 shows the locations of auto-ignition sources for both injection cases. Comparing the distribution of the CNG blend (Fig. 4), it is possible to see the premature ignition of the mixture in the fuel-rich areas of the combustion chamber compared to the rest of the mixture. This causes an uneven fuel ignition throughout the combustion chamber. Nevertheless, combustion is more stratified due to better insulation of the cylinder walls and a longer diffusion period. The advantage of CNG injection, by forming a homogeneous mixture, is the reduction of soot formation which occurs in poor combustion areas [7][8].

DOI: 10.1051/itmconf/20171507016

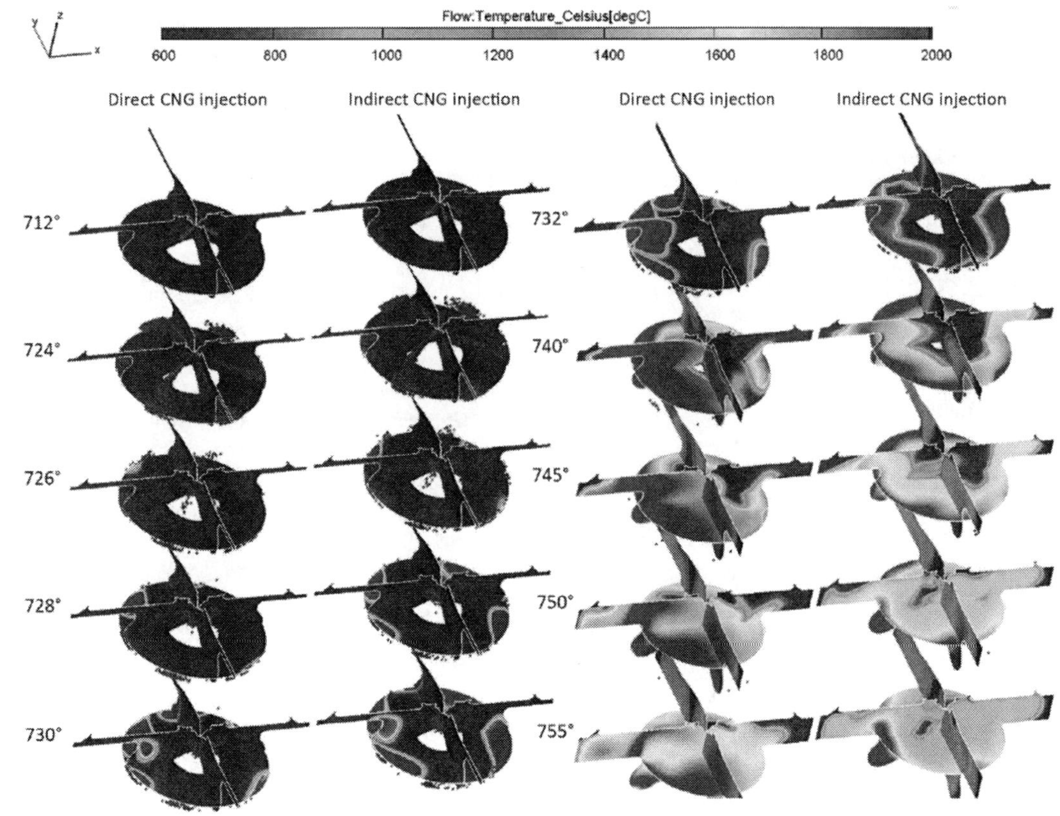

Fig. 6 Temperature distribution and diesel droplets for direct and indirect injection case in different crank angle position.

Figure 7 shows the average temperature in the combustion chamber for different crankshaft rotation angles. Higher temperatures do not mean greater heat losses because the wall temperature is lower than the average.

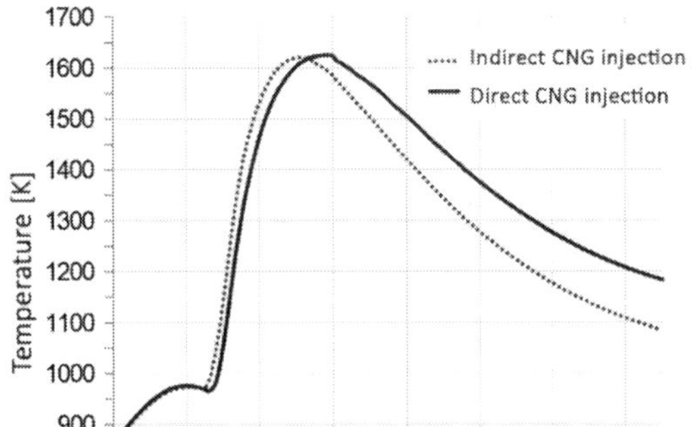

Fig. 7 Average in-cylinder temperature for direct and indirect gas injection.

In spite of similar maximal temperature, the stratification of the mixture results in a polytrophic expansion with the polytrophic exponent more closely related to the adiabatic factor (Fig. 5), thus reducing the heat loss. An early reignition pilot is favourable for faster ignition and combustion process [9]. A dual fuel CNG injection decreases the thermal efficiency slightly when compared to the full diesel injection as shown is study [10][11].

4 Conclusions

The full diesel fuel combustion process starts with the auto-ignition of the pre-blended fuel / air mixture, with the kinetic ignition at the beginning. In the majority of the injection, diffusion combustion dominates, leading to the evaporation rate and fuel burning rate equilibrium. Diffusion of the heterogeneous mixture allows to control the combustion rate, respectively extending and increasing the average effective pressure when compared to the combustion of the homogeneous mixture. Long fuel injection reduces the maximum temperature and pressure values, reducing the risk of excessive emissions of harmful nitrogen oxides. Through indirect injection, CNG produces a homogeneous mixture in the combustion chamber. The later the injection, the less homogeneous the mixture at the time of diesel injection. However, it is important to remember that there is a limitation of direct injection of CNG, as the counter-pressure in the combustion chamber increases. The advantages of heterogeneous

DOI: 10.1051/itmconf/20171507016

mixtures are: better thermal efficiency (less heat is lost through the cylinder walls during the combustion and expansion phase) and a higher degree of fuel burnout. Proper direct injection is more efficient than indirect injection in terms of engine economy (higher cycle efficiency) and reduced emissions of harmful substances. Improvements are also occurring intransient states, where direct injection is more reliable as the entire dose is applied to the combustion chamber during the same cycle. In the indirect injection, a part of the injected dose is taken from subsequent cycles. The advantage of CNG injection, by forming a homogeneous mixture, is the reduction soot formation which occurs in poor combustion areas.

References

1. K. U. Leuven, F. Ingenieurswetenschappen, D. Werktuigkunde, A. T. Mechanica, E. N. Energieconversie, *Katholieke Universiteit Leuven Thesis* (2008).
2. M. J. Staś, J. Kones, **7**, 485–496, (2000).
3. K. Ryu, Energy Convers. Manag., **76**, 506–516, (2013).
4. A. Ambrozik, T. Ambrozik, D. Kurczyński, P. Łagowski, Proc. Inst. Veh., **2**, no 8, 5–12, (2012).
5. A. I. Ramírez, S. Som, T. P. Rutter, D. E. Longman, and S. K. Aggarwal, *Spring Technical Meeting of the Central States Section of the Combustion Institute*, 22–24, (2012)
6. Y. Shi and R. D. Reitz, *Modelling Diesel Combustion-Multi-dimensional CFD modelling of diesel combustion:Overview*. (2009).
7. A. Paykani, A. H. Kakaee, P. Rahnama, and R. D. Reitz, Energy, **90**, 814–826, (2015).
8. R. G. Papagiannakis and D. T. Hountalas, Appl. Therm. Eng., **23**, no. 3, pp. 353–365, (2003).
9. S. Imran, D. R. Emberson, A. Diez, D. S. Wen, R. J. Crookes, and T. Korakianitis, Appl. Therm. Eng., **67**, no. 1–2, 354–365, (2014).
10. S. Imran, D. R. Emberson, B. Ihracska, D. S. Wen, R. J. Crookes, and T. Korakianitis, Int. J. Hydrogen Energy, **39**, no. 10, pp. 5163–5175, (2014).
11. M. Xu, W. Cheng, Z. Li, H. Zhang, T. An, and Z. Meng, Appl. Energy, **179**, 1185–1193, (2016)

AUTHOR INDEX

AUTHOR INDEX